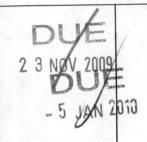

WELDING: PRINCIPLES AND PRACTICE

WELDING: PRINCIPLES AND PRACTICE

Henry Horwitz, P.E.

DUTCHESS COMMUNITY COLLEGE,
State University of New York

HOUGHTON MIFFLIN COMPANY Boston
Dallas Geneva, Illinois Hopewell, New Jersey
Palo Alto London

Illustrations by CHARTHOUSE.

PRINTED IN THE U.S.A.

Library of Congress Catalog Card Number: 77-76341

ISBN: 0-395-24473-0

CONTENTS

PREFACE

This book has been prepared for all students of welding, in or out of the classroom environment, and for all those who want a better understanding of what happens to a weld during the welding operation and after the weld has been made.

All of the industrially important welding processes—including adhesive bonding—and their applications are discussed. However, the greatest emphasis has been placed on the "bread-and-butter" skills. To this end, the book includes practice exercises on oxyacetylene welding, brazing, hard surfacing and the various electric–arc welding processes.

The book also includes separate chapters on such topics as welding consumables, welding metallurgy, metal properties of importance to the welder, inspection of welds, and welder certification. Reference tables and a glossary of key terms are among the informative items included at the end of the book.

To gain the most benefit from this book, it is suggested that students devote a minimum of five hours per week to studying the text, in addition to time spent in the classroom and laboratory.

I wish to express my thanks to the American Society for Metals, the American Welding Society, and many others who freely gave permission to use their materials as part of this textbook, as well as to Robert Sysum (Los Medanos College) and Melvin F. Taylor (Sam Houston State University) who provided valuable reviews of the manuscript.

It would be appreciated if all users of this book, both students and instructors, would send comments and suggestions on how to improve future editions to the author, care of the publisher.

H. H.

PART 1
INTRODUCTION TO WELDING

CHAPTER 1
WELDING: YESTERDAY AND TODAY

It is said that about 2500 years ago, a Greek blacksmith named Glaukos, living in the city of Khios, invented the *welding* of iron (Figure 1-1). In this process, pieces of iron were heated in a furnace, or forge, until they were soft. Then, with the aid of hammering, they were fused together as a unit. Prior to that time, metals were joined by riveting or by soldering, a nonfusion process often employing gold as a solder. The practice of forge welding, however, continued almost unchanged until about 80 years ago, when the invention of modern welding processes provided increasingly efficient means of joining metal plates or shapes, castings, forgings, or forgings to castings. First to be developed was *arc welding*, followed very quickly by *oxyacetylene welding*. These first welding processes were used primarily to repair damaged or worn metal parts.

Modern welding of metals, like ancient forge welding, is the joining of metals by fusion. However, with the development of welding technology and the improvement of testing methods, it became apparent that a complete and permanent fusion could be attained between two (or more) metals, with the weld area stronger than either of the pieces being joined. With the proper welding materials and techniques, almost any two pieces of metal can be fused into a single unit. Overlapping of pieces to be joined is not necessary, and thickness at the weld need be no greater than the thickness of either member.

Figure **1-1** Welding may have looked something like this during the 1600s.

All metals are weldable provided the proper process and technique are used. Occasionally an attempt to weld metals fails because one of these two factors—the proper process or the proper technique—has been overlooked. However, if the engineer and the welder understand the composition, structure, and properties of a metal, they will be able to design and make better welds. This underscores the close relationship between the metallurgy of a metal and its *weldability,* which we shall examine later.

1-1 NATURE OF THE WORK

Generally, the job of the welder or welding machine operator is to join (weld) two pieces of metal by applying intense heat, pressure, or both to melt the edges of the metal so that they fuse permanently. During this process, the worker can use various types of devices to obtain the necessary heat, with or without the aid of pressure—or the necessary pressure, with or without the aid of heat—to fuse the edges of the metal in a controlled fashion. These welding procedures are used in the manufacturing and repair of many different products ranging from water faucets, refrigerators, cars, and trains to electronic equipment, airplanes, ships, and missiles.

In the most common welding processes, there are several different sources of heat and a number of methods for controlling and focusing it. In fact, more than 40 different heat-based welding processes have been developed. However, these various processes can be grouped into three categories: the *arc welding* process, which obtains heat from an electric arc and maintains it between two *electrodes* or between an electrode and the work; the *gas-welding* process, which obtains heat, in the form of a flame, through the mixture of oxygen and some other combustible gas (fuel), usually acetylene; and the *resistance-welding* process, which obtains heat from resistance of the workpiece to an electric current. Two of the processes used to weld metal, the arc and the gas methods, can also be used to cut and gouge metal.

Since the 1940s, welding technology has improved so rapidly that the old concepts and definitions of welding are no longer completely accurate. However, it is accurate to say that most welding today is accomplished by one of the above processes. Increasingly, welding is being seen as the joining of metals and plastics by any method that does not use fastening devices. Therefore, cold welding, or *solid state welding*, is an important subject for our examination, as is welding with sound or with light (Figure 1-2). Soldering, brazing, the welding of plastics, and adhesive bonding are all areas that will be examined later.

1-2 ARC WELDING

Arc welding is widely accepted as the best, most economical, most natural, and most practical process of joining metals. In the most commonly used manual arc welding process, the arc welder obtains a suitable electrode, attaches the ground cable to the workpiece, and adjusts the electric current to "strike an arc," that is, to create an intense current that jumps between the electrode and the metal. Next, the electrode is moved along the seams of the metal to be welded, allowing sufficient time for the arc heat to melt the metal. The molten metal from the electrode is deposited in the joint and, together with the molten metal edges, solidifies to form a solid connection. The welder selects the electrode (filler metal) used to produce the arc according to job specifications (Figure 1-3).

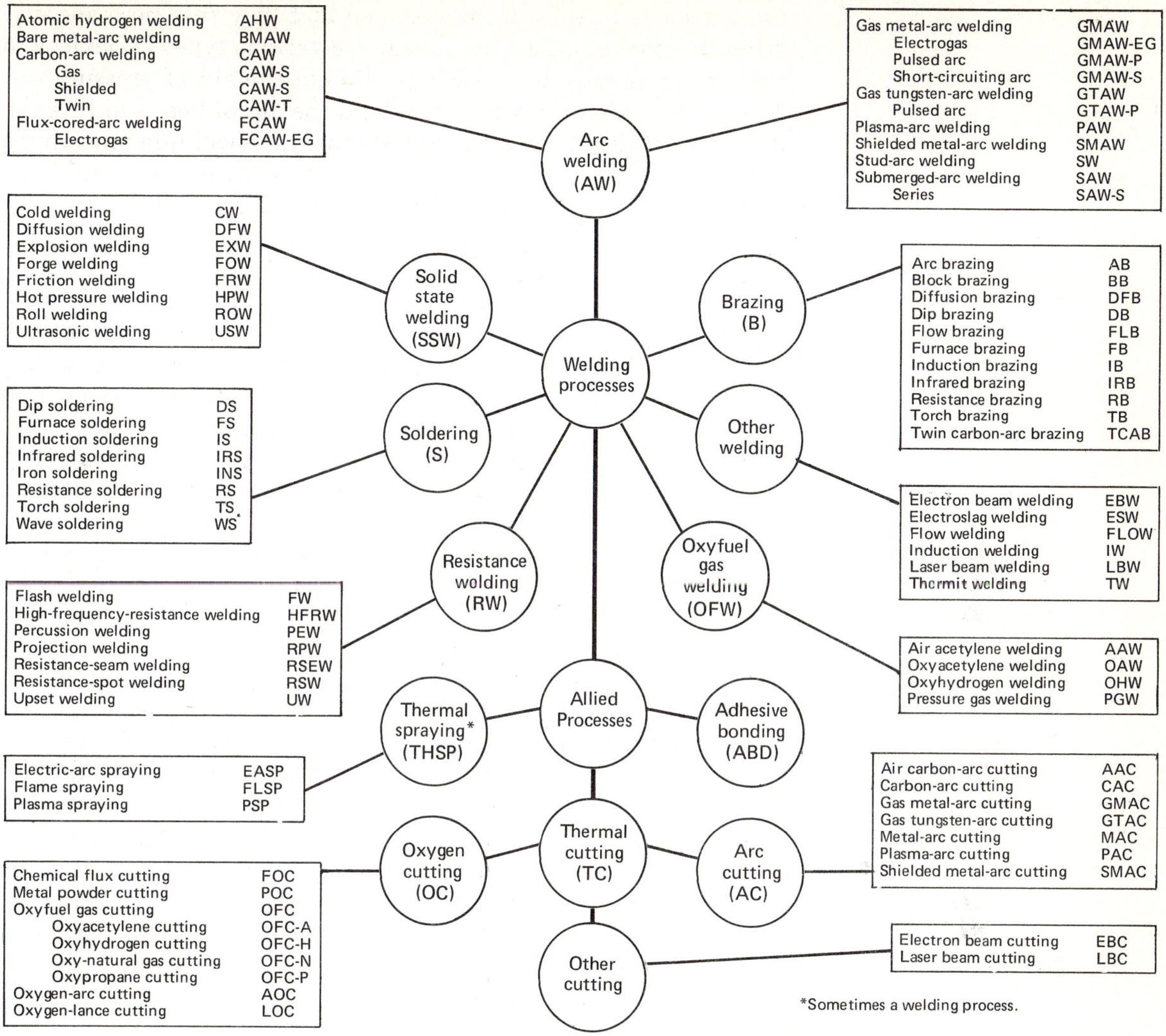
Welding processes
Arc welding (AW)
Atomic hydrogen welding AHW
Bare metal-arc welding BMAW
Carbon-arc welding CAW
Gas CAW-S
Shielded CAW-S
Twin CAW-T
Flux-cored-arc welding FCAW
Electrogas FCAW-EG
Gas metal-arc welding GMAW
Electrogas GMAW-EG
Pulsed arc GMAW-P
Short-circuiting arc GMAW-S
Gas tungsten-arc welding GTAW
Pulsed arc GTAW-P
Plasma-arc welding PAW
Shielded metal-arc welding SMAW
Stud-arc welding SW
Submerged-arc welding SAW
Series SAW-S
Solid state welding (SSW)
Cold welding CW
Diffusion welding DFW
Explosion welding EXW
Forge welding FOW
Friction welding FRW
Hot pressure welding HPW
Roll welding ROW
Ultrasonic welding USW
Brazing (B)
Arc brazing AB
Block brazing BB
Diffusion brazing DFB
Dip brazing DB
Flow brazing FLB
Furnace brazing FB
Induction brazing IB
Infrared brazing IRB
Resistance brazing RB
Torch brazing TB
Twin carbon-arc brazing TCAB
Soldering (S)
Dip soldering DS
Furnace soldering FS
Induction soldering IS
Infrared soldering IRS
Iron soldering INS
Resistance soldering RS
Torch soldering TS
Wave soldering WS
Other welding
Electron beam welding EBW
Electroslag welding ESW
Flow welding FLOW
Induction welding IW
Laser beam welding LBW
Thermit welding TW
Resistance welding (RW)
Flash welding FW
High-frequency-resistance welding HFRW
Percussion welding PEW
Projection welding RPW
Resistance-seam welding RSEW
Resistance-spot welding RSW
Upset welding UW
Oxyfuel gas welding (OFW)
Air acetylene welding AAW
Oxyacetylene welding OAW
Oxyhydrogen welding OHW
Pressure gas welding PGW
Allied Processes
Thermal spraying* (THSP)
Electric-arc spraying EASP
Flame spraying FLSP
Plasma spraying PSP
Adhesive bonding (ABD)
Thermal cutting (TC)
Oxygen cutting (OC)
Chemical flux cutting FOC
Metal powder cutting POC
Oxyfuel gas cutting OFC
Oxyacetylene cutting OFC-A
Oxyhydrogen cutting OFC-H
Oxy-natural gas cutting OFC-N
Oxypropane cutting OFC-P
Oxygen-arc cutting AOC
Oxygen-lance cutting LOC
Arc cutting (AC)
Air carbon-arc cutting AAC
Carbon-arc cutting CAC
Gas metal-arc cutting GMAC
Gas tungsten-arc cutting GTAC
Metal-arc cutting MAC
Plasma-arc cutting PAC
Shielded metal-arc cutting SMAC
Other cutting
Electron beam cutting EBC
Laser beam cutting LBC
*Sometimes a welding process.

Figure **1-2** Master chart of welding and allied processes. (Redrawn by permission from American Welding Society, *Welding Handbook*, 6th ed., 1969, section 2.)

There are several arc welding processes. *Carbon-arc welding* is the earliest modern welding technique. In this process, an arc is struck between a pure carbon electrode and the grounded workpiece, or between two carbon electrodes joined near the weld surface. The carbon electrodes are not consumed in the process. If filler metal is needed to complete the weld, welding rods must be used. Today, however, the carbon-arc process is used primarily to cut or gouge metals.

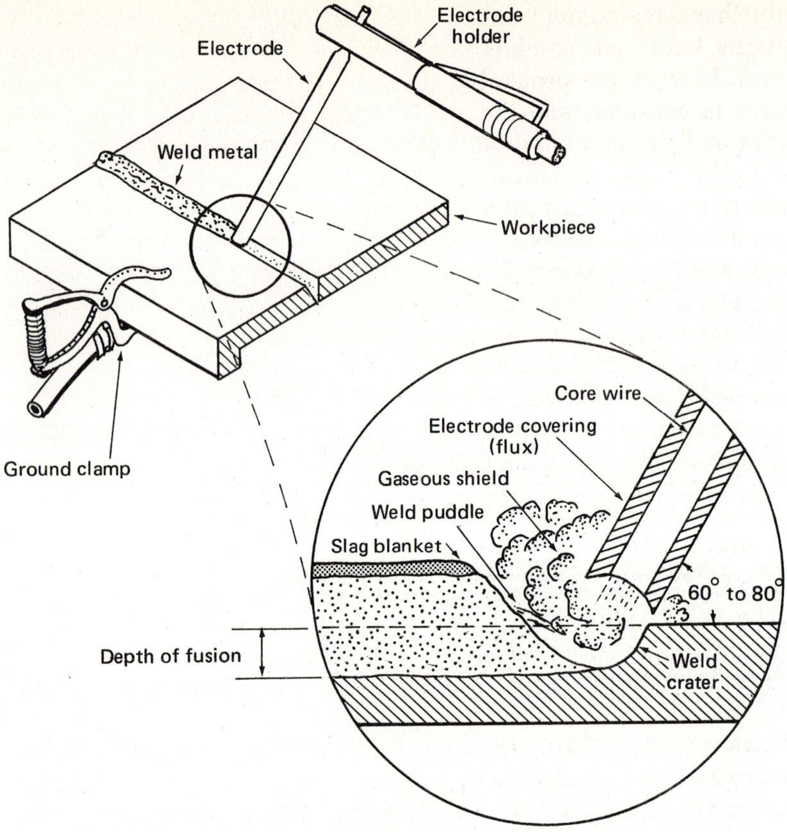
Electrode
Electrode holder
Weld metal
Workpiece
Ground clamp
Core wire
Electrode covering (flux)
Gaseous shield
Weld puddle
Slag blanket
60° to 80°
Depth of fusion
Weld crater

Figure **1-3** The arc welding process. (By permission, from *Metals Handbook* Volume 6, Copyright American Society for Metals, 1971.)

The carbon-arc process was followed very quickly by the development of *metal-arc welding,* in which a consumable metal rod was used as an electrode. At first, the electrodes were bare metal rods, which caused significant arc stabilization problems. The development of electrode coatings, commonly called *flux,* greatly resolved arc stabilization problems and resulted in what is called *shielded metal–arc welding,* the most widely used of the electric processes. On heating, the flux vaporizes, forming a gaseous shield around the arc and the weld. The shielding gas prevents oxygen and nitrogen in the air from forming weakening oxides or nitrides with the welded metal. The development of the manual shielded metal–arc process was soon extended to semiautomatic and automatic welding machines.

Further development of the concepts behind shielded metal–arc welding led to *gas shielded–arc welding*. There are two such processes. In both, the protective gases are brought in from a separate source (a cylinder) and the arc is struck between bare metal electrodes and the grounded workpiece. The gases flow from a protective collar out of and around the electrode, forming the protective atmosphere. In *gas tungsten–arc welding*, the electrodes are made of nonconsumable tungsten. Shielding is from externally supplied gases, and the necessary filler metal is supplied by welding rods. In *gas metal–arc welding*, the electrode is a continuous filler metal that is protected by externally supplied gases.

There are several other arc welding processes, such as submerged-arc welding, atomic hydrogen welding, and plasma-arc welding. But the processes mentioned in the preceding paragraph are the most common and will be discussed more extensively later.

1-3 GAS WELDING

Gas welding, or flame welding, uses an intensely hot flame produced by the combination of a fuel gas with air or oxygen. The most commonly used fuel gases are acetylene, natural gas, propane, and butane. Very often the fuels are burned with oxygen, which allows for a much higher combustion temperature.

Oxyacetylene welding (Figure 1-4) is the most common gas-welding process. Oxygen and acetylene combined in a mixing chamber and burned at the torch tip produce the highest flame temperature (about 6000°F, which is well beyond the melting point of most metals). Therefore, welding is possible with or without a filler metal. Parts may be fused by bringing them into contact as they are melted by the torch; when the torch is removed, the metal parts fuse as they cool. If filler metal is needed to complete a weld, welding rods are selected according to job specifications and melted by the torch heat. The selection of proper welding rods, torch tips, regulator adjustments for both the oxygen and the acetylene supplies, and welding position are all a matter of experience and knowledge of the process. Much of this will be examined in greater detail later.

The disadvantages of fuel-gas welding revolve around the fact that a number of metals react unfavorably, or even violently, in the presence of carbon, hydrogen, or oxygen, all of which are present in the fuel-gas process. Gas welding is also cooler, slower, and more distorting than arc welding. However, for welding in hard-to-reach places or with metals that have lower melting points, such as lead or thin sheet metals, gas welding is often more efficient than other processes. In combination with a stream of oxygen or air, the oxyacetylene torch is also an excellent cutting or gouging torch.

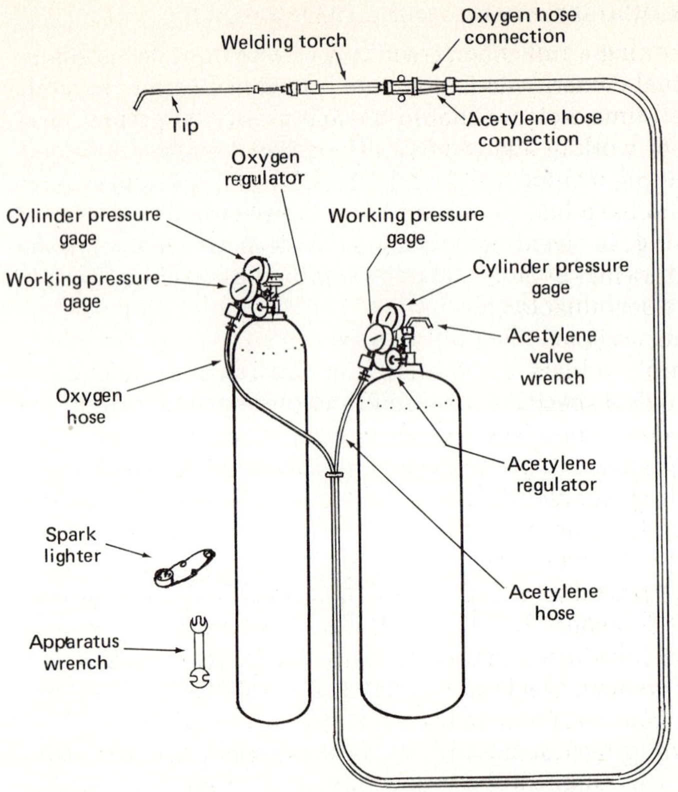

Oxygen hose connection
Welding torch
Tip
Acetylene hose connection
Oxygen regulator
Cylinder pressure gage
Working pressure gage
Working pressure gage
Cylinder pressure gage
Acetylene valve wrench
Oxygen hose
Acetylene regulator
Spark lighter
Acetylene hose
Apparatus wrench

Figure **1-4** A complete oxyacetylene welding outfit. (Redrawn by permission from Airco Welding Products, A Division of Airco, Inc., *Oxyacetylene Welding and Oxygen Cutting Instruction Course*, 1966, p. 28.)

1-4 RESISTANCE WELDING

Resistance welding is a machine process used primarily in the mass production of parts requiring relatively simple welding operations. The weld is made by heat generated by the resistance of the work-pieces to the flow of electricity at the intended location, and fused by the pressure of contacting electrodes. The resistance welding machine operator makes the necessary adjustments in the machine to control the current and the pressure and then feeds and aligns the work. After the welding operation is completed, the operator removes the work from the machine. Some types of resistance welding are spot, projection, flash, and upset welding, all of which will be examined in detail later.

1-5 WELDER QUALIFICATIONS AND ADVANCEMENT

A person planning a full-time career as a welder or cutter needs to develop manual dexterity and good hand-eye coordination. In addition, most full-time welders should be able to lift 100 pounds, and bend, stoop, or work in awkward positions fairly easily.

For entry-level manual welding jobs, most employers prefer to hire those who have had high school, industrial arts, or vocational school training in welding methods. Courses in mathematics, mechanical drawing, blueprint reading, and/or other kinds of metalworking are also valuable.

Before being assigned jobs in which strength and/or quality of the weld are critical, welders may have to pass certain tests to determine their ability to do a specific kind of work. The ability of the welder to do work meeting the requirements of the American Society of Mechanical Engineers (ASME) *Boiler and Pressure Vessel Code* and others is the basis for certification of welders. However, certification under these codes is not a general certification, but rather a certification for a particular type of work or employer, and is valid only for a specified time period (see Chapter 23). Occasionally, a welder may be promoted to inspector, who checks welds for general conformance with specifications and quality of workmanship. Welders may also become foremen. Welders with two years' training at a technical institute or a community or junior college may qualify for employment as welding technicians. Some welders, after obtaining sufficient experience, establish their own welding and repair shops.

1-6 PLACES OF EMPLOYMENT AND POSITION TITLES

According to the U.S. Department of Labor, Bureau of Labor Statistics, an estimated 555,000 welders and oxygen and arc cutters were employed throughout this country in 1972. About 385,000 of these workers were employed in manufacturing industries, mostly in those making such durable goods as transportation equipment and fabricated metal products. Of the 150,000 welders and cutters in other industries, the greatest number were employed in construction firms and establishments performing miscellaneous repair services. The remaining 20,000 were widely scattered among other nonmanufacturing industries, such as retail chain stores and food-processing plants. Workers employed in the field of welding are known by such position-descriptive titles as arc welder, combination welder, fitter welder, resistance-welding welder, arc welding machine operator, gas welder, inspector, and others.

The widespread use of the welding and cutting processes enables these workers to find jobs in every state. Most of the 22,000 new job openings expected each year during the 1970s will be found in the seven major metalworking areas of the United States, namely, Pennsylvania, California, Ohio, Michigan, Illinois, Texas, and New York.

1-7 EARNINGS AND WORKING CONDITIONS

A welder's earnings depend to a large extent on the skill requirements of the particular job and on the industry or activity in which he or she is employed.

Skilled manual welders in the fabricated steel industry averaged $8.00 per hour in the early 1970s. The highest hourly rate currently being paid is in the San Francisco–Oakland area—$12.00 per hour. Resistance welding machine operators and oxygen and arc cutters generally earn somewhat less—the average hourly wage in these occupations is about $4.50.

Welders may work inside in a well-ventilated and well-lighted shop, outside on a construction site, or in the confined space of an underground tunnel. The various trade and safety organizations have developed a series of rules and requirements (see Chapter 13) regulating various welding procedures and health precautions.

Working conditions of resistance welding operators are usually safer than those for other welders, since they are largely free of the hazards associated with manual welding operations.

1-8 PROFESSIONAL AND LABOR ORGANIZATIONS REPRESENTING WELDERS

Many welders and cutters are members of the American Welding Society and various labor unions. Among the unions representing welders are:

- The International Brotherhood of Boilermakers, Iron Workers, Shipbuilders, Blacksmiths, Forgers, and Helpers; 8th at State Avenue, Kansas City, Kansas 66101.
- The International Association of Machinists and Aerospace Workers; 1300 Connecticut Avenue NW, Washington, D.C. 20036.
- The International Union, United Automobile, Aerospace, and Agricultural Implement Workers of America; 800 East Jefferson Avenue, Detroit, Michigan 48214.
- The United Association of Journeymen and Apprentices of the Plumbing and Pipe Fitting Industry of the United States and Canada; 901 Massachusetts Avenue NW, Washington, D.C. 20001.

REVIEW QUESTIONS

1. Was welding discovered during the B.C. or the A.D. period?

2. Name the categories into which the more than 40 heat-based welding processes can be grouped.

3. Describe generally the process of arc welding. Name some of the arc welding processes.

4. Describe the process of oxyacetylene welding.

5. What determines the selection of the filler metal to be used for welding a given joint?

6. Name and describe some of the welding processes other than those mentioned above.

7. Differentiate between a welding operator and a welder.

8. What do employers usually require as the minimum qualifications for a person applying for a manual welding job?

9. What is meant by the term *certified welder*?

10. What kind of training is recommended for a person seeking employment as a welding technician?

11. According to the U.S. Department of Labor, Bureau of Labor Statistics, how many welders and cutters were employed in the United States during 1972?

12. In what industries were most welders and cutters employed during 1972?

13. What were the average hourly wages of welders in the fabricated steel industry during 1972?

14. Name some of the professional and labor organizations representing welders.

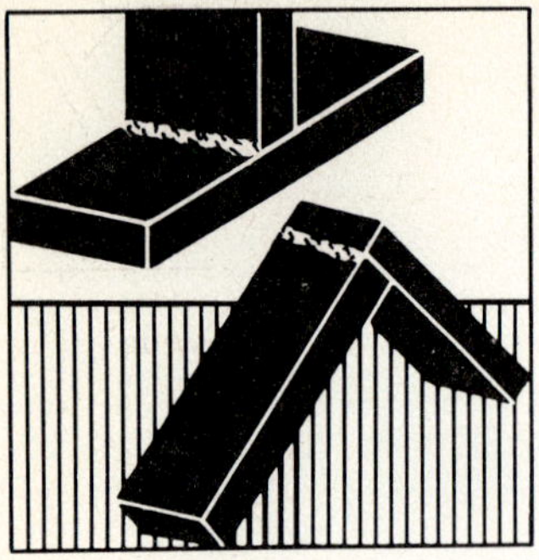

CHAPTER 2
TYPES OF WELDS AND WELDED JOINTS

Welded joints, regardless of the joining process, are designed primarily for the strength and safety required by services they must perform. The manner in which stress will be applied in service—whether tension, shear, bending, or torsion—must be considered. Different designs may be required, depending on whether the loading is static or dynamic. Different designs may also be needed where fatigue is involved. Joints may be designed to reduce or eliminate stress raisers and to obtain an acceptable pattern of residual stresses. Joints that will be subjected to corrosion or erosion must be made so that they are free of irregularities, crevices, and other defects which will render them susceptible to such forms of attack. The design must also reflect consideration of *joint efficiency*, which is defined as the ratio of the joint strength to that of the base metal and is generally expressed as a percentage.

In addition, joints are designed for economy and accessibility during construction. Among the factors involved in construction are control of distortion and shrinkage cracking, the facilitation of good workmanship, and the production of sound welds. Accessibility during construction not only ensures lower costs, but also provides an opportunity for better workmanship, reduction of flaws, and control of distortion and residual stresses.

2-1 TYPES OF WELDS

One aspect of joint design is the type of weld employed in the joining. There are five basic weld types: bead, weave, fillet, plug, and groove. The selection of weld type is as integral to joint efficiency as is joint design. One type of weld is selected over another because of its specific relation to the joint efficiency.

Bead welds are made by a single pass with the filler metal with *no* side-to-side motion. Bead welding is used primarily to build up worn surfaces and is not often used for joints. An example of a bead weld is shown in Figure 2-1.

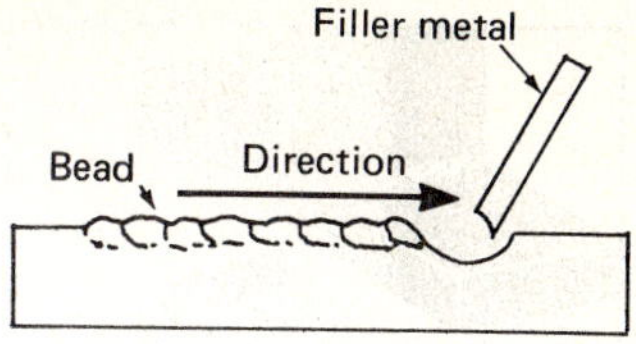
Filler metal
Bead
Direction

Figure **2-1** Bead weld

Weave welds are accomplished by running a bead with some amount of side-to-side motion (Figure 2-2). The width of the bead is a matter of design or necessity. There are also several weave patterns, such as zigzag, circular, oscillating, and others. Weave welds are also used primarily for build-up of surfaces.

Fillet welds are similar to, but faster to make than groove welds and are often preferred in similar situations because of economy. However, single-fillet welds sometimes are not as strong as groove welds, while a double fillet compares favorably for strength (see Figure 2-3). Fillet-welded joints are simple to prepare from the standpoint of edge preparation and fit-up, although sometimes more welding is required than with groove-welded joints. Fillet welds are often combined with other welds to improve stress distributions, for example, at a tee joint. Concave fillets are most effective when the direction of stress is transverse to the joint.

Plug and *slot welds* primarily take the place of rivets. They are used to fuse two pieces of metal whose edges, for some reason, cannot be fused. Either an interior circle (plug) or a slot is welded, leaving the edges free (see Figure 2-4).

Groove welds are made in the groove between two pieces of metal. They are used in many combinations, depending on accessibility,

Figure **2-2** Weave welding motions

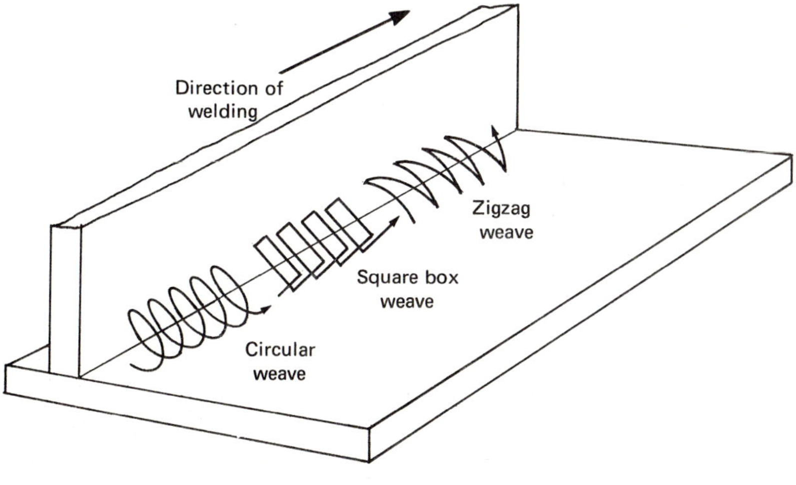
Direction of
welding
Zigzag
weave
Square box
weave
Circular
weave

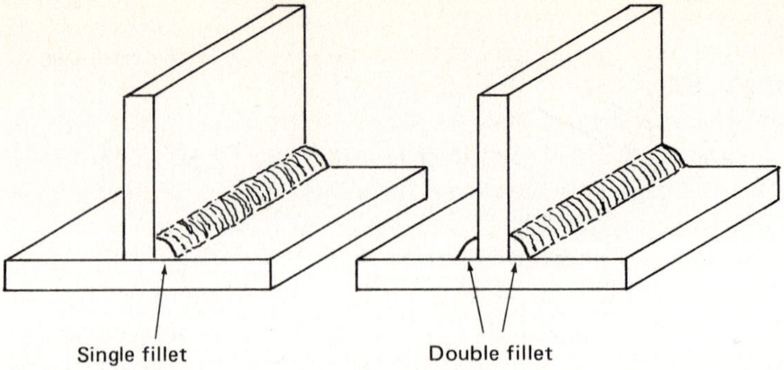
Single fillet
Double fillet

Figure **2-3** Single- and double-fillet welds

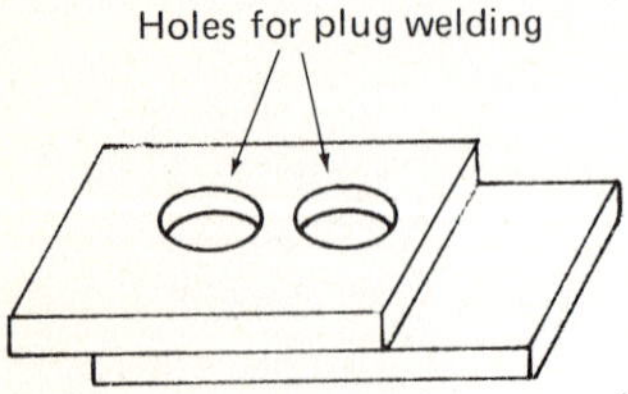
Holes for plug welding

Figure **2-4** Prepared plates for plug welding

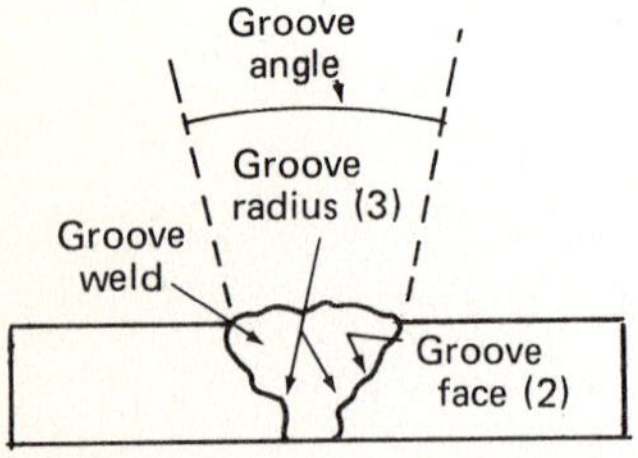
Groove angle
Groove radius (3)
Groove weld
Groove face (2)

Figure **2-5** Groove weld in cross-section

economy, design, and the type of welding process used. An example of a groove weld is illustrated in Figure 2-5. The groove includes (1) the groove angle, (2) the groove face, and (3) the groove radius.

You will attempt many of these types of welds in the practice section (Part 3) of this book. However, a welder must be prepared to do them in any of the usual welding positions: flat, horizontal, vertical, and overhead (Figure 2-6). Of course, the flat position is the easiest. The molten metal is held in position (until it begins to solidify) by the force of gravity. This position also enables maximum deposition rates. Next in ease of welding is the horizontal weld, in which the force of gravity also helps to some extent.

Figure **2-6** The four standard welding positions

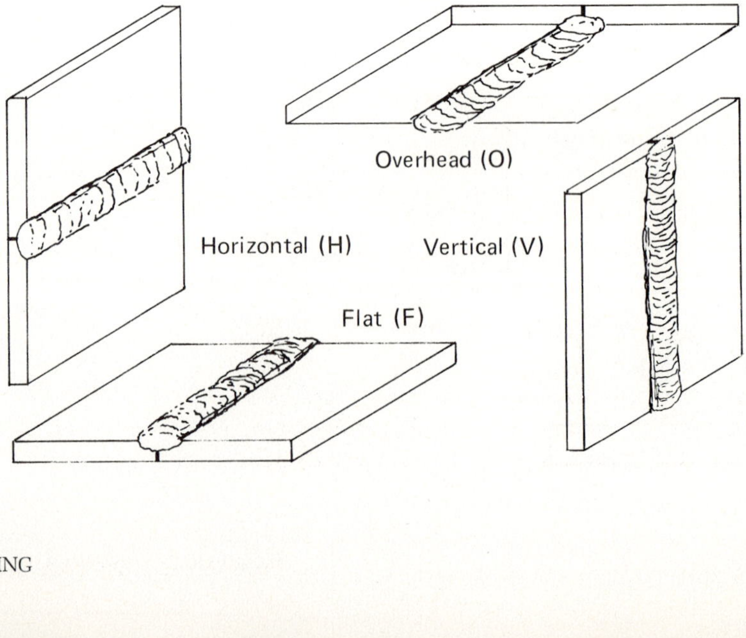
Overhead (O)
Horizontal (H)
Vertical (V)
Flat (F)

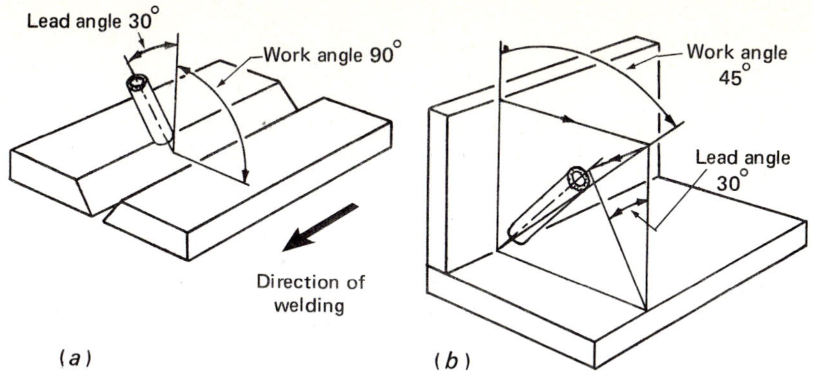
Lead angle 30°
Work angle 90°
Direction of welding
(a)
Work angle 45°
Lead angle 30°
(b)

Figure **2-7** (*a*) Electrode position for flat-position arc welding. (*b*) Electrode position for horizontal-fillet arc welding. (By permission, from *Metals Handbook* Volume 6, Copyright American Society for Metals, 1971.)

Welding in positions other than flat (and occasionally horizontal) is called *out-of-position welding* and often requires the use of manipulative techniques (Figures 2-7 and 2-8) and electrodes that result in faster freezing of molten metal and slag to counteract the effect of gravity (see Table 2-1). Another way to deal with the problem of out-of-position welds is to make use of one of the many kinds of welding positioners (see Figure 2-9, page 18).

Figure **2-8** Usual electrode clearances for avoiding interference and allowing greater welder visibility. (By permission, from *Metals Handbook* Volume 6, Copyright American Society for Metals, 1971.)

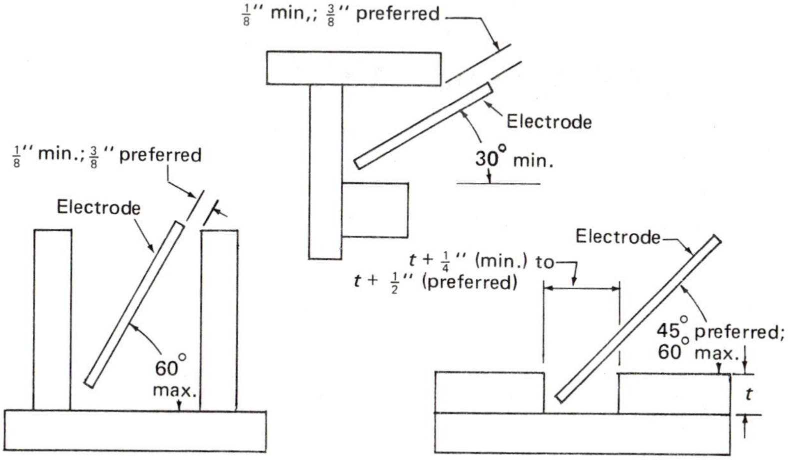
1/8" min,; 3/8" preferred
Electrode
30° min.
1/8" min.; 3/8" preferred
Electrode
60° max.
Electrode
t + 1/4" (min.) to
t + 1/2" (preferred)
45° preferred;
60° max.
t

Table **2-1** Recommended* electrodes for out-of-position welding

AWS–ASTM ELECTRODE CLASSIFICATION*	WELDING CATEGORY	GENERAL CHARACTERISTICS
	60,000-psi MINIMUM TENSILE STRENGTH	
E6010	Freeze†	Molten weld metal freezes quickly; suitable for welding in all positions with DC reverse-polarity power; has a low-deposition rate and deeply penetrating arc; can be used to weld all types of joints.
E6011	Freeze†	Similar to E6010, except can be used with AC as well as DC power.
E6012	Follow	Faster travel speed and smaller welds than E6010; AC or DC, straight-polarity power; penetration less than E6010. Primary use is for single-pass welding of thin-gage sheet metal in flat, horizontal, and vertical-down positions.
E6013	Follow	Similar to E6012, except can be used with DC (either polarity) or AC power.
E6027	Fill	Deposition rate high since covering contains about 50% iron powder; primary use is for multipass, deep-groove, and fillet welding in the flat position or horizontal fillets, using DC (either polarity) or AC power.

2-2 JOINT DESIGNS

There are five basic joint styles: lap, butt, corner, flange, and T. *Lap joints* are essentially two overlapping pieces of metal fused together by spot, fillet, plug, or slot welds. The weld of a *butt joint* is between the surface planes of the two parts. Butt joints can be plain, square, beveled, V, single-J grooves, single-U grooves, or doubled. *Corner joints* are what the name implies: welds between two parts at a 90-degree angle. They are either half-lap, corner-to-corner, or full inside, and can be prepared to form single-bevel, single-V, or single-U grooves. *Flange joints,* or *edge joints,* result from fusing the adjoining surface of each part so that the weld is within the surface

Table **2-1** *(cont.)*

AWS–ASTM ELECTRODE CLASSIFICATION*	WELDING CATEGORY	GENERAL CHARACTERISTICS
	70,000-psi MINIMUM TENSILE STRENGTH	
E7014	Fill-freeze	Higher deposition rate than E6010; usable with DC (either polarity) or AC power; primary use is for inclined and short, horizontal fillet welds.
E7018	Fill-freeze	Suitable for welding low and medium-carbon steels (0.55% C max) in all positions and types of joints. Weld-metal quality and mechanical properties highest of all mild-steel electrodes; usable with DC reverse polarity or AC power.
E7024	Fill	Higher deposition rate than E7014; suitable for flat-position welding and horizontal fillets.
E7028	Fill	Similar to type E7018; used for welding horizontal fillets and grooved fillet welds in flat position.

*E6020, E7015, and E7016 are not included because of their limited usage. Only electrodes up to 3/16-in. diameter can be used in all welding positions (flat, horizontal, vertical, and overhead).

†When used for welding sheet metal, these electrodes have follow-freeze characteristics.

Source: *The Procedure Handbook of Arc Welding*, 12th ed. (Cleveland, Ohio: Lincoln Electric Company), p. 6.2-12.

planes of both parts. They can be single- or double-flanged joints. *T joints* are just that. But they, too, can be single-bevel, double-bevel, single-J, and double-J. The effects of poor fit-up (gap between plate edges) due to distortion and cracking are shown in Figure 2-10 (page 19).

The proportions of grooves for butt, corner, flange, and T joints, as well as plug welds, as recommended by the American Welding Society, are shown in Figure 2-11 (pages 20–23), which shows the typical designs and dimensions of joints used for welding with the submerged-arc, shielded metal–arc, gas tungsten–arc, gas metal–arc, flux-cored arc, and gas welding processes (except pressure-gas

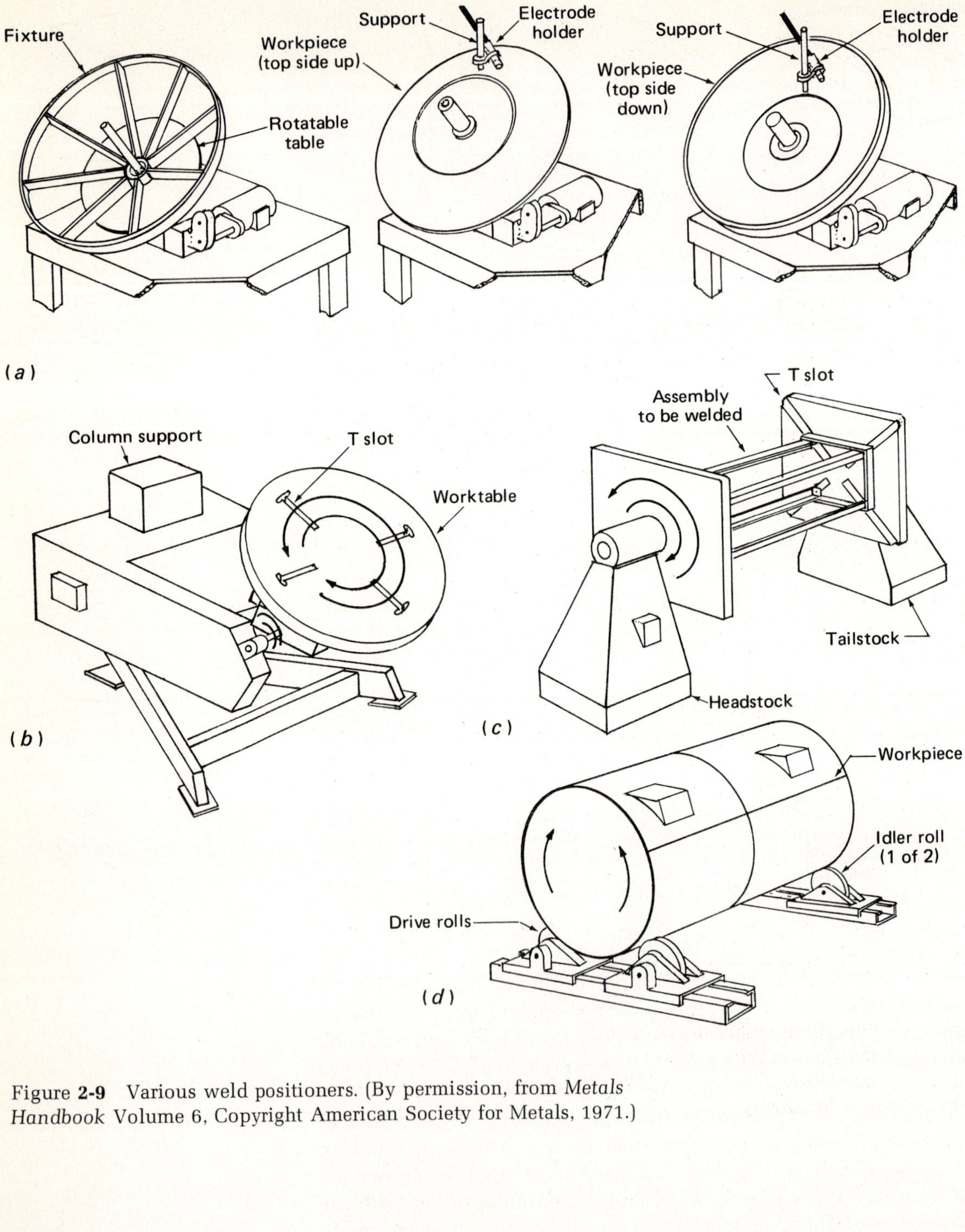
Fixture
Rotatable table
Workpiece (top side up)
Support
Electrode holder
Workpiece (top side down)
Support
Electrode holder
(a)
Column support
T slot
Worktable
(b)
T slot
Assembly to be welded
Tailstock
Headstock
(c)
Workpiece
Idler roll (1 of 2)
Drive rolls
(d)

Figure **2-9** Various weld positioners. (By permission, from *Metals Handbook* Volume 6, Copyright American Society for Metals, 1971.)

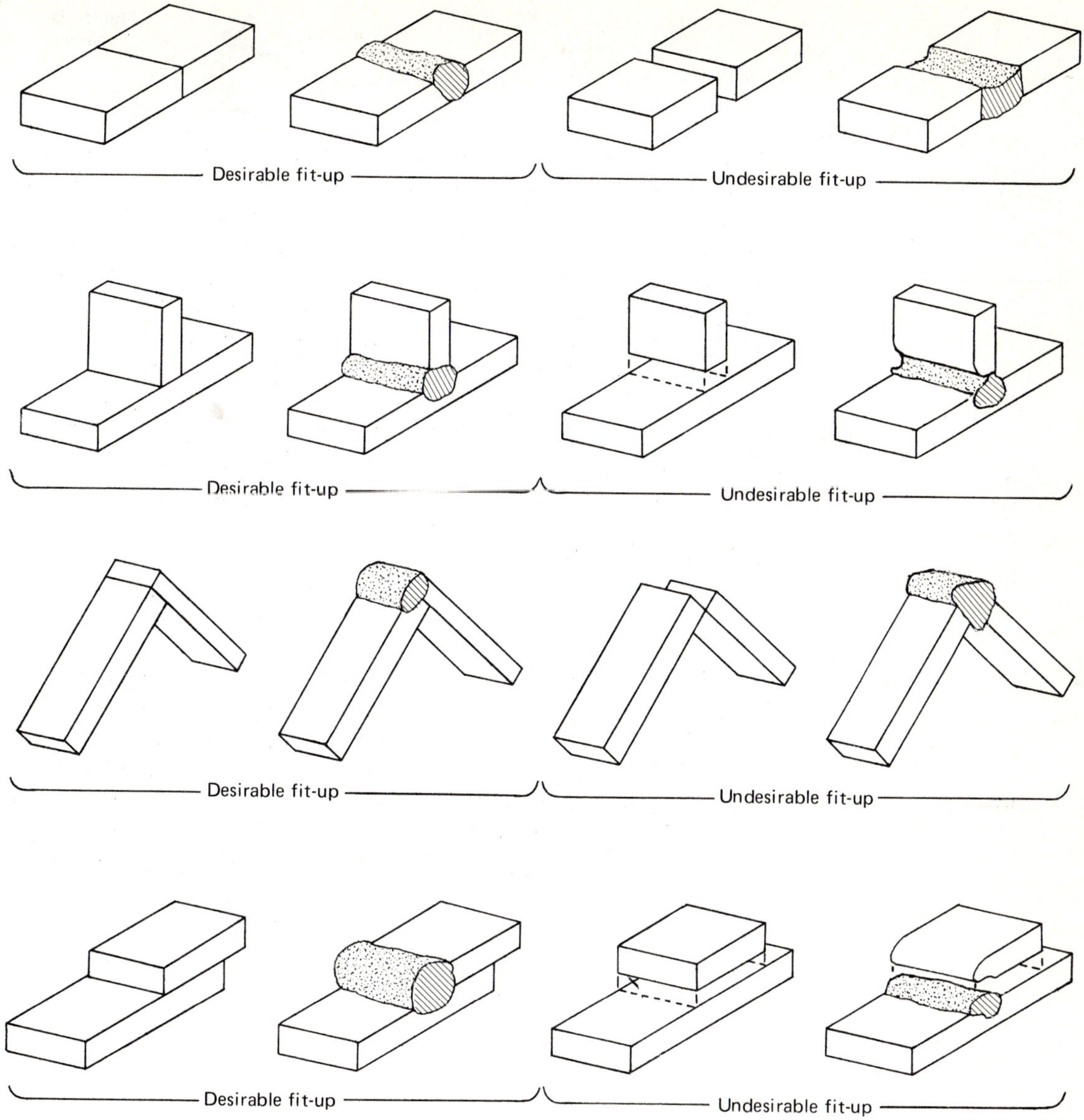
Desirable fit-up
Undesirable fit-up
Desirable fit-up
Undesirable fit-up
Desirable fit-up
Undesirable fit-up
Desirable fit-up
Undesirable fit-up

Figure **2-10** Effect of desirable and undesirable fit-up on weld soundness. Poor fit-up in the joints in the top three rows can result in complete melt-through. (By permission, from *Metals Handbook* Volume 6, Copyright American Society for Metals, 1971.)

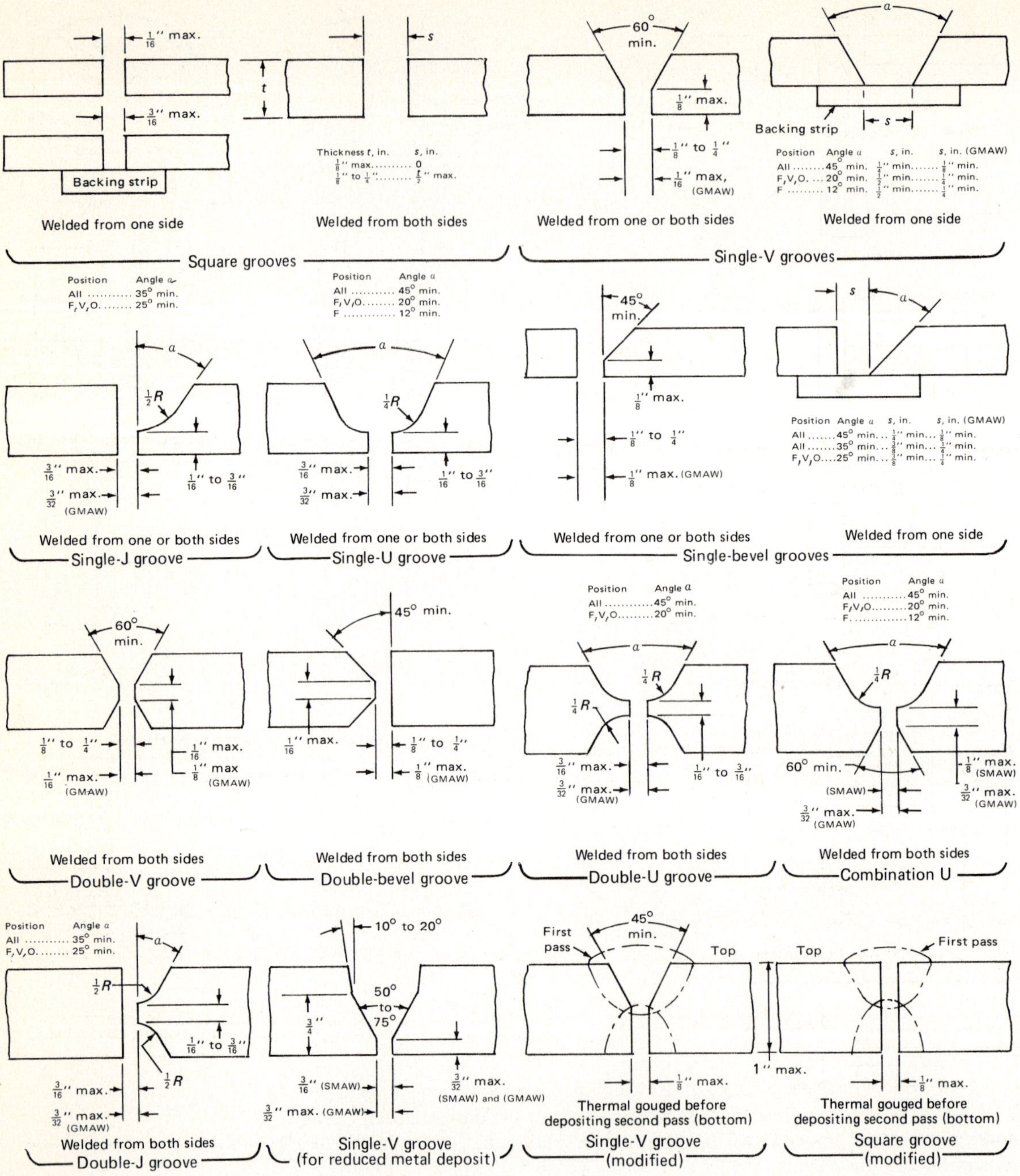

1/16'' max.
3/16'' max.
Backing strip
Welded from one side
s
t
Thickness t, in. s, in.
1/8'' max. 0
1/8'' to 1/4'' .... t/2'' max.
Welded from both sides
Square grooves
60° min.
1/8'' max.
1/8'' to 1/4''
1/16'' max, (GMAW)
Welded from one or both sides
a
Backing strip
Position Angle a s, in. s, in. (GMAW)
All .... 45° min. 1/4'' min. ...... 1/8'' min.
F,V,O. ... 20° min. 1/2'' min. ...... 1/4'' min.
F .... 12° min. 1/2'' min. ...... 1/4'' min.
Welded from one side
Single-V grooves
Position Angle a
All .... 35° min.
F,V,O. .... 25° min.
1/2R
3/16'' max.
3/32'' max. (GMAW)
1/16'' to 3/16''
Welded from one or both sides
Single-J groove
Position Angle a
All .... 45° min.
F,V,O. .... 20° min.
F .... 12° min.
1/4R
3/16'' max.
3/32'' max.
1/16'' to 3/16''
Welded from one or both sides
Single-U groove
45° min.
1/8'' max.
1/8'' to 1/4''
1/8'' max. (GMAW)
Welded from one or both sides
Position Angle a s, in. s, in. (GMAW)
All .... 45° min. .. 1/4'' min. .. 1/8'' min.
All .... 35° min. .. 3/8'' min. .. 1/4'' min.
F,V,O .... 25° min. .. 1/2'' min. .. 1/4'' min.
Welded from one side
Single-bevel grooves
60° min.
1/8'' to 1/4''
1/16'' max. (GMAW)
1/16'' max.
1/8'' max (GMAW)
Welded from both sides
Double-V groove
45° min.
1/16'' max.
1/8'' to 1/4''
1/8'' max. (GMAW)
Welded from both sides
Double-bevel groove
Position Angle a
All .... 45° min.
F,V,O .... 20° min.
1/4R
3/16'' max.
3/32'' max. (GMAW)
1/16'' to 3/16''
Welded from both sides
Double-U groove
Position Angle a
All .... 45° min.
F,V,O .... 20° min.
F. .... 12° min.
1/4R
60° min.
1/8'' max. (SMAW)
3/32'' max. (GMAW)
(SMAW)
3/32'' max. (GMAW)
Welded from both sides
Combination U
Position Angle a
All .... 35° min.
F,V,O. .... 25° min.
1/2R
1/16'' to 3/16''
3/16'' max.
3/32'' max. (GMAW)
Welded from both sides
Double-J groove
10° to 20°
50° to 75°
3/4''
3/16'' (SMAW)
3/32'' max. (GMAW)
3/32'' max. (SMAW) and (GMAW)
Single-V groove (for reduced metal deposit)
45° min.
First pass
Top
1'' max.
1/8'' max.
Thermal gouged before depositing second pass (bottom)
Single-V groove (modified)
Top
First pass
1/8'' max.
Thermal gouged before depositing second pass (bottom)
Square groove (modified)

Figure **2-11** Recommended proportions for arc- and gas-welded grooves: (*a*) butt joints (By permission, from *Metals Handbook* Volume 6, Copyright American Society for Metals, 1971.)

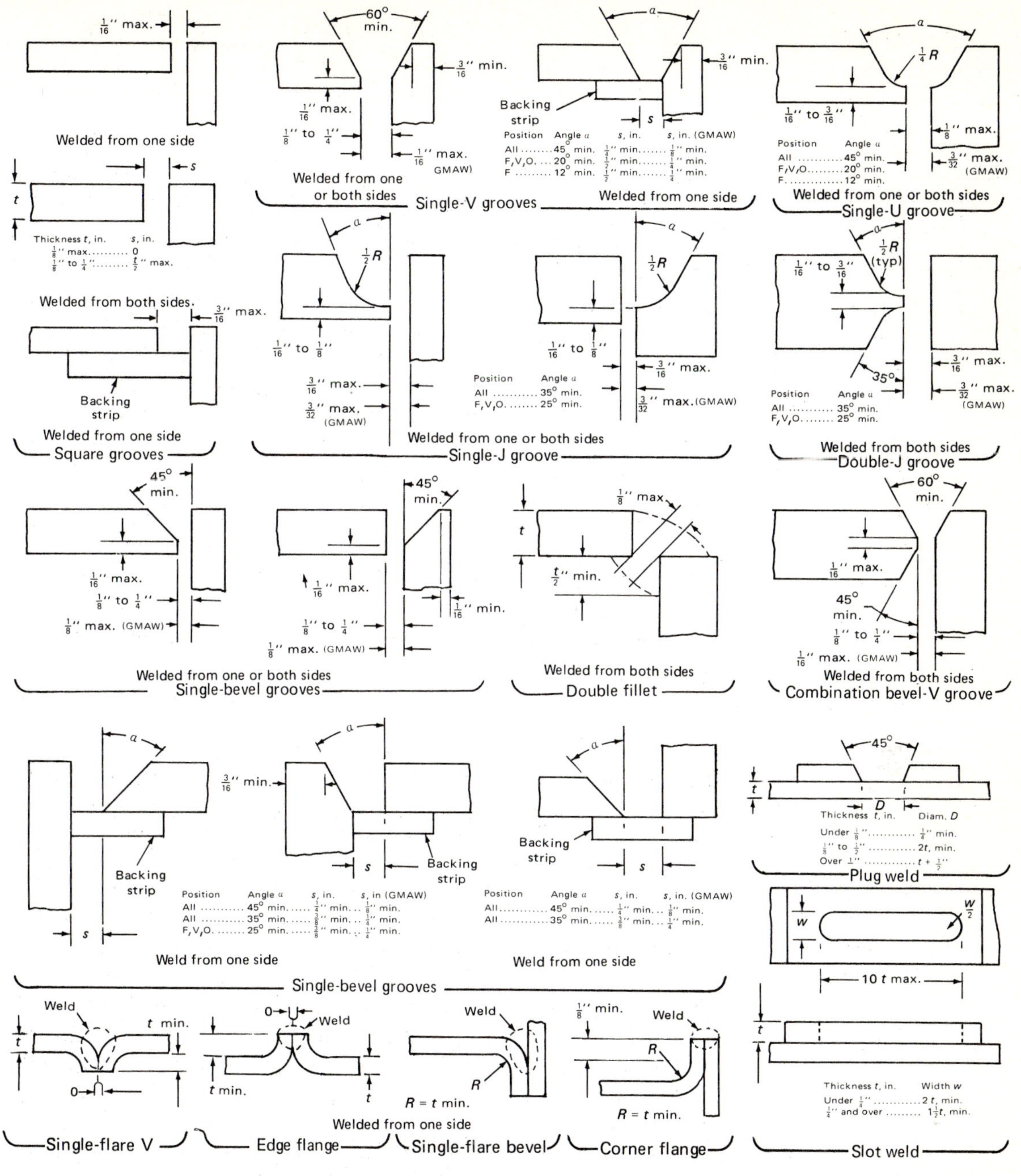
1/16" max.
Welded from one side
s
t
Thickness t, in. s, in.
1/8" max......... 0
1/8" to 1/4"........ t/2" max.
Welded from both sides.
3/16" max.
Backing strip
Welded from one side
Square grooves
60° min.
3/16" min.
1/16" max.
1/8" to 1/4"
1/16" max. (GMAW)
Welded from one or both sides
Single-V grooves
a
3/16" min.
Backing strip
s
Position Angle a s, in. s, in. (GMAW)
All........45° min. 1/4" min....... 1/8" min.
F,V,O.... 20° min. 1/2" min....... 1/4" min.
F .... 12° min. 1/2" min....... 1/4" min.
Welded from one side
a
1/4 R
1/16" to 3/16"
1/8" max.
Position Angle a
All ....45° min.
F,V,O.........20° min.
F..............12° min.
3/32" max. (GMAW)
Welded from one or both sides
Single-U groove
a
1/2 R
1/16" to 1/8"
3/16" max.
3/32" max. (GMAW)
a
1/2 R
1/16" to 1/8"
3/16" max.
Position Angle a
All ....35° min.
F,V,O........25° min.
3/32" max. (GMAW)
Welded from one or both sides
Single-J groove
a
1/2 R (typ)
1/16" to 3/16"
35°
3/16" max.
Position Angle a
All .... 35° min.
F,V,O........ 25° min.
3/32" max. (GMAW)
Welded from both sides
Double-J groove
45° min.
1/16" max.
1/8" to 1/4"
1/8" max. (GMAW)
45° min.
1/16" max.
1/16" min.
1/8" to 1/4"
1/8" max. (GMAW)
Welded from one or both sides
Single-bevel grooves
1/8" max.
t
t/2" min.
Welded from both sides
Double fillet
60° min.
1/16" max.
45° min.
1/8" to 1/4"
1/16" max. (GMAW)
Welded from both sides
Combination bevel-V groove
a
3/16" min.
Backing strip
s
Position Angle a s, in. s, in (GMAW)
All .... 45° min...... 1/4" min... 1/8" min.
All .... 35° min...... 3/8" min... 1/4" min.
F,V,O. .... 25° min...... 3/8" min... 1/4" min.
Weld from one side
a
Backing strip
s
Position Angle a s, in. s, in. (GMAW)
All........... 45° min...... 1/4" min... 1/8" min.
All............35° min...... 3/8" min... 1/4" min.
Weld from one side
Single-bevel grooves
45°
t
D
Thickness t, in. Diam. D
Under 1/8"........... 1/4" min.
1/8" to 1/2" .... 2t, min.
Over 1/2" .... t + 1/2"
Plug weld
w
w/2
10 t max.
t
Thickness t, in. Width w
Under 1/4" .... 2 t, min.
1/4" and over .... 1 1/2 t, min.
Slot weld
Weld
t min.
t
0
Weld
t min.
t
R
R = t min.
Weld
1/8" min.
Weld
R
R = t min.
Welded from one side
Single-flare V
Edge flange
Single-flare bevel
Corner flange

Figure **2-11** (*b*) (*cont.*) corner and flange joints and plug welds

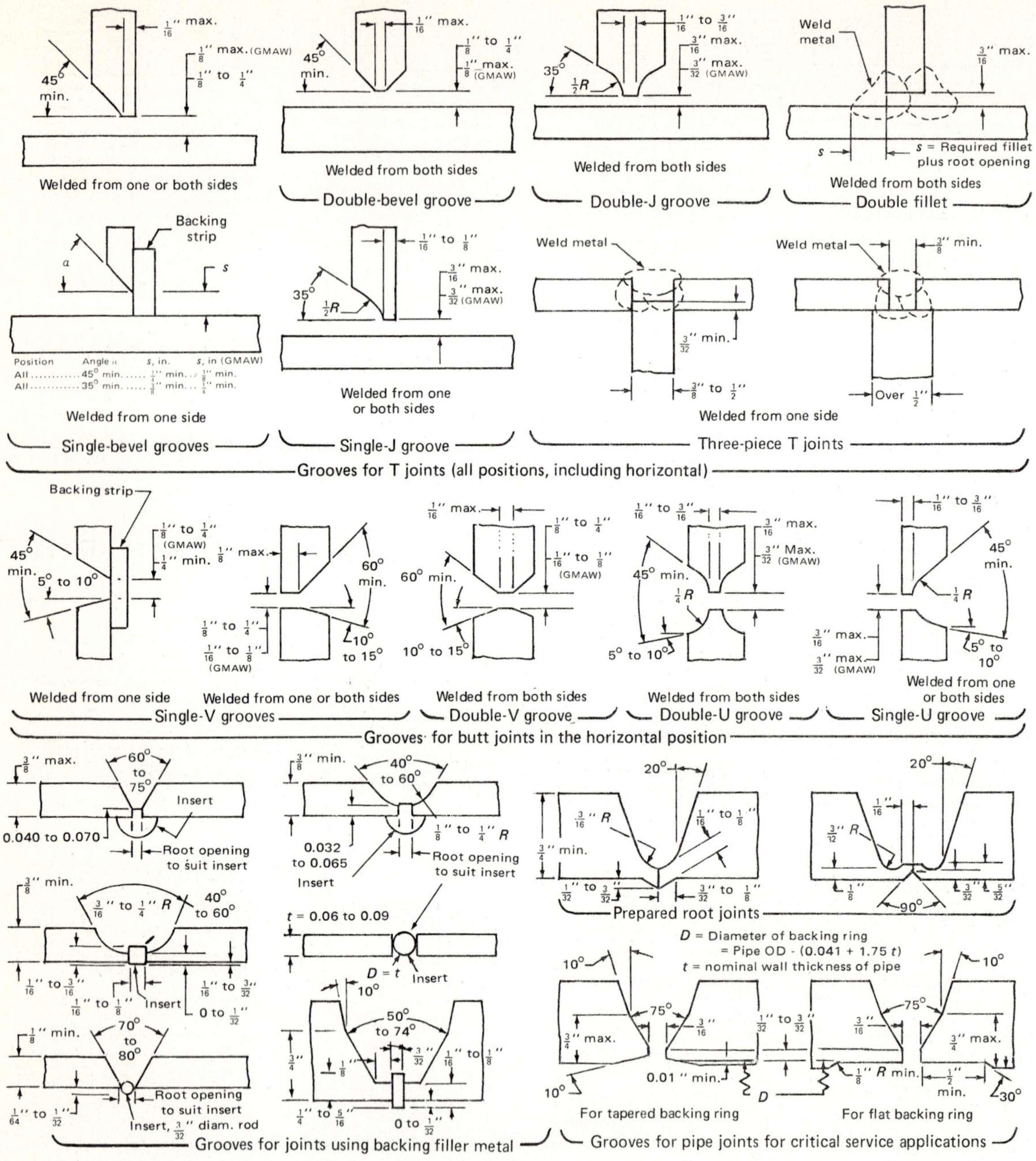
Welded from one or both sides
Welded from both sides
Double-bevel groove
Welded from both sides
Double-J groove
Weld metal
s = Required fillet plus root opening
Welded from both sides
Double fillet
Backing strip
Welded from one side
Single-bevel grooves
Welded from one or both sides
Single-J groove
Weld metal
Welded from one side
Three-piece T joints
Grooves for T joints (all positions, including horizontal)
Backing strip
Welded from one side
Welded from one or both sides
Single-V grooves
Welded from both sides
Double-V groove
Welded from both sides
Double-U groove
Welded from one or both sides
Single-U groove
Grooves for butt joints in the horizontal position
Insert
Root opening to suit insert
Root opening to suit insert
Insert
Prepared root joints
D = Diameter of backing ring = Pipe OD - (0.041 + 1.75 t)
t = nominal wall thickness of pipe
Root opening to suit insert
Insert, 3/32" diam. rod
Grooves for joints using backing filler metal
For tapered backing ring
For flat backing ring
Grooves for pipe joints for critical service applications

Figure **2-11** (*c*) (*cont.*) T joints and horizontal butt joints

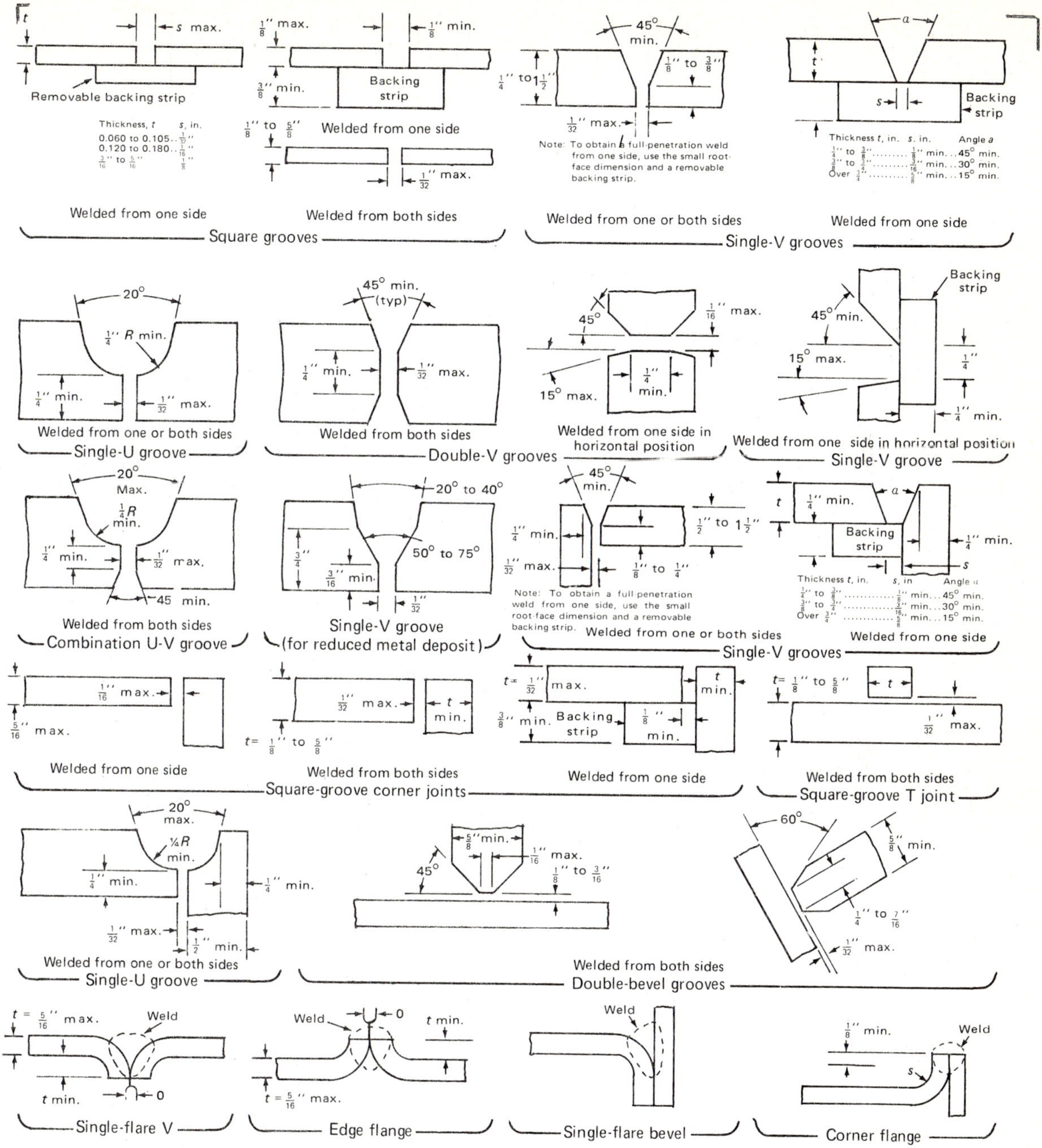

Removable backing strip
Welded from one side
Welded from both sides
Square grooves
Backing strip
Welded from one side
Note: To obtain a full-penetration weld from one side, use the small root-face dimension and a removable backing strip.
Welded from one or both sides
Welded from one side
Single-V grooves
Welded from one or both sides
Single-U groove
Welded from both sides
Welded from one side in horizontal position
Double-V grooves
Welded from one side in horizontal position
Single-V groove
Welded from both sides
Combination U-V groove
Single-V groove (for reduced metal deposit)
Welded from one or both sides
Welded from one side
Single-V grooves
Welded from one side
Welded from both sides
Welded from one side
Square-groove corner joints
Welded from both sides
Square-groove T joint
Welded from one or both sides
Single-U groove
Welded from both sides
Double-bevel grooves
Single-flare V
Edge flange
Single-flare bevel
Corner flange

Figure **2-11** (*d*) (*cont.*) submerged arc–welded grooves

welding). The proportions and limitations shown are those generally found to be necessary, with suitable welding procedures, to obtain complete joint penetration and acceptable reinforcement and shape of the weld metal area.

The grooves themselves, depending on final use (normal maintenance and repair; high-pressure, leak-tight use; and so on) may be prepared by any one of the following methods: flame cutting, grinding, sawing, milling, or shearing.

The selection of root openings and groove angles is also greatly influenced by the materials to be joined, location of the joint in the weldment, and the performance required. J- and U-groove joints may be used to minimize the amount of weld metal required when the savings are sufficient to justify the more difficult and costly chamfering operations. These joints are particularly useful in the welding of heavy-thickness pieces. One disadvantage of J- and bevel-groove joints is that they are difficult to weld soundly, due to the common problem of slag entrapment along their one straight side.

The most important criterion for the strength in a groove weld is the degree of joint penetration. Since welded joints are usually designed so that they will be equal in strength to the base metal, groove-welded joint designs that result in welds extending completely through the members being joined are most commonly used. One of the principles of design is the selection of joint designs that will result in the desired degree of joint penetration.

REVIEW QUESTIONS

1. State the reasons for designing, instead of randomly selecting, the shape of a welded joint.

2. What is meant by the expression "efficiency of a welded joint"? In what terms and relative to what factors is it expressed?

3. When and why are fillet welds used in preference to groove welds?

4. Referring to Figure 2-11, proportion the following joint grooves for welding from one side only in the horizontal position:

a. *Butt joints.* single-V, double-V, double-bevel

b. *Flange joints.* single-V, double-V, single-bevel

c. *T joints.* single-bevel, double-fillet, three-piece

and for the following material thicknesses: ⅛ inch, ¼ inch, ⅜ inch, ½ inch, ⅝ inch.

5. Name the methods by which welding grooves may be prepared.

6. Which of the methods named in question 5 would be used to prepare a joint for normal maintenance? Which method would be used to prepare a high-pressure leak-tight joint? Why should the same methods not be used for both applications?

7. Which of the four welding positions is the easiest for welding? Why?

8. What equipment is commonly used to keep the joint in the flat and the horizontal positions?

9. Referring to Figure 2-8, calculate the minimum length of the electrode stick-out and electrode lead angle needed for making a fillet weld inside a box having the following inside dimensions: 9 × 4 × 6 inches deep. *Note:* As part of your answer, indicate whether your answer applies to the 4-inch or the 9-inch dimension.

10. Name the four welding positions.

11. What is meant by *out-of-position* welding?

12. What are the angular limits for flat-position welding? horizontal-position welding?

13. Why are fillet welds often combined with other welds?

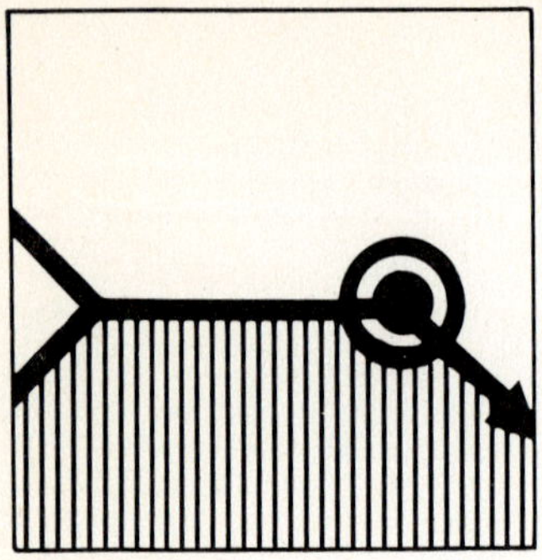

CHAPTER 3

WELDING SYMBOLS AND NONDESTRUCTIVE WELD-TESTING SYMBOLS

To facilitate presentation, this chapter has been divided into two sections. The first section deals with welding symbols—the concise language that designers and engineers use to convey the precise methodology necessary for a welder to produce a weld appropriate to the workpiece. The second section deals with nondestructive testing symbols, which illustrate how and where completed welds are to be tested, undamaged, to ensure their appropriateness.

WELDING SYMBOLS

In order for welding to have attained the position it has in construction and manufacturing, it has had to prove adequate to design needs. Once the processes proved adequate for design purposes, it was necessary to provide a means for designers to communicate to welders exactly what kind of, and to what extent, welding was necessary in each case. To neglect this could be dangerous as well as costly. For example, to write, "To be welded throughout," or, "To be welded completely," on a drawing may generally indicate the extent of welding, but not the necessary strength. If strength were essential in a design, a dangerous situation could result if the wrong technique were used. If strength were not the essential factor, a less extensive weld might work just as well and be much less expensive.

Certain shops, in their desire for safety, use much more welding than is necessary.

To combat problems of this sort, a welding language—that is, a set of welding symbols—has been developed which briefly gives the welder or supervisor all the information necessary for a proper weld. In practice, many companies need only a few of the symbols, but as long as the symbols are from the universal set, everyone will be speaking the same language.

For example, in the past the use of the terms *far side* and *near side* has led to confusion, because in drawings where joints are shown in cross-section, all sides are equidistant from the reader. In the AWS system, the *joint* is the basis of reference. Any joint whose welding is indicated by a symbol will always have an *arrow side* and an *other side*. Accordingly, the terms *arrow side, other side,* and *both sides* are used to locate the weld with respect to the joint.

The tail of the symbol is used for designating the welding specifications, procedures, or supplementary information to be used. If a welder knows the size and type of a weld, he or she has only part of the information necessary for completing that weld. The process to be used, identification of filler metal, whether or not peening or root chipping is required, and other pertinent data must also be known. The notation indicating these data, to be placed in the tail of the symbol, will usually be established by each user. If notations are not used, the tail of the symbol may be omitted.

3-1 ELEMENTS OF A WELDING SYMBOL

The AWS makes a distinction between the terms *weld symbol* and *welding symbol*. The *weld symbol* is the ideograph used to indicate the desired type of weld. The assembled *welding symbol* consists of the following elements, or such of these elements as are necessary: reference line with arrow, basic weld symbols, dimensions, and other data; supplementary symbols; finish symbols; and tail, which contains specifications, process, or other references. The information conveyed by the welding symbol [Figure 3-1(*a*)] is thus easily and accurately read and long descriptive notes are not necessary.

The *reference line* of a welding symbol is the line [Figure 3-1(*b*)] depicted on a horizontal plane and joined to a tail and an arrow. The reference line is the basis of each simplified symbol and provides for the orientation and standard location of the elements of a welding symbol. The positions of the tail and the arrow may be interchanged, but the elements of the symbol are always in the same position on the reference line.

To show the location of a weld, an *arrow* is drawn with the head pointing directly to the joint where the weld is to be made. The

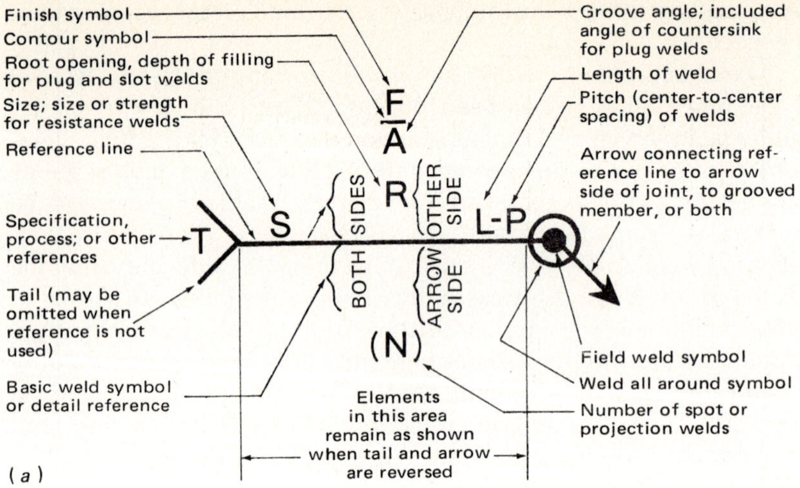

Finish symbol
Contour symbol
Root opening, depth of filling for plug and slot welds
Size; size or strength for resistance welds
Reference line
Specification, process; or other references
Tail (may be omitted when reference is not used)
Basic weld symbol or detail reference
F
A
R
S
T
L-P
(N)
BOTH SIDES
OTHER SIDE
ARROW SIDE
Groove angle; included angle of countersink for plug welds
Length of weld
Pitch (center-to-center spacing) of welds
Arrow connecting reference line to arrow side of joint, to grooved member, or both
Field weld symbol
Weld all around symbol
Number of spot or projection welds
Elements in this area remain as shown when tail and arrow are reversed
(a)

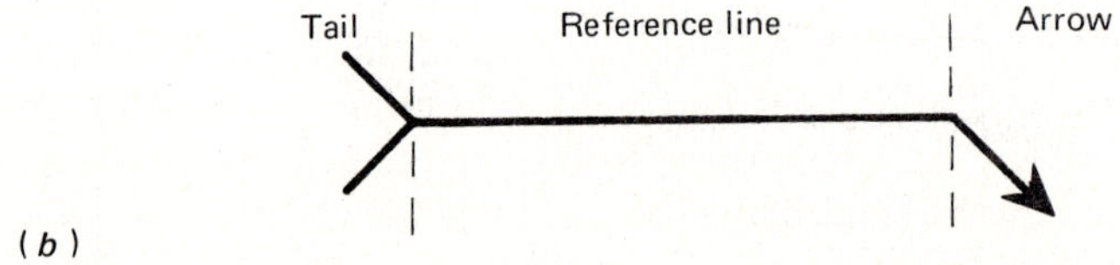

Tail
Reference line
Arrow
(b)

Figure **3-1** (*a*) Complete welding symbol. (*b*) The tail, reference line, and arrow: the basic welding symbol. (Redrawn by premission from American Welding Society, *Welding Handbook,* 6th ed., 1968, section 1, p. 1.8.)

placement of the weld symbol (Figure 3-2) can be used to indicate the arrow side, other side, or both sides of the joint. The weld symbols will be illustrated later in this chapter.

The *dimensions* shown in a welding symbol indicate the size, groove angle, root opening, length of weld, pitch (center-to-center spacing) of welds; the depth of filling of plug or slot welds; and the included angle of countersink for plug welds (Figure 3-3). One or more of these may be specified, depending on the type of joint and instruction required. When welds on both sides of a joint have the same dimensions, one or both may be dimensioned on the welding symbol [Figure 3-3(*b*)].

The size of a fillet weld is determined by the length of its largest leg [see Figure 3-3(*d*)]. This dimension is shown to the left of the weld symbol on the same side of the reference line. When fillet

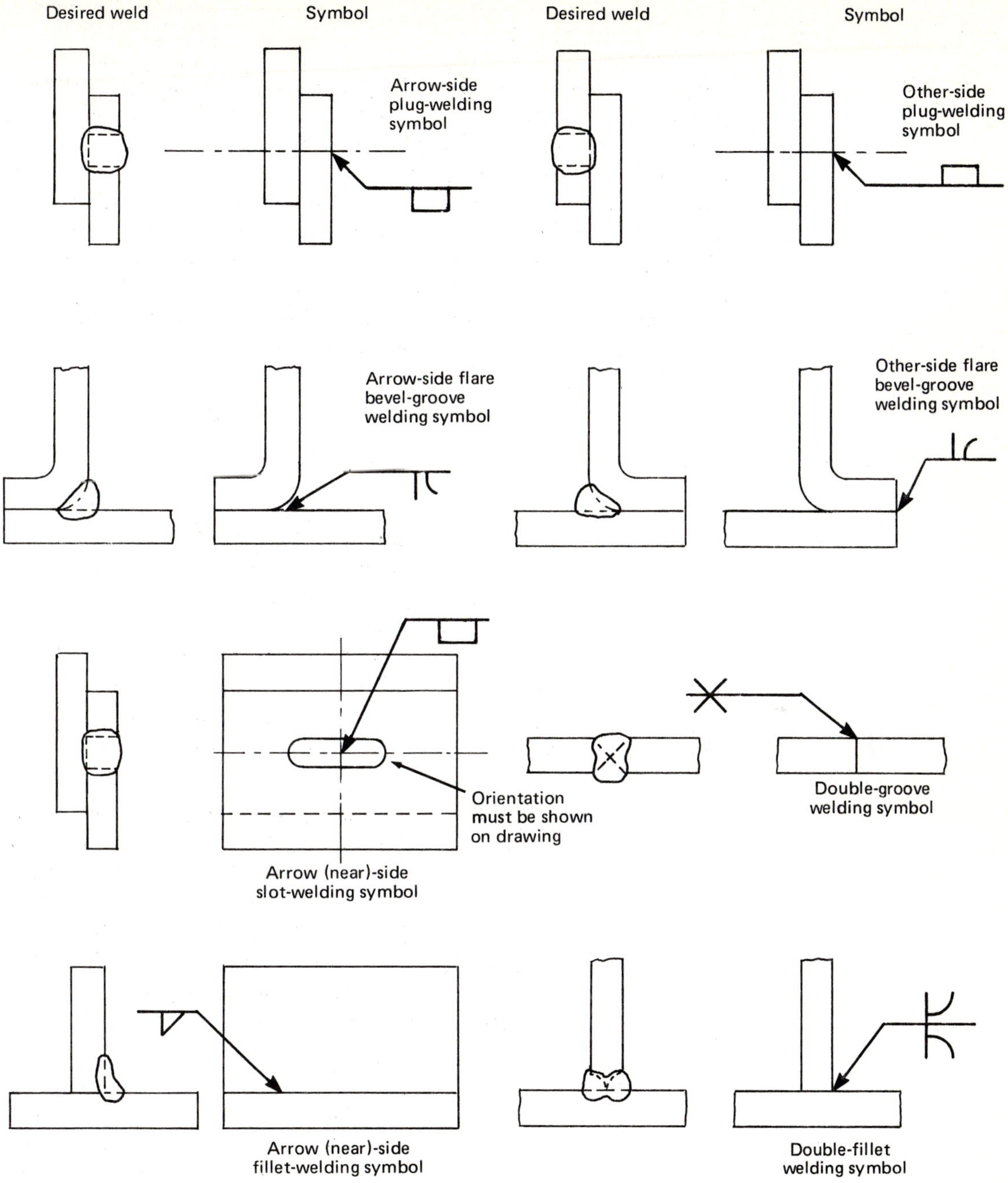
Desired weld
Symbol
Desired weld
Symbol
Arrow-side plug-welding symbol
Other-side plug-welding symbol
Arrow-side flare bevel-groove welding symbol
Other-side flare bevel-groove welding symbol
Orientation must be shown on drawing
Arrow (near)-side slot-welding symbol
Double-groove welding symbol
Arrow (near)-side fillet-welding symbol
Double-fillet welding symbol

Figure **3-2** Location and significance of the arrow in welding symbols

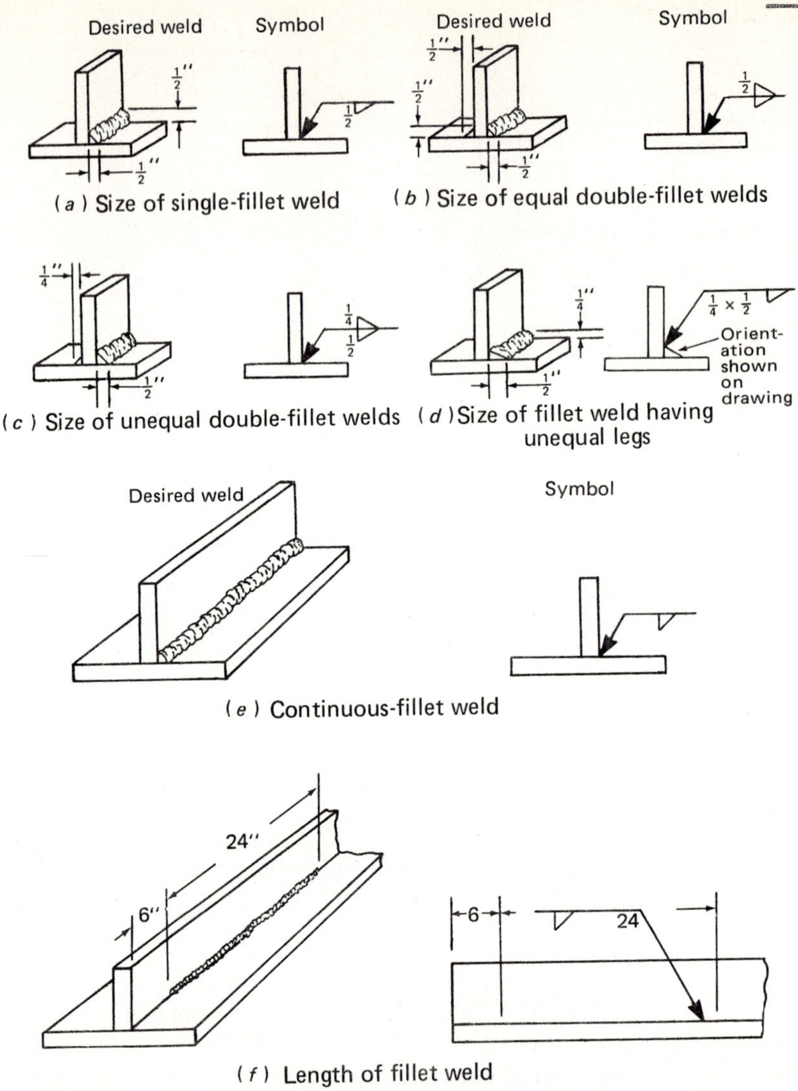
Desired weld
Symbol
Desired weld
Symbol
(a) Size of single-fillet weld
(b) Size of equal double-fillet welds
(c) Size of unequal double-fillet welds
1/4 × 1/2
Orientation shown on drawing
(d) Size of fillet weld having unequal legs
Desired weld
Symbol
(e) Continuous-fillet weld
24"
6"
6
24
(f) Length of fillet weld

Figure **3-3** Some welding symbols and the welds represented by them. (Redrawn by permission from American Welding Society, *Welding Handbook*, 6th ed., 1968, section 1, p. 1.50.)

welds differ in size, both are dimensioned [see Figure 3-3(c)]. Further illustrations of fillet-welding symbols can be found in the rest of Figure 3-3. The size of a groove weld is the joint penetration (the depth of chamfering plus the root penetration, when specified).

In Figure 3-4, root penetration is *not* specified. However, in Figure 3-5, root penetration *is* specified exactly by the ⅛ added to the

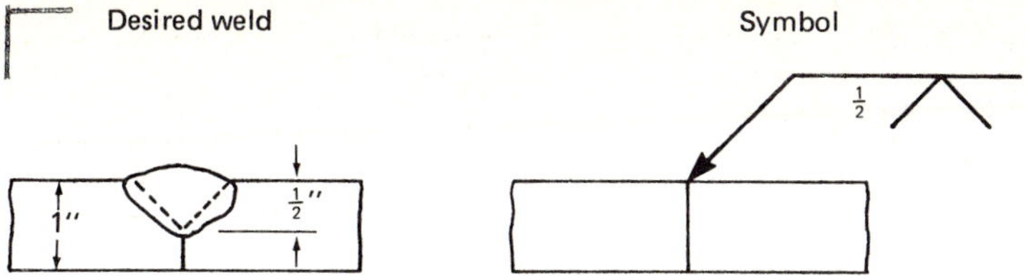
Desired weld
Symbol
1"
1/2"
1/2

Figure **3-4** Welding symbol *not* specifying penetration

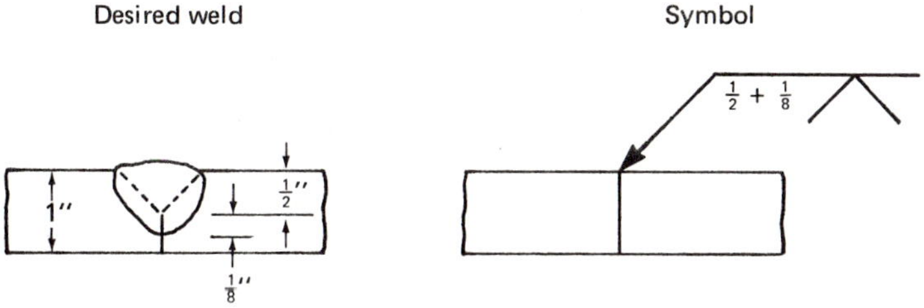
Desired weld
Symbol
1"
1/2"
1/8"
1/2 + 1/8

Figure **3-5** Welding symbol with root penetration specified

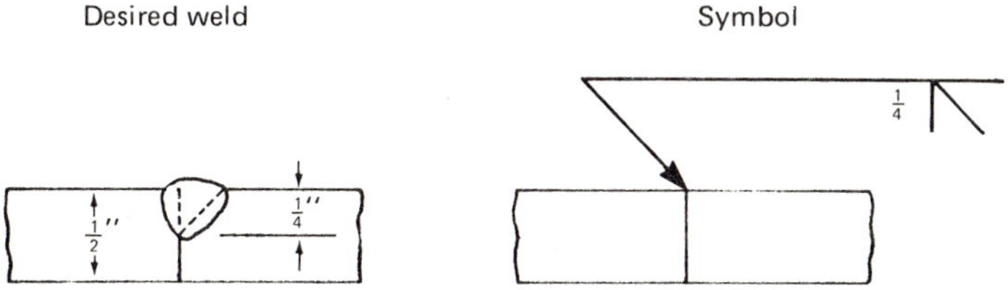
Desired weld
Symbol
1/2"
1/4"
1/4

Figure **3-6** Welding symbol specifying depth of groove weld

groove size. The size of a groove is shown to the left of the weld symbol. The arrow points to the plate to be beveled when a single bevel is specified. Groove dimension *is* shown when (1) the groove weld extends only partly through the parts being joined, as in Figure 3-6, or (2) the root penetration is specified in addition to the depth of chamfer, as in Figures 3-5 and 3-7.

The groove dimension is *not* shown when (1) the single-groove weld extends completely through the parts being joined. If 100% penetration is desired, either the melt-through or back weld symbol is added, as in Figure 3-8; or (2) both sides of a double-groove weld are the same and the weld extends completely through the parts being joined, as in Figure 3-9.

The size of a flare-groove weld is considered to extend only to the tangent points of the members (see Figure 3-10). For flange welds,

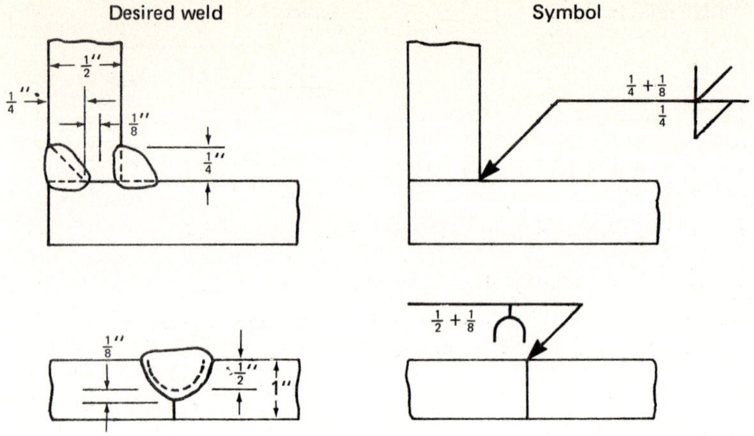

Desired weld
Symbol
1"

Figure **3-7** Welding symbols specifying depth of grooves and depth of penetration

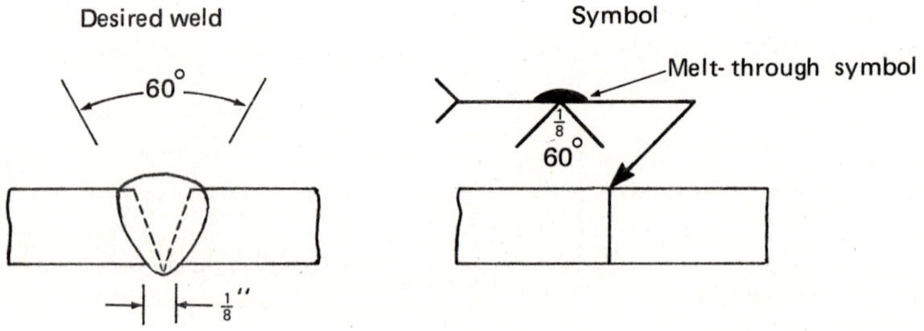

Desired weld
Symbol
60°
Melt-through symbol
60°

Figure **3-8** Welding symbol specifying 100% penetration

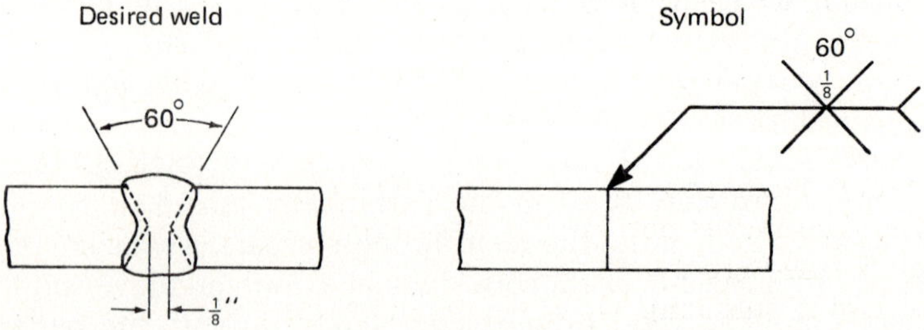

Desired weld
Symbol
60°
60°

Figure **3-9** Welding symbol specifying a double-groove weld with both grooves having the same dimensions

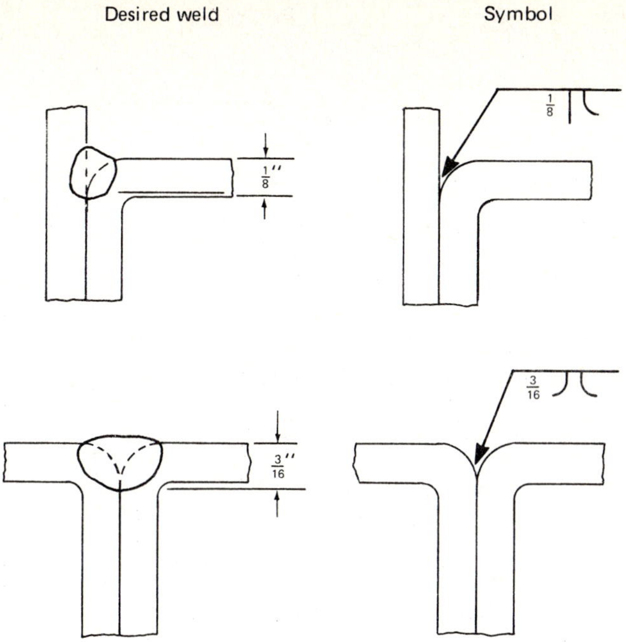

Desired weld
Symbol
1/8"
1/8
3/16"
3/16

Figure **3-10** Welding symbols specifying the size of flare-groove welds

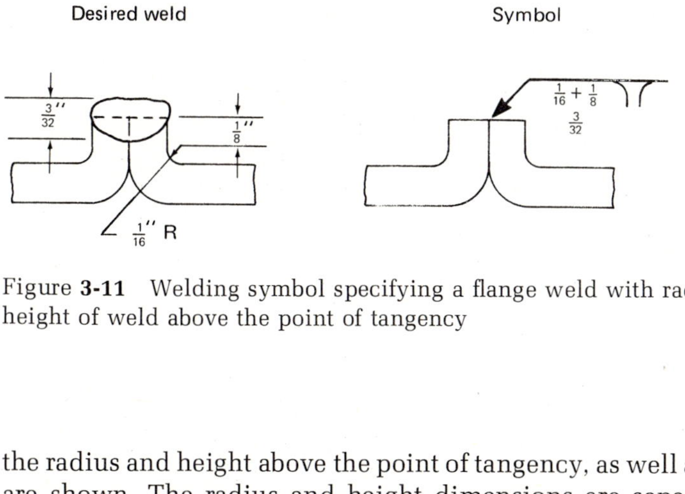

Desired weld
Symbol
3/32"
1/8"
1/16" R
1/16 + 1/8
3/32

Figure **3-11** Welding symbol specifying a flange weld with radius and height of weld above the point of tangency

the radius and height above the point of tangency, as well as the size, are shown. The radius and height dimensions are separated by a "plus" mark (see Figure 3-11).

The size of a surfacing weld is shown to the left of the weld symbol and indicates the minimum height of the build-up. (The length and width of surface to be covered are shown by specific dimensions on the drawing (see Figure 3-12).

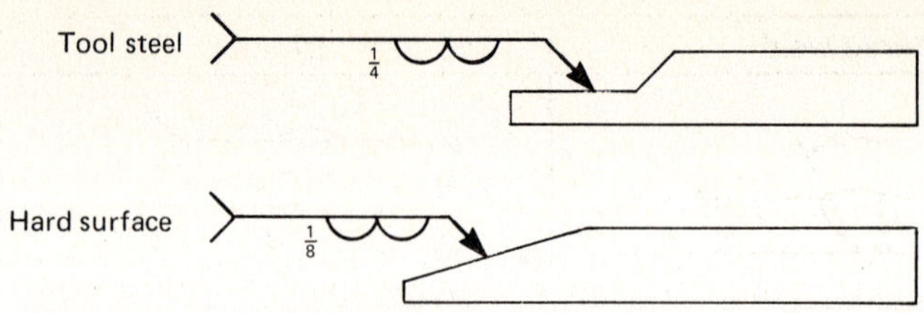
Tool steel
1/4
Hard surface
1/8

Figure **3-12** Welding symbols specifying maximum build-up height of a surface to be built up

Groove angles and root openings, if in accordance with established shop standards, are not shown on the welding symbol. If not in accordance with shop standards, they are shown as illustrated in Figure 3-13 (Figures 3-13–3-19 are on pages 35–39).

The length and pitch of welds are shown in symbols as illustrated in Figure 3-14. When the length of the weld is not given, the symbols apply between abrupt changes in the direction of welding, except when the "weld-all-around" symbol is used.

The weld-all-around symbol is a supplementary weld symbol used to show a weld that extends completely around a joint. Its use is illustrated in Figure 3-15(*a*). The field weld symbol is also a supplementary symbol, and is used to show welds not made in the shop or place of initial construction, as in Figure 3-15(*b*).

Finish symbols are added to indicate what the weld surface should look like at the end of the process. Welds that are to be welded approximately flat-faced without recourse to any method of finishing are shown by adding the flush-contour symbol to the weld symbol, observing the usual location significance (see Figure 3-16). Where normal build-up from standard welding procedures is acceptable, the contour symbol is omitted.

Welds that are to be made flat-faced by mechanical means are shown by adding to the weld symbol both the flush-contour symbol and the user's standard finish symbol, observing the usual location significance (see Figure 3-17).

Welds that are to be mechanically finished to a convex contour are shown by adding to the weld symbol both the convex-contour symbol and the user's standard finish symbol, observing the usual location significance (see Figure 3-18).

When a specification, process, or other reference is used with a welding symbol, the reference is placed in the tail, as shown in Figure 3-19. When the use of a definite process is required, it may be indicated in the tail by one or more of the letter designations listed in Tables 3-1 and 3-2, on pages 40–41.

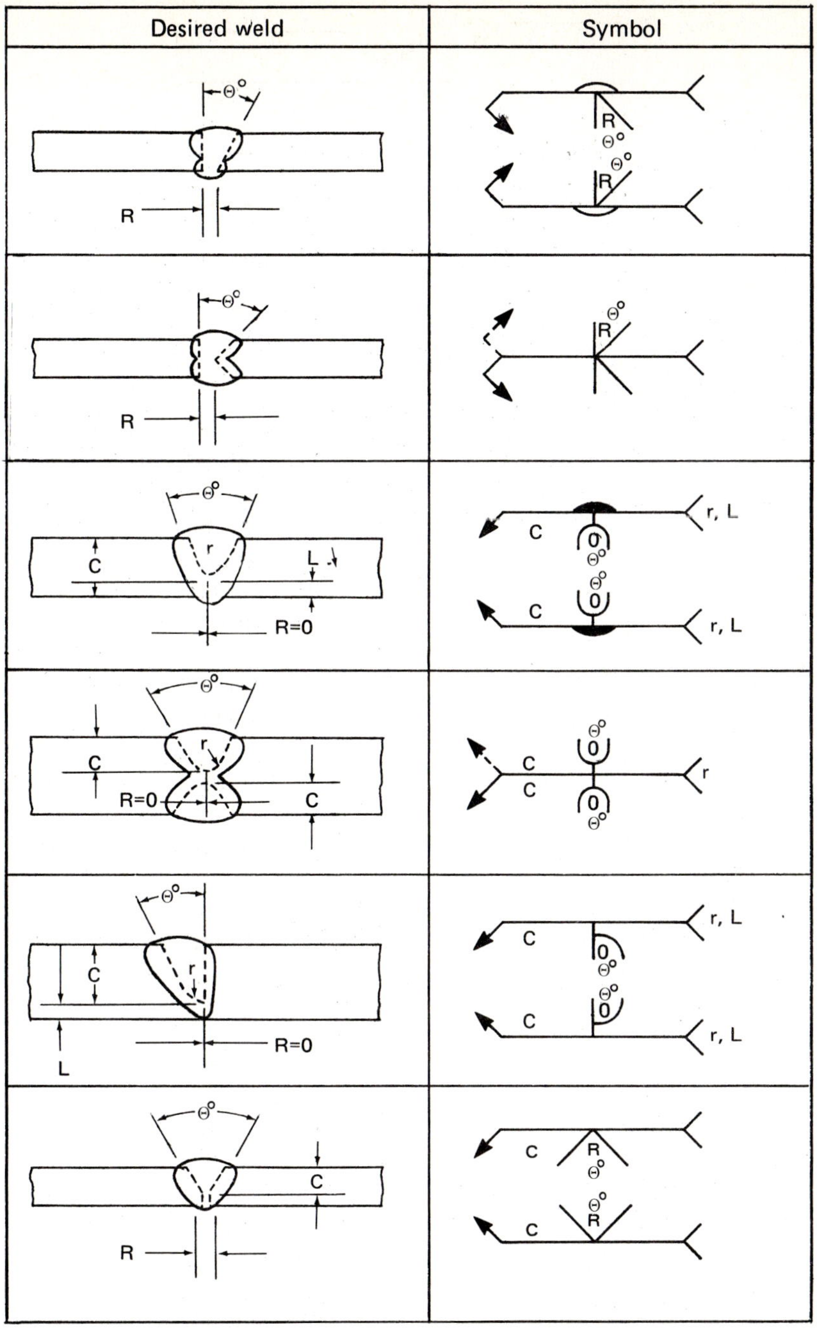
Desired weld
Symbol
θ°
R
C
L
r
R=0
r, L
0

Figure **3-13** How to specify groove angles and root openings that are not standard in the shop where the weld is made. In groove joints, the root face (land dimension (L) is sometimes specified instead of chamfer (C). Root, opening, and angle may be omitted if they are user's standard.

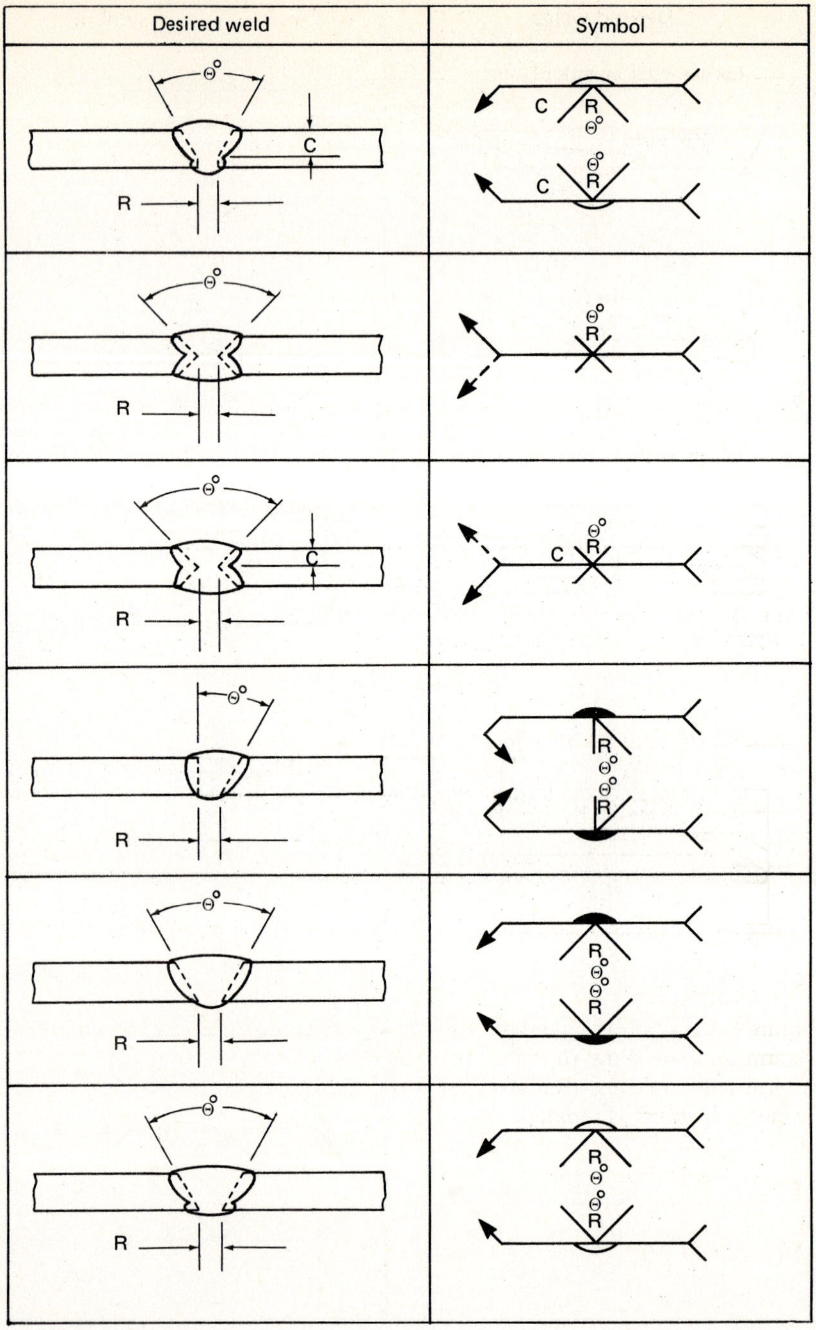
Desired weld
Symbol

Figure **3-13** *(cont.)*

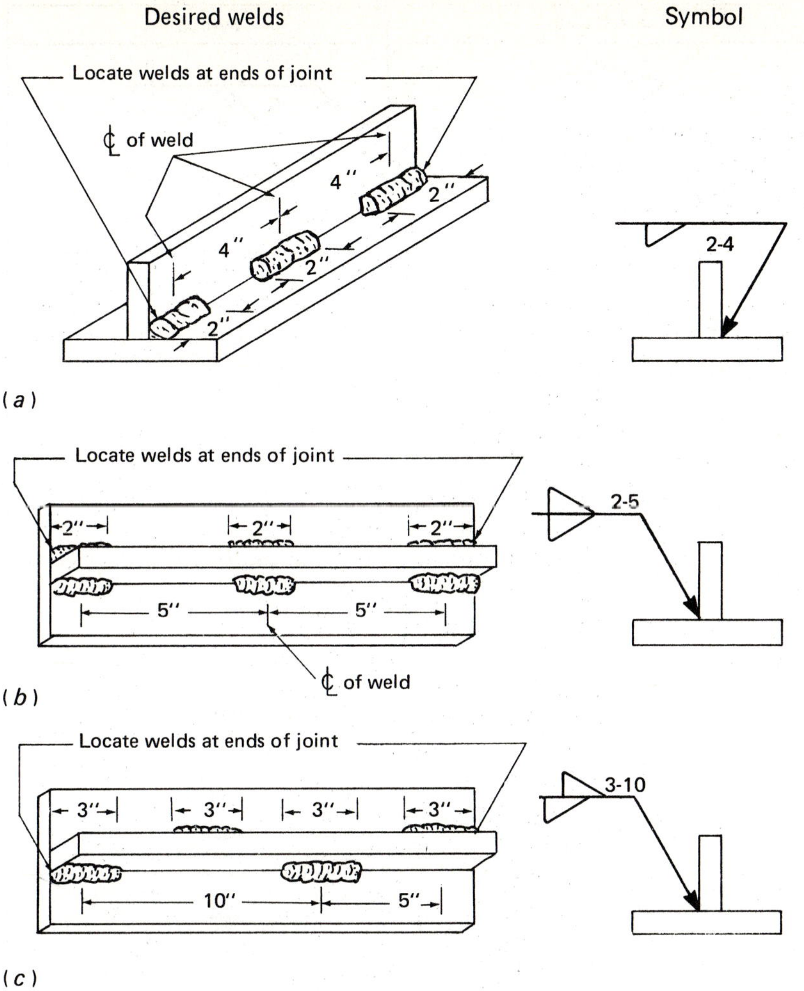

Desired welds
Symbol
Locate welds at ends of joint
℄ of weld
4″
4″
2″
2″
2″
2-4
(a)
Locate welds at ends of joint
2″
2″
2″
5″
5″
℄ of weld
2-5
(b)
Locate welds at ends of joint
3″
3″
3″
3″
10″
5″
3-10
(c)

Figure **3-14** Welding symbols specifying length and pitch of welds: (*a*) intermittent welding; (*b*) chain intermittent welding; (*c*) staggered intermittent welding. (Redrawn by permission from American Welding Society, *Welding Handbook*, 6th ed., 1968, section 1, p. 1.50.)

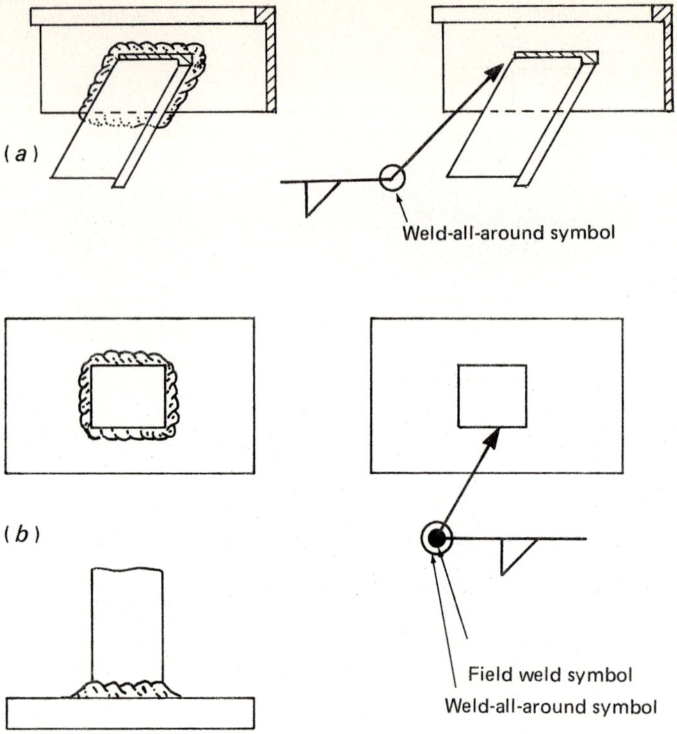
(a)
Weld-all-around symbol
(b)
Field weld symbol
Weld-all-around symbol

Figure **3-15** The weld-all-around symbol and the field weld symbol

Figure **3-16** Finish symbols for flat-faced welds with no recourse to machining. (Courtesy of the American Welding Society.)

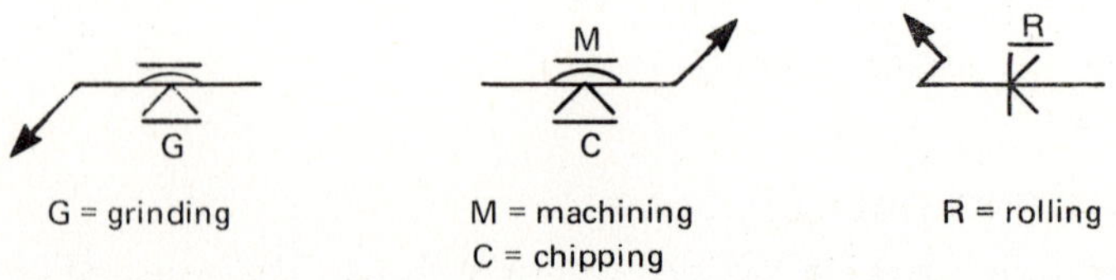
G
M
C
R
G = grinding
M = machining
C = chipping
R = rolling

Figure **3-17** Finish symbols for mechanically finished, flat-faced welds. (Courtesy of the American Welding Society.)

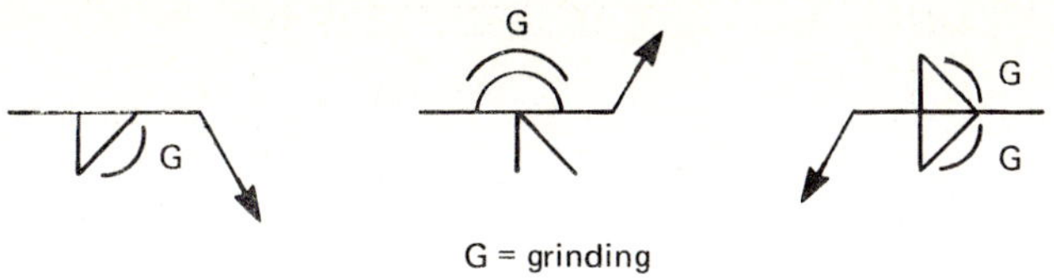
G
G
G
G
G
G = grinding

Figure **3-18** Finish symbols for mechanically finished, convex welds. (Courtesy of the American Welding Society.)

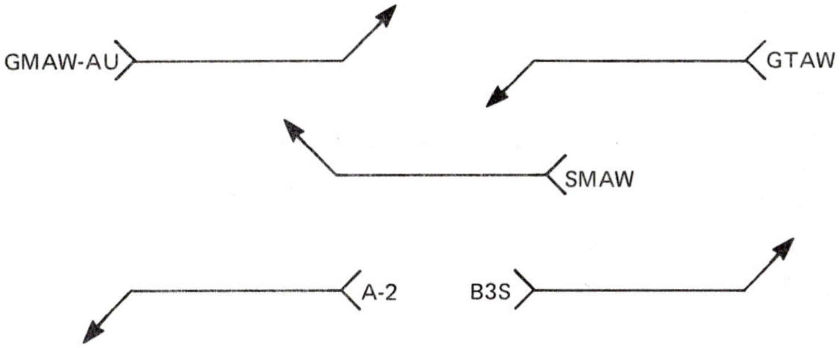
GMAW-AU
GTAW
SMAW
A-2
B3S

Figure **3-19** Placement of specification, process, or other references in welding symbol tail. (Courtesy of the American Welding Society.)

3-2 EXTENT OF WELDING DENOTED BY SYMBOLS

Except when the weld-all-around symbol is used, the extent of welding is usually designated by the applicable weld symbol at those locations of abrupt changes in the direction of welding, as shown in Figures 3-20 to 3-23. Welding symbols denoting welding on the hidden joints may be indicated as shown in Figure 3-24. (Figures 3-20–3-31 are on pages 41–47.)

3-3 BASIC WELDING SYMBOLS

The basic welding symbols are shown in Figure 3-25. How the symbols are used on drawings and the welds they indicate are shown in Figure 3-26.

Table **3-1** Designation of welding processes by letters*

	WELDING PROCESS	LETTER DESIGNATION
Brazing	Torch brazing	TB
	Twin-carbon–arc brazing	TCAB
	Furnace brazing	FB
	Induction brazing	IB
	Resistance brazing	RB
	Dip brazing	DB
	Block brazing	BB
	Flow brazing	FLB
Flow welding	Flow welding	FLOW
Resistance welding	Flash welding	FW
	Upset welding	UW
	Percussion welding	PEW
Induction welding	Induction welding	IW
Arc welding	Bare metal–arc welding	BMAW
	Stud welding	SW
	Gas-shielded stud welding	GSSW
	Submerged arc welding	SAW
	Gas tungsten–arc welding	GTAW
	Gas metal–arc welding	GMAW
	Atomic hydrogen welding	AHW
	Shielded metal–arc welding	SMAW
	Twin-carbon–arc welding	TCAW
	Carbon-arc welding	CAW
	Gas carbon–arc welding	GCAW
	Shielded carbon–arc welding	SCAW
Thermit welding	Nonpressure thermit welding	NTW
	Pressure thermit welding	PTW
Gas welding	Pressure gas welding	PGW
	Oxyhydrogen welding	OHW
	Oxyacetylene welding	OAW
	Air-acetylene welding	AAW
Forge welding	Roll welding	RW
	Die welding	DW
	Hammer welding	HW

Note: Letter designations have not been assigned to arc-spot, resistance-spot, arc-seam and resistance-seam welding or to projection welding since the weld symbols used are adequate.

*The following suffixes may be used if desired to indicate the method of applying the above processes:

Automatic welding	—AU
Machine welding	—ME
Manual welding	—MA
Semiautomatic welding	—SA

Source: Amercian Welding Society, *Welding Handbook,* 6th ed., section 1, p. 1.16.

Table **3-2** Designation of cutting processes by letters*

CUTTING PROCESS	LETTER DESIGNATION	CUTTING PROCESS	LETTER DESIGNATION
Arc cutting	AC	Oxygen cutting	OC
Air-carbon–arc cutting	AAC	Chemical flux cutting	FOC
Carbon-arc cutting	CAC	Metal powder cutting	POC
Metal-arc cutting	MAC	Oxygen-arc cutting	AOC

*The following suffixes may be used if desired to indicate the methods of applying the above processes:

Automatic cutting	—AU
Machine cutting	—ME
Manual cutting	—MA
Semiautomatic cutting	—SA

Source: American Welding Society, *Welding Handbook,* 6th ed., section 1, p. 1.11.

NONDESTRUCTIVE TESTING SYMBOLS

The assembled testing symbol (Figure 3-27) consists of the reference line with arrow, basic testing symbols, test-all-around symbol, number of tests (N), and tail, with extent of test, specification, process, or other references. The arrow connects the reference line to the part to be tested. The side of the part to be tested, to which the arrow points, is considered the *arrow side.* The other side is referred to simply as the *other side.*

The location of the abbreviation for the nondestructive test(s) to which the part is to be submitted is as shown in Figure 3-28. A list of nondestructive tests, together with the standard abbreviations used to represent the test method on the reference line, is given in Table 3-3, on page 46.

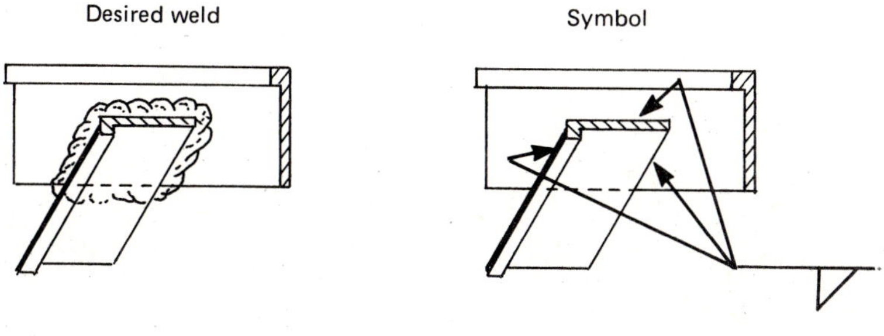
Desired weld
Symbol

Figure **3-20** Extent-of-welding symbol. (Redrawn by permission from American Welding Society, *Welding Handbook,* 6th ed., 1968, section 1, p. 1.51.)

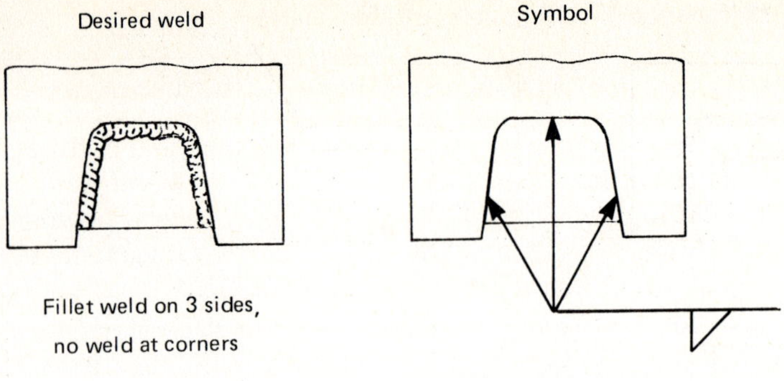
Desired weld
Symbol
Fillet weld on 3 sides,
no weld at corners

Figure **3-21** Extent-of-welding symbol. (Redrawn by permission from American Welding Society, *Welding Handbook*, 6th ed., 1968, section 1, p. 1.51.)

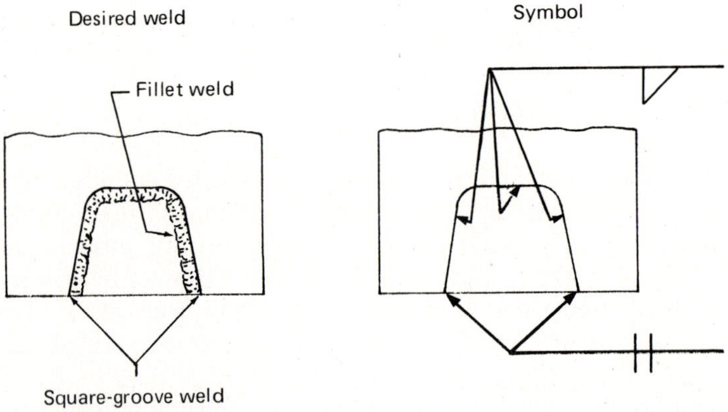
Desired weld
Symbol
Fillet weld
Square-groove weld

Figure **3-22** Extent-of-welding symbol. (Redrawn by permission from American Welding Society, *Welding Handbook*, 6th ed., 1968, section 1, p. 1.51)

Sometimes it is necessary to specify the location and direction of the x-ray source in relation to the inspected weld. When this is necessary, it is usually done as shown in Figure 3-29. The x-ray source is to be located 24 inches from the joint at a 30-degree angle to the vertical arm.

When only a certain length of a weld is to be inspected, the length (in inches) is placed to the right of the abbreviation used to call out the test, as shown in Figure 3-30. At other times, especially when the acceptance of a weld is based on statistical results, the number of tests to be performed at random is specified in parentheses under the abbreviation of the test to be performed, as shown in Figure 3-31.

Nondestructive testing symbols may be combined with welding symbols or used independently.

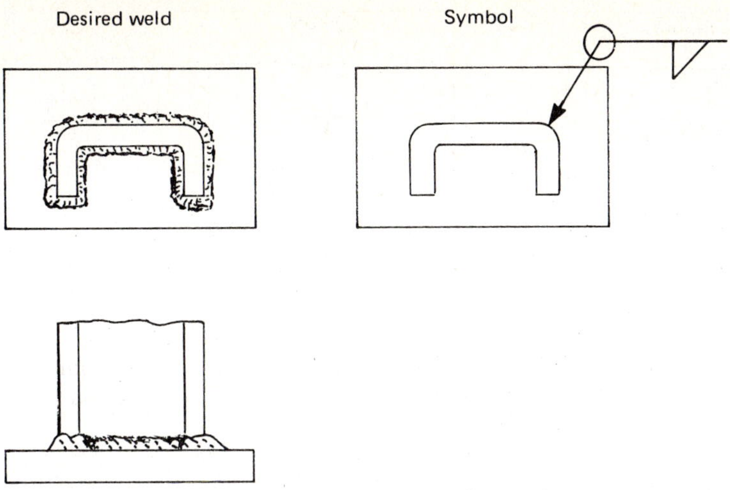
Desired weld
Symbol

Figure **3-23** Extent-of-welding symbol. (Redrawn by permission from American Welding Society, *Welding Handbook*, 6th ed., 1968, section 1, p. 1.51.)

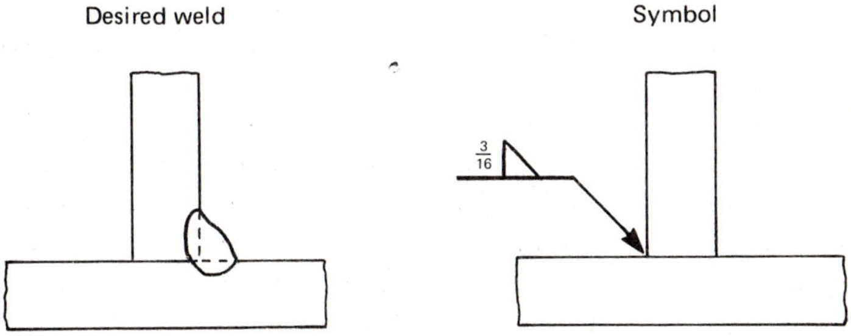
Desired weld
Symbol
3/16

Figure **3-24** Welding symbol indicating the welding of a hidden joint

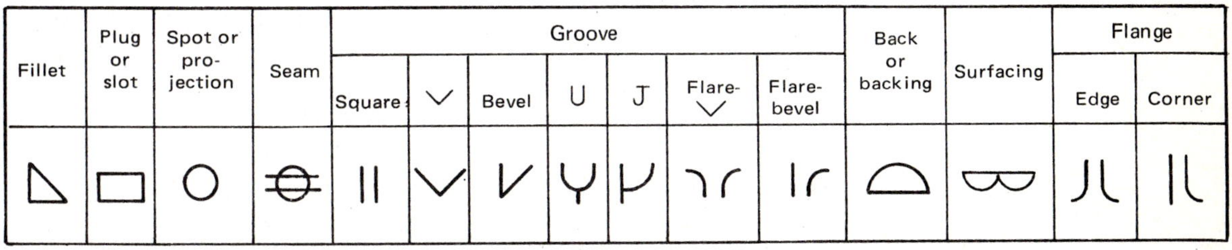
Fillet
Plug or slot
Spot or projection
Seam
Groove
Square
V
Bevel
U
J
Flare-V
Flare-bevel
Back or backing
Surfacing
Flange
Edge
Corner

Figure **3-25** The basic welding symbols. (Redrawn by permission from American Welding Society, *Welding Handbook*, 6th ed., 1968, section 1, p. 1.6.)

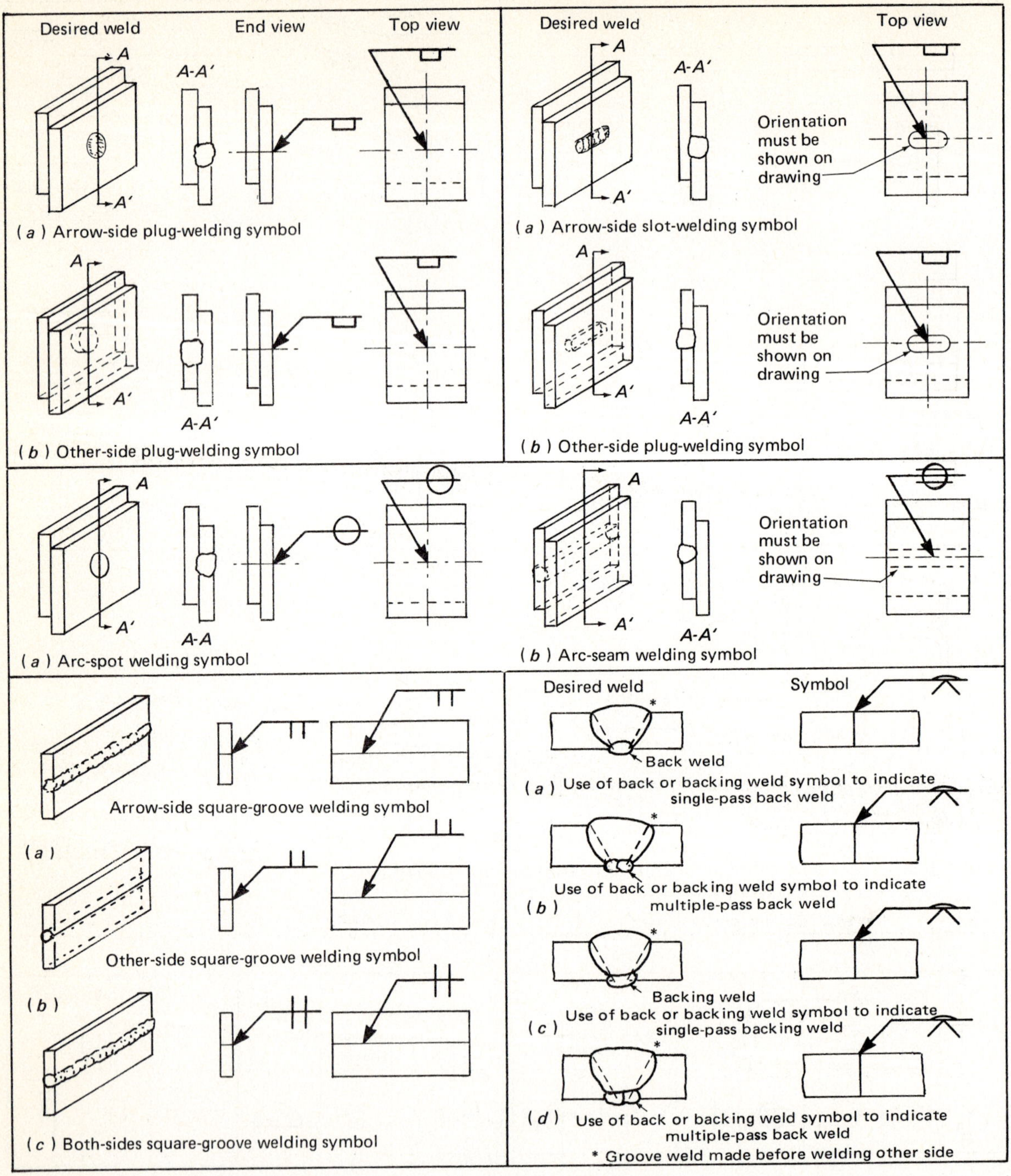
Desired weld
End view
Top view
A
A-A'
A'
(a) Arrow-side plug-welding symbol
(b) Other-side plug-welding symbol
Desired weld
Top view
Orientation must be shown on drawing
(a) Arrow-side slot-welding symbol
(b) Other-side plug-welding symbol
A-A
(a) Arc-spot welding symbol
(b) Arc-seam welding symbol
Arrow-side square-groove welding symbol
(a)
Other-side square-groove welding symbol
(b)
(c) Both-sides square-groove welding symbol
Desired weld
Symbol
Back weld
(a) Use of back or backing weld symbol to indicate single-pass back weld
(b) Use of back or backing weld symbol to indicate multiple-pass back weld
Backing weld
(c) Use of back or backing weld symbol to indicate single-pass backing weld
(d) Use of back or backing weld symbol to indicate multiple-pass back weld
* Groove weld made before welding other side

Figure **3-26** Examples of welding symbols and the welds they indicate. (Redrawn by permission from American Welding Society, *Welding Handbook*, 6th ed., 1968, section 1, pp. 1.45–1.49.)

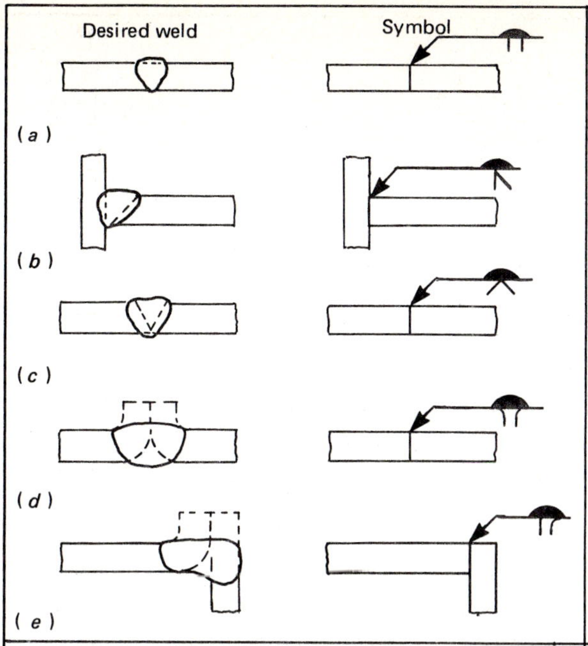
Desired weld
Symbol
(a)
(b)
(c)
(d)
(e)

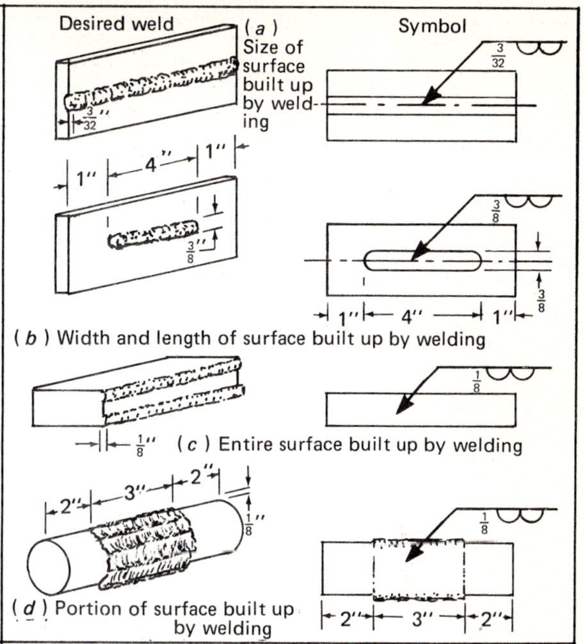
Desired weld
Symbol
(a) Size of surface built up by welding
3/32
3/32"
1"
4"
1"
3/8
3/8"
1" 4" 1"
(b) Width and length of surface built up by welding
1/8
1/8"
(c) Entire surface built up by welding
2"
3"
2"
1/8"
1/8
(d) Portion of surface built up by welding
2" 3" 2"

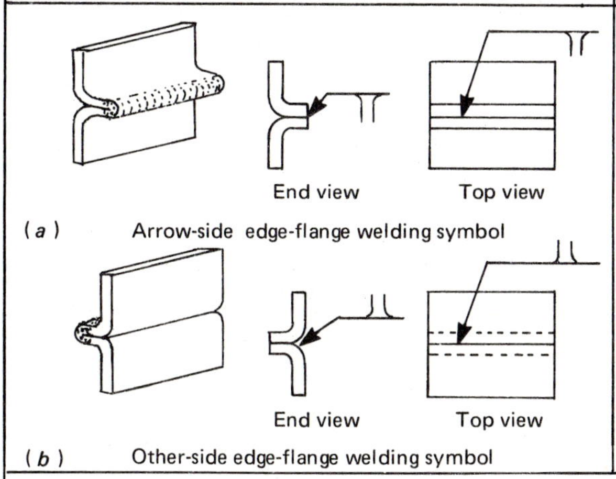
End view
Top view
(a) Arrow-side edge-flange welding symbol
End view
Top view
(b) Other-side edge-flange welding symbol

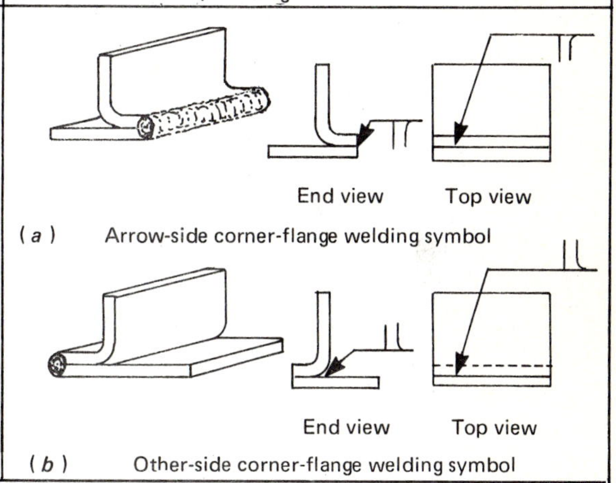
End view
Top view
(a) Arrow-side corner-flange welding symbol
End view
Top view
(b) Other-side corner-flange welding symbol

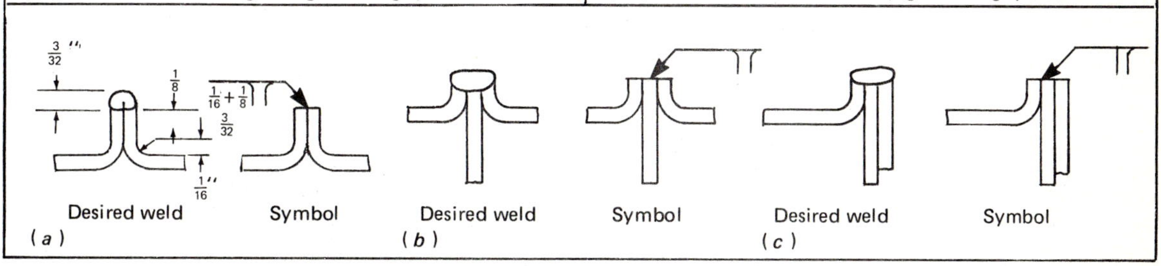
3/32"
1/8
1/16 + 1/8
3/32
1/16"
Desired weld
Symbol
(a)
Desired weld
Symbol
(b)
Desired weld
Symbol
(c)

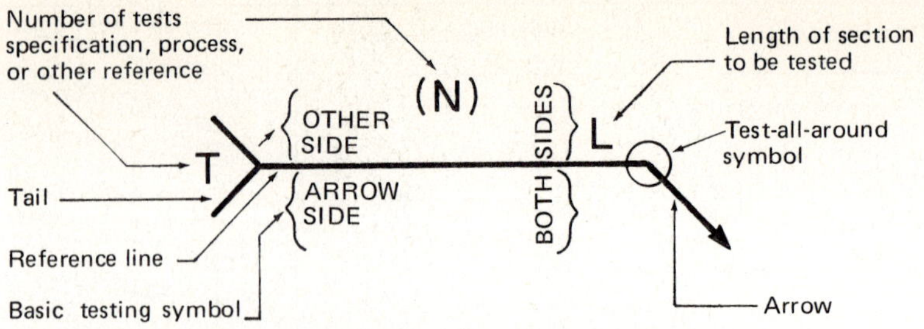
Number of tests
specification, process,
or other reference
(N)
Length of section
to be tested
OTHER
SIDE
BOTH SIDES
L
Test-all-around
symbol
T
Tail
ARROW
SIDE
Reference line
Basic testing symbol
Arrow

Figure **3-27** The assembled testing symbol. (Redrawn by permission from American Welding Society. *Welding Handbook,* 6th ed., 1968, section 1, p. 1.39.)

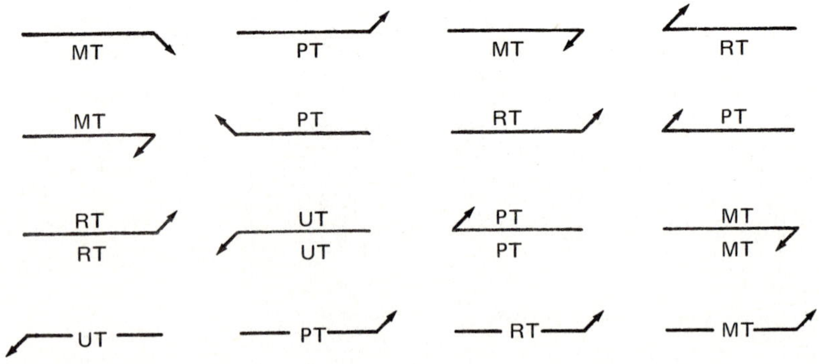
MT
PT
MT
RT
MT
PT
RT
PT
RT
RT
UT
UT
PT
PT
MT
MT
UT
PT
RT
MT

Figure **3-28** Location of test abbreviations on the reference line. (Redrawn by permission from American Welding Society, *Welding Handbook,* 6th ed., 1968, section 1, pp. 1.39, 1.40.)

Table **3-3** Nondestructive tests and their abbreviations

TYPE OF TEST	ABBREVIATION
X-ray	RT
Magnetic particle	MT
Liquid penetrant	PT
Ultrasonic	UT

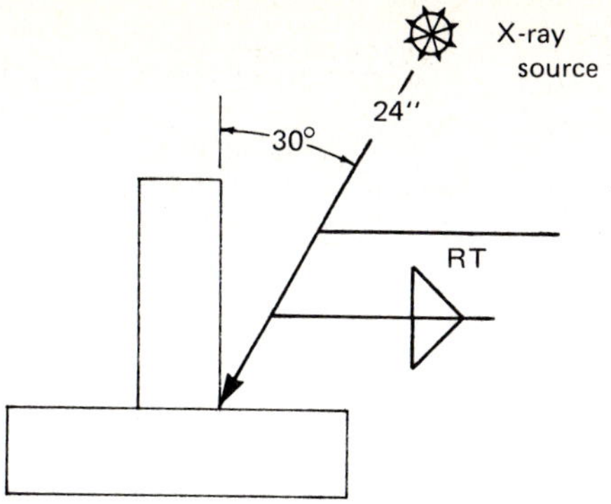
X-ray
source
24''
30°
RT

Figure **3-29** Symbol showing location of x-ray source in relation to the part to be tested. (Redrawn by permission from American Welding Society, *Welding Handbook,* 6th ed., 1968, section 1, p. 1.40.)

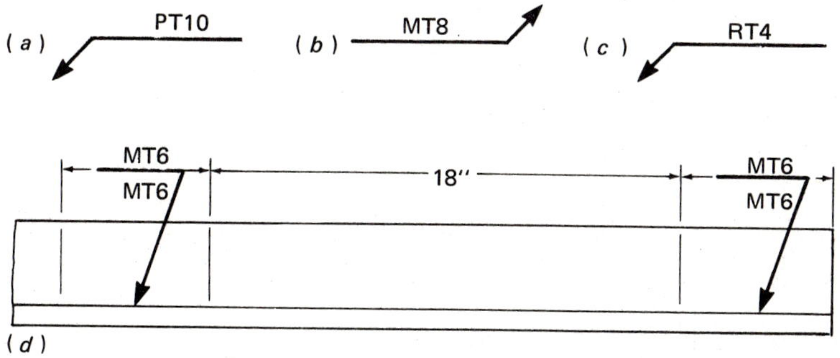
(a)
PT10
(b)
MT8
(c)
RT4
MT6
MT6
18''
MT6
MT6
(d)

Figure **3-30** Test symbols indicating specific lengths of the weld to be tested. (Redrawn by permission from American Welding Society, *Welding Handbook,* 6th ed., 1968, section 1, p. 1.41.)

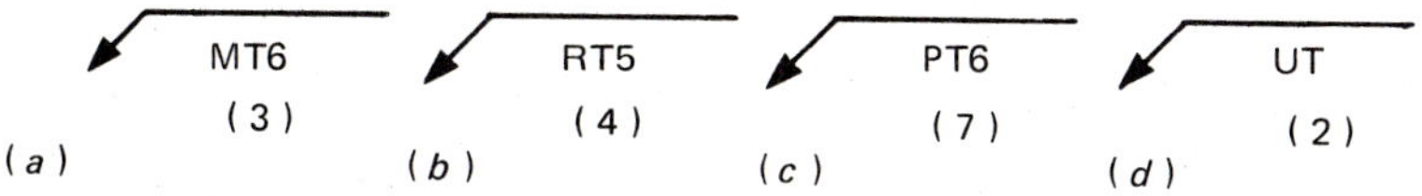
MT6
(3)
(a)
RT5
(4)
(b)
PT6
(7)
(c)
UT
(2)
(d)

Figure **3-31** Testing symbols indicating (in parentheses) the number of tests to be made. (Redrawn by permission from American Welding Society, *Welding Handbook,* 6th ed., 1968, section 1, p. 1.42.)

REVIEW QUESTIONS

1. Differentiate the terms *weld symbol* and *welding symbol.*

2. What is the maximum number of elements comprising an assembled welding symbol?

3. Draw a welding symbol and identify each of its elements by name.

4. Is it always necessary for each welding symbol to have the maximum number of elements? Explain your answer.

5. The term *boxing* is used in reference to what type of weld?

6. Give the meaning of each of the following terms:
   a. Intermittent weld
   b. Plug weld
   c. Groove weld
   d. Field weld

7. What is the name of the small, short bead used as a temporary fastener?

8. Why are welding symbols used?

9. Name the welding processes whose abbreviations are:
   a. AHW
   b. UW
   c. IB
   d. TW
   e. SMAW

10. Figure 3-29 shows a combination weld and nondestructive testing symbol. Describe:
   a. the test method to be used.
   b. where and how far from the joint the test device is to be positioned.
   c. the type of weld to be inspected.

CHAPTER 4
WELDING CONSUMABLES

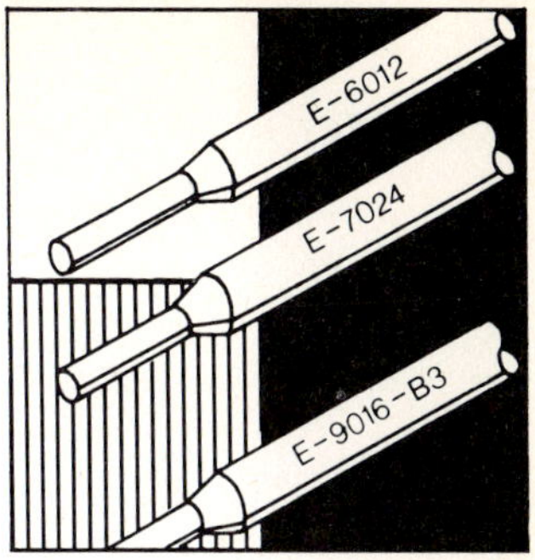
E-6012
E-7024
E-9016-B3

Welding consumables are the materials used up during welding, such as electrodes, welding rods, fluxes, fuel gases, and externally applied shielding gases.

FILLER METALS

Essentially all filler metals discussed in this chapter are classified according to one of the specifications developed jointly by the AWS–ASTM (American Welding Society–American Society for Testing and Materials) Subcommittee on Filler Metals. The filler metals and their corresponding AWS–ASTM specifications are:

- Mild steel–covered arc welding electrodes: A5.1
- Iron and steel gas-welding rods: A5.2
- Aluminum and aluminum-alloy arc welding electrodes: A5.3
- Corrosion-resisting chromium and chromium–nickel steel–covered welding electrodes: A5.4
- Low-alloy steel–covered arc welding electrodes: A5.5
- Copper and copper-alloy arc welding electrodes: A5.6
- Copper and copper-alloy welding rods: A5.7
- Brazing filler metal: A5.8

- Corrosion-resisting chromium and chromium–nickel steel welding rods and bare electrodes: A5.9
- Aluminum and aluminum-alloy welding rods and bare electrodes: A5.10
- Nickel and nickel alloy–covered welding electrodes: A5.11
- Surfacing welding rods and electrodes: A5.13
- Nickel and nickel-alloy bare welding rods and electrodes: A5.14
- Welding rods and covered electrodes for welding cast iron: A5.15
- Titanium and titanium-alloy bare welding rods and electrodes: A5.16
- Bare mild-steel electrodes and fluxes for submerged-arc welding: A5.17
- Mild-steel electrodes for gas metal–arc welding: A5.18
- Magnesium-alloy welding rods and bare electrodes: A5.19
- Mild-steel electrodes for flux-cored arc welding: A5.20
- Composite-surfacing welding rods and electrodes: A5.21

Through the filler metal specifications, the user is informed that a certain electrode or welding rod can produce a weld metal having specific mechanical properties. At the same time, the specification system classifies electrodes for various positions of welding, for their ability to penetrate adequately into the root of a joint, and for power supply (alternating or direct current).

4-1 ARC WELDING ELECTRODES

The development of flux-coated electrodes, capable of making welds with physical properties that equal or exceed those of the parent metal, has made arc welding into the most widely used welding process.

Prior to the development of the coated electrode, the atmospheric gases in the high-temperature welding zone formed oxides and nitrides with the weld metal. In general, oxides are low in tensile strength and ductility and tend to reduce the normal properties of the base metals. Electrode-coating materials provide an automatic cleansing and deoxidizing action in the molten crater. As the coating burns in the arc, it releases a gaseous, inert atmosphere that protects the molten end of the electrode as well as the molten weld pool. This atmosphere excludes harmful oxygen and nitrogen from the molten weld area, while the burning residue of coating forms a slag to cover the deposited weld metal. This slag also excludes oxygen and nitrogen from the weld until it has cooled to such a point that oxides and nitrides no longer form. In addition, the slag slows the cooling, producing a more ductile weld.

Besides these benefits, other advantages are gained by coating electrodes. The coating improves weld appearance, provides easier arc striking, helps maintain the arc, regulates the depth of penetration, reduces spatter, improves the x-ray quality of the weld, and sometimes adds alloying agents to the weld metal or restores lost elements. The slag from the coating not only protects the weld bead but also assists in shaping it. In addition, powdered iron has been added to the coating of many of the basic electrode types. In the intense heat of the arc, the iron powder is converted into steel and contributes metal to the weld deposit. When added in relatively large amounts, the speed of welding is appreciably increased and weld appearance is improved. The electrode coating also serves as an insulator for the core wire of the electrode. It affects the arc length and welding voltage and controls the welding position in which the electrode can be used.

The composition of the electrode coating is extremely important. The blending of the proper ingredients is virtually an art. Besides correctly balancing the previously listed performance characteristics, the coating should have a melting point somewhat lower than that of the core wire or base metal. The resultant slag must have a lower density in order to be quickly and thoroughly expelled from the cooling weld metal. Where the electrode is to be used for overhead or vertical welding, the slag formed from the melted coating must solidify quickly to help hold the molten metal against the force of gravity.

In most cases, the differences in the operational characteristics of an electrode can be attributed to the coating. The core wire is generally from the same wire stock. For the common E-60XX series of electrodes, the core wire is SAE 1010 carbon steel, having a carbon range of 0.05–0.15%.

4-2 AWS–ASTM CLASSIFICATION OF CARBON-STEEL ELECTRODES

This classification consists of a series of four- or five-digit numbers (Figure 4-1) prefixed with the letter *E*. The *E* indicates electric welding use. The numerals to the left of the last two digits multiplied by 1000 give the minimum tensile strength of the (stress-relieved) deposited metal; the next-to-last digit tells the power supply, type of slag, type of arc, penetration, and presence of iron powder. For data regarding further interpretation of these classification numbers, see Tables 4-1 and 4-2.

Some of these electrodes are for direct current (dc) and some, for alternating current (ac). Some dc electrodes are for straight polarity (electrode holder connected to the negative pole, as in Figure 4-2), and some are for reverse polarity (electrode holder attached to the positive pole, as in Figure 4-3).

Table **4-1** Interpretation of last digit in AWS electrode classification

	LAST DIGIT								
	0	1	2	3	4	5	6	7	8
Power supply	[a]	AC or DC rev. polarity	AC or DC	AC or DC	AC or DC	DC rev. polarity	AC or DC rev. polarity	AC or DC	AC or DC rev. polarity
Type of slag	[b]	Organic	Rutile	Rutile	Rutile	Low hydrogen	Low hydrogen	Mineral	Low hydrogen
Type of arc	Digging	Digging	Medium	Soft	Soft	Medium	Medium	Soft	Medium
Penetration	[c]	Deep	Medium	Light	Light	Medium	Medium	Medium	Medium
Iron powder in coating	0–10%	None	0–10%	0–10%	30–50%	None	None	50%	30–50%

[a]E-6010 is DC reverse polarity; E-6020 is AC or DC
[b]E-6010 is organic; E-6020 is mineral
[c]E-6010 is deep penetration; E-6020 is medium penetration

Source: *Metals and How to Weld Them* (Cleveland, Ohio: James F. Lincoln Arc Welding Foundation), p. 94.

Table **4-2** AWS electrode classification system

DIGIT	SIGNIFICANCE	EXAMPLE
1st two or 1st three	Min. tensile strength (stress relieved)	E-60XX = 60,000 psi (min.) E-110XX = 110,000 psi (min.)
2nd last	Welding position	E-XX1X = all positions E-XX2X = horizontal and E-XX3X = flat
Last	Power supply, type of slag, type of arc, amount of penetration, presence of iron powder in coating	

Note: Prefix "E" (to left of a 4- or 5-digit number) signifies arc-welding electrode.

Source: *Metals and How to Weld Them* (Cleveland, Ohio: James F. Lincoln Arc Welding Foundation), p. 94.

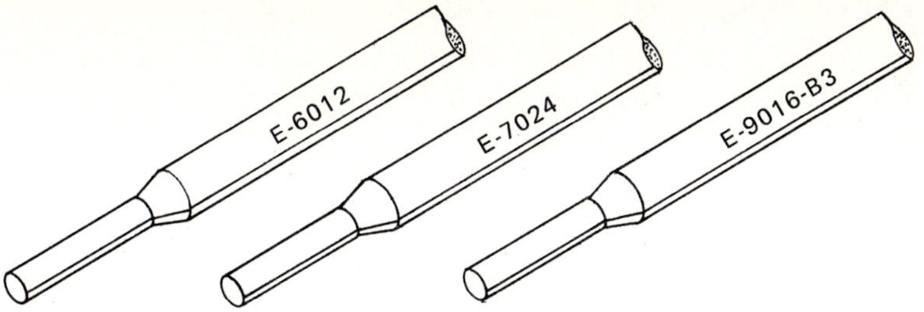

E-6012
E-7024
E-9016-B3

Figure **4-1** AWS–ASTM flux-coated carbon-steel electrode identification markings. (Redrawn by permission from *Fundamentals of Service Welding,* John Deere Service Publications, 1971, Fig. 47.)

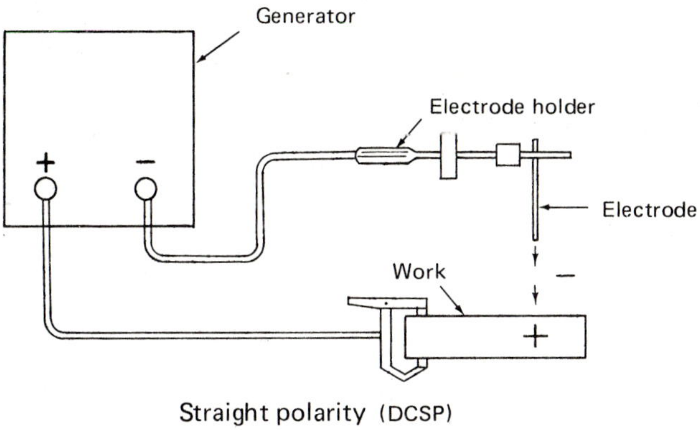

Generator
Electrode holder
+
−
Electrode
Work
−
+
Straight polarity (DCSP)

Figure **4-2** A straight-polarity arc welder (negative electrode). (Redrawn by permission from *Fundamentals of Service Welding,* John Deere Service Publications, 1971, Fig. 27.)

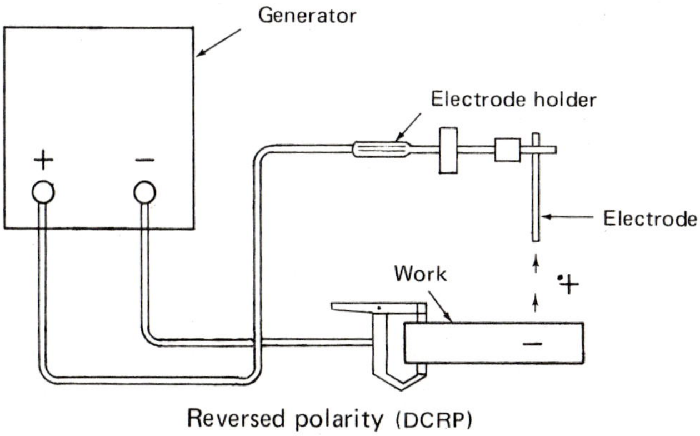

Generator
Electrode holder
+
−
Electrode
Work
+
−
Reversed polarity (DCRP)

Figure **4-3** A reverse-polarity arc welder (positive electrode). (Redrawn by permission from *Fundamentals of Service Welding,* John Deere Service Publications, 1971, Fig. 27.)

4-3 USE OF CARBON-STEEL ARC WELDING ELECTRODES

The arc welding electrodes for welding low- and medium-carbon steels carry AWS classification numbers E-4510 and E-6010, E-6011, E-6012, E-6013, E-7014, E-7015, E-7016, E-7018, E-6020, E-6024, E-6027, and E-7028. E-4510 is a bare electrode; the others, all in the E-6000 series, are coated electrodes.

AWS E-45XX electrodes are bare and have a tensile strength of 45,000 psi in the stress-relieved condition. Because bare electrodes are seldom used, our principal concern is with the E-60XX series.

AWS E-6010 electrodes are coated with high-cellulose sodium. They are designed for all-position welding with reverse-polarity direct current. They are best suited for vertical and overhead welding and some sheet metal applications.

The thickness of the coating is held at a minimum to facilitate welding in the vertical and overhead positions, but is sufficient to develop the shielding necessary for a high-quality deposit. Some coatings have a small amount (less than 10% by coating weight) of iron powder to improve arc characteristics. The arc has a digging characteristic that results in deep penetration. This calls for skilled manipulation of the electrode by the operator to minimize spatter and the tendency to undercut. The slag formed is light and easily removed. The profile of fillet welds is more or less convex on horizontal and vertical deposits. Beads deposited by E-6010 electrodes have a rather coarse ripple.

The E-6010 electrode is excellent for temporary tacking because of its ductility and deep-penetration qualities. Its physical properties are excellent and, when properly applied, its deposits will meet the most exacting inspection standards.

AWS E-6011 electrodes are coated with high-cellulose potassium. They are sometimes described as the counterpart of the E-6010 type. Performance characteristics of the two electrodes are quite similar; however, the E-6011 electrodes perform equally well with either an ac or a dc power supply. These electrodes have a forceful digging arc, resulting in deep penetration. While the coating on E-6011 electrodes is slightly heavier than the coating on E-6010, the resultant slag and weld profiles are similar.

AWS E-6012 electrodes are coated with high-titania sodium. They are designed for general-purpose welding in all positions, with either dc or ac power. They are specifically recommended for horizontal and most downhill welding applications. An E-6012 electrode has a rather quiet arc, with medium penetration and no spatter. Good build-up and no excess penetration make the electrode excellent for welding under poor fit-up conditions. Since the arc is high stabilizing, the welds have a good appearance and are relatively free from undercut. Fillet welds usually have a convex profile of a smooth, even ripple in the horizontal or the vertical down position. The slag coverage is complete, and the slag may easily be removed.

When used with a dc power supply, straight polarity is preferred. It is used extensively where appearance and high rates of deposition are more important than maximum ductility. For example, this electrode is particularly suited to making highly satisfactory welds on sheet metal, where single-pass welds must pass radiographic inspection.

Some proprietary E-6012 type electrodes have a small amount of iron powder in the coating to improve arc characteristics.

AWS E-6013 electrodes are coated with high-titania potassium and can be used in all positions, with ac or dc. These electrodes are similar to the E-6012 electrodes but produce less spatter and tend to undercut less. The beads are fine rippled and superior in appearance to the beads produced with the E-6012 electrodes.

Slag removal is easier with the E-6013, and the arc is very stable. This facilitates striking and maintaining the arc, even with extremely small (1/16 and 5/64 inch) electrodes and makes E-6013 ideal for welding thin metals. The arc is soft and the penetration very light. The mechanical properties of the E-6013 are slightly better than the E-6012. The same may be said for its radiographic quality.

Changing from one manufacturer's to another's E-6013 electrode may result in a change in the nature of the molten metal transfer in the arc stream. Some manufacturers compound their coating, so that a globular transfer is obtained, while others produce a fine-spray transfer. Ordinarily, the spray transfer is preferred for vertical or overhead deposits. The amount of spatter from this electrode varies with different brands, too. Some manufacturers have also introduced small amounts of iron powder into the E-6013 electrode coatings.

AWS E-7014 electrodes have a coating similar to that of the E-6012 and E-6013 types. However, the coating of this electrode type is considerably thicker, since it contains a substantial amount of iron powder (30% of the coating weight). The presence of iron powder permits higher welding currents, which means higher deposition rates and welding speeds. The thicker coating does not make it as ideally suited to out-of-position production welding on thin-gage material; however, it will perform adequately when the occasional job demands it. Its performance characteristics do make this electrode particularly well suited to production welding of irregularly shaped products, where some out-of-position welding is encountered.

Mechanical properties of E-7014 weld metal compare favorably with those of the E-6012 and E-6013 metals. Fillet-weld contour ranges from flat to slightly convex. Slag removal is very easy and sometimes self-cleaning. Shallow penetration and rapid solidification are characteristics that make this electrode type well suited to handling poor fit-up conditions.

AWS E-7015 electrodes are coated with low-hydrogen sodium. They were the first dc reverse-polarity, all-position electrodes designed for the welding of high-sulfur and high-carbon steels which tend to develop weld porosity and crack under the weld bead. Metallurgists found that the presence of hydrogen in molten metal tends to promote underbead cracking and the formation of porosity during solidification. The E-7015 electrode's coating was designed to have a very low moisture content, thereby preventing the introduction of hydrogen into the weld. The successful performance of this electrode led to the later development of the E-6016 and E-6018 types, which also have very low moisture contents in the coatings. These electrodes are commonly known as *low-hydrogen electrodes.*

AWS E-7016 electrodes have potassium silicate or other potassium salts added to the coating, to make the electrode suitable for use with ac as well as dc reverse polarity.

AWS E-7018 electrodes are of low-hydrogen design, have a 30% powdered iron coating. Like the E-7016 electrodes, this type operates on either ac or dc reverse polarity. They have all the desirable low-hydrogen characteristics that produce sound welds on troublesome steels, such as high-sulfur, high-carbon, and low-alloy grades. Their slightly thicker coating and powdered iron content make them generally easier to use than the other low-hydrogen types. For these reasons, they are the most widely used electrodes.

The minerals of the low-hydrogen electrodes' coatings are limited to inorganic compounds, such as calcium fluoride, calcium carbonate, magnesium-aluminum silicate, ferrous alloys, and such binding agents as sodium and potassium silicate. These electrodes are referred to as *lime-ferritic* because of the general use of lime-type coatings. (Lime is a decomposition product of such compounds as calcium carbonate.)

Since the coating of these electrodes is heavier than normal, vertical and overhead welding is usually limited to the small-diameter electrodes. The current used is somewhat higher than for E-6010 electrodes of corresponding size.

The mechanical properties (including impact strength) of low-hydrogen electrodes are superior to those of E-6010 electrodes that deposit weld metal of similar composition. The use of low-hydrogen electrodes reduces the preheat and postheat of welds, thus making for better welding conditions and lowering or eliminating of preheating costs.

AWS E-6020 electrodes are coated with high-iron oxide and are designed to produce high-quality horizontal fillet welds at high welding speeds, using either ac or dc straight polarity. In the flat position, these electrodes can be used with alternating or direct current of either polarity. E-6020 electrodes are characterized by a forceful, spraying type of arc and heavy slag, which completely covers the deposit but is quite easily removed. Penetration is medium at normal welding currents, but high currents and high travel speeds will result in deep penetration. The deposits of this electrode are usually flat or even slightly concave in profile and have a smooth, even ripple. The radiographic qualities are excellent, and the weld beads show medium spatter and a slight tendency to undercut.

AWS E-6024 electrodes are ideally suited for production fillet welding. Their 50% powdered iron coating assists in producing a high deposition rate and welding speeds that are considerably higher than those of the E-6012, E-6013, and E-7014 types, which have similar performance characteristics. Operating characteristics include a soft, quiet arc that produces practically no spatter. The ability to drag the electrode coating on the parent metal while welding produces a very smooth bead appearance. Physical properties of the weld deposit compare favorably with the E-6012, E-6013, and E-7014 types. As in the case of the E-6020, the welding position is limited to flat and horizontal.

AWS E-6027 electrodes are of 50% iron design, for either ac or dc operation. The arc characteristics of this type closely approximate the E-6020 type. Having a very high deposition rate and a slag that crumbles for easy removal, the E-6027 electrode is particularly well suited to multiple-pass, deep-groove welding.

The E-6027 electrodes produce high-quality weld metal, having physical properties that closely resemble those of E-6010 electrodes. Operating characteristics make this electrode slightly harder to handle than the E-7024 type; however, properly deposited weld beads can have a smoother appearance.

AWS E-7028 electrodes have a low-hydrogen coating containing 50% powdered iron. Designed for ac or dc reverse-polarity operation, the heavy coating gives this electrode a very high deposition rate. Although capable of producing the physical properties and weld quality typical of low-hydrogen designs, these electrodes are suitable for only flat and horizontal welding positions.

The approximate range (in amperes) and the typical storage and rebake conditions for carbon-steel electrodes are shown in Figure 4-4 and Table 4-3, respectively.

4-4 ALLOY-STEEL ELECTRODES

The expanding use of high-strength alloy steels has initiated development of coated electrodes capable of producing weld deposits having tensile strength exceeding 100,000 psi. Mechanical properties of this magnitude are achieved through the use of alloy steel as the core wire of the electrode. In most electrode designs, the electrode coating is of a lime-ferritic nature typical of the low-hydrogen designs and frequently contains powdered iron. For this reason, these high-tensile-strength electrodes usually have an E-XX15, E-XX16, or E-XX18 classification. Operation characteristics parallel those of typical 60,000-psi low-hydrogen designs.

The flexible system of electrode identification established by the AWS readily catalogs these electrodes in groups established for the E-60XX series. For example, the E-11018 electrode has a 110,000-psi tensile strength, and like the E-7018 type will weld in all positions, uses ac or dc reverse-polarity power, has a low-hydrogen slag, has medium arc force and penetration, and has 30% iron powder in its coating. Table 4-4 lists the AWS standard suffixes that indicate specific additions of alloying elements. In a complete electrode designation, these later symbols appear after the four- or five-digit basic number (see Figure 4-1).

4-5 AWS–ASTM SPECIFICATIONS FOR GAS-WELDING RODS

Gas-welding rods or wires are steel rods having no coverings. The welding operation is determined solely by the composition of the rods and the welding flame used. The various classes of gas-welding rods are briefly described below.

Figure **4-4** Chart for determining the current electrodes will carry. Small variations may be found when comparing different makes and types. (Redrawn by permission from *The Procedure Handbook of Arc Welding*, 11th ed., Lincoln Electric Company, Cleveland, Ohio, p. 2-32.)

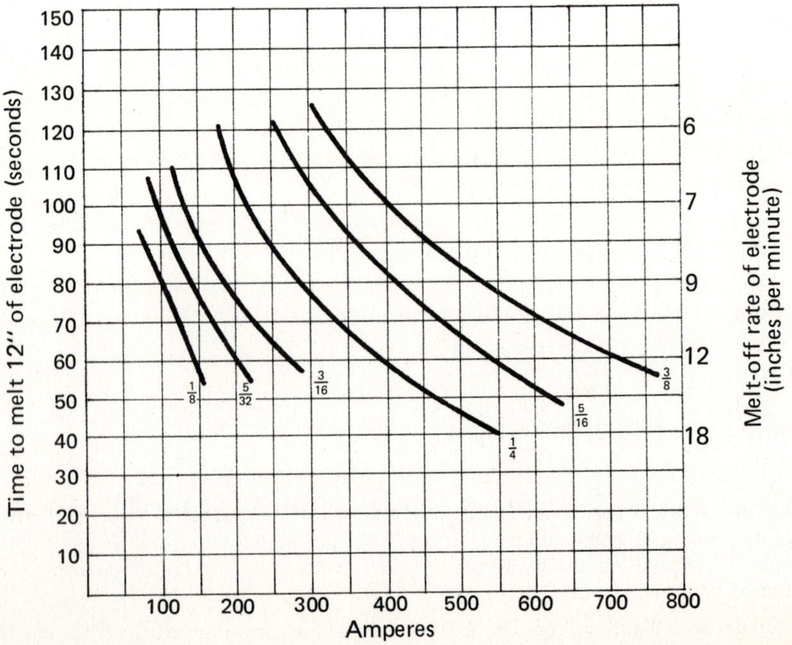

Time to melt 12" of electrode (seconds)
150
140
130
120
110
100
90
80
70
60
50
40
30
20
10
1/8
5/32
3/16
1/4
5/16
3/8
6
7
9
12
18
Melt-off rate of electrode (inches per minute)
100
200
300
400
500
600
700
800
Amperes

Table **4-3** Electrode reconditioning guide

AIRCO TYPE	AWS–ASTM CLASS	STORAGE WHEN SHIPMENT ARRIVES	STORAGE WHEN BOXES ARE OPENED	RECONDITION TEMP. °F AND TIME AT TEMP.
LOW HYDROGEN AND LOW HYDROGEN LOW ALLOY IRON POWDER COATED ELECTRODES				
7016	E-7016	68°F to 110°F at 50% relative humidity	Store in oven at 350°F to 450°F	600°F for 30–60 minutes at temp.
7016-M	E-7016	"	"	"
8016-C1	E-8016-C1	"	"	"
8016-B2	E-8016-B2			
9016-B3	E-9016-B3	"	"	800°F for 30–60 minutes at temp.
10016-G	E-10016-G	"	"	"
10016-D2	E-10016-D2	"	"	"
Easyarc 7018	E-7018	"	"	"
Easyarc 7028	E-7028	"	"	"
CODE-ARC 7018	E-7018	"	"	"
CODE-ARC 7018-A1	E-7018-A1	"	"	"
CODE-ARC 8018-C3	E-8018-C3	"	"	"
CODE-ARC 11018-M	E-11018-M	"	"	"
CODE-ARC 12018-M	E-12018-M	"	"	"
CONVENTIONAL COATED AND IRON POWDER COATED ELECTRODES				
6010	E-6010	Between conditions 110°F at 68% relative humidity and 68°F at 50% rel. humidity	175°F or 68°F to 110°F at 50% relative humidity	250°F for 30 minutes at temp.
6011	E-6011	"	"	"
6011C				
6012	E-6012	"	"	"
6012C				
6013	E-6013	"	"	"
6013C				
6020	E-6020	"	"	"
Easyarc 6027	E-6027	"	"	375°F for 60 minutes at temp.
Easyarc 7014	E-7014	"	"	"
Easyarc 7024	E-7024	"	"	"
7010A1	E-7010A1	"	"	250°F for 30 minutes at temp.
7020A1	E-7020A1	"	"	
10013-G	E-10013-G	"	"	"

Caution: If electrodes have become wet and are able to be reconditioned, they should be heated at 180°F for 1 hr. before reconditioning at temperatures as recommended under Recondition Temp. column. If electrodes become too wet they cannot be reconditioned. Operating and weld metal characteristics should be checked to make sure that electrodes are usable after reconditioning.

Source: Airco Welding Products, *Airco Electrode Pocket Guide,* ADC 1107A, pp. 149–150.

Table **4-4** AWS designation of major alloying elements in arc welding electrodes

SUFFIX TO AWS ELECTRODE NO.	ALLOY ELEMENT, %				
	Mo (*Molybdenum*)	Cr (*Chromium*)	Ni (*Nickel*)	Mn (*Manganese*)	Va (*Vanadium*)
A_1	0.5				
B_1	0.5	0.5			
B_2	0.5	1.25			
B_3	1.0	2.25			
B_4	0.5	2.0			
C_1			2.5		
C_2			3.5		
C_3			1.0		
D_1	0.3			1.5	
D_2	0.3			1.75	
G*	0.2	0.3	0.5	1.0	0.1

*Need have minimum content of one element only.

Source: *Metals and How to Weld Them* (Cleveland, Ohio: James F. Lincoln Arc Welding Foundation), p. 99.

Class RG65 welding rods are used for the oxyacetylene welding of carbon and low-alloy steels that exhibit strengths in the range of 65,000 to 75,000 psi. They are used on sheet, plate, tubes, and pipes. When a base-metal alloy analysis is used for some specific property, such as creep resistance or corrosion resistance, then the filler metal analysis should match the base-metal alloy analysis. Class RG65 welding rods are of low-alloy steel analysis.

Class RG60 welding rods are used for the oxyacetylene welding of carbon steels in the strength range of 50,000 to 65,000 psi and for welding wrought iron. They may also be used for low-alloy steels that fall in this range. These are general-purpose gas-welding rods, of medium strength and good ductility, which are most commonly used for the welding of carbon-steel pipes for power plants, process piping, and other conditions of severe service.

Class RG45 welding rods are of a simple, low-carbon–steel analysis. Most rods of this class are of the following nominal composition: carbon, 0.07% maximum; manganese, 0.25% maximum; phosphorus and sulfur, each 0.04% maximum; silicon, 0.08% maximum. These welding rods are general-purpose rods and may be used to join wrought iron.

4-6 STAINLESS STEEL ELECTRODES AND WELDING RODS

A variety of stainless steel electrodes (prefix E) and welding rods (prefix R, or prefix ER if the filler metal can be used as either an electrode or as a welding rod) are being manufactured. These filler metals can produce a weld metal similar in composition to that of most base metals.

Manganese and silicon are included in the electrode coating to reduce oxidation; titanium is included to promote arc stability, produce an easily removable slag, and prevent carbide precipitation. Lime is an extremely important ingredient in the coating, since it tends to eliminate hydrogen, the formation of which leads to underbead cracking. Any material that is high in carbon is excluded because of the affinity of chromium for carbon, especially at welding temperatures. The low-hydrogen type of coating used on stainless steel electrodes is similar to that employed on certain carbon-steel welding electrodes: *E*-7015 and *E*-7016.

Filler metals comparable in composition to the 300, 400, and 500 series of stainless steels have been developed. The 200 series of steels (low nickel, high manganese) are welded using the corresponding 300-type electrode or welding rod.

Filler metals suitable for welding similar stainless steels are listed in Tables 4-5*a* and 4-5*b*, while those suitable for welding stainless steels to other alloys are listed in Tables 4-6 and 4-7.

Table **4-5*a*** Filler metals recommended for commonly used austenitic chromium-nickel stainless steels

AISI BASE METAL TYPES	AWS–ASTM FILLER METAL DESIGNATION
201	ER308
202	ER308
301, 302, 304, 308	ER308
304L	ER308L
309	ER309
310	ER310
316	ER316
316L	ER316L
317	ER317
330	ER330
321	ER321
347	ER347
17-7PH	W17-7PH
PH15-7 Mo	WPH15-7 Mo
17-4PH	17-4PH
AM350	AM350
AM355	AM355
A286	A286

Table **4-5*b*** Recommended current ranges

ELECTRODE SIZE, in.	AC OR DC CURRENT, AMP		MAXIMUM ARC VOLTAGE
	FLAT HORIZONTAL AND OVERHEAD	*VERTICAL*	
3/64	15–25	15–25	23
1/16	20–40	25–40	24
5/64	30–60	35–55	24
3/32	45–90	45–65	24
1/8	70–120	70–95	25
5/32	100–160	100–125	26
3/16	130–190	130–145	27
1/4	210–300	—	28
5/16	250–400	—	29

Source: American Welding Society, *Welding Handbook*, 6th ed., section 4, pp. 65.16, 65.18.

Table **4-6** Covered electrodes recommended for welds between stainless, heat resisting, and carbon steels and other alloys

BASE METALS → ↓	CARBON STEEL	1¼ Cr- ½ Mo	2¼ Cr- 1 Mo	NICKEL	INCONEL	MONEL	COPPER-NICKEL ALLOYS
201	[d]; E309	[d]; E309	[d]; E309	ENiCrFe-3	ENiCrFe-3	ENiCrFe-3; ENiCu-2	[o]; ENiCrFe-3
202	[d]; E309	[d]; E309	[d]; E309	ENiCrFe-3	ENiCrFe-3	ENiCrFe-3; ENiCu-2	[o]; ENiCrFe-3
301	[d]; E309	[d]; E309	[d]; E309	ENiCrFe-3	ENiCrFe-3	ENiCrFe-3; ENiCu-2	[o]; ENiCrFe-3
302	[d]; E309	[d]; E309	[d]; E309	ENiCrFe-3	ENiCrFe-3	ENiCrFe-3; ENiCu-2	[o]; ENiCrFe-3
302B	[d]; E309	[d]; E309	[d]; E309	ENiCrFe-3	ENiCrFe-3	ENiCrFe-3; ENiCu-2	[o]; ENiCrFe-3
303[a]	[g]; E309	[g]; E309	[g]; E310	[p]	[p]	[p]	[q]
304	[d]; E309	[d]; E309	[d]; E309	ENiCrFe-3	ENiCrFe-3	ENiCrFe-3; ENiCu-2	[o]; ENiCrFe-3
304L	[d]; E309	[d]; E309	[d]; E309	ENiCrFe-3	ENiCrFe-3	ENiCrFe-3; ENiCu-2	[o]; ENiCrFe-3
305	[d]; E309	[d]; E309	[d]; E309	ENiCrFe-3	ENiCrFe-3	ENiCrFe-3; ENiCu-2	[o]; ENiCrFe-3
308	[d]; E309	[d]; E309	[d]; E309	ENiCrFe-3	ENiCrFe-3	ENiCrFe-3; ENiCu-2	[o]; ENiCrFe-3
309	[d]; E309; [e]	[d]; E309; [e]	[d]; E309; [e]	ENiCrFe-3	ENiCrFe-3	ENiCrFe-3; ENiCu-2	[o]; ENiCrFe-3
309S	[d]; E309; [e]	[d]; E309; [e]	[d]; E309; [e]	ENiCrFe-3	ENiCrFe-3	ENiCrFe-3; ENiCu-2	[o]; ENiCrFe-3
310	E310; [e]	E310; [e]	E310; [e]	ENiCrFe-3	ENiCrFe-3	ENiCrFe-3; ENiCu-2	[o]; ENiCrFe-3
310S	E310; [e]	E310; [e]	E310; [e]	ENiCrFe-3	ENiCrFe-3	ENiCrFe-3; ENiCu-2	[o]; ENiCrFe-3
314	[d]; E309; [e]	[d]; E309; [e]	[d]; E309; [e]	ENiCrFe-3	ENiCrFe-3	ENiCrFe-3; ENiCu-2	[o]; ENiCrFe-3
316	[d]; E309; [e]	[d]; E309; [e]	[d]; E309; [e]	ENiCrFe-3	ENiCrFe-3	ENiCrFe-3; ENiCu-2	[o]; ENiCrFe-3
316L	[d]; E309; [e]	[d]; E309; [e]	[d]; E309; [e]	ENiCrFe-3	ENiCrFe-3	ENiCrFe-3; ENiCu-2	[o]; ENiCrFe-3
317	[d]; E309; [e]	[d]; E309; [e]	[d]; E309; [e]	ENiCrFe-3	ENiCrFe-3	ENiCrFe-3; ENiCu-2	[o]; ENiCrFe-3
317L[b]	[d]; E309; [e]	[d]; E309; [e]	[d]; E309; [e]	ENiCrFe-3	ENiCrFe-3	ENiCrFe-3; ENiCu-2	[o]; ENiCrFe-3
321	[d]; E309; [e]	[d]; E309; [e]	[d]; E309; [e]	ENiCrFe-3	ENiCrFe-3	ENiCrFe-3; ENiCu-2	[o]; ENiCrFe-3
330[b]	[g]; E312; [e]	[g]; E312; [e]	[g]; E312; [e]	ENiCrFe-3	ENiCrFe-3	ENiCrFe-3; ENiCu-2	[o]; ENiCrFe-3
347	[g]; E312; [e]	[g]; E312; [e]	[g]; E312; [e]	ENiCrFe-3	ENiCrFe-3	ENiCrFe-3; ENiCu-2	[o]; ENiCrFe-3
348	[g]; E312; [e]	[g]; E312; [e]	[g]; E312; [e]	ENiCrFe-3	ENiCrFe-3	ENiCrFe-3; ENiCu-2	[o]; ENiCrFe-3
403	[r]; [l]	[s]; [l]	[t]; [l]	ENiCrFe-3	ENiCrFe-3	ENiCrFe-3; ENiCu-2	[o]; ENiCrFe-3
405	[r]; [l]	[s]; [l]	[t]; [l]	ENiCrFe-3	ENiCrFe-3	ENiCrFe-3; ENiCu-2	[o]; ENiCrFe-3
410	[r]; [l]	[s]; [l]	[t]; [l]	ENiCrFe-3	ENiCrFe-3	ENiCrFe-3; ENiCu-2	[o]; ENiCrFe-3
414	[r]; [l]	[s]; [l]	[t]; [l]	ENiCrFe-3	ENiCrFe-3	ENiCrFe-3; ENiCu-2	[o]; ENiCrFe-3
416[a]	E309	E309	E309	[p]	[p]	[p]	[q]
420	[r]; [l]	[s]; [l]	[t]; [l]	ENiCrFe-3	ENiCrFe-3	ENiCrFe-3; ENiCu-2	[o]
430	[r]; [l]	[s]; [l]	[t]; [l]	ENiCrFe-3	ENiCrFe-3	ENiCrFe-3; ENiCu-2	[o]
430F[a]	E309	E309	E309	[p]	[p]	[p]	[q]
431	[r]; [l]	[s]; [l]	[t]; [l]	ENiCrFe-3	ENiCrFe-3	ENiCrFe-3; ENiCu-2	[o]

440A	[r, l]	[s, l]	[t, l]	ENiCrFe-3	ENiCrFe-3	ENiCrFe-3; ENiCu-2	[o]
440B	[r, l]	[s, l]	[t, l]	ENiCrFe-3	ENiCrFe-3	ENiCrFe-3; ENiCu-2	[o]
440C	[r, l]	[s, l]	[t, l]	ENiCrFe-3	ENiCrFe-3	ENiCrFe-3; ENiCu-2	[o]
446	[r, l, e]	[s, l, e]	[t, l, e]	ENiCrFe-3	ENiCrFe-3	ENiCrFe-3; ENiCu-2	[o]
501	[r, l]	[s, l]	[t, l]	ENiCrFe-3	ENiCrFe-3	ENiCrFe-3; ENiCu-2	[o]
502	[r, l]	[s, l]	[t, l]	ENiCrFe-3	ENiCrFe-3	ENiCrFe-3; ENiCu-2	[o]
505	[r, l]	[s, l]	[t, l]	ENiCrFe-3	ENiCrFe-3	ENiCrFe-3; ENiCu-2	[o]
Carbon steel	[r]	[u]	[u]	ENiCrFe-3	ENiCrFe-3	ENiCrFe-3; ENiCu-2	[o]
¼Cr-½Mo	[s]	[s]	[s]	ENiCrFe-3	ENiCrFe-3	ENiCrFe-3; ENiCu-2	[o]
2¼Cr-1Mo			[t]	ENiCrFe-3	ENiCrFe-3	ENiCrFe-3; ENiCu-2	[o]
Nickel				ENi-1	ENiCrFe-3	ENiCu-2	ECuNi;ENiCu-2
Inconel					ENiCrFe-3	ENiCu-2; ENiCrFe-3	ECuNi; ENiCu-2
Monel						ENiCu-2	ECuNi; ENiCu-2
Copper-nickel alloys							ECuNi

When two or more electrodes are listed for a given combination of base metals, the first is suitable for most applications. Others have been listed as alternates, or as recommendations for more demanding applications, or for use where cracking might occur with particular lots or heats of material welded with the first electrode. The service requirements at the joint will dictate the most suitable electrode. Recommended electrodes were selected because they are adequate for most applications, and because they are commonly available. The recommendations do not imply that other electrodes would be unsuitable for many applications. In those instances the electrodes shown were selected on the basis of material costs.

Suppliers may be contacted regarding preheat and postheat treatments for best serviceability.

Electrode designations are in accordance with the following AWS–ASTM Filler Metal Specifications: Stainless steels, AWS A5.4, ASTM A298, nickel-base alloys. AWS A5.11, ASTM B295, mild steels, AWS A5.1, ASTM A233; low alloy steels, AWS A5.5, ASTM A316; copper alloys, AWS A5.6, ASTM B225.

[a]Includes selenium-bearing grade, although welding usually is not recommended for any of the free-machining steels when high quality joints are required.
[b]Not a standard AISI grade designation.
[d]Butter chromium steel with E309; complete joint with E308.
[e]ENiCrFe-3 preferred, especially for joints for high temperature service in low sulfur atmospheres, or when one member is a copper-bearing alloy.
[g]Butter chromium steel with E309, and chromium-nickel steel with E312; complete joint with E308.
[l]E309 or E310 may be used when matching composition weld metal is not required.
[o]Butter copper-nickel member with two or more layers of ENi-CrFe-3; complete joint with ENiCrFe-3.
[p]Butter free-machining steel; complete joint with ENiCrFe-3.
[q]Butter free-matching steel with two layers of E309 (312 for chromium-nickel free-machining steel); and butter nickel-copper member with one or two layers of ENiCrFe-3, complete joint with ENiCrFe-3.
[r]Any E60XX or E70XX mild steel electrode.
[s]E8015-B2L, E8016-B2, or E8018-B2 (low alloy steel electrodes).
[t]E9015-B3L, E9015-B3, E9016-B3, or E9018-B3 (low alloy steel electrodes).
[u]E7015, E7016, E7018 or E7028 (mild steel electrodes).

Source: American Welding Society, *Welding Handbook*, 6th ed., section 4, p. 64.12.

Table **4-7** Covered electrodes recommended for welds between stainless and heat-resisting steels

BASE METALS→ ↓	201	202	301	302	302B	303[a]	304	304L	305
201	E308[c]	E308[c]	E308[c]	E308	E308	E308; E312	E308	E308	E308
202		E308[c]	E308[c]	E308	E308	E308; E312	E308	E308	E308
301			E308[c]	E308	E308	E308; E312	E308	E308	E308
302				E308	E308	E308; E312	E308	E308	E308
302B					E308	E308; E312	E308	E308	E308
303[a]						E308; E312	E308; E312	E308; E312	E308; E312
304							E308[h]	E308	E308
304L								E308L	E308
305									E308; E310
308									
309									
309S									
310									
310S									
314									
316									
316L									
317									
317L[b]									
321									

When two or more electrodes are listed for a given combination of base metals, the first is suitable for most applications. Others have been listed as alternates, or as recommendations for more demanding applications, or for use where cracking might occur with particular lots or heats of material welded with the first electrode. The service requirements of the joint will dictate the most suitable electrode. Recommended electrodes were selected because they are adequate for most applications, and because they are commonly available. The recommendations do not imply that other electrodes would be

308	309	309S	310	310S	314	316	316L	317	317L[b]	321
E308	E308	E308	E308	E308	E308; E312	E308	E308	E308	E308	E308
E308	E308	E308	E308	E308	E308; E312	E308	E308	E308	E308	E308
E308	E308	E308	E308	E308	E308; E312	E308	E308	E308	E308	E308
E308	E308	E308	E308	E308	E308; E312	E308	E308	E308	E308	E308
E308	E308	E308	E308	E308	E308; E312	E308	E308	E308	E308	E308
E308; E312	E308; E312	E308; E312	E312	E312	E312	E308; E312	E308; E312	E308; E312	E308; E312	E308; E312
E308	E308	E308	E308	E308	E308; E312	E308	E308	E308	E308	E308
E308	E308	E308	E308	E308	E308; E312	E308	E308L	E308	E308L	E308L
E308	E308	E308	E310; E308	E310; E308	E312	E308	E308	E308	E308	E308
E308	E308	E308	E308	E308	E308; E312	E308	E308	E308	E308	E308
	E309	E309	E309	E309	E309; E312	E309	E309; E316	E309	E309	E308; E347
		E309S[i]; E309	E309	E309S[i]; E309	E309 E312	E309S[i]; E316L	E316	E316	E316	E308; E347
			E310	E310	E310; E312	E316	E316	E317	E317L[j]	E308
				E310	E310; E312	E316	E316	E317	E317L[j]	E308
					E310; E312	E316	E316	E317	E317	E308
						E316[k]	E316	E316	E316	E308
							E316L	E316	E316L	E308L
								E317	E317	E308
									E317L[j]	E308L; E347
										E347

unsuitable for many applications. In those instances the electrodes shown were selected on the basis of material costs.

Suppliers may be contacted regarding preheat and postheat treatments for best serviceability.

Electrode designations are in accordance with the following AWS–ASTM Filler Metal Specifications: Stainless steels, AWS A5.4, ASTM A298; nickel-base alloys, AWS A5.11, ASTM B295; mild steels, AWS A5.1, ASTM A233; low alloy steels, AWS A5.5, ASTM A316; copper alloys, AWS A5.6, ASTM B225.

Table **4-7** *(cont.)*

330[b]	347	348	403	405	410	414	416[a]	420	430
E312; E309	E308	E308	[d]; E309	[d]; E309	[d]; E309	[d]; E309	[d]; E309	[d]; E309	[d]; E309
E312; E309	E308	E308	[d]; E309	[d]; E309	[d]; E309	[d]; E309	[d]; E309	[d]; E309	[d]; E309
E312; E309	E308	E308	[d]; E309	[d]; E309	[d]; E309	[d]; E309	[d]; E309	[d]; E309	[d]; E309
E312; E309	E308	E308	[d]; E309	[d]; E309	[d]; E309	[d]; E309	[d]; E309	[d]; E309	[d]; E309
E312; E309	E308	E308	[d]; E309	[d]; E309	[d]; E309	[d]; E309	[d]; E309	[d]; E309	[d]; E309
E312; E309	E308; E312	E308; E312	[d]; E309	[d]; E309	[d]; E309	[d]; E309	[d]; E309	[d]; E309	[d]; E309
E312; E309	E308	E308	[d]; E309	[d]; E309	[d]; E309	[d]; E309	[d]; E309	[d]; E309	[d]; E309
E312; E309	E308L	E308L	[d]; E309	[d]; E309	[d]; E309	[d]; E309	[d]; E309	[d]; E309	[d]; E309
E312; E309	E308	E308	[d]; E309	[d]; E309	[d]; E309	[d]; E309	[d]; E309	[d]; E309	[d]; E309
E312; E309	E308	E308	[d]; E309	[d]; E309	[d]; E309	[d]; E309	[d]; E309	[d]; E309	[d]; E309
E312; E309	E308; E347	E308; E347	[d]; E309	[d]; E309	[d]; E309	[d]; E309	[d]; E309	[d]; E309	[d]; E309
E312; E309	E308; E347	E308; E347	[d]; E309	[d]; E309	[d]; E309	[d]; E309	[d]; E309	[d]; E309	[d]; E309
E312; E310	E308; E347	E308; E347	[d]; E309	[d]; E309	[d]; E309	[d]; E309	[d]; E309	[d]; E309	[d]; E309
E312; E310	E308; E347	E308; E347	[d]; E309	[d]; E309	[d]; E309	[d]; E309	[d]; E309	[d]; E309	[d]; E309
E312; E310	E308	E308	[d]; E309	[d]; E309	[d]; E309	[d]; E309	[d]; E309	[d]; E309	[d]; E309
E312; E309	E308;[k]	E308; E316	[d]; E309	[d]; E309	[d]; E309	[d]; E309	[d]; E309	[d]; E309	[d]; E309
E312; E309	E316L	E316L	[d]; E309	[d]; E309	[d]; E309	[d]; E309	[d]; E309	[d]; E309	[d]; E309

[a]Includes selenium-bearing grade, although welding usually is not recommended for any of the free-machining steels when high quality joints are required.
[b]Not a standard AISI grade designation.
[c]Weld metal will be somewhat lower in strength than base metal.
[d]Butter chromium steel with E309, complete joint with E308.
[e]ENiCrFe-3 preferred, especially for joints for high temperature service in low sulfur atmospheres, or when one member is a copper-bearing alloy.
[f]Butter chromium-nickel steel with E312 complete joint with E310.

430F[a]	431	440A	440B	440C	446	501	502	505[b]	←BASE METALS ↓
[d]; E309	[d]; E309	[d]; E309	[d]; E309	[d]; E309	E310;[e]	[d]; E309	[d]; E309	[d]; E309	201
[d]; E309	[d]; E309	[d]; E309	[d]; E309	[d]; E309	E310;[e]	[d]; E309	[d]; E309	[d]; E309	202
[d]; E309	[d]; E309	[d]; E309	[d]; E309	[d]; E309	E310;[e]	[d]; E309	[d]; E309	[d]; E309	301
[d]; E309	[d]; E309	[d]; E309	[d]; E309	[d]; E309	E310;[e]	[d]; E309	[d]; E309	[d]; E309	302
[d]; E309	[d]; E309	[d]; E309	[d]; E309	[d]; E309	E310;[e]	[d]; E309	[d]; E309	[d]; E309	302B
[d]; E309	[d]; E309	[d]; E309	[d]; E309	[d]; E309	[f]	[g]; E309	[g]; E309	[g]; E309	303[a]
[d]; E309	[d]; E309	[d]; E309	[d] E309	[d]; E309	E310;[e]	[d]; E309	[d]; E309	[d]; E309	304
[d]; E309	[d]; E309	[d]; E309	[d]; E309	[d]; E309	E310;[e]	[d]; E309	[d]; E309	[d]; E309	304L
[d]; E309	[d]; E309	[d]; E309	[d]; E309	[d]; E309	E310;[e]	[d]; E309	[d]; E309	[d]; E309	305
[d]; E309	[d]; E309	[d]; E309	[d]; E309	[d]; E309	E310;[e]	[d]; E309	[d]; E309	[d]; E309	308
[d]; E309	[d]; E309	[d]; E309	[d]; E309	[d]; E309	E310;[e]	[d]; E309;[e]	[d]; E309;[e]	[d]; E309;[e]	309
[d]; E309	[d]; E309	[d]; E309	[d]; E309	[d]; E309	E310;[e]	[d]; E309;[e]	[d]; E309;[e]	[d]; E309;[e]	309S
[d]; E309	[d]; E309	[d]; E309	[d]; E309	[d]; E309	E309; E310;[e]	E310;[e]	E310;[e]	E310;[e]	310
[d]; E309	[d]; E309	[d]; E309	[d]; E309	[d]; E309	E309; E310;[e]	E310;[e]	E310;[e]	E310;[e]	310S
[d]; E309	[d]; E309	[d]; E309	[d]; E309	[d]; E309	E309; E310;[e]	[d]; E309;[e]	[d]; E309;[e]	[d]; E309;[e]	314
[d]; E309	[d]; E309	[d]; E309	[d]; E309	[d]; E309	E310;[e]	[d]; E309;[e]	[d]; E309;[e]	[d]; E309;[e]	316
[d]; E309	[d]; E309	[d]; E309	[d]; E309	[d]; E309	E310;[e]	[d]; E309;[e]	[d]; E309;[e]	[d]; E309;[e]	316L

[g]Butter chromium steel with E309 and chromium-nickel steel with E312; complete joint with E308.

[h]E308L recommended for cryogenic applications.

[i]E309S is not a standard electrode designation. The "S" implies 0.08% C max, compared with a 0.15% C max for E309.

[j]E317L is not a standard electrode designation. The "L" implies 0.04% C max, compared with 0.08% C max for E317.

[k]E16-8-2 recommended for less embrittlement during long-time high temperature service.

[l]E309 or E310 may be used when matching composition weld metal is not required.

[m]Special electrode with matching composition (including carbon content) required when hardness of quenched and tempered weld metal must equal that of base metal.

Table **4-7** *(cont.)*

330[b]	347	348	403	405	410	414	416[a]	420	430
E312; E309	E308L;[k]	E308L;[k]	[d]; E309	[d]; E309	[d]; E309	[d]; E309	[d]; E309	[d]; E309	[d]; E309
E312; E309	E308L	E308L	[d]; E309	[d]; E309	[d]; E309	[d]; E309	[d]; E309	[d]; E309	[d]; E309
E312; E309	E347	E347	[d]; E309	[d]; E309	[d]; E309	[d]; E309	[d]; E309	[d]; E309	[d]; E309
E330	E312; E309	E312; E309	[g]; E312	[g]; E312	[g]; E312	[g]; E312	[g]; E312	[g]; E312	[g]; E312
	E347	E347	[d]; E309	[d]; E309	[d]; E309	[d]; E309	[d]; E309	[d]; E309	[d]; E309
		E348	[d]; E309	[d]; E309	[d]; E309	[d]; E309	[d]; E309	[d]; E309	[d]; E309
			E410;[l]	E410;[l]	E410;[l]	E410;[l]	E410;[d]	E410;[l]	E430;[l]
				E410;[l]	E410;[l]	E410;[l]	E410;[d]	E410;[l]	E430;[l]
					E410;[l]	E410;[l]	E410;[d]	E410;[l]	E430;[l]
						[m]; E309	E410;[d]	E410;[l]	E430;[l]
							E410; E309	E309	E410; E309
								E420; E410;[l]	E410;[l]
									E430;[l]

[n]28 Cr is not a standard electrode designation. It signifies an electrode which has a typical as-deposited analysis of 0.10 C, 1.0 Mn, 0.50 Si, 29 Cr and 0.15 N.
[o]Butter copper-nickel member with two or more layers of ENiCrFe-3; complete joint with ENiCrFe-3.
[p]Butter free-machining steel with two layers of E309 (E312 for chromium-nickel free-machining steel), complete joint with ENiCrFe-3.
[q]Butter free-machining steel with two layers of E309 (312 for chromium-nickel free-machining steel); and butter nickel-copper member with one or two layers of ENiCrFe-3, complete joint with ENiCrFe-3.

430F[a]	431	440A	440B	440C	446	501	502	505[b]	←BASE METALS ↓
[d]; E309	[d]; E309	[d]; E309	[d]; E309	[d]; E309	E310;[e]	[d]; E309;[e]	[d]; E309;[e]	[d]; E309;[e]	317
[d]; E309	[d]; E309	[d]; E309	[d]; E309	[d]; E309	E310;[e]	[d]; E309;[e]	[d]; E309;[e]	[d]; E309;[e]	317L[b]
[d]; E309	[d]; E309	[d]; E309	[d]; E309	[d]; E309	E310;[e]	[d]; E309;[e]	[d]; E309;[e]	[d]; E309;[e]	321
[g]; E312	[g]; E312	[g]; E312	[g]; E312	[g]; E312	[g]; E312	[g]; E312;[e]	[g]; E312;[e]	[g]; E312;[e]	330[b]
[d]; E309	[d]; E309	[d]; E309	[d]; E309	[d]; E309	[d]; E309	[d]; E309;[e]	[d]; E309;[e]	[d]; E309;[e]	347
[d]; E309	[d]; E309	[d]; E309	[d]; E309	[d]; E309	[d]; E309	[d]; E309;[e]	[d]; E309;[e]	[d]; E309;[e]	348
E309	E410;[l]	E410;[l]	E410;[l]	E410;[l]	E410;[l]	E502;[l]	E502;[l]	E505;[l]	403
E309	E410;[l]	E410;[l]	E410;[l]	E410;[l]	E410;[l]	E502;[l]	E502;[l]	E505;[l]	405
E309	E410;[l]	E410;[l]	E410;[l]	E410;[l]	E410;[l]	E502;[l]	E502;[l]	E505;[l]	410
E309	E410;[l]	E410;[l]	E410;[l]	E410;[l]	E410;[l]	E502;[l]	E502;[l]	E505;[l]	414
E309	E410; E309	E309	E309	E309	E309	E309	E309	E309	416[a]
E309	E410;[l]	E420;[l]	E420;[l]	E420;[l]	E430;[l]	E502;[l]	E502;[l]	E505;[l]	420
E309	E430;[l]	E430;[l]	E430;[l]	E430;[l]	E430;[l]	E502;[l]	E502;[l]	E505;[l]	430
E309	E309	E309	E309	E309	E309; E310	E309	E309	E309	430F[a]
	[m]; E309	[m]; E309	[m]; E309	[m]; E309	E309; E310	E502;[l]	E502;[l]	E505;[l]	431
		[m]; E309	[m]; E309	[m]; E309	E309; E310	E502;[l]	E502;[l]	E505;[l]	440A
			[m]; E309	[m]; E309	E309; E310	E502;[l]	E502;[l]	E505;[l]	440B
				[m]; E309	E309; E310	E502;[l]	E502;[l]	E505;[l]	440C
					28 Cr[n]; E310	E502;[l];[e]	E502;[l];[e]	E505;[l];[e]	446
						[m]; E502;[l]	E502;[l]	E502;[l]	501
							E502;[l]	E502;[l]	502
								E505;[l]	505[b]

[r]Any E60XX or E70XX mild steel electrode.
[s]E8015-B2L, or E8016-B2, or E8018-B2 (low alloy steel electrodes).
[t]E9015-B3L, E9015-B3, E9016-B3, or E9018-B3 (low alloy steel electrodes).
[u]E7015, E7016, E7018 or E7028 (mild steel electrodes).

Source: American Welding Society, *Welding Handbook*, 6th ed., section 4, p. 64.30.

4-7 FILLER METALS FOR ALUMINUM ALLOYS

Aluminum is the second most widely fabricated metal. It is extensively welded by both gas metal–arc and gas tungsten–arc welding processes. Because of the relatively poor operability and the necessity for complete flux removal after welding, covered filler rods are seldom used for welding aluminum. The filler metals that have been found to be suitable for general-purpose arc welding of various combinations of aluminum alloys are listed in Tables 4-8 to 4-10.

The standard sizes of bare aluminum welding electrodes on expendable spools are 0.030, 3/64, 1/16, 3/32, and 1/8 inch. These are available on 10-, 12½-, and 15-pound spools. Wire diameters up to 1/16 inch are also available on 1-pound spools.

Bare welding rods in straight lengths and coils are supplied in 1/16-, 3/32-, 1/8-, 5/32-, 3/16-, and 1/4-inch diameters. The standard-length aluminum rod is 36 inches, and the most common package contains five pounds. Other lengths and sizes of packages available are described in AWS A5.10-69.

Covered aluminum electrodes used for arc welding of aluminum are either 14 inches long, produced in diameters of 3/32, 1/8, 5/32, 3/16, and 1/4 inch; or 18 inches long, in diameters of 5/16 and 3/8 inch. Standard packages containing one, five and ten pounds are available.

4-8 FILLER METALS FOR NICKEL ALLOYS

Prior to selecting a nickel-alloy filler metal, the composition of the base metal to be welded should be checked. The analysis of the metal to be welded should match the composition of the filler metal as closely as possible. The filler metal will, however, have a percentage of other elements that have been added to satisfy usability requirements, for example, to control porosity or hot cracking tendencies. Two of the most important guidelines for the welder to follow in using nickel-alloy filler metals are (1) to operate at low heat inputs, and (2) to minimize puddling or wide weaving.

The uses of the various types of nickel and nickel-alloy filler metals are shown in Table 4-11. The joining processes applicable to nickel alloys are shown in Table 4-12.

4-9 COPPER AND COPPER-ALLOY FILLER METALS

The specifications for filler metals available for welding copper and copper alloys are covered in AWS specifications A5.6-69 and A5.7-69 for arc and gas welding, respectively.

Generally, it is not advisable to use bare copper-alloy rods as electrodes, because the deposited metal may contain minute particles of oxide, which results in a loss of ductility within the weld joint. Most copper and copper alloy–covered electrodes are designed to operate with dc reverse polarity.

The filler metals used in arc welding of coppers and copper alloys to dissimilar metals are listed in Table 4-13 (*a* and *b*).

4-10 MAGNESIUM AND MAGNESIUM-ALLOY FILLER METALS

Magnesium welding rods and electrodes are classified on the basis of filler metal chemistry. The classification used in AWS A5.19 and MIL-R-6944 is based on the standard nomenclature established in *ASTM Recommended Practices B275, Codification of Light Metals and Alloys, Cast and Wrought.*

The composition of the four most commonly used electrode wires for gas metal–arc welding are given in Table 4-14. The choice of filler metal is governed by the composition of the base metal.

Filler metals ERAZ61A and ERAZ92A are considered satisfactory for welding the following magnesium alloys in cast or wrought form to themselves or to each other: AZ10A, AZ31B, AZ31C, AZ61A, AZ80A, ZE10A, ZK21A, HK31A, HM21A, and HM31A. It should be noted that ERAZ61A is preferred over ERAZ92A because it is less costly, and that EREZ33A is recommended for joining these two alloys over any other cast magnesium alloys.

Filler metals for welding magnesium are available on expendable spools containing $\frac{3}{4}$ or 10 pounds of wire and in straight-length rods. The spooled electrode is available in standard diameter sizes of 0.040, $\frac{3}{64}$, $\frac{1}{16}$, $\frac{3}{32}$, and $\frac{1}{8}$ inch. Uncoated straight-length rods are 36 inches long and are furnished in diameters of $\frac{1}{16}$, $\frac{3}{32}$, $\frac{1}{8}$, $\frac{5}{32}$, and $\frac{1}{4}$ inch. The range of metal thicknesses that can be welded with the various filler metals, amperages, processes, and modes of metal transfer to be used is shown in Table 4-15, parts *a* and *b*.

4-11 FILLER METALS FOR TITANIUM AND TITANIUM ALLOYS

Titanium and titanium alloys have a great affinity for oxygen, nitrogen, carbon, hydrogen, and other weld contaminants. Because of this, the selection of the correct filler metal depends on the application, type of base metal, and expected pick-up of impurities during welding.

The 14 titanium and titanium-alloy filler metals are classified in AWS A5.16, in accordance with the AWS classification system, being used as either electrodes or gas-welding rods.

ERTi-1 to ERTi-4 electrodes are essentially unalloyed titanium and are used to weld unalloyed titanium-base metals. The higher the number of the ERTi filler is, the higher the corresponding tensile strength will be.

(Text continues on page 88)

Table **4-8** Filler metals commonly used in arc welding combinations of aluminum alloys

	EASE OF WELDING					STRENGTH OF WELDED JOINT (AS WELDED)						CORROSION RESISTANCE					
ALLOYS TO BE WELDED	1100	4043	5654	5356	5554	5556	1100	4043	5654	5356	5554	5556	1100	4043	5654	5356	5554	5556
To weld alloy 1100 to:																		
1100	B	A	—	C	—	C	B	A	—	A	—	A	A	A	—	—	—	—
3003, alclad 3003	A	A	—	B	—	B	B	A	—	A	—	A	A	A	—	—	—	—
3004, alclad 3004	C	A	—	B	—	B	B	A	—	A	—	A	A	A	—	—	—	—
5005, 5050	B	A	—	B	—	B	B	A	—	A	—	A	A	A	—	—	—	—
5052, 5154, 5454	—	—	—	—	—	—	—	—	—	—	—	—	—	—	—	—	—	—
5083, 5086, 5456	—	—	—	—	—	—	—	—	—	—	—	—	—	—	—	—	—	—
6063, 6101	—	A	—	B	—	B	—	A	—	A	—	A	—	A	—	—	—	—
6061	—	A	—	B	—	B	—	A	—	A	—	A	—	A	—	—	—	—
To weld alloy 3003 to:																		
3003, alclad 3003	A	A	—	B	—	B	C	B	—	A	—	A	A	A	—	—	—	—
3004, alclad 3004	—	A	—	B	—	B	—	B	—	A	—	A	—	A	—	—	—	—
5005, 5050	B	A	—	B	—	B	C	B	—	A	—	A	A	A	—	—	—	—
5052	—	A	—	B	—	B	—	B	—	A	—	A	—	A	—	—	—	—
5154	—	A	C	B	C	B	—	B	A	A	A	A	—	C	A	B	A	B
5454	—	A	—	B	C	B	—	B	—	A	A	A	—	C	—	B	A	B
5083, 5086, 5456	—	A	—	A	—	A	—	B	—	A	—	A	—	B	—	A	—	A
6063, 6101	—	A	—	B	—	B	—	B	—	A	—	A	—	A	—	—	—	—
6061	—	A	—	B	—	B	—	B	—	A	—	A	—	A	—	—	—	—
To weld alclad 3003 to:																		
Alclad 3003	A	A	—	B	—	B	C	B	—	A	—	A	A	A	—	B	—	B
3004, alclad 3004	—	A	—	B	—	B	—	B	—	A	—	A	—	A	—	B	—	B
5005, 5050	B	A	—	B	—	B	C	B	—	A	—	A	A	A	—	B	—	B
5052	—	A	—	B	—	B	—	B	—	A	—	A	—	A	—	B	—	B
5154	—	A	C	B	C	B	—	B	A	A	A	A	—	C	A	B	A	B
5454	—	A	—	B	C	B	—	B	—	A	A	A	—	C	—	B	A	B
5083, 5086, 5456	—	A	—	A	—	A	—	B	—	A	—	A	—	B	—	A	—	A
6063, 6101	—	A	—	B	—	B	—	B	—	A	—	A	—	A	—	B	—	B
6061	—	A	—	B	—	B	—	B	—	A	—	A	—	A	—	B	—	B

To weld alloy 3004 to:																		
3004, alclad 3004	—	A	C	B	C	B	—	D	C	B	C	A	—	A	B	—	B	—
5005, 5050	—	A	—	B	—	B	—	B	—	A	—	A	—	A	—	—	—	—
5052	—	A	—	B	—	B	—	C	—	B	—	A	—	A	—	—	—	—
5154	—	A	C	B	C	B	—	D	C	B	C	A	—	C	A	B	A	B
5454	—	A	—	B	C	B	—	D	—	B	C	A	—	C	—	B	A	B
5083, 5086, 5456	—	A	—	A	—	A	—	C	—	B	—	A	—	B	—	A	—	A
6063, 6101	—	A	—	B	—	B	—	C	—	B	—	A	—	A	—	—	—	—
6061	—	A	—	B	—	B	—	C	—	B	—	A	—	A	—	—	—	—
To weld alclad 3004 to:																		
Alclad 3004	—	A	C	B	C	B	—	D	C	B	C	A	—	A	B	C	B	C
5005, 5050	—	A	—	B	—	B	—	B	—	A	—	A	—	A	—	B	—	B
5052	—	A	—	B	—	B	—	C	—	B	—	A	—	A	—	B	—	B
5154	—	A	C	B	C	B	—	D	C	B	C	A	—	C	A	B	A	B
5454	—	A	—	B	C	B	—	D	—	B	C	A	—	C	—	B	A	B
5083, 5086, 5456	—	A	—	A	—	A	—	C	—	B	—	A	—	B	—	A	—	A
6063, 6101	—	A	—	B	—	B	—	C	—	B	—	A	—	A	—	B	—	B
6061	—	A	—	B	—	B	—	C	—	B	—	A	—	A	—	B	—	B
To weld alloy 5005 or 5050 to:																		
5005, 5050	C	A	—	B	—	B	—	B	—	A	—	A	A	A	—	—	—	—
5052	—	A	—	B	—	B	—	B	—	A	—	A	—	A	—	—	—	—
5154	—	A	C	B	C	B	—	B	A	A	A	A	—	C	A	B	A	B
5454	—	A	—	B	C	B	—	B	—	A	A	A	—	C	—	B	A	B
5083, 5086, 5456	—	A	—	A	—	A	—	B	—	A	—	A	—	B	—	A	—	A
6063, 6101	—	A	—	B	—	B	—	B	—	A	—	A	—	A	—	—	—	—
6061	—	A	—	B	—	B	—	B	—	A	—	A	—	A	—	—	—	—
To weld alloy 5052 to:																		
5052	—	A	B	A	C	A	—	D	C	B	C	A	—	C	B	—	B	—
5154	—	A	B	A	C	A	—	D	C	B	C	A	—	C	A	B	A	B
5454	—	A	B	A	C	A	—	D	C	B	C	A	—	C	B	B	A	B
5083, 5086, 5456	—	—	—	A	—	A	—	—	—	B	—	A	—	—	—	A	—	A
6063, 6101	—	A	C	B	C	B	—	B	A	A	A	A	—	A	B	—	B	—
6061	—	A	C	B	C	B	—	D	C	B	C	A	—	A	B	—	B	—
To weld alloy 5083 or 5456 to:																		
5154	—	—	B	A	B	A	—	—	C	B	C	A	—	—	A	A	A	A
5454	—	—	—	A	B	A	—	—	—	B	C	A	—	—	—	B	A	B

Table **4-8** *(cont.)*

ALLOYS TO BE WELDED	EASE OF WELDING						STRENGTH OF WELDED JOINT (AS WELDED)						CORROSION RESISTANCE					
	1100	4043	5654	5356	5554	5556	1100	4043	5654	5356	5554	5556	1100	4043	5654	5356	5554	5556
To weld alloy 5083 or 5456 to:																		
5083, 5086, 5456	—	—	—	A	—	A	—	—	—	B	—	A	—	—	—	A	—	A
6063, 6101	—	A	B	A	B	A	—	B	A	A	A	A	—	A	A	A	A	A
6061	—	A	B	A	B	A	—	D	C	B	C	A	—	A	A	A	A	A
To weld alloy 5086 to:																		
5154	—	—	B	A	B	A	—	—	C	B	C	A	—	—	A	A	A	A
5454	—	—	—	A	B	A	—	—	—	B	C	A	—	—	—	B	A	B
5086	—	—	—	A	—	A	—	—	—	B	—	A	—	—	—	A	—	A
6063, 6101	—	A	B	A	B	A	—	B	A	A	A	A	—	A	A	A	A	A
6061	—	A	B	A	B	A	—	D	C	B	C	A	—	A	A	A	A	A
To weld alloy 5154 to:																		
5154	—	—	B	A	B	A	—	—	C	B	C	A	—	—	A	—	A	—
5454	—	—	B	B	B	A	—	—	C	B	C	A	—	—	A	B	A	B
6063, 6101	—	A	C	B	C	B	—	B	A	A	A	A	—	A	B	—	B	—
6061	—	A	C	B	C	B	—	D	C	B	C	A	—	A	B	—	B	—
To weld alloy 5454 to:																		
5454	—	—	B	A	B	A	—	—	C	B	C	A	—	—	B	B	A	B
6063, 6101	—	A	C	B	C	B	—	B	A	A	A	A	—	B	B	—	A	—
6061	—	A	C	B	C	B	—	D	C	B	C	A	—	B	B	—	A	—
To weld alloy 6061 to:																		
6063, 6101	—	A	C	B	C	B	—	B	A	A	A	A	—	A	B	C	B	C
6061	—	A	C	B	C	B	—	D	C	B	C	A	—	A	B	C	B	C
To weld alloy 6063 or 6101 to:																		
6063, 6101	—	A	C	B	C	B	—	B	A	A	A	A	—	A	B	C	B	C

ALLOYS TO BE WELDED	SERVICE AT SUSTAINED TEMP ABOVE 150°F						COLOR MATCH AFTER ANODIZING						DUCTILITY					
	1100	4043	5654	5356	5554	5556	1100	4043	5654	5356	5554	5556	1100	4043	5654	5356	5554	5556
To weld alloy 1100 to:																		
1100	A	A	—	—	—	—	A	—	—	B	—	B	A	D	—	B	—	C
3003, alclad 3003	A	A	—	—	—	—	A	—	—	B	—	B	A	D	—	B	—	C
3004, alclad 3004	A	A	—	—	—	—	A	—	—	B	—	B	A	D	—	B	—	C
5005, 5050	A	A	—	—	—	—	A	—	—	B	—	B	A	D	—	B	—	C
5052, 5154, 5454	—	—	—	—	—	—	—	—	—	—	—	—	—	—	—	—	—	—
5083, 5086, 5456	—	—	—	—	—	—	—	—	—	—	—	—	—	—	—	—	—	—
6063, 6101	—	A	—	—	—	—	—	—	—	A	—	A	—	C	—	A	—	B
6061	—	A	—	—	—	—	—	—	—	A	—	A	—	C	—	A	—	B
To weld alloy 3003 to:																		
3003, alclad 3003	A	A	—	—	—	—	A	—	—	B	—	B	A	D	—	B	—	C
3004, alclad 3004	—	A	—	—	—	—	—	—	—	A	—	A	—	C	—	A	—	B
5005, 5050	A	A	—	—	—	—	A	—	—	B	—	B	A	D	—	B	—	C
5052	—	A	—	—	—	—	—	—	—	A	—	A	—	C	—	A	—	B
5154	—	—	—	—	—	—	—	—	B	A	A	A	—	C	A	A	A	B
5454	—	A	—	—	A	—	—	—	—	A	A	A	—	C	—	A	A	B
5083, 5086, 5456	—	—	—	—	—	—	—	—	—	A	—	A	—	C	—	A	—	B
6063, 6101	—	A	—	—	—	—	—	—	—	A	—	A	—	C	—	A	—	B
6061	—	A	—	—	—	—	—	—	—	A	—	A	—	C	—	A	—	B
To weld alclad 3003 to:																		
Alclad 3003	A	A	—	—	—	—	A	—	—	B	—	B	A	D	—	B	—	C
3004, alclad 3004	—	A	—	—	—	—	—	—	—	A	—	A	—	C	—	A	—	B
5005, 5050	A	A	—	—	—	—	A	—	—	B	—	B	A	D	—	B	—	C
5052	—	A	—	—	—	—	—	—	—	A	—	A	—	C	—	A	—	B
5154	—	—	—	—	—	—	—	—	B	A	A	A	—	C	A	A	A	B
5454	—	A	—	—	A	—	—	—	—	A	A	A	—	C	—	A	A	B
5083, 5086, 5456	—	—	—	—	—	—	—	—	—	A	—	A	—	C	—	A	—	B
6063, 6101	—	A	—	—	—	—	—	—	—	A	—	A	—	C	—	A	—	B
6061	—	A	—	—	—	—	—	—	—	A	—	A	—	C	—	A	—	B
To weld alloy 3004 to:																		
3004, alclad 3004	—	A	—	—	A	—	—	—	B	A	A	A	—	C	A	A	A	B

Table **4-8** (*cont.*)

ALLOYS TO BE WELDED	SERVICE AT SUSTAINED TEMP ABOVE 150°F						COLOR MATCH AFTER ANODIZING						DUCTILITY					
	1100	4043	5654	5356	5554	5556	1100	4043	5654	5356	5554	5556	1100	4043	5654	5356	5554	5556
To weld alloy 3004 to:																		
5005, 5050	—	A	—	—	—	—	—	—	—	A	—	A	—	C	—	A	—	B
5052	—	A	—	—	—	—	—	—	—	A	—	A	—	C	—	A	—	B
5154	—	—	—	—	—	—	—	—	B	A	A	A	—	C	A	A	A	B
5454	—	A	—	—	A	—	—	—	—	A	A	A	—	C	—	A	A	B
5083, 5086, 5456	—	—	—	—	—	—	—	—	—	A	—	A	—	C	—	A	—	B
6063, 6101	—	A	—	—	—	—	—	—	—	A	—	A	—	C	—	A	—	B
6061	—	A	—	—	—	—	—	—	—	A	—	A	—	C	—	A	—	B
To weld alclad 3004 to:																		
Alclad 3004	—	A	—	—	A	—	—	—	B	A	A	A	—	C	A	A	A	B
5005, 5050	—	A	—	—	—	—	—	—	—	A	—	A	—	C	—	A	—	B
5052	—	A	—	—	—	—	—	—	—	A	—	A	—	C	—	A	—	B
5154	—	—	—	—	—	—	—	—	B	A	A	A	—	C	A	A	A	B
5454	—	A	—	—	A	—	—	—	—	A	A	A	—	C	—	A	A	B
5083, 5086, 5456	—	—	—	—	—	—	—	—	—	A	—	A	—	C	—	A	—	B
6063, 6101	—	A	—	—	—	—	—	—	—	A	—	A	—	C	—	A	—	B
6061	—	A	—	—	—	—	—	—	—	A	—	A	—	C	—	A	—	B
To weld alloy 5005 or 5050 to:																		
5005, 5050	A	A	—	—	—	—	A	—	—	B	—	B	A	D	—	B	—	C
5052	—	A	—	—	—	—	—	—	—	A	—	A	—	C	—	A	—	B
5154	—	—	—	—	—	—	—	—	B	A	A	A	—	C	A	A	A	B
5454	—	A	—	—	A	—	—	—	—	A	A	A	—	C	—	A	A	B
5083, 5086, 5456	—	—	—	—	—	—	—	—	—	A	—	A	—	C	—	A	—	B
6063, 6101	—	A	—	—	—	—	—	—	—	A	—	A	—	C	—	A	—	B
6061	—	A	—	—	—	—	—	—	—	A	—	A	—	C	—	A	—	B
To weld alloy 5052 to:																		
5052	—	A	—	—	B	—	—	—	A	A	B	B	—	C	A	A	A	B
5154	—	—	—	—	—	—	—	—	A	A	B	B	—	C	A	A	A	B
5454	—	A	—	—	A	—	—	—	B	A	A	A	—	C	A	A	A	B
5083, 5086, 5456	—	—	—	—	—	—	—	—	—	A	—	A	—	—	—	A	—	B

6063, 6101	—	A	—	—	A	—	—	—	B	A	A	A	—	C	A	A	A	B
6061	—	A	—	—	A	—	—	—	A	A	B	B	—	C	A	A	A	B
To weld alloy 5083 or 5456 to:																		
5154	—	—	—	—	—	—	—	—	B	A	A	A	—	—	A	A	A	B
5454	—	—	—	—	—	—	—	—	—	A	A	A	—	—	—	A	A	B
5083, 5086, 5456	—	—	—	—	—	—	—	—	—	A	—	A	—	—	—	A	—	B
6063, 6101	—	—	—	—	—	—	—	—	B	A	A	A	—	C	A	A	A	B
6061	—	—	—	—	—	—	—	—	B	A	A	A	—	C	A	A	A	B
To weld alloy 5086 to:																		
5154	—	—	—	—	—	—	—	—	B	A	A	A	—	—	A	A	A	B
5454	—	—	—	—	—	—	—	—	—	A	A	A	—	—	—	A	A	B
5086	—	—	—	—	—	—	—	—	—	A	—	A	—	—	—	A	—	B
6063, 6101	—	—	—	—	—	—	—	—	B	A	A	A	—	C	A	A	A	B
6061	—	—	—	—	—	—	—	—	B	A	A	A	—	C	A	A	A	B
To weld alloy 5154 to:																		
5154	—	—	—	—	—	—	—	—	A	A	B	B	—	—	A	A	A	B
5454	—	—	—	—	—	—	—	—	B	A	A	A	—	—	A	A	A	B
6063, 6101	—	—	—	—	—	—	—	—	B	A	A	A	—	C	A	A	A	B
6061	—	—	—	—	—	—	—	—	A	A	B	B	—	C	A	A	A	B
To weld alloy 5454 to:																		
5454	—	—	—	—	A	—	—	—	B	A	A	A	—	—	A	A	A	B
6063, 6101	—	A	—	—	A	—	—	—	B	A	A	A	—	C	A	A	A	B
6061	—	A	—	—	A	—	—	—	B	A	A	A	—	C	A	A	A	B
To weld alloy 6061 to:																		
6063, 6101	—	A	—	—	B	—	—	—	B	A	A	A	—	C	A	A	A	B
6061	—	A	—	—	B	—	—	—	B	A	B	B	—	C	A	A	A	B
To weld alloy 6063 or 6101 to:																		
6063, 6101	—	A	—	—	B	—	—	—	B	A	A	A	—	C	A	A	A	B

(Ratings are relative, in decreasing order of merit, and apply only within a given block. Combinations having no rating are not recommended.)

Source: By permission, from *Metals Handbook* Volume 6, Copyright American Society for Metals, 1971.

Table **4-9** Guide to selection of filler-metal alloys for arc welding various combinations of heat-treatable aluminum alloys

Ratings are relative, in decreasing order of merit, and apply to a given base-metal combination and postweld condition. The use of base metals as filler metals, or of combinations indicated here by dashes as having no ratings, is not recommended.

ALLOYS TO BE WELDED[a]	POSTWELD CONDITION[b]	EASE OF WELDING 2319	4043	4145	5039	5556[f]	5554[g]	STRENGTH[c] 2319	4043	4145	5039	5556[f]	5554[g]
To weld 2014 or 2024 to:													
2014, 2024 or 2219	X	C	B	A	—	—	—	A	B	A	—	—	—
	Y	C	B	A	—	—	—	A	C	B	—	—	—
To weld 2219 to:													
2219	X	A	A	A	—	—	—	A	B	B	—	—	—
	Y or Z	A	A	A	—	—	—	A	C	B	—	—	—
To weld 6061, 6063 or 6101 to:													
1100	X	—	A	—	—	B	—	—	A	—	—	A	—
2014 or 2024	X	—	B	A	—	—	—	—	A	A	—	—	—
2219	X	—	A	A	—	—	—	—	A	A	—	—	—
3003, 3004, 5005 or 5050	X	—	A	—	—	B	—	—	B	—	—	A	—
5052, 5154 or 5454	X	—	A	—	—	B	C	—	C	—	—	A	B
5083, 5086 or 5456	X	—	—	—	—	A	B	—	—	—	—	A	B
6061, 6063 or 6101	X	—	A	—	—	B	C	—	C	—	—	A	B
	Y or Z	—	A	—	—	h	B	—	A	—	—	h	B
To weld 7005 or 7039 to:													
5052, 5154 or 5454	X	—	A	—	A	A	B	—	D	—	A	B	C
5083, 5086 or 5456	X	—	—	—	A	A	—	—	—	—	A	B	—
6061 or 6063	X	—	A	—	A	A	B	—	D	—	A	B	C
	Y or Z	—	A	—	A	h	B	—	C	—	A	h	B
7005 or 7039	X	—	—	—	A	A	—	—	—	—	A	B	—
	Y or Z	—	—	—	A	h	—	—	—	—	A	h	—
To weld 7075 or 7178 to:													
7075 or 7178	X	—	A	A	B	B	—	—	C	C	A	B	—
	Y or Z	—	A	A	B	h	—	—	B	B	A	h	—

[a]Ratings for both bare and alclad materials are the same. [b]X = naturally aged for 30 days or longer; Y = postweld solution heat treated and artificially aged; Z = postweld artificially aged. [c]Ultimate strength from cross-weld tensile test. [d]Ratings based on free-bend elongation of weld. [e]Ratings based on continuous or alternate immersion in fresh or salt water. [f]5183 and 5356 have the same ratings as 5556. [g]Filler alloy 5554 is suitable for

	DUCTILITY[d]						CORROSION RESISTANCE[e]					
	2319	4043	4145	5039	5556[f]	5554[g]	2319	4043	4145	5039	5556[f]	5554[g]
To weld 2014 or 2024 to:												
2014, 2024	A	A	B	—	—	—	A	B	B	—	—	—
or 2219	A	B	B	—	—	—	A	B	B	—	—	—
To weld 2219 to:												
2219	A	B	B	—	—	—	A	B	B	—	—	—
	A	B	B	—	—	—	A	B	B	—	—	—
To weld 6061, 6063 or 6101 to:												
1100	—	B	—	—	A	—	—	A	—	—	B	—
2014 or 2024	—	A	B	—	—	—	—	A	A	—	—	—
2219	—	A	B	—	—	—	—	A	A	—	—	—
3003, 3004, 5005 or 5050	—	B	—	—	A	—	—	A	—	—	B	—
5052, 5154 or 5454	—	B	—	—	A	A	—	A	—	—	B	A
5083, 5086 or 5456	—	—	—	—	B	A	—	—	—	—	A	A
6061, 6063	—	B	—	—	A	A	—	A	—	—	C	B
or 6101	—	B	—	—	[h]	A	—	A	—	—	[h]	B
To weld 7005 or 7039 to:												
5052, 5154 or 5454	—	B	—	A	A	A	—	B	—	A	A	A
5083, 5086 or 5456	—	—	—	A	A	—	—	—	—	A	A	—
6061 or 6063	—	B	—	A	A	A	—	A	—	A	A	A
	—	B	—	A	[h]	A	—	A	—	A	[h]	A
7005 or 7039	—	—	—	A	A	—	—	—	—	A	A	—
	—	—	—	A	[h]	—	—	—	—	A	[h]	—
To weld 7075 or 7178 to:												
7075 or 7178	—	B	B	A	A	—	—	B	B	A	A	—
	—	B	B	A	[h]	—	—	A	A	B	[h]	—

welding 6061, 6063 and 7005 prior to brazing. [h]Filler alloy not recommended because of possible susceptibility to stress-corrosion cracking when postweld heat treated.

Source: By permission from *Metals Handbook* Volume 6, Copyright American Society for Metals, 1971.

Table **4-10** Filler metals suitable for general-purpose gas shielded–arc welding of various combinations of aluminum alloy base metals

BASE METALS TO BE WELDED (IN COLUMN BELOW, AND IN COLUMN HEADS AT RIGHT)	319, 333, 355, C355	13, 43, 356	214, A214, B214, F214	7005, 7039, A612, C612, D612	6061, 6063, 6101, 6151	5456	5454	5154, 5254[a]	5086	5083	5052, 5652[a]	5005, 5050	3004, ALCLAD 3004	2219	2014, 2024	1100, 3003, ALCLAD 3003	1060, EC
1060, EC	4145[b,c]	4043[c,d]	4043[c,e]	4043[c]	4043[c]	5356[b]	4043[c,e]	4043[c,e]	5356[b]	5356[b]	4043[c]	1100[b]	4043	4145	4145	1100[b]	1260[b,f]
1100, 3003, alclad 3003	4145[b,c]	4043[c,d]	4043[c,e]	4043[c]	4043[c]	5356[b]	4043[c,e]	4043[c,e]	5356[b]	5356[b]	4043[c,e]	4043[e]	4043[e]	4145	4145	1100[b]	
2014, 2024	4145[g]	4145	—	—	4145	—	—	—	—	—	—	—	—	4145[g]	4145[g]		
2219	4145[b,c,g]	4145[b,c]	4043[c]	4043[c]	4043[c,d]	4043	4043[c]	4043[c]	4043	4043	4043[c]	4043	4043	2319[b,c,d]			
3004, alclad 3004	4043[c]	4043[c]	5654[h]	5356[e]	4043[h]	5356[e]	5654[h]	5654[h]	5356[e]	5356[e]	4043[c,e]	4043[e]	4043[e]				
5005, 5050	4043[c]	4043[c]	5654[h]	5356[e]	4043[h]	5356[e]	5654[h]	5654[h]	5356[e]	5356[e]	4043[c,e]	4043[e,j]					
5052, 5652[a]	4043[c]	4043[c,h]	5654[h]	5356[e,k]	5356[b,h]	5356[h]	5654[h]	5654[a,h]	5356[e]	5356[e]	5654[a,b,h]						
5083	—	5356[b,c,e]	5356[e]	5183[e,k]	5356[e]	5183[e]	5356[e]	5356[e]	5356[e]	5183[e]							
5086	—	5356[b,c,e]	5356[e]	5356[e,k]	5356[e]	5356[e]	5356[h]	5356[h]	5356[e]								
5154, 5254[a]	—	4043[c,h]	5654[h]	5356[h,k]	5356[b,h]	5356[h]	5654[h]	5654[a,h]									
5454	4043[c]	4043[c,h]	5654[h]	5356[h,k]	5356[b,h]	5356[h]	5554[b,e]										
5456	—	5356[b,c,e]	5356[e]	5556[e,k]	5356[e]	5556[e]											
6061, 6063, 6101, 6151	4145[b,c]	4043[c,h]	5356[b,h]	5356[b,c,h,k]	4043[c,h]												
7005, 7039, A612, C612, D612	4043[c]	4043[c,h,k]	5356[h,k]	5039[e]													
214, A214, B214, F214	—	4043[c,h]	5654[h,j]														
13, 43, 356	4145[b,c]	4043[c,j]															
319, 333, 355, C355	4145[b,c,j]																

Note: All filler metals shown here are covered by AWS specification A5.10-69, prefixed by the letters "ER". Throughout this table, the prefix has been omitted, to conserve space. Filler metals 5356, 5556, and 5654 are not suitable for sustained service at temperatures higher than 150°F. Other service conditions, such as immersion in fresh or salt water or exposure to specific chemicals, may also limit the choice of filler metal. Where no filler metal is listed, the base-metal combination is not recommended for welding. [a]Base metals 5254 and 5652 are used for hydrogen peroxide service. Filler metal 5654 is used for welding both alloys for service at temperatures of 150°F and below. [b]4043 may be used for some jobs. [c]4047 may be used for some jobs. [d]4145 may be used for some jobs. [e]5183, 5356 or 5556 may be used. [f]1100 may be used for some jobs. [g]2319 may be used for some jobs. [h]5183, 5356, 5554, 5556 and 5654 may be used. In some cases they provide improved color match after anodizing treatment, highest weld ductility, and higher weld strength. Filler metal 5554 is suitable for service at elevated temperature. [j]Filler metal of the same composition as the base metal is sometimes used. [k]5039 may be used for some jobs.

Source: By permission, from *Metals Handbook* Volume 6, Copyright American Society for Metals, 1971.

Table **4-11** Nickel and nickel-alloy filler metals

AWS DESIGNATION	USE
ENi-1	Covered electrode for welding nickel to itself and to steel. It may also be used for surfacing steel.
RNi-2	Bare rods used without flux for oxyacetylene welding of nickel.
ERNi-3	Bare wire or rods for gas-shielded arc welding of nickel to itself and to steel. These may also be used for surfacing steel.
TYPICAL NICKEL-COPPER FILLER METALS	
ENiCu-1	Covered electrode, used for welding Monel to steel and for surfacing steel, as well as the steel on the clad side of Monel clad steel.
ENiCu-2	Same as above but meets more rigid bend requirements.
ENiCu-4	Covered electrode used where Cb is undesirable.
RNiCu-5	Bare wire for oxyacetylene welding of Monel with flux.
ERNiCu-7	Bare wire or rod for gas-shielded arc and submerged arc welding of Monel to itself and for surfacing steel and the alloy side of the clad steel.
NICKEL-CHROMIUM-IRON FILLER METALS	
ENiCrFe-1	Covered electrode, e.g., Inconel 132, for welding Inconel 600.
ENiCr-1	Covered electrode, e.g., Chromend[1] 20/80, for welding high-nickel alloys to carbon steel.
ENiCrFe-3	Covered electrode, e.g., Inconel 182, for more rigid use than ENiCrFe-1; it is used widely for high-temperature applications involving cast base alloys.
RNiCr-4	Bare rod, e.g., Inconel 42, for oxyacetylene welding Inconel 600.
ERNiCrFe-5	Bare rod and wire, e.g., Inconel 62, for gas-shielded arc welding Inconel 600.
ERNiCr-3	Bare wire and rod, e.g., Inconel 82, for gas-shielded arc welding Inconel 600 and related alloys, e.g., cast alloys used for high temperatures.

Table **4-11** (*cont.*)

AWS DESIGNATION	USE
	NICKEL-CHROMIUM-IRON FILLER METALS
ERNiCrFe-6	Bare wire and rod for gas-shielded arc welding high-nickel alloys to carbon and austenitic stainless steels.
ERNiCrFe-7	Bare wire and rod for gas-shielded arc welding Inconel X-750 and 722 which are age-hardenable alloys.
	TYPICAL NICKEL-BASE FILLER METALS FOR WELDING DISSIMILAR METALS
ENiCrFe-2	Covered electrode, e.g., Inco weld A[2].
ENiMo-3	Covered electrode, e.g., Hastelloy W[3].
ERNiCrFe-6	Bare wire and rod for gas-shielded arc welding, e.g., Inconel 92[2].
ERNiMo-6	Bare wire and rod for gas-shielded arc welding, e.g., Hastelloy[3].
	NICKEL-MOLYBDENUM FILLER METALS
ENiMo-1	Covered electrode Hastelloy B for welding the alloy to itself, to steel, and to other metals.
ENiMoCr-1	Covered electrode Hastelloy C for welding the alloy to itself.
ENiMo-3	Covered electrode Hastelloy W for welding the alloy to itself, to steel, and to other metals.
ERNiMo-4	Bare filler rod or wire for gas-shielded arc welding Hastelloy B as above.
ERNiMo-5	Bare filler rod or wire for gas-shielded arc welding Hastelloy C as above.
ERNiMo-6	Bare filler rod or wire for gas-shielded arc welding Hastelloy W.

[1]Registered trademark of the Arcos Corp.
[2]Registered trademark of International Nickel Co., Inc.
[3]Registered trademark of Stellite Div., Cabot Corp.

Source: American Welding Society, *Welding Handbook*, 6th ed., section 5, Tables 94.9–94.13.

Table **4-12** Joining processes applicable to nickel alloys

ALLOY	SHIELDED METAL-ARC	GAS TUNGSTEN ARC	GAS METAL ARC	SUBMERGED ARC	ELECTRON BEAM	OXYACETYLENE	BRAZING
Nickel 200	X	X	X	X	X	X	X
Nickel 201	X	X	X	X	X	—	X
MONEL alloy 400	X	X	X	X	X	X	X
MONEL alloy 401	—	X	—	—	X	—	X
MONEL alloy 404	—	X	—	—	—	X	X
MONEL alloy R-405	X	X	X	—	X	X	X
MONEL alloy K-500	X	X	X	—	X	X	X
MONEL alloy 502	X	X	X	—	X	X	X
INCONEL alloy 600	X	X	X	X	X	X	X
INCONEL alloy 601	X	X	X	X	X	X	X
INCONEL alloy 625	X	X	X	X	X	—	X
INCONEL alloy 706		X	—	—	X	—	X
INCONEL alloy 718	—	X	X	—	X	—	X
INCONEL alloy X-750	—	X	—	—	X	—	X
INCOLOY alloy 800	X	X	X	X	X	X	X
INCOLOY alloy 825	X	X	X	—	X	—	X
INCOLOY alloy 901	—	X	—	—	X	—	X
Alloy 713C	—	—	—	—	—	—	X
UDIMET 500	X	X	—	—	X	—	X
UDIMET 700	—	X	—	—	X	—	X
RENE 41	—	X	—	—	X	—	X
ASTROLOY	—	X	—	—	X	—	X
WASPALOY	—	X	—	—	X	—	X
CARPENTER 20Cb3	X	X	X	—	X	—	X
HASTELLOY alloy B	X	X	X	—	X	—	X
HASTELLOY alloy C	X	X	X	—	X	—	X
HASTELLOY alloy C-276	X	X	X	—	X	—	X
HASTELLOY alloy D	—	—	—	—	—	X	X
HASTELLOY alloy F	X	X	—	—	X	—	X
HASTELLOY alloy G	X	X	X	—	—	—	X
HASTELLOY alloy N	X	X	—	—	X	—	X
HASTELLOY alloy R-235	X	X	X	—	X	—	X
HASTELLOY alloy X	X	X	X	—	X	—	X
Alloy IN100	—	—	—	—	—	—	X

X = Weldable by this method.
— = Not weldable by this method or no information available; consult manufacturer.
Source: American Welding Society, *Welding Handbook*, 6th ed., section 4, p. 67.8.

Table **4-13*a*** Recommended aluminum bronze and phosphor bronze electrodes for welding like and dissimilar metals

BASE METAL	STAINLESS STEEL	LOW-NICKEL STEEL	MALLEABLE IRON	CAST IRON	HIGH-CARBON STEEL	MED.-CARBON STEEL	LOW-CARBON STEEL	ALUMINUM BRONZE 92/8	CUPRO-NICKEL 90/10
Copper (deoxidized)	1G	3G	3G	3G	3G	3G	3G	3F	—
Phosphor bronze (5.0 Sn)	3C	3C	3G	3C	3C	3C	3C	3C	—
Phosphor bronze (10.0 Sn)	3C	3C	3C	3C	3C	3C	3C	3C	—
Silicon bronze (3.0 Si)	1A	1A	1A	1B	1A	1A	1A	1A	1A
Naval (yellow) brass (39.0 Zn, 0.8 Sn)	1E	3, 1D	3C	3, 1E	3, 1E	3, 1D	3, 1D	1, 3C	1B
Manganese bronze (40.0 Zn, 1.0 Al, 1.0 Fe)	1E	3, 1F	1, 3F	3, 1E	3, 1D	3, 1D	3, 1D	1, 3E	1B
Manganese bronze (25.0 Zn, 4.0 Al, 3.0 Fe, 3.5 Mn)	1E	3, 1F	1, 3F	3, 1E	3, 1E	3, 1D	3, 1D	1, 3E	1B
Manganese bronze (25.0 Zn, 6.0 Al, 3.0 Fe, 3.5 Mn)	1E	3, 1F	1, 3F	3, 1E	3, 1E	3, 1D	3, 1D	1, 3E	1B
Muntz metal (40.0 Zn)	1E	3, 1F	1, 3F	3, 1E	3, 1E	3, 1D	3, 1D	1, 3E	1B
Admiralty bronze (28.0 Zn, 1.0 Sn)	1E	3, 1F	1, 3F	3, 1E	3, 1E	3, 1D	3, 1D	1, 3E	1B
Navy G (8.0 Sn, 4.0 Zn)	1E	3C	3D	3C	3D	3C	3C	3C	—
Cartridge brass (30.0 Zn)	—	3, 1F	3, 1D	3, 1E	3, 1E	3, 1D	3, 1D	1E	1B
Cupro-nickel 70/30	1A	1A	1A	1A	1A	1A	1A	1A	1A
Cupro-nickel 90/10	1A	1B	1A	1A	1A	1A	1A	1A	1A
Aluminum bronze 92/8	1A	1B	1B	1B	1B	1B	1B	1A	
Low-carbon steel	—	1B	1B	1B	1D				
Med.-carbon steel	—	1B	1B	1B					
High-carbon steel	—	1C	1C	1C					
Cast iron	1C	1C	1C	1C					
Malleable iron	1C	1C	3, 1B	3, 1C					

Source: Airco Electrode Pocket Guide, ADC 1107A, p. 123. Courtesy of Airco Welding Products.

CUPRO-NICKEL 70/30	CARTRIDGE BRASS	NAVY G	ADMIRALTY BRONZE	MUNTZ METAL	MANGANESE BRONZE (HIGH TENSILE) 66	MANGANESE BRONZE (MED. TENSILE) 64	MANGANESE BRONZE (LOW TENSILE) 62	NAVAL (YELLOW) BRASS	SILICON BRONZE	PHOSPHOR BRONZE (C)	PHOSPHOR BRONZE (A)	COPPER (DEOXIDIZED)
—	3G	3G	3G	3G	3G	3G	3G	3G	3G	3G	3G	3G
—	3D	3D	3D	3D	3D	3D	3D	3D	3A	3C	3C	
—	3D	3D	3D	3D	3D	3D	3D	3D	3A	3C		
1A	1A	3C	1A	3, 1D	3, 1D	3, 1D	3, 1, 2C	4, 1C	3, 1A			
1B	1, 3F	3, 1F	3, 1F	3, 1F	3, 1F	3, 1F	3E	3F				
1B	1, 3F	3F	3, 1F	3, 1F	3, 2, 1F	1E	3, 1F					
1B	1, 3F	3F	3, 1F	3, 1F	1E	1, 2F						
1B	1, 3F	3F	3, 1F	3, 1F	1, 2F							
1B	1, 3F	3F	3, 1F	3, 1F								
1B	3, 1F	3, 1F	3, 1F									
—	3, 1F	3D										
1C	3, 1F											
1A												

Bronze electrode selection and preheat chart

AIRCO ELECTRODE DESIGNATION	PREHEAT and INTERPASS TEMPERATURE
1—E Cu Al-A2	A—150°F
2—E Cu Al-B	B—300°F
3—E Cu Sn-A	C—400°F
	D—500°F
	E—600°F
	F—700°F
	G—600–1,000°F

Table **4-13*b*** Recommended current ranges

STANDARD SIZE OF ELECTRODE (DIAMETER × LENGTH)	AMPERAGE RANGE
5/16″ × 12″	40–60
3/22″ × 12″	50–90
1/8″ × 14″	90–130
5/32″ × 14″	130–150
3/16″ × 14″	150–210
1/4″ × 18″	210–275
Except for ECuSn-A:	
1/8″ × 14″	80–120
3/22″ × 14″	130–190

Source: *Airco Electrode Pocket Guide*, ADC 1107A, p. 123. Courtesy of Airco Welding Products.

Table **4-15*a*** Typical joint designs used for gas—shielded arc welding of various thicknesses of magnesium alloy sheet and plate

WELD AND JOINT TYPE (SEE ILLUSTRATION)	APPLICABLE RANGE OF WORK-METAL THICKNESS, IN.[a]					
	GAS TUNGSTEN-ARC WELDING[b]			*GAS METAL-ARC WELDING*[c]		
	AC	DCSP	DCRP	SHORT-CIRCUITING ARC	PULSED ARC	SPRAY ARC
A[d]	0.025 to 1/4	0.025 to 1/2	0.025 to 3/16	0.025 to 3/16	0.090 to 1/4	3/16 to 3/8
B[e]	1/4 to 3/8	1/4 to 3/8	3/16 to 3/8	f	3/16 to 1/4	1/4 to 1/2
C[g]	3/8[h]	3/8[h]	3/8[h]	f	f	1/2[h]
D[j]	0.040 to 1/4	0.040 to 1/4	0.040 to 1/4	1/16 to 3/16	1/16 to 1/4	3/16 to 1/2
E[k]	3/16[h]	3/16[h]	3/16[h]	f	1/8 to 1/4	1/4[h]
F[m]	0.025 to 1/4	0.025 to 1/2	0.025 to 5/32	1/16 to 5/32	0.090 to 3/16	5/32 to 3/8
G[n]	1/16 to 3/16	1/16 to 3/8	1/16 to 1/8	1/16 to 5/32	0.090 to 1/4	5/32 to 3/4
H[p]	3/16[h]	3/8[h]	1/8[h]	f	1/4 to 3/8	3/8[h]
J[q]	0.040[h]	0.040[h]	0.025[h]	0.040 to 5/32	0.090 to 1/4	5/32[h]

[a]Suggested minimum and maximum thickness limits. [b]Using 300-amp ac or dcsp, or 125-amp dcrp. [c]Using 400-amp dcrp. [d]Single-pass complete-penetration weld. Suitable for thin material. [e]Complete-penetration weld. Suitable for thick material. On material thicker than suggested maximum, use double-V-groove butt joint to minimize distortion. [f]Not recommended because spray-arc welding is more practical or economical, or both. [g]Complete-penetration weld. Used on thick material. Minimizes distortion by equalizing shrinkage stress on both sides of joint. [h]No maximum. Thickest material in commercial use could be welded in this type of joint. [j]Single-pass complete-penetration weld. For material thicker than suggested maximum, use single-bevel-groove corner joint because it requires less welding, especially if a square corner is required. [k]Single-pass or multiple-pass complete-penetration weld. Used on thick material to minimize welding. Produces square joint corners. [m]Single-weld T-joint. Thickness limits are based on 40% joint penetration. [n]Double-weld T-joint. Suggested thickness limits based on 100% joint penetration. [p]Double-weld T-joint. Used on thick material requiring 100% joint penetration. [q]Single or double-weld joint. Strength dependent on size of fillet. Maximum strength in tension on double-weld joints is obtained when lap equals five times the thickness of the thinner member.

Source: By permission, from *Metals Handbook* Volume 6, Copyright American Society for Metals, 1971.

Table **4-14** Compositions of electrodes and filler metals used in gas shielded–arc welding of magnesium alloys (AWS A5.19-69)

ELEMENT	ER AZ61A	ER AZ101A	ER AZ92A	ER EZ33A
Aluminum	5.8 to 7.2	9.5 to 10.5	8.3 to 9.7	—
Beryllium	0.0002 to 0.0008	0.0002 to 0.0008	0.0002 to 0.0008	—
Manganese	0.15 min	0.13 min	0.15 min	—
Zinc	0.40 to 1.5	0.75 to 1.25	1.7 to 2.3	2.0 to 3.1
Zirconium	—	—	—	0.45 to 1.0
Rare earth	—	—	—	2.5 to 4.0
Copper	0.05 max	0.05 max	0.05 max	—
Iron	0.005 max	0.005 max	0.005 max	—
Nickel	0.005 max	0.005 max	0.005 max	—
Silicon	0.05 max	0.05 max	0.05 max	—
Others (total)	0.30 max	0.30 max	0.30 max	0.30 max
Magnesium	Remainder	Remainder	Remainder	Remainder

Source: By permission, from *Metals Handbook* Volume 6, Copyright American Society for Metals, 1971.

Table **4-15*b*** Illustration of typical joint designs.

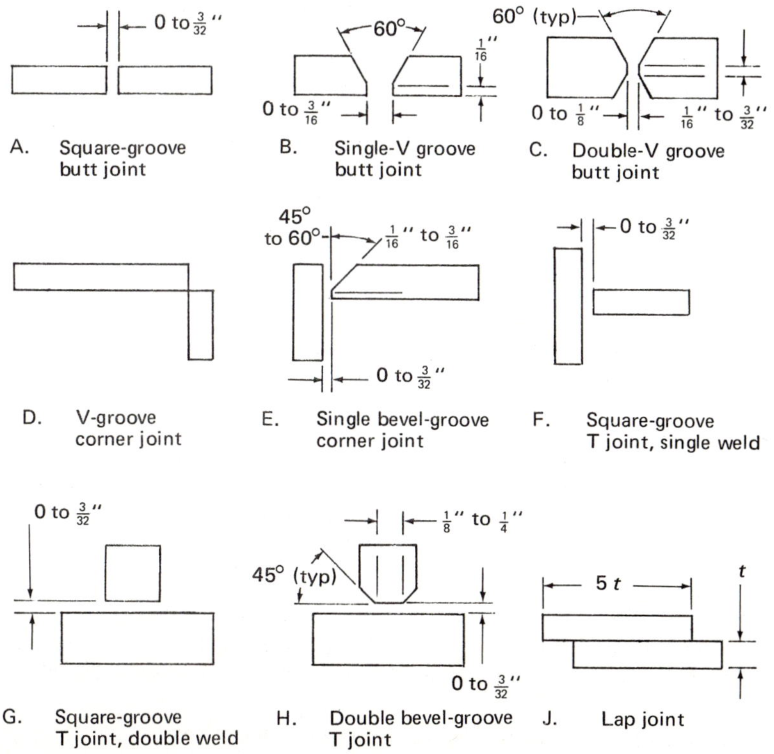

0 to 3/32"
60°
1/16"
0 to 3/16"
60° (typ)
0 to 1/8"
1/16" to 3/32"
A. Square-groove butt joint
B. Single-V groove butt joint
C. Double-V groove butt joint
45° to 60°
1/16" to 3/16"
0 to 3/32"
0 to 3/32"
D. V-groove corner joint
E. Single bevel-groove corner joint
F. Square-groove T joint, single weld
0 to 3/32"
1/8" to 1/4"
45° (typ)
0 to 3/32"
5 t
t
G. Square-groove T joint, double weld
H. Double bevel-groove T joint
J. Lap joint

The alpha-alloy filler metals ERTi-5A1-2.5Sn, ERTi-8A1-1Mo-1V, ERTi-3A1-2.5V, and ERTi-6A1-2Cb-1Ta-1Mo are used to weld base metal of the same designation (composition).

Similarly, the weldable alpha-beta alloy Ti-6A1-4V is welded by use of the ERTi-6A1-4V filler metal. The various compositions of beta alloys are welded with beta-alloy filler metal ERTi-13V-11Cr-3A1.

Because titanium and titanium alloys have a great affinity to impurities that may affect the welded joint, some procedures require that the filler metal be cleaned just before use. Cleaning can be done by hand wiping, using a cellulose sponge or lint-free cloth and a chlorine-free solvent, such as methyl-ethyl-ketone, toluene, or acetone.

It is very important that plastic gloves be worn during the cleaning process. This eliminates contamination caused by handprints and the plasticizer present in rubber gloves.

Filler metals for welding titanium and its alloys are available in straight lengths, in coils without support, and on expendable spools. Standard lengths of 36 inches (914 mm) are available in 5-pound (2.27 kg), 10-, 25-, and 50-pound quantities. The spooled wire is available in the 1-pound (0.45 kg), 10-, and 20-pound sizes. Spooled wire in the 1-pound size is available in standard-diameter size of 0.030 inch (0.76 mm), 0.035, or 0.045 inch. Wires of 1/16- and 1/8-inch diameters are available in spools and coils of 25 pounds (11.34 kg) and 50 pounds, as well as in the standard straight lengths.

4-12 SURFACING FILLER METALS

Surfacing filler metals are usually available as bare-cast or tubular rod, or as covered-solid or tubular wire, and are produced in bare or covered form. The availability of a particular alloy in a given form depends on the alloy's ability to be cast, shaped into wire or tubing, or produced in proper form. Because surfacing alloys are available in so many forms, they are covered by two AWS specifications: bare electrodes are covered by AWS A5.21 and covered electrodes, by AWS A5.13.

Before continuing, it should be pointed out that, although many surfacing alloys are covered by the AWS specifications, many more are not, because they are proprietary or have limited availability.

The system for classifying surfacing electrodes and welding rods used in the specifications (AWS A5.13 and AWS A5.21) follows the standard pattern used in other AWS–ASTM filler metal specifications. The letter *E* at the beginning of each classification indicates *electrode;* the letter *R* indicates *welding rod.* The letters immediately after the *E* or *R* are the chemical symbols for the principal elements in the classification. Thus, CoCr is a cobalt-chromium alloy, CuZn is a copper-zinc alloy, and so on.

When the limits of the percentage content of the principal filler-metal alloying elements result in two or more distinct range limits, the individual range groupings are identified by the letters *A*, *B*, *C*, and so on, as in ECuSn-A.

Further subdividing of groups A and B is done only to indicate the percentage limits of carbon contained in the alloy. For example, the 2 following the B in group 1B2 indicates that the alloys with this designation contain an excess of 2% carbon, and alloys in groups 1B1 contain less than 2% carbon. Similarly, alloys in group 2A2 contain more than 1.75% carbon, and alloys in group 2A1 contain less than 1.75% carbon.

To select the correct surfacing alloy for any particular application, knowledge of the causes of deterioration of the part should be established. One or more of the following factors will probably be involved:

- Sliding metal to metal contact, with or without lubrication.
- Rolling contact against a metallic or nonmetallic surface.
- Heavy shock or impact loading without deformation or serious cracking.
- Varying earth abrasion or other types of wear, causing metal loss.
- Corrosion (atmospheric or otherwise).
- Resistance to hot deformation.
- Machinability.
- Freedom from deposit porosity.
- Ease of application to a specific welding process.

In practice, the choice of a surfacing alloy is usually based on the experience of the user or the supplier. Some applications for which specific classes of surfacing alloys have proven to be satisfactory are listed below.

- *High-speed steel* (modified): EFe-A, EFe-B, EFe-C. These are used when hardness is required at temperatures up to 1100°F and when good wear resistance and toughness is needed. They are generally interchangeable, except that A and B grades have higher carbon and are better for edge-holding applications. C grade is best for hot and tough work.
- *Austenitic manganese steel:* EFeMn-A (Ni-Mn). These classes are approximately the same, except that B deposits have slightly higher yield strength. They are best used for metal-to-metal wear and impact applications, mainly because of their work-hardening characteristics.
- *Austenitic high-chromium iron:* EFeCr-A. This is used for facing agricultural machinery parts (large areas, heavy materials). Because

of associated impact, it is not suitable for the repair of machinery used in very rocky soil. Primarily, it resists erosion or low-stress scratching abrasion.

- *Cobalt-base:* ECoCr-A. This is used mainly for exhaust-valve contact surfaces because of its resistance to heat, corrosion, and erosion service. ECoCr-B and ECoCr-C have higher carbon and are therefore used on applications where greater hardness and abrasion resistance, but not impact resistance, are needed.
- *Copper-base alloy:* ECuA1-A2. This is used for surfacing bearing surfaces within 130 to 190 Bhn hardness and for corrision resistance. ECuA1-B and ECuA1-C are used for bearing surfaces and 140 to 290 Bhn hardness. ECuA1-D and ECuA1-E are for bearing and wear-resistant surfaces requiring even higher hardness, 230 to 390 Bhn. ECuSi-A is essentially for corrosion-resisting, not bearing, service. ECuSn (copper-tin) is for lower-hardness bearing surfaces, corrosion resistance, and sometimes wear-resistance applications. ECuZn-E (leaded bronze) produces a porous deposit to retain oil in locomotive journal-box overlays.
- *Nickel-chromium-boron:* ENiCr-A, ENiCr-B, ENiCr-C. These are for good metal-to-metal wear resistance, good low-stress, scratch abrasion resistance, corrosion resistance, and retention of hardness when hot. Hardness increases from A to C, while machinability and toughness decrease.

Surfacing alloys used on various base metals are discussed further in the chapter on hard surfacing (Chapter 17).

4-13 SUBMERGED-ARC WELDING ELECTRODES AND FLUXES

The mild-steel electrodes and fluxes to be used with the submerged-arc welding process are tabulated in Tables 4-16 (*a* and *b*) and 4-17, respectively, and are classified in AWS A5.17. In the classification system, the letter *E* indicates an electrode, as in the other classifying systems, but here the similarity stops. The next letter, *L*, *M*, or *H*, indicates low-, medium-, or high-manganese, respectively. The following number or numbers indicate the approximate carbon content in hundredths of one percent. If there is a suffix, K, this indicates a silicon-killed steel.

Fluxes are classified on the basis of the mechanical properties of the weld deposit made with a particular electrode. The classification designation given to a flux consists of a prefix, F (indicating a flux), followed by a two-digit number representative of the tensile-strength and impact requirements for tests made in accordance with

Table **4-16*a*** AWS classifications and composition limits for electrodes for submerged-arc welding (AWS A5.17)

AWS CLASSIFICATION	COMPOSITION, %[a] C	Mn	Si
CARBON STEEL (LOW MANGANESE) ELECTRODES			
EL8	0.10 max	0.30–0.55	0.05 max
EL8K	0.10 max	0.30–0.55	0.10–0.20
EL12	0.07–0.15	0.35–0.60	0.05 max
CARBON STEEL (MEDIUM MANGANESE) ELECTRODES			
EM5K[b]	0.06 max	0.90–1.40	0.40–0.70
EM12	0.07–0.15	0.85–1.25	0.05 max
EM12K	0.07–0.15	0.85–1.25	0.15–0.35
EM13K	0.07–0.19	0.90–1.40	0.45–0.70
EM15K	0.12–0.20	0.85–1.25	0.15–0.35
2% MANGANESE STEEL ELECTRODE			
EH14	0.10–0.18	1.75–2.25	0.05 max

[a]Electrodes of all classes also contain maximums of 0.035 S, 0.03 P. 0.15 Cu (independent of coating), 0.50 total other elements.
[b]Also contains 0.05 to 0.15 Ti, 0.02 to 0.12 Zr, 0.05 to 0.15 Al—exclusive of the 0.50% content of "total other elements."

Table **4-16*b*** Current ranges for electrode wires used in submerged-arc welding

WIRE DIAMETER, IN.	CURRENT RANGE, AMP[a]	WIRE DIAMETER, IN.	CURRENT RANGE, AMP[a]
0.045	100 to 350	5/32	340 to 1100
1/16	115 to 500	3/16	400 to 1300
5/64	125 to 600	7/32	500 to 1400
3/32	150 to 700	1/4	600 to 1600
1/8	220 to 1000	5/16	1000 to 2500
		3/8	1500 to 4000

[a]Upper and lower limits of ranges are extremes and are rarely used.

Source: By permission, from *Metals Handbook* Volume 6, Copyright American Society for Metals, 1971.

Table **4-17** Mechanical-property requirements for flux classification (AWS A5.17)[a]

Tensile strength:	
Classes F60 thru F64	62,000 to 80,000 psi
Classes F70 thru F74	72,000 to 95,000 psi
Yield strength (0.2% offset) min:	
Classes F60 thru F64	50,000 psi
Classes F70 thru F74	60,000 psi
Elongation in 2 in., min (all classes)	22%[b]
Charpy V-notch impact strength:	
Classes F60 and F70	Not required
Classes F61 and F71	20 ft-lb at 0 F
Classes F62 and F72	20 ft-lb at −20 F
Classes F63 and F73	20 ft-lb at −40 F
Classes F64 and F74	20 ft-lb at −60 F

[a]Mechanical properties are those of the weld metal, produced by one of the classes of flux in combination with an electrode of one of the classes shown in Table 4-16*a*. A flux designation consists of the class number, as shown in the present table, followed by the designation of the electrode used in combination with it (for example, F60-EL8). [b]For each increase of one percentage point, the tensile strength or the yield strength, or both, may decrease 1000 psi to minimums as follows: tensile strength, 60,000 psi (F60 through F64) and 70,000 psi (F70 through F74); yield strength, 48,000 psi (F60 through F64) and 58,000 psi (F70 through F74).

Source: By permission, from *Metals Handbook* Volume 6, Copyright American Society for Metals, 1971.

the specification. This is then followed by a set of letters and numbers corresponding to the classification of the electrode used with the flux.

Special proprietary fluxes and alloys are available that adapt this process to the production of sound welds on high-strength alloy steels, stainless steels, austenitic-manganese steels, and the depositing of hard-surfacing alloys with specific abrasion- or impact-resisting properties.

4-14 FILLER METALS FOR GAS TUNGSTEN–ARC (TIG) WELDING

As in the selection of filler metals used in the shielded metal–arc and oxyacetylene processes, the selection of a filler metal for the TIG welding process depends primarily on the metal being joined. The actual filler metals used for welding with the oxyacetylene welding process and gas metal–arc (MIG) process are also used for welding with this process.

4-15 FACTORS AFFECTING WELD METAL PROPERTIES

The mechanical properties (tensile strength and hardness) of the weld depend on the analysis (chemical composition) of the weld deposit. Since the base metal and the filler metal are both brought to the molten state and mixed within the weld crater, it is apparent that the final analysis of the weld is related to the analysis of the base metal, the analysis of the filler metal, and their mixing ratio.

In shielded metal–arc welding, the chemical analysis of the deposited metal depends upon the analysis of the core wire and on the analysis and action of the electrode coating. The composite design can be arranged to deposit practically any carbon or alloy steel. Electrode coatings can be made to add carbon to the deposit, to add alloying metals, and also to burn objectionable ingredients out of the weld.

One class of electrode coating reacts with the sulfur in the core wire to give a deposit that is appreciably lower (20–30%) in sulfur than the original electrode.

Some shielded metal–arc electrodes deposit metal as very fine droplets or as a spray, and these particles must be heated to a very high temperature when they are in the arc. In fact, some of the iron may be vaporized. The intense heating and the cleaning action of the slag results in a purified weld metal.

Penetration controls the mixing ratio of filler and parent metals. It may be increased or decreased by manipulation of the electrode. In gas, carbon-arc, and submerged-arc welding, the penetration may be changed by alternating the welding procedure. Inclining the work and welding downhill decreases the penetration, whereas welding uphill ensures deep penetration.

In shielded metal–arc welding, an E-6010 electrode used with straight polarity (negative electrode) gives a spray type of arc that will deposit only a little metal, most of it flying off into the air as fine droplets. This type of arc is useful to fuse the flanged edges of sheet steel (vertical-edge welds).

By using a slow travel speed and keeping the arc stream directed on the deposited metal, the penetration may be kept to a minimum. The penetration may even be reduced to zero, resulting in no weld. In contrast, fast travel and directing the arc stream on the base metal ahead of the deposited metal result in more penetration. Figure 4-5 gives the approximate amount of filler metal that is melted and mixed with the base metal in the weld.

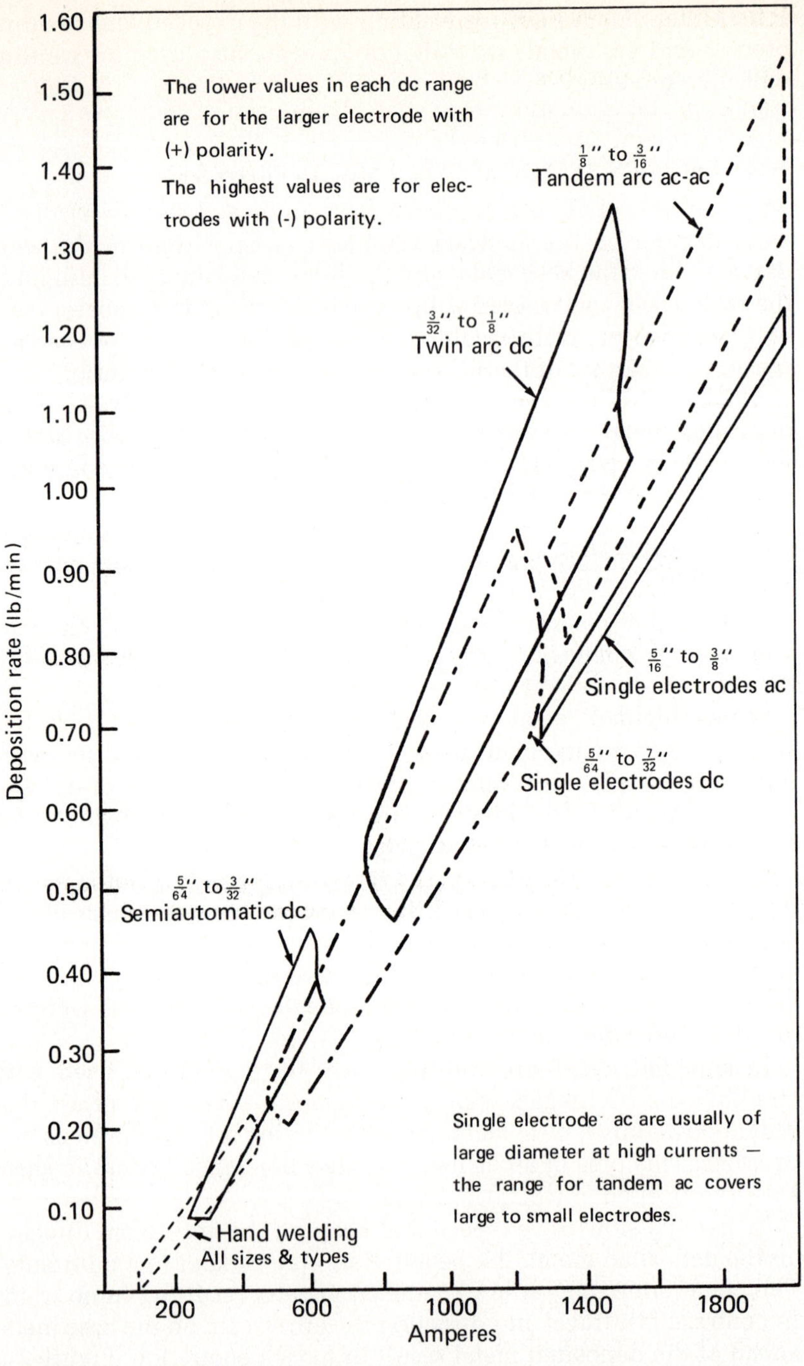

The lower values in each dc range are for the larger electrode with (+) polarity.
The highest values are for electrodes with (-) polarity.
1/8" to 3/16"
Tandem arc ac-ac
3/32" to 1/8"
Twin arc dc
5/16" to 3/8"
Single electrodes ac
5/64" to 7/32"
Single electrodes dc
5/64" to 3/32"
Semiautomatic dc
Hand welding
All sizes & types
Single electrode ac are usually of large diameter at high currents — the range for tandem ac covers large to small electrodes.
Deposition rate (lb/min)
1.60
1.50
1.40
1.30
1.20
1.10
1.00
0.90
0.80
0.70
0.60
0.50
0.40
0.30
0.20
0.10
200
600
1000
1400
1800
Amperes

Figure **4-5** Approximate deposition rate of submerged-arc processes on mild steel. (Redrawn by permission from *The Procedure Handbook of Arc Welding*, 11th ed., Lincoln Electric Company, Cleveland, Ohio, p. 2-94.)

SHIELDING AND FUEL GASES

The principal purpose of a shielding gas is to protect the molten metal from contamination by oxygen, nitrogen, and hydrogen in the air, while the primary purpose of a fuel gas is to create a combustion temperature sufficient to melt the metal to be welded.

4-16 SHIELDING GASES

Shielding gases are consumables used with the gas tungsten–arc and gas metal–arc welding processes. Although, in theory, any of the inert gases—helium, argon, neon, xenon, or krypton—could be used, the only ones plentiful enough for practical uses in welding are helium and argon. These gases provide satisfactory shielding for the more reactive metals, such as aluminum, magnesium, beryllium, columbium, tantalum, titanium, and zirconium.

Although pure inert gases protect metal at any temperature from reaction with constituents of the atmosphere (air), they are not suitable for all welding applications.

4-17 SHIELDING GASES FOR GAS TUNGSTEN–ARC (TIG) WELDING

Either argon, helium, or a mixture of the two (see Table 4-18) is commonly used in gas tungsten–arc welding.

4-18 SHIELDING GASES FOR GAS METAL–ARC (MIG) WELDING

The most commonly used gases for gas metal–arc welding are listed in Table 4-19.

4-19 FUEL GASES

The commonly used fuel gases are listed in Table 4-20 (page 98). All fuel gases have one common property: they all require oxygen to support combustion. However, to achieve good combustion and be used for welding, a fuel gas, when burned with oxygen, must have (1) a high flame temperature, (2) a high rate of flame propagation, (3) adequate heat content, and (4) the minimum chemical reaction of the flame with base and filler metals.

Most commercially used oxygen is obtained from air by the liquid-air process. In the liquid-air process, the various gases in the air (oxygen, nitrogen, argon, neon, helium, carbon dioxide, and water vapor) are separated from each other by bringing the atmosphere

Table **4-18** Shielding gases for TIG welding of various metals

METAL	THICKNESS	MANUAL	MECHANIZED
Aluminum	Under ⅛ in.	Ar (ACHF)	Ar (ACHF) or He (DCSP)
	Over ⅛ in.	Ar (ACHF)	Ar-He (He-75) (ACHF) or He (DCSP)
Steels	Under ⅛ in.	Ar (DCSP)	Ar (DCSP)
	Over ⅛ in.	Ar (DCSP)	Ar-He (He-75) or He (DCSP)
Copper	Under ⅛ in.	Ar-He (DCSP) (He-75)	Ar-He (DCSP) (He-75)
	Over ⅛ in.	He (DCSP)	Ar-He (He-90) or He (DCSP)
Stainless steels	Under ⅛ in.	Ar (DCSP)	Ar-He or Ar-H_2 (DCSP) (H-1 to H-15) (He-75)
	Over ⅛ in.	Ar-He (DCSP) (He-75)	He (DCSP) (He-H_2)
Nickel alloys	Under ⅛ in.	Ar (DCSP)	Ar-He (He-75) He (DCSP)
	Over ⅛ in.	Ar-He (DCSP) (He-75)	He (DCSP)

Source: By permission, from Union Carbide Corporation, Linde Division, *How To Do TIG (Heliarc) Welding*, F-1607-D 5/70.

collected in special tanks to the boiling point (temperature) of each of the gases to be separated from the mixture.

Oxygen (boiling point 183.0°C) is removed from the mixture by boiling, and nitrogen (boiling point 195.8°C) is removed with it. Then the oxygen is separated from the nitrogen by the rectification process. The 99.5% pure oxygen obtained from the rectification process is placed into large storage tanks, from which it is compressed into the steel cylinders that are familar to most welders.

With few exceptions, oxygen cylinders, like acetylene cylinders, are not sold, but remain the property of the oxygen manufacturer. Oxygen cylinders are charged with oxygen at a pressure of 2200 lb/in^2 at 70°F. If the temperature of the cylinder is allowed to rise above 70°F, the pressure in a full cylinder will rise above 2200 lb/in^2. To prevent explosions arising from dangerously excessive pressures, each cylinder has a safety device to release the oxygen much before there is any danger of the cylinder rupturing.

Facilities consuming large volumes of oxygen generally obtain the oxygen in liquid form. The liquid oxygen, which by federal law must be stored in facilities located at some distance from the point of use, is discharged into an automatic converter that changes it to a gas and distributes it through pipes to the point of use.

Acetylene and hydrogen are the only fuel gases commercially available that have all the desired properties. Other gases, such as propane and natural gas, have sufficiently high flame temperature

Table **4-19** Gas metal–arc shielding gas selection chart

SPRAY-ARC MODE OF METAL TRANSFER

METAL	ARGON	HELIUM	ARGON-OXYGEN	[1]Ar-CO_2	ARGON-HELIUM GAS MIXTURES	[2]Ar-O_2-CO_2	[3]CO_2
Aluminum	X	X			X (90% He-10% Ar) or (He-75)		
Carbon steels			X (O_2-5)	X (C-25)		X (6% CO_2) (5% O_2)	X
Low-alloy steels			X (O_2-2)				
Copper	X	X	X (O_2-1)		X (90% He-10% Ar) or (He-75)		
Stainless steels			X (O_2-1) (O_2-2)				
Nickel alloys	X	X			X (He-75)		
Reactive metals	X	X					

SHORT-CIRCUITING MODE OF METAL TRANSFER

METAL	ARGON	HELIUM	ARGON-HELIUM MIXTURES	ARGON-CO_2 MIXTURES	ARGON-HELIUM CO_2 MIXTURES	CO_2
Aluminum	X	X	X (He-75)			
Carbon steel				X (C-25) or (C-50)		X Wire designed for CO_2 req'd
High-strength steel					X (A-415)	
Copper			X (He-75)			
Stainless steels				X (C-25)	X (A-1025)	
Nickel alloy	X		X (90% He-10% Ar) or (He-75)		X (A-1025)	
Reactive metals	X	X	X (He-75)			

[1]Suitable for high current > 300 amperes spray arc operation
[2]Higher quality on heavy mill scale plate when used with Linde 83 and 85 wires
[3]Used with flux cored wire and for high speed solid wire welding

Source: By permission, from Union Carbide Corporation, Linde Division, *How To Do Manual Metal Inert Gas Welding*, F-51-110-A 5/71.

Table **4-20** Commonly used fuel gases

GAS	HEAT VALUE (BTU PER CU FT)	FLAME TEMPERATURE WITH OXYGEN (DEGREES F)
Acetylene	1433	6300
Butane	2999	5300
MAPP®	2406	6000
Methane	914	5000
Natural gas	1200	4600
Propane	2309	5300

but exhibit low flame-propagation rates. These other gas flames are excessively oxidizing at oxyfuel gas ratios high enough to produce usable heat-transfer rates. Flame-holding devices, such as counterbores on the tips, are necessary for stable operation and good heat transfer, even at the higher oxyfuel gas ratios. These gases, however, can be used for brazing, soldering, and similar operations, where the demands upon the flame characteristics and heat transfer rates are not excessive.

The most widely used fuel gas is acetylene. Acetylene is a compound of carbon and hydrogen (C_2H_2), containing 92.3% hydrogen by weight. It is colorless, lighter than air, and has a sweet, easily identifiable odor similar to garlic. Acetylene contained in cylinders has an odor slightly different from that of pure generated acetylene. This difference is due to the acetone vapor contained in the cylinder.

Acetylene is a highly combustible compound that liberates heat upon decomposition. Acetylene contains a higher percentage of carbon than any of the numerous other hydrocarbons. This combination gives the oxyacetylene mixture the hottest flame of any of the commercially used gases.

Warning: At temperatures above 1435°F (780°C) or at pressures above 30 psi, acetylene gas is unstable, and an explosive decomposition may result even without the presence of oxygen.

Note: It has become an accepted safe practice never to use acetylene at pressures exceeding 15 psi in generators, cylinders, pipelines, or hoses.

When used with hydrogen, the oxyhydrogen flame is capable of producing a maximum flame temperature of 5500°F, considerably below the temperature produced by the oxyacetylene flame, but much higher than that produced by any other fuel except for MAPP. The oxyhydrogen flame is used mainly to weld metals with low welding points, such as lead, aluminum, and magnesium. A decided disadvantage of the oxyhydrogen flame is that it is very difficult to see, making it most difficult (impossible for all practical purposes) to adjust the flame by sight. Because of this, the adjustment of the flame is controlled through pressure regulators. MAPP is a liquified acetylene compound (stabilized methylacetylene and propadiene) patented by the Dow Chemical company.

Natural gas and liquefied petroleum gas are used principally for applications such as preheating, soldering, and brazing. Because it is difficult to obtain a nonoxidizing flame with propane and butane, these gases are used primarily for brazing.

REVIEW QUESTIONS

1. What is meant by the term *welding consumables* as used in this text?

2. Name the societies that have developed the specifications covering filler metals.

3. What purpose does the atmosphere formed by the burning electrode serve?

4. Name three important characteristics of the electrode coating.

5. Describe completely the following AWS electrode designations: E-6010; E-6016; E-6018; E-7010A_1; E-6012C_2.

6. Differentiate between *straight* and *reverse* polarity.

7. Name the AWS classification for gas welding rods used to weld carbon and low-alloy steels.

8. What are the similarities between an E-6015 electrode and the electrodes used for stainless steels?

9. Referring to Table 4-11, select the correct filler metal to weld (a) AISI 310 to monel; (b) AISI 501 to carbon steel; (c) AISI 303 to AISI 310; (d) AISI 348 to AISI 440A.

10. Referring to Tables 4-8 and 4-9, select the best filler metal for the following conditions:

a. Ease of welding.
b. Ductility.
c. Color match after anodizing.
d. Corrosion resistance as applied when joining: aluminum alloy 1100 to aluminum alloy 6061; aluminum alloy 3004 to aluminum alloy 6061; aluminum alloy 6061 to aluminum alloy 2024.

11. List three welding processes commonly used to join Inconel alloy X-750.

12. Why is it recommended that rubber gloves *not* be worn during preweld cleaning of titanium and titanium alloys?

13. Why aren't all surfacing filler metals covered by AWS specifications?

14. Differentiate between the classification system for submerged-arc electrodes and the system used to classify filler metals used in the TIG and MIG processes.

15. What is the principal purpose of a shielding gas?

16. Name the inert gases commonly used as shielding gases.

17. What is the purpose of a fuel gas?

18. Name the commonly used fuel gases and the welding processes in which they are used.

PART 2
WELDING PROCESSES

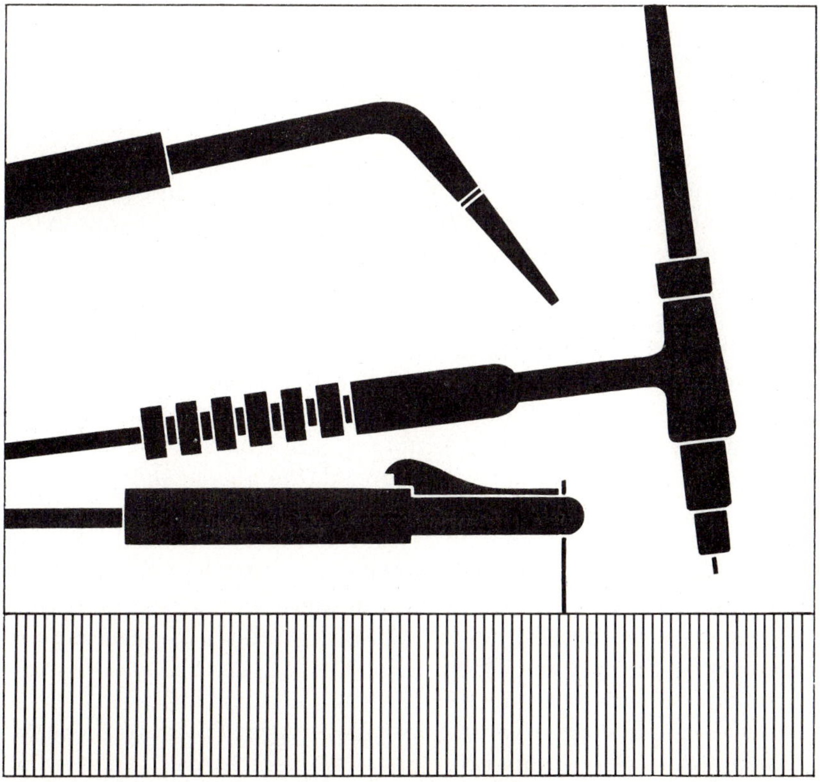

CHAPTER 5
ARC WELDING PROCESSES AND EQUIPMENT

Of all welding processes, arc welding is the one most widely used. In arc welding, the heat needed to melt the electrode and the metal of the workpiece is generated by their resistance (friction) to the flow of electricity (current).

When electricity is conducted through a wire, the movement of the electric energy through the wire causes friction, and the friction heats the wire. Since the wire is heated as a result of resistance to the flow of electricity, it follows that the greater the flow of electricity (current) through a wire of a given diameter, the more friction will result. The increase in friction will then result in an increase of heat.

It might help to compare the flow of electricity through a wire to the flow of water through a pipe. The rate of water flow through a pipe is expressed in gallons per minute. The rate of electricity flow through a wire is expressed in amperes. Since the flow of electricity determines the amount of heat produced, the greater the amperage, the greater the heat in the arc.

Because aluminum and copper offer less resistance to the flow of electricity than do other metals, they will heat less than will an equal-diameter wire of steel when an equal quantity of electrical current is passed through them. Metals that are not large enough in diameter for the current carried will get very hot and may even melt.

Industry has been using this heating-due-to-electrical-resistance principle for a long time and in many ways. Probably the best-known application of this principle is an electric fuse.

A fuse is intended to be the weakest point in a circuit. A small piece of metal within the fuse—the conductor—is designed to carry only a limited amount of current. If more than this amount of current goes through the fuse, the metal heats up and melts. This breaks the circuit. Another application of this principle is the electric-arc furnace, shown in Figure 5-1, used for producing alloy and stainless steels.

A vessel to hold the metal, usually called a *crucible,* is connected to one terminal of a high-ampere power source. An electrode, usually carbon, is connected to the other terminal. The metal to be melted is placed in the crucible and the electrode is brought into contact with it, thus completing the circuit. The resistance to the flow of current through the metal generates enough heat to melt it.

5-1 THE ARC WELDING CIRCUIT

To create a welding circuit, one must have a source of electric power. In most arc welding processes, it is the welding machine. Two cables are used. One connects the electrode holder to one terminal of the welder, and is called the *electrode cable* or *electrode lead.* The other cable connects the ground clamp to the other terminal and is called the *ground* or the *work lead* (see Figure 5-2). Both

Figure **5-1** An electric-arc furnace. (Redrawn by permission from James F. Lincoln Arc Welding Foundation, *Metals and How to Weld Them,* 2d ed., 1954, 1962, p. 7.)

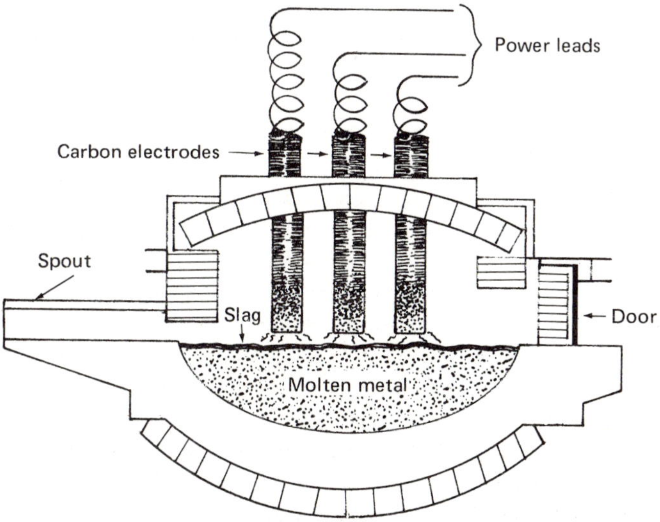
Power leads
Carbon electrodes
Spout
Slag
Door
Molten metal

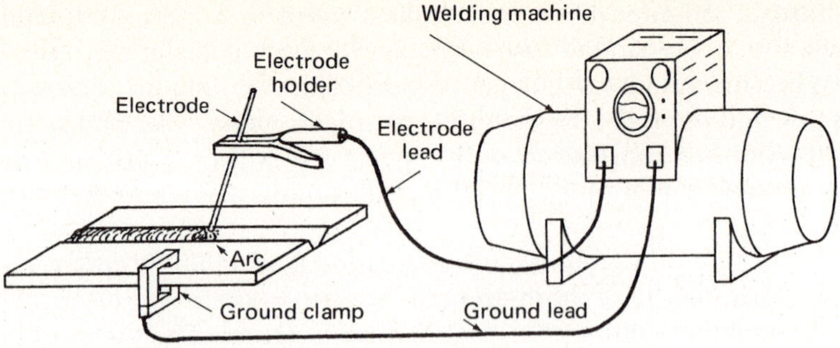
Welding machine
Electrode holder
Electrode
Electrode lead
Arc
Ground clamp
Ground lead

Figure **5-2** The basic arc welding set-up

cables should be of a size, length, and material that will carry the required current with little resistance; otherwise power will be wasted in the form of heat in the cable.

The important part of the welding circuit is the hook-up of the electrode and the ground or work lead. In dc welding, the electrode and ground lead can be hooked up in two different ways. One way is referred to as straight polarity (SP), the other is reverse polarity (RP). In the RP hook-up, the ground or work lead must be connected to the negative (−) terminal and the electrode lead to the positive (+) terminal. In the SP hook-up, the ground or work lead must be connected to the (+) terminal and the electrode lead to the minus (−) terminal.

On some welding machines, the polarity may be changed by means of a polarity selection switch provided for that purpose. On other machines, the welder must disconnect both the electrode and ground leads and reconnect them to the opposite terminals (ground cable where electrode cable was and electrode cable where ground cable was) to reverse polarity.

In ac welding, there is only one way the ground or work lead and electrode lead can be hooked up to the terminals. The Union Carbide Corporation, in their *Welding Power Handbook,* states that "alternating current arc welders are just like dc arc welders, except they have two halves. One for each half cycle of ac." Study of Figure 5-3 shows that an ac arc behaves like a reverse-polarity dc arc for one half-cycle, then like a straight-polarity dc arc for the other half-cycle.

Regardless of hook-up used, the welder causes the electric current to flow by touching the tip of the electrode to the work and creating an arc. The flow of electricity is stopped when the tip of the electrode is moved away from the work and the circuit is broken.

If the welder closes the circuit by touching the electrode to the work and holds it in direct contact there, the electrode will soon become red hot. The resistance to the flow of electricity through the electrode, which is small in cross-section and a poor conductor, causes this heating. With the electrode held against the workpiece, enough heat will usually be generated to fuse the two together (and sometimes to fuse the electrode to the electrode holder as well). If the workpiece is thin and light, it, too, will become very hot; because of its small size it will offer resistance to the flow of current.

5-2 THE ELECTRIC ARC

Dry air is a poor conductor of electricity—it can almost be classified as a nonconductor. Electricity really doesn't flow through air. But under certain conditions it will jump across an air gap in an arc. The welding current flowing through this high-resistance air gap generates an intense arc heat, which may be from 6000 to 10,000°F. As a result, the base metal melts at the point where the arc touches it, and the electrode melts (and becomes filler metal) at the point where the arc touches the tip of the electrode.

To keep the arc stable and consistent, certain chemicals are put into the electrode coatings to help contain and direct the arc, and also to protect the molten filler metal from the air while it transfers across the arc (see Chapter 4).

The point to understand here is that the resistance across an electric arc creates the heat that a welder uses to melt metal. The welding machine or power source can be adjusted to deliver

Figure **5-3** Simplified schematic of an arc welding power supply

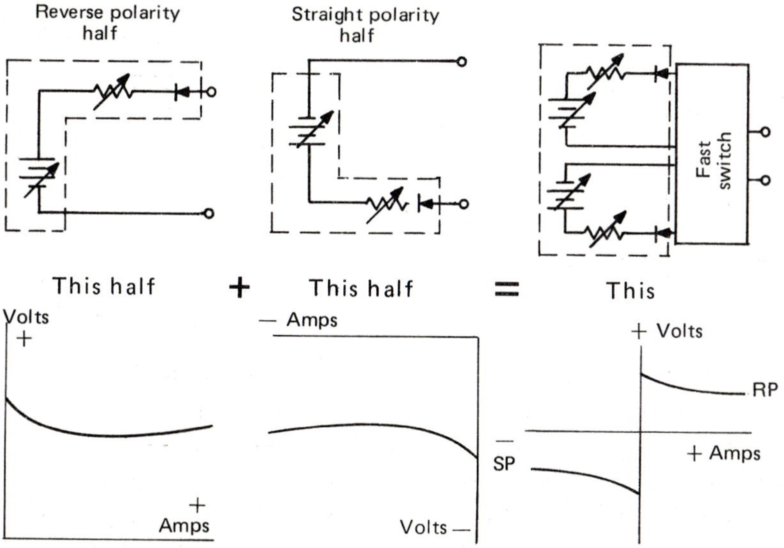

Reverse polarity half
Straight polarity half
Fast switch
This half
+
This half
=
This
Volts
+
+
Amps
— Amps
Volts —
+ Volts
RP
+ Amps
—
SP

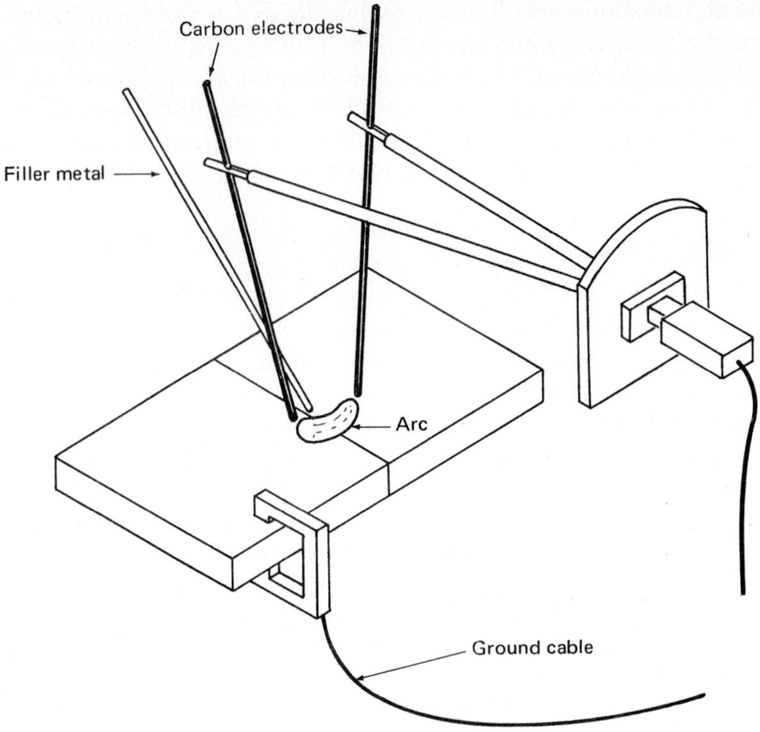
Carbon electrodes
Filler metal
Arc
Ground cable

Figure **5-4** Twin carbon–arc welding set-up

the right amount and type of power to the end of the electrode. The welding-machine operator must direct the arc to the right spot on the work, hold the arc at the correct length, and move it at the proper rate of speed to get the desired results.

5-3 CARBON-ARC WELDING

As mentioned in Chapter 1, carbon-arc welding is one of the earliest forms of welding. In this method, the arc is struck between the carbon electrode and the parent metal or, in twin carbon-arc, between two carbon electrodes. Filler metal, usually with the aid of flux, is fed into the weld pool in much the same manner as with oxyacetylene welding (see Figure 5-4).

One of the major problems with carbon-arc welding is arc stabilization. The use of flux helps with this problem, as does reverse polarity with the single carbon electrode method. However, the best method of arc stabilization is the incorporation of a solenoid in one of the carbon electrode holders. The magnetic force of the solenoid helps to compress the arc and maintain the welding point.

With the development of other, more efficient arc welding methods, the carbon-arc method has fallen into disuse. However, often a single carbon-arc device is combined with a high-pressure air line for what is called an *air carbon-arc cutting method*. The intense heat of the arc melts the metal, and the air blows it free, producing the cut. In fact, today most carbon-arc use is for cutting or gouging.

5-4 SHIELDED METAL–ARC WELDING

The principal welding process throughout the world is still shielded metal–arc welding, using flux coated electrodes. Like the other electric processes, the heat of the arc is used to bring the workpiece and a consumable electrode to the molten state. The circuit is usually set up as shown in Figure 5-5.

In this process, the arc actually carries tiny globules of molten metal from the tip of the electrode to the weld puddle on the work surface. The key principle in this process, however, is the shielding, which is obtained from decomposition of the electrode coating in the arc. The coating serves one or all of the following three functions:

1. The creation of an inert atmosphere that shields the molten metal from oxygen and nitrogen (or other contaminants) in the air.

2. The addition of deoxidizers or scavengers to refine the grain structure of the weld metal.

3. The formation of a fast-hardening slag blanket which protects the weld puddle.

Figure **5-5** Basic shielded metal–arc welding set-up

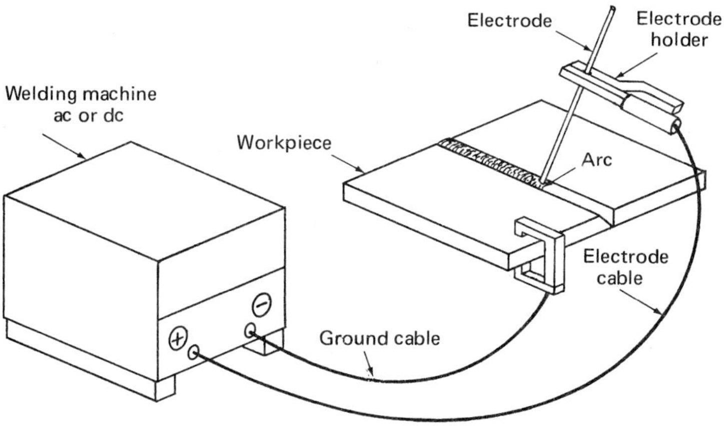
Electrode
Electrode holder
Welding machine ac or dc
Workpiece
Arc
Electrode cable
Ground cable

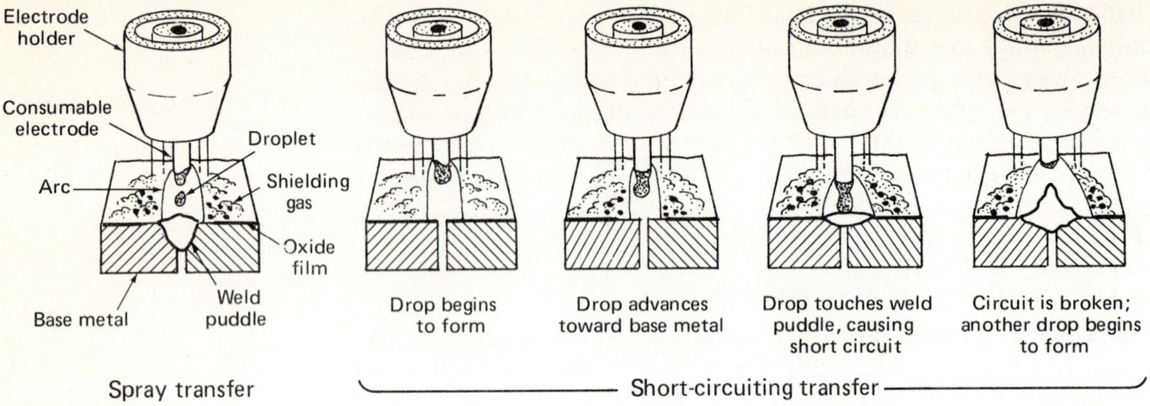
Electrode holder
Consumable electrode
Droplet
Arc
Shielding gas
Oxide film
Base metal
Weld puddle
Drop begins to form
Drop advances toward base metal
Drop touches weld puddle, causing short circuit
Circuit is broken; another drop begins to form
Spray transfer
Short-circuiting transfer

Figure **5-6** Methods of metal transfer in gas metal–arc welding. (By permission, from *Metals Handbook* Volume 6, Copyright American Society for Metals, 1971.)

As noted in the previous chapter, various electrodes and coatings have been developed to address specific welding problems. Therefore, as far as the welder is concerned, shielded metal–arc welding consists of selecting the correct electrode, setting the correct amperage, striking and holding the arc, and being able to weld in the proper position for the job.

Flux-cored metal–arc welding is a specialized version of this process in which the electrode is a continuously fed wire with a flux-filled core. The composition and function of the flux is essentially the same as with a coated electrode. The advantage of this process is its adaptability to semiautomatic and automatic methods of application.

5-5 GAS METAL–ARC WELDING (GMAW)

The GMAW (also known as MIG, metal inert gas) process is essentially a dc reverse-polarity process in which the solid, bare consumable electrode is shielded from the atmosphere by an externally supplied protective atmosphere, usually carbon dioxide, argon–carbon dioxide mixtures, or helium-base gases. There are two means of applying this process. An *all-position method,* using a manually manipulated gun, and an *automatic-head method,* which is used primarily for flat-position welding.

Metal transfer with the MIG process is by one of two methods: the *spray-arc method* and the *short-circuiting method* (Figure 5-6). The electrodes used in the spray-arc method are larger in diameter—

0.045 to 0.125 inch versus 0.020 to 0.45 inch—than those used in the short-circuiting method; the arc is on all the time. Because of this, the spray-arc method produces a heavy deposition of filler metal. Therefore, the spray-arc method should be restricted to single- or multipass welding in the flat or horizontal position in ⅛-inch-thick or thicker weldments. The short-circuit method is exceptionally well suited to welding thin sections in any welding position.

Flux shielded–arc welding is a variation of this process, in which a continuously supplied flux-cored electrode is used as well as the shielding of carbon dioxide gas. This double protection affords safer, stronger welds in the semiautomatic and automatic applications.

5-6 GAS TUNGSTEN–ARC WELDING (GTAW)

The GTAW (also known as TIG, tungsten inert gas) process is an arc process which uses a virtually nonconsumable tungsten electrode and an externally supplied protective inert atmosphere, usually helium, argon, or a mixture of both (see Figure 5-7). The manipulative techniques necessary for welding with this process are similar to those required for fuel-gas welding: one hand is used to manipulate the torch and the other is used to feed the filler metal.

Figure **5-7** Essentials of the TIG welding process. (With permission from Union Carbide Corporation, Linde Division, *How to Do TIG (Heliarc) Welding*, F-1607-D 5/70, p. 3.)

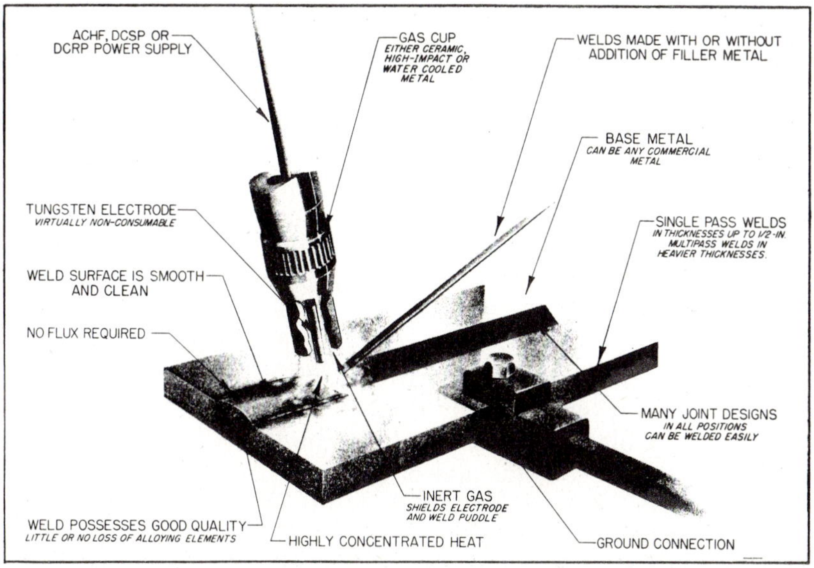
ACHF, DCSP OR
DCRP POWER SUPPLY
GAS CUP
EITHER CERAMIC, HIGH-IMPACT OR WATER COOLED METAL
WELDS MADE WITH OR WITHOUT
ADDITION OF FILLER METAL
BASE METAL
CAN BE ANY COMMERCIAL METAL
TUNGSTEN ELECTRODE
VIRTUALLY NON-CONSUMABLE
SINGLE PASS WELDS
IN THICKNESSES UP TO 1/2-IN. MULTIPASS WELDS IN HEAVIER THICKNESSES.
WELD SURFACE IS SMOOTH
AND CLEAN
NO FLUX REQUIRED
MANY JOINT DESIGNS
IN ALL POSITIONS CAN BE WELDED EASILY
INERT GAS
SHIELDS ELECTRODE AND WELD PUDDLE
WELD POSSESSES GOOD QUALITY
LITTLE OR NO LOSS OF ALLOYING ELEMENTS
HIGHLY CONCENTRATED HEAT
GROUND CONNECTION

The high–current density electric arc produced by this process makes it possible to weld at higher speeds and obtain greater penetration than with fuel-gas or shielded metal–arc welding. Extremely high quality welds can be made with this process, but everything depends on the set-up of the equipment and proper preparation of the base metal (cleaning). This process may also be manual, semiautomatic, or automatic.

5-7 SUBMERGED-ARC WELDING

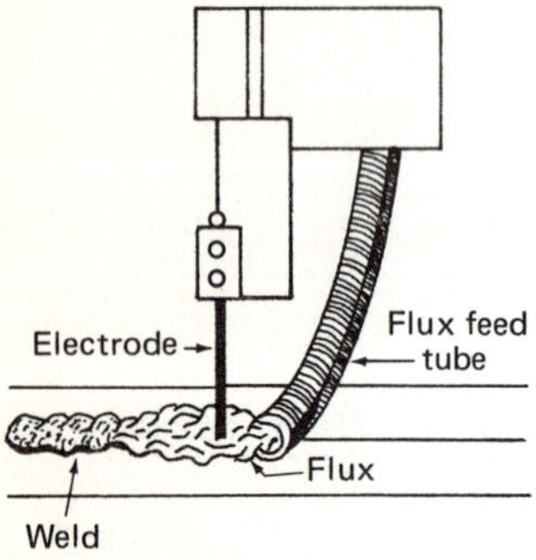
Electrode
Flux feed tube
Flux
Weld
Figure **5-8** The submerged-arc welding process

Submerged-arc welding is either a semiautomatic or an automatic process. A bare metal electrode or electrodes are used and the arc is shielded by a separately supplied blanket of granular, fusible flux. There is no visible evidence of the arc in this method. The arc, molten electrode, and weld puddle are completely submerged in the conductive, high-resistance flux.

A specially designed welding head (Figure 5-8) feeds the continuous electrode and the flux separately. By varying the chemical composition of the flux, a variety of metals and alloys can be welded in several joint designs. However, submerged-arc welding is primarily a production process used for straight-line welds, especially in box framing.

5-8 SOURCES OF POWER FOR ARC WELDING

The AWS classifies commercially available welding machines (power supplies) according to type of power source and type of control. Table 5-1 relates the process requirements to the type of power supply.

The rotating electric-motor generator and transformer-rectifier dc-type welding machines are of the constant-current and constant-voltage types. They have either a dc generator or an ac generator with a rectifier and a way of automatically adjusting the voltage and current output to meet the demands of the arc load at all times. Welding machines of this class are driven by an engine (Figure 5-9, page 119) or an electric motor.

Direct-current welding machines are more versatile than alternating-current machines. They have a wide current range, can be used to weld with all types of electrodes, and are usually preferred for out-of-position (vertical and overhead) welding, pipe welding, and anywhere reverse polarity (Figure 5-10, page 119) is needed. However, arc blow is occasionally encountered with dc straight polarity. When it is, welds may be poor.

Generator dc and generator-rectifier dc welding machines must have the output ratings shown in Table 5-2 to qualify under National Electrical Manufacturers Association (NEMA) *Standard EW 1-24*.

According to NEMA, constant current-transformer ac arc welding machines are classified as shown in Tables 5-3 to 5-5.

1. *Industrial environment.* Machines so classified must have the ability to deliver rated output at high duty cycles and must withstand heavy production service in industrial environments. Single-operator machines in this class have the output ratings (under load) shown in Table 5-3.

2. *Limited-service.* These transformer-type ac machines deliver rated output at lower duty cycles and work under less rigorous conditions than do industrial units. They frequently have two or more open-circuit voltage ranges. The low range normally has approximately 80 volts open circuit. The machine is rated for a maximum current output at a specified voltage on a specified duty cycle. Limited-service machines have output ratings as shown in Table 5-4.

3. *Limited-input.* As with limited-service welding machines, these also deliver rated output at lower duty cycles and are designed for use on single-phase power distribution systems of limited capacity, such as those used in home workshops. Output ratings (under load) for these machines are shown in Table 5-5.

Transformer-rectifier ac arc welders are classified in a manner similar to the transformer-type ac arc welders. The NEMA output ratings for each type of power supply are shown in Tables 5-6 to 5-8.

The NEMA output ratings for transformer-rectifier ac/dc arc welding machines are shown in Tables 5-9 to 5-11. It should be pointed out that these machines (unlike the transformer type) are basically single-phase ac transformers with rectifiers that convert alternating current to direct current. They can be operated as either ac or dc welding machines, depending on the welding conditions. The machine may be a standard unit for shielded metal–arc welding, or it may be a unit equipped with automatic gas and water controls for tungsten inert gas–arc (TIG) welding (Figure 5-11, page 120).

Another type of transformer-rectifier power supply is the high-frequency stabilized arc welding machine. These machines are classified according to their output currents: ac only, dc only, and ac/dc, and they come in the following types: transformer ac, transformer-rectifier dc, and transformer-rectifier ac/dc.

The Miller Electric Manufacturing Company of Appleton, Wisconsin, states that although conventional ac power sources may be used for ac TIG welding, their duty cycles must be lowered (Table 5-12), because partial rectification occurring at the arc introduces a dc component that causes overheating of the main transformer.

The welding machine required for the tungsten-arc (TIG) welding process may be an ac/dc rectifier or a dc generator, which can be either motor or engine driven.

(Text continues on page 121)

Table **5-1** Power source classification

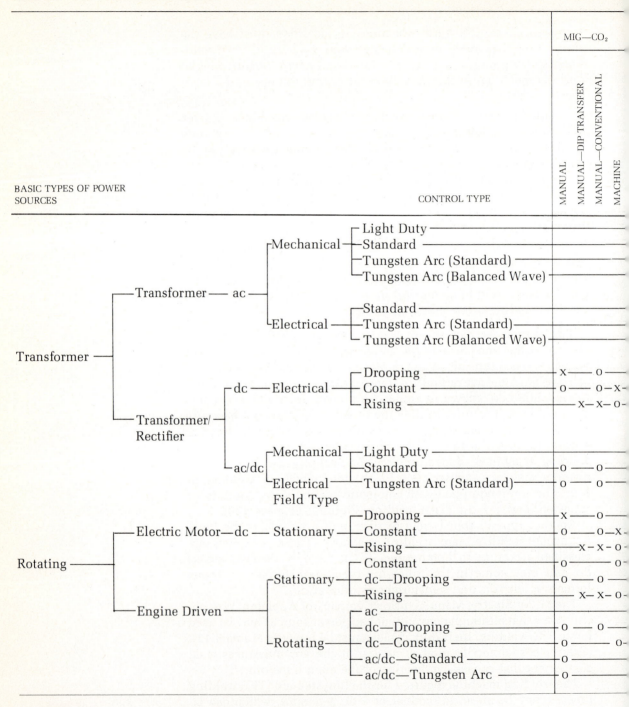

BASIC TYPES OF POWER SOURCES				CONTROL TYPE	MIG—CO_2 MANUAL	MIG—CO_2 MANUAL—DIP TRANSFER	MIG—CO_2 MANUAL—CONVENTIONAL	MIG—CO_2 MACHINE
Transformer	Transformer	ac	Mechanical	Light Duty				
				Standard				
				Tungsten Arc (Standard)				
				Tungsten Arc (Balanced Wave)				
			Electrical	Standard				
				Tungsten Arc (Standard)				
				Tungsten Arc (Balanced Wave)				
	Transformer/ Rectifier	dc	Electrical	Drooping	x		o	
				Constant	o		o	x
				Rising		x	x	o
		ac/dc	Mechanical	Light Duty				
				Standard	o		o	
			Electrical Field Type	Tungsten Arc (Standard)	o		o	
Rotating	Electric Motor	dc	Stationary	Drooping	x		o	
				Constant	o		o	x
				Rising		x	x	o
	Engine Driven		Stationary	Constant	o			o
				dc—Drooping	o		o	
				Rising		x	x	o
			Rotating	ac				
				dc—Drooping	o		o	
				dc—Constant	o			o
				ac/dc—Standard	o			
				ac/dc—Tungsten Arc	o			

x = preferred
o = also usable
√ = stick electrodes

Source: Courtesy of Air Reduction Sales Company (Airco, Inc.).

MIG OTHER			MIG CUTTING		TIG—A.C. ALUMINUM			TIG—A.C. OTHER			TIG-D.C.S.P.			TIG-SPOT		
MANUAL	MANUAL—PUSH & PULL GUNS	MACHINE	MANUAL	MACHINE	MANUAL	MACHINE HOLDER	AUTOMATIC HEAD	MANUAL	MACHINE HOLDER	AUTOMATIC HEAD	MANUAL	MACHINE HOLDER	AUTOMATIC HEAD	STANDARD	WITH CRATER ELIM.	STICK ELECTRODES
																√
										O						√
					X	O										
							X									
					O	O		X	X	O						√
					X	X	X		O							√
X	O		X								O	O	X	X	X	√
O	O	X	O	X												
	X	O														
																√
O	O		O							X	O	O	X	X	X	√
O	O		O		O	O		X	X	O	X	X	O	O	O	√
X	O		X								O	O		O		√
O	O	X	O	X												
	X	O														
O		O														
O	O										O	O				√
	X	O														
																√
O	O										O	O				√
O		O														
O											O	O				√
O					X	X		X	X		X	X				√

Table **5-2** Generator dc and generator-rectifier dc arc welders

OUTPUT RATINGS—60 PERCENT DUTY CYCLE AT RATED OUTPUT								
RATED			MINIMUM			MAXIMUM		
AMPERES	*AT*	*LOAD VOLTS**	*AMPERES*	*AT*	*LOAD VOLTS**	*AMPERES*	*AT*	*LOAD VOLTS**
150		26	20		20	185		27
200		28	30		21	250		30
250		30	40		22	310		32
300		32	60		22	375		35
400		36	80		23	500		40
500		40	100		24	625		44
600		44	120		25	750		44

*The above load voltages are based on the formula $E = 20 + 0.04\ I$, where E is the load voltage and I is the load current. For currents above 600 amperes, the voltage shall remain constant at 44 volts. Where the output current indicator is calibrated in amperes, the calibration points shall be based on the load voltages determined for each current setting by the formula $E = 20 + 0.04\ I$, where E is the load voltage and I is the load current. For currents above 600 amperes, the voltage shall remain constant at 44 volts.

Source: Reprinted by permission from the NEMA Standards Publication for Electric Arc-welding Apparatus, EW 1-1971, © 1971. Information is based on ANSI C87.1-1976.

Table **5-4** Limited-service ac transformers

OUTPUT RATINGS—20 PERCENT DUTY CYCLE AT RATED OUTPUT								
RATED			MINIMUM			MAXIMUM		
AMPERES	*AT*	*LOAD VOLTS*	*AMPERES*	*AT*	*LOAD VOLTS*	*AMPERES*	*AT*	*LOAD VOLTS*
295		30	50		22	295		30

Source: Reprinted by permission from the NEMA Standards Publication for Electric Arc-welding Apparatus, EW 1-1971, © 1971. Information is based on ANSI C87.1-1976.

Table **5-3** Industrial transformer ac arc welders (constant current)

OUTPUT RATINGS—60 PERCENT DUTY CYCLE AT RATED OUTPUT								
RATED			MINIMUM			MAXIMUM (AT 35 PERCENT DUTY CYCLE)		
AMPERES	*AT*	*LOAD VOLTS**	*AMPERES*	*AT*	*LOAD VOLTS**	*AMPERES*	*AT*	*LOAD VOLTS**
200		28	40		22	250		30
300		32	60		22	375		35
400		36	80		23	500		40
500		40	100		24	625		44
600		44	120		25	750		44
HIGH CURRENT RATINGS† (ONE-HOUR DUTY RATING)								
750		44	187		28	935		44
1000		44	250		30	1250		44
1500		44	450		38	1875		44

*The above load voltages are based on the formula $E = 20 + 0.04\,I$, where E is the load voltage and I is the load current. For currents above 600 amperes, the voltage shall remain constant at 44 volts.
†These sizes are rated for service with automatic machine arc welders. In testing to determine the rating, temperature rise and other characteristics, rated current at rated load voltage shall be applied for one hour immediately followed by the application of 75 percent rated current at rated load voltage for three hours. The temperature rise shall be measured at the end of the one-hour period and at the completion of the three-hour period at 75 percent rated current and load voltage.

Source: Reprinted by permission from the NEMA Standards Publication for Electric Arc-welding Apparatus, EW 1-1971, © 1971. Information is based on ANSI C87.1-1976.

Table **5-5** Limited-input ac transformers

OUTPUT RATINGS—20 PERCENT DUTY CYCLE AT RATED OUTPUT										
RATED			MINIMUM			MAXIMUM			RATED INPUT AMPERES	
AMPERES	*AT*	*LOAD VOLTS*	*AMPERES*	*AT*	*LOAD VOLTS*	*AMPERES*	*AT*	*LOAD VOLTS*	*WITH POWER FACTOR CORRECTION**	*WITHOUT POWER FACTOR CORRECTION*
180		25	20		20	180		25	37	46

*The no-load input current of a power-factor-corrected arc welder shall not exceed 50 percent of its rated input current.

Source: Reprinted by permission from the NEMA Standards Publication for Electric Arc-welding Apparatus. EW 1-1971 © 1971. Information is based on ANSI C87.1-1976.

Table **5-6** Industrial dc transformer rectifiers

OUTPUT RATINGS—60 PERCENT DUTY CYCLE AT RATED OUTPUT								
RATED			MINIMUM			MAXIMUM (AT 35 PERCENT DUTY CYCLE)		
AMPERES	*AT*	*LOAD VOLTS**	*AMPERES*	*AT*	*LOAD VOLTS**	*AMPERES*	*AT*	*LOAD VOLTS**
200		28	40		22	250		30
300		32	60		22	375		35
400		36	80		23	500		40
500		40	100		24	625		44
600		44	120		25	750		44
800		44	160		26	1000		44

*The above load voltages are based on the formula $E = 20 + 0.04\,I$, where E is the load voltage and I is the load current. For currents above 600 amperes, the voltage shall remain constant at 44 volts.

Source: Reprinted by permission from the NEMA Standards Publication for Electric Arc-welding Apparatus, EW 1-1971, © 1971. Information is based on ANSI C87.1-1976.

Table **5-8** Limited-input dc transformer rectifiers

OUTPUT RATINGS—20 PERCENT DUTY CYCLE AT RATED OUTPUT										
RATED			MINIMUM			MAXIMUM			RATED INPUT AMPERES	
AMPERES	*AT*	*LOAD VOLTS*	*AMPERES*	*AT*	*LOAD VOLTS*	*AMPERES*	*AT*	*LOAD VOLTS*	*WITH POWER FACTOR CORRECTION*	*WITHOUT POWER FACTOR CORRECTION**
150		25	20		20	150		25	37	46

*The no-load input current of a power-factor-corrected arc welder shall not exceed 50 percent of its maximum input current.

Source: Reprinted by permission from the NEMA Standards Publication for Electric Arc-welding Apparatus, EW 1-1971, © 1971. Information is based on ANSI C87.1-1976.

Table **5-7** Limited-service dc transformer rectifiers

OUTPUT RATINGS—20 PERCENT DUTY CYCLE AT RATED OUTPUT								
RATED			MINIMUM			MAXIMUM		
AMPERES	*AT*	*LOAD VOLTS*	*AMPERES*	*AT*	*LOAD VOLTS*	*AMPERES*	*AT*	*LOAD VOLTS*
240		30	50		22	240		30

Source: Reprinted by permission from the NEMA Standards Publication for Electric Arc-welding Apparatus, EW 1-1971, © 1971. Information is based on ANSI C87.1-1976.

Table **5-9** Industrial ac/dc transformer rectifiers

OUTPUT RATINGS—60 PERCENT DUTY CYCLE AT RATED OUTPUT											
RATED			MINIMUM			MAXIMUM					
						AC AMPERES	*AT*	*LOAD VOLTS**	*DC AMPERES*	*AT*	*LOAD VOLTS**
AC/DC AMPERES	*AT*	*LOAD VOLTS**	*AC/DC AMPERES*	*AT*	*LOAD VOLTS**	(AT 35 PERCENT DUTY CYCLE)			(AT 60 PERCENT DUTY CYCLE)		
200		28	40		22	250		30	200		28
300		32	60		22	375		35	300		32
400		36	80		23	500		40	400		36
500		40	100		24	625		44	500		40

*The above load voltages are based on the formula $E = 20 + 0.04\ I$, where E is the load voltage and I is the load current. For currents above 600 amperes, the voltage shall remain constant at 44 volts. Where the output current indicator is calibrated in amperes, the calibration points shall be based on the load voltages determined for each current setting by the formula $E = 20 + 0.04\ I$, where E is the load voltage and I is the load current. For currents above 600 amperes, the voltage shall remain constant at 44 volts.

Source: Reprinted by permission from the NEMA Standards Publication for Electric Arc-welding Apparatus, EW 1-1971, © 1971. Information is based on ANSI C87.1-1976.

Table **5-10** Limited-service ac/dc transformer rectifiers

OUTPUT RATINGS—20 PERCENT DUTY CYCLE AT RATED OUTPUT											
RATED			MINIMUM			MAXIMUM					
AC/DC AMPERES	*AT*	*LOAD VOLTS*	*AC/DC AMPERES*	*AT*	*LOAD VOLTS*	*AC AMPERES*	*AT*	*LOAD VOLTS*	*DC AMPERES*	*AT*	*LOAD VOLTS*
295		30	50		22	295		30	240		30

Source: Reprinted by permission from the NEMA Standards Publication for Electric Arc-welding Apparatus, EW 1-1971, © 1971. Information is based on ANSI C87.1-1976.

Table **5-11** Limited-input ac/dc transformer rectifiers

OUTPUT RATINGS—20 PERCENT DUTY CYCLE AT RATED OUTPUT													
RATED			MINIMUM			MAXIMUM						RATED INPUT AMPERES	
AC/DC AMP	*AT*	*LOAD VOLTS*	*AC/DC AMP*	*AT*	*LOAD VOLTS*	*AC AMP*	*AT*	*LOAD VOLTS*	*DC AMP*	*AT*	*LOAD VOLTS*	*WITH POWER FACTOR CORRECTION**	*WITHOUT POWER FACTOR CORRECTION*
180		25	20		20	180		25	150		25	37	46

*The no-load input current of a power-factor-corrected arc welder shall not exceed 50 percent of its maximum input current.

Source: Reprinted by permission from the NEMA Standards Publication for Electric Arc-welding Apparatus, EW 1-1971, © 1971. Information is based on ANSI C87.1-1976.

Table **5-12** Conversion factors to obtain 100% duty cycle rating for ac power supplies used for TIG welding

POWER SUPPLY PRESENT DUTY CYCLE	POWER SUPPLY RATED AMPS TIMES:		DUTY CYCLE	DERATED DUTY CYCLE AC TIG WELDING DERATED AMPS TIMES:
60%	75%	=	100%	70%
50%	70%	=	100%	70%
40%	55%	=	100%	70%
30%	50%	=	100%	70%
20%	45%	=	100%	70%

Source: *The Gas Tungsten-Arc Welding Process*, Form GTA-1-74, Miller Electric Manufacturing Company, Appleton, Wisconsin.

Figure **5-9** Portable gasoline motor–driven welding power source. (With permission from Miller Electric Manufacturing Company, Publication FLP-1 7/73, p. 24.)

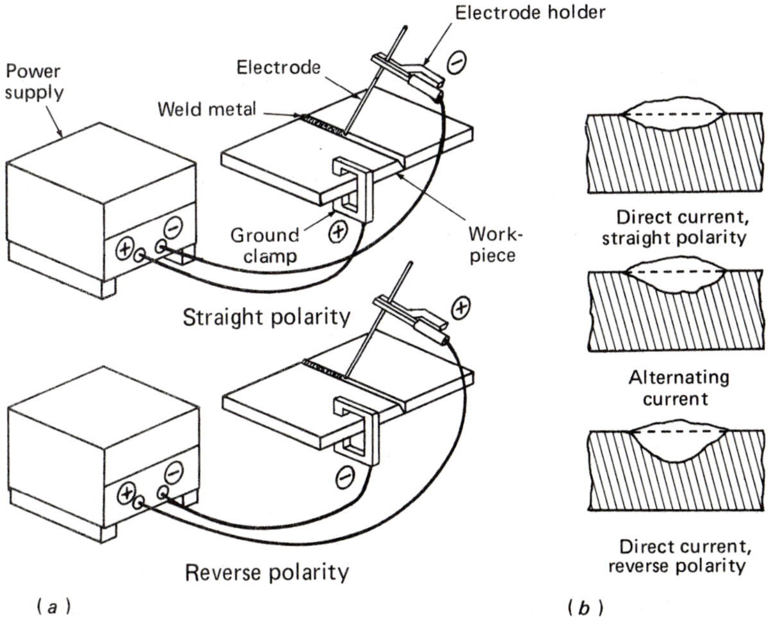
Electrode holder
Power supply
Electrode
Weld metal
Ground clamp
Work-piece
Straight polarity
Reverse polarity
Direct current, straight polarity
Alternating current
Direct current, reverse polarity
(a)
(b)

Figure **5-10** (*a*) Straight polarity versus reverse polarity. (*b*) Relative depths of penetration for different current characteristics. (By permission, from *Metals Handbook* Volume 6, Copyright American Society for Metals, 1971.)

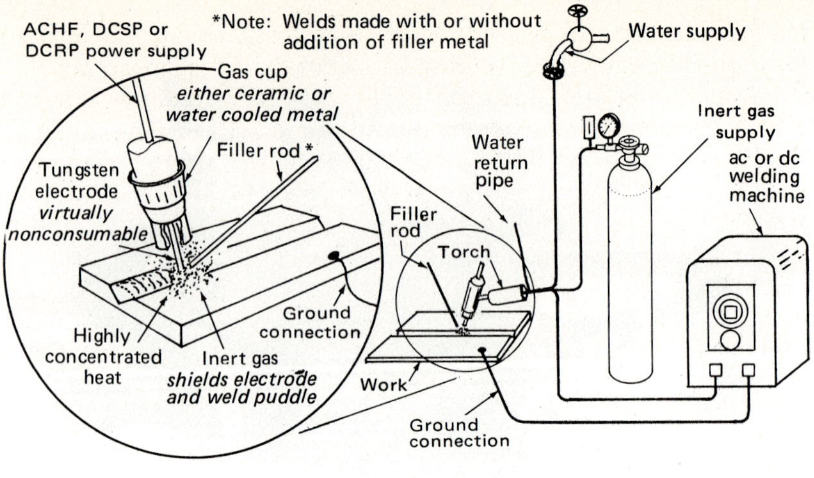
ACHF, DCSP or
DCRP power supply
*Note: Welds made with or without
addition of filler metal
Water supply
Gas cup
either ceramic or
water cooled metal
Filler rod *
Tungsten
electrode
virtually
nonconsumable
Highly
concentrated
heat
Inert gas
shields electrode
and weld puddle
Ground
connection
Water
return
pipe
Inert gas
supply
ac or dc
welding
machine
Filler
rod
Torch
Work
Ground
connection

Figure **5-11** Welding unit equipped with a water supply for TIG welding. (Adapted by permission from Union Carbide Corporation, Linde Division, *How To Do TIG (Heliarc) Welding*, F-1607-D 5/70, p. 3; and Hobart Brothers Company.)

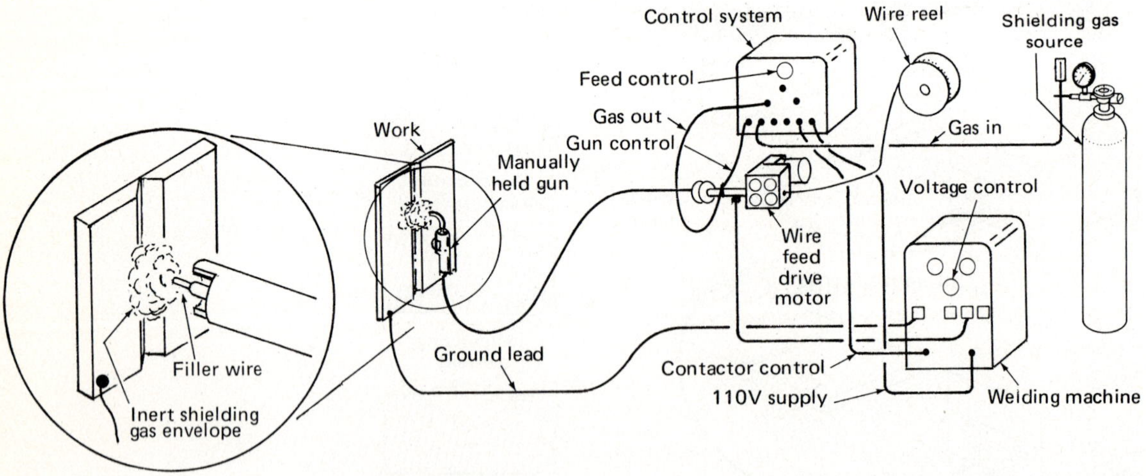
Control system
Wire reel
Shielding gas
source
Feed control
Gas out
Gas in
Work
Gun control
Manually
held gun
Voltage control
Wire
feed
drive
motor
Filler wire
Inert shielding
gas envelope
Ground lead
Contactor control
110V supply
Welding machine

Figure **5-12** Component parts of a typical MIG welding set-up. (Adapted by permission from James F. Lincoln Arc Welding Foundation, *Metals and How To Weld Them*, 2d ed., 1954, 1962; and Hobart Brothers Company.)

If a dc generator only is used, the welding machine may not be used for high-frequency (HF) welding of aluminum or magnesium.

Power supplies designed specifically for TIG welding incorporate a high-frequency stabilizer in the ac circuit. The high-frequency oscillator is used to assist in starting the arc in welding carbon or stainless steel, as well as for breaking down the oxide coating on aluminum.

The power supply used for the gas metal–arc (MIG) welding process is called a constant-voltage (CV) type of power source (Figure 5-12). It can be a dc rectifier or a motor- or engine-driven generator. The output power of a CV machine supplies the same voltage no matter what amperage is used.

The power supply used for the flux shielded–arc welding process is essentially the same as that used for the MIG process. When using a spray-type arc (Figure 5-13) with this process, an auxiliary gas shield of either CO_2 or argon (as applicable) is required.

Power supplies used for submerged-arc welding are motor-generators and transformer-rectifiers for dc output and transformers for ac output (Figure 5-14). Either ac or dc produces acceptable results in submerged-arc welding, although each has distinct advantages in specific applications, depending on the amperage range, diameter of the electrode wire, and travel speed.

Either of two general types of dc power supplies are used for the flux-cored (vapor-shielded) metal–arc welding process. When a constant-current power supply is employed, it is usually matched to an arc voltage–sensing electrode-wire feeder (Figure 5-15).

Constant-voltage transformer-rectifiers are used extensively to supply the power for electroslag welding, because they are easily controlled and are standard in many plants, being used for gas metal–arc and flux-cored–arc welding. The power rating of a machine is usually 750 amps and 60 volts at 100% duty cycle. A power supply of this rating is required for each wire fed; thus, total capacity for welding with three electrode wires would be 2250 amps.

Constant-current transformer-rectifiers have been used successfully for electroslag welding, but they are more difficult to control. Their ready availability has usually been the principal reason for their use.

Motor-generators have also proved satisfactory for electroslag welding. However, motor-generator sets having a capacity of 750 amps at 100% duty cycle are far more costly than transformer-rectifiers of the same capacity.

A motor-generator or a storage battery is used to supply the dc power needed for arc-stud welding. Although the power supplies used for shielded metal–arc (stick) welding can be used, power supplies designed specifically for arc-stud welding are recommended for greater efficiency and the most appropriate voltage and amperage. Sometimes it is possible to obtain the high amperages needed by connecting two or more regular arc welding power supplies in series.

Plasma-arc welding (Figures 5-16 and 5-17) is generally considered to be a refinement of the gas tungsten–arc (TIG) process. It uses the same power supplies; but, instead of an inert gas (argon or helium), it uses a plasma-forming gas, such as nitrogen or hydrogen.

5-9 SOME TERMS THAT NEED EXPLANATION

Some terms have cropped up in this chapter that need to be explained. The following definitions attempt to do just that.

INPUT-OUTPUT VOLTAGE This refers to the relationship of input voltage to the output voltage of a transformer. This ratio can also be described as the ratio of the number of turns of wire on the input or primary side of the transformer to the number of turns on the secondary or output side.

There are two types of transformers, *step-up transformers* and *step-down transformers*. Welding power supplies are usually step-down transformers, because the voltage supplied by power companies for industrial consumption is too high for welding purposes. Therefore, a transformer is used to step down a high-input voltage to

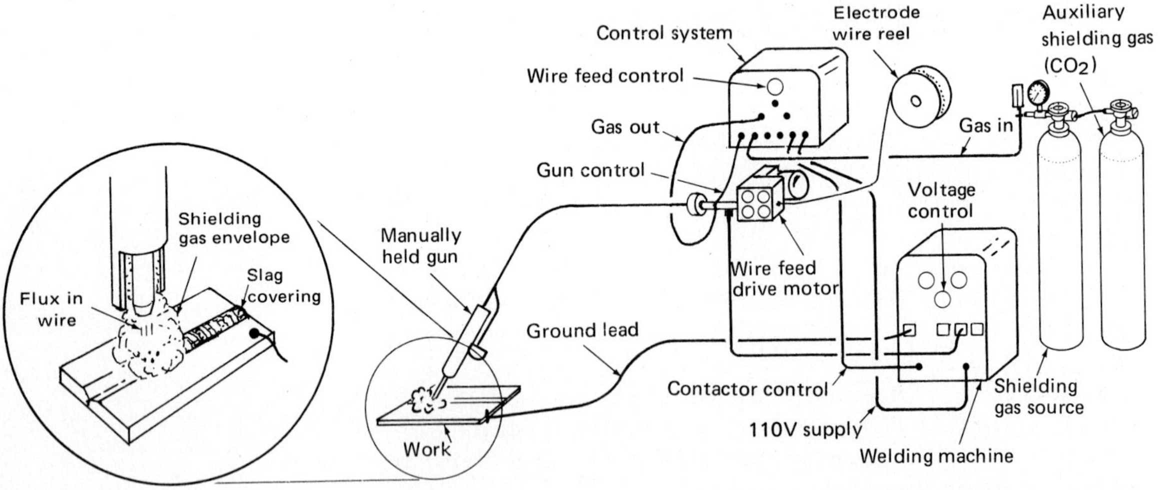
Control system
Electrode wire reel
Auxiliary shielding gas (CO_2)
Wire feed control
Gas out
Gas in
Gun control
Voltage control
Shielding gas envelope
Manually held gun
Slag covering
Flux in wire
Wire feed drive motor
Ground lead
Contactor control
Shielding gas source
110V supply
Work
Welding machine

Figure **5-13** MIG welding set-up modified for an auxiliary gas shield. (Adapted by permission from James F. Lincoln Arc Welding Foundation, *Metals and How to Weld Them,* 2d ed., 1954, 1962; and Hobart Brothers Company.)

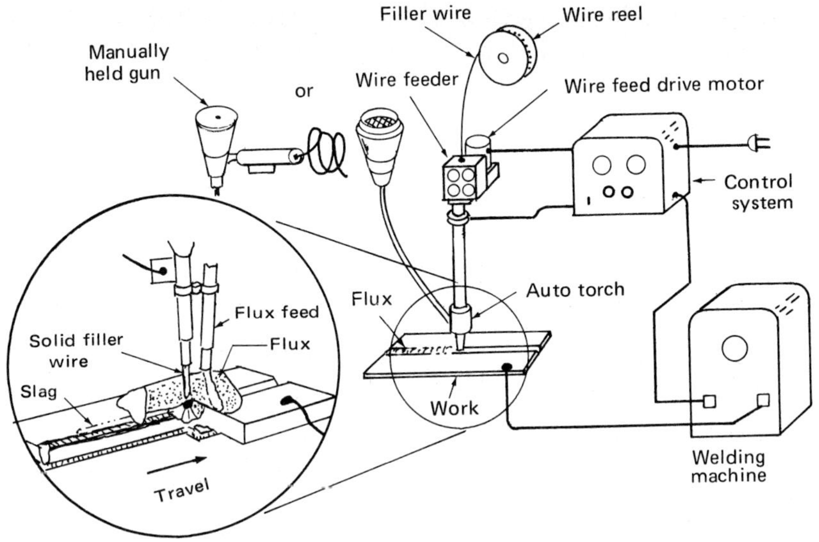

Filler wire
Wire reel
Manually held gun
or
Wire feeder
Wire feed drive motor
Control system
Flux feed
Solid filler wire
Flux
Slag
Flux
Auto torch
Work
Travel
Welding machine

Figure **5-14** Component parts of a submerged-arc welding set-up. (Adapted by permission from James F. Lincoln Arc Welding Foundation, *Metals and How to Weld Them,* 2d ed., 1954, 1962; and Hobart Brothers Company.)

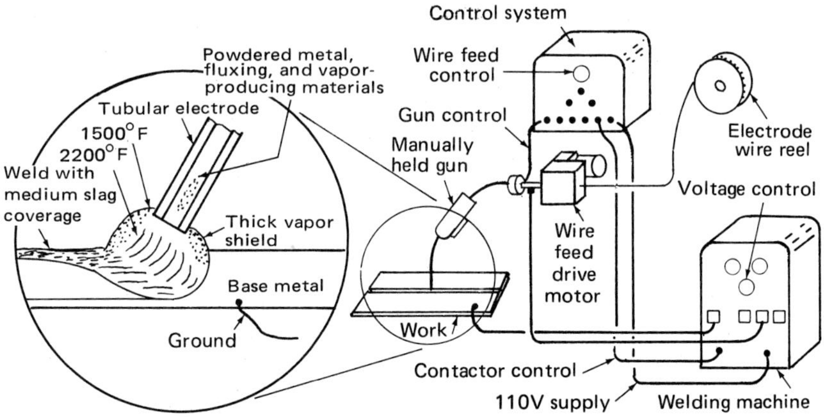

Control system
Powdered metal, fluxing, and vapor-producing materials
Wire feed control
Tubular electrode
Gun control
1500° F
2200° F
Manually held gun
Electrode wire reel
Weld with medium slag coverage
Voltage control
Thick vapor shield
Wire feed drive motor
Base metal
Ground
Work
Contactor control
110V supply
Welding machine

Figure **5-15** Component parts of a flux-cored metal–arc welding set-up. (Adapted by permission from James F. Lincoln Arc Welding Foundation, *Metals and How to Weld Them,* 2d ed., 1954, 1962; and Hobart Brothers Company.)

a lower, more acceptable voltage for arc welding. At the same time, the transformer provides a means of delivery for high amperage (50 to 500 amps) from the relatively low-current power lines.

In step-down transformers, the primary winding has more turns of wire than the secondary winding. For example, in the transformer shown in Figure 5-18, the input voltage is 100 and the input amperage is 0.25 amps. The ratio of primary turns to secondary turns is 4:1 (100:25), so the input voltage is reduced by one quarter while input amperage is increased four times. The result is an output voltage of 25 and an output amperage of 1. The power output is the same as the input (25 watts), but voltage has been decreased with a commensurate increase in amperage.

The ratio of primary to secondary turns to the step-down of voltage can be expressed as

$$\frac{N_1}{N_2} = \frac{E_1}{E_2}$$

where N_1 is the number of turns on the primary winding, N_2 is the number of turns on the secondary winding, E_1 is the input voltage, and E_2 is the output voltage.

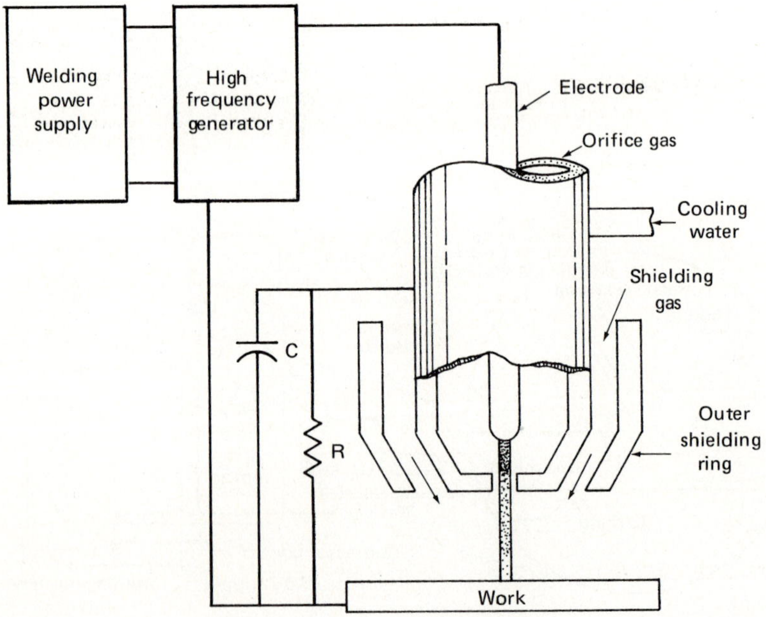
Welding power supply
High frequency generator
Electrode
Orifice gas
Cooling water
Shielding gas
C
R
Outer shielding ring
Work

Figure **5-16** Simplified diagram of the basic plasma-arc welding process. (Redrawn by permission from Miller Electric Manufacturing Company, *Fundamentals of Gas Metal-Arc (MIG) Welding,* Form: MTS-6A 3/71.)

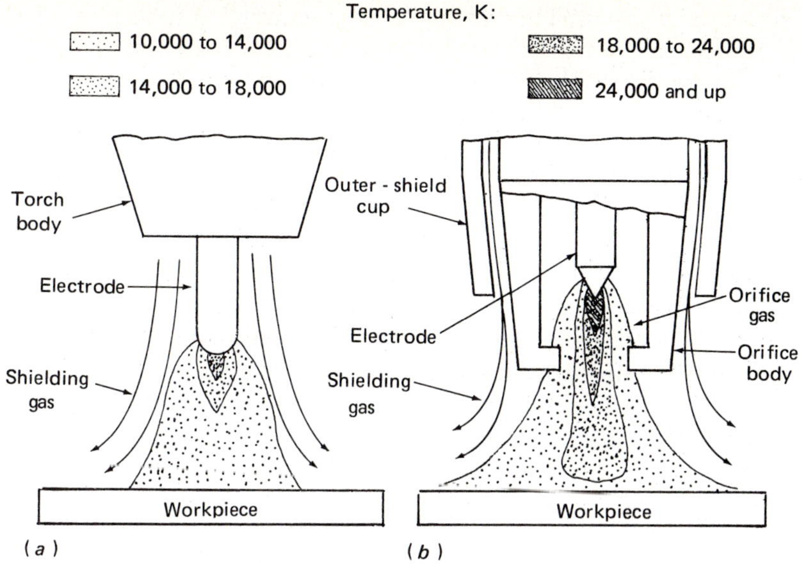
Temperature, K:
10,000 to 14,000
14,000 to 18,000
18,000 to 24,000
24,000 and up
Torch body
Electrode
Shielding gas
Workpiece
(a)
Outer - shield cup
Electrode
Shielding gas
Orifice gas
Orifice body
Workpiece
(b)

Figure **5-17** Comparison of (*a*) a nonconstricted arc used for TIG welding and (b) a constricted arc used for plasma-arc welding. (By permission, from *Metals Handbook* Volume 6, Copyright American Society for Metals, 1971.)

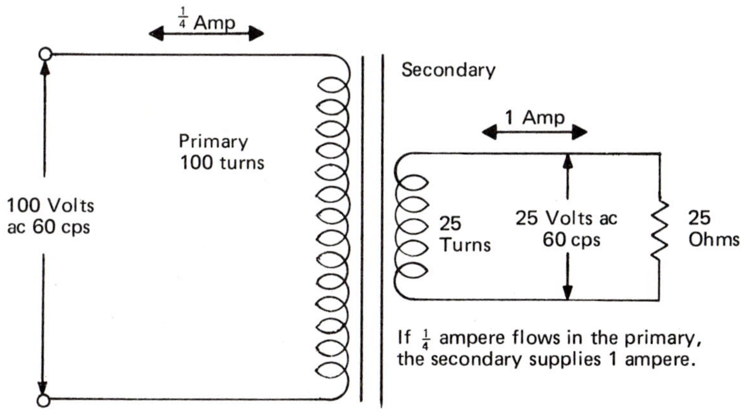
¼ Amp
Primary
100 turns
100 Volts
ac 60 cps
Secondary
1 Amp
25
Turns
25 Volts ac
60 cps
25
Ohms
If ¼ ampere flows in the primary,
the secondary supplies 1 ampere.

Figure **5-18** Schematic of a step-down transformer

The current of transformer-type welding machines may be controlled in several different ways. There may be taps on the secondary coil; that is, an output lead may be plugged in at, say, 25, 50, or 75 turns, and so on, depending on the desired output. Other methods of adjustment involve moving the core of the transformer or moving the coils within the transformer; and a more advanced method involves introducing another electric circuit into the transformer by means of a reactor. Each of these methods provides the control needed to obtain the proper volt-ampere characteristics for good welding.

DUTY CYCLE The *duty cycle* is one of the most important rating points of a welding power supply. It expresses (as a percentage) the portion of the time that the power supply must deliver its rated output in each one of successive 10-minute intervals. Thus a 60% duty cycle (the standard industrial rating) means that the power supply can deliver its rated load output for 6 out of every 10 minutes. (Steady operation at rated load for 36 minutes out of one hour is not a 60% duty cycle. The rating is based on successive 10-minute intervals.) Power supplies rated at 100% duty cycle produce their rated output continuously without exceeding the established temperature limits. They are commonly used for automatic and semiautomatic welding applications.

Duty cycle is the most important factor when deciding the type of power supply to be acquired or used for any specified application. Industrial units for manual welding are rated at 60% duty cycle. NEMA rates limited-service power supplies at 20% duty cycle. Individual manufacturers have sometimes rated power supplies at duty-cycle values other than those above; but the rating method using 10-minute intervals is standard.

In any duty-cycle rating, the maximum allowable temperature of the components in the unit is the determining factor. These maximum temperatures are specified by various organizations and agencies interested in insulation standards.

An important point to remember is that the duty cycle of a power supply is based on the output current, not on the kVA or kW rating. Two useful and approximate formulas are given below for determining a new duty cycle at other than rated output, or for determining an output other than rated at a new specified duty cycle.

$$T_a = \left(\frac{I}{I_a}\right)^2 T$$

$$I_a = I\sqrt{\frac{T}{T_a}}$$

where T is the given duty cycle (in percent), T_a is the required duty cycle (in percent), I is the rated current at the given duty cycle, and I_a is current at the required duty cycle.

For very high current output supplies (750 amps and higher), another duty cycle rating is usually used. This is the *one-hour duty rating*. These power supplies are designed for the demanding service of semiautomatic or automatic welding systems. In determining the rated output of these machines, each is loaded for one hour at the rated output and then tested. Then the output is reduced immediately to 75% of the rated current value and operation is continued for an additional three hours. At this time, the test period is terminated. Temperatures are measured at the end of the first hour and at the conclusion of the test. These temperatures must be within established allowable limits.

ARC BLOW *Arc blow* is a phenomenon caused by a magnetic disturbance close to the arc when welding with dc current. The conditions under which arc blow can occur are shown in Figure 5-19. The weld extends from end 1 to end 4, with the ground connection being made to the backing bar at 1. As the weld progresses from 1 to 4, the arc and arc flame will show a decidedly forward pull for about one-eighth to one-fourth the length of the seam. This pull is maximum at the start and becomes almost zero at 2. Between 2 and 3 the arc is most stable. At 3 the arc shows a backward pull that increases as welding proceeds to 4. This tendency to pull back may be enough to extinguish the arc. An extremely unstable arc (Figure 5-20) results. The magnetic field is set up in the plane of the parts being joined and

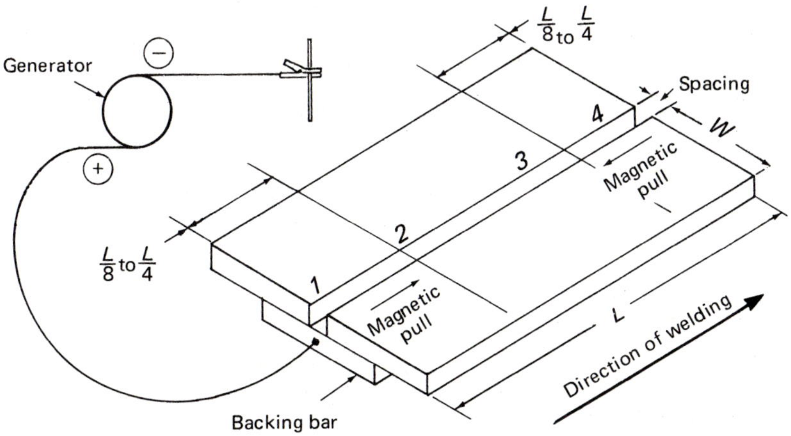

Generator
−
+
L/8 to L/4
L/8 to L/4
Spacing
W
1
2
3
4
Magnetic pull
Magnetic pull
L
Direction of welding
Backing bar

Figure **5-19** Conditions under which arc blow can occur. (Adapted by permission from *The Procedure Handbook of Arc Welding,* 11th ed., Lincoln Electric Company, Cleveland, Ohio, Fig. 3-46.)

circumferentially around the electrode and the plates, as shown in Figure 5-21. Field F_1 is set up around the electrode, field F_2 around the plates being joined, and field F_3 in the plates adjacent to the arc and in a direction similar to that of field F_1. Of the three fields, field F_3 is the one that apparently causes the arc to draw away from either end of the weld. If arc blow is noticeable, and it is not convenient to change to ac welding, where arc blow does not exist, the following remedies may be tried:

- Reduce current.
- Use the backstepping technique on welds exceeding six inches in length.
- Weld toward a heavy tack weld from ground direction.
- When nearing the end of the weld, affix the ground connection at the end of the weld and weld toward it.
- Use two ground connections and place them as far away from the weld (and each other) as possible.
- Coil the ground cable around the workpiece. (This technique requires experimentation on the part of the welder as to how many turns of the cable are to be used and where to place them.)

RECTIFIERS Rectifiers are devices used to change (rectify) ac into dc. Some ac/dc welders are the dry-plate selenium and silicon types. The *silicon rectifier* has a shape and size somewhat resembling a spark plug. In the *selenium rectifier*, a layer of selenium about 0.002–0.003 inch thick is deposited on a support plate of steel or aluminum. This plate also has the function of dissipating the heat generated by the current flow. A front electrode contacts the other side of the selenium layer. These individual modular rectifier elements are mounted in stacks. Current will flow from the front electrode to the selenium and its support plate, but the backward resistance from the other direction of flow is about 1000 times higher, making reverse current very small. Another type of rectifier used in ac/dc welders is the *semiconductor type*. This type of rectifier is arranged in a bridge circuit. By the use of rectifiers in the bridge, it can be seen that current flow through the meter is always in one direction. When the voltage being measured has a wave form, as shown in Figure 5-22, the path of current flow will be from the lower input terminal through rectifier 3, through the instrument, and then through rectifier 2, thus completing its path back to the source's upper terminal. The next half-cycle of the input voltage (indicated by the dashed sine wave) will cause the current to pass through rectifier 1, through the instrument, and through rectifier 4, completing the path back to the source.

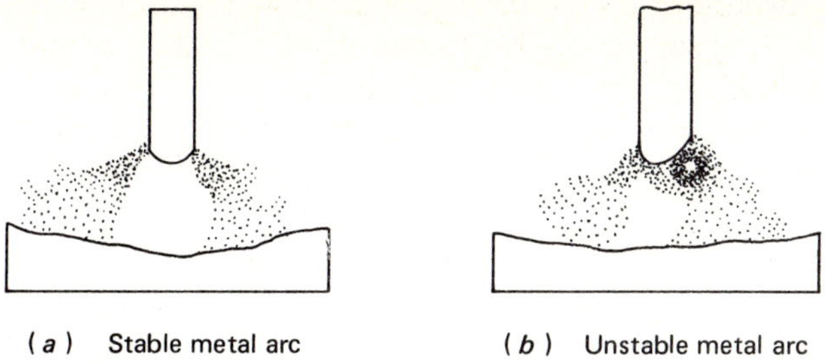
(a) Stable metal arc
(b) Unstable metal arc

Figure **5-20** Stable versus unstable arc. (Adapted by permission from *The Procedure Handbook of Arc Welding*, 11th ed., Lincoln Electric Company, Cleveland, Ohio, Fig. 3-44.)

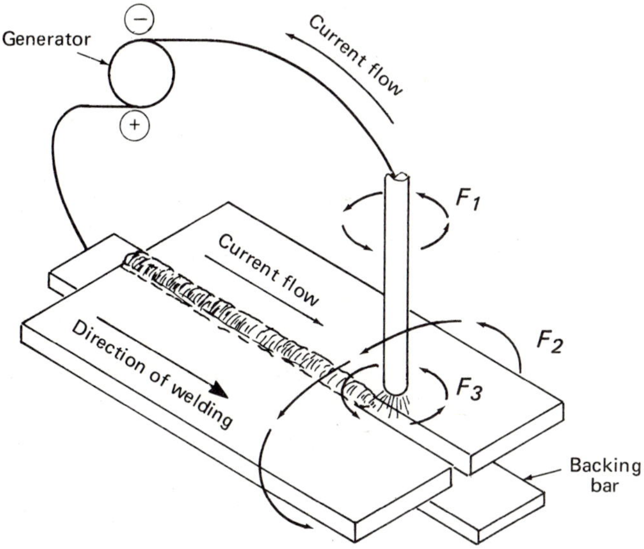
Generator
−
+
Current flow
F1
Current flow
Direction of welding
F2
F3
Backing bar

Figure **5-21** Magnetic field set up in workpiece and around arc. (Adapted by permission from *The Procedure Handbook of Arc Welding*, 11th ed., Lincoln Electric Company, Cleveland, Ohio, Fig. 2-85.)

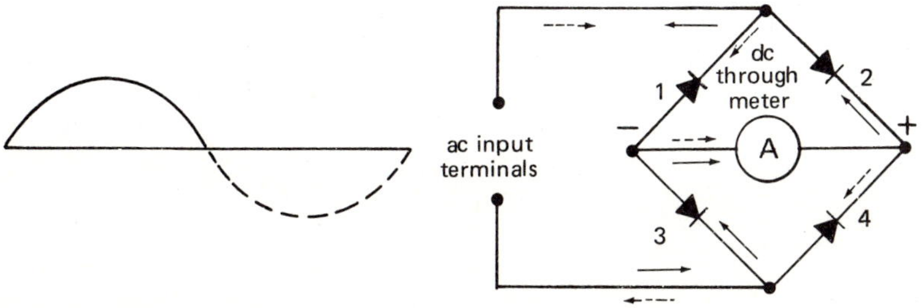
dc through meter
1
2
−
+
A
ac input terminals
3
4

Figure **5-22** Simple connection of a full-wave rectifier

REVIEW QUESTIONS

1. Name the unit used to express the rate of flow of electricity through a wire.

2. How is heat needed to melt the electrode and base metal in arc welding generated?

3. List and describe the major arc welding processes.

4. How has the AWS classified commercially available welding machines?

5. Why are dc welding machines more versatile than ac machines?

6. How does the NEMA classify constant-current transformer ac welding machines?

7. Name the principal components needed for welding with the following processes: GTAW, SMAW, GMAW.

8. When hooking up a welding machine for straight-polarity welding, which terminal has the negative polarity?

9. What is meant by *input-output voltage*?

10. What is meant by *duty cycle*?

11. What is arc blow?

12. Does ac reduce arc blow? Explain.

13. What would be the most appropriate NEMA welder classification of a welding machine to be purchased for use in a general welding shop? an automobile body and fender shop? a home workshop?

14. According to the Miller Electric Manufacturing Company, what causes overheating of the main transformer of a conventional ac power supply when it is used for TIG welding?

15. What is the purpose of the high-frequency oscillator found in power supplies specifically designed for TIG welding?

16. Can a constant-current transformer ac arc welder be used for MIG welding? If it can, must it be modified? If it must be modified, what must be added to or removed from it and why?

CHAPTER 6
FUEL-GAS WELDING

Fuel-gas welding, or flame welding, was the second modern welding process to be developed. In this process, workpieces are fused by the heat of a flame, without electricity. The flame results from combustion of a fuel gas with air or oxygen.

The most commonly used fuel gases are acetylene, hydrogen, natural gas, propane, butane, and a newly developed gas called methylacetylene propadiene (MAPP). Usually these gases are burned with oxygen rather than air because the large nitrogen component of air (which contributes nothing to the combustion) results in a low-temperature flame that is below the fusion temperature of most metals.

All the fuel gases used in welding are composed of both carbon and hydrogen, and are generally burned with pure oxygen. As a result, fuel-gas welding cannot be performed on metals (for example, titanium) that are harmed by these elements. Further, the combustion of these fuels with oxygen produces carbon dioxide and water, which can also be damaging to certain metals. Therefore, the welder must be certain that the metals being cut or welded using the fuel-gas process are not reactive with the resulting compounds. There are many such metals.

The necessary flame temperature usually determines which fuel gas is to be used in this process. The oxyacetylene mixture gives the highest temperature, approximately 6300°F at the cone. An oxyhydrogen flame provides about 4000°F maximum. The other gases are

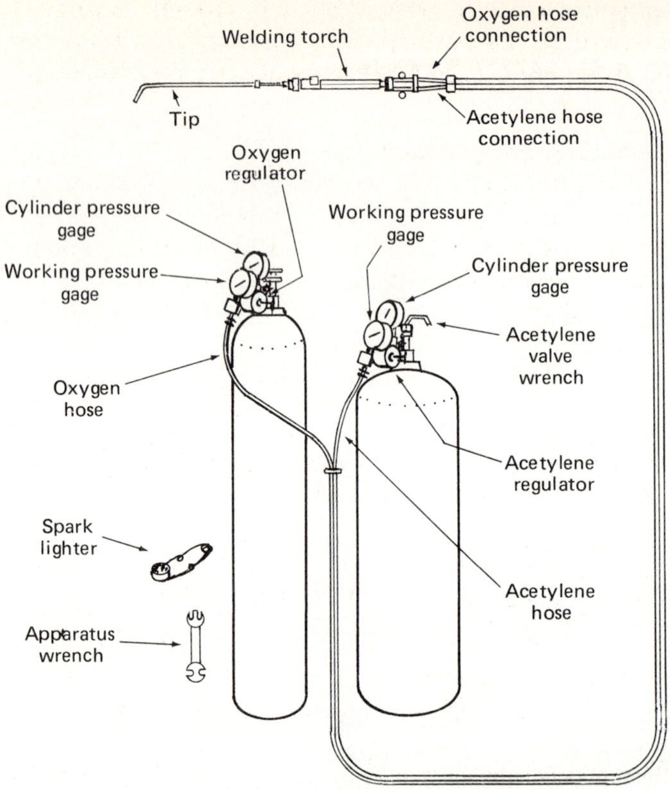

Oxygen hose connection
Welding torch
Tip
Acetylene hose connection
Oxygen regulator
Cylinder pressure gage
Working pressure gage
Working pressure gage
Cylinder pressure gage
Acetylene valve wrench
Oxygen hose
Acetylene regulator
Spark lighter
Acetylene hose
Apparatus wrench

Figure **6-1** A complete oxyacetylene welding outfit. (Redrawn by permission from Airco Welding Products, A Division of Airco, Inc., *Oxyacetylene Welding and Oxygen Cutting Instruction Course,* 1966, p. 28.)

lower, with the exception of MAPP, which is comparable to acetylene in temperature and has shown itself to be generally safer to use.

In flame cutting, when the oxyfuel gas flame heats the metal red hot, a stream of oxygen or air is blown into the hot metal, causing it to separate or cut. This is the general principle of flame cutting.

6-1 MATERIALS AND EQUIPMENT FOR FUEL-GAS WELDING

Fuel-gas welding equipment (Figure 6-1) consists of a gas supply, regulators for gas pressure control, hoses, torches, a torch igniter, goggles, and welding rods.

From the standpoint of safety, it is important that all torches, regulators or reducing valves, and acetylene generators be examined, tested, and found to have all practicable safeguards. Most insurance companies or local authorities will accept equipment listed by Underwriters' Laboratory, Inc., Northbrook, Illinois, or approved and listed by Factory Mutual Laboratories, Norwood, Massachusetts.

TORCHES A welding torch (Figure 6-2), mixer, and *welding tip* should be selected according to the recommendations of the equipment manufacturer, usually supplied in a booklet. (When the instruction booklet is not available, Tables 16-1 and 16-2, on pages 402 and 417, should be consulted for guidance in the selection of the correct welding-tip size and corresponding gas pressure.)

Figure **6-2** Typical welding torch. (By permission, from *Metals Handbook* Volume 6, American Society for Metals, 1971.)

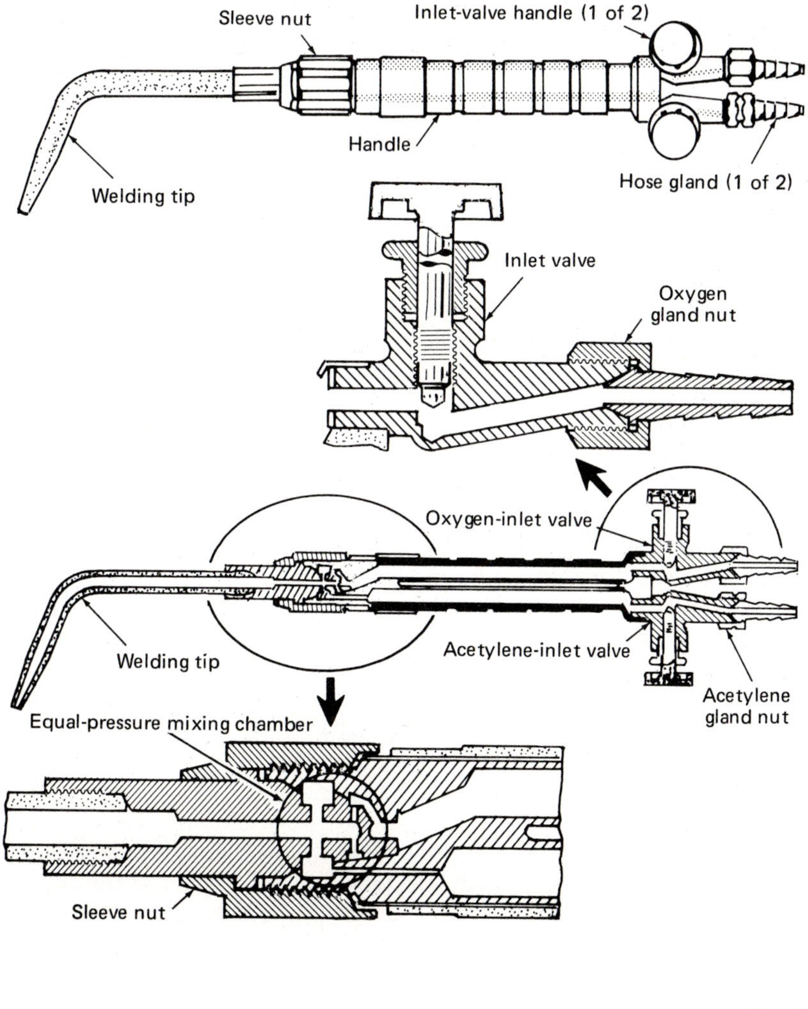

Sleeve nut
Inlet-valve handle (1 of 2)
Handle
Hose gland (1 of 2)
Welding tip
Inlet valve
Oxygen
gland nut
Oxygen-inlet valve
Welding tip
Acetylene-inlet valve
Acetylene
gland nut
Equal-pressure mixing chamber
Sleeve nut

The torch is the most important piece of gas-welding equipment. The torch mixes and controls the flow of gases to produce the required flame. A torch consists of a body with two inlet valves, a mixing chamber, and a tip (Figure 6-2). One end of the green oxygen hose is connected to one of the inlet valves with right-hand connections and the other end is connected to the oxygen regulator with right-hand connections. One end of the red fuel-gas hose is connected to the other inlet valve with left-hand connections (usually with one groove around it), the other end connecting in the same manner to the fuel-gas regulator. Oxygen and fuel-gas hoses *always* work in this manner.

The *torch*, or blowpipe, as it is sometimes called, provides a means of mixing the oxygen and fuel gas to provide the correct mixture at the tip. Tips attach to the blowpipe at the end of the mixing chamber by means of a special nut and come in various sizes for different welding and cutting jobs. Tip size is measured at the inside diameter of the outlet. Tips should be kept free and clean of accumulated metal during welding to ensure proper flame size and control and should be cleaned frequently with special wire-bristle brushes and/or tip cleaners.

There are two basic types of torches: the *balanced-pressure* type and the *injector* type. In balanced-pressure torches, the mixture nozzle has a central orifice around which are several smaller holes. One or the other of the gases (depending on torch manufacturer) enters by the central orifice under 1 to 15 psi pressure. The other gas enters through the smaller holes under the same pressure. In the injector type of torch, the oxygen passes through an injector nozzle and creates a suction that draws the fuel gas into the mixing chamber. Not much gas pressure is necessary with this type of torch.

REGULATORS Regulators (Figure 6-3), or automatic reducing valves, should be used only with the gases for which they are designed and marked. They should be used only for the pressure and flow ranges indicated in the manufacturer's literature. *Caution*: Never force a regulator-to-cylinder connection that does not make up readily. If it does not make up without forcing, it is probably the wrong connection or in need of cleaning or repair.

Regulators serve two basic functions: (1) they reduce cylinder pressure to a level acceptable to torches, and (2) they maintain constant pressure to the torch. The most common regulators are oxygen and acetylene regulators. Oxygen regulators are often green (like oxygen hoses) and have right-hand threads. Acetylene regulators are often colored red and have left-hand threads.

CYLINDERS Oxygen is available in steel cylinders of 20 to 300 cubic feet capacity, with pressures up to 2200 psi. The cylinders are usually painted green and have a green protective cap that should

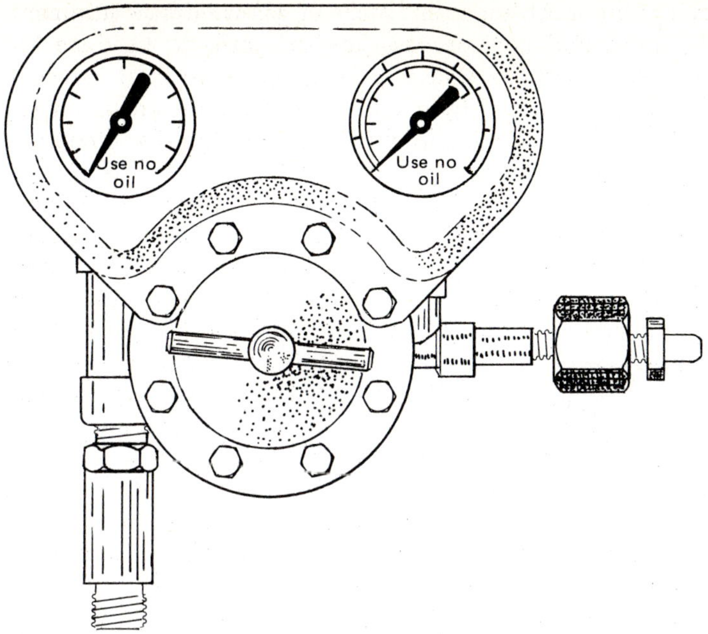
Use no oil
Use no oil

Figure **6-3** A typical regulator

always be screwed on securely when regulators and hoses have been removed. Acetylene cylinders are usually red and are pressurized to 250 psi. Most acetylene tanks also contain an absorbent material soaked with a dissolved chemical to stabilize the acetylene. This allows the use of pressures for storage above 15 psi (Figure 6-4).

Only cylinders that comply with the specifications of the Interstate Commerce Commission (ICC) effective at the date of their manufacture should be used. The cylinders must be maintained and charged with gas in accordance with ICC regulations.

A compressed gas cylinder must legibly show the chemical name, or a commonly accepted name, of the gas that it contains.

Cylinders should always be stored and used in an upright position (Figure 6-5). If acetylene cylinders are used in a horizontal position, it is possible for acetone to be drawn into the regulator, creating a fire hazard by interfering with the proper regulation and burning of the oxyacetylene gases. During storage and when in use, all cylinders should be fastened securely to a rigid object. If the valve is broken off by a cylinder's falling, this may cause the cylinder to be propelled violently.

Acetylene should never be drawn from a cylinder at an hourly rate exceeding one-seventh the cylinder capacity. Manifolding should be used for greater flows.

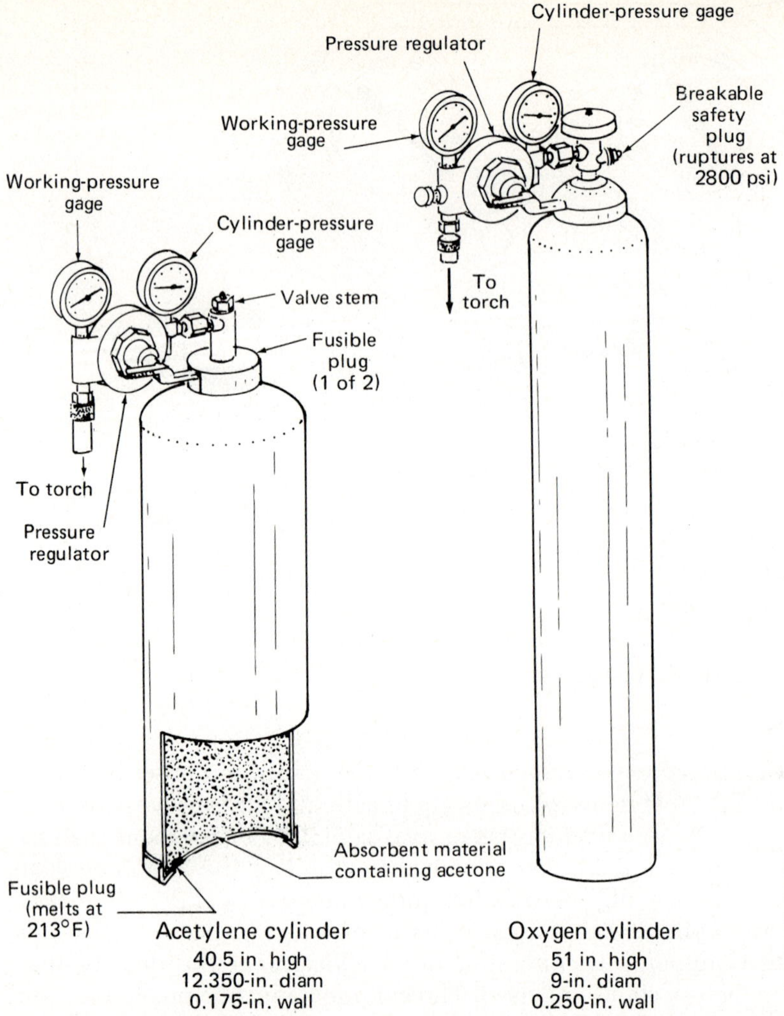
Cylinder-pressure gage
Pressure regulator
Breakable safety plug (ruptures at 2800 psi)
Working-pressure gage
Working-pressure gage
Cylinder-pressure gage
Valve stem
To torch
Fusible plug (1 of 2)
To torch
Pressure regulator
Absorbent material containing acetone
Fusible plug (melts at 213°F)
Acetylene cylinder
40.5 in. high
12.350-in. diam
0.175-in. wall
Oxygen cylinder
51 in. high
9-in. diam
0.250-in. wall

Figure **6-4** Portable gas cylinders and regulators used in gas welding. (By permission, from *Metals Handbook* Volume 6, American Society for Metals, 1971.)

MANIFOLDING CYLINDERS Cylinders are manifolded (Figure 6-6) to centralize the gas supply to provide a continuous supply of gas at a rate (volume in ft^3/hr) greater than could be supplied by an individual cylinder of the same size. Manifolds should be designed and constructed of material suitable for the particular gas and service for which they are to be used. The wide range of pressures encountered in the gases used, from the relatively low pressures of acetylene to

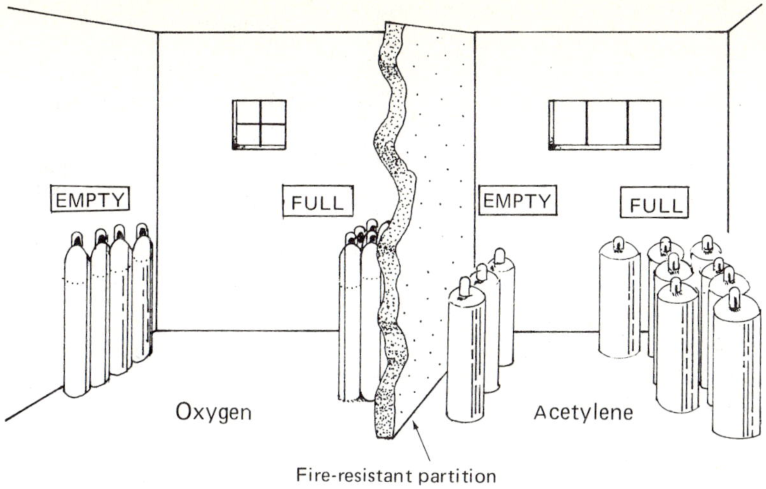
EMPTY
FULL
EMPTY
FULL
Oxygen
Acetylene
Fire-resistant partition

Figure **6-5** Safe storage for oxygen and acetylene cylinders. (Redrawn by permission from Union Carbide Corporation, Linde Division, *The Oxy-Acetylene Handbook,* 2d ed., 1960, p. 27.)

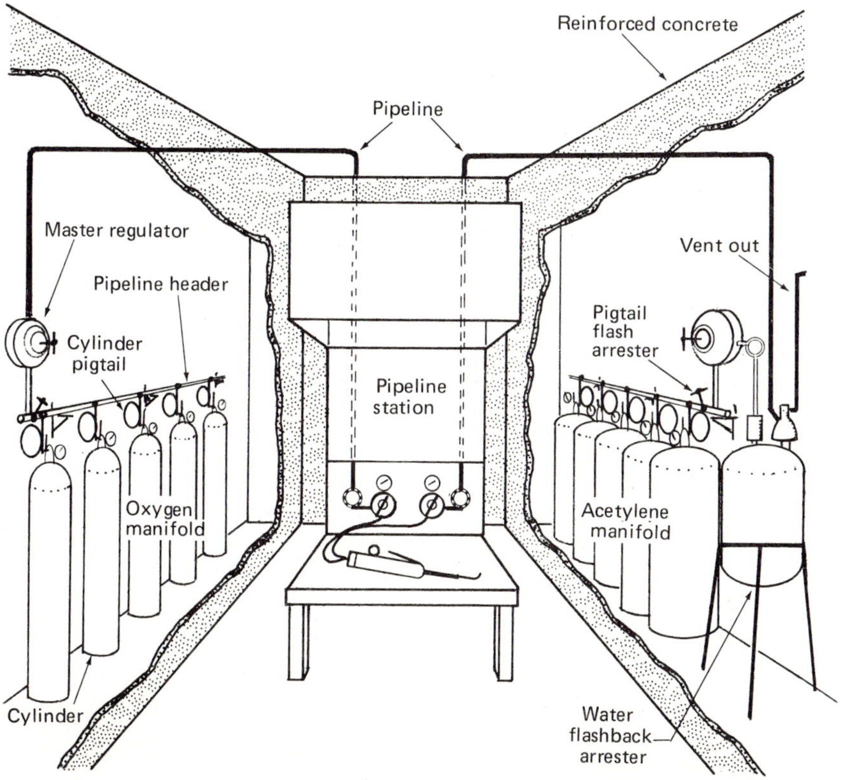
Reinforced concrete
Pipeline
Master regulator
Pipeline header
Cylinder
pigtail
Vent out
Pigtail
flash
arrester
Pipeline
station
Oxygen
manifold
Acetylene
manifold
Cylinder
Water
flashback
arrester

Figure **6-6** Basic manifolding system. For clarity, only one pipeline station is shown.

the high pressures of oxygen, makes it necessary that care be employed in the construction, installation, and maintenance of manifolds. Manifolds should be obtained from and installed under the supervision of persons thoroughly familiar with proper manifolding practices.

Warning: Do not, under any circumstances, use copper pipes in acetylene manifolds. Acetylene forms copper acetylide when placed in contact with copper. This is a highly unstable compound known to explode spontaneously.

ACETYLENE GENERATORS When used, acetylene generators should be marked plainly with the maximum hourly production (in cubic feet) for which they are designed, the weight and size of the carbide necessary for a single charge, the manufacturer's name and address, and the name or number of the type of generator (Figure 6-7). Double-rated generators are designed to generate acetylene at twice the maximum hourly rate of a single-rated generator of the same carbide capacity. The total hourly output of a generator must not exceed the rate for which it is approved and marked.

6-2 FUEL-GAS WELDING FLAMES

Two types of flames are seen in the welding industry: *premix* and *nozzle-mix*. In premix, the one most commonly used in manual welding, the fuel gas and the oxygen mix in the torch chamber and are usually totally mixed before combustion occurs at the tip. This type of flame is blue or nearly invisible. Nozzle-mix flames are produced by having separate passages for the fuel and the oxygen, no mixing chamber, and combustion and mixing occurring immediately outside the tip. These flames are usually long and yellowish. Nozzle-mix flames are heat-radiating flames often used in industrial furnaces, while premix flames concentrate heat into a cone of the very high temperature necessary for welding.

Figure 6-8 shows the basic premix oxyacetylene flame with cone and envelope. The *cone* is a premix flame, while the *envelope* may draw air from the atmosphere, thus becoming a nozzle-mix flame. As seen in the figure, an *oxidizing flame,* which results from a mixture of more oxygen than acetylene, is the hottest. However, the excess oxygen oxidizes the weld metal and may not be practical in some instances. *Carburizing flames* result from a mixture of more acetylene than oxygen and are sometimes used to add carbon to the weld metal. *Neutral flames* result from equal proportions of oxygen and acetylene and are the most commonly used in general welding.

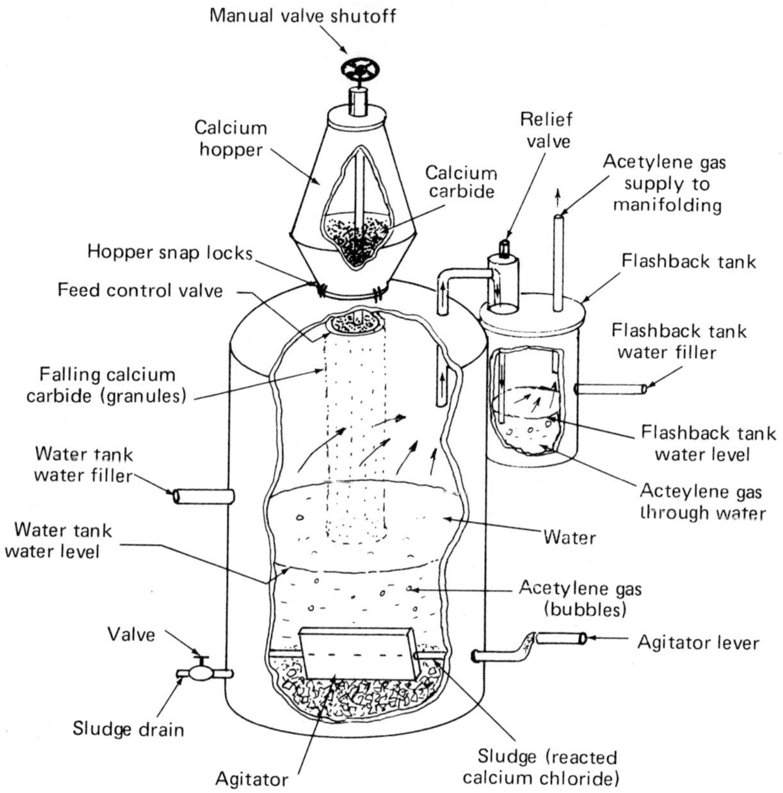
Manual valve shutoff
Calcium hopper
Relief valve
Calcium carbide
Acetylene gas supply to manifolding
Hopper snap locks
Flashback tank
Feed control valve
Flashback tank water filler
Falling calcium carbide (granules)
Flashback tank water level
Water tank water filler
Acteylene gas through water
Water tank water level
Water
Acetylene gas (bubbles)
Valve
Agitator lever
Sludge drain
Sludge (reacted calcium chloride)
Agitator

Figure **6-7** Simplified diagram of an acetylene generator. (Redrawn by permission from American Welding Society, *Welding Handbook,* 6th ed., 1970, section 3A, p. 41.10.)

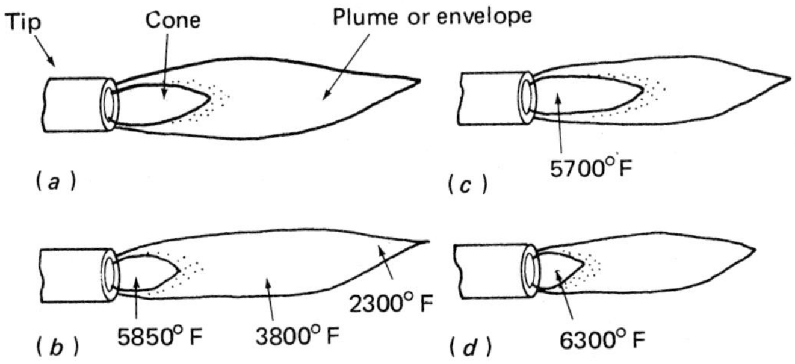
Tip
Cone
Plume or envelope
(a)
(c)
5700°F
2300° F
(b)
5850° F
3800° F
(d)
6300° F

Figure **6-8** (*a*) Basic premix flame. (*b*) Neutral oxyacetylene flame. (*c*) Carburizing flame. (*d*) Oxidizing flame.

6-3 FUEL-GAS WELDING TECHNIQUES

There are two common techniques employed in oxyacetylene (or any other fuel-gas) welding: forehand and backhand. *Forehand welding* is the most popular. In this technique, the top of the torch is held at an angle of 15 degrees from the vertical, so that the flame points in the direction of travel. This way, the torch preheats the work to be welded. As this occurs, the welding rod, which is held in the free hand, is dipped into the puddle at 15 degrees from the vertical in the direction opposite to the torch angle. The rod is then withdrawn from the puddle and moved along with the weld. Of course, experience will allow a welder to vary this technique to obtain greater or lesser penetration, or to respond to other specific aspects of the job.

Backhand welding is not as popular as forehand welding, but must be used on thick metals to ensure good penetration and weld strength. In the backhand technique, the torch is tilted in a direction opposite to the direction of travel. The torch angle varies with the metal thickness, up to vertical for very thick metals (the flame pointing directly on the metal surface). The filler rod is held in the molten puddle all the time and often moved side to side in a weaving motion. The welding tip for backhand welding should always be one size larger than that used for forehand welding, to ensure that heat loss due to the continual presence of the welding rod is counteracted.

6-4 SPECIAL USES OF FUEL-GAS FLAMES

The effects of welding on metal parts are described in detail in Chapters 18 and 19. Most of the time the welding heat, unless carefully controlled, has detrimental effects on metal. Sometimes, however, it is necessary to bend a part that is straight, straighten a part that is bent, harden a part that is soft, or soften a part that is hard. Because of the intense heat (from 2300°F at the plume to 6300°F at the cone) and because it is so easy to control and manipulate, oxyacetylene is commonly used for such processes as flame hardening, flame annealing, flame priming, flame tempering, forming and bending, and straightening. These applications are described in the following sections and depend on the intense heat and ease of manipulation of the oxyacetylene flame. Conceivably, however, other fuel gases could be used.

FLAME HARDENING *Flame hardening* is used for localized (spot) hardening of gray cast irons, nodular cast irons, and pearlitic cast irons having combined carbon contents ranging from 0.35% to 0.80%. Cast and wrought plain-carbon steels ranging from 0.37% to 0.55% carbon are also commonly flame hardened.

Although alloy steels are not usually flame hardened, those alloys that have been found suitable to flame hardening are listed in Table 6-1. Four flame-hardening methods are in common use: spot or stationary, progressive, spinning, and combination progressive–spinning.

Table **6-1** Response of steels and cast irons to flame hardening

MATERIAL	TYPICAL ROCKWELL C HARDNESS AS AFFECTED BY QUENCHANT		
	AIR[a]	*OIL*[b]	*WATER*[b]
PLAIN CARBON STEELS			
1025 to 1035	—	—	33 to 50
1040 to 1050	—	52 to 58	55 to 60
1055 to 1075	50 to 60	58 to 62	60 to 63
1080 to 1095	55 to 62	58 to 62	62 to 65
1125 to 1137	—	—	45 to 55
1138 to 1144	45 to 55	52 to 57[c]	55 to 62
1146 to 1151	50 to 55	55 to 60	58 to 64
CARBURIZED GRADES OF PLAIN CARBON STEELS[d]			
1010 to 1020	50 to 60	58 to 62	62 to 65
1108 to 1120	50 to 60	60 to 63	62 to 65
ALLOY STEELS			
1340 to 1345	45 to 55	52 to 57[c]	55 to 62
3140 to 3145	50 to 60	55 to 60	60 to 64
3350	55 to 60	58 to 62	63 to 65
4063	55 to 60	61 to 63	63 to 65
4130 to 4135	—	50 to 55	55 to 60
4140 to 4145	52 to 56	52 to 56	55 to 60
4147 to 4150	58 to 62	58 to 62	62 to 65
4337 to 4340	53 to 57	53 to 57	60 to 63
4347	56 to 60	56 to 60	62 to 65
4640	52 to 56	52 to 56	60 to 63
52100	55 to 60	55 to 60	62 to 64
6150	—	52 to 60	55 to 60
8630 to 8640	48 to 53	52 to 57	58 to 62
8642 to 8660	55 to 63	55 to 63	62 to 64
CARBURIZED GRADES OF ALLOY STEELS[d]			
3310	55 to 60	58 to 62	63 to 65
4615 to 4620	58 to 62	62 to 65	64 to 66
8615 to 8620	—	58 to 62	62 to 65
MARTENSITIC STAINLESS STEELS			
410 and 416	41 to 44	41 to 44	—
414 and 431	42 to 47	42 to 47	—
420	49 to 56	49 to 56	—
440 (typical)	55 to 59	55 to 59	—
CAST IRONS (ASTM CLASSES)			
Class 30	—	43 to 48	43 to 48
Class 40	—	48 to 52	48 to 52
Class 45010	—	35 to 43	35 to 45
50007, 53004, 60003	—	52 to 56	55 to 60
Class 80002	52 to 56	56 to 59	56 to 61
Class 60-45-15	—	—	35 to 45
Class 80-60-03	—	52 to 56	55 to 60

[a]To obtain the hardness results indicated, those areas not directly heated must be kept relatively cool during the heating process. [b]Thin sections are susceptible to cracking when quenched with oil or water. [c]Hardness is slightly lower for material heated by spinning and combination progressive-spinning methods than for material heated by progressive or stationary methods. [d]Hardness values of carburized cases containing 0.90 to 1.10% C.

Source: By permission, from *Metals Handbook* Volume 2, Copyright American Society for Metals, 1961–

The *spot* (stationary) *method,* illustrated in Figure 6-9, consists of locally heating selected areas with a suitable flame head and subsequently *quenching.* The heating head may be of either single- or multiple-orifice design, depending on the extent of the area to be hardened. The heat input must be balanced to obtain a uniform temperature over the entire selected area. After heating, the parts are usually immersion quenched; however, in some mechanized operations, a spray quench may be used.

Basically, the spot method requires no elaborate equipment (except, perhaps, fixtures and timing devices to assure uniform processing of each piece). However, the operation may be automated by indexing the heated parts into either a spray quench or a suitable quench bath.

The *progressive method,* illustrated in Figure 6-10, is used to harden large areas that are beyond the scope of the spot method. The size and shape of the workpiece, as well as the volume of oxygen and fuel gas required to heat an area, are factors in the selection of this method. In progressive hardening, the flame head is usually of the multiple-orifice type, and quenching facilities may be either integrated with the flame head or separate from it. The flame head progressively heats a narrow band of area that is subsequently quenched as the head and quench traverse the workpiece.

The equipment needed for flame hardening by the progressive method consists of one or more flame heads and a quenching means, mounted on a movable carriage that runs on a track at a regulated speed (flame cutting machines are adaptable to this type of flame hardening). Workpieces mounted on a turntable or in a lathe can be hardened readily by the progressive method; either the flame head or the workpiece may move. There is no practical limit to the length of parts that can be hardened by this method, because it is easy to lengthen the track over which the flame head travels. Single passes as wide as 60 inches can be made; wider areas must be hardened in more than one pass.

When more than one pass is required to cover a flat surface, or when cylindrical surfaces are hardened progressively, such surfaces will exhibit soft bands because of overlapping or underlapping of the heated zones. These soft bands can be minimized, however, by closely controlling the extent of the overlapping. (Wherever overlapping occurs, the possibility of severe thermal upset and cracking should be anticipated. Tests should be conducted to determine whether overlapping will cause cracking or other harmful effects.) Simple curved surfaces may be hardened progressively by means of contoured flame heads, and some irregular surfaces may be traversed by the use of tracer-template methods.

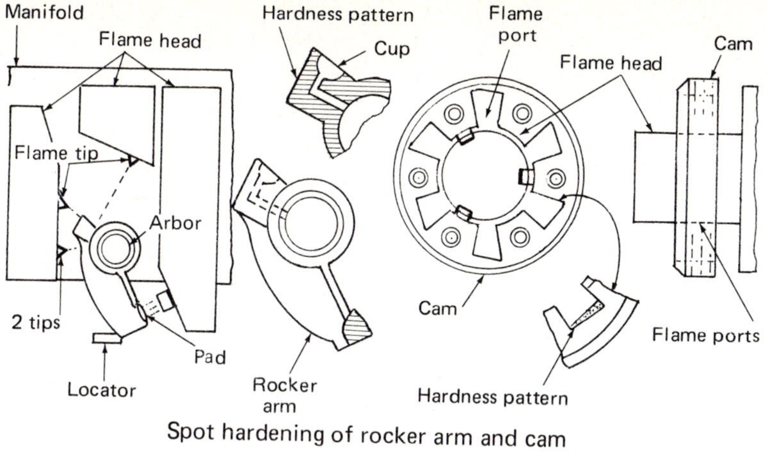
Manifold
Flame head
Flame tip
Arbor
2 tips
Pad
Locator
Hardness pattern
Cup
Rocker
arm
Flame
port
Flame head
Cam
Cam
Flame ports
Hardness pattern
Spot hardening of rocker arm and cam

Figure **6-9** Spot method of flame hardening. (By permission, from *Metals Handbook* Volume 2, Copyright American Society for Metals, 1971.)

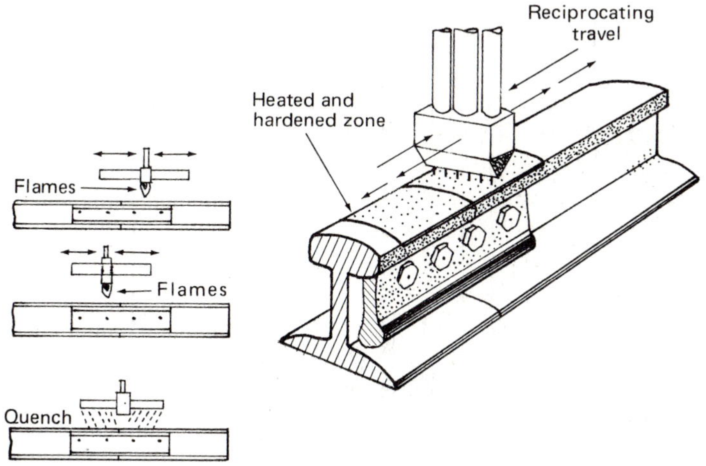
Reciprocating
travel
Heated and
hardened zone
Flames
Flames
Quench

Figure **6-10** Progressive method of flame hardening. (Redrawn by permission from Airco Welding Products, A Division of Airco, Inc., *Oxyacetylene Welding and Oxygen Cutting Instruction Course*, 1966, p. 99.)

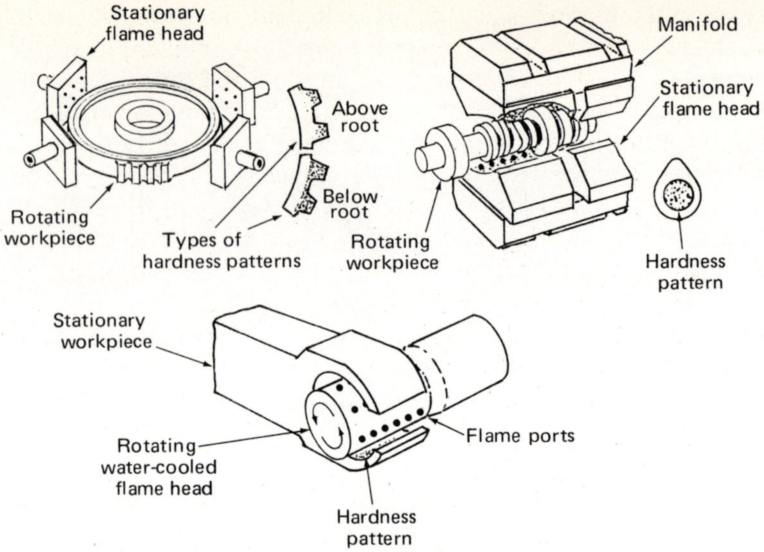

Stationary
flame head
Rotating
workpiece
Above
root
Below
root
Types of
hardness patterns
Manifold
Stationary
flame head
Rotating
workpiece
Hardness
pattern
Stationary
workpiece
Rotating
water-cooled
flame head
Flame ports
Hardness
pattern

Figure **6-11** Spinning method of flame hardening. (By permission, from *Metals Handbook* Volume 2, Copyright American Society for Metals, 1971.)

The rate of travel of the flame head over the surface is governed mainly by the heating capacity of the head, the depth of case required, the composition and shape of the work, and the type of quench employed. Speeds in the range of 2 to 12 in./min are typical with oxyacetylene heating heads. Ordinarily, water at ambient temperature is used as a quench, although air is sometimes used when a less severe quench is indicated. Under special conditions, warm or hot water, or a solution of water and soluble oil, also may be employed.

The *spinning method* (Figure 6-11) is applied to round or semiround parts, such as wheels, cams, or small gears. In its simplest form, the spinning method employs a mechanism for rotating or spinning the workpiece, in either a horizontal or vertical plane, while the surface is being heated by the flame head. One or more water-cooled heating heads, equal in width to the surface to be heated, are employed. The speed of rotation is relatively unimportant, provided uniform heating is obtained. After the surface has been heated to the desired temperature, the flame is extinguished or withdrawn and the work is quenched by immersion or spray, or by a combination of both.

The spinning method is particularly adaptable to extensive mechanization and automation; this makes it possible, for example, for all the cams on a camshaft to be hardened at the same time.

Commercially built machines are available that provide automatic control of timing, temperature, and quenching, as well as accurate control of gas flow, so that close metallurgical specifications can be met consistently. Frequently, when production is sufficient, the spinning method can be set up so that parts are either loaded manually and automatically unloaded, or both loaded and unloaded automatically.

A recent development utilizes a rotating flame head for the internal spin hardening of oddly shaped parts that would present handling problems if the parts themselves were rotated. Each part is positioned by a simple handling device, and the flame head rotates inside the part to be hardened.

In contrast to the progressive method, in which acetylene usually is used (because of its high flame temperature and rapid heating rates), satisfactory results can be obtained in spin hardening with natural gas, propane, or manufactured gas. The choice of gas depends on the shape, size, and composition of the workpiece, and on the depth of case required, as well as on the relative cost and availability of each gas.

A wide choice of quenchants is also possible in the spin-hardening method. Because the flame is extinguished or withdrawn before the part is quenched, any appropriate quenchant may be used for immersion quenching. In spray quenching, the quenchant is usually water or a water-base liquid; air has also been used.

The *combination progressive-spinning method* (Figure 6-12), as the name implies, combines the progressive and spinning methods for hardening long parts, such as shafts and rolls. The workpiece is rotated, as in the spinning method, but in addition, the heating heads traverse the roll or shaft from one end to the other. Only a narrow circumferential band is heated progressively as the flame head moves from one end of the work to the other. The quench follows immediately behind the heating head, either as an integral part of the head or as a separate quench ring. This method provides a means of hardening large surface areas with relatively low gas flows. Progressive-spinning units designed to handle a broad range of diameters and lengths are available commercially.

The cost of oxyfuel gas required to heat one square inch of steel to 1500°F to a depth of ⅛ inch is the same for all gases except for oxypropane, which is approximately twice as expensive. The ratio of oxygen to fuel gas is 1:1 for acetylene, 0:5 for city gas, 1:75 for natural gas, and 4:0 for propane.

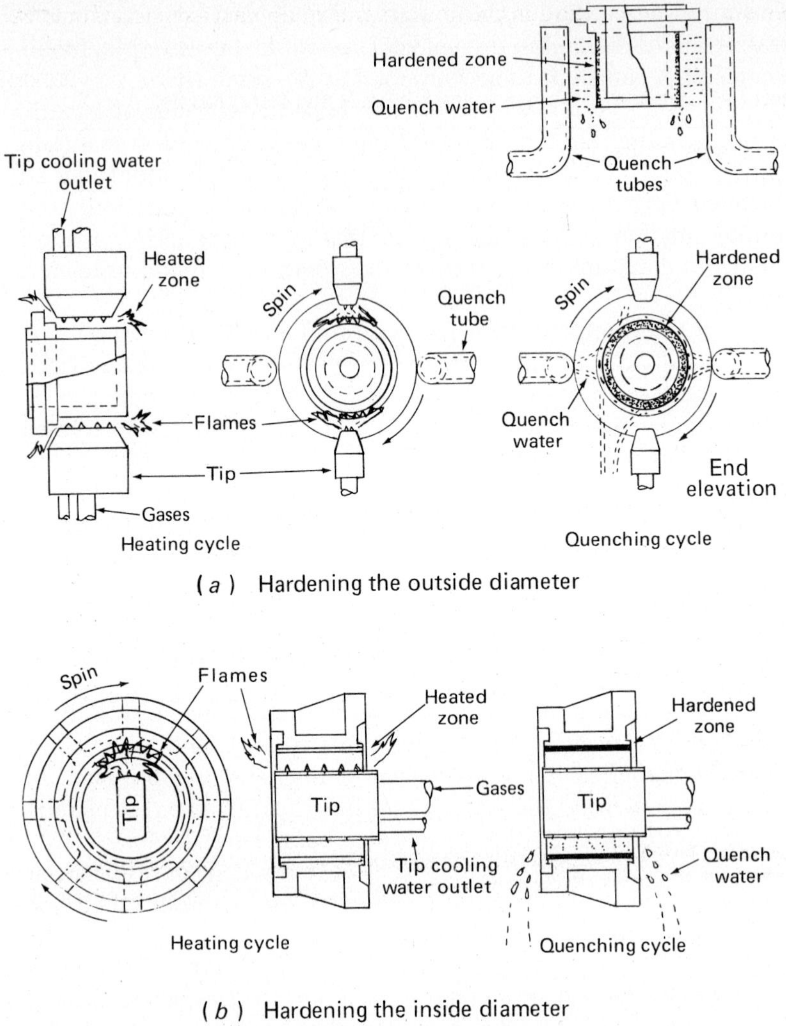
Hardened zone
Quench water
Quench
tubes
Tip cooling water
outlet
Heated
zone
Spin
Quench
tube
Hardened
zone
Flames
Tip
Gases
Quench
water
End
elevation
Heating cycle
Quenching cycle
(a) Hardening the outside diameter
Spin
Flames
Heated
zone
Hardened
zone
Gases
Tip
Tip
Tip
Tip cooling
water outlet
Quench
water
Heating cycle
Quenching cycle
(b) Hardening the inside diameter

Figure **6-12** Combination progressive-spinning method of flame hardening. (Redrawn by permission from Airco Welding Products, A Division of Airco, Inc., *Oxyacetylene Welding and Oxygen Cutting Instruction Course*, 1966, p. 100.)

To minimize cracking of prehardened areas during flame hardening, it is advisable to preheat uniformly throughout to a temperature below the lower transformation point prior to flame hardening. Surface conditions that have proven detrimental to flame hardening are listed in Table 6-2.

Table **6-2** Surface conditions of steel parts that are detrimental to flame hardening

DEFECT OR CONDITION	PROBABLE ORIGIN OF CONDITION	DETRIMENTAL EFFECTS TO BE EXPECTED ON FLAME-HARDENED AREAS
Laps, seams, folds, fins (wrought parts)	Rolling mill or forging operations	Localized overheating (or, at worst, surface melting), with consequent grain growth, brittleness, and greater hazard of cracking
Scale (adherent)[a]	Rolling or forging; prior heat treatment; flame cutting	1. Insulating action against heating, with resulting underheated areas and soft spots 2. Localized retardation of quench, causing soft spots
Rust, dirt[a]	Storage and handling of material or parts	1. Similar to scale as noted above 2. Severe rusting may result in pitted surfaces that will remain after hardening
Decarburization	Present in as-received steel bar stock; from heating for forging or prior heat treatment of parts or stock	In severely decarburized work, no hardening response will be found when parts are tested by file or other superficial means[b]
Pinholes, shrinkage (castings)	Casting defects	Localized overheating (or, at worst, surface melting), with consequent grain growth, brittleness, and greater hazard of cracking
Coarse-grained gate areas (castings)	Casting gates located on areas to be flame hardened. (Avoid, if possible.)	Increased cracking hazard during quench, compared to nongated areas. Shrinkage defects also are likely in these areas

[a]In addition to detrimental effects on flame hardened surfaces, scale, rust, and dirt may pop loose into the path of the flame and cause fouling of oxyfuel gas burners or react chemically with ceramic air-fuel gas burner parts (causing their rapid deterioration). When these materials enter a closed quenching system, they may clog strainers, plug quench orifices, and cause excessive wear of pumps. [b]Partial decarburization lowers surface hardness as a direct function of actual carbon content of stock at surface, provided steel was adequately heated and quenched.

Source: By permission, from *Metals Handbook* Volume 2, Copyright American Society for Metals, 1961–.

FLAME ANNEALING When the carbon or alloy content of a ferrous metal is high enough, hardening results when the heated zone is cooled rapidly (this is discussed in more detail in Chapter 19).

Carbon steels containing 0.30% carbon or less do not harden sufficiently to prevent the use of flame-cut pieces in structures or in subsequent fabrication where bending is involved. When cut by

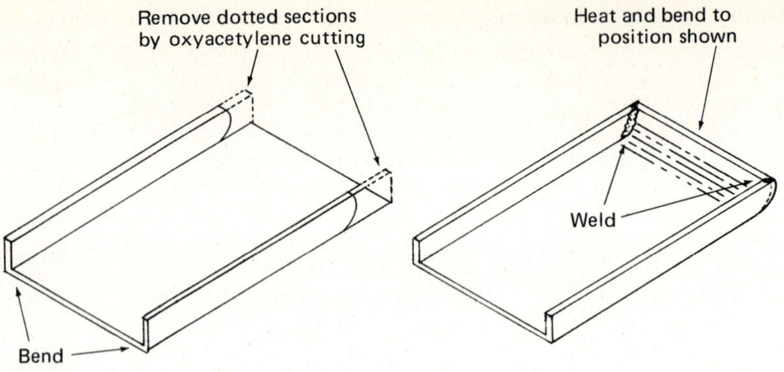
Remove dotted sections
by oxyacetylene cutting
Heat and bend to
position shown
Weld
Bend

Figure **6-13** Cutting and bending with oxyacetylene torch. (Redrawn by permission from Union Carbide Corporation, Linde Division, *The Oxy-Acetylene Handbook,* 2d ed., 1960, p. 399.)

flame, steels containing alloying elements and steels containing more than 0.30% carbon may harden sufficiently to prevent their use for some purposes.

When hardening is likely to occur along a cut edge, oxyacetylene flames, applied by suitable equipment, can be used either to prevent hardening or to soften an already hardened cut surface. The term *flame annealing* is applied to this process and to the selective flame softening of areas of a hardened steel part.

Tool steels and certain high-alloy steels will crack during flame cutting, or very soon after, unless they are heated to 400 to 800°F before being cut.

TEMPERING The theory of tempering is detailed in Chapter 19. For our purposes here, it is important only to note that quench-hardened steel sometimes fails in use because of residual internal stresses. Tempering makes a tougher, less brittle steel with lower internal stresses. It is accomplished by reheating the hardened metal, preferably before it has cooled to room temperature. Time at temperature is the key element in tempering. This is determined in part by the composition of the steel and in part by the experience of the welder. The important point to remember is that tempering toughens a hard steel, reducing brittleness and, to some extent, hardness.

FORMING, BENDING, AND STRAIGHTENING The oxyfuel gas flame is frequently used to heat such metal parts as ornamental ironwork, truck bodies, tanks, angle iron, and similar metal parts prior to forming and/or fabrication (Figure 6-13).

When it is necessary to bend pipe or other metal shapes without loss of strength, as would result in cold bending, the oxyacetylene flame can be used to heat the pipe in preparation for what is known as *wrinkle bending*. Wrinkle bending requires no special skill on the part of the welder, but it does require the use of specially predesigned bending fixtures (Figure 6-14). Any designed degree of bend can be obtained by this method. Sharp bends are made by spacing the wrinkles (bends) closely and making each wrinkle provide a change of direction of from 8 to 12 degrees. Long, smooth bends may be made by having each wrinkle provide a change of direction from 3 to 6 degrees and by spacing the bends farther apart.

The oxyfuel-gas flame is also used to preheat parts that are to be straightened (Figure 6-15), upset (Figure 6-16), or stretched (Figure 6-17).

FLAME PRIMING Flame priming is a method in which specialized oxyacetylene equipment is used to prepare surfaces of large structures (such as bridges, storage tanks, ships, and others) for painting by scrubbing them with high-velocity, high-temperature oxyacetylene flames in such a way as to loosen scale, foreign material, and contaminants, leaving the surface warm and dry. After passage of the flames, the surfaces must be swept clean of all loosened debris, preferably with a wire brush, and painted immediately.

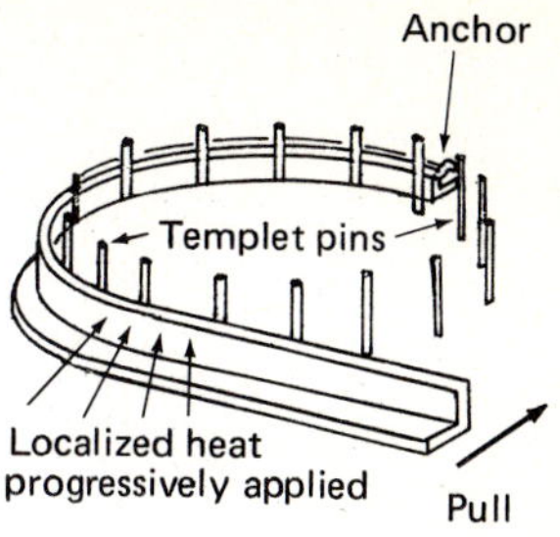
Anchor
Templet pins
Localized heat
progressively applied
Pull

Figure **6-14** Special fixture for wrinkle bending. (Redrawn by permission from Union Carbide Corporation, Linde Division, *The Oxy-Acetylene Handbook,* 2d ed., 1960, p. 401.)

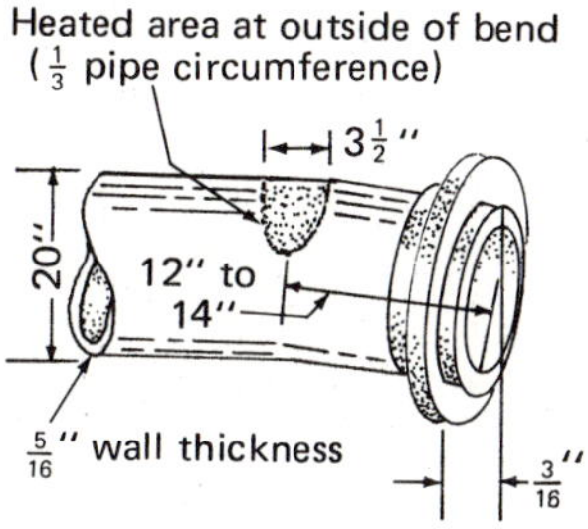
Heated area at outside of bend
($\frac{1}{3}$ pipe circumference)
$3\frac{1}{2}''$
20''
12'' to
14''
$\frac{5}{16}''$ wall thickness
$\frac{3}{16}''$

Figure **6-15** Straightening with the oxyacetylene flame. (Redrawn by permission from Union Carbide Corporation, Linde Division. *The Oxy-Acetylene Handbook,* 2d ed., 1960, p. 400.)

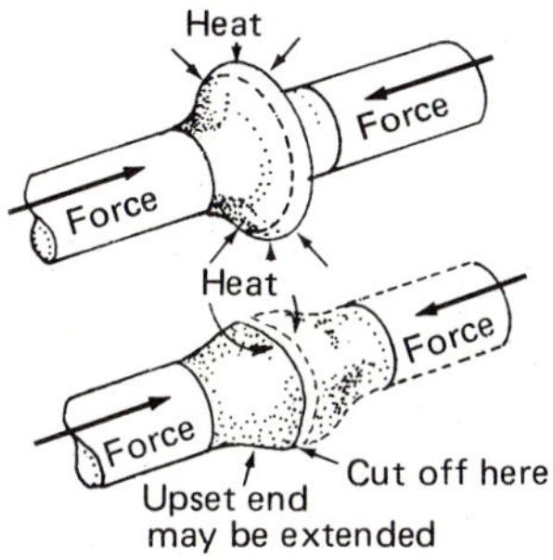
Heat
Force
Force
Heat
Force
Force
Cut off here
Upset end
may be extended

Figure **6-16** Upset welding tube ends by use of the oxyacetylene flame and forging. (Redrawn by permission from Union Carbide Corporation, Linde Division, *The Oxy-Acetylene Handbook,* 2d ed., 1960, p. 402.)

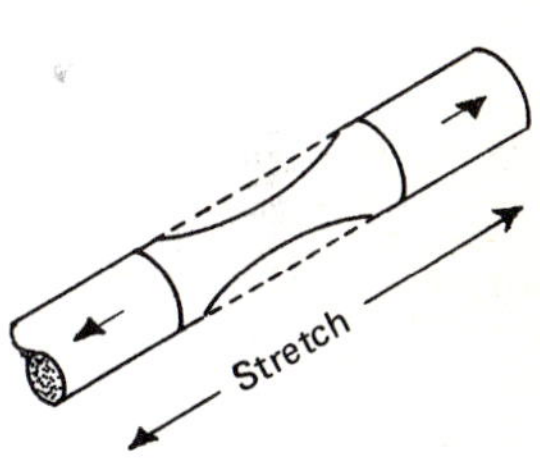
Stretch

Figure **6-17** Stretching and "necking in" pipe with the oxyacetylene torch. (Redrawn by permission from Union Carbide Corporation, Linde Division, *The Oxy-Acetylene Handbook,* 2d ed., 1960, p. 402.)

REVIEW QUESTIONS

1. Name the most commonly used fuel gases.

2. Why are fuel gases for welding usually burned with commercially pure oxygen?

3. How is flame cutting accomplished?

4. List the equipment necessary for fuel-gas welding or cutting.

5. Which hoses connect to the torch and regulator with left-hand connections?

6. Name the two common types of torches and explain how they work.

7. What are the two basic functions of regulators?

8. Why should acetylene cylinders be used only in an upright position?

9. Why are cylinders manifolded?

10. Name and explain the two types of flames used in the welding industry.

11. Name and explain the three kinds of premix oxyacetylene flames.

12. Explain forehand and backhand welding.

13. Sketch, name, and identify the temperature zones of the oxyacetylene flame.

14. Name three special uses of the oxyacetylene flame.

15. Differentiate between the following flame hardening processes: (a) spot and progressive; (b) progressive and spinning.

16. What could happen to a partially prehardened part if it is not uniformly preheated prior to flame hardening?

17. List three surface conditions that have proven to be detrimental to flame hardening.

18. What is meant by the term *wrinkle bending*?

19. Why are parts wrinkle bent?

CHAPTER 7
RESISTANCE-WELDING PROCESSES

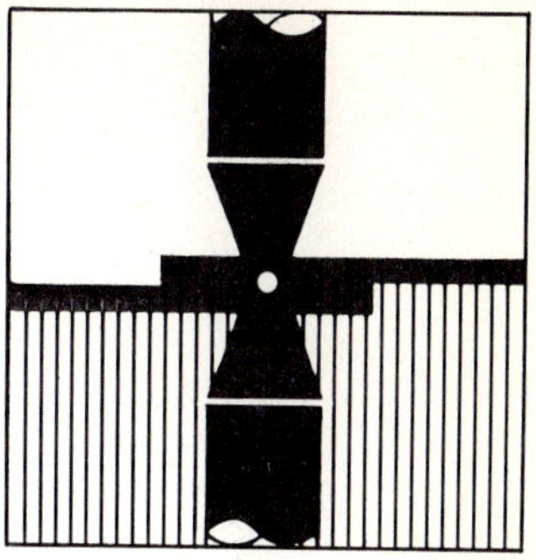

Resistance welding consists of a group of processes (Figure 7-1) in which the heat required for welding is generated by the resistance of the parts to the passage of an electric current. It differs from the fusion-welding processes in that it requires, in addition to heat, the application of mechanical pressure to forge the parts together. The pressure refines the grain structure and produces a weld with physical properties that in most cases are equal, and sometimes even superior, to those of the base metal.

Resistance-welding equipment is classified according to its electrical operation—as direct-energy or as stored-energy type. Although both single-phase and three-phase welding machines are commercially available, the single-phase, direct-energy machine is the most commonly used, because it is the simplest and least expensive in terms of acquisition cost, installation, and maintenance.

The electrical system of a single-phase ac resistance-welding machine (Figure 7-2) consists of transformers, top switch, and secondary circuit, including electrodes. The electrical energy used by the electrodes is taken directly from the power line, like water from a fire hydrant.

Sufficient voltage must be available in the electrical system to provide the current required to furnish sufficient heat to produce a weld (Figure 7-3) of the required size.

The voltage needed to produce the required current is determined by the geometry of the secondary loop (Figure 7-4), the electrical

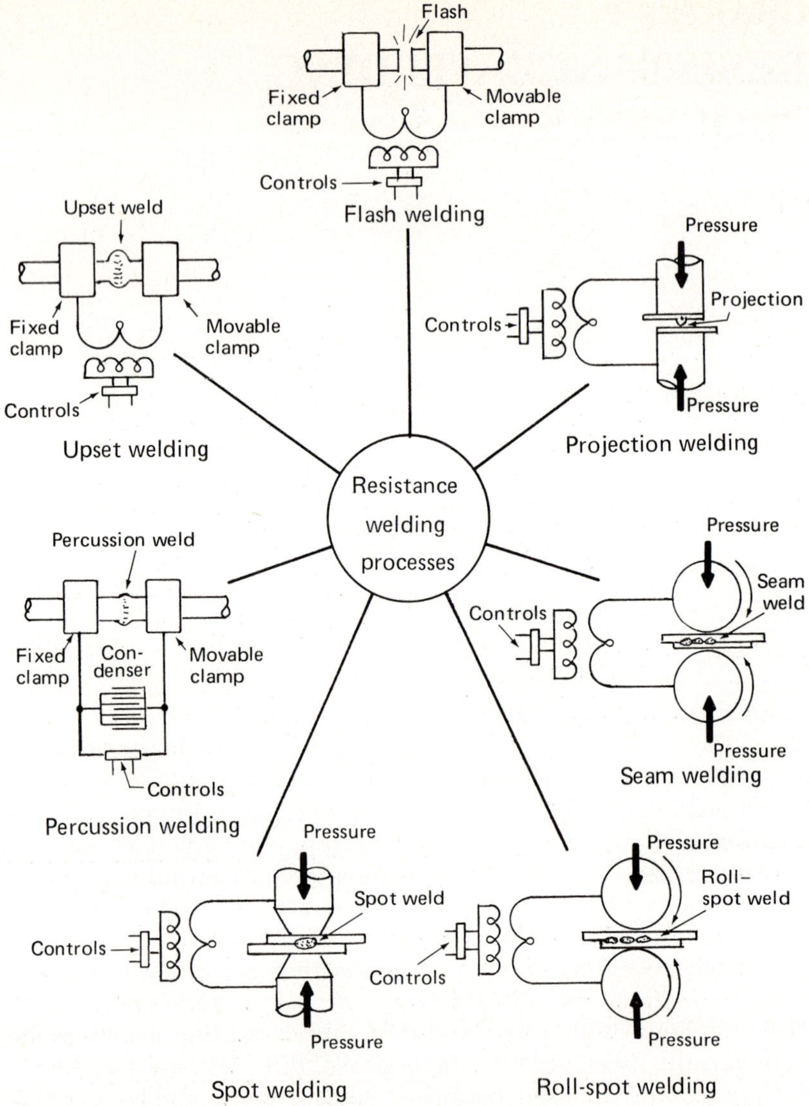
Flash
Fixed clamp
Movable clamp
Controls
Flash welding
Upset weld
Fixed clamp
Movable clamp
Controls
Upset welding
Pressure
Projection
Controls
Pressure
Projection welding
Resistance welding processes
Percussion weld
Fixed clamp
Con-denser
Movable clamp
Controls
Percussion welding
Pressure
Seam weld
Controls
Pressure
Seam welding
Pressure
Spot weld
Controls
Pressure
Spot welding
Pressure
Roll-spot weld
Controls
Pressure
Roll-spot welding

Figure **7-1** The resistance-welding processes. (Redrawn from *Welding Engineering* by Boniface E. Rossi. Copyright 1954 McGraw-Hill Book Company. Used with permission of McGraw-Hill Book Company.)

conductivity of the conductor in the circuit, the contact joint resistance, and the resistance of the work. The required voltage is referred to as the *secondary open-circuit voltage* and, when multiplied by the welding current and divided by 1000, gives the demand kVA of the welding transformer. The term *kVA* is used in the resistance-welding industry to express the work capability of the

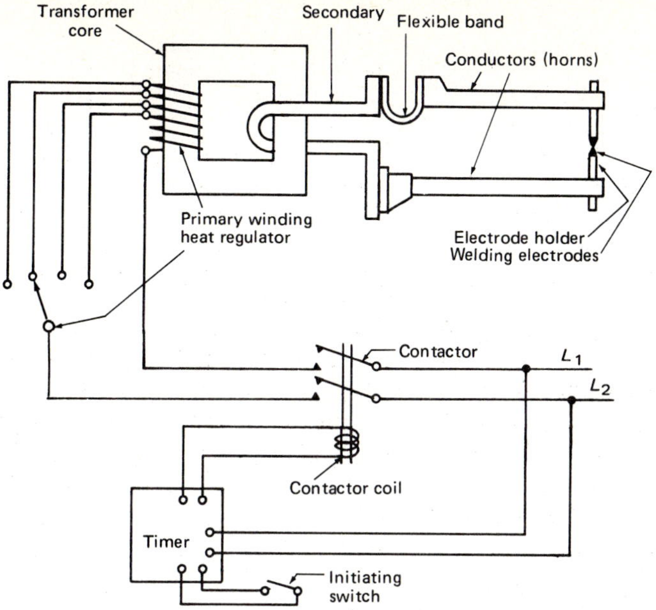
Transformer core
Secondary
Flexible band
Conductors (horns)
Primary winding heat regulator
Electrode holder
Welding electrodes
Contactor
L_1
L_2
Contactor coil
Timer
Initiating switch

Figure **7-2** Schematic of a single-phase ac resistance-welding machine. (Redrawn from *Welding Engineering* by Boniface E. Rossi. Copyright 1954 McGraw-Hill Book Company. Used with permission of McGraw-Hill Book Company.)

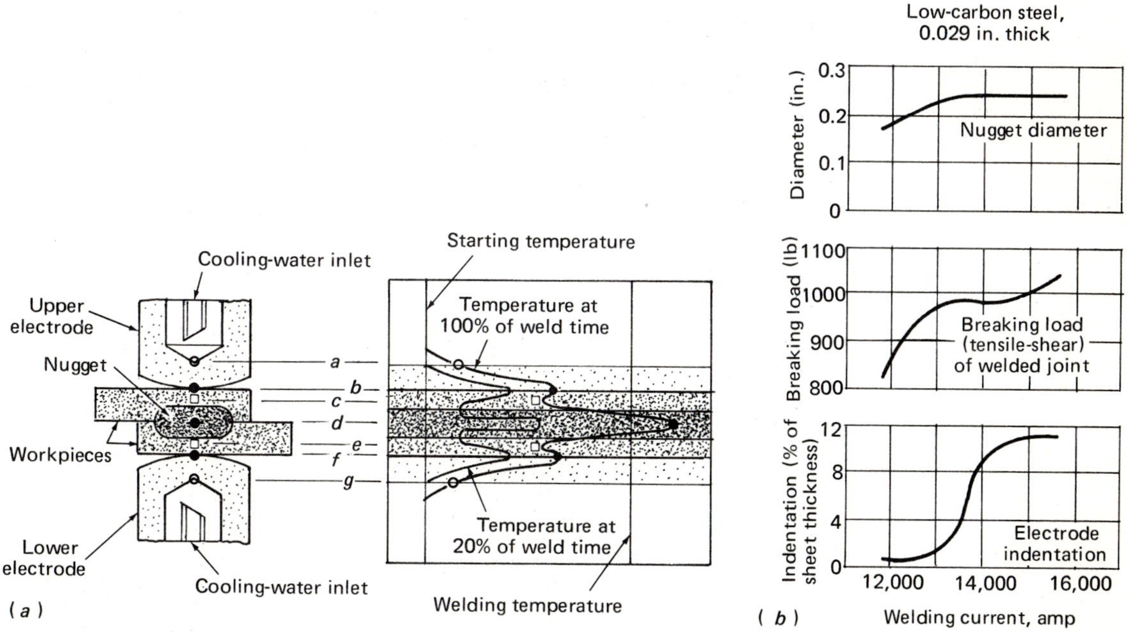
Cooling-water inlet
Upper electrode
Nugget
Workpieces
Lower electrode
Cooling-water inlet
a
b
c
d
e
f
g
(a)
Starting temperature
Temperature at 100% of weld time
Temperature at 20% of weld time
Welding temperature
Low-carbon steel, 0.029 in. thick
Diameter (in.)
0.3
0.2
0.1
0
Nugget diameter
Breaking load (lb)
1100
1000
900
800
Breaking load (tensile-shear) of welded joint
Indentation (% of sheet thickness)
12
8
4
0
Electrode indentation
12,000
14,000
16,000
Welding current, amp
(b)

Figure **7-3** (*a*) Major points of heat generation in resistance spot-welding. (*b*) Effects of welding current on nugget diameter, tensile shear strength, and electrode indentation in resistance spot-welding. (By permission, from *Metals Handbook* Volume 6, Copyright American Society for Metals, 1971.)

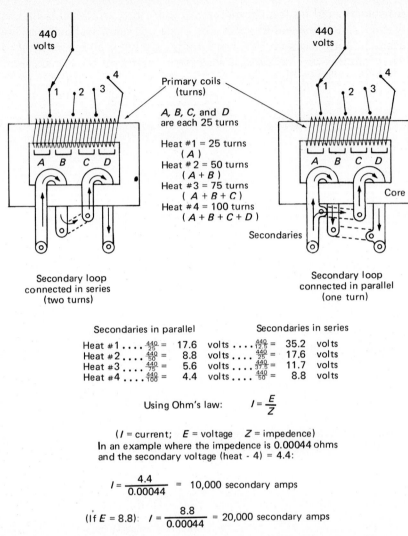

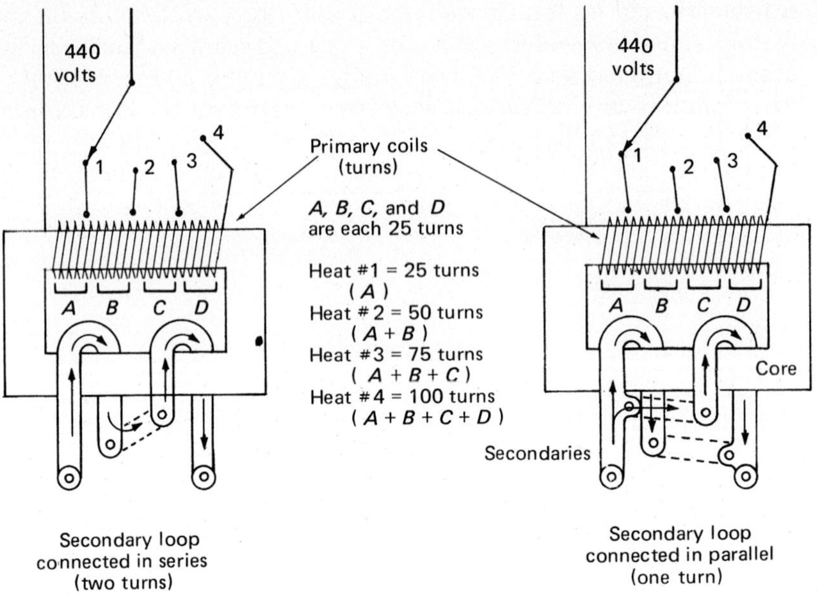
440
volts
1
2
3
4
Primary coils
(turns)
A, B, C, and D
are each 25 turns
Heat #1 = 25 turns
(A)
Heat #2 = 50 turns
(A + B)
Heat #3 = 75 turns
(A + B + C)
Heat #4 = 100 turns
(A + B + C + D)
A B C D
Core
Secondaries
Secondary loop
connected in series
(two turns)
Secondary loop
connected in parallel
(one turn)

	Secondaries in parallel		Secondaries in series	
Heat #1	.... $\frac{440}{25}$ =	17.6 volts	.... $\frac{440}{12.5}$ =	35.2 volts
Heat #2	.... $\frac{440}{50}$ =	8.8 volts	.... $\frac{440}{25}$ =	17.6 volts
Heat #3	.... $\frac{440}{75}$ =	5.6 volts	.... $\frac{440}{37.5}$ =	11.7 volts
Heat #4	.... $\frac{440}{100}$ =	4.4 volts	.... $\frac{440}{50}$ =	8.8 volts

Using Ohm's law: $I = \frac{E}{Z}$

(I = current; E = voltage Z = impedence)
In an example where the impedence is 0.00044 ohms
and the secondary voltage (heat - 4) = 4.4:

$$I = \frac{4.4}{0.00044} = 10{,}000 \text{ secondary amps}$$

$$(\text{If } E = 8.8)\text{:} \quad I = \frac{8.8}{0.00044} = 20{,}000 \text{ secondary amps}$$

Figure **7-4** Effect of the geometry of the secondary loop on heat generation. (Redrawn from *Resistance Welding, Designing, Tooling and Applications* by W. A. Stanley. Copyright 1950 McGraw-Hill Book Company. Used with permission of McGraw-Hill Book Company.)

welding machine, or duty cycle. The *work capability* involves not only how big a weld the machine can produce, but also how often it can produce such welds without "burning out" its transformer or related equipment.

The duty cycle of a resistance-welding transformer is defined as the percentage of time in each one-minute period that the transformer is actually carrying current. It is standard practice to rate resistance-welding transformers on a 50% duty cycle. This means that all resistance-welding machines (except seam welders) can be used at their maximum kVA rating indefinitely for 30 seconds out of every minute without causing excessive heating of the transformer.

Dirt, rust, and other foreign matter on the surfaces of the parts to be joined increases the resistance to electric current and must therefore be removed. Because the mechanical (dirt) and chemical (rust) films are in no way similar, different methods must be employed to remove them. Metal manufacturers can usually provide thorough information on the correct cleaning method to be used for a particular metal.

7-1 SPOT WELDING

Spot welding is the most widely used form of resistance welding. The three variations of the basic process are shown in Figure 7-5.

In its simplest application [Figure 7-5(*a*)], spot welding consists merely of clamping two or more pieces of sheet metal between two copper or copper-alloy welding electrodes and passing electric current of sufficient strength through the pieces to cause welding or bonding of the pieces.

A pictorial sequence of the making of a spot weld and the resultant spot is given in Figure 7-6. The sequence consists of:

- *Squeeze time.* The time between the initial application of the electrode pressure on the work, and the first application of current in making spot and seam welds by resistance welding and in projection or upset welding.
- *Weld time.* The time that the welding current flows through the parts being welded, usually expressed in cycles. For example, in welding a 0.125-inch-thick assembly of SAE 1010 with a frequency of 60 Hz (hertz), the weld time would be 10 cycles or 10 c/60 Hz (c/s) = ⅙ s.

Figure **7-5** Three arrangements of workpiece and electrodes in resistance spot-welding. (By permission, from *Metals Handbook* Volume 6, Copyright American Society for Metals, 1971.)

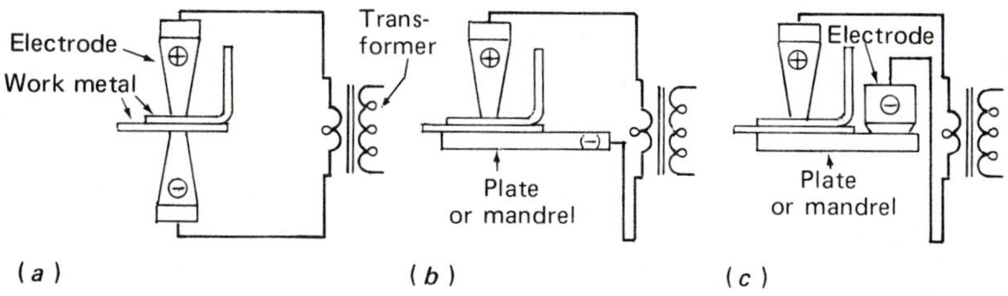
Electrode
Work metal
Trans-
former
Plate
or mandrel
Electrode
Plate
or mandrel
(a)
(b)
(c)

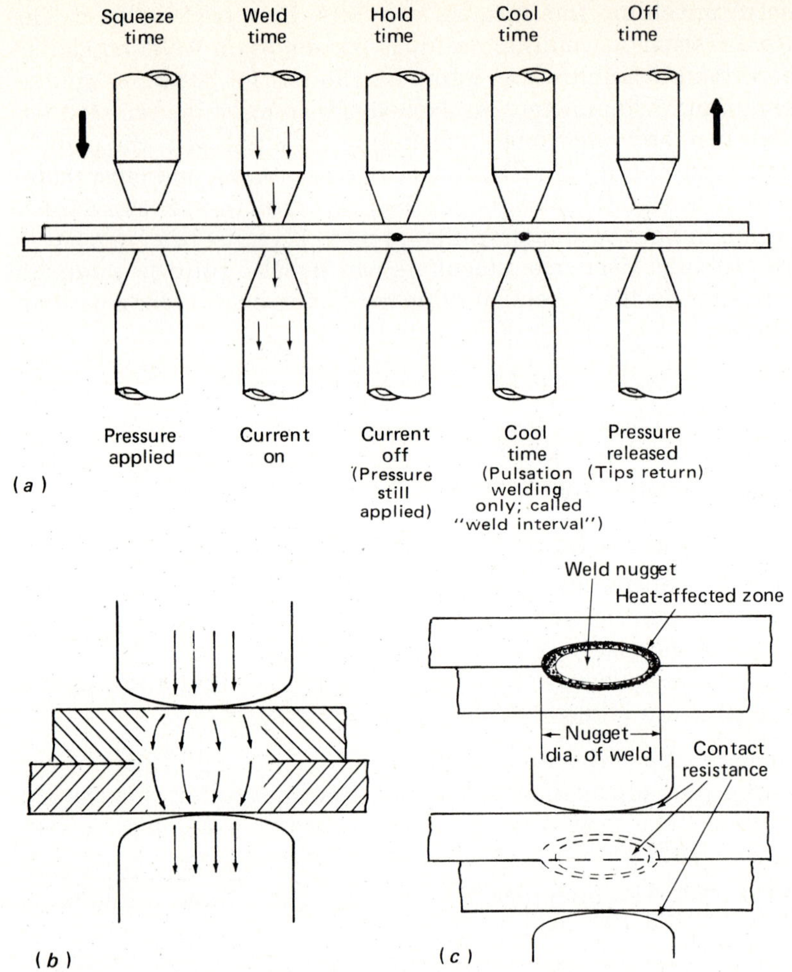

Squeeze time
Weld time
Hold time
Cool time
Off time
Pressure applied
Current on
Current off (Pressure still applied)
Cool time (Pulsation welding only; called "weld interval")
Pressure released (Tips return)
(a)
Weld nugget
Heat-affected zone
Nugget dia. of weld
Contact resistance
(b)
(c)

Figure **7-6** (*a*) Pictorial sequence of the making of a spot weld. (*b*) Path of the welding current in a spot weld. (*c*) Heat-affected zone in relation to weld nugget. (Redrawn from *Resistance Welding, Designing, Tooling and Applications* by W. A. Stanley. Copyright 1950 McGraw-Hill Book Company. Used with permission of McGraw-Hill Book Company.)

- *Hold time.* The time during which pressure is applied at the point of welding after the welding current has ceased to flow. The hold time is used to allow the plastic weld nugget to cool or harden, after which the pressure is released and the tip is retracted. Figure 7-7 is a graphical representation of the same sequence.

To make a weld with three-phase power supplies, some metals, such as alloy steel, magnesium, and Ni-Cr-Fe-Ti (Inconel), require several impulses of current while the electrodes are still closed. This is called *multiple-impulse* or *pulsation* welding when applied to the spot- or projection-welding process, and *interrupted timing* when applied to the seam-welding process.

Standardized, general-purpose spot-welding machines include rocker-arm machines, press-type spot machines, and projection-welding machines.

The *rocker-arm machine* (Figure 7-8) is the simplest and most commonly used type. It is readily adaptable for ordinary spot welding on most weldable metals and is generally available in three types of operation: air, foot, and mechanical (motor or cam). It has a capacity of two pieces of approximately 13-gage (0.089-inch) SAE 1020 steel, a throat depth of from 12 to 36 inches, and a transformer capacity from 5 to 50 kVA.

Foot-operated machines (Figure 7-9) are best suited for miscellaneous sheet metal fabrication, particularly where the runs are not

Figure **7-7** Graphical representation of various resistance spot-welding cycles, showing relation of time for each segment and relative values of welding current and electrode force for each segment. (By permission, from *Metals Handbook* Volume 6, Copyright American Society for Metals, 1971.)

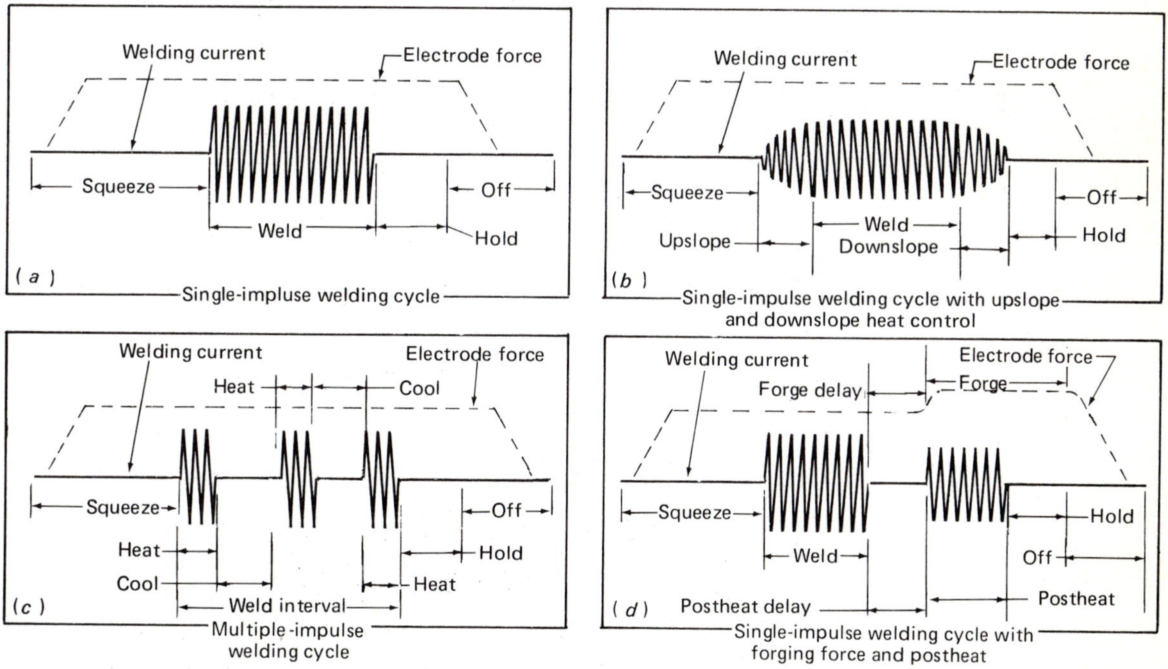

Welding current
Electrode force
Squeeze
Off
Weld
Hold
(a)
Single-impluse welding cycle
Welding current
Electrode force
Squeeze
Off
Upslope
Weld
Downslope
Hold
(b)
Single-impulse welding cycle with upslope and downslope heat control
Welding current
Electrode force
Heat
Cool
Squeeze
Off
Heat
Hold
Cool
Heat
(c)
Weld interval
Multiple-impulse welding cycle
Welding current
Electrode force
Forge delay
Forge
Squeeze
Hold
Weld
Off
(d)
Postheat delay
Postheat
Single-impulse welding cycle with forging force and postheat

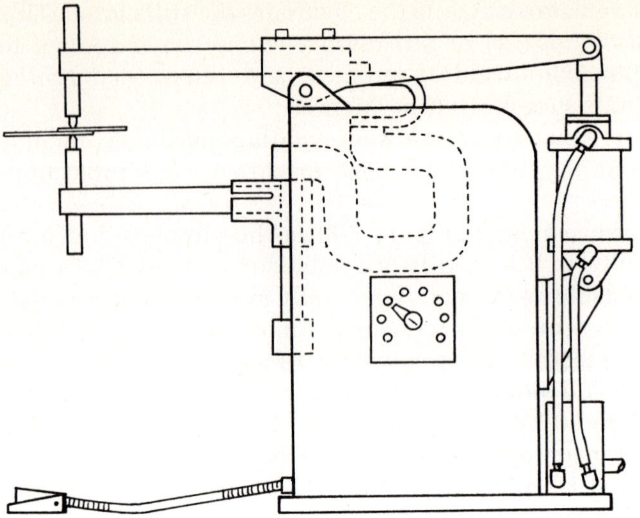

Figure **7-8** Standard air-operated rocker-arm spot-welding machine. (Redrawn by permission from American Welding Society, *Welding Handbook*, 6th ed., 1970, section 2, p. 28.5.)

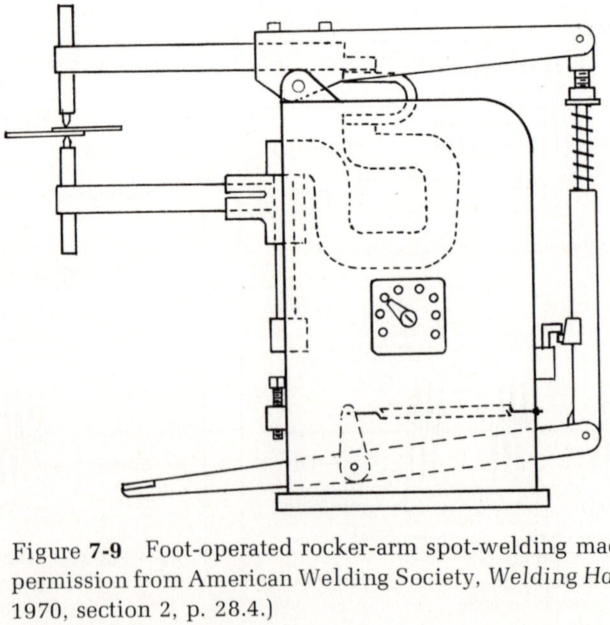

Figure **7-9** Foot-operated rocker-arm spot-welding machine. (Redrawn by permission from American Welding Society, *Welding Handbook*, 6th ed., 1970, section 2, p. 28.4.)

long. Motor-operated machines (Figure 7-10) are ideal for long production runs, where compressed air is at a premium or not readily available. Air-operated machines are the most popular type of rocker-arm machine and are not limited in operation by operator fatigue.

7-2 ROLL-SEAM WELDING

Roll-seam welding, or *seam welding*, as it is commonly called, consists of making a series of overlapping spot welds. Such a weld is normally gas and liquid tight. Two rotating circular electrodes (electrode wheels), or one rotating and one bar-type electrode (Figure 7-11) are used for transmitting the current. In this latter arrangement, the process is referred to as *butt-seam welding*. All roll-seam welds (Figure 7-12) are lap welds. There are two general types of seams: longitudinal and circular.

A longitudinal machine [Figure 7-13(*a*)] has its electrodes arranged to feed the work into the throat of the machine. Circular machines [Figure 7-13(*b*)] have the electrodes arranged to feed the work across the throat of the machine. A universal machine is one in which the electrodes may quickly be changed over from longitudinal to circular or vice versa.

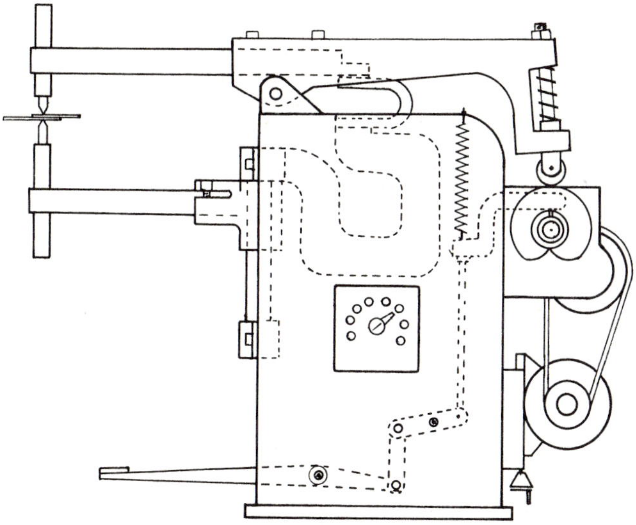

Figure **7-10** Motor-operated rocker-arm spot-welding machine. (Redrawn by permission from American Welding Society, *Welding Handbook*, 6th ed., 1970, section 2, p. 28.7.)

Two-wheel machines, whether longitudinal or circular, may have either or both wheels power driven, thus pulling the work through them, or both may be idling, rotated by the friction of the stock, which is pulled through by an outside force. Single-wheel machines are almost invariably of the idling type, with the work moved through by the mandrel, or fixture.

The electrical principles described in the previous section on spot welding also apply to seam welding. Clean welding surfaces are more important in seam welding than in any other resistance welding process. In spot welding, the detrimental effects of scale and surface dirt can be compensated for by adjustments of time, pressure, and current or by other means, but this cannot be done in seam welding. The best results are obtained with cold-finished steel properly cleaned of dirt and rust. The next best is hot-rolled steel properly pickled, oiled, and wiped clean just before welding. Material covered with mill scale, such as may be found on blue or black annealed stock, is considered unweldable. Materials that can be welded after they are properly cleaned include high-carbon, stainless, and coated steels, as well as alloys of aluminum, nickel, and magnesium. Seam welding of copper and high-copper alloys is not recommended.

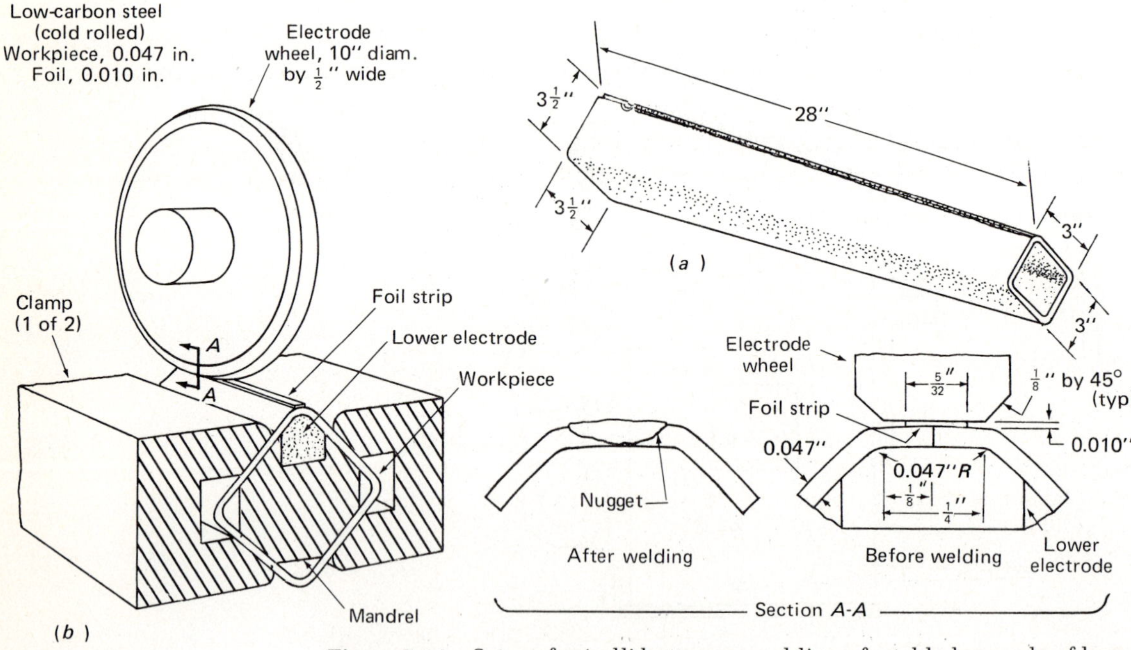

Low-carbon steel (cold rolled)
Workpiece, 0.047 in.
Foil, 0.010 in.
Electrode wheel, 10" diam. by 1/2" wide
Clamp (1 of 2)
A
A
Foil strip
Lower electrode
Workpiece
Mandrel
(b)
3 1/2"
3 1/2"
28"
3"
3"
(a)
Electrode wheel
5/32"
1/8" by 45° (typ)
Foil strip
0.010"
0.047"
0.047"R
1/8"
1/4"
Nugget
After welding
Before welding
Lower electrode
Section A-A

Figure **7-11** Set-up for 'roll' butt-seam welding of a table leg made of low-carbon steel. (By permission, from *Metals Handbook* Volume 6, Copyright American Society for Metals, 1971.)

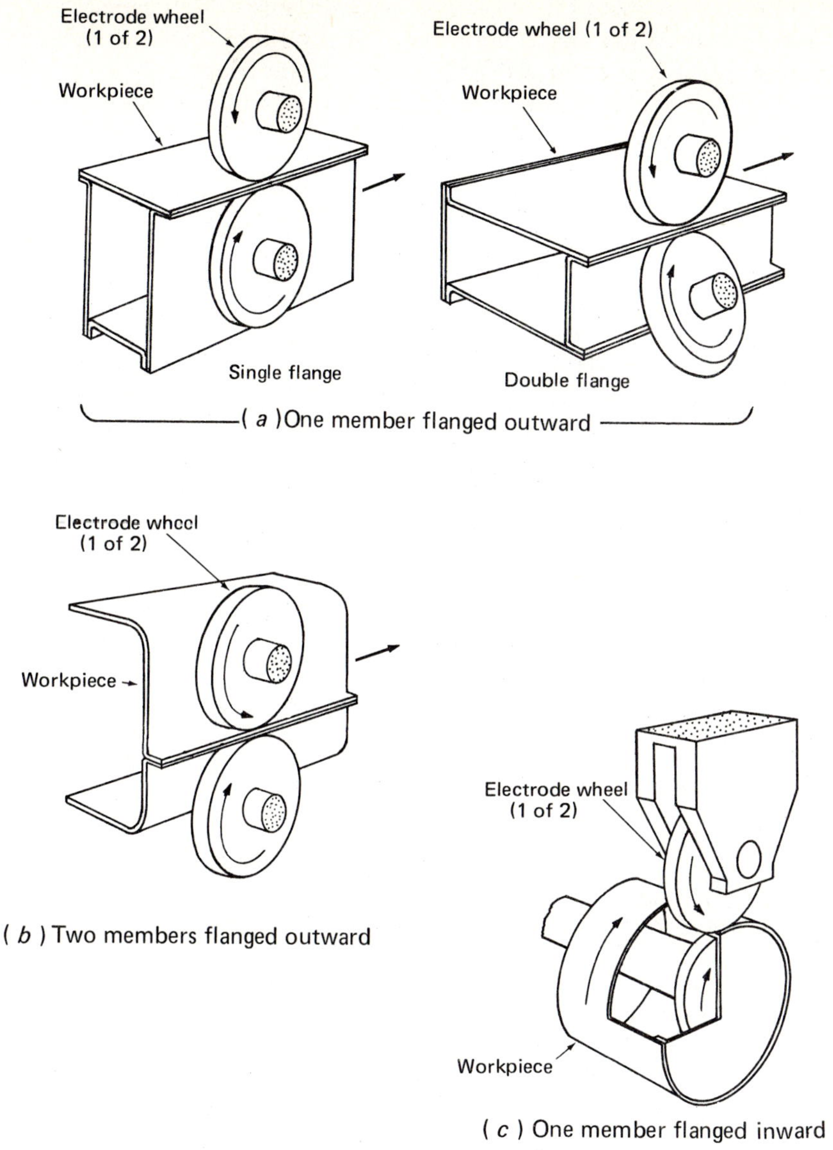
Electrode wheel
(1 of 2)
Workpiece
Electrode wheel (1 of 2)
Workpiece
Single flange
Double flange
(a)One member flanged outward
Electrode wheel
(1 of 2)
Workpiece
(b) Two members flanged outward
Electrode wheel
(1 of 2)
Workpiece
(c) One member flanged inward

Figure **7-12** Various arrangements of electrode wheels for making flange-joint lap-seam welds. (By permission, from *Metals Handbook* Volume 6, Copyright American Society for Metals, 1971.)

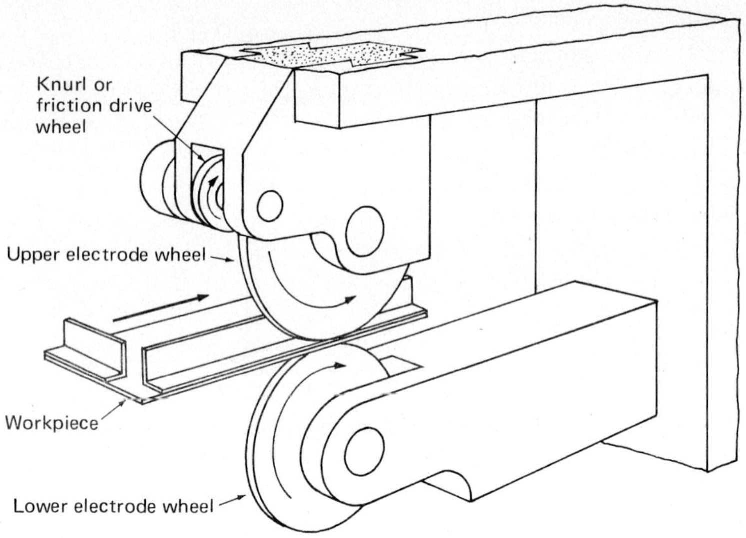
Knurl or
friction drive
wheel
Upper electrode wheel
Workpiece
Lower electrode wheel

(*a*) Longitudinal machine

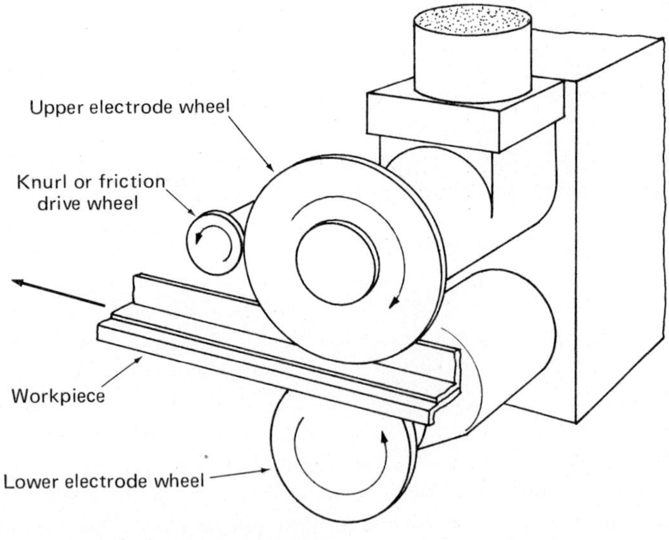
Upper electrode wheel
Knurl or friction
drive wheel
Workpiece
Lower electrode wheel

(*b*) Circular machine

Figure **7-13** Longitudinal and circular resistance seam-welding machines. (By permission, from *Metals Handbook* Volume 6, Copyright American Society for Metals, 1971.)

7-3 PROJECTION WELDING

In the projection welding process (Figure 7-1), current and heat flow are localized at a point or points predetermined by the design or configuration of one or both of the two parts to be welded (Figure 7-14). Spherical projections are used for welding assemblies made of steel sheet and plate. Projections can also be coined or forged on the ends or faces of screws, nuts, and similar fasteners (Figure 7-15).

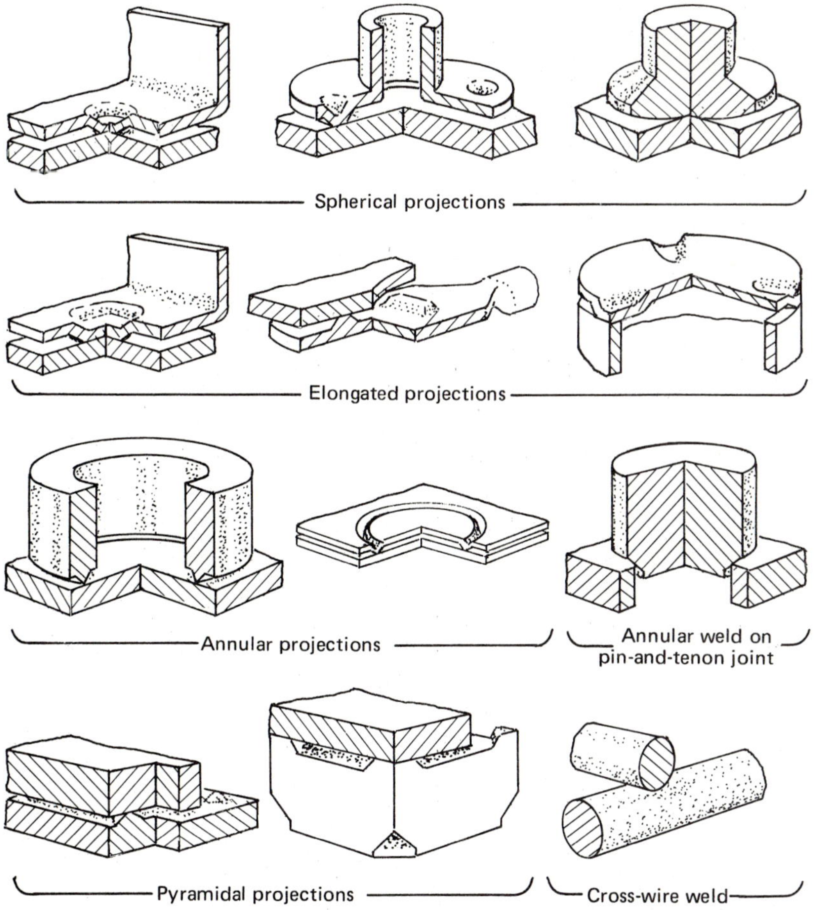

Spherical projections
Elongated projections
Annular projections
Annular weld on pin-and-tenon joint
Pyramidal projections
Cross-wire weld

Figure **7-14** Variations on basic types of projections in projection welding. (By permission, from *Metals Handbook* Volume 6, Copyright American Society for Metals, 1971.)

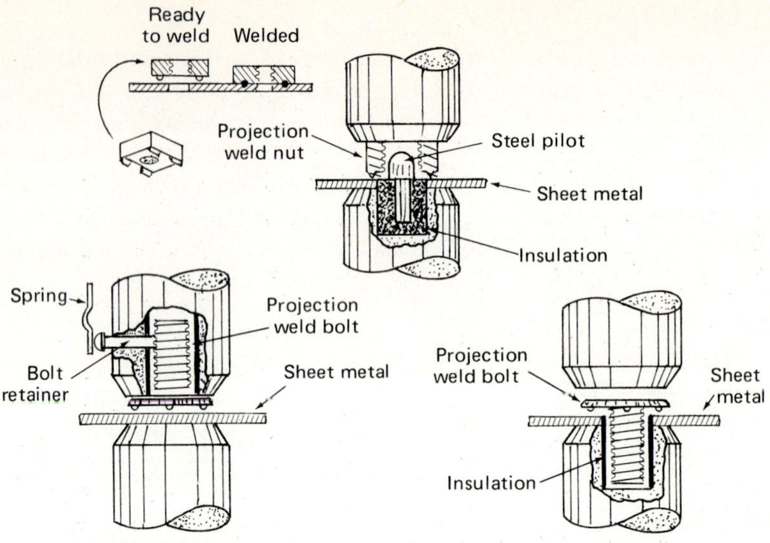
Ready
to weld
Welded
Projection
weld nut
Steel pilot
Sheet metal
Insulation
Spring
Projection
weld bolt
Bolt
retainer
Sheet metal
Projection
weld bolt
Sheet
metal
Insulation

Figure **7-15** Projection-welded square nuts and studs. (Redrawn from *Resistance Welding, Designing, Tooling and Applications* by W. A. Stanley. Copyright 1950 McGraw-Hill Book Company. Used with permission of McGraw-Hill Book Company.)

Elongated projections are often used instead of spherical projections, where the shape of the parts makes an elongated weld more suitable and where welds made with spherical projections will not meet strength requirements. Annular projections are used for welding tubing to sheet metal, for making liquid-tight or gas-tight connections, as in attaching mounting studs to liquid reservoirs, and for joining thin sheet metal parts in cases where spherical projections may collapse during welding. Pyramidal projections are coined or forged on the face of nuts. Cross-wire welds (Figure 7-16) are used in making wire projects, such as cylindrical wire baskets and circuit boards. Projection-welding machines (Figure 7-17) are similar in principle to the press-type spot-welding machines discussed in section 7-1.

The metal that can be projection welded most satisfactorily is low-carbon steel (0.20% carbon maximum) with a section thickness between 0.010 and 0.250 inch. Naval brass, monel (nickel-copper), and austenitic stainless steels in any combination of two can also be successfully welded. Although coated metals, such as terneplate, tin plate, and aluminized steel, can also be welded, electrode pick-up necessitates frequent cleaning and redressing of the electrodes.

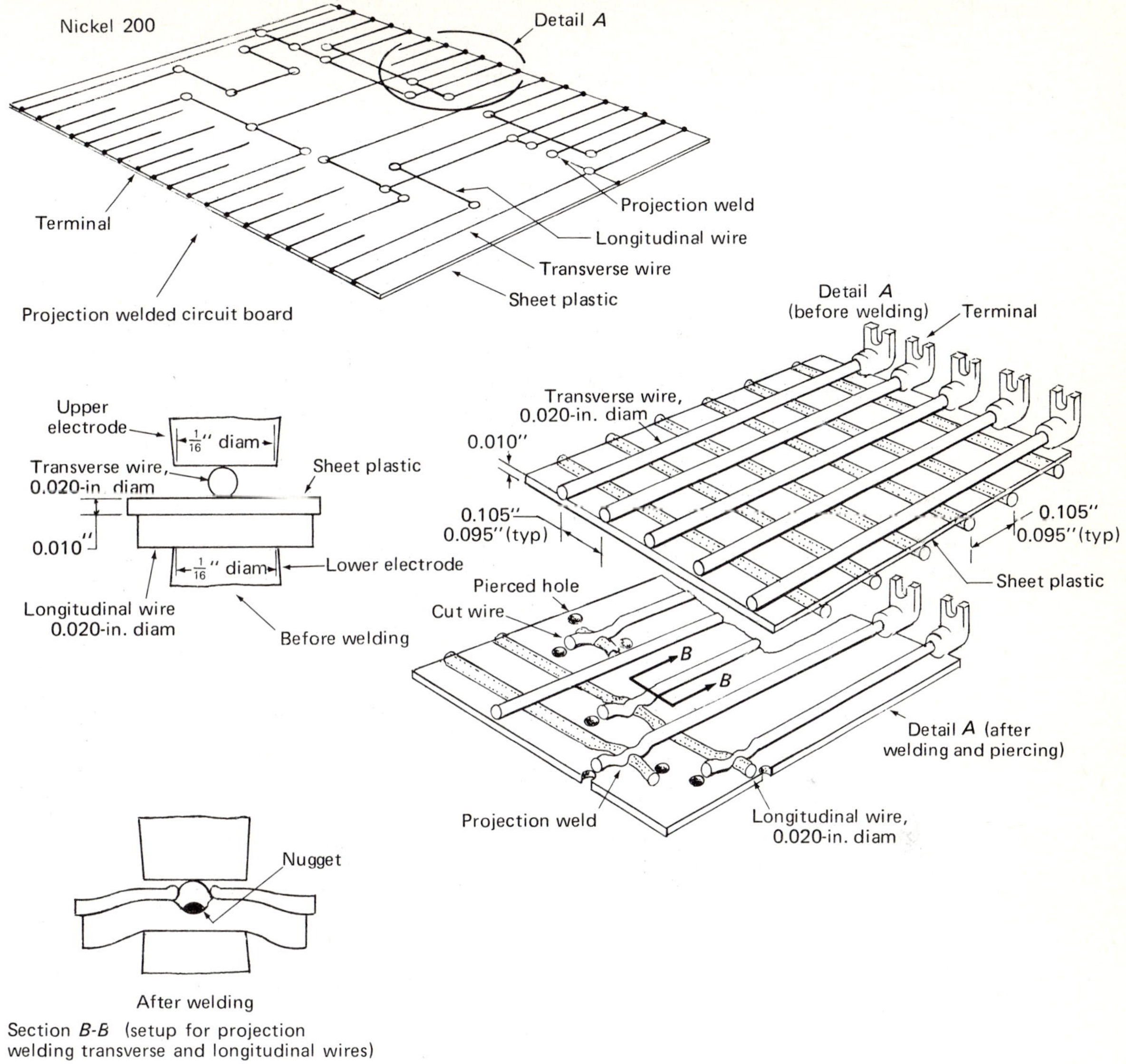

Nickel 200
Detail A
Terminal
Projection weld
Longitudinal wire
Transverse wire
Sheet plastic
Projection welded circuit board
Detail A
(before welding)
Terminal
Upper electrode
1/16" diam
Transverse wire, 0.020-in. diam
Sheet plastic
0.010"
Lower electrode
Longitudinal wire 0.020-in. diam
Before welding
Transverse wire, 0.020-in. diam
0.010"
0.105"
0.095" (typ)
0.105"
0.095" (typ)
Sheet plastic
Pierced hole
Cut wire
B
B
Detail A (after welding and piercing)
Projection weld
Longitudinal wire, 0.020-in. diam
Nugget
After welding
Section B-B (setup for projection welding transverse and longitudinal wires)

Figure **7-16** Projection-welded circuit board. (By permission, from *Metals Handbook* Volume 6, Copyright American Society for Metals, 1971.)

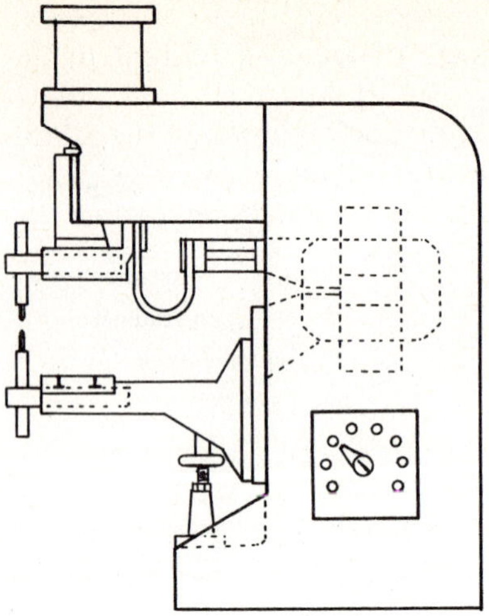

Figure **7-17** Typical press-type combination spot- and projection-welding machine. (Redrawn by permission from American Welding Society, *Welding Handbook,* 6th ed., 1969, section 2.)

7-4 ELECTRODES FOR SPOT, SEAM, AND PROJECTION WELDING

The electrodes used for these processes perform three important functions:

1. They conduct the welding current to the work.

2. They transmit the proper pressure or force to the weld area to produce a satisfactory weld.

3. They dissipate the heat from the weld zone more or less rapidly, depending on the process being employed and the necessity for removal of heat.

The first of these functions is electrical. If the application of pressure did not have to be considered, suitable electrode selection could be made almost entirely on the basis of electrical and thermal conductivity, taking into account the resistance of the electrode itself and the resistance between the electrode and the work surface at the area of contact.

The second function is mechanical. During welding operations, the electrodes are often subjected to considerable stresses, and they must withstand these stresses at high temperatures without excessive deformation. This is so because the current must not only be conducted to the work but must also be localized within a fixed area.

The transmitted pressure not only forges the heated workpieces together, but also restricts the passage of welding current to the localized area.

The conductivity of the electrodes must be higher than that of the metals being welded because, in addition to conducting electricity, the electrodes also conduct heat away from the exterior surfaces of the weld material. This function is very important because it is used in reverse when dissimilar metals are being welded to obtain a heat balance. The heat balance is obtained by using electrodes with lower heat conductivity than the weld metal to prevent too rapid heat dissipation from one of the metals of a dissimilar combination. This procedure is also useful when heat-treatable parts are to be welded.

Successful performance of any resistance welding operation, however simple or complicated it may be, is dependent upon the use of correct electrodes. The RWMA (Resistance Welder Manufacturers' Association) has classified electrode materials into two composition groups: copper-base alloys and refractory metal compositions. The alloys are available in the form of bars, forgings, castings, and inserts. They are sold under a number of trade names (Table 7-1). The general recommendations of the RWMA for the selection of the proper electrode material for spot welding similar and dissimilar materials are given in Table 7-2.

Electrode face shapes have been standardized (Figure 7-18) and are identified by the code developed by the RWMA for that purpose. The minimum face diameter for type A, B, D, and E electrodes can be determined by using the formula

$$\text{Face diameter} = 0.10 + 2t$$

where t = the thickness (in inches) of the base metal contacting the electrode.

Electrodes designed for spot welding can be used for projection welding if the electrode face is large enough to cover the projection being welded or the pattern of projections being welded simultaneously by the electrode.

Electrode wheels are available with standard face contours (Figure 7-19) and, depending on the machine size, they have the common widths and diameters in inches listed in Table 7-3.

7-5 FLASH WELDING

Flash welding is a resistance butt-welding process in which two workpieces are clamped in suitable current-carrying fixtures that hold them end to end in very light contact (Figure 7-20). Electric current is made to flow through the workpiece to produce flashing (arcing) which, in combination with electrical resistance, heats the abutting ends to the fusion point. When the abutting ends reach the

Table **7-1** Resistance-weldable similar and dissimilar metal combinations

	GROUP A: COPPER-BASE ALLOYS				
MANUFACTURER	CLASS 1	CLASS 2	CLASS 3	CLASS 4	CLASS 5
Acme Electric Welder Co.	Acmeloy	Superloy	Stainaloy	Hardaloy	—
American Metal Climax, Inc.	Amzirc	Amzirc; Amcron	—	—	—
Ampco Metal, Inc.	Ampcoloy 99	Ampcoloy 97	Ampcoloy 95	Ampcoloy 83	Ampcoloy 92
Eisler Engineering Co., Inc.	Eisler E	Eisler 3	Eisler 100	Eisler 73	Eisler ED
Electroloy Co., Inc.	Electroloy A	Electroloy XX	Electroloy TX	Electroloy B	Electroloy Molin #2
Hercules Welding Products Co.	H-1	H-2	H-3	H-4	H-5
Mallory Metallurgical Co.	Elkonite A	Mallory 3	Mallory 100	Mallory 73	Elkaloy D
Tipaloy, Incorporated	Tipaloy 100	Tipaloy 130	Tipaloy 200	Tipaloy T-4	Tipaloy T-5
Tuffaloy Products, Inc.	Tuffaloy 88	Tuffaloy 77	Tuffaloy 55	Tuffaloy 44	Tuffaloy 66
Weldaloy Products Co.	Weldaloy 10	Weldaloy 20	Weldaloy 30	Weldaloy 40	Weldaloy 50
Availability	Bars, Forgings	Bars, Forgings, Castings	Bars, Forgings, Castings	Bars, Forgings, Castings	Castings

GROUP B: REFRACTORY METALS COMPOSITIONS

MANUFACTURER	CLASS 10	CLASS 11	CLASS 12	CLASS 13	CLASS 14
Acme Electric Welder Co.	Tungstenite 40	Tungstenite 42	Tungstenite 43	Tungstenite 80	Tungstenite 90
Ampco Metal, Inc.	Ampcoloy 3W1	Ampcoloy 3W10	Ampcoloy 3W20	Ampcoloy 3W100	Ampcoloy 3M100
Dobbins Mfg. Company	Cletaloy 1CT, 3CT, 10CT	20CT	30CT, 10BC, 30BC, 10CC, 20CC, 30CC	—	—
Eisler Engineering Co., Inc.	E-1W3	E-10W3	E-20W3	E100W	—
Electroloy Co., Inc.	Electroloy 1	Electroloy 10	Electroloy 20	Electroloy 100	Electroloy 500
Hercules Welding Products Co.	H-1W	H-10W	H-20W	H-100W	H-14M
Mallory Metallurgical Co.	Elkonite 1W3	Elkonite 10W3	Elkonite 20W3	Elkonite 100W	Elkonite 100M
Tipaloy, Incorporated	T-1W	T-10W	T-20W	T-100W	T-100M
Tuffaloy Products, Inc.	Tuffaloy 1W	Tuffaloy 10W	Tuffaloy 20W	Tuffaloy 100W	Tuffaloy 100M
Weldaloy Products Co.	Weldtung 10CT8	Weldtung 20CT	Weldtung 30CT	Weldaloy W	Weldaloy M
Availability	Bars, Inserts	Bars, Inserts	Bars, Inserts	Bars, Inserts	Bars, Inserts

Source: *1964–1965 Welding Data Book*, pp. 186–187. Courtesy of *Welding Data Book/Welding Design and Fabrication Magazine.*

Table **7-2** Recommended electrode materials for spot-welding similar and dissimilar materials using conventional spot-welding methods

TO WELD SIMILAR METALS

FERROUS

To weld similar metals read block under metal to be welded	TIN PLATE STEEL		TERNE PLATE STEEL		GALVANIZED IRON ZINC PLATE		CADMIUM PLATE STEEL		CHROME PLATE STEEL		STAINLESS STEEL 18-8 TYPE		SCALY H.R. STEEL		C.R. STEEL, H.R. STEEL CLEAN	
	B	I	A	I	A	I (II)	B	I	A	II	A	III (II)	B	I (II)	A	II
	I	3	I	3	I (II)	3	I	3	II	3	III (II)		I (II)	2	II	

TO WELD DISSIMILAR METALS

FERROUS ALLOYS	STAINLESS STEEL 18-8 TYPE		CHROME PLATE STEEL		CADMIUM PLATE STEEL		GALVANIZED IRON		TERNE PLATE STEEL		TIN PLATE STEEL	
Cold rolled steel	A	II (III)	A	II	B	II	B	I	A	I (II)	B	I
hot rolled steel, clean[a]	II		II	3	II	3	II	3	II	3	II	3
Tin plate			B	II	B	I (II)	B	I	B	I (II)		
steel			I	3	I	3	I	3	I	3		
Terne plate	B	II	B	II	B	I (II)	B	I				
steel	I	3	I	3	I	3	I	3				
Galvanized iron	B	II	B	II	B	I						
zinc plate	I	3	I	3	I	3						
Cadmium	B	II	B	II								
plate steel	I	3	I	3								
Chrome	A	III (II)										
plate steel	II	3										

LEGEND

BLOCK INTERPRETATION

Weld-ability	Electrode against
Electrode against	Special information

WELDABILITY
A-Excellent
B-Good

ELECTRODES
R.W.M.A. SPECIFICATIONS
I. Group A Class 1
II. Group A Class 2
III. Group A Class 3
Note: Electrode materials in circles are second choice. Example (II)

SPECIAL INFORMATION
1. Special conditioned required.
2. Good practice recommends cleaning before welding.
3. If plating is heavy, weld strength is questionable.

TO WELD SIMILAR METALS

NONFERROUS

ALUMINUM	ALUMINUM ALLOYS DURALUMINUM		CUPRO NICKEL		NICKEL SILVER		NICKEL		NICKEL ALLOYS MONEL NICHROME (HIGH RES.)		BRASS YELLOW 25%–40% ZINC		PHOSPHOR BRONZE GRADE A, C AND D		SILICON BRONZE EVERDUR OLYMPIC DURONZE HERCULOY	
B I(II)	B	I(II)	A	II	B	II	A	II	A	II	B	II	A	II	A	II
I(II) 2	I(II)	2	II		II		II		II		II		II		II	

TO WELD DISSIMILAR METALS

NONFERROUS ALLOYS	NICKEL ALLOYS		NICKEL		PHOSPHOR BRONZE		SILICON BRONZE		YELLOW BRASS		NICKEL SILVER	
Cupro nickel	B	II			B	II	B	II			B	II
	II				II		II				II	
Silicon bronze everdur-olympic duronze-herculoy					B	II	A	II	B	II		
					II		II		II			
Nickel silver	B	II			B	II	B	II				
	II	I			II	I	II					
Nickel alloys	A	II	A	II								
	II		II									
Stainless steel 18-8 type	B	II	B	II								
	III(II)	I	II(III)									

	ALUMINUM	
Aluminum alloys duraluminum	B	I(I)
	I(II)	2

This chart adopted by the Resistance Welder Manufacturers' Association embodies the best general recommendations that can be made for the selection of the proper electrode materials for spot welding similar and dissimilar metals.

These recommendations for electrodes can be effective only when the other essential factors of the correct time, current and pressure are properly controlled.

[a] Plain carbon steel

Source: Resistance Welder Manufacturers' Association.

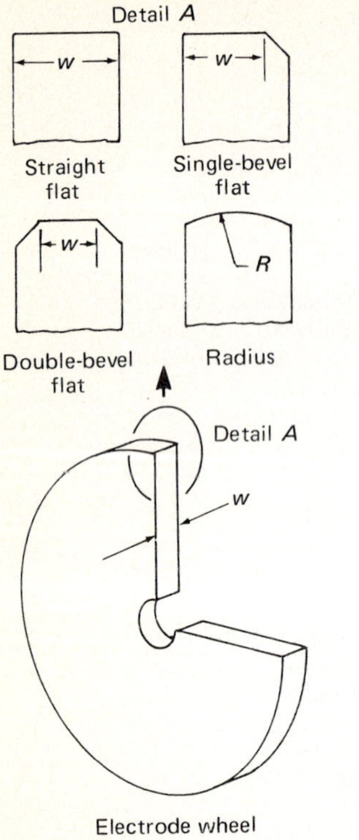
Detail A
Straight flat
Single-bevel flat
Double-bevel flat
Radius
Detail A
Electrode wheel

Figure **7-19** Standard face contours of seam-welding electrode wheels. (By permission, from *Metals Handbook* Volume 6, Copyright American Society for Metals, 1971.)

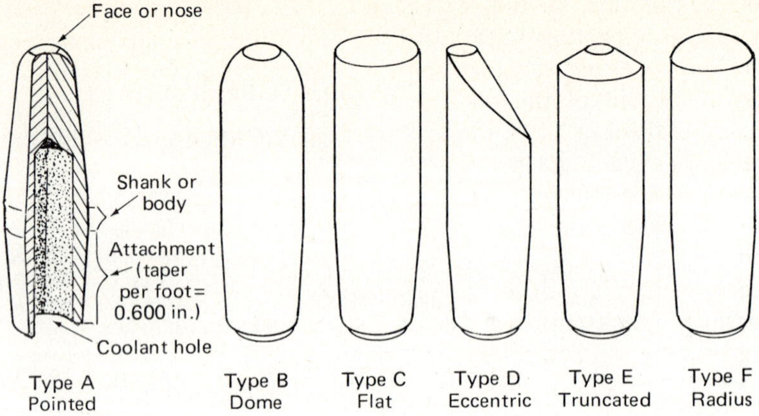
Face or nose
Shank or body
Attachment (taper per foot= 0.600 in.)
Coolant hole
Type A Pointed
Type B Dome
Type C Flat
Type D Eccentric
Type E Truncated
Type F Radius

Figure **7-18** Standard types of electrode face or nose shapes. (By permission, from *Metals Handbook* Volume 6, Copyright American Society for Metals, 1971.)

Table **7-3** Common sizes of electrode wheels used for resistance seam-welding

MACHINE SIZE	WHEEL DIAMETER, IN.	WHEEL WIDTH, IN.
Small	7	3/8
Medium	8	3/8–1/2
Large	10–12	3/8–3/4

Source: By permission, from *Metals Handbook* Volume 6, Copyright American Society for Metals, 1971.

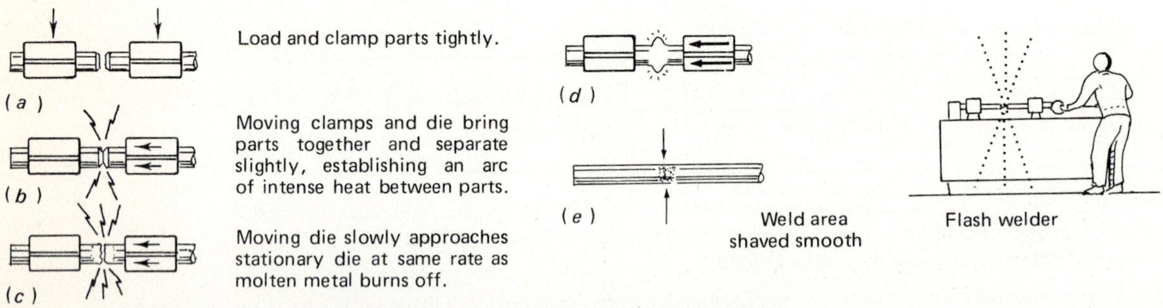
Load and clamp parts tightly.
(a)
Moving clamps and die bring parts together and separate slightly, establishing an arc of intense heat between parts.
(b)
Moving die slowly approaches stationary die at same rate as molten metal burns off.
(c)
(d)
(e)
Weld area shaved smooth
Flash welder

Figure **7-20** Typical flash-welding operation. (Redrawn from *Resistance Welding, Designing, Tooling and Applications* by W. A. Stanley. Copyright 1950 McGraw-Hill Book Company. Used with permission of McGraw-Hill Book Company.)

proper temperature for the correct depth, the workpieces are suddenly brought together with sufficient force to cause an upsetting action. The upsetting action pushes the molten metal and part of the plastic metal out of the weld area into a weld upset. In many applications (Figure 7-21), the upset must be removed after welding by scarfing or machining.

Flash welding can be used for joining many ferrous and nonferrous alloys (see Table 7-4), except for cast iron, lead, tin, zinc, bismuth, and antimony alloys.

Some specific applications of flash welding are: miter joints between window frame extrusions, extremely large crankshafts (14-ft long × 8-in. diameter at the main bearings), jet engine mounts, steel bands for contour roll forming of automotive wheel rims, and others (Figure 7-21).

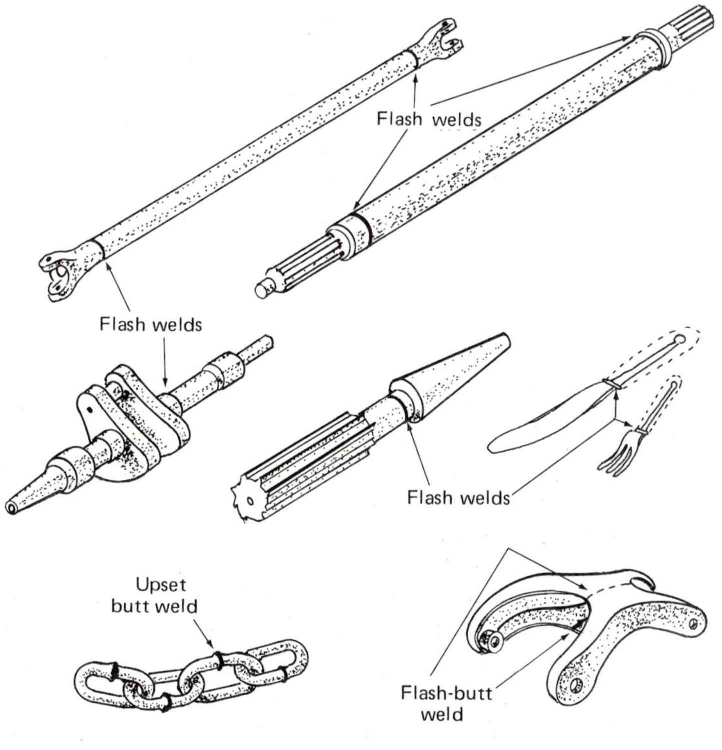
Flash welds
Flash welds
Flash welds
Upset
butt weld
Flash-butt
weld

Figure **7-21** Examples of flash butt-welding applications. (Redrawn from *Resistance Welding, Designing, Tooling and Applications* by W. A. Stanley. Copyright 1950 McGraw-Hill Book Company. Used with permission of McGraw-Hill Book Company.)

Table **7-4** Some combinations of metals that have been flash welded

BASE METAL	METALS THAT HAVE BEEN FLASH WELDED TO BASE METALS LISTED IN THE FIRST COLUMN										
	ALUMINUM ALLOYS	COPPER ALLOYS	MAGNESIUM ALLOYS	MOLYBDENUM	NICKEL ALLOYS	STEELS, CARBON AND ALLOY	STEELS, STAINLESS	STEELS, TOOL	TANTALUM	TITANIUM ALLOYS	TUNGSTEN
Aluminum alloys	X	X	X	—	X	—	—	—	—	—	—
Copper alloys	X	X	X	—	X	X	X	X	—	X	—
Magnesium alloys	X	X	X	—	—	—	—	—	—	—	—
Molybdenum	—	X	—	X	X	X	X	X	—	—	—
Nickel alloys	X	X	—	X	X	X	X	X	X	—	X
Steels, carbon and alloy	—	X	—	X	X	X	X	X	X	X	X
Steels, stainless	—	X	—	X	X	X	X	X	X	X	X
Steels, tool	—	X	—	X	X	X	X	X	X	—	X
Tantalum	—	X	—	—	X	X	X	X	X	—	—
Titanium alloys	—	X	—	—	—	X	X	—	—	X	—
Tungsten	—	—	—	—	X	X	X	X	—	—	X

Source: By permission, from *Metals Handbook* Volume 6, Copyright American Society for Metals, 1971.

A standard flash-welding machine consists of a low-impedance transformer; a stationary platen; a movable platen on which the clamping disk, electrodes, and other tools needed to position and hold the workpieces are mounted (Figure 7-22); flashing and upsetting mechanisms; and the necessary electrical, air, or hydraulic controls.

7-6 UPSET-BUTT WELDING

Upset-butt welding was the earliest form of resistance welding. In this process (Figure 7-23) fusion can be produced simultaneously over the entire area of the abutting surfaces, or progressively along a joint, by the heat obtained from the resistance to the current flow through the area of contact of those surfaces. Welding force is applied before heating is started and is maintained throughout the heating period.

Upset-butt welding processes include:

- Upset-butt welding of parts end to end (wire, bars, strip, and others).

Figure **7-22** Fixture used to position and hold formed tubing for flash welding. (By permission, from *Metals Handbook* Volume 6, Copyright American Society for Metals, 1971.)

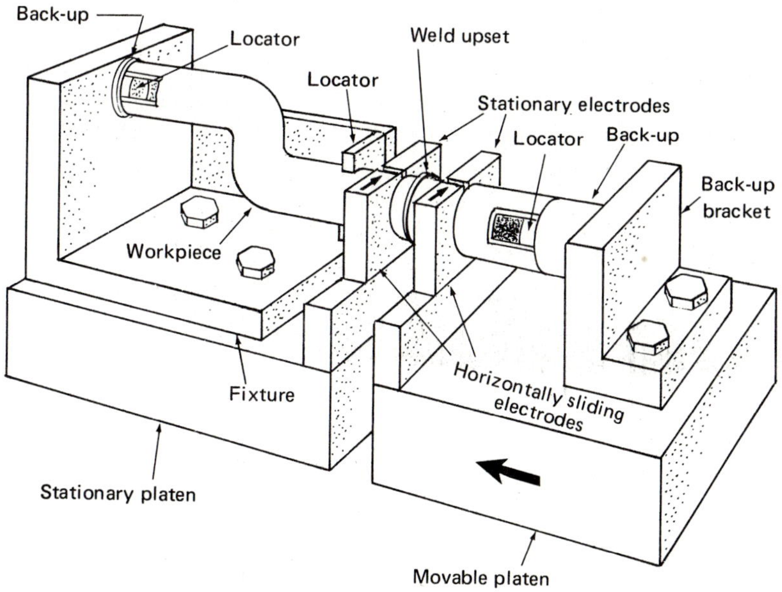
Back-up
Locator
Weld upset
Locator
Stationary electrodes
Locator
Back-up
Back-up
bracket
Workpiece
Fixture
Horizontally sliding
electrodes
Stationary platen
Movable platen

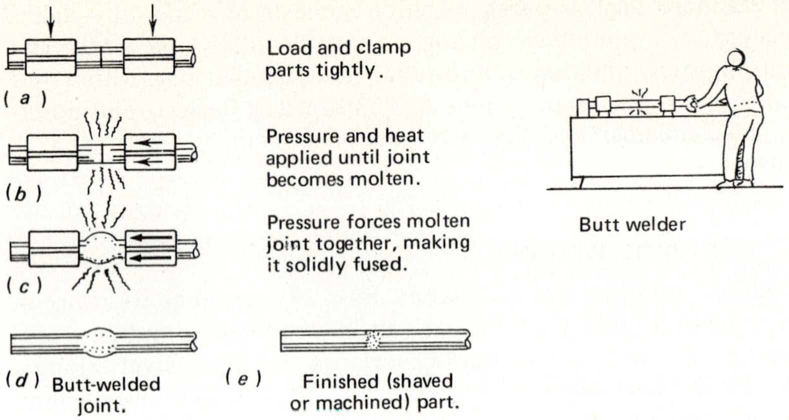

Load and clamp parts tightly.
(a)
Pressure and heat applied until joint becomes molten.
(b)
Pressure forces molten joint together, making it solidly fused.
(c)
Butt welder
(d) Butt-welded joint.
(e) Finished (shaved or machined) part.

Figure **7-23** Typical upset butt-welding operation. (Redrawn from *Resistance Welding, Designing, Tooling and Applications* by W. A. Stanley. Copyright 1950 McGraw-Hill Book Company. Used with permission of McGraw-Hill Book Company.)

- Continuous butt-seam welding with low-frequency current.
- Continuous butt-seam welding with high-frequency current.

Upset-butt welding is widely used in wire mills and in the welding of products made from wire. Continuous butt-seam low-frequency welding is widely used to make joints such as the longitudinal joint in tubing or pipe. In this process, the formed tube passes under the electrodes, which transfer the welding current into the material being welded, and through a set of rolls, which provide the welding force. The amount of upset is regulated by the relative position of the welding electrodes and the rolls applying the upset force. The required welding heat is governed by the current passing through the work and the speed at which the tube goes through the rolls.

Continuous butt-seam high-frequency welding is also used to make joints like the longitudinal joint in tubing or pipe, but in terms of welding speed per kVA, this process is not as efficient as the low-frequency process. A wide variety of materials in wire, bar, strip, and tubing forms can be joined by these methods. The materials include, among others: aluminum alloys, brass, copper, gold, nickel alloys, stainless steels, and low-carbon and high-carbon steels.

Although aluminum alloys are weldable by these processes, a specially designed machine is usually required for the proper control of upsetting pressure during completion of the weld.

7-7 PERCUSSION WELDING

Percussion welding is a resistance-welding process in which the heat is obtained from an arc produced by rapid discharge of electrical energy. The force is applied percussively during or immediately after the electrical discharge. A shallow layer of metal on the contact surfaces of the workpiece is melted by the heat of the arc produced between them, and one of the workpieces is impacted against the other, extinguishing the arc, expelling molten metal, and completing the weld. The heat input is intense but extremely brief and localized, enabling the percussion welding of one small component (Figure 7-24) to another, or a small component (Figure 7-25) to a larger one.

There are two percussion welding methods: the *capacitor-discharge method* and the *magnetic-force method*. Percussion welding is similar to stud welding (an arc welding process) in three important respects:

1. Welding heat is obtained from an arc.

Figure **7-24** Simplified diagram of the sequence of steps in capacitor-discharge percussion welding: (*a*) contactor is momentarily closed to charge the capacitor; (*b*) one of the parts to be welded is advanced toward the other and rapidly accelerated; (*c*) the arc forms across the gap just before the parts meet, melting the work surface of each part, and exploding the nib (arc starter), if one is used; (*d*) the arc is extinguished as one part is impacted against the other, expelling molten metal as flash and forging the parts together to complete the weld. (By permission, from *Metals Handbook* Volume 6, Copyright American Society for Metals, 1971.)

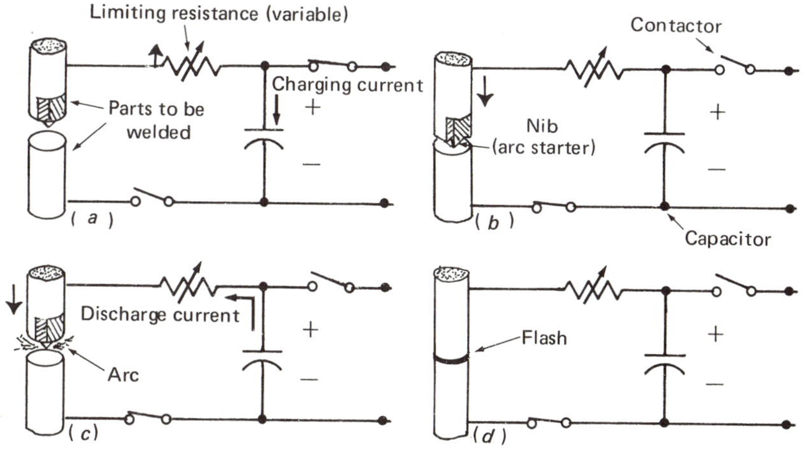

Limiting resistance (variable)
Contactor
Charging current
Parts to be welded
Nib
(arc starter)
(a)
(b)
Capacitor
Discharge current
Arc
Flash
(c)
(d)

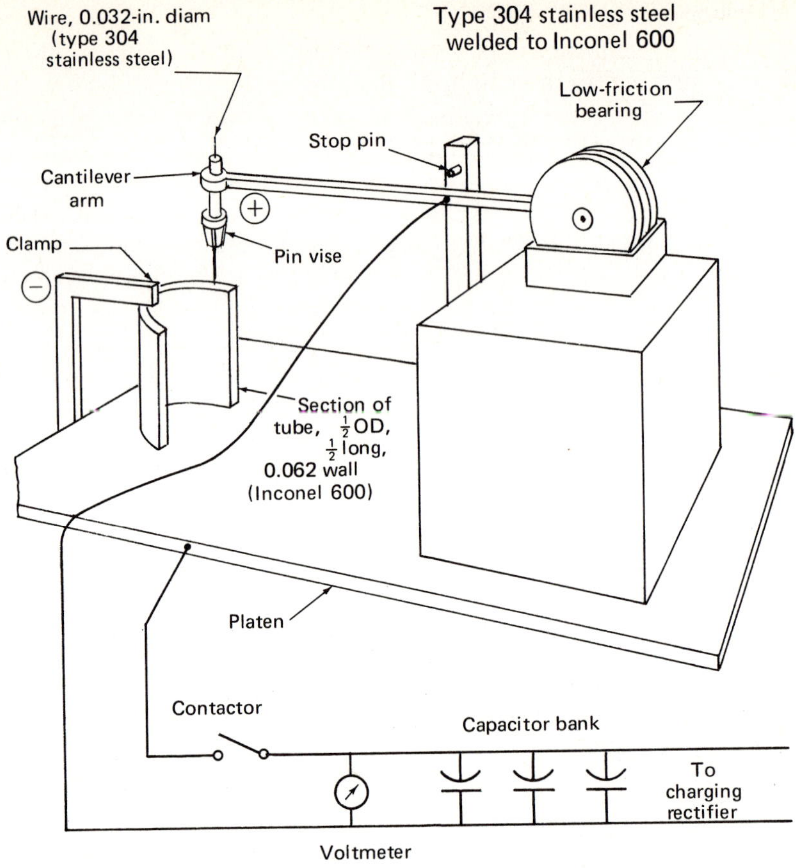
Wire, 0.032-in. diam
(type 304
stainless steel)
Type 304 stainless steel
welded to Inconel 600
Low-friction
bearing
Stop pin
Cantilever
arm
Clamp
Pin vise
Section of
tube, ½ OD,
½ long,
0.062 wall
(Inconel 600)
Platen
Contactor
Capacitor bank
To
charging
rectifier
Voltmeter

Figure **7-25** Set-up for capacitor-discharge percussion welding of a stainless steel wire to a larger Inconel 600 specimen. (By permission, from *Metals Handbook* Volume 6, Copyright American Society for Metals, 1971.)

2. Force is applied percussively.

3. Arc-starting methods used in the several variations of the two processes are closely similar.

However, percussion and stud welding differ in various aspects of equipment (Figure 7-26), technique, and process variables. Recommended applications for percussion (capacitor discharge) and stud (electric-arc) welding are listed in Table 7-5.

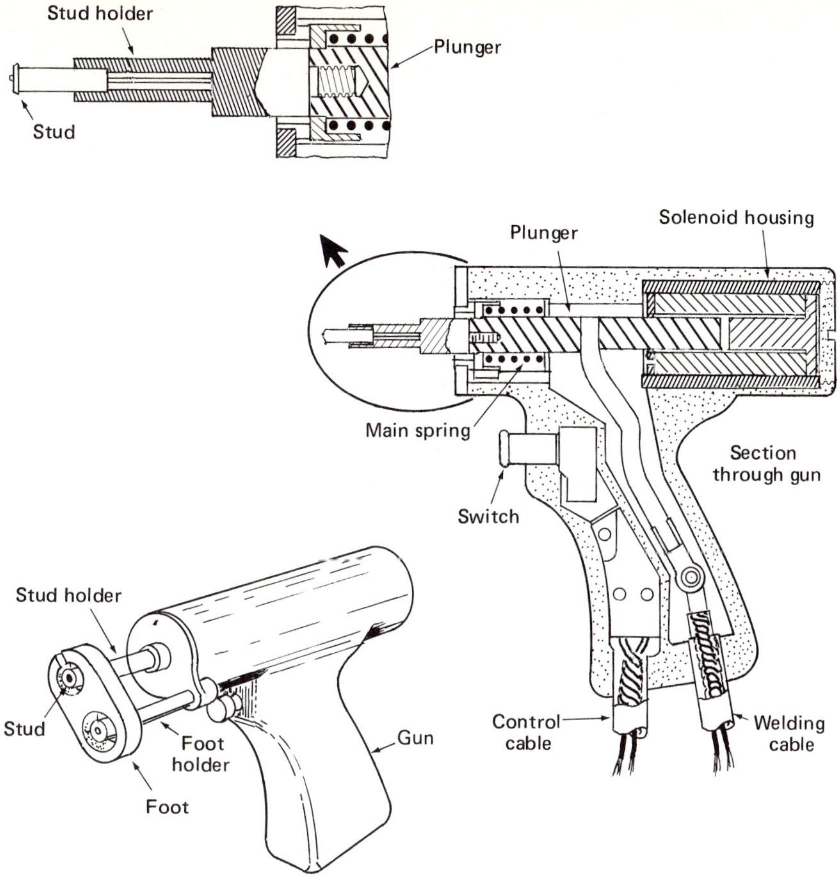
Stud holder
Plunger
Stud
Solenoid housing
Plunger
Main spring
Section
through gun
Switch
Stud holder
Stud
Foot
holder
Gun
Foot
Control
cable
Welding
cable

Figure **7-26** Guns for capacitor-discharge stud welding. Arc stud-welding guns are similar but have a ferrule holder in the foot. (By permission, from *Metals Handbook* Volume 6, Copyright American Society for Metals, 1971.)

Almost any like or unlike metals or alloys can be joined by percussion welding. Metals joined include copper alloys, aluminum alloys, nickel alloys, low-carbon steels, medium-carbon steels, and stainless steels. Gold, silver, copper-tungsten, silver-tungsten, and silver-cadmium oxide are percussion welded to copper alloys for commonly used assemblies for electrical contacts. Copper, thermocouple, and low-expansion alloys are percussion welded to molybdenum.

Table **7-5** Percussion-welding process selection chart

COMPARISON OF PROCESS VARIABLES IN PERCUSSION AND STUD WELDING

ITEM	PERCUSSION WELDING				STUD WELDING			
	CAPACITOR-DISCHARGE METHOD			*MAGNETIC-FORCE METHOD*	*ARC METHOD*[a]	*CAPACITOR-DISCHARGE METHOD*		
	LOW-VOLTAGE		HIGH-VOLTAGE					
Power supply	Capacitor	Capacitor	Capacitor	Transformer[b]	Rectifier[c]	Capacitor	Capacitor	Capacitor
Current supplied to arc	dc	dc	dc	ac	dc	dc	dc	dc
Voltage, V	50 to 150	12 to 120	1000 to 3000	10 to 35	60 to 100	100 to 200[d]	100 to 200[d]	100 to 200[d]
Arc-starting method	Nib plus dc voltage (initial gap)	High-frequency ac pulse plus dc voltage[e]	Dc voltage (initial gap)	Nib plus first half-cycle of ac (initial contact)	Nib plus dc voltage (draw arc after contact)	Nib plus dc voltage (initial gap)	Nib plus dc voltage (initial contact, no retraction)	Dc voltage (draw arc by retracting after contact)
Arc time, milliseconds	0.15 to 1	1 max	1 max	8 max	100 to 1000	3 to 6	3 to 6	6 to 12

[a]A ferrule is used to confine the molten metal in this method; flux may be used, depending on workpiece and size. [b]Resistance-welding transformer. [c]Arc-welding type of rectifier or motor-generator; no energy storage. [d]Approximate. [e]Initial gap.

CONDITIONS FOR AND RESULTS OF PERCUSSION WELDING OF SEVEN COMBINATIONS OF WORK METALS[a]

WORKPIECES WELDED	WIRE SIZE	WELDING VOLTAGE, V	RESISTANCE, OHMS[b]	INITIAL GAP, IN.[c]	TENSILE STRENGTH, psi	LOCATION OF FAILURE
Chromel wire to Alumel wire	0.015-in. diam	130	1.0	3/8	77,300	Alumel
Copper wire to Nichrome wire	0.015-in. diam	160	1.5	3/8	39,100	Copper
Copper wire to stainless plate	0.015-in. diam	150	1.0	1 1/4	40,000	Copper
Nichrome wire to stainless plate	0.015-in. diam	350	1.0	2 1/4	145,500	Weld
Chromel-Alumel wire to stainless plate	0.150-in. diam	350	1.0	2 1/4	65,000	Weld
Thorium wire to thorium wire	0.040-in. square	350	1.0	2 1/4	Not measured	Wire
Thorium wire to Zircaloy-2 wire	0.040-in. square	350	1.0	2 3/4	Not measured	Zircaloy-2

[a]The welding machine used had a capacitor bank had a variable output rated at 20 to 400 microfarads and 600 volts max. The series resistor was a 0 to 7.5-ohm, 50-watt stepless potentiometer. The effective weight of the cantilever was about 6 oz, but additional weight could be added, if needed. [b]Setting of potentiometer. [c]Distance of fall of wire (workpiece) attached to pivoted arm.

Source: By permission, from *Metals Handbook* Volume 6, Copyright American Society for Metals, 1971.

REVIEW QUESTIONS

1. Name and describe five resistance-welding processes.

2. Name and describe the three kinds of time referred to in the spot-welding process.

3. Describe the similarities between the projection-type and the press-type resistance-welding machines.

4. Describe the functions of the electrodes used in resistance welding.

5. What does the abbreviation RWMA represent?

6. Draw a sketch of, and calculate the minimum face diameter of, a standard RWMA type B electrode to be used for spot welding a 0.0325-inch-thick piece of carbon steel to a 0.0612-inch-thick piece of carbon steel, with the 0.0325-inch-thick piece in contact with the electrode.

7. Name the metals that are not joined by flash welding and stud welding.

8. Describe the upset-welding process.

9. List the similarities between percussion and stud welding.

10. Describe the weld sequence involved in spot welding.

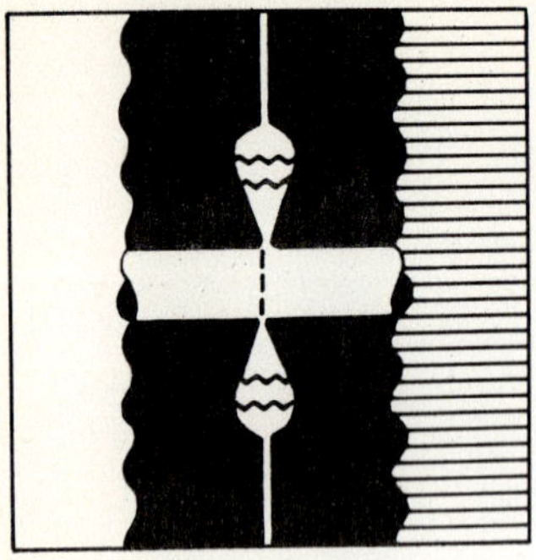

CHAPTER 8
SOLID STATE WELDING PROCESSES

Solid state welding consists of a group of welding processes in which fusion is produced essentially at temperatures below the melting point of the base metals being joined, without the addition of a brazing filler metal. Pressure may or may not be used.

To join metals in the solid state it is necessary to achieve mechanical intimacy of contact by careful preparation of the surfaces to be joined. Surface preparation includes the generation of an acceptable surface smoothness and the removal of rust, dirt, oil, and other contaminants and all absorbed moisture or gas. This is accomplished by machining, abrading, grinding, polishing, or chemical etching followed by degreasing with alcohol, trichlorethene, acetone, detergents, or by means of a vacuum bake-out.

8-1 COLD-PRESSURE WELDING

Cold-pressure welding produces welds without the application of external heat, but rather by subjecting the weld metals to sufficient pressure to cause them to deform plastically at room temperature.

The cold-pressure welding process is best suited to the joining of high-purity and commercially pure aluminum; other nonferrous metals, such as aluminum alloys, cadmium, lead, copper, nickel, zinc, and silver; or combinations of nonferrous metals of differing hardness.

The welding pressure, which may be applied manually or by power tool, can be a slow squeeze or an impact in a range from 20,000 psi for aluminum to 160,000 psi for copper (Figure 8-1).

In addition to the square-butt or seam welds, some other welds made by this process are:

- *Trap welds.* Designed to permit the welding of inserts into similar or dissimilar nonferrous metals by flowing metal around the insert, somewhat like plastic molding.
- *Wave welds.* Used on flat stock where the standard straight-line weld is considered undesirable because of structural or aesthetic considerations.
- *Stagger welds.* Used to join thin metal sheets to heavy bar stock. This type of weld places dots or short-line welds along two or more parallel lines.
- *Sandwich welds.* Best suited for applications where indentation of the work surfaces produced by the dies is objectionable. With this method, a third piece of metal, such as a suitable piece of wire, is sandwiched between the two workpieces and all three are welded together in a single operation. The insert flattens out between the two sheets, producing a weld without indentation on either side of the joint.

8-2 FRICTION WELDING

Friction welding is a process in which the welding heat is produced by direct conversion of mechanical energy to thermal energy at the interface of the workpieces, without the application of heat from

Figure **8-1** Cold welding: (*a*) the dies advance toward each other until the joint surfaces touch; (*b*) additional pressure causes the metal to upset and flow into the flash recesses; (*c*) completing the weld. (Redrawn by permission from American Welding Society, *Welding Handbook*, 6th ed., 1971, section 3B.)

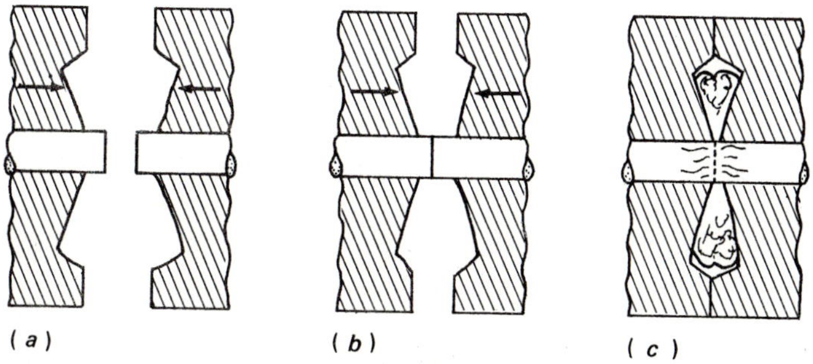

(a)
(b)
(c)

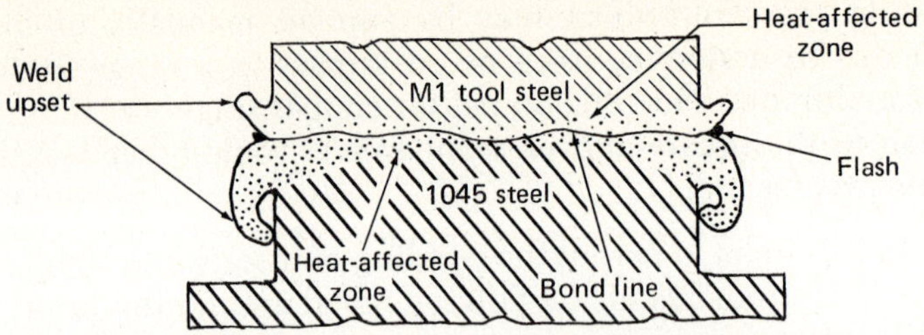
Heat-affected zone
Weld upset
M1 tool steel
1045 steel
Flash
Heat-affected zone
Bond line

Figure **8-2** Cross-section of a friction weld between two dissimilar steels. The greater amount of weld upset occurred in the 1045 steel because of its greater forgeability. When workpieces made of the same or closely similar metal are welded, approximately equal amounts of weld upset are formed on both sides of the bond line. (By permission, from *Metals Handbook* Volume 6, Copyright American Society for Metals, 1971.)

external sources. Friction welds are made by holding a nonrotating workpiece in contact with a rotating one under constant or gradually increasing pressure. The interface (bond line) of the two pieces reaches welding temperature and then the rotation is stopped to complete the weld. Welding occurs under the influence of pressure that is applied while the heated zone is in the plastic (mushy) temperature range.

A cross-section through a friction weld joining two dissimilar steels is shown in Figure 8-2. Steel with the greater forgeability has the greater amount of weld upset. When similar metals are welded, the amount of upset is about the same on both sides of the bond line.

There are three methods of joining workpieces by friction welding: (1) conventional friction welding, (2) inertia welding, and (3) flywheel friction welding.

In *conventional friction welding,* mechanical energy is converted to heat energy by rotating one workpiece while pressing it against a nonrotating workpiece. After a specified period of time, rotation is suddenly stopped and the pressure is increased and held for another specified period of time, producing a weld. The machine required for conventional friction welding resembles an engine lathe equipped with an efficient spindle braking system; a means of applying and controlling axial pressure; and a weld cycle timer to compensate for such workpiece variables as melting point, thermal conductivity of the metal, and metallurgical changes that occur during the heating cycle, particularly when dissimilar metals are being welded.

In *inertia welding,* the workpiece component that is to be rotated is held in a collet chuck–flywheel assembly. The other workpiece is clamped in a nonrotating vise. The rotating assembly is then accelerated to a predetermined speed, at which time the flywheel is disconnected from the power supply and the workpieces are brought into contact under a constant force. Flywheel energy is rapidly converted to heat at the interface, and welding occurs as rotation ceases.

The inertia-welding machine is constructed with a horizontal bed and overhead tie bars to contain the axial pressure and torque reactions, and to ensure accurate spindle-to-bed alignment. The spindle is driven by a hydrostatic motor through a change-gear transmission. A hydraulically actuated tailstock retracts adjustably for loading and unloading. A self-centering vise can be attached to the tailstock for clamping cylindrical parts, and fixtures are used for holding asymmetrical parts for which the vise is not suitable. The spindle has a means for mounting a collet chuck, and a draw bar is used for opening and closing the collet.

Flywheel size (moment of inertia of the flywheel or spindle) is adjusted by adding or removing flywheel disks. The spindle speed and axial pressure are adjusted by dials on the control panel.

Two characteristics of inertia welding of 1-inch-diameter low-carbon steel bars (continuously decreasing the surface velocity of the workpiece and continuously changing the torque at the weld interface) are illustrated in Figure 8-3. Surface velocity begins at some

Figure **8-3** Variations of surface velocity, torque, axial force, and weld upset in a one-second inertia weld. (By permission, from *Metals Handbook* Volume 6, Copyright American Society for Metals, 1971.)

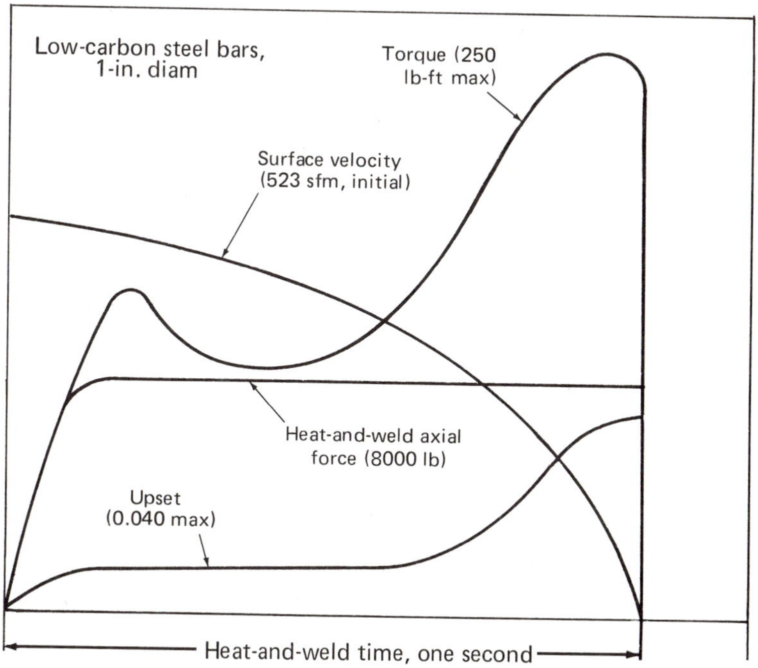
Low-carbon steel bars,
1-in. diam
Torque (250
lb-ft max)
Surface velocity
(523 sfm, initial)
Heat-and-weld axial
force (8000 lb)
Upset
(0.040 max)
Heat-and-weld time, one second

initial value and decreases along an essentially parabolic curve to zero, at which time the weld is completed. Heating and welding time is usually 0.2 to 4 seconds. Torque has a peak value of short duration early in the cycle, gradually decreases, and then increases until the velocity has decreased to the value at which welding begins, at which time the torque rises sharply. This high torque is accompanied by the forging in the weld zone and is responsible for much of the upsetting. The high-torque phase, present only in inertia welding, refines the grain structure and expells any oxides at the weld interface. The gradually decreasing and increasing part of the torque curve is essential to the formation of good welds. The second (low-torque) phase generally will not develop if the initial velocity is too low. Conditions for inertia welding of 1-inch-diameter bars in combinations of similar and dissimilar metals are listed in Table 8-1.

Other differences between inertia welding and conventional friction welding are input power at the weld interface and heating time. The power needed for the weld itself is not important in inertia welding, because whatever power is required can always be supplied by deceleration of the flywheel at the required rate. In conventional friction welding, power is limited by the size of the drive motor.

The high power used in inertia welding is a result of rather rapidly applied axial pressure. Power demands in conventional friction welding are controlled and limited to motor capacity by applying the axial pressure slowly: usually 2 to 4 seconds elapse before the full pressure is applied. The lower heating rates of conventional friction welding require more energy because much of the heat is conducted away from the weld interface. By rapid applications of small amounts of energy, inertia welding produces narrower heat-affected zones than those produced in conventional friction welding.

In inertia welding, intensely hot working of the weld zone followed immediately by rapid cooling results in a very small grain size in the as-welded condition. Subsequent heat treatment will restore the grains to their normal size.

Flywheel friction welding is done with a machine in which mechanical energy is stored in, and released by, a flywheel in amounts predetermined and gaged by flywheel speed. The amount of energy released by the flywheel is determined by its speed when axial pressure is first applied, and by the speed at which the clutch disengages the spindle from the motor.

Flywheel friction welding incorporates features of both the conventional and the inertia processes. Flywheels are connected to the

Table **8-1** Conditions for inertia welding of 1-inch-diameter bars in combinations of similar and dissimilar metals.

	WELDING CONDITIONS			RESULTANT WELD CONDITIONS		
WORK METAL	*SPINDLE SPEED, rpm*	*AXIAL FORCE, lb*	*FLYWHEEL SIZE, lb-ft²*[a]	*WELD ENERGY, ft-lb*	*METAL LOST, in*[b]	*TOTAL TIME, sec*[c]
METALS WELDED TO THEMSELVES						
1018 steel	4600	12,000	6.7	24,000	0.10	2.0
1045 steel	4600	14,000	7.8	28,000	0.10	2.0
4140 steel	4600	15,000	8.3	30,000	0.10	2.0
Inconel 718	1500	50,000	130.0	50,000	0.15	3.0
Maraging steel	3000	20,000	20.0	30,000	0.10	2.5
Type 410 stainless steel	3000	18,000	20.0	30,000	0.10	2.5
Type 302 stainless steel	3500	18,000	14.0	30,000	0.10	2.5
Copper, commercially pure	8000	5,000	1.0	10,000	0.15	0.5
Copper alloy 260 (cartridge brass, 70%)	7000	5,000	1.2	10,000	0.15	0.7
Titanium alloy Ti-6Al-4V	6000	8,000	1.7	16,000	0.10	2.0
Aluminum alloy 1100	5700	6,000	2.7	15,000	1.15	1.0
Aluminum alloy 6061	5700	7,000	3.0	17,000	0.15	1.0
DISSIMILAR-METAL COMBINATIONS						
Copper to 1018 steel	8000	5,000	1.4	15,000	0.15	1.0
M2 tool steel to 1045 steel	3000	40,000	27.0	40,000	0.10	3.0
Nickel alloy 718 to 1045 steel	1500	40,000	130.0	50,000	0.15	2.5
Type 302 stainless to 1020 steel	3000	18,000	20.0	30,000	0.10	2.5
Sintered high-carbon steel to 1018	4600	12,000	8.3	30,000	0.10	2.5
Aluminum 6061 to type 302 stainless	5500	5000 & 15,000[d]	3.9	20,000	0.20	3.0
Copper to aluminum alloy 1100	2000	7,500	11.0	7,500	0.20	1.0

[a]Moment of inertia of the flywheel [b]Total axial shortening of workpieces during welding. [c]Includes heat time and weld time. [d]The 5000-lb force is applied during the heating stage of the weld; force is increased to 15,000 lb near the end of the weld.

Source: By permission, from *Metals Handbook* Volume 6, Copyright American Society for Metals, 1971.

drive motor and to the spindle, and are coupled through an integral clutch. The drive motor–flywheel system rotates continuously and is coupled to the flywheel-spindle system to bring the rotating workpiece to the proper speed. The motor flywheel is disengaged from the spindle flywheel after the desired energy has been extracted. The spindle flywheel, having a low moment of inertia, comes to rest quickly, without braking, to complete the weld.

In operation, one workpiece is clamped in a collet chuck that is mounted to the spindle. Then the clutch is engaged, which causes the workpiece to be rotated at a predetermined speed. The mating workpiece is clamped in the tailstock and then brought into contact with the rotating workpiece. Pressure is then applied to heat the workpieces. At a predetermined time or spindle speed, the clutch is disengaged and a welding pressure is applied that stops rotation and completes the weld. After the flywheel is disengaged, the heating pressure can be continued or the welding pressure can be applied immediately. Thus, both kinetic and direct mechanical energy can be used to heat the workpieces to welding temperature, although the machine is designed so that all the heating is derived from the kinetic energy of the freely rotating flywheel, and none from direct mechanical energy.

Almost any combination (other than babbits and cast irons) of ferrous and nonferrous and sintered metals can be joined by friction welding. The weld upset usually reaches a high hardness resulting from rapid quenching, so the upset is sometimes removed by hot shearing (instead of grinding or machining) before the weld has cooled.

8-3 DIFFUSION WELDING

In the diffusion welding process, the weld is made by the application of pressure and elevated temperature after the surfaces to be joined have been properly prepared. Once true metal-to-metal contact is established, the atoms are within the attractive force fields of each other and produce a high-strength joint (see Table 8-2). The joint at this time resembles a grain boundary (Figure 8-4), because the metal crystal lattices at each side of the joint have different orientations. However, when observed under a metallurgical microscope, the boundary may differ from a grain boundary in that it may contain more impurities, inclusions, and voids, which remain at a weld interface if full asperity deformation has not occurred.

Interlayers (Figure 8-5) in common use, ranging in thickness from 0.001 to more than 0.010 inch, can be applied by electroplating, evaporation, or as powdered fillers. They serve one or more of the following purposes:

Table **8-2** Shear strength of single lap-joint specimens welded with Ni-Be intermediate alloys

BASE METAL	INTERMEDIATE ALLOY	WELDING TEMPERATURE °F	WELDING TIME MINUTES	SHEAR STRENGTH, psi	
				ROOM TEMPERATURE	*1500°F*
AISI 410	A	2190	2	Not welded	
AISI 410	B	2110	5	31,500	
AISI 410	C	2110	5	46,550	
AISI 347	A	2175	1	Not welded	
AISI 347	B	2110	5	33,300	
AISI 347	C	2110	5	39,600	
AISI 410	D	2075	4		7400
AISI 410	D	2010	10		8800
AISI 410	E	1960	10		8500
AISI 410	E	2100	½		7750
AISI 347	D	2080	10		15,000
AISI 347	D	2100	10		20,000
AISI 347	E	2080	10		18,800
AISI 347	E	2100	10		15,300
Inconel X	B	2110	5	35,850	
Inconel X	C	2110	5	72,900	
Haynes 25	B	2110	5	43,350	
Haynes 25	C	2100	5	31,250	
Inconel X	D	2100	10		50,950
Inconel X	D	2075	10		32,750
Inconel X	D	2010	10		44,400
Inconel X	E	2030	½		37,250
Inconel X	E	1995	10		30,000
Inconel X	E	2010	10		36,500
Haynes 25	D	1940	10		30,250
Haynes 25	D	2055	7		25,850
Haynes 25	E	2100	10		31,000

Intermediate alloy composition
A Ni-0.28% Be
B Ni-1.51% Be
C Ni-3.02% Be
D Ni-20% Cr-0.3% Mn-3.0% Be
E Ni-5.8% Be-13.6% Cr-0.1% C-0.3% Mn-3.9% Be

Source: American Welding Society, *Welding Handbook*, 6th ed., section 3B, p. 52.21.

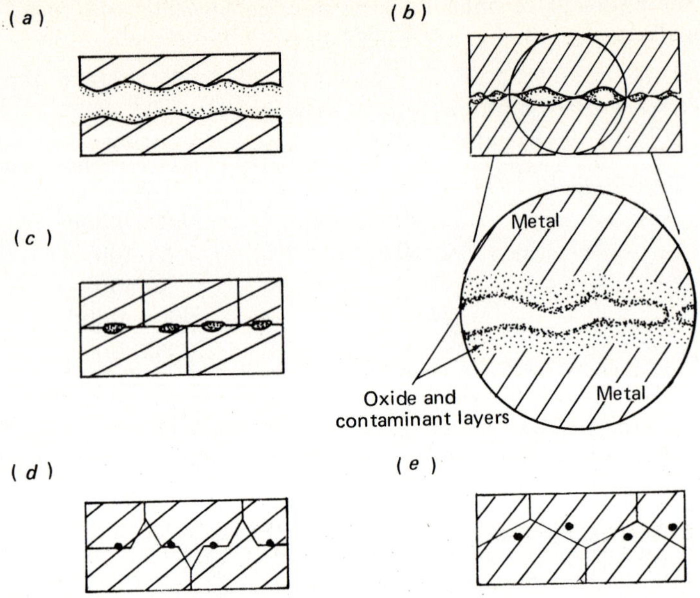

(a)
(b)
(c)
Metal
Oxide and contaminant layers
Metal
(d)
(e)

Figure **8-4** Sequential diagram of a two-step diffusion-welding process. (Redrawn by permission from American Welding Society, *Welding Handbook,* 6th ed., 1971, section 3B, pp. 52.4, 52.5.)

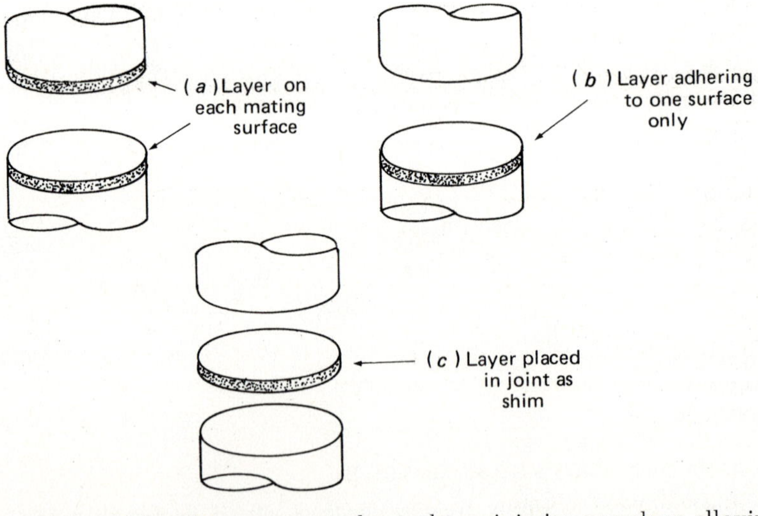

(a) Layer on each mating surface
(b) Layer adhering to one surface only
(c) Layer placed in joint as shim

Figure **8-5** Interlayers commonly used to minimize or reduce alloying compatibility problems between metals that are difficult to alloy. (Redrawn by permission from American Welding Society, *Welding Handbook,* 6th ed., 1971, section 3B, p. 52.13.)

- They enable the diffusion welding of normally nonweldable metal combinations by using a lower-strength intermediate metal.
- They allow the modification of surface conditions by the use of electroplate or intermediate foil to minimize oxide-film problems.
- They minimize or solve alloying compatibility problems (Table 8-2) when joining dissimilar metals.
- They minimize distortion with a soft interlayer by confining deformation to the low-strength intermediate metal.

Because the most important diffusion-welding parameters—time, temperature, and pressure—are similar to resistance-welding parameters, standard resistance-welding equipment, with the addition of an inert-gas atmospheric control device at the point of welding, is used for this process. The biggest advantage in using this type of equipment for diffusion welding is the speed at which joints can be made. Cycle times are measured in seconds rather than hours (as with the other diffusion-welding approaches), but the preparation of large weld areas becomes time consuming and requires numerous overlaps to achieve welding over an area larger than the face of the electrode.

Other types of standarized diffusion-welding equipment include high-pressure isostatic equipment (Figure 8-6) and mechanical or hydraulic presses of some sort.

The high-pressure isostatic arrangement consists basically of a hot pressing operation (pressures up to 150,000 psi with simultaneous temperature in excess of 3000°F) performed in a high-pressure autoclave. The inert-gas working fluid provides true isostatic pressure to any part within the pressure chamber. If a standard mechanical or hydraulic press is to be used, it must be of sufficient load and size capacity and have an adequate heating chamber and pressure-time control.

Although a number of materials in similar and dissimilar combinations have been joined, most applications of this process have been with titanium alloys, zirconium alloys, and nickel-base alloys. Plain-carbon steels can also be welded by this process, but the application of this process to carbon steel is limited, since carbon steel is more easily welded by other processes. To date, diffusion welding has found most of its applications in the atomic energy and aerospace industires.

8-4 EXPLOSION WELDING

In this process, the metals to be joined are metallurgically bonded (welded) by a high-velocity movement (jetting) produced by the controlled detonation of an explosive (Figure 8-7). The explosive,

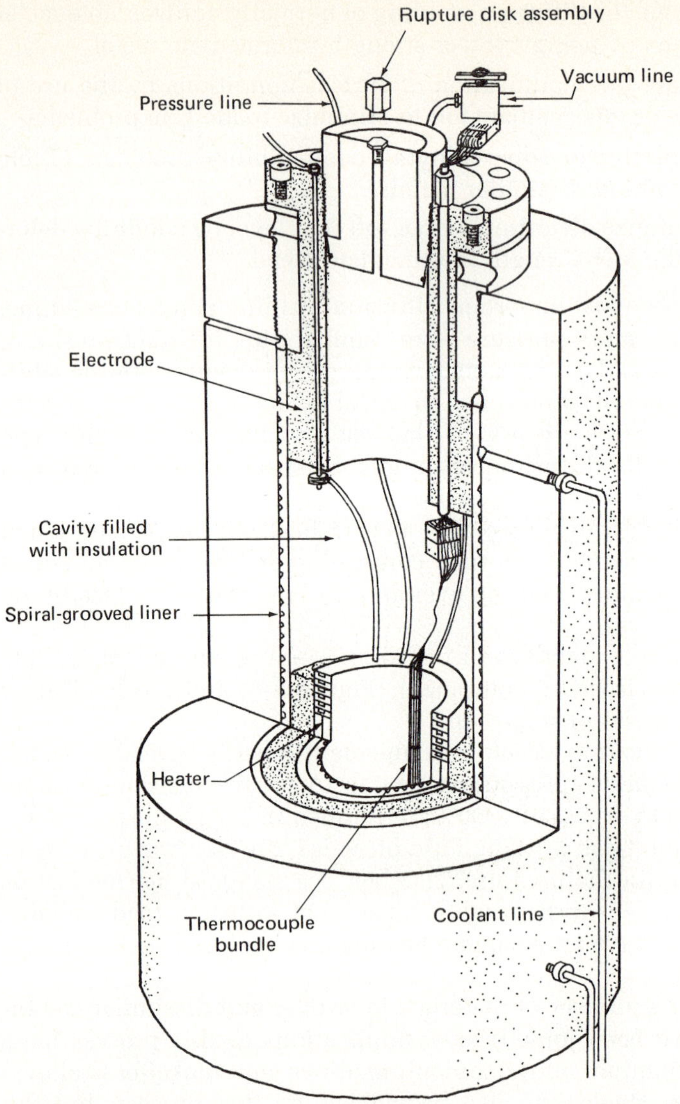
Rupture disk assembly
Vacuum line
Pressure line
Electrode
Cavity filled
with insulation
Spiral-grooved liner
Heater
Thermocouple
bundle
Coolant line

Figure **8-6** Cross-section of a high-pressure isostatic diffusion welder. (Redrawn by permission from American Welding Society, *Welding Handbook*, 6th ed., 1971, section 3B, p. 52.25.)

which may be in plastic, liquid, or granular form, is evenly placed over one of the metals to be joined, and the other piece of metal is placed on an anvil which, depending upon the thickness of the metal to be welded, could consist of a bed of sand for thick metals, or steel or reinforced concrete for thinner metals. If the *prime* (the metal to which the explosive charge is attached) is so thin that it

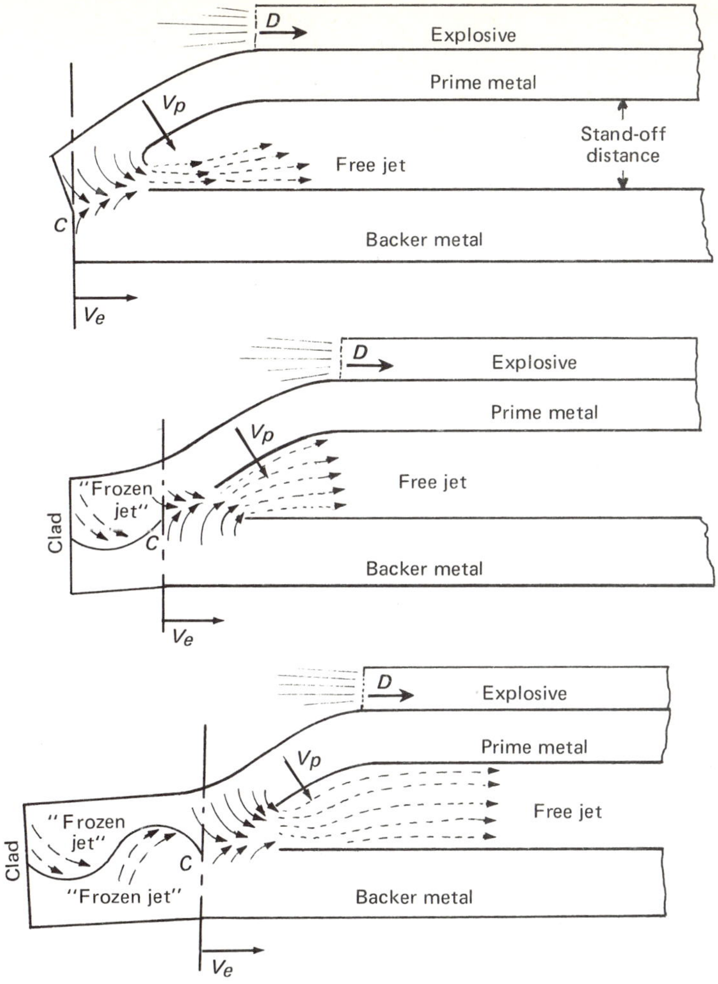
D
Explosive
Prime metal
Vp
Stand-off
distance
Free jet
C
Backer metal
Ve
D
Explosive
Prime metal
Vp
Free jet
"Frozen
jet"
Clad
C
Backer metal
Ve
D
Explosive
Prime metal
Vp
Free jet
"Frozen
jet"
Clad
C
"Frozen jet"
Backer metal
Ve

Figure **8-7** The explosion-welding process. (Redrawn by permission from American Welding Society, *Welding Handbook*, 6th ed., 1971, section 3B, p. 51.9.)

tends to sag, internal supports in the form of thin, rectangular metal ribbons are positioned between the two metals.

The important variables in this welding process are the minimum collision velocity and the minimum collision angle. The three types of metallurgical bonds resulting from variations in collision velocity and collision angle are shown in Figure 8-8.

Although the process is routinely used to produce clads of aluminum to steel, titanium to steel, or other combinations of metals forming brittle intermetallics measuring more than 300 square feet,

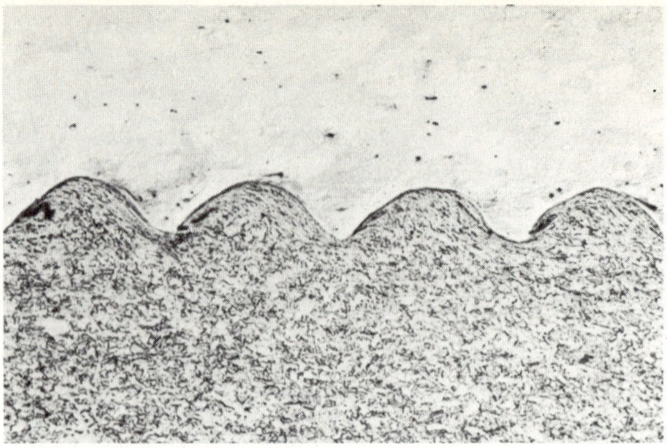

(a)

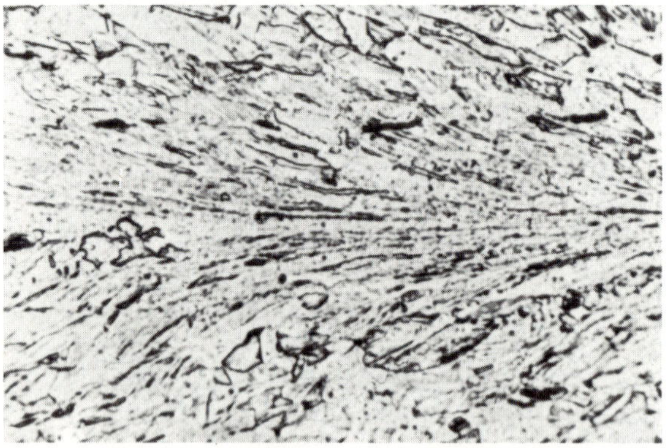

(b)

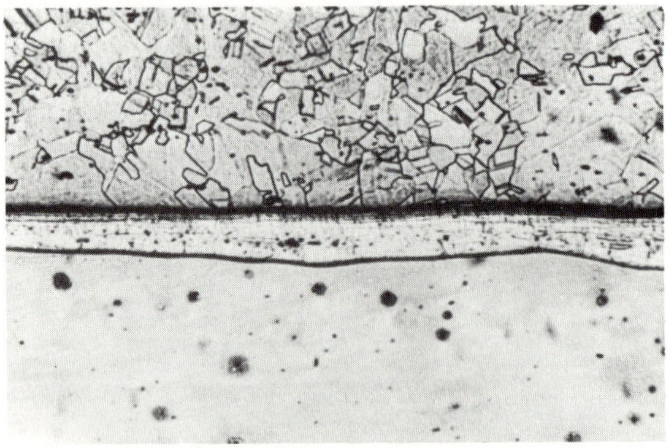

(c)

Figure **8-8** Three types of metallurgical bonds resulting from variations in collision velocity and collision angle: (*a*) Cu-Ni bond with a wavy interface (×50); (*b*) Direct metal-to-metal copper bond; (*c*) Cu-Ni bond showing continuous layer zone. (By permission from American Welding Society, *Welding Handbook*, 6th ed., 1971, section 3B, p. 51.7.)

with prime plates up to 1.3 inches thick and no upper limit for the backer plate, its major application (by tonnage) has been the cladding of cupro-nickel billets for the production of U.S. coinage.

As a rule of thumb, metals with 5% or greater elongation in a two-inch-gage length and a charpy V-notch impact resistance of 10 foot-pounds or greater (such as those used in the fabrication of chemical process vessels, anode rods used in primary aluminum reduction, superstructures for ocean-going vessels, or leak-free tubular joints for cryogenic process piping), can be welded by this process.

Warning: Explosives must be handled only by personnel well equipped and well trained in their safe use.

8-5 ULTRASONIC WELDING

This process welds metal by the local application of high-frequency (between 10,000 and 175,000 Hz) vibratory energy, while the parts are held together under pressure. The pressures vary with the size of the welding machine used (Figure 8-9). The clamping force (Table 8-3) depends on the wattage required to weld the assembly. The wattage can be obtained by reference to Figure 8-10, or can be calculated by use of the following equation:

$$E = 150(1.5H)(1.5t)$$

where E equals the energy in watt-seconds; H is the Vickers microhardness number; and t equals the thickness of sheet in contact with powered sonotrode, in inches. The metals that are commonly welded by this process are tabulated in Table 8-4.

Although it is possible to weld aluminum having a thickness of up to 0.10 inch, the thickness of most materials welded has an upper limit in the range of 0.015 to 0.040 inch. There appears to be no lower limit to weldable thickness, since fine wires of less than 0.0005-inch diameter and thin foils of 0.00017-inch thickness have been welded satisfactority.

Whenever possible, the thinnest or the softest member should be placed in contact with the sonotrode. Inert atmospheres or special surface preparations are not required for making welds by this process.

Representative applications of ultrasonic welding equipment are to be found in:

- The *electronics industry,* where it is used to attach fine (0.005- to 0.020-inch-diameter) aluminum or gold lead wires to transistors, diodes, and other semiconductor devices, or to ceramic or glass substrates. It has also been used successfully to hermetically encapsulate microcircuits and other electronic components, such as transistor or diode cans.

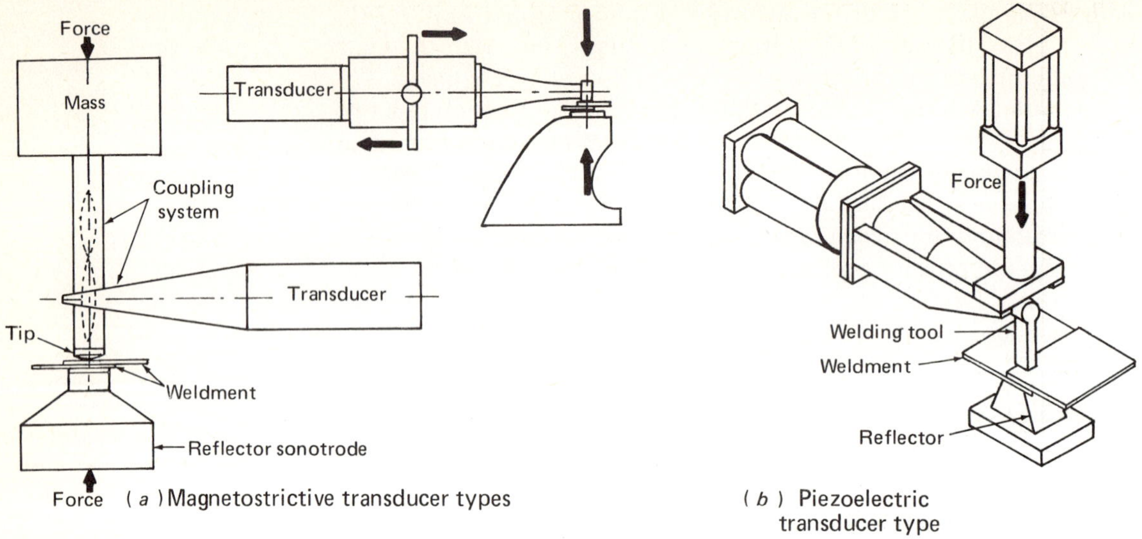
Force
Mass
Coupling system
Transducer
Transducer
Tip
Weldment
Reflector sonotrode
Force
(a) Magnetostrictive transducer types
Force
Welding tool
Weldment
Reflector
(b) Piezoelectric transducer type

Figure **8-9** Typical force systems used for ultrasonic welding. (Redrawn by permission from *Welding Data Book, 1964–1965*, Welding Data Book/ Welding Design and Fabrication Magazine, p. E/102.)

Table **8-3** Clamping force ranges for ultrasonic spot-type welding machines of various power capacities (smooth-surface tips)

WELDING MACHINE: ELECTRICAL POWER TO MAGNETOSTRICTIVE TRANSDUCER (WATTS)	APPROXIMATE CLAMPING FORCE RANGE
20	4–175 grams
100	0.5–15 pounds
300-lateral drive	50–100 pounds
300-wedge-reed	50–300 pounds
600	50–550 pounds
2000	50–1000 pounds
4000	300–1200 pounds

Source: American Welding Society, *Welding Handbook*, 6th ed., section 3B, 59.15.

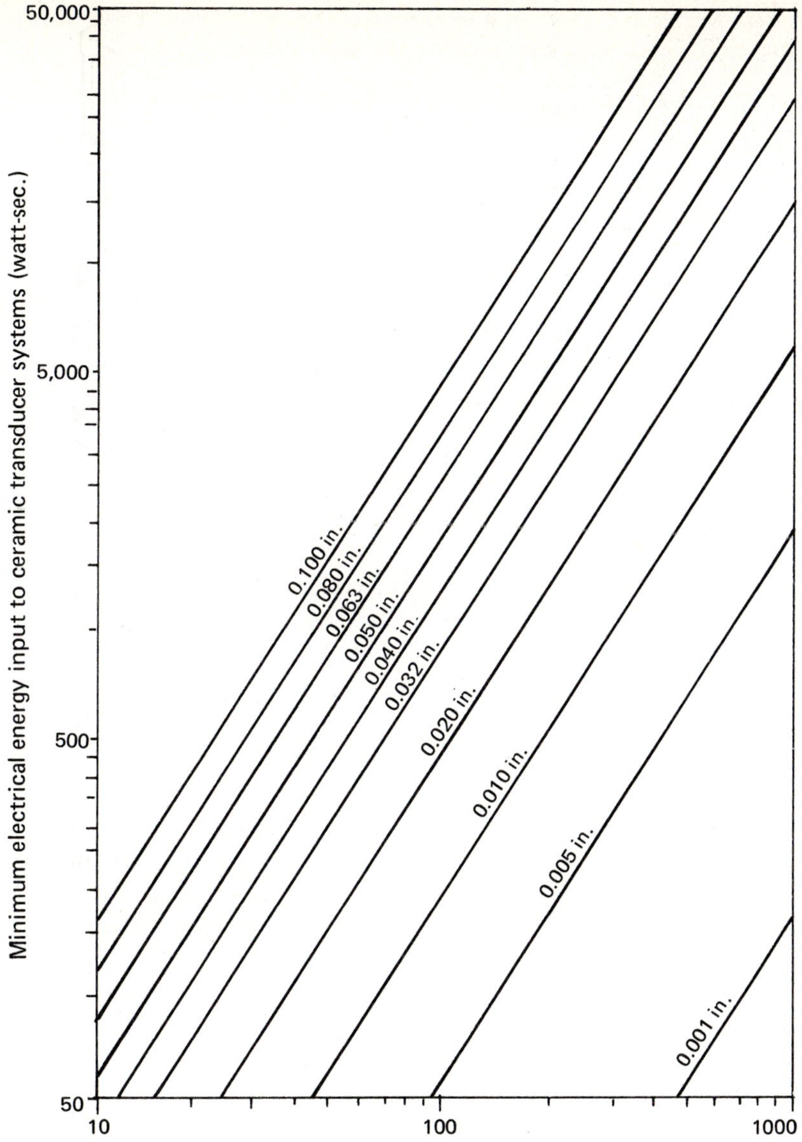
Minimum electrical energy input to ceramic transducer systems (watt-sec.)
50,000
5,000
500
50
10
100
1000
0.100 in.
0.080 in.
0.063 in.
0.050 in.
0.040 in.
0.032 in.
0.020 in.
0.010 in.
0.005 in.
0.001 in.

Figure **8-10** Minimum electrical energy required for ultrasonic spot welding as a function of material hardness and thickness. (Redrawn by permission from American Welding Society, *Welding Handbook*, 6th ed., 1971, section 3B, p. 59.7.)

Table **8-4** Metals and alloys that have been successfully joined by ultrasonic welding or in which welding feasibility has been demonstrated

	Al	Be	Cu	Ge	Au	Fe	Mg	Mo	Ni	Pd	Pt	Si	Ag	Ta	Sn	Ti	W	Zr
Al & alloys	•	•	•	•	•	•	•	•	•	•	•	•	•	•	•	•	•	•
Be & alloys		•														•		
Cu & brass			•		•	•		•	•		•		•			•		•
Germanium					•						•							
Gold					•	•			•		•	•						
Iron & steel						•		•	•		•		•			•	•	•
Mg & alloys							•						•					
Mo & alloys								•	•					•		•	•	•
Ni & alloys									•		•					•		
Pd & alloys										•			•					
Pt & alloys												•						
Silicon														•				
Ag & alloys													•	•				•
Ta & alloys														•				
Tin															•			
Ti & alloys																•		
W & alloys																	•	
Zr & alloys																		•

Source: American Welding Society, *Welding Handbook*, 6th ed., section 3B, p. 59.11.

• The *sterile-packaging industry,* where it permits packaging of hospital supplies, precision instrument parts, ball bearings, primary explosives, slow-burning propellants, pyrotechnics, living-tissue cultures, and chemicals (such as phosphorus, lithium, aluminum hydride, and nitrogen perchlorate) that would react in air.

• The *aerospace industry,* where it is used for the fabrication of leak-tight joints in a bellows consisting of 6061-T6 aluminum welded to AISI 321 stainless steel, both materials 0.031 inch thick. It is also used for the fabrication of insulated aluminum heater ducts up to 40 inches in diameter and consisting of 0.003-inch-thick aluminum foil wrapped around a 1-inch-thick layer of insulation.

• The *metal-fabrication industry,* where 0.001-inch-thick beryllium foil windows have been assembled to AISI 310 stainless steel frames to provide a helium leak-tight board with a clearview area of ¾ inch, and for helium leak-tight bonds in pinch-off weld closures used in capillary tubes for refrigeration, air conditioning, and similar applications.

REVIEW QUESTIONS

1. How is fusion produced in solid state welding?
2. Name and describe three friction-welding processes.
3. Differentiate between the terms *upset* and *flash*.
4. What kind of metal cannot be joined by friction welding?
5. Why are interlayers used for diffusion welding of certain metals?
6. Why is the diffusion-welding process not commonly used to weld carbon steels?
7. Describe the principles of the explosion-welding process.
8. What is the rule of thumb commonly used to determine which metals can be joined by explosion welding?
9. List five metals commonly welded by the ultrasonic welding process.
10. For what purposes are the following welds used?
    a. Trap weld.
    b. Wave weld.
    c. Stagger weld.
    d. Sandwich weld.
11. Differentiate between the conventional and the inertia friction-welding methods.
12. In what industry is diffusion welding most widely used?
13. How does the ultrasonic process weld metals?
14. Calculate the number of watt-seconds required to weld a 0.020-inch-thick aluminum plate to a 0.010-inch-thick aluminum plate.

CHAPTER 9

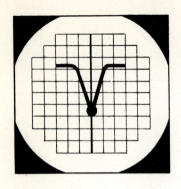

OTHER WELDING PROCESSES

The welding processes to be discussed in this chapter are electron-beam welding, electroslag welding, electrogas welding, laser-beam welding, thermit welding, the welding of plastics, and industrially obsolete welding processes.

9-1 ELECTRON-BEAM WELDING

The electron beam of the electron-beam welding process has been compared to a hot wire passing through butter. As the wire moves along, the butter melted by it flows past and solidifies behind it. During welding (Figure 9-1), the beam first punches a hole in the metal, then the metal flows around it in much the same way. The three classes of electron-beam welding equipment in common use are high vacuum, medium vacuum, and nonvacuum.

The high-vacuum system is the most widely used electron-beam welding process. In this system, the electron beam is produced in a high-vacuum environment by an electron gun, which usually consists of a tungsten or tantalum cathode. Electrons are emitted from the cathode, which is heated to about 4600°F or higher. The electrons are gathered, accelerated to a high velocity, and shaped into a beam by electrical fields between the cathode, grid, and anode. The beam is collimated and focused by passing through the field of an electromagnetic focusing coil, or magnetic lens. Electron beams can be deflected from their normal path by magnetic deflection coils,

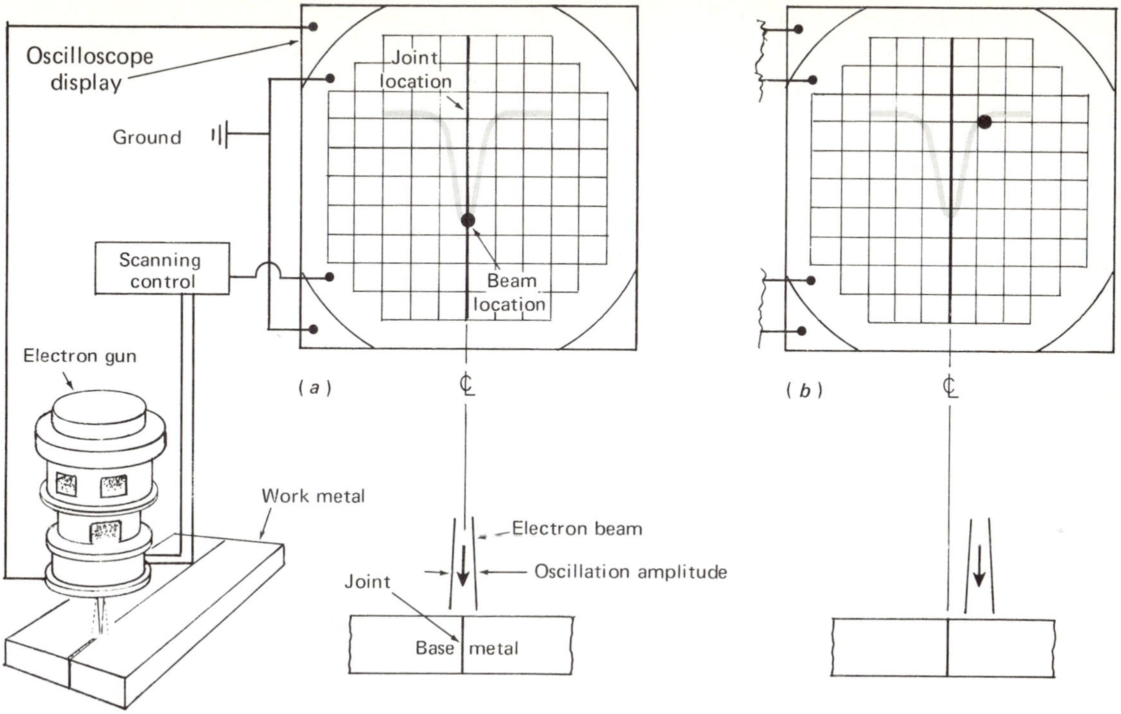
Oscilloscope
display
Ground
Joint
location
Beam
location
Scanning
control
Electron gun
(a)
(b)
Work metal
Electron beam
Oscillation amplitude
Joint
Base
metal

Figure **9-1** Electronic scanning of a joint, showing equipment, joint to be welded, and oscilloscope display: (*a*) beam centered on the joint; (*b*) beam displaced to the right. (By permission, from *Metals Handbook* Volume 6, Copyright American Society for Metals, 1971.)

usually located below the focusing coil. Beams are focused to about 0.010 to 0.030 inches in diameter and have a power density of about 10^6 watts per square inch, which is sufficient to vaporize any metal.

The medium-vacuum system operates with a vacuum of from 2×10^{-1} to 10^{-3} torr in the work chamber, while the gun chamber in which the beam is generated, collimated, and focused is held at from 10^{-4} to 10^{-5} torr, as in high-vacuum welding.

The equipment (Figure 9-2) used for nonvacuum electron-beam welding consists of an optical column and orifice system. The electron gun and beam focusing system are similar to those used for the other types of electron-beam welding.

With electron-beam welding, as with other welding processes, it is important that the relative positions of the heat source and the weld joint be established accurately prior to initiation of the welding cycle. This requirement is more complicated with electron-beam welding because, in general, the weldments produced are relatively

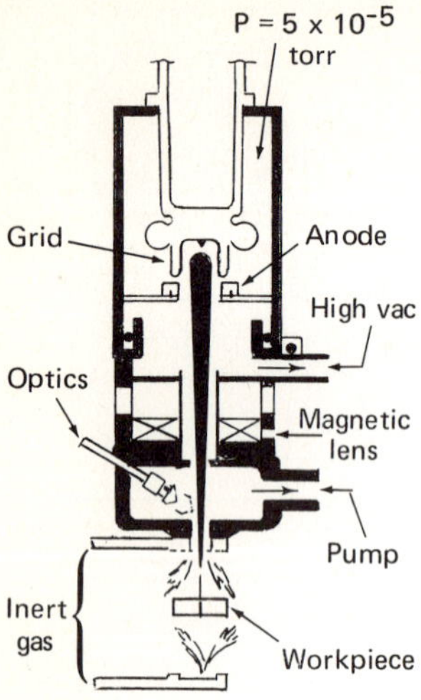
P = 5 x 10^{-5}
torr
Grid
Anode
High vac
Optics
Magnetic
lens
Pump
Inert
gas
Workpiece

Figure **9-2** Simplified cross-section of a nonvacuum electron-beam optical column and orifice system. (Redrawn by permission from American Welding Society, *Welding Handbook,* 6th ed., 1970, section 3A, p. 47.25.)

narrow (Figure 9-3) and the workpiece is usually contained in a vacuum chamber, making visual alignment of the workpiece and heat source difficult. Two scanning methods [Figures 9-1(*a*) and 9-1(*b*)] have been developed to keep the beam on a straight seam.

Joint tracking methods include:

- *Manual joint tracking.* In this method, the operator must closely observe the joint at the point where the weld is being made to anticipate any change in direction of the beam.
- *Automatic joint tracking.* Two methods may be used to track a joint automatically. In one, the joint is tracked using a stylus or follower wheel as a probe. The probe can be made to ride in a groove or parallel to the joint, or on an edge that duplicates the contour of the joint. Probes can induce translation mechanically or electromechanically. The second method can be used only if the parts have been machined precisely to some given contour and are in position with extreme accuracy, involving continuous-path numerical control.

Commercially pure metals or alloys that can be welded by the electron-beam method are listed in Table 9-1. To enable the welding of two incompatible metals, a filler metal (Table 9-2) can be used, or each of the incompatible metals can be welded to a transition piece that is compatible with each. With the exception of alloy and tool steels, all metals listed in Table 9-1 can be welded without preweld or postweld heat treatment.

High-vacuum electron-beam welding, as mentioned above, is the most widely used electron-beam welding method. It has found wide application in the nuclear power, electronics, and microelectronics industries. The medium and nonvacuum methods are used widely by the aerospace and automotive industries. The last few years saw the introduction of electron-beam welders into job shops serving industry.

In addition to the safe practices outlined in Chapter 13, electron-beam welders should study *Recommended Safe Practices for Electron Beam Welding,* AWS A6.4.

9-2 ELECTROSLAG WELDING

There are two systems, illustrated in Figures 9-4 and 9-5, of electroslag welding: conventional electroslag and the consumable–guide tube system. In the conventional electroslag system, the consumable electrode is fed downward through a nonconsumable guide tube, which requires vertical travel of the welding head; while in the consumable–guide tube system, the filler metal is supplied by the electrode and its guiding member. The electrode wire is fed to

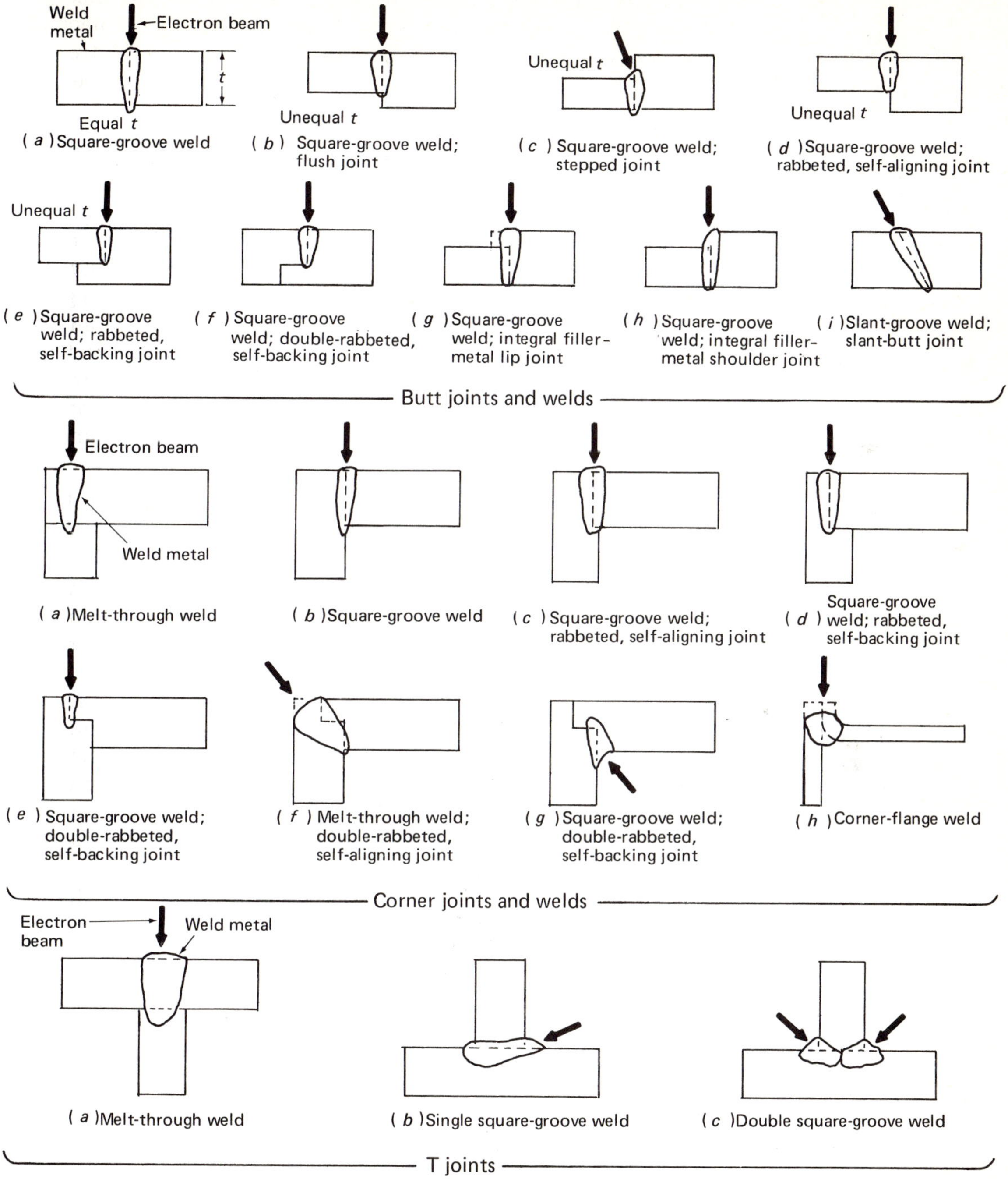
Weld metal
Electron beam
t
Equal t
(a) Square-groove weld
Unequal t
(b) Square-groove weld; flush joint
Unequal t
(c) Square-groove weld; stepped joint
Unequal t
(d) Square-groove weld; rabbeted, self-aligning joint
Unequal t
(e) Square-groove weld; rabbeted, self-backing joint
(f) Square-groove weld; double-rabbeted, self-backing joint
(g) Square-groove weld; integral filler-metal lip joint
(h) Square-groove weld; integral filler-metal shoulder joint
(i) Slant-groove weld; slant-butt joint
Butt joints and welds
Electron beam
Weld metal
(a) Melt-through weld
(b) Square-groove weld
(c) Square-groove weld; rabbeted, self-aligning joint
(d) Square-groove weld; rabbeted, self-backing joint
(e) Square-groove weld; double-rabbeted, self-backing joint
(f) Melt-through weld; double-rabbeted, self-aligning joint
(g) Square-groove weld; double-rabbeted, self-backing joint
(h) Corner-flange weld
Corner joints and welds
Electron beam
Weld metal
(a) Melt-through weld
(b) Single square-groove weld
(c) Double square-groove weld
T joints

Figure **9-3** Types of joints and welds used in electron-beam welding. (By permission, from *Metals Handbook* Volume 6, Copyright American Society for Metals, 1971.)

Table **9-1** Metals that can be welded with the electron beam process

	Silver	Aluminum	Gold	Beryllium	Cadmium	Cobalt	Chromium	Copper	Iron	Magnesium	Manganese	Molybdenum	Columbium	Nickel	Lead	Platinum	Rhenium	Tin	Tantalum	Titanium	Vanadium	Tungsten
Aluminum	2																					
Gold	1	5																				
Beryllium	5	2	5																			
Cadmium	2	5	5	4																		
Cobalt	3	5	2	5	3																	
Chromium	2	5	3	5	3	2																
Copper	2	2	1	5	5	2	2															
Iron	3	5	2	5	3	2	2	2														
Magnesium	5	2	5	5	1	5	5	5	3													
Manganese	2	5	5	5	3	2	2	1	2	5												
Molybdenum	3	5	2	5	4	5	1	3	2	3	3											
Columbium	4	5	4	5	4	5	5	2	5	4	5	1										
Nickel	2	5	1	5	3	1	2	1	2	5	2	5	5									
Lead	2	2	5	4	2	2	2	2	2	5	2	3	4	2								
Platinum	1	5	1	5	5	1	2	1	1	5	5	2	5	1	5							
Rhenium	3	4	4	5	4	1	1	3	5	4	4	5	5	3	4	2						
Tin	2	2	5	3	2	5	2	2	5	5	5	3	5	5	2	5	3					
Tantalum	5	5	4	5	4	5	5	3	5	4	5	1	1	5	4	5	5	5				
Titanium	2	5	5	5	5	5	1	5	5	3	5	1	1	5	5	5	5	5	1			
Vanadium	3	5	3	5	4	5	3	3	1	4	5	1	1	5	4	5	3	5	5	1		
Tungsten	3	5	4	5	4	5	1	3	5	3	3	1	1	5	3	1	5	3	1	2	1	
Zirconium	5	5	5	5	3	5	5	5	5	3	5	5	1	5	5	5	5	5	2	1	5	5

1 VERY DESIRABLE
(Solid Solubility in all Combinations)
2 PROBABLY ACCEPTABLE
(Complex Structures May Exist)
3 USE WITH CAUTION
(Insufficient Data for Proper Evaluation)
4 USE WITH EXTREME CAUTION
(No Data Available)
5 UNDESIRABLE COMBINATIONS
(Intermetallic Compounds Formed)

Source: American Welding Society, *Welding Handbook*, 6th ed., section 3A, p. 47.44.

Table **9-2** Filler metals or transition shims for incompatible metals to be welded by the electron beam process

MATERIAL #1	MATERIAL #2	FILLER SHIM
Tough Pitch Copper	Tough Pitch Copper	Nickel
Tough Pitch Copper	Mild Steel	Nickel
Hastelloy X	SAE 8620 Steel	1/32″ 321 Stainless Steel
304 Stainless Steel	Monel	1/32″ Hastelloy B
A-286 Stainless Steel	SAE 4140 Steel	1/32″ Hastelloy B
Inconel 713C	Inconel 713C	Udimet 500
Rimmed Steel	Rimmed Steel	Aluminum

Source: American Welding Society, *Welding Handbook*, 6th ed., section 3A, p. 47.42.

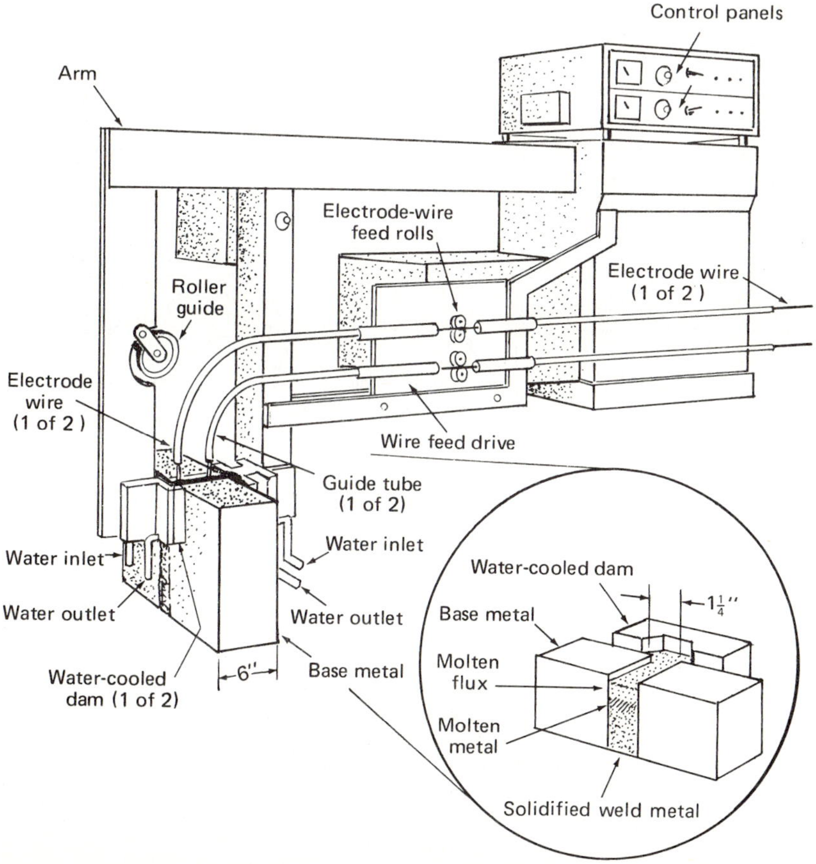
Control panels
Arm
Electrode-wire
feed rolls
Roller
guide
Electrode wire
(1 of 2)
Electrode
wire
(1 of 2)
Wire feed drive
Guide tube
(1 of 2)
Water inlet
Water inlet
Water outlet
Water outlet
Water-cooled
dam (1 of 2)
6″
Base metal
Water-cooled dam
1 1/4″
Base metal
Molten
flux
Molten
metal
Solidified weld metal

Figure **9-4** Major components and equipment set-up for conventional electroslag welding. (By permission, from *Metals Handbook* Volume 6, Copyright American Society for Metals, 1971.)

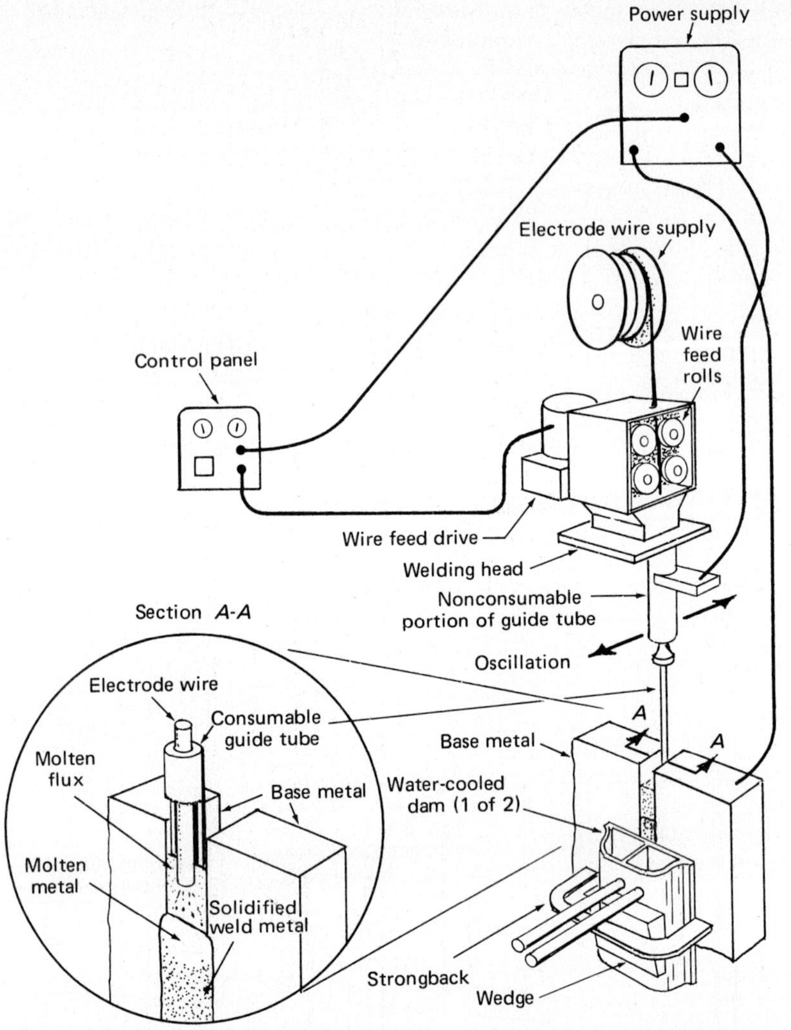
Power supply
Electrode wire supply
Wire
feed
rolls
Control panel
Wire feed drive
Welding head
Nonconsumable
portion of guide tube
Section A-A
Oscillation
Electrode wire
Consumable
guide tube
Molten
flux
Base metal
Molten
metal
Solidified
weld metal
A
A
Base metal
Water-cooled
dam (1 of 2)
Strongback
Wedge

Figure **9-5** Electroslag-welding equipment with the consumable–guide tube system. (By permission, from *Metals Handbook* Volume 6, Copyright American Society for Metals, 1971.)

the weld puddle through a tube that is also consumable. The use of a consumable guide eliminates the need for vertical travel of the welding head, thereby simplifying the equipment.

In both systems, the heat from the layer of molten flux (slag) melts a consumable electrode and the surface of the base metal to produce a weld puddle. Filler metal is fed into a molten pool contained in a pocket formed by copper shoes or dams (water cooled or not) that bridge the gap between the weldments. The process utilizes the electrical resistivity of the molten flux to continuously produce the heat needed to melt the filler metal and adjacent base metal. During welding, the molten flux maintains a protective cover over the joint. Although the axis of the weld joint is vertical, the process is actually flat-position welding with vertical travel.

Among the many base materials that have been electroslag welded in production, the largest tonnages have been in the following steels: ASTM A515 and A516; ASTM A302 and A533; ASTM A387 and A542; ASTM A543 and HY 580; and ASTM A36 and A441. In addition to the above materials, SAE 1020 and SAE 1045 wrought and cast stainless steels and such high-strength alloy steels as D-6ac have been successfully welded by these methods.

Both of the electroslag welding systems are most often used for welding plate from 1¼ inches thick upward. The thickest section that has been successfully welded to date has been a 36-inch-thick steel section, while the thinnest has been a ⅜-inch-thick steel plate, and the longest successfully welded joint measured 20 feet.

The quantity, composition, and form of the electrodes used varies with the thickness of the job and the composition of the base material. Although all three types of electrodes (solid, flux-cored, and braided) have been successfully used, the solid electrode is preferred.

The flux used must have several important characteristics. It must have an electrical resistivity that is sufficiently high to generate the heat required for welding. It must also have a viscosity that is fluid enough for good circulation to occur, for even distribution of heat, and to permit it to rise easily above the molten metal to prevent slag inclusions, but not so fluid as to leak out of small openings between the work and the cooling shoes. The composition of a typical flux that has performed satisfactorily is: 35% SiO_2, 40% MnO, 5% Al_2O_3, 7% CaO, 6% CaF_2, 3% FeO, 3% TiO_2, and 1% Na_2O.

Electroslag welding is used, in both shop and field applications, in machine building when it is not possible to buy the thickness of plate required in the width or length necessary. Also, since acceptance of the *ASME Boiler and Pressure Vessel Code* (Code Case 1355) in March 1965, the process has been used extensively for fabrication of heavy-wall pressure vessels for the chemical, petroleum, marine, and power-generating industries. All electroslag welds in pressure vessels fabricated to the requirements of the ASME Code must meet both the radiographic and ultrasonic inspection standards in Code Case 1355, after (in most cases) the recommended postweld heat treatment to achieve grain refinement in the weld zone and in the base metal heat-affected zone.

9-3 ELECTROGAS WELDING

In its mechanical aspects and its application to welding practice, electrogas welding resembles conventional electroslag welding, from which it was developed. Electrically, electrogas welding differs from electroslag welding in two ways: (1) the heat is produced by an electric arc and not by the electrical resistance of a slag, and (2) only direct current can be used, whereas either alternating or direct current can be used for electroslag welding.

The equipment used for electrogas welding (Figure 9-6) closely resembles that for conventional electroslag welding. Therefore, a change from one process to the other requires only a change from shielding gas to flux, or from flux to shielding gas (80% argon + 20% carbon dioxide). Thus selection between processes is based on cost and application requirements, not on capital expenditure.

For work from ¾ to 3 inches thick, the electrogas and conventional electroslag systems are closely competitive. However, for sections thicker than 3 inches, electroslag welding is usually more practical.

9-4 LASER-BEAM WELDING

Laser-beam welding is accomplished by focusing a xenon light beam through a ruby (aluminum oxide with a small concentration of chromium oxide in solution). During exposure, some of the chromium atoms are excited to a high energy level, causing the ruby to emit a red light. Some of this red light escapes from the end of the crystal in the form of an almost perfectly monochromatic and nondivergent beam of red light. This beam can be manipulated with simple optical systems to obtain localized heating, thus bringing about fusion at the point of contact of two workpieces, making a weld. The only other heat source that rivals the laser in heat output (W/cm^2) is the electron beam.

As in electron-beam welding, there is no need for mechanical contact of any kind with the workpiece and no requirement that the worked material be a conductor of electricity.

Welds involving different metals and joint designs have been made satisfactorily with the laser. These involve similar and dissimilar metals, including materials such as copper, nickel, tantalum, stainless steel, Dumet, Kovar, aluminum, tungsten, titanium, columbium, zirconium, and superalloys. The welds generally fall into the categories of wire-to-wire, sheet-to-sheet, wire-to-sheet, tube-to-sheet, and stud welds.

Different configurations can be used for wire-to-wire welds. With the butt configuration, two wires, preferably square ended, are placed end to end, and the laser beam is absorbed by the metal, creating a weld nugget. In the lap joint, the two pieces of wire are placed side by side and the beam is directed at the area where they

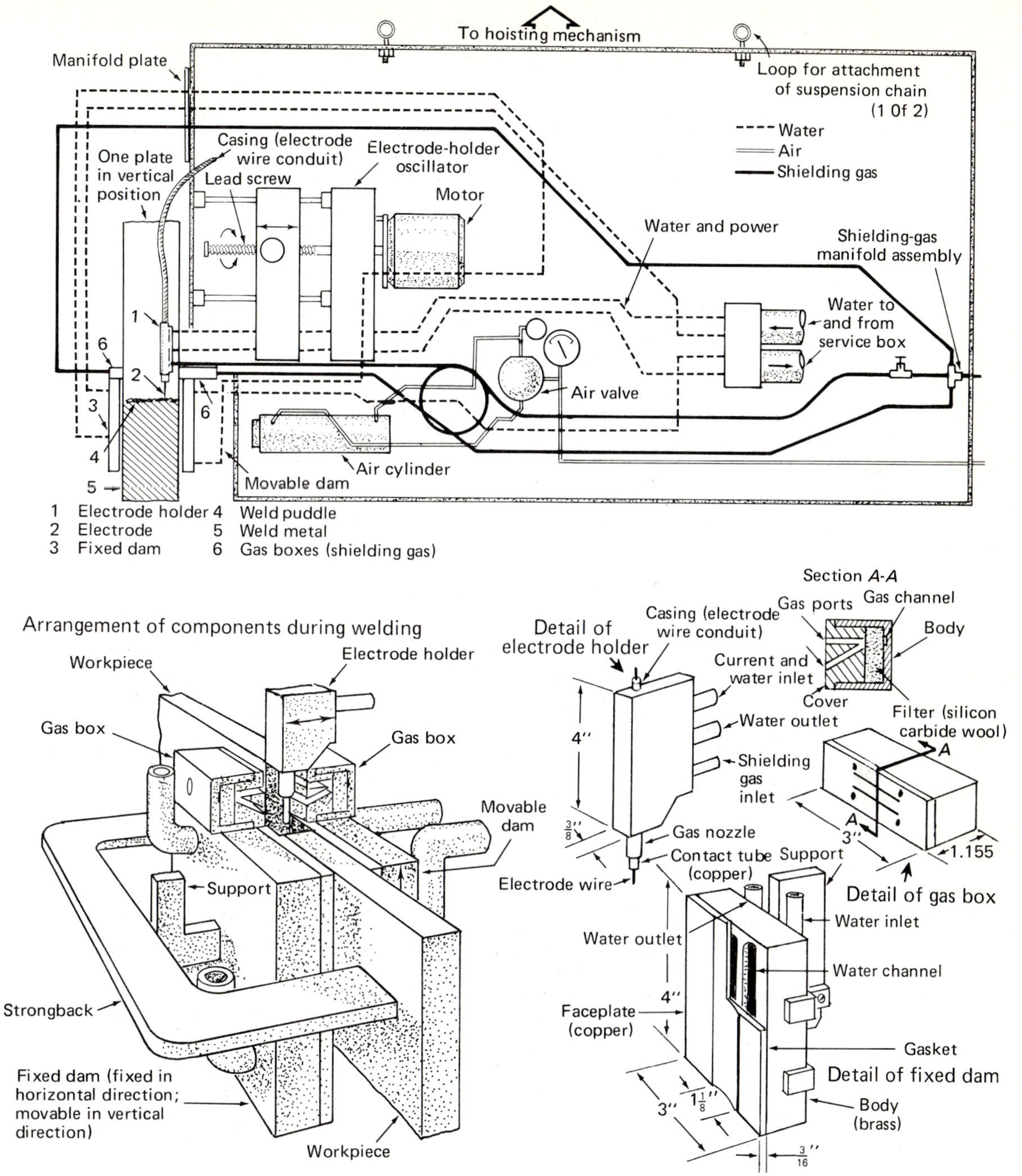

To hoisting mechanism
Manifold plate
Loop for attachment of suspension chain (1 Of 2)
Water
Air
Shielding gas
One plate in vertical position
Casing (electrode wire conduit)
Lead screw
Electrode-holder oscillator
Motor
Water and power
Shielding-gas manifold assembly
Water to and from service box
Air valve
Air cylinder
Movable dam
1 Electrode holder
2 Electrode
3 Fixed dam
4 Weld puddle
5 Weld metal
6 Gas boxes (shielding gas)
Arrangement of components during welding
Workpiece
Electrode holder
Gas box
Gas box
Movable dam
Support
Strongback
Fixed dam (fixed in horizontal direction; movable in vertical direction)
Workpiece
Detail of electrode holder
Casing (electrode wire conduit)
Current and water inlet
Water outlet
Shielding gas inlet
4″
3/8″
Gas nozzle
Contact tube (copper)
Electrode wire
Section A-A
Gas ports
Gas channel
Body
Cover
Filter (silicon carbide wool)
A
A
3″
1.155
Detail of gas box
Support
Water inlet
Water outlet
Water channel
4″
Faceplate (copper)
Gasket
Detail of fixed dam
Body (brass)
3″
1 1/8″
3/16″

Figure **9-6** Typical unit for electrogas welding (power supply, electrode wire-feed system, suspension chains, and hoisting mechanism not shown). (By permission, from *Metals Handbook* Volume 6, Copyright American Society for Metals, 1971.)

touch. The T joint is a variation in which one wire is at right angles to the other. Seam welds between sheets can be made with overlapping spot welds. A microsection of a weld between overlapping sheets is shown in Figure 9-7. Welds are shown with specific "long-pulse" welding conditions in Figure 9-8. A tube-to-fin seam weld in zircalloy is shown in Figure 9-9.

The preferred joint configuration for laser-welding wires is probably the lap joint, although sound welds can be produced in all cases. By comparison, the cross configuration is preferable for resistance welding, so that pressure can be easily applied to the wires, and the weld bead forms at the joint interface. With the lap joint, the laser energy is delivered to the precise spot where the weld nugget is needed without stringent requirements for lead length, alignment, and so on. Multiple welds can be made as desired for added strength. Only a simple fixture or clamping arrangement that holds the wires together is necessary. When the cross configuration is used, it is better to direct the laser beam at an angle to the joint, so that it strikes the interface between the wires. This eliminates the need to melt through the top wire in order to join it to the bottom one. The strength of wire weld joints is equivalent to that of the parent metal in the annealed condition.

Good seam welds can be made between thin sheets with a laser-welding machine. Generally, the required laser energy output for welding a given thickness of sheet is approximately 50% higher than that required for welding the same diameter wire. Thus, if 10 joules produce a fully penetrated weld on 0.015-inch-diameter wire, 15 joules will be required for 0.015-inch-thick sheet. At present, seam

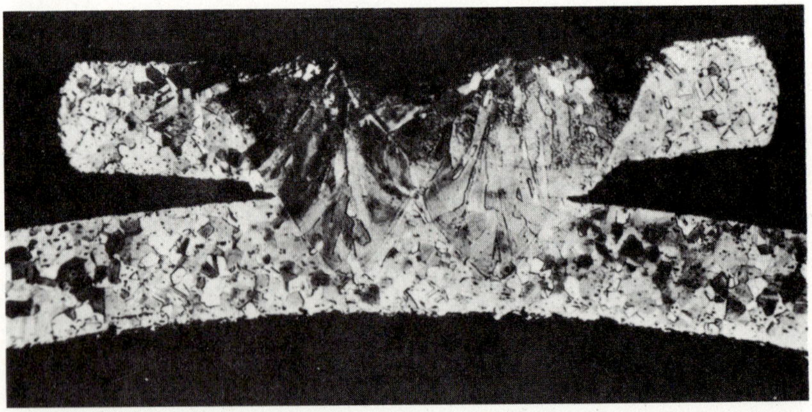

Figure **9-7** Microsection of a weld made in overlapping sheets with a laser beam. (By permission from American Welding Society, *Welding Handbook*, 6th ed., 1970, section 3A, p. 55.9.)

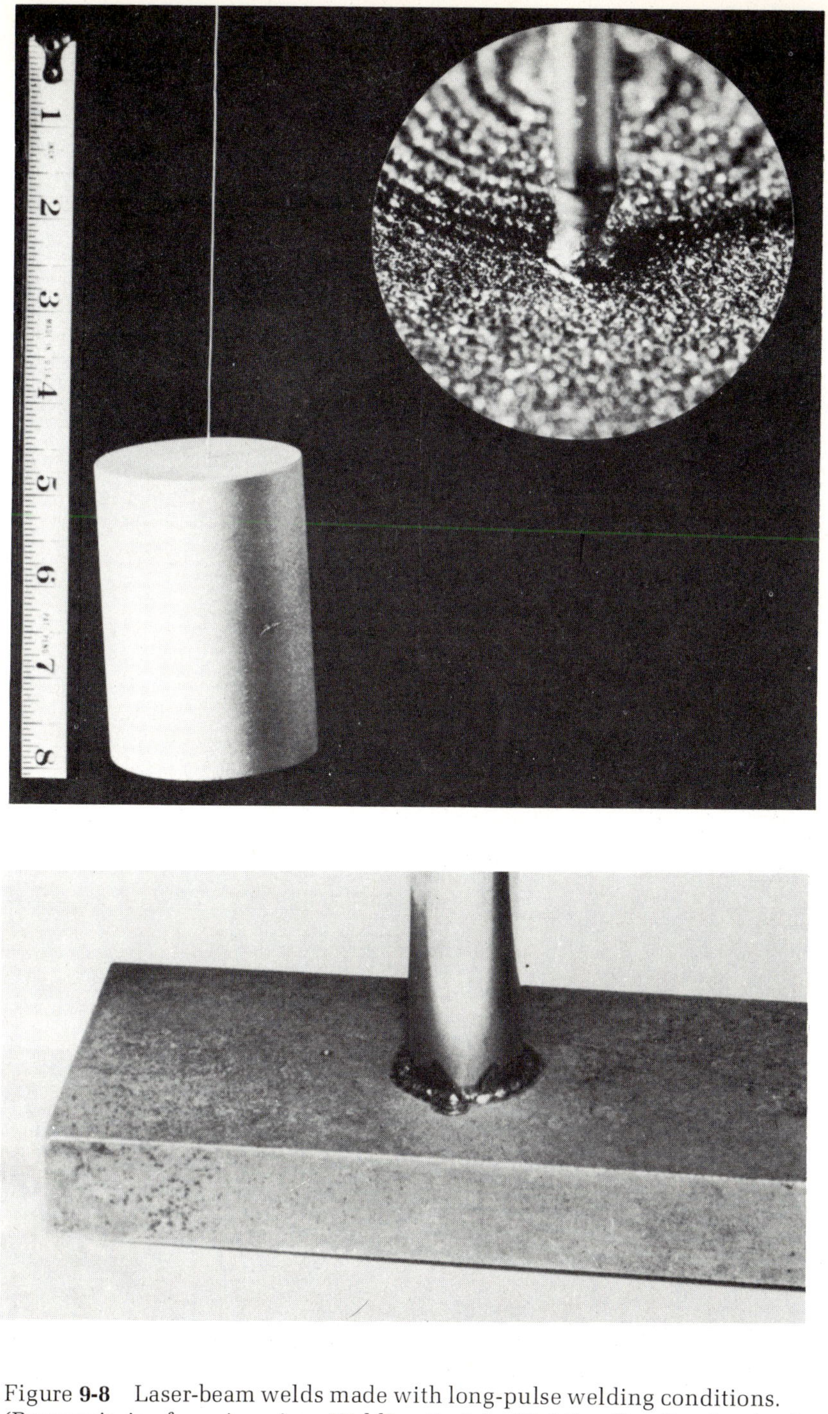

Figure **9-8** Laser-beam welds made with long-pulse welding conditions. (By permission from American Welding Society, *Welding Handbook*, 6th ed., 1970, section 3A, p. 55.11.)

Figure **9-9** Laser-beam tube-to-fin seam weld in zircalloy. (By permission from American Welding Society, *Welding Handbook,* 6th ed., 1970, section 3A, p. 55.10.)

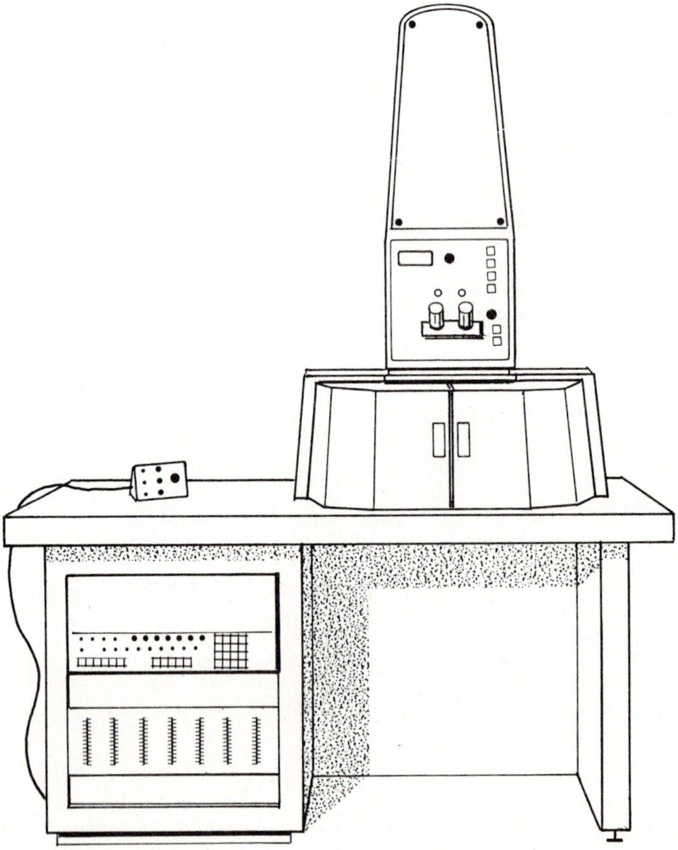

Figure **9-10** Front view of a laser-welding system

welds are limited to low welding speeds and, consequently, short lengths. With the repetition rate of current lasers, it would take about an hour to produce a weld a few feet long. However, it has been shown that good welds can be produced with the laser and, as advances are made in the laser power output and repetition rate, welding speeds should increase many times over. The front view of a laser welding system is shown in Figure 9-10.

9-5 THERMIT WELDING

Thermit welding is a process in which the heat for welding is obtained by surrounding the parts to be joined with superheated liquid metal and slag. The superheated metal and slag are produced by igniting the thermit (a mixture consisting mostly of aluminum and iron oxide), which is contained in a special crucible (Figure 9-11). During the combustion time (about 35 seconds), the alumi-

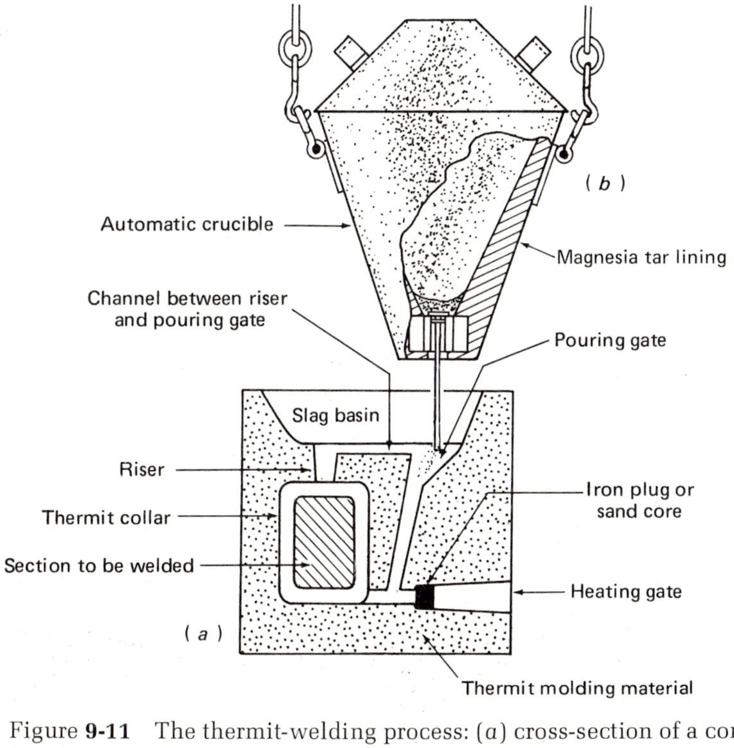

(b)
Automatic crucible
Magnesia tar lining
Channel between riser and pouring gate
Pouring gate
Slag basin
Riser
Iron plug or sand core
Thermit collar
Section to be welded
Heating gate
(a)
Thermit molding material

Figure **9-11** The thermit-welding process: (*a*) cross-section of a completed mold; (*b*) cross-section of the crucible. (Redrawn by permission from American Welding Society, *Welding Handbook*, 6th ed., 1971, section 3B, pp. 57.10, 57.14.)

num combines with the oxygen of the iron oxide and forms aluminum oxide (slag), leaving a pure form of superheated (4500°F) iron or thermit steel.

Thermit welding can be done, with or without the application of pressure, in the original fabrication and repair of large iron and steel parts such as railway and street-car rails (Figure 9-12), frogs and switches, crank shafts, rolling-mill rolls, stern posts and rudder posts of ships, and concrete reinforcing bars.

A modified thermit-welding process, called *cadwelding,* was developed by Erico Products, Inc., of Cleveland, Ohio. The method can be used to weld copper to copper or copper to steel, and is principally employed to make electrical connections and to weld signal bonds to rails. The thermit mixture used for this process consists of copper oxide and aluminum. The heat developed during the melting of the mixture is much less than would be required to braze or solder the same conductors, so the procedure may be performed in installations where the high-intensity magnetic fields associated with other welding methods cannot be tolerated.

9-6 WELDING OF PLASTICS

A convenient way to join plastics is to weld them. The methods most commonly used are: hot-gas welding, friction welding, heated-tool welding, and ultrasonic welding.

Hot-gas welding may be done manually or semiautomatically. The equipment required for the hot-gas welding of plastics is similar to that used for oxyacetylene welding (see Figure 9-13). Welding is accomplished by applying pressure on the welding rod, held in one hand, while supplying heat to the base material and the rod with the hot gas from the welding torch, held in the other hand. Both pressure and heat must be held constant, otherwise the material may melt, char, or become distorted.

The welding rod should be cut at angle of 60 degrees (to provide a thin wedge that is readily heated, facilitating the start of the weld) and held at an angle of 70 degrees to the base material.

In the semiautomatic welding of plastics, the welding rod is fed through a tube in the hot-gas torch, thereby preheating it and achieving a higher welding speed than is possible with manual welding, where both the base and filler materials must be preheated separately (Figure 9-14). The preparation and types of joints used in hot-gas welding are shown in Figure 9-15.

The AWS states that friction welding is a term that describes a joint resulting from frictional heat generated by rubbing two surfaces together. The heat resulting from the friction melts the plastic

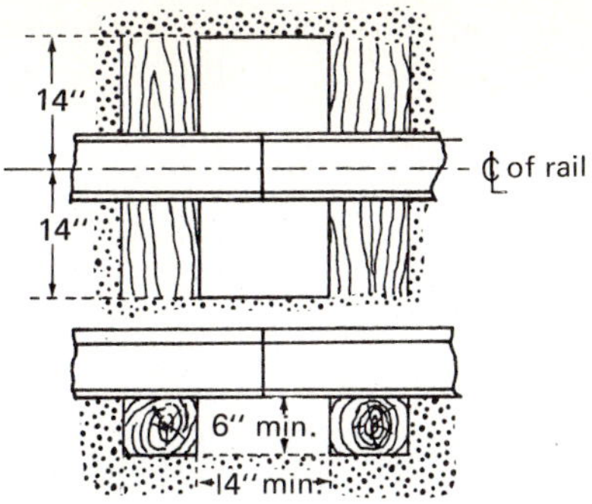
14"
℄ of rail
14"
6" min.
14" min.

1 Open up pavement as shown.

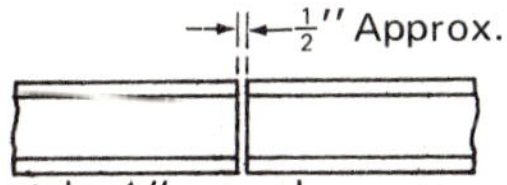
$\frac{1}{2}$" Approx.

2 Cut approximately $\frac{1}{2}$" gap where necessary between rail ends, cutting through head with a saw and through the web and base with a torch.

3 Clean rail ends thoroughly removing all dirt and rust at least 3" either side of joint.

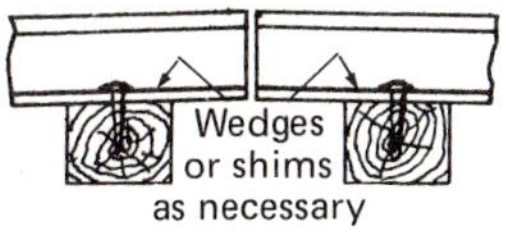
Wedges
or shims
as necessary

4 Tighten rail ends to ties adjacent to joint by shimming between rail and tie if necessary, at same time raising joint to required height to insure correct surface after welding and grinding.

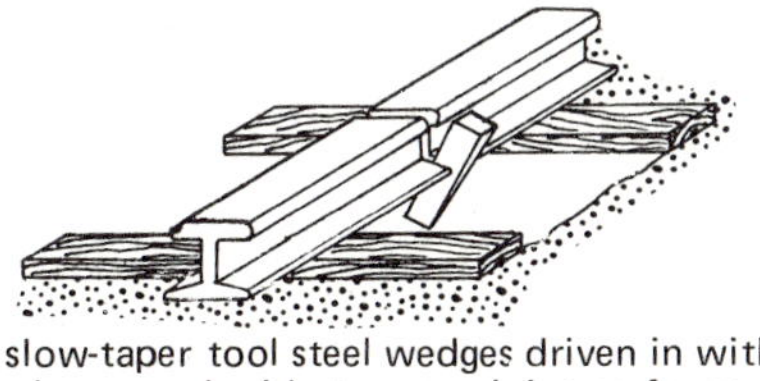

5 Special slow-taper tool steel wedges driven in with sledges between bases each side to open joint as far as possible.

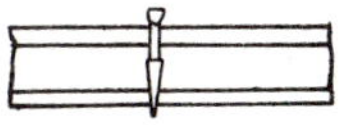

6 Slightly oversize insert selected, bevelled a little with a hammer along bottom edges, and driven between heads.

7 Base wedges knocked out. The compression on insert if work has been done properly should be sufficient to hold weight of car.

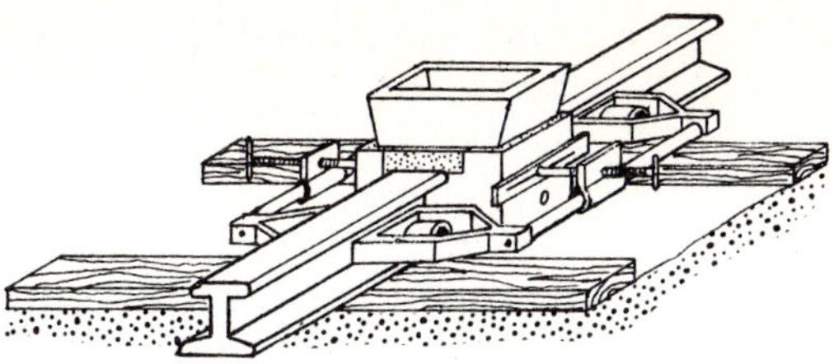

Welding loose rails under traffic, the necessary pressure that prevents the relative motion of the rail ends under load is exerted by the use of compression clamps as shown by above illustration.

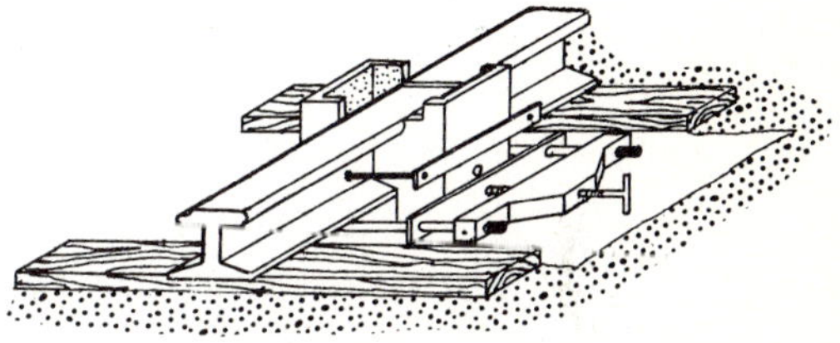

8 Special molds applied which do not interfere with operation of cars. Molds must be carefully luted. Molds secured with special clamp through standard bolt holes and by regular square clamp at bottom.

9 Rail preheated from outside to avoid interference with traffic. Iron cap placed over mold tops between cars to assist in heating heads of rails.

10 When rail is heated, burner is withdrawn, vent hole plugged and strips of asbestos laid in groove of rail at joint, and molding material rammed and trowelled into groove to the level of head and lip; then heating burner is replaced.

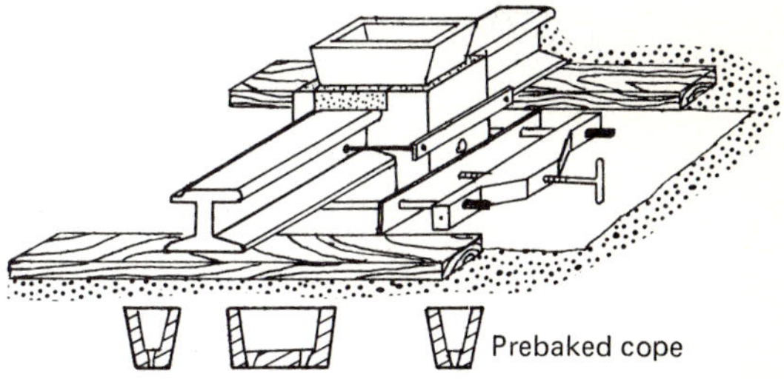
Prebaked cope

11 A prebaked cope made on a special pattern is next put into place and molding material packed around the base of the cope. At the same time the fully charged crucible is mounted in position.

12 The preheating hole is then plugged and the weld poured in the usual manner. In two or three minutes the cope can be pried off and a car permitted to pass. The flange of the wheel automatically cleaning the sand from the groove.

Figure **9-12** Making a thermit rail weld under traffic. (Courtesy of the American Welding Society.)

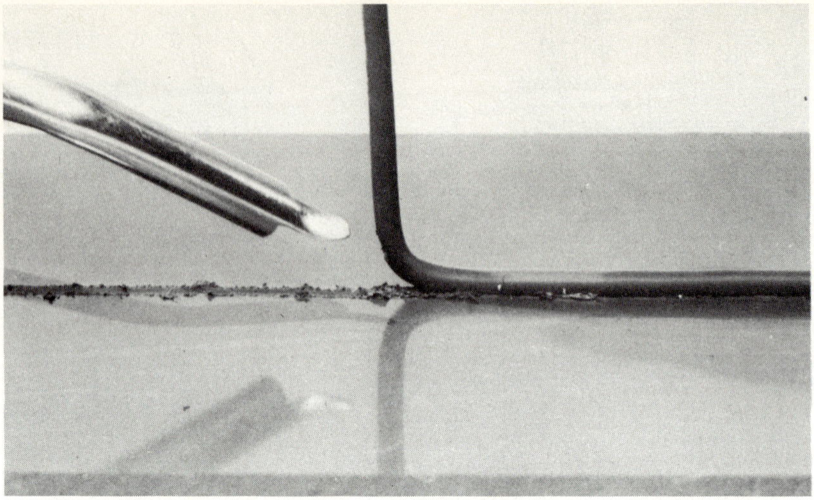

Figure **9-13** Manual hot-gas welding of plastics. (By permission from American Welding Society, *Welding Handbook,* 6th ed., 1970, section 3A, p. 56.12.)

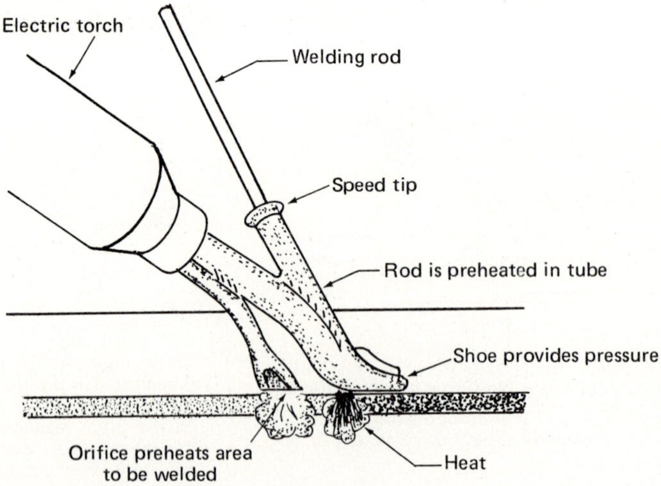

Electric torch
Welding rod
Speed tip
Rod is preheated in tube
Shoe provides pressure
Orifice preheats area
to be welded
Heat

Figure **9-14** Semiautomatic hot-gas method for welding plastics. (Redrawn by permission from American Welding Society, *Welding Handbook,* 6th ed., 1970, section 3A, p. 56.14.)

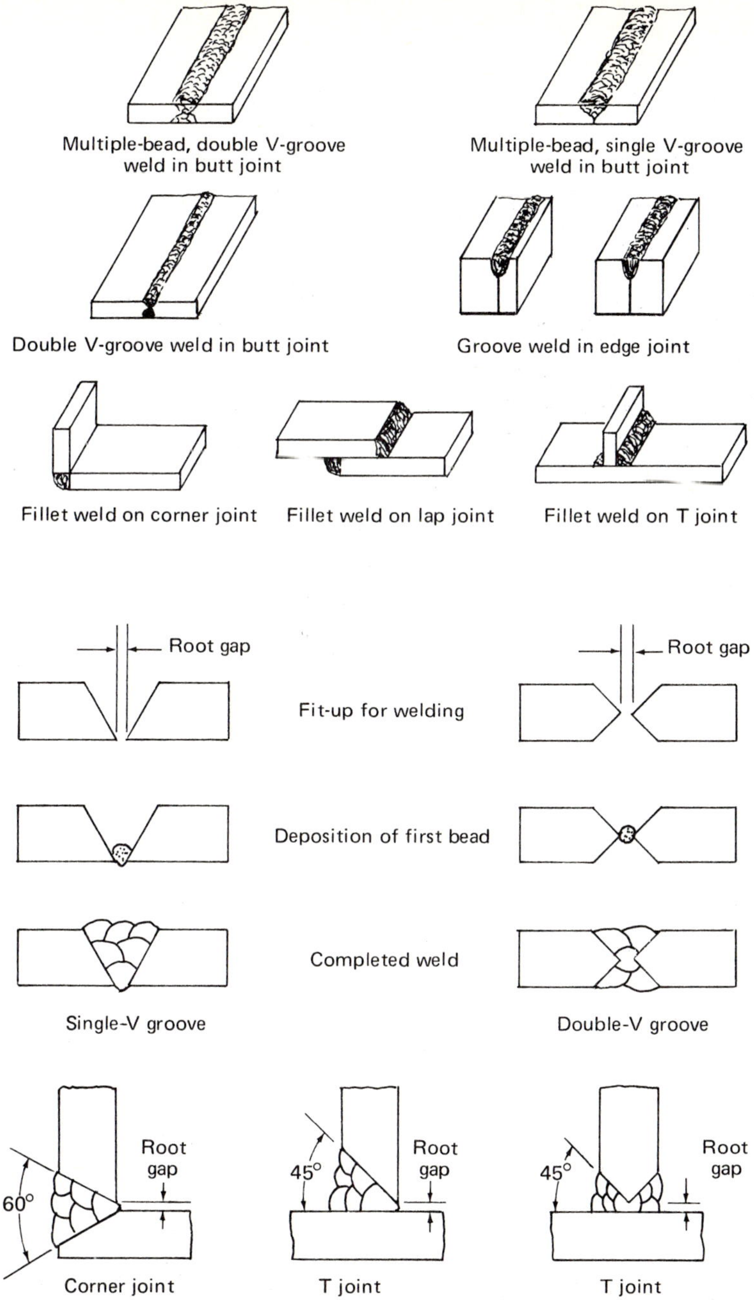
Multiple-bead, double V-groove weld in butt joint
Multiple-bead, single V-groove weld in butt joint
Double V-groove weld in butt joint
Groove weld in edge joint
Fillet weld on corner joint
Fillet weld on lap joint
Fillet weld on T joint
Root gap
Root gap
Fit-up for welding
Deposition of first bead
Completed weld
Single-V groove
Double-V groove
Root gap
60°
Corner joint
45°
Root gap
T joint
45°
Root gap
T joint

Figure **9-15** Preparation of, and completed joints in plastic. (Redrawn by permission from American Welding Society, *Welding Handbook*, 6th ed., 1970, section 3A, pp. 56.10, 56.11.)

surfaces to be welded, pressure is then applied, and movement of the assembly is stopped. After cooling has taken place, the assembly is removed from the machine. At present, spin welding is the most common friction-welding technique in use.

Typical operating velocities, weld strengths, and joint designs are shown in Tables 9-3 and 9-4, and Figure 9-16, respectively.

Heated-tool welding is accomplished by bringing the parts either directly into contact with, or within ⅛ inch of, the heated tool and increasing the temperature of the parts until the area to be joined becomes molten (Figure 9-17). As soon as the area is molten, the parts are held in contact with a pressure that, depending on the materials to be joined, ranges from 5 to 300 psi.

The theory and equipment used for the *ultrasonic welding* of plastics is similar to that used for the ultrasonic welding of metal (Figure 9-18).

Correct joint design is very important for ultrasonic welding. The key to good design is the use of an energy director (Figure 9-19). The energy director localizes vibrations so that a restricted volume of plastic is quickly brought to its melting temperature and flows over the width of the joint to produce uniform coverage between the joining surfaces. Some typical joints used in ultrasonic welding are shown in Figure 9-20. The plastics most commonly welded are listed in Table 9-5. They are available in sheet, block, rod, tubular, and extruded forms.

Although it is sometimes possible to identify thermoplastic materials by feel, because the polyolefins feel slippery and slightly greasy to the touch and usually have a very slippery surface, a more reliable field test for identifying plastics for welding is to burn a ⅛-inch-thick by 3-inch-long sample of the material to be welded with a flame and compare the results with those listed in Table 9-6, on page 224.

9-7 OBSOLETE WELDING PROCESSES

Because some processes are no longer used on a large industrial scale, the AWS considers them to be obsolete. These processes include: forge welding, carbon-electode welding, bare metal–arc welding, impregnated-tape metal–arc welding, and atomic hydrogen welding.

Some of these processes have been discussed briefly in the early chapters of this book. However, if the reader wishes to obtain more information on these processes, he or she is referred to Chapter 61, section 3B, of the *Welding Handbook*, 6th ed., or to the author.

Table **9-3** Operating velocities and pressures for plastic welding

MATERIAL	AVERAGE POINT VELOCITY, FT/SEC	INITIAL* PRESSURE, LB/IN²
Nylon	5 to 50	25 to 150
Acetal	5 to 35	25 to 150
Acrylic resin	10 to 35	15 to 125
Polyethylene	5 to 60	10 to 100

*In some applications a secondary high pressure, or squeezing, after initial melting has proved beneficial in improving weld quality. This secondary pressure may be many times greater than that used to obtain initial melting, and is generally limited only by material stiffness and strength.

Source: American Welding Society, *Welding Handbook*, 6th ed., section 3B, p. 56.16.

Table **9-4** Strengths in plastic welding

WELDED TO	POLYETHYLENE	ACRYLIC	NYLON	ACETAL
Polyethylene	70–95*	—	—	—
Acrylic	—	75–85*	—	30–35*
Nylon	—	—	50–70*	25–55
Acetal	—	30–35	25–55	30–70

*Strengths indicated as percent of base material strength

Source: American Welding Society, *Welding Handbook*, 6th ed., section 3B, p. 56.16.

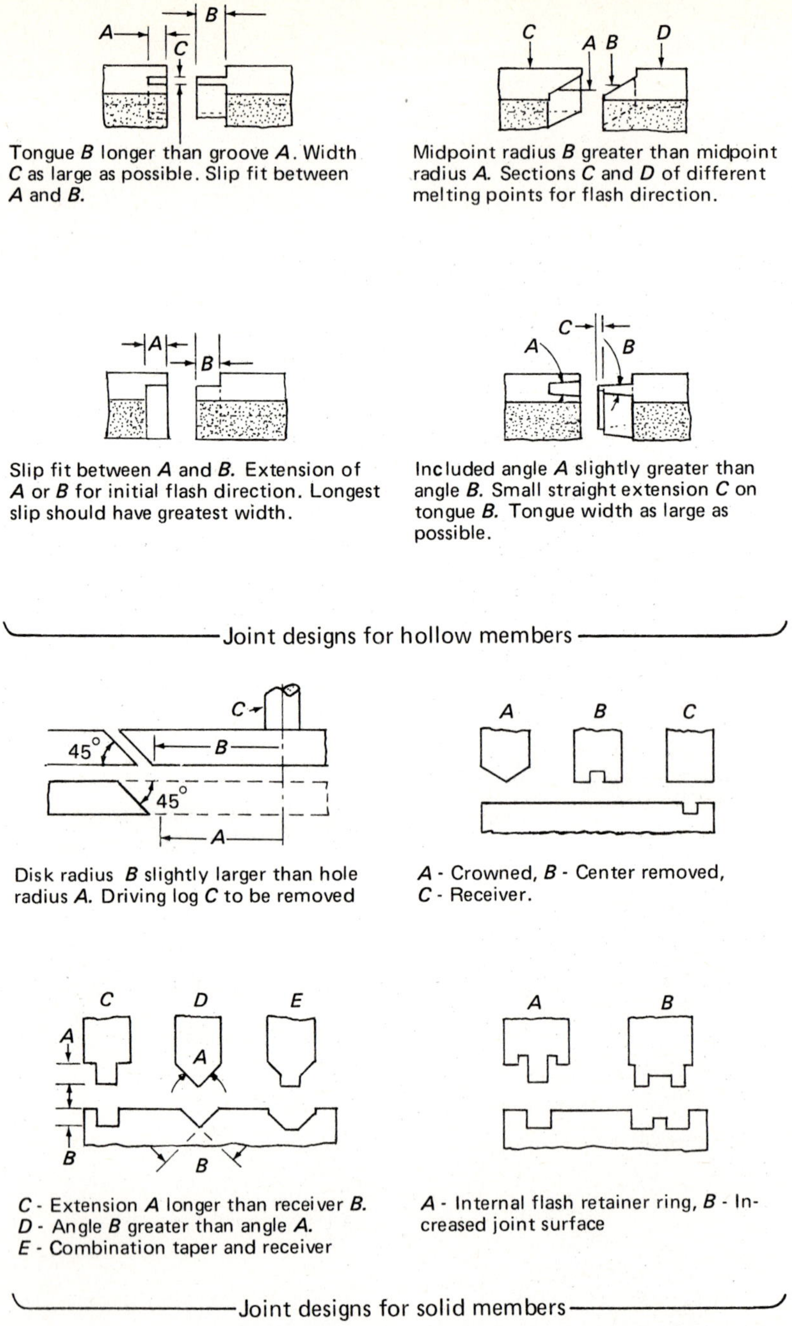
Tongue B longer than groove A. Width C as large as possible. Slip fit between A and B.
Midpoint radius B greater than midpoint radius A. Sections C and D of different melting points for flash direction.
Slip fit between A and B. Extension of A or B for initial flash direction. Longest slip should have greatest width.
Included angle A slightly greater than angle B. Small straight extension C on tongue B. Tongue width as large as possible.
Joint designs for hollow members
45°
45°
Disk radius B slightly larger than hole radius A. Driving log C to be removed
A - Crowned, B - Center removed, C - Receiver.
C - Extension A longer than receiver B.
D - Angle B greater than angle A.
E - Combination taper and receiver
A - Internal flash retainer ring, B - Increased joint surface
Joint designs for solid members

Figure **9-16** Typical joint designs for friction welding hollow and solid plastic members. (Redrawn by permission from American Welding Society, *Welding Handbook*, 6th ed., 1970, section 3A, pp. 56.17, 56.18.)

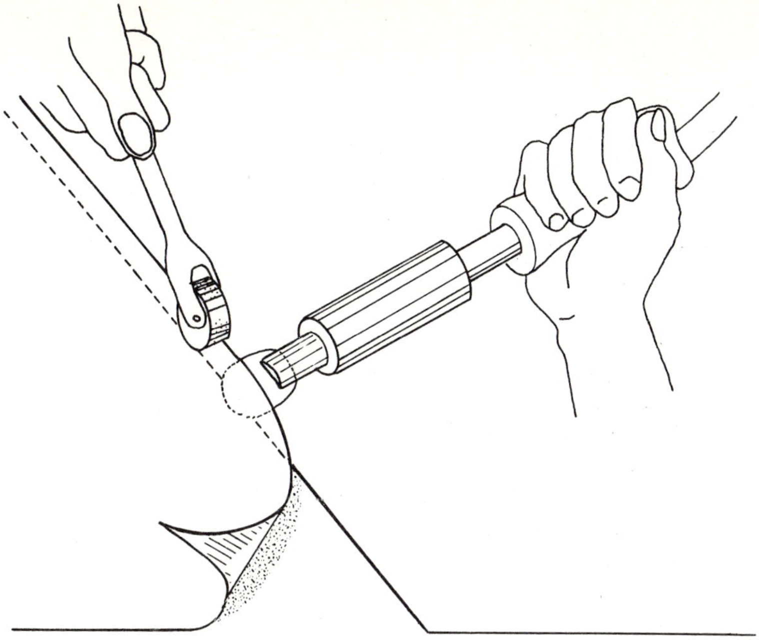

Figure **9-17** Heated-shoe seam-welding tool in use. (Redrawn by permission from American Welding Society, *Welding Handbook,* 6th ed., 1970, section 3A, p. 56.20.)

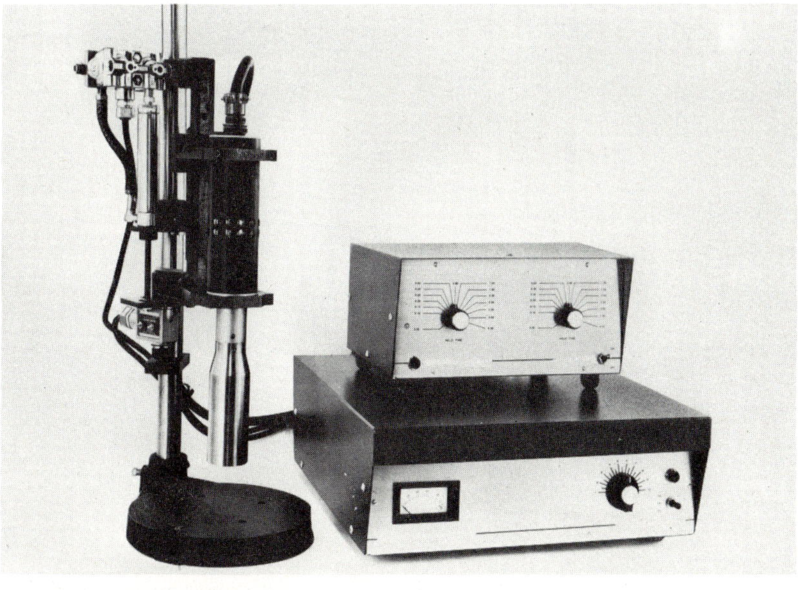

Figure **9-18** Equipment for ultrasonic welding of plastics. (Courtesy of the American Welding Society.)

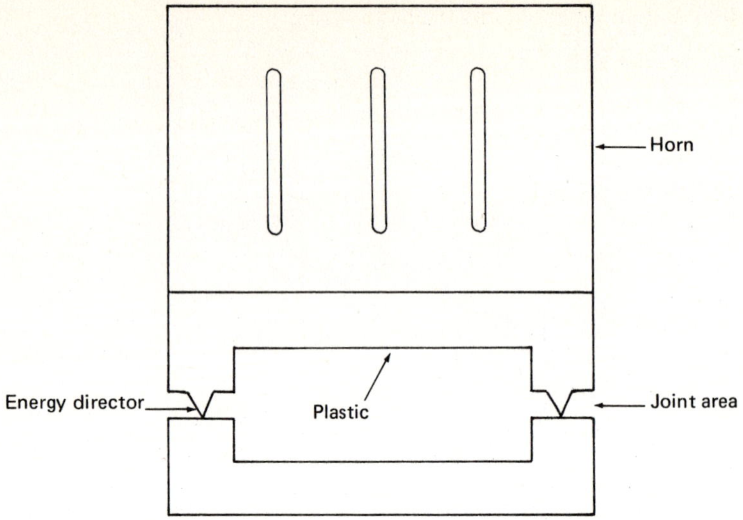
Horn
Energy director
Plastic
Joint area

Figure **9-19** End view of plastic parts to be welded, showing the energy director

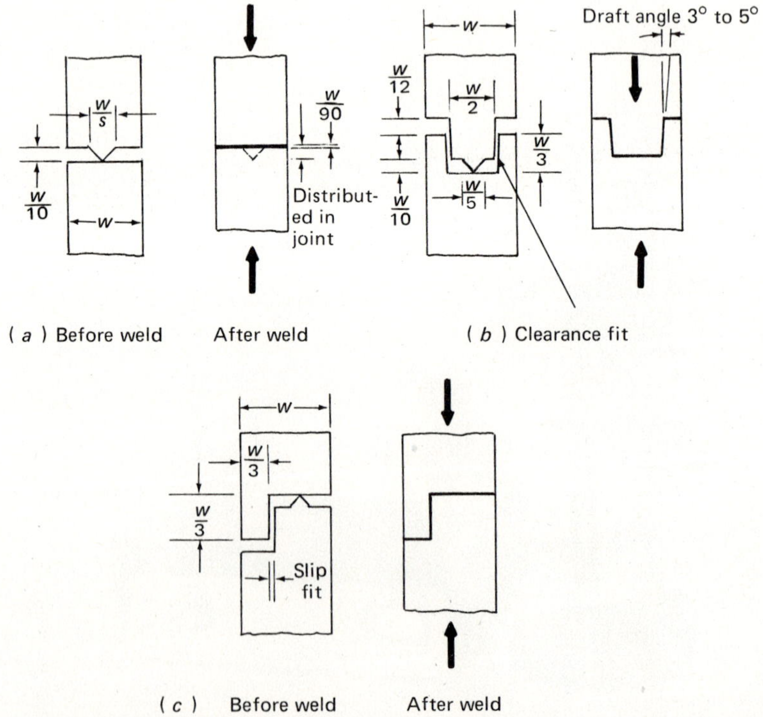
W/S
W/10
W
W/90
Distributed in joint
W
W/12
W/2
W/3
W/10
W/5
Draft angle 3° to 5°
(a) Before weld
After weld
(b) Clearance fit
W
W/3
W/3
Slip fit
(c) Before weld
After weld

Figure **9-20** **Common joint designs for ultrasonic welding of plastics:** (*a*) butt joint with energy director; (*b*) tongue-and-groove joint with energy director; (*c*) step joint with energy director. (Redrawn by permission from American Welding Society, *Welding Handbook*, 6th ed., 1971, section 3B, pp. 56.24, 56.25.)

Table **9-5** The most commonly welded plastics

	Material	Characteristics*			Remarks
		Staking and Inserting	Welding: Near Field	Welding: Far Field	
General Purpose Toys—Appliances Housewares	Polystyrene unfilled	E	E	E	Excellent acoustical properties; produces strong, smooth joints.
	Rubber-modified	E	E	G–P	Welding characteristics depend on degree of impact resistance.
	Glass-filled	E	E	E	Weldable with filler content up to 30%.
	SAN	E	E	E	Particularly good as glass-filled compounds.
	ABS	E	E	G	Can be bonded to other polymers such as SAN, styrene, and acrylics.
Engineering Plastics Automotive—Appliance Electronic	Polycarbonate	E	E	E	High melting temperature requires high energy levels. Oven dried or "as molded" parts perform best due to hygroscopic nature of the material.
	Nylon	E	G	F	
	Polysulfone	E	G	G–F	
	Acetal	E	G	G	Requires high energy and long ultrasonic exposure because of low coefficient of friction.
	Acrylics	E	E	G	Weldable to ABS and SAN; applications include dials, radio cases, and meter housing; in sheet form, joints must be machined.
	Polyphenylene oxide	G	G	G–F	High melting temperatures require high energy levels.
	Noryl	E	G	E–G	
	Phenoxy	E	G	G–F	
High Volume Low Cost Applications	Polypropylene (Olefins)	E	G–P	F–P	Horn design for welding is particularly critical; filled compounds usually better, but need individual testing.
	Polyethylene (Olefins)	E	G–P	F–P	
	Butyrates	G–F	P	P	Weldability varies with formulation and part configuration; however, these materials usually perform well in staking and inserting applications. Decomposition of some formulations may occur.
	Cellulosics	G–F	P	P	
	Acetates	G–F	P	P	
	Vinyls	E–F	F–P	F–P	

*Near-field welding refers to joint 1/4″ or less from area of horn contact; far-field welding to joint more than 1/4″ from contact area. E = Excellent; G = Good; F = Fair; P = Poor.

Source: American Welding Society, *Welding Handbook*, 6th ed., section 3B, p. 56.28.

Table **9-6** Thermoplastic burning test

MATERIAL	SPECIFIC GRAVITY	BURNING CHARACTERISTICS			
		*RATE**	*ODOR*	*FLAME*	*EFFECT ON MATERIAL*
Cellulose acetate butyrate (CAB)	1.15–1.22	0.5–1.5	Butyric acid (rancid butter)	Dark yellow with slight blue edges; some black smoke (not sooty)	Melts; drips; drippings continue to burn
Polyamide (nylon)	1.09–1.14	Self-extinguishing	Burning wood or hair	Blue with yellow top	Melts, drips, and froths
Polyethylene					
1. Low density	0.910–0.925	1.0–1.1	Burning parafin	Bottom blue, top yellow	Melts and drips
2. Medium density	0.926–0.940	1.0–1.1	Burning parafin	Bottom blue, top yellow	Melts and drips
3. High density	0.941–0.965	1.0–1.1	Burning parafin	Bottom blue, top yellow; white smoke	Melts and drips
Polypropylene	0.90–0.91		Burning parafin	Bottom blue, top yellow; white smoke	Melts and drips
Polyvinyl chloride (PVC)					
Type I	1.35–1.45	Self-extinguishing	Hydrochloric acid	Yellow, green on bottom edges;	Softens; chars
Type II	1.35–1.45	Self-extinguishing	Hydrochloric acid	Spurts green and yellow; white smoke	Softens; chars
Styrene copolymers (ABS)					
Type I	1.00–1.08	1.3	Artificial illuminating gas	Yellow; black smoke	Softens; chars
Type II	1.06–1.08	1.3	Artificial illuminating gas	Yellow; black smoke	Softens; chars

*ASTM Test No. D635-44 (rate in inches per minute)

Source: American Welding Society, *Welding Handbook*, 6th ed., section 3B, p. 56.8.

REVIEW QUESTIONS

1. Describe the electron-beam welding process.

2. By what name is the focusing coil of an electron-beam welding machine known?

3. Which AWS safety publication should an electron-beam welding-machine operator be familiar with?

4. Name the two electroslag welding systems and differentiate between them.

5. What methods are commonly used to inspect electroslag-welded pressure vessels fabricated to the ASME Boiler and Pressure Vessel Code?

6. How is laser-beam welding done?

7. Name some of the metals that are commonly joined by laser-beam welding.

8. Describe the thermit-welding process and name some of its applications.

9. Name and describe three methods commonly used to weld plastics.

10. Name the welding processes considered to be obsolete by the AWS.

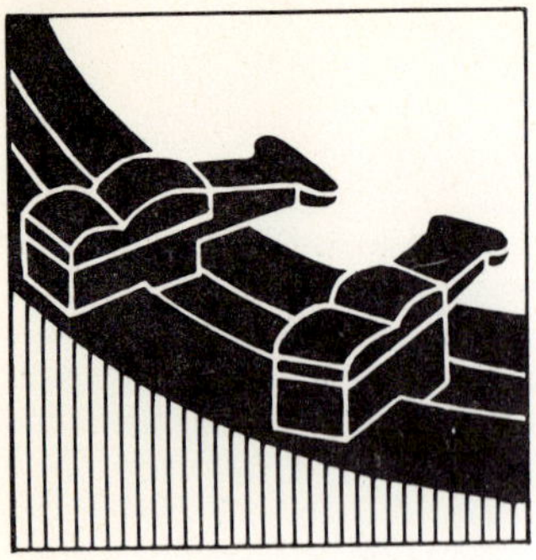

CHAPTER 10
BRAZING

This chapter and Chapters 11 and 12 are concerned with what might be called *low-temperature welding processes:* brazing, soldering, and adhesive bonding. In each process, the weld is attained at a temperature below the melting point of the base metal. There is no fusion; therefore, the metallurgical problems associated with fusion are avoided. Distortion is kept to a minimum, and exceedingly thin metals can be joined or sealed against leaks with relative ease. Perhaps the best advantage of these processes is their applicability in joining unlike materials.

The American Welding Society defines *brazing* as a process "wherein coalescence is produced by heating to suitable temperatures above 800°F and by using a nonferrous filler metal having a melting point below that of the base metal, the filler metal being distributed between the closely fitted surfaces of the joint by capillary attraction."

Brazing, like soldering, is dependent for its success upon the fact that a molten metal of low surface tension will flow easily and evenly over the surface of a properly heated and chemically clean base metal, just as water flows over a clean glass plate.

10-1 BRAZING METHODS

Of the five brazing methods described in the following paragraphs and compared in Figure 10-1 and Table 10-1, the manual torch brazing method is the method most widely used.

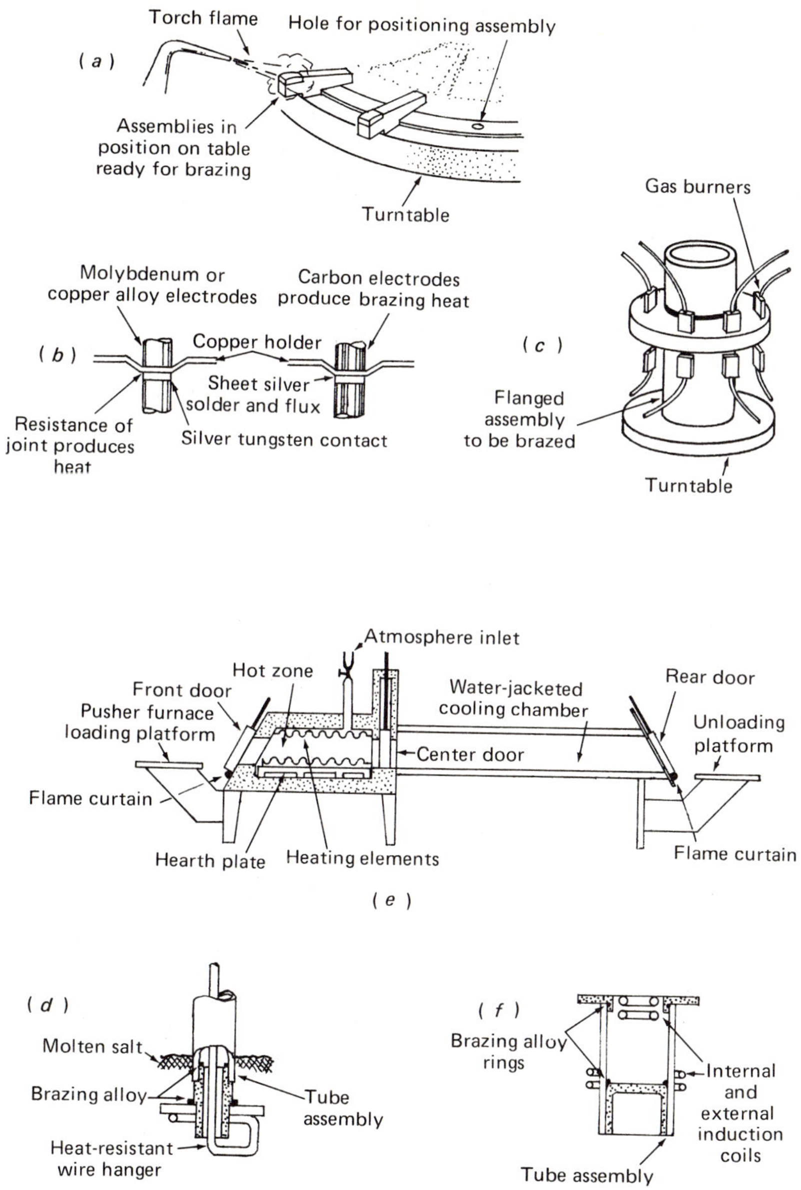

(a)
Torch flame
Hole for positioning assembly
Assemblies in position on table ready for brazing
Turntable
(b)
Molybdenum or copper alloy electrodes
Carbon electrodes produce brazing heat
Copper holder
Sheet silver solder and flux
Resistance of joint produces heat
Silver tungsten contact
(c)
Gas burners
Flanged assembly to be brazed
Turntable
Atmosphere inlet
Hot zone
Front door
Pusher furnace loading platform
Flame curtain
Water-jacketed cooling chamber
Rear door
Unloading platform
Center door
Hearth plate
Heating elements
Flame curtain
(e)
(d)
Molten salt
Brazing alloy
Tube assembly
Heat-resistant wire hanger
(f)
Brazing alloy rings
Internal and external induction coils
Tube assembly

Figure **10-1** Comparison of the common brazing methods. (Redrawn from V. H. Laughner and A. D. Hargan, *Handbook of Fastening and Joining of Metal Parts,* McGraw-Hill, New York, 1956, p. 317.)

Table **10-1** The basic brazing processes (to be used with Figure 10-1)

METHOD AND DESCRIPTION	ADVANTAGES	LIMITATIONS
Hand torch: Gas welding torch burning acetylene, hydrogen, or methane, together with air or oxygen [see Fig. 10-1(*a*)]	1. Flexible method, applicable to a great many assemblies. 2. Equipment is low-cost. 3. Heat can be localized.	1. Work will oxidize. 2. Skilled operator required. 3. Relatively slow. 4. Base metals limited. 5. Multiple or inaccessible joints cannot be brazed. 6. Localized heating will cause distortion.
Resistance: Heat is supplied to the work by conduction of heat from hot carbon electrodes or by resistance of work to the passage of current [see Fig. 10-1(*b*)]	1. Short operator-training period. 2. Equipment is low-cost and simple. 3. Localized rapid heating. 4. Moderate production rate. 5. Effective for long-seam brazes.	1. Individual handling necessary. 2. Distortion is possible. 3. Size of assembly is limited.
Radiant gas burner: Ceramic burners, using gas-air mixture under pressure, apply concentrated heat to work on conveyors or automatic indexing tables [see Fig. 10-1(*c*)]	1. Unskilled operators can be used. 2. Localized, rapid heating. 3. Controlled heat is ideal for mass-production work.	1. Costly equipment. 2. Some distortion. 3. Work oxidizes. 4. Exact temperature control difficult.
Salt bath: Also called dip brazing. Work is dipped into molten salt bath to heat. Bath can have fluxing action [see Fig. 10-1(*d*)]	1. Accurate temperature control. 2. Rapid heating. 3. Copper brazing possible. 4. No flux needed for some applications.	1. Fixtures required. 2. Danger of explosion with wet parts. 3. Salt carryover. 4. Parts must be washed. 5. Some oxidation during cooling period.
Furnace: Work is heated in a furnace in presence of reducing atmosphere. Filler material is in the form of foil, paste, slugs, powder, or molten spray [see Fig. 10-1(*e*)]	1. No residual stresses. 2. Good for mass production. 3. Uniform heating reduces distortion. 4. Control of carburization or decarburization. 5. Bright surfaces. 6. Accurate control of temperature.	1. Requires fixturing or self-positioning of components. 2. Localized heating not possible. 3. Expensive equipment for small lots.

FILLER RODS OR MATERIALS	BASE METALS THAT CAN BE BRAZED	APPLICATIONS
1. Naval brass 2. Brass brazing alloy 3. Manganese bronze 4. Low-fuming bronze 5. Phosphor bronze 6. Silver alloys (with or without phosphorus) 7. Copper silicon 8. Nickel silver	1. Steel 2. Copper 3. Copper alloys 4. Stainless steel 5. Nickel 6. Nickel alloys	1. In assemblies where self-jigging is impractical. 2. In assemblies that require extensive fixturing. 3. Small production runs, where inexpensive equipment will offset inherent disadvantages.
1. Copper phosphorus 2. Silver alloys (with or without phosphorus)	1. Steel 2. Copper 3. Copper alloys 4. Stainless steel 5. Nickel 6. Nickel alloys	1. Where combination of heat and pressure are needed to ensure good joint. 2. Small parts, especially electrical components, brazed with silver and phos-copper.
1. Naval brass 2. Brass brazing alloy 3. Manganese bronze 4. Low fuming bronze 5. Phosphor bronze 6. Silver alloys 7. Copper silver 8. Nickel silver	1. Steel 2. Copper 3. Copper alloys 4. Stainless steel 5. Nickel 6. Nickel alloys	1. Mass-production parts where large parts need not be heated. 2. Silver and phos-copper brazing where parts are on production line basis.
1. Brass brazing alloys 2. Silver alloys (with or without phosphorus)	1. Steel 2. Copper 3. Copper alloys 4. Stainless steel 5. Nickel 6. Nickel alloys	1. With assemblies that lend themselves to suspension in fixtures. 2. For parts that require rapid heating.
1. Electrolytic copper 2. Deoxidized copper 3. Brass brazing alloys 4. Naval brass 5. Copper phosphorus 6. Silver alloys	1. Steel 2. Copper 3. High-copper alloys 4. Stainless steel 5. Nickel alloys 6. Nickel	1. Widely used for copper brazing. 2. With multiple or inaccessible joints. 3. Mass production where assembly permits fixturing or self-positioning.

Table **10-1** *(cont.)*

METHOD AND DESCRIPTION	ADVANTAGES	LIMITATIONS
Induction: High-frequency currents induce eddy currents in the work and produce heat [see Fig. 10-1(*f*)]	1. Rapid, localized heating. 2. Adapted to mass production.	1. Possible distortion due to local heating. 2. Copper brazing not usually practical. 3. Temperature difficult to control.
Molten metal dip: Parts are immersed in bath of molten brazing alloy [see Fig. 10-1(*d*)].	1. Brazes great many joints at once.	1. Applications extremely limited. 2. Parts must be clean and well fluxed. 3. Bath must be large to keep down temperature drop on immersion.

Source: V. H. Laughner and A. D. Hargan, *Handbook of Fastening and Joining of Metal Parts* (New York: McGraw-Hill Book Company, 1956), p. 323.

TORCH BRAZING Torch brazing is a brazing process in which the heat is obtained from a gas flame or flames impinging on or near the joint to be brazed. Torches used in this process may be of the hand-held type or may consist of fixed burners having one or many flames. Several types of fuel gas are available for combustion with oxygen or air. Torch brazing can be done as a completely manual, partly mechanized, or completely automatic process.

Torch brazing can be used on carbon, low-alloy, and stainless steels; aluminum alloys; copper and copper alloys; and magnesium alloys. Highly alloyed steels, heat-resisting alloys, and reactive metals are usually brazed by other methods because they require special atmospheres or closer control of the thermal cycle, or both.

Manual torch brazing with the oxyfuel-gas torch is the brazing process with which all welders must be familiar. The equipment used for manual torch brazing is similar to, or the same as, the equipment used in oxyfuel-gas welding.

Manual torch brazing is best applied where production quantities are small, since equipment cost is low. It is also used in applications where physical size, joint configuration, or other considerations make it difficult or impossible to braze by other methods. The main drawbacks are the labor and the relative skill needed for efficient production. Where production quantities are larger, automatic torch

FILLER RODS OR MATERIALS	BASE METALS THAT CAN BE BRAZED	APPLICATIONS
1. Nickel silver 2. Silver alloys 3. Copper phosphorus	1. Steel 2. Copper alloys 3. Stainless steel 4. Nickel alloys 5. Nickel	1. Applications where the joint is not too deep and components not too heavy. 2. Mass-production silver brazing.
1. Brass brazing alloy	1. Steel	1. Wire basket. 2. Assemblies made of narrow metal strips.

brazing is used, often producing brazed assemblies at the rate of 400 to 1400 per hour.

Acetylene, natural gas, propane, and proprietary gas mixtures are the types of fuel gas most often used in the torch brazing of steel. Hydrogen, butane, and producer (city) gas are seldom used. In manual torch brazing, commercially pure oxygen is chiefly used as the combustion agent because of its fast heating rate. Increased production of oxygen for steelmaking and other purposes has resulted in larger supplies and lower prices. This has favored the use of natural gas and propane. These gases require larger oxygen-to-fuel ratios than acetylene, but are less costly than acetylene.

Compressed air involves only the cost of compression and advantage can be taken of this in automatic operations, where the investment in pumping and mixing equipment can be amortized with the rest of the equipment. Another factor favoring compressed air–natural gas combustion in automatic applications is that if conveyor malfunction occurs, the workpieces will not be destroyed by overheating.

FURNACE BRAZING Furnace brazing is a mass-production process for joining the components of small assemblies by metallurgical bonds, using a nonferrous filler metal as the bonding material and a furnace as the heat source. Furnace brazing is feasible only if the filler metal can be preplaced on the joint before brazing and retained in position during brazing.

The process requires the use of a suitable inert-gas (argon or helium) atmosphere to protect the steel assemblies against oxidation, or oxidation and decarburization, during brazing and cooling (which is accomplished in chambers adjacent to the brazing furnace). The proper brazing atmosphere also makes possible the proper wetting of the joint surfaces by the molten copper filler metal, usually without the use of brazing flux.

Although filler metals other than copper can be used in furnace brazing carbon and low-alloy steels, copper is generally preferred because of its low cost and the high strength of the joints produced. The high brazing temperature necessary when copper filler metals are used (2000 to 2100°F) is also advantageous when steel assemblies are to be heat treated after brazing.

INDUCTION BRAZING Induction brazing is a process in which the surfaces of components to be joined are selectively heated to brazing temperature by electrical energy supplied from an induction heating unit. The energy is transmitted to the workpiece by induction, rather than by an electrical connection, using an inductor or work coil. Heating is the result of eddy currents in the work metal, which, by virtue of electrical resistivity and the flow of induced alternating current, generate heat. When the work metal being heated is ferromagnetic, as are most steels, some slight additional heating results from hysteresis. However, all heating due to hysteresis ceases when the temperature of the work metal is raised to the Curie point (about 1420°F). Above this temperature, heating by electrical resistance continues at a reduced rate as the temperature rises.

Most of the heat induced in the workpiece by electric current is limited to a thin surface layer close to the inductor. Distribution of heat to other areas of the workpiece depends on conduction. In general, heat flow by conduction, although fairly rapid, is minimized by the rapidity at which induction heating takes place. The depth of heating by induction depends mainly on the frequency of the alternating current. As the frequency is increased, both the theoretical depth of current penetration and the depth of the heated zone in the workpiece decrease.

The brazing filler metal (Figure 10-2) is usually preplaced in a carefully designed joint (Figure 10-3) and coil set-up (Figure 10-4) to assure that the surfaces of all members of the joint reach the brazing temperature at the same time. Flux is usually employed, except when an atmosphere is specifically introduced to perform the same functions.

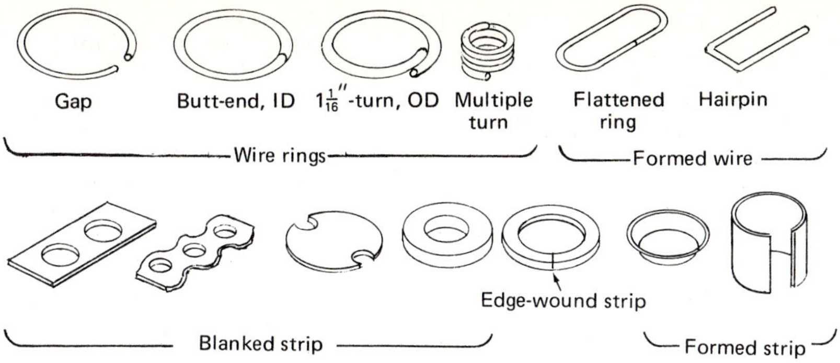
Gap
Butt-end, ID
$1\frac{1}{16}''$-turn, OD
Multiple turn
Flattened ring
Hairpin
Wire rings
Formed wire
Edge-wound strip
Blanked strip
Formed strip

Figure **10-2** Common filler metal preforms used in induction brazing. (By permission, from *Metals Handbook* Volume 6, Copyright American Society for Metals, 1971.)

RESISTANCE BRAZING Resistance brazing is a resistance joining process in which the workpieces are heated locally and the filler metal preplaced between them is melted by the heat obtained from resistance to the flow of electric current through the electrodes and the work (Figure 10-5). In the usual application of the process, the heating current is passed through the joint itself. Resistance-welding equipment is used, and the pressure needed for establishing electrical contact across the joint is ordinarily applied through the electrodes. The electrode pressure is also the usual means for providing the tight fit needed for capillary behavior in the joint. The heat for resistance brazing can be generated mainly in the workpieces themselves, in the electodes, or in both, depending on their electrical resistivity and dimensions.

DIP BRAZING Dip brazing in molten salt, also called *salt bath brazing* and *molten chemical-bath dip brazing,* is a process in which the assembly to be brazed is immersed in a bath of molten salt, which provides the heat and may provide the fluxing action for brazing. The bath temperature is maintained above the melting point of the filler metal but below the melting range of the base metal. This method is largely confined to brazing small parts such as wires or narrow strips of metal.

10-2 BRAZING FILLER METALS

Brazing filler metals are metals that are added when making a braze. Filler metals used in brazing have melting points of about 800°F, but below that of the metal being brazed, and have properties suitable

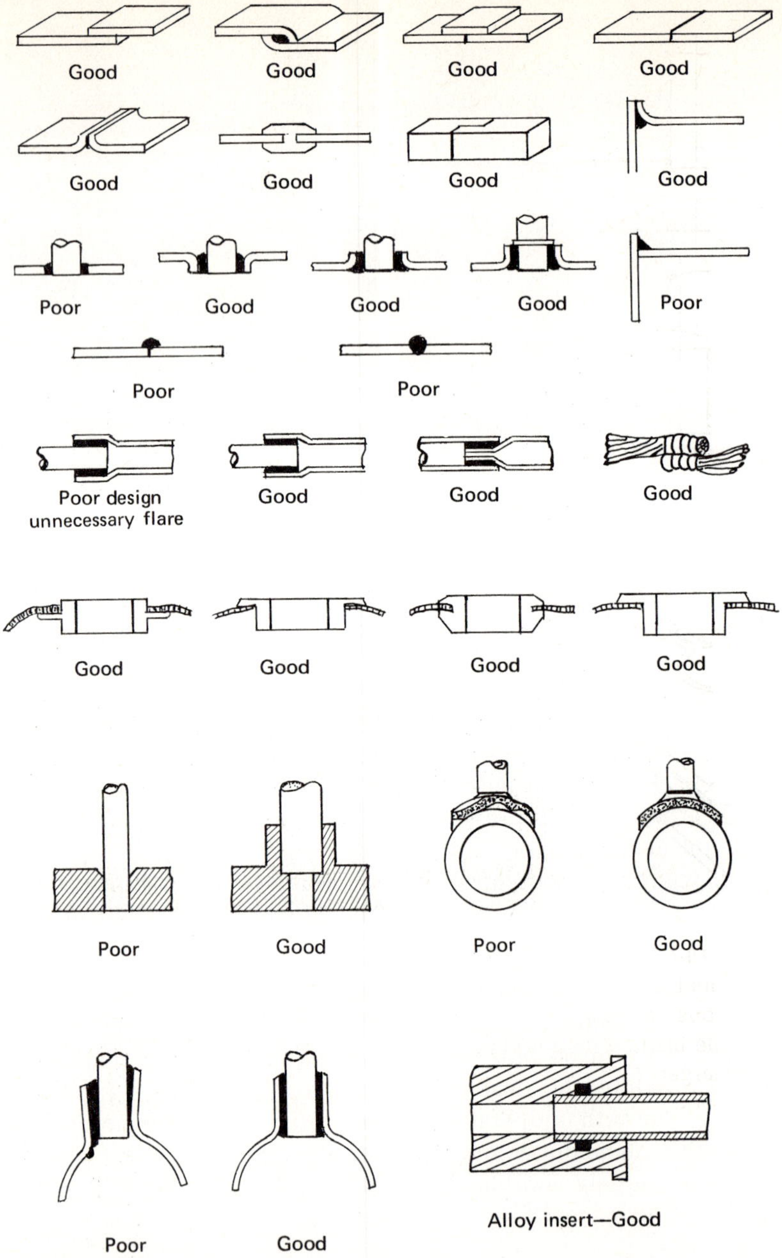
Good
Good
Good
Good
Good
Good
Good
Good
Poor
Good
Good
Good
Poor
Poor
Poor
Poor design
unnecessary flare
Good
Good
Good
Good
Good
Good
Good
Poor
Good
Poor
Good
Alloy insert—Good
Poor
Good

Figure **10-3** Good and bad practice in brazing joint design. (Redrawn from V. H. Laughner and A. D. Hargan, *Handbook of Fastening and Joining of Metal Parts*, McGraw-Hill, New York, 1956, p. 462.)

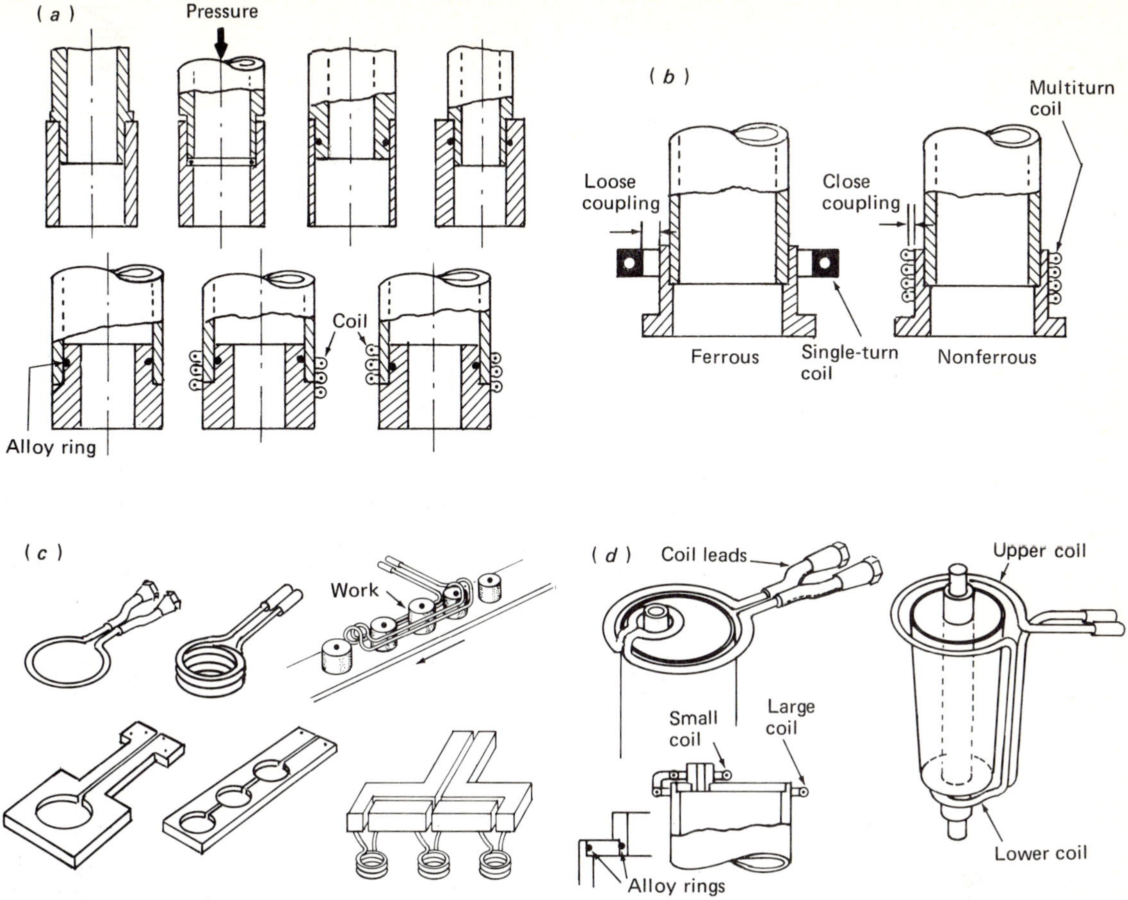
(a)
Pressure
Coil
Alloy ring
(b)
Multiturn coil
Loose coupling
Close coupling
Ferrous
Single-turn coil
Nonferrous
(c)
Work
(d)
Coil leads
Upper coil
Small coil
Large coil
Alloy rings
Lower coil

Figure **10-4** Placement and design of induction-brazing coils: (*a*) proper placement of silver alloy ring and coil determines flow; (*b*) coil arrangement for ferrous and nonferrous parts; (*c*) commonly used coils; (*d*) designs for multiple brazing of single assemblies. (Redrawn from V. H. Laughner and A. D. Hargan, *Handbook of Fastening and Joining of Metal Parts,* McGraw-Hill, New York, 1956, pp. 462, 464.)

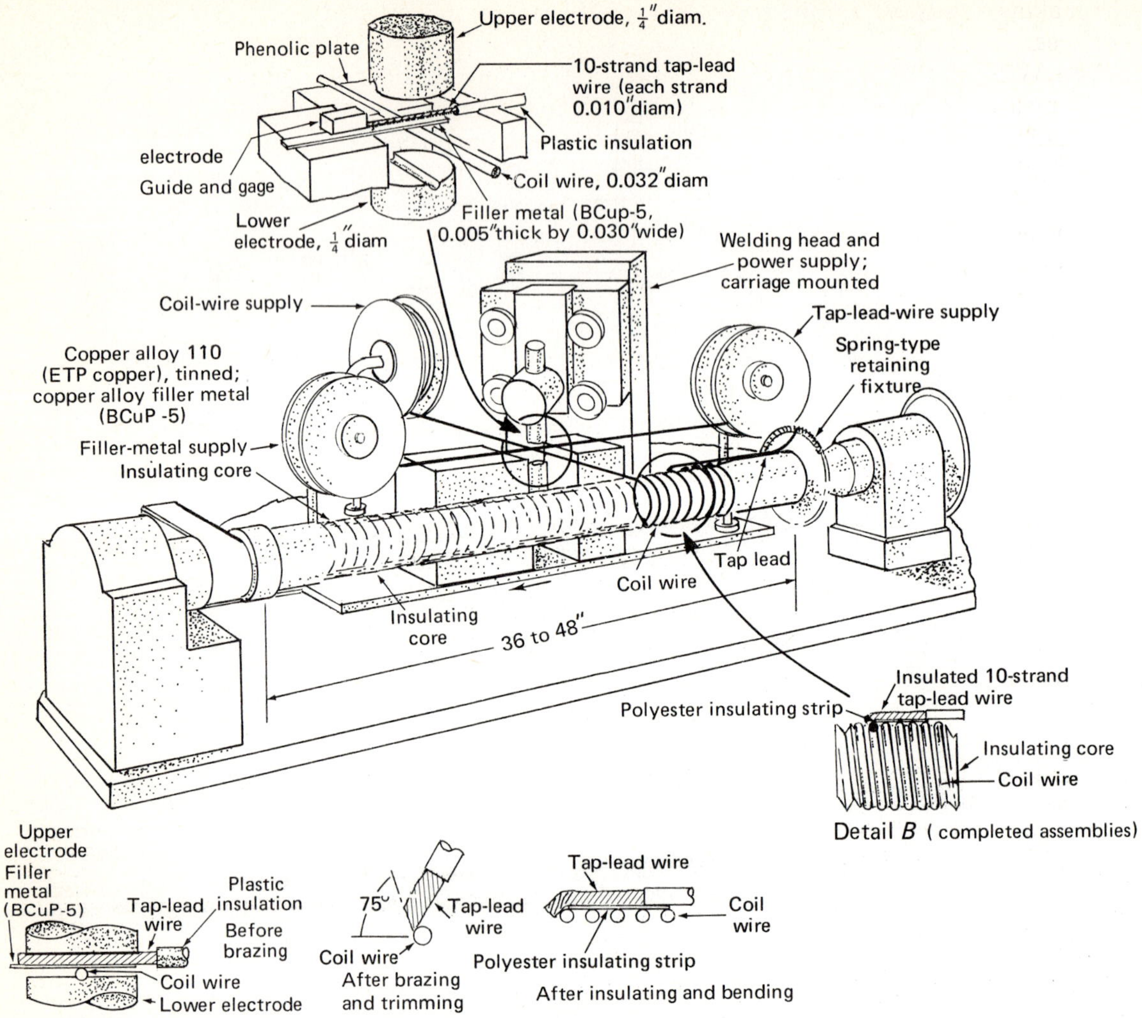

Upper electrode, 1/4" diam.
Phenolic plate
10-strand tap-lead wire (each strand 0.010" diam)
Plastic insulation
electrode
Guide and gage
Coil wire, 0.032" diam
Lower electrode, 1/4" diam
Filler metal (BCup-5, 0.005" thick by 0.030" wide)
Welding head and power supply; carriage mounted
Coil-wire supply
Tap-lead-wire supply
Spring-type retaining fixture
Copper alloy 110 (ETP copper), tinned; copper alloy filler metal (BCuP -5)
Filler-metal supply
Insulating core
Tap lead
Coil wire
Insulating core
36 to 48"
Insulated 10-strand tap-lead wire
Polyester insulating strip
Insulating core
Coil wire
Detail B (completed assemblies)
Upper electrode
Filler metal (BCuP-5)
Tap-lead wire
Plastic insulation
Before brazing
Coil wire
Lower electrode
75°
Tap-lead wire
Coil wire
After brazing and trimming
Tap-lead wire
Coil wire
Polyester insulating strip
After insulating and bending

Figure **10-5** Arrangement of cross-wire resistance brazing of copper-wire tap leads to solid wire while the solid wire is being wound into a coil. (By permission, from *Metals Handbook* Volume 6, Copyright American Society for Metals, 1971.)

for making joints by capillary attraction between closely filleted surfaces.

The AWS classification system described in AWS A5.8 for brazing filler metals is based on chemical composition rather than on mechanical property requirements. The mechanical properties of a brazed joint depend, among other things, on the base metal and filler metal used. Therefore, a classification system based on mechanical properties would be misleading; it would only apply if the brazing filler metal were used on a given base metal. If it is necessary to determine the mechanical properties of a given base metal and filler metal combination, the joint should be tested as specified in AWS C3.2, *Standard Method of Evaluating the Strength of Brazed Joints.*

The seven basic filler metal groups, listed in Table 10-2, are identified by the principal element or elements in their chemical composition. For example, in the filler metal designation BCuP-2, the *B* identifies it as a brazing filler metal (like a prefix *E* identifies electrodes and *R* designates welding rods in other AWS specifications). The *RB* in RBCuZn-A indicates that the filler metal is suitable as a welding rod and as a brazing filler metal. CuZn means copper-zinc, the two principal elements in this particular brazing filler metal. Similarly, in other brazing filler metals, Si is for silicon, Ag is for silver, and so on, using standard chemical symbols. The numeral or letter suffix denotes one particular chemical analysis within a group (1 is one analysis within the same group and 2, another, similar to the classification of filler metals used for surfacing).

In addition to the 14 classes of filler metals described in AWS A5.8, there are a number of proprietary filler metals available with uncommon metals that are not, as of this writing, in such wide use as to have become industrial standards.

Brazing filler metals are commonly available in the form of rings, foil, strips, slugs, powder, and paste.

10-3 BRAZING FLUXES

Brazing fluxes (Table 10-3) are commercially available in the form of paste, liquid, or powder. The so-called general purpose fluxes consist mainly of borax or mixtures of borax and boric acid.

For brazing aluminum, fluxes containing fluorides and chlorides are generally specified and can be obtained from such companies as Eutectic Welding Alloys Corporation, Air Reduction Sales Corporation, Linde Air Products Company, and Aluminum Company of America, to name a few. Each composition may be best suited to one particular brazing method. For example, Alcoa No. 33 is a hand-brazing flux, Alcoa No. 34 is a furnace-brazing compostion that is slightly less active than No. 33, and Alcoa No. 52 is recommended for furnace brazing.

Table **10-2** The seven basic brazing filler-metal groups

AWS CLASSIFICATION	ALUMINUM SILICON			COPPER PHOSPHORUS				
	BAISi-2	BAISi-3	BAISi-4	BCuP-1	BCuP-2	BCuP-3	BCuP-4	BCuP-5
Air Products & Chemicals, Inc., K.G.M. Division					KGM PCO	KGM PCO-5		KGM PCO-15
Air Reduction Sales Co.			Airco 718			Aircosil 5		Aircosil 15
Alloy Specialties Co.				Alloy No. 1300-B	Alloy No. 1300	Alloy No. 1220	Alloy No. 1200	Alloy No. 1180
All-State Welding Alloys Co., Inc.		All-State No. 716	All-State No. 718		Silflo 0 All-State No. 21	Silflo 5	All-State No. 29	Silflo 15
Aluminum Co. of America		716 Altigweld, & coiled	718 Altigweld, & coiled					
Aluminum Wire Products Co., Inc.			AWP 718					
American Brazing Alloys Corp.		Ambraze 716	Ambraze 718	Phos-Sil(1)	Phos-Sil 0	Phos-Sil 6	Phos-Sil 6F	Phos-Sil 15
Arcos Corp.		Alumar-716	Alumar-718					
Belmont Smelting & Refining Wks., Inc.				Belmont BCuP-1	Belmont BCuP-2			
Dalweld Company, Inc.		Surebraze 716	Surebraze 718	Phosphorus Copper-0	Phosphorus Copper-0	Phosphorus Silver-6	Phosphorus Silver-6F	Phosphorus Silver-15

Engelhard Industries, Inc., American Platinum & Silver Div.						Silvaloy 5		Silvaloy 15
Eutectic Welding Alloys Corp.			EutecRod 900X		EutecRod 187			EutecRod 1803
Fusion Engineering Co.				P-1-1300	P-2-1300	SP-3-1200	SP-4-1200	SP-5-1300
Goldsmith Bros Div., National Lead Co.					G. B. No. 1300	G. B. No. 6		G.B. No. 15
Handy & Harman			Handy Alumibraze			Sil-Fos-5		Sil-Fos
Marquette Mfg. Co.		No. 716			No. 1301		No. 1306	No. 1300
National Cylinder Gas Div. of Chemetron Corp.			NCG 718					
Pacific Welding Alloys Mfg. Co.	Pacific 20	Pacific 30	Pacific 40	Pacific Cu 1	Pacific Cu 2			
United Wire & Supply Corp.					Phoson-0	Phoson-6, 5		Phoson-15
United States Welding Alloys Corp.		US 716	US 718					
Welco Alloys Corp.		Welco 10	Welco 120	Welco 20			Welco 602	
Westinghouse Elec. Corp.				Phos Copper 5	Phos Copper	Phos Silver 5	Phos Silver 6	Phos Silver 15

Source: 1964–1965 *Welding Data Book*, pp. E132–133. Courtesy of *Welding Data Book/Welding Design and Fabrication Magazine.*

Table **10-2** *(cont.)*

AWS CLASSIFICATION	SILVER												
	B Ag-1	B Ag-1a	B Ag-2	B Ag-3	B Ag-4	B Ag-5	B Ag-6	B Ag-7	B Ag-8	B Ag-8a	BAg-13	BAg-18	BAg-19
Air Products & Chemicals, Inc., K.G.M. Division													
Air Reduction Sales Co.	Aircosil 45		Aircosil 35	Aircosil 3	Aircosil E	Aircosil G	Aircosil H	Aircosil J	Aircosil M				
Alloy Specialites Co.	Silver Solder No. 45	Silver Solder No. 50	Silver Solder No. 35	Silver Solder No. 350		Silver Solder No. 145							
All-State Welding Alloys Co., Inc.	All-State No. 101				All-State No. 100			All-State No. 155					
Aluminum Co. of America													
Aluminum Wire Products Co., Inc.													
American Brazing Alloys Corp.	Ambraze 45	Ambraze 50	Ambraze 35	Ambraze 350	Ambraze 24	Ambraze 14	Ambraze 25	Ambraze 56	Ambraze 72				
Arcos Corp.													
Belmont Smelting & Refining Wks., Inc.													
Dalweld Company, Inc.	Surebraze 45	Surebraze 50	Surebraze 35	Surebraze 530	Surebraze 420	Surebraze 541	Surebraze 520	Surebraze 560	Surebraze 721				

Engelhard Industries, Inc., American Platinum & Silver Div.	Silvaloy 45	Silvaloy 50	Silvaloy 35	Silvaloy 503	Silvaloy 250	Silvaloy A-18	Silvaloy A-25	Silvaloy A-355	Silvaloy 301				
Eutectic Welding Alloys Corp.					EutecRod 200				EutecRod 1806				
Fusion Engineering Co.	S-4-1000	S-4-1050	S-4-1100	S-4-1200	S-4-1240	S-4-1250	S-4-1275	S-5-1150	Eut. 1400				
Goldsmith Bros Div., National Lead Co.	G.B. No.45	G.B. No. 50	G.B. No. 35	G.B. No. 350	G.B. No. 240	G.B. No. 145	G.B. No. 250	G.B. No. 72					
Handy & Harman	Easy-Flo 45	Easy-Flo	Easy-Flo 35	Easy-Flo 3	Braze SS	Braze DE	Braze ETX	Braze 560	Braze BT	Lithobraze BT	Braze 541	Braze 603	Lithobraze 925
Marquette Mfg. Co.	No. 1175	1050	1335	1350	1240	1145	1250	1356	1372				
National Cylinder Gas Div. of Chemetron Corp.													
Pacific Welding Alloys Mfg. Co.		Pacific 1800	Pacific 1801	Pacific 1802	Pacific 1803								
United Wire & Supply Corp.	SilBond 45	SilBond 50	SilBond 35	SilBond 40N	Sil 40N	Sil 45	Sil 50	Sil 56T	Sil 72				
United States Welding Alloys Corp.													
Welco Alloys Corp.													
Westinghouse Elec. Corp.	Co-Silver 45C	Co-Silver 50C	Co-Silver 35C	Co-Silver 50N	Co-Silver 40N	Co-Silver 45	Co-Silver 50	Co-Silver 56T	Co-Silver 72-28				

Table **10-2** (*cont.*)

AWS CLASSIFICATION	PRECIOUS METALS				COPPER & COPPER ZINC			
	BAu-1	BAu-2	BAu-3	BAu-4	BCu-1	BCu-1a	RBCuZn-A	RBCuZn-D
Air Reduction Sales Co.							Airco 20	
All-State Welding Alloys Co., Inc.								All-State No. 11 & 13
American Brazing Alloys Corp.					Ambraze Copper		Naval Bronze	Nickel Silver
Dalweld Company, Inc.							Dalweld #48	Nickel Silver
Fusion Engineering Co.							B-1600	B-1700
Handy & Harman	Premabraze 401	Premabraze 403	Premabraze 129	Premabraze 130				
Pacific Welding Alloys Mfg. Co.					Pacific Cu 100	Pacific Cu 200	Pacific CuZn A	Pacific CuZn D
United States Welding Alloys Corp.								US Nickel Silver

	MAGNESIUM			NICKEL						
	BMg-1	BMg-2	BMg-2a	BNi-1	BNi-2	BNi-3	BNi-4	BNi-5	BNi-6	BNi-7
All-State Welding Alloys Co., Inc.	All-State No. 61 Spoolarc AZ92A	All-State No. 63 Spoolarc AZ61A								
American Brazing Alloys Corp.	Ambraze 61A									
Dalweld Company, Inc.	Dalweld 61 A									
Pacific Welding Alloys Mfg. Co.	Pacific Mg 101	Pacific Mg 102	Pacific Mg 103	Pacific Ni 1001	Pacific Ni 1002	Pacific Ni 1003				
Wall Colmonoy Corp.				Nicrobraz 120, Standard	LM Nicrobraz	Nicrobraz 130	Nicrobraz 135	Nicrobraz 30	Nicrobraz 10	Nicrobraz 50
Welco Alloys Corp.	Welco 60									

Table **10-3** General data on applications for commercially available brazing fluxes*

AWS BRAZING FLUX TYPE NO.	METAL COMBINATIONS FOR WHICH VARIOUS FLUXES ARE SUITABLE		EFFECTIVE TEMPERATURE RANGE OF FLUX, °F	MAJOR CONSTITUENTS OF FLUX	PHYSICAL FORM	METHODS OF APPLICATION†
	BASE METALS	*FILLER METALS*				
1	Aluminum and aluminum alloys	BAISi	700–1190	Fluorides; Chlorides	Powder	1,2,3,4,
2	Magnesium alloys	BMg	900–1200	Fluorides; Chlorides	Powder	3,4,
3A	Copper and copper-base alloys (except those with aluminum) iron-base alloys; cast iron; carbon and alloy steel; nickel and nickel-base alloys; stainless steels; precious metals (gold, silver, palladium, etc.)‡	BCuP BAg	1050–1600	Boric Acid, Borates, Fluorides, Fluoborate Wetting Agent	Powder Paste Liquid	1,2,3,
3B	Copper and copper-base alloys (except those with aluminum); iron-base alloys; cast iron; carbon and alloy steel; nickel and nickel-base alloys; stainless steels; precious metals (gold, silver, palladium, etc.)	BCu BCuP BAg BAu RBCuZn BNi	1350–2100	Boric Acid Borates Fluorides Fluoborate Wetting Agent	Powder Paste Liquid	1,2,3,
4	Aluminum-bronze; aluminum-brass§	BAg, BCuZn, BCuP	1050–1600	Borates Fluorides Chlorides	Powder Paste	1,2,3,
5	Copper and copper-base alloys (except those with aluminum) nickel and nickel-base alloys; stainless steels; carbon and alloy steels; cast iron and miscellaneous iron-base alloys; precious metals (except gold and silver)	BCu, BCuP BAg-(8–19) BAu, BCuZn BNi	1400–2200	Borax Boric Acid Borates	Powder Paste Liquid	1,2,3,

*This table provides a guide for classification of most of the proprietary fluxes available commercially.
For additional data consult AWS specification for brazing filler metal A5.8; consult also AWS Brazing Manual, 1963 Ed.
†1—Sprinkle dry powder on joint; 2—dip heated filler metal rod in powder or paste; 3—mix to paste consistency with water, alcohol, monochlorobenzene, etc.; 4—molten flux bath.
‡Some Type 3A fluxes are specifically recommended for base metals listed under Type 4.
§In some cases Type 1 flux may be used on base metals listed under Type 4.

Source: American Welding Society, *Welding Handbook*, 6th ed., section 3B, p. 60.59.

10-4 PREPARING THE PARTS TO BE BRAZED

For proper brazing results, cleanliness and proper positioning of the parts of an assembly are essential. Removal of oxides, dirt, grease, moisture, and oil is necessary for complete cohesion and penetration of the brazing material throughout the joint. When flux protects against oxidation during heating, it must effectively prevent the oxidizing elements in the air from reaching the surfaces of the joint. But when an inert gas like argon or helium furnishes protection, little additional protection of the joint area is required.

The cleaning processes in common use include acids, solvents, vapor degreasing, power brushing, grinding, sandblasting, and others. Of these methods, only power brushing, solvents, and grinding are used individually or in combination for production- and repair-type brazing.

Prior to using solvents:

- Make certain that the area is properly ventilated.
- Wear gloves or use special skin ointments to prevent extraction of natural oils from the skin and minimize the risk of getting dermatitis.
- Do not use solvents close to open flames or in areas where the temperature is close to, or in excess of, the flashpoint of the solvent.

Caution: It must also be pointed out that solvents must be disposed of in accordance with federal, state, county, and local environmental regulations. Solvent wastes can usually be disposed of by pouring on dry ground, well away from buildings. In no case should solvent wastes be burned in furnaces or heaters or discharged into sewers or streams.

Removal of oxides and tarnish before brazing can be done mechanically by filing, grinding, tumbling, or shot blasting. Sandblasting or wire brushing are not recommended. With sandblasting, any particles of sand retained by the surface interfere with the flow of brazing material. Wire brushing sometimes smears metal over the oxide rather than removing it. Pickling is used when there is no clean, machined surface.

Grease and oil are best removed by degreasing. Their removal is important, because even during the rapid heating of the joint area they are usually not burned away entirely. They leave a carbonaceous residue that may seriously interfere with the flow of the brazing alloy.

Parts to be production brazed are usually assembled by self-jigging. *Self-jigging* is a method of assembly in which the component parts incorporate design features that will ensure that when assembled, they will remain in proper relationship throughout the brazing cycle without the aid of auxiliary fixtures. Self-jigging is the pre-

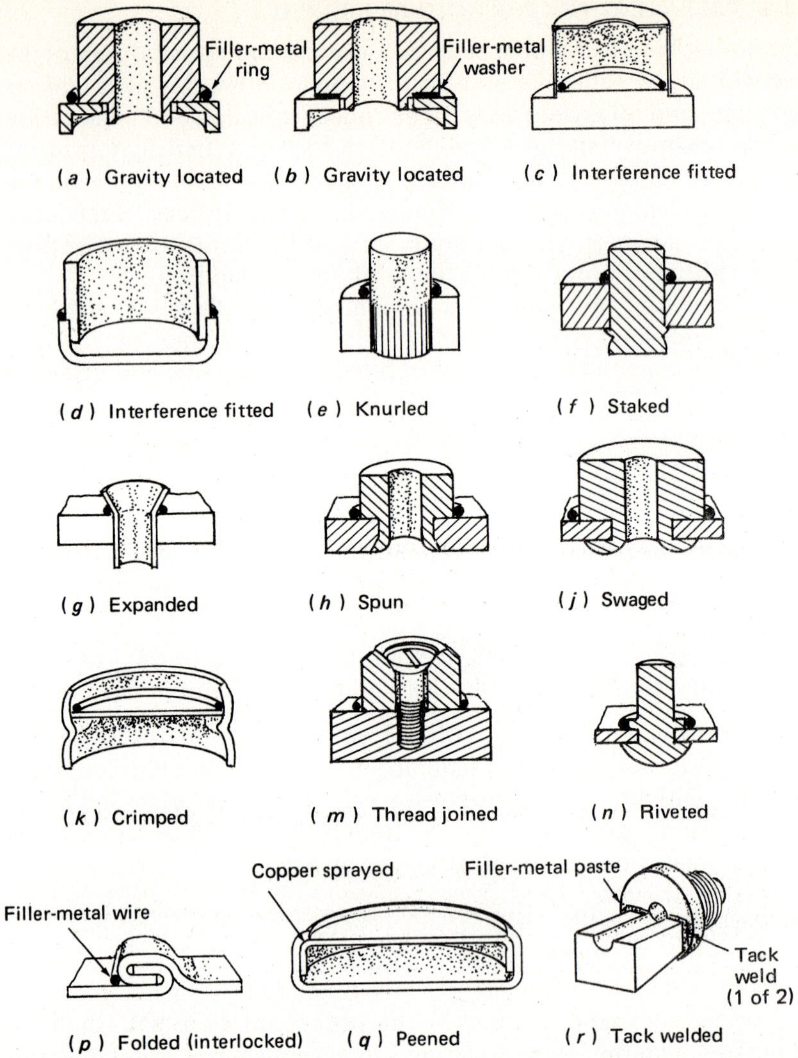
Filler-metal ring
Filler-metal washer
(a) Gravity located
(b) Gravity located
(c) Interference fitted
(d) Interference fitted
(e) Knurled
(f) Staked
(g) Expanded
(h) Spun
(j) Swaged
(k) Crimped
(m) Thread joined
(n) Riveted
Copper sprayed
Filler-metal paste
Filler-metal wire
Tack weld (1 of 2)
(p) Folded (interlocked)
(q) Peened
(r) Tack welded

Figure **10-6** Common methods of self-jigging. (By permission, from *Metals Handbook* Volume 6, Copyright American Society for Metals, 1971.)

ferred method of assembly, as it eliminates the initial and replacement cost of auxiliary fixtures and the cost of heating them during brazing. It is usually the better method of holding components. The various methods by which self-jigging can be accomplished (see Figure 10-6) are discussed below.

GRAVITY LOCATING The simplest method of assembling two components is to rest one on top of the other with the brazing filler metal either wrapped around one component near the joint [Figure

10-6(*a*)] or placed between components [Figure 10-6(*b*)]. The principal disadvantage of gravity locating may be the lack of a dependable means of orienting the components or keeping them from moving in relation to one another. Nevertheless, some production components are assembled in this manner, especially those in which the upper component is relatively heavy.

INTERFERENCE OR PRESS FITTING This requires the expansion or contraction of mating component surfaces. In Figure 10-6(*c*), the cup has an inside diameter that is smaller than that of the mating projection of the underlying plate. The extent of interference seldom exceeds about 0.001 inch, per inch of diameter, up to about 3-inch diameters. Nevertheless, most interference fits require considerable force to achieve assembly. This force is generally provided by an arbor press or similar tool. Thus an interference fit is a press fit. Lighter interference fits, such as that shown in Figure 10-6(*d*), may provide zero clearance or a very slight gap between the mating surfaces of components. These, too, require some external force, such as that provided by an arbor press, to achieve assembly. Fits with zero clearance are referred to as *size-to-size fits*. Some method is used to prevent slippage when the components are heated in the furnace, particularly if the joint is a vertical axis. In Figure 10-6(*d*), a shoulder on one of the components is used to ensure stability.

KNURLING In high-production manufacturing, there will be considerable variation in joint clearance among the assemblies being brazed. Typical brazed assemblies, in which a round male member is fitted to a female member, are subject to either of two conditions: (1) the male part is off-center, thus allowing all the clearance on one side, or (2) the male member is out-of-round, so that all of the clearance will be on two opposite sides with no clearance (or even interference) on the other two sides.

Knurling the end of the male member [Figure 10-6(*e*)] is sometimes a way of correcting the conditions described above and obtaining uniformity among brazed joints. Often, knurling can be done during machining of the part, thus adding very little to the cost. If knurling must be done in a secondary operation, the extra cost can often be balanced against the cost of the rejects that would be encountered if knurling were not done.

When the male member is tubular, prick punching may be substituted for knurling. Usually two rows of prick punch marks near the end of the male member are sufficient. Prick punching is easily done, but since it involves a secondary operation, the cost must be justified.

STAKING Figure 10-6(*f*) shows how staking will effectively lock two components in position. Burrs are turned up on the shaft by driving a punch into it. This method is commonly used to retain the orientation of such assemblies as cams, levers, and gears on shafts or on a common tube. It is sometimes used as a substitute for tack welding, knurling, or interference fitting.

EXPANDING This method is commonly used for the assembly of tubes to tube sheets. The tubular component is pressed into a header sheet and expanded in the hole to lock the assembly, as shown in Figure 10-6(g). Rings of brazing filler metal can be placed over the tube before or after the expanding operation. To avoid obtaining a mere line contact in expanding, a leader can be placed on the expanding tool to project into the tube and support the tube wall while the end of the tube is being flared.

SPINNING When the diameter of a hole in an assembly cannot be altered during assembly, as when a hub is fastened to a lever, the assembly can be locked together [as shown in Figure 10-6(*h*)] by spinning in a riveting machine. The same result can be obtained by flaring the tenon in a press. Tolerances for the punched hole and tenon must be held closely to ensure the close joint clearances required for brazing. The punched hole must be chamfered to allow room for the spun or pressed end of the tenon. The spinning method of assembly is used for parts of various types of business machines, many of which were formerly assembled by cross-drilling and pinning of the hubs.

SWAGING An inexpensive and effective method of assembling a stud in the hole in a hollow body is to *swage* it in place, as shown in Figure 10-6(*j*). This method is acceptable when it is not necessary to maintain accuracy of the diameter of the hole in the hub, and when the projection on the flange can be tolerated. The principal advantage of swaging is that close tolerances do not have to be held on the tenon or the punched hole, because the swaging operation forces the components into intimate contact. In addition to other applications, swaging has been used to assemble a valve body in the float chamber of refrigerators. The resulting bond after furnace copper brazing is strong, tight, leak-proof, and capable of withstanding high pressure.

CRIMPING Figure 10-6(*k*) shows the assembly of a disk, shell, and copper filler-metal ring in which the disk and ring are held in place by crimping the end of the shell. Also shown is an inexpensive method of forming "stoppers," or indentations around the shell, against which the disk rests.

In general, it is preferable to set an assembly of this type on end in the furnace so that the filler metal will flow down through the joints. However, if the tubular component is long, the assembly must be laid on its side to clear through the furnace; so an oversize ring of hard copper wire filler metal can be sprung in place close to the joint. If the diameter of the tube is 2 inches or more, the filler-metal wire and adjoining steel surfaces should be coated with copper powder paste, which will harden and prevent the wire from sagging away from the joint at the top as the assembly is heated. The paste also provides an auxiliary supply of filler metal.

THREAD JOINING This has been used for assembling components of replacement punch holders for die sets used in punch presses. As shown in Figure 10-6(*m*), the shank is held in place on the punch-holder plate by a screw. Because drilling and tapping are required, this method of assembly is generally limited to small production quantities.

RIVETING Riveting is illustrated in Figure 10-6(*n*), and is a modification of the spinning and swaging methods that use a rivet as part of the assembly. It is widely used to assemble the vanes for the outer disks of fan wheels before furnace copper brazing. The combination of riveting and copper brazing markedly extends the service life of the assembly.

FOLDING OR INTERLOCKING Several methods and designs of folding and interlocking can be used to secure joints, such as that shown in Figure 10-6(*p*). These methods are widely used in the manufacture of brazed tubing or tubular assemblies. Copper filler metal is supplied either in the form of copper plating or by welding a copper wire at the base of the joint. Capillary action will draw the filler metal to all areas of the joint.

PEENING Assembly of two hollow shells by the peening method is shown in Figure 10-6(*q*). The stamped components are pressed together and the outer shell is peened with an air hammer along the periphery. To apply filler metal to an assembly of this type, copper can be sprayed on the joint interfaces with an oxyacetylene spray gun before assembly.

TACK WELDING Prior to copper brazing, the tip and shank of the electrode holder for an atomic-hydrogen welding torch are assembled by tack welding, as shown in Figure 10-6(*r*). The filler metal consists of a small amount of copper powder paste daubed around the joint. Any oxide formed during tack welding is reduced by the protective atmosphere during furnace brazing.

The tack-welding method of assembly usually requires careful investigation to determine the most strategic point or points for placing the weld. For economy, the number of tack welds per assembly should be held to a minimum.

When it is not possible or practical to self-jig a joint for brazing, the lap joint (Figure 10-7) is the preferred type of joint. These joints depend for their strength on penetration between close, conforming surfaces, rather than on external fillets; the joints are usually intended to be stressed in shear. A rule of thumb is to make the length of the joint at least three times the thickness of the thinnest section. However, when more precise work must be performed, the length of the joint of lap area can be designed for any desired safety factor by use of the following formulas. For tubular joints:

$$X = \frac{W(D - W)Y \cdot T}{LD}$$

For flat joints:

$$X = \frac{Y \cdot T \cdot W}{L}$$

where:

X = length of lap in inches

W = wall thickness of weaker member in inches

D = diameter of shear area in inches

Y = safety factor (usually 4 or 5)

L = shear strength of brazing alloy in lb/in.^{2}

T = Tensile strength of weaker member in lb/in.^{2}

The value of shear strength is taken as a minimum value of 25,000 psi for all brazing alloys containing appreciable amounts of silver.

In joining a 0.050-inch annealed monel sheet of 70,000-psi tensile strength to a metal of equal or greater strength, using a brazing alloy

Figure **10-7** Standard method for calculating required minimum overlap for brazed tubular and lap joints. (Redrawn by permission from American Welding Society, *Welding Handbook*, 6th ed., 1971, section 3B, p. 60.19.)

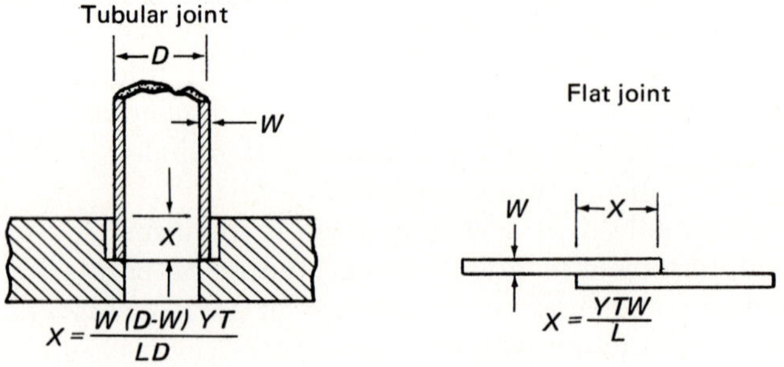

Tubular joint
D
W
X
X = W (D-W) YT / LD
Flat joint
W
X
X = YTW / L

with a shear strength of 25,000 psi, the depth of shear area, assuming a safety factor of 4, should be

$$X = \frac{4 \times 70{,}000 \times 0.050}{25{,}000} = 0.56\text{-inch lap}$$

In joining a ¾-inch copper tube having 0.064-inch wall thickness to a ¾-inch steel tube sheet (header), the depth of shear should be as follows (when $T = 33{,}000$ psi and $Y = 5$):

$$X = \frac{0.64 \times (0.75 - 0.064) \times 5 \times 33{,}000}{25{,}000 \times 0.75} = 0.486\text{-inch lap}$$

The maximum shear strength of a joint depends on the ductility of the filler metal and base metal, and is attained by maintaining proper clearance on the joint. A good rule of thumb is to limit filler-metal stretch to 50% on cooling. Thus a starting clearance can be selected (Figure 10-8) based on the change of gap, which will prevent the alloy from stretching more than 50% (base-metal stretch is not compensated for).

A serious condition results when the female member of a tube-socket joint has a lower coefficient of thermal expansion than the male: the gap spreads on cooling and puts tension on the brazing alloy.

It takes two steps to select a room temperature gap that will limit a stretch to 50%. First, find the gap change, on nomograph, that occurs during heating to brazing temperature; then multiply by 3. The nomograph is based on the equation:

$$\Delta C_D = D\Delta T(\alpha_2 - \alpha_1)$$

where:

ΔC_D = gap change between room temperature and brazing temperature (in inches)

D = nominal diameter of the joint (in inches)

ΔT = net difference between room temperature and solidus temperature of the brazing alloy (°F)

α_2 = mean value of expansion coefficient for base metal of *female* member over the temperature range (inches/inches/°F)

α_1 = mean value of expansion coefficient for base metal of *male* member over the temperature range T (inches/inches/°F)

In cases where the nomograph cannot be read with sufficient accuracy or one of the variables is off the scale, the value of ΔC_D can be calculated by using the equation on which the nomograph is based.

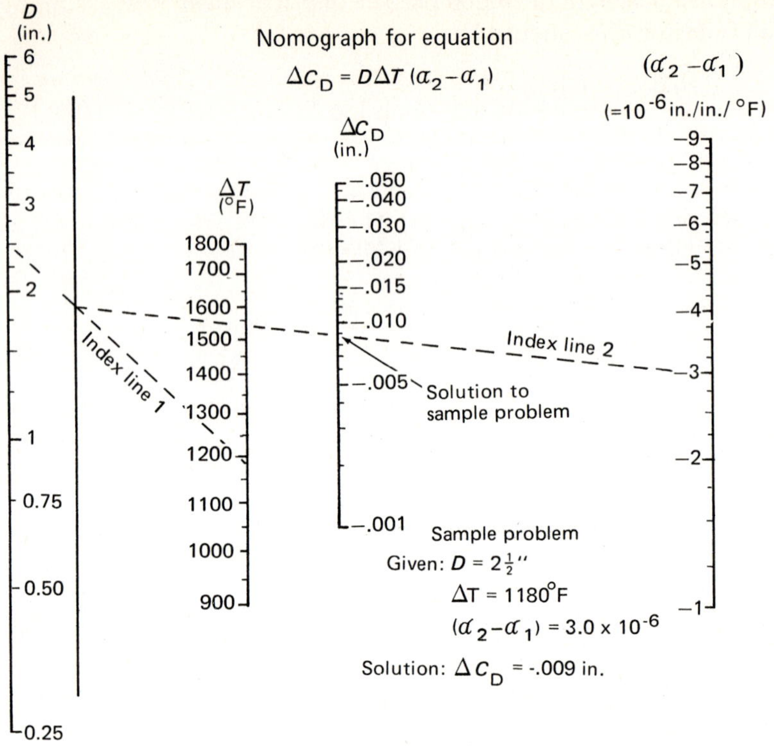

D (in.)
Nomograph for equation
ΔC_D = DΔT (α2−α1)
(α2 −α1)
(=10^-6 in./in./°F)
ΔC_D (in.)
ΔT (°F)
6
5
4
3
2
1
0.75
0.50
0.25
1800
1700
1600
1500
1400
1300
1200
1100
1000
900
−.050
−.040
−.030
−.020
−.015
−.010
−.005
−.001
−9
−8
−7
−6
−5
−4
−3
−2
−1
Index line 1
Index line 2
Solution to sample problem
Sample problem
Given: D = 2½"
ΔT = 1180°F
(α2−α1) = 3.0 x 10^-6
Solution: ΔC_D = -.009 in.

Notes:

1 This nomograph gives change in diameter caused by heating. Clearance to promote brazing filler metal flow must be provided at brazing temperature.

2 D = nominal diameter of joint, inches
C_D = change in clearance, inches
T = brazing temperature minus room temperature, °F
α_1 = mean coefficient of thermal expansion, male member, in./ in./ °F
α_2 = mean coefficient of thermal expansion, female member, in./ in./ °F

3 This nomograph assumes a case where α_1 exceeds α_2, so that scale value for $(\alpha_2-\alpha_1)$ is negative. Resultant values for ΔC_D are therefore also negative, signifying that the joint gap reduces upon heating. Where $(\alpha_2-\alpha_1)$ is positive, values of ΔC_D are read as positive, signifying enlargement of the joint gap upon heating.

Figure **10-8** Standard nomograph for calculating minimum clearance for brazed tubular joints. (Redrawn by permission from American Welding Society, *Welding Handbook*, 6th ed., 1971, section 3B, p. 60.19.)

10-5 BASE AND FILLER-METAL EFFECTS REQUIRING SPECIAL TREATMENTS

Unless special treatments are employed with certain base and filler metals, they will result in poor-quality brazed joints. These phenomena and the treatments found useful to keep them to a minimum are discussed in the following paragraphs.

CARBIDE PRECIPITATION Carbide precipitation means that the carbon combines preferentially with the chromium and is rejected as chromium carbide, usually at the grain boundary. Carbide precipitation occurs when alloys containing chromium are heated to temperatures ranging from approximately 800 to 1500°F (450 to 820°C). Carbide precipitation decreases the corrosion resistance of the heat-affected zone. Carbide precipitation can be kept to a minimum by postbraze heating the brazed part to a temperature anywhere from 1850 to 2050°F (1000 to 1100°C), maintaining it at that temperature for 2 hours, and then allowing it to cool in the atmosphere.

RESIDUAL OXIDES Residual oxides—oxides remaining on the surface of the base metal after it has been cleaned by the usual cleaning procedures—of aluminum, titanium, silicon, magnesium, manganese, and beryllium can be reduced by special fluxes. Chromium oxides may be removed with fluoride-bearing fluxes.

HYDROGEN EMBRITTLEMENT Hydrogen embrittlement occurs when metals that have not completely deoxidized during casting and solidification processes are brazed in hydrogen-containing atmospheres. It is due to the fact that the molecules of water vapor (which form when the trapped oxygen is combined with diffused hydrogen) are too large to diffuse to the surface. The trapped water vapor creates a large amount of pressure inside the metal structure and tears it apart, thereby lowering its tensile strength. Hydrogen embrittlement of nonferrous metals can be minimized by avoiding the use of hydrogen brazing atmospheres. Hydrogen embrittlement of ferrous metals can be minimized by heating the metal throughout to 300°F and maintaining it at that temperature until all the hydrogen has diffused. The holding or baking time can be anywhere from a few hours to a few days.

SULFUR EMBRITTLEMENT Sulfur embrittlement occurs when nickel and nickel-copper alloys are brazed or preheated in a sulfur-type atmosphere. It results from a brittle and weak, low-melting nickel-sulfide compound that forms at the grain boundaries of the base metal, causing it to crack. Sulfur embrittled material cannot be salvaged and is usually scrapped.

PHOSPHORUS EMBRITTLEMENT Phosphorus embrittlement occurs when phosphorus combines with other metals to form phosphides. Phosphorus embrittlement can be minimized by avoiding the use of copper-phosphorus filler metals with iron or nickel-base alloys.

VAPOR PRESSURE Vapor pressure is the pressure generated by the vaporization of some elements in the filler metal when the filler metal is heated to high temperatures in a vacuum. Great care must be taken when selecting filler metals for brazing parts that must operate at high temperatures in a vacuum.

STRESS CRACKING Stress cracking is known to occur frequently when age-hardenable materials that have high annealing temperatures are brazed. Stress cracks occur almost instantaneously during brazing and are usually visible because the filler metal tends to flow into the cracks. The occurence of stress cracking can be reduced by:

- Using annealed-temper rather than hard-temper material.
- Annealing cold-worked parts prior to brazing.
- Removing the source of externally applied stress, such as parts that do not fit properly, jigs that exert stress on the parts, or overhanging unsupported weights.
- Redesigning parts or revising joint design.
- Heating at slower rates. Heavy parts can be heated so rapidly that stresses are set up by steep thermal gradients.
- Heating the fluxed and assembled parts in a torch-brazing application to a high enough temperature to affect stress relief, cooling to the brazing temperature, and then hand feeding the filler metal.
- Selecting a brazing filler metal that is less likely to induce this type of damage.

POSTBRAZE HEAT TREATING Postbraze heat treatments are often used to improve the mechanical properties of a brazed part. Care must be taken that the postbraze temperatures be kept lower than the melting-point temperature of the filler metal.

ALLOYING The term *alloying* as used here refers to the interaction between filler and base metal by diffusion and dissolution. Both diffusion and dissolution occur when brazing nickel, monel, or cupro-nickels, using copper as a filler metal. Dissolution often occurs when brazing base metals having relatively thin (0.032-inch) cross-sections, and when brazing aluminum and magnesium. Diffusion is known to occur when brazing high-temperature, ferrous-base alloys with filler metals containing boron. Diffusion can also occur when brazing gold to base (nonprecious) metals at a slow pace.

REVIEW QUESTIONS

1. What is the AWS definition of brazing?

2. How many brazing methods are there? Name them and give a short description of each.

3. Furnace brazing is best for brazing what type of joint(s)?

4. All welders should be able to braze with what process?

5. What atmosphere should be used for brazing nickel and high-nickel alloys?

6. In which of the brazing processes are filler-metal preforms commonly used?

7. Describe the principles of (a) resistance brazing; (b) salt bath brazing.

8. What is the AWS type number of the flux used for brazing carbon and alloy steels when using AWS type BCu filler metal?

9. How can the waste of the solvent used for cleaning the parts to be brazed be disposed of?

10. What is the name of the simplest method commonly used to assemble two components to be brazed? What are the disadvantages of this method? When is it commonly used?

11. What is meant by the term *self-jigging*?

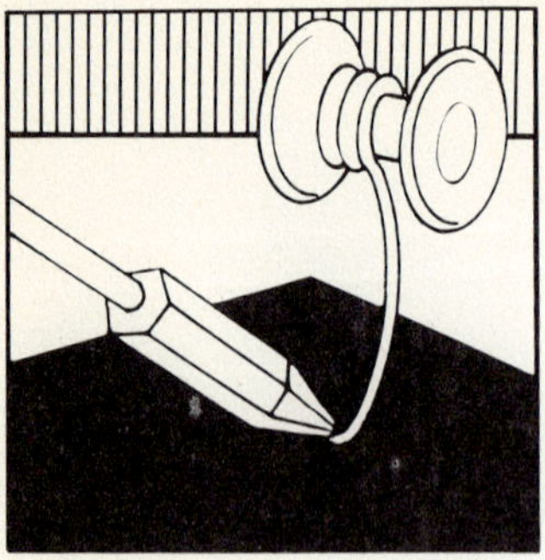

CHAPTER 11
SOLDERING

The soldering process is another of the low-temperature joining processes and involves a metallurgical or solvent action between the solder and the metal being joined (Figure 11-1). The joint is chemical in character rather than purely physical, because the attachment is formed in part by chemical (capillary) action rather than by mere physical attraction. The properties of a solder joint are different from those of the original solder, because during the soldering process the solder is partly converted to a new and different alloy due to a solvent action between the respective metals. The physical properties of this new alloy are not necessarily the same as those of the original solder. The physical properties of a soldered connection depend, therefore, on the extent to which alloy formation has taken place during soldering, and are subject to wide variation due to the variables listed below:

- Inadequate type and amount of flux and solder for metal being joined.
- Inadequate preparation of joint.
- Insufficient amount of heat applied to the seam.
- Insufficient cooling time.

11-1 KINDS OF SOLDER

Most solders are alloys of tin and lead. The percentage composition of tin and lead determines the physical and mechanical properties (Table 11-1). Solder is available in many forms—bar, stick, foil, wire, strip, or powder. It also can be obtained in circular or semicircular rings, although these forms are generally reserved for brazing alloys.

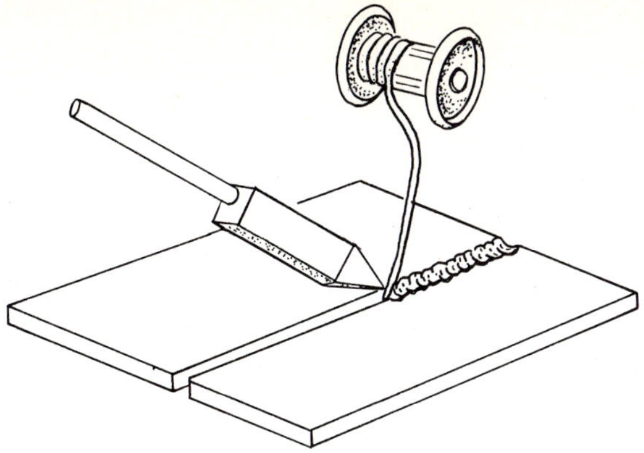

Figure **11-1** The soldering process. (Redrawn by permission from *Fundamentals of Service Welding*, John Deere Service Publications, Fig. 86.)

Table **11-1** Melting range of some tin-lead solders

NOMINAL COMPOSITION, PERCENT			
TIN	*LEAD*	MELTING RANGE, °F	TYPICAL USES
4	96	560–600	For coating metals and differential soldering
10	90	515–575	For coating metals and differential soldering
15	85	435–555	For coating and joining metals
30	70	360–495	General-use solder
33	67	360–485	General-use solder
38	62	360–465	General-use solder
40	60	360–460	General-use solder
45	55	360–440	Hermetic sealing
50	50	360–415	Special soldering applications
60	40	360–370	For low-temperature soldering
62	38	360–360	Eutectic solder of fixed melting point
75	25	360–380	Special soldering applications

11-2 FLUXES

The function of the flux (Table 11-2) is to remove the nonmetallic oxide film from the metal surface and keep it removed during the soldering operation so that the clean, free metals may make mutual metallic contact. The flux does not consitute a part of the soldered joint. After soldering, the flux residue, still retaining its quota of captured oxides, lies inert on the surface of the soldered joint.

Table **11-2** Solder fluxes for various metals

BASE METAL, ALLOY, OR APPLIED FINISH	FLUX REQUIREMENTS			
	NONCORROSIVE	*CORROSIVE*	*SPECIAL FLUX AND/OR SOLDER*	*SOLDERING NOT RECOMMENDED*
Aluminum			X	
Aluminum-bronze			X	
Beryllium				X
Beryllium copper		X		
Brass	X	X		
Cadmium	X	X		
Cast iron			X	
Chromium				X
Copper	X	X		
Copper-chromium		X		
Copper-nickel		X		
Copper-silicon		X		
Gold	X			
Inconel			X	
Lead	X	X		
Magnesium			X	
Manganese-bronze (high tensile)				X
Monel		X		
Nickel		X		
Nichrome			X	
Palladium	X			
Platinum	X			
Rhodium		X		
Silver	X	X		
Stainless steel			X	
Steel		X		
Tin	X	X		
Tin-bronze	X	X		
Tin-lead	X	X		
Tin-nickel	X	X		
Tin-zinc	X	X		
Titanium				X
Zinc		X		
Zinc-die castings			X	

Source: *Fundamentals of Service Welding*, (John Deere Service Publications, 1971), p. 56.

Soldering fluxes can be divided into four groups: (1) the chloride or acid type, (2) the organic type, (3) the rosin or resin type, and (4) fluxes for aluminum.

11-3 THE APPLICATION OF SOLDER

It is essential that the metal being soldered be free of nonmetallic impurities. Embedded graphite and slag in castings; magnetic oxide on drawn steel; enamel, shellac, or rubber insulation on wire; and paint, rust, sulphides, or heavy oxides must all be removed from a metal prior to soldering, so that the molten solder may make metallic contact with the metal.

Electrodeposited platings must be removed either before soldering or by the alloy action of the solder during soldering; otherwise the solder will be attached to the plating but not to the base metal.

Since soldering is an operation involving an action between two metals, it is essential that the metal being soldered be as hot as the molten solder during the entire soldering operation. If it is not, the solder will only freeze to produce a *cold joint,* and no alloy action will take place. It is particularly important that the soldering iron or other source of heat be of high enough temperature (Table 11-3) to heat the metal to solder-melting and solder-alloying temperatures.

Soldering cannot be done without a flux. Even if a metal is clean, it rapidly acquires an oxide film of submicroscopic thickness due to the heat of soldering. Because this film insulates the metal from the solder, it must be broken up and removed by the soldering flux.

For successful soldering, the soldering iron must be *tinned* during use (Figure 11-2). The purpose of tinning a soldering iron is to provide a completely metallic surface through which the heat may flow readily from the iron to the metal being soldered; if there were no tinning, the hot iron would oxidize and the heat could not flow through the surface oxide film into the solder.

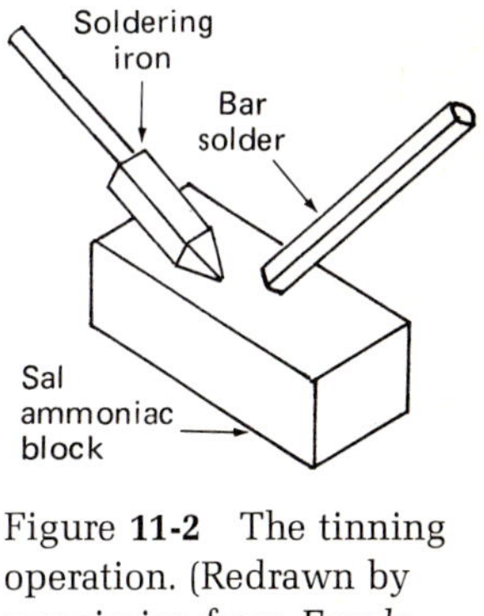
Soldering iron
Bar solder
Sal ammoniac block

Figure **11-2** The tinning operation. (Redrawn by permission from *Fundamentals of Service Welding,* John Deere Service Publications, Fig. 84.)

Because one hears a great deal about the tensile strength of solder and the strength of soldered joints, it is important to realize that the primary purpose of soldering is not to secure strength, but to secure a permanently sound, nonporous, continuous metallic connection that is not affected by temperature change and lends itself to minor torsional strain and stress without rupture. Whenever possible, it is advisable to make sure that a joint is mechanically secure before soldering, then to solder the connection and allow it to remain undisturbed until it is completely solid. The alloy attachment lies in the thin film of solder (approximately 0.004-inch thick) between the two metals being joined. To add solder after this alloy film has formed is superfluous and wasteful.

Table **11-3** Thermal capacities of soldering irons and the jobs for which they are used

WORK TO BE DONE	ELECTRICALLY HEATED IRONS: CHOICE OF BIT DIAM. (IN.)[1]	ELECTRICALLY HEATED IRONS: HEAT RATING (WATTS)	EXTERNALLY HEATED IRONS[2]: SIZE OF BIT (LB)
Printed circuit soldering.	Up to 1/4	35 Max.	1/4
Very light soldering on radios, telephones, electrical appliances, fuses, fine instruments, and for home use.	1/4–7/16	44–52	1/2
Medium soldering on switchboards, telephones, radios, electrical appliances and light manufacturing.	3/8–1/2	60–70	1
Fast soldering on radios, telephones, electrical appliances, jewelry, etc. For light medium jobs in home, factory and schools.	3/8–9/16	85–100	1 1/2
High-speed soldering on radios and telephones. Medium light soldering on tinware, toy motors, type bars, fuses, etc., tinsmithing, plumbing and wiring.	5/8–7/8	130–150	2
High-speed soldering on light tinware, art glass, toys, small metal patterns, organ pipes, etc.	5/8–1	170–200	2 1/2
Medium tinware, light roofing, gutters, ventilating flues, electrical, airplane and other medium manufacturing, shipboard repairs.	7/8–1 3/8	225–250	3
Roofing, refrigerators, copper and galvanized iron, heavy tinware, metal patterns, ship, auto and airplane building.	1 1/8–1 3/8	300–350	4
Heavy roofing and cornices, vats, tanks, ventilating flues, auto radiators, armatures, plumbing and shipbuilding.	1 3/8–1 7/8	350–650	5
Extra heavy duty work.	2	1250	10

[1]The bit diameters vary with the manufacturer of electrically heated irons and sizes used depend somewhat upon the working space available.

[2]Bits for externally heated irons are usually furnished in pairs. Thus a 1/2-lb bit is identified as "1 lb per pair."

Source: American Welding Society, *Soldering Manual*, p. 63.

Some metals solder easily, some solder with difficulty, and some do not solder at all. Copper, tin plate, and clean carbon steel solder easily with any of the three major types of fluxes. Brass, particularly if it is oxidized, does not solder well with conventional rosin, but responds well to the activated resins and some of the stabilized resins. Nickel plate, the nonferrous nickel alloys, and beryllium copper are not soldered with rosin flux, but solder very well with the activated resins. Cadmium plate and silver plate solder fairly well with rosin if the plating is clean and of good quality, and if the base metal is properly prepared prior to plating; otherwise these metals require an activated resin, particularly silver plate, which rapidly acquires a tarnish film of silver sulphide in sulfur atmospheres. Zinc, if clean or in a moderate state of oxidation, can be soldered with a good activated resin, but, if badly oxidized, requires a chloride- or organic-type flux. The nickel ferrous and chrome-nickel ferrous alloys require a chloride-type flux or, depending on the chromium content, a stainless steel flux, which is a special chloride flux with a high mineral acidity. Such metals as aluminum, tantalum, magnesium, and molybdenum can be soldered satisfactorily only with proprietary fluxes.

11-4 SOLDERING WITH THE SOLDERING IRON

Various soldering devices are illustrated in Figure 11-3. When using flux-core solder with an iron it is pertinent to remember that this product consists of two substances (solder and flux) that are physically and chemically very dissimilar. For instance, the flux, in most cases, is liquid or semiliquid at room temperatures, with a tendency to vaporize at 212°F, while the solder, depending upon composition, does not become liquid below 361–620°F. For this reason, the soldering flux, which is contained within the solder core, has a tendency to vaporize or decompose while the solder is being melted. Flux-core solder must be applied at the exact junction between the metal assembly and the soldering iron, so that solder and flux may be liberated simultaneously at the specific point where solder is desired and so that the flux may exert its activity at the point where it is needed.

In the application of any solder (Figure 11-4), it should be remembered that the sole function of the soldering iron is to transmit heat, and that the metal being soldered must be heated to solder-melting and solder-alloying temperature before actual soldering will take place. The flat face of the iron should be held directly against the assembly, so that the heat may be most effectively transmitted to the

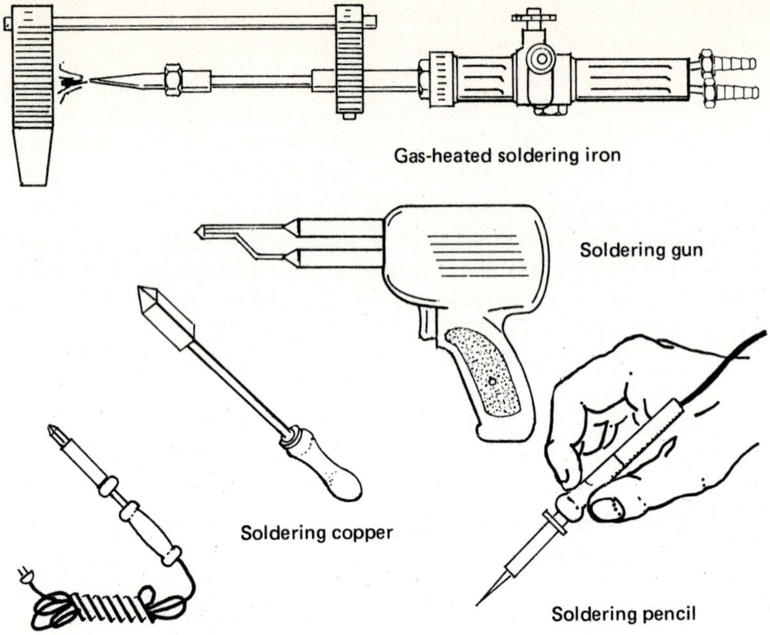
Gas-heated soldering iron
Soldering gun
Soldering copper
Soldering pencil
Electric soldering iron

Figure **11-3** Various soldering devices. (Redrawn by permission from *Fundamentals of Service Welding*, John Deere Service Publications, Fig. 82.)

metal being soldered. In general, there will be an insulating oxide barrier between the hot iron and the work, but as soon as cored solder is applied, the flux quickly breaks down the oxide film and metal-to-metal contact is established through the medium of molten solder.

Note: Make sure that the thermal capacity of the soldering iron is adequate for the job intended. Very often solder is condemned simply because the capacity of the iron is inadequate to heat the mass of metal that is to be soldered.

To obtain good results apply the flat face of the adequately heated soldering iron directly against the assembly and simultaneously apply the cored solder strand at the exact point of iron contact (but only after the metal is hot).

11-5 DIP-POT SOLDERING

In the case of dip-pot soldering, one often hears the complaint that the pot needs periodic additions of tin because the tin boils out of

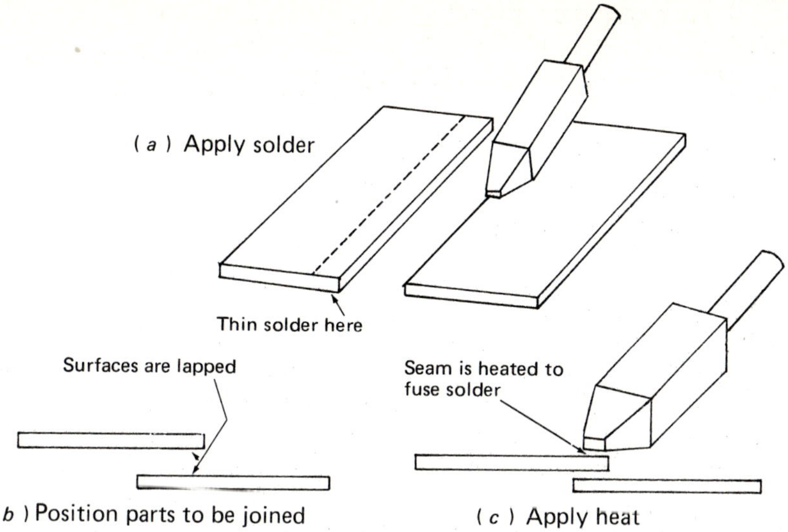
(a) Apply solder
Thin solder here
Surfaces are lapped
Seam is heated to
fuse solder
(b) Position parts to be joined
(c) Apply heat

Figure **11-4** How to solder a simple lap joint. (Redrawn by permission from *Fundamentals of Service Welding*, John Deere Service Publications, Fig. 85.)

the solder pot, or because the tin, which is lighter, comes to the top and has to be replaced. These are, of course, erroneous explanations. It is impossible to alter the tin-lead ratio of solder as long as it is completely liquid. What happens is that when the solder cools at the close of a day's work, there is a eutectic segregation with a precipitation of lead-rich crystals; when the solder is subsequently remelted, the low-melting, tin-rich eutectic constituent melts first and, if the solder is used while in this partly melted condition, there will be selective abstraction of the liquid portion, which is relatively high in tin. After solder has once solidified, it should not be used until it is again completely liquid and stirred.

Metallic contamination of the solder due to solvent action of the molten solder on the metal being dipped is common to dip-pot soldering. This solvent action is a normal and necessary reaction without which soldering could not take place. In the case of steel parts or assemblies, solder contamination due to solvent action is generally inconsequential because steel is relatively insoluble in solder. When the molten solder finally becomes sluggish and unworkable due to metallic contamination, it should be replaced with fresh, pure solder. It is fruitless and wasteful to add pure tin or pure solder to the bath in an attempt to compensate for metallic contamination.

Surface oxidation of the solder pot is always a problem. Slag removal by periodic additions of a cover flux is not very effective, because cover fluxes generally contain objectionable chlorides and because the gradual accumulation of flux residue is equally as objectionable as the original slag. Periodic skimming of the solder pot with a flat steel or porcelain spatula, or even a piece of cardboard, is generally preferable and more practical.

11-6 SOLDERING JOINTS

The form of the joint used in soldering will be an important factor in determining the maximum load it can support without failure due to fracture or deformation. Figure 11-5 shows a number of common types of joints. *Butt joints* (*a*) are weakest; that is, they only support loads corresponding to the tensile strength of the solder. However, as the solder exists only as thin film, its strength will exceed that of the solder in wire or strip form.

When properly made, *lap joints* (*b*) or *scarfjoints* (*d*) will withstand loads corresponding to tensile stresses several times the strength of the solder. A lap joint often cannot be used where there are strict joint-thickness limitations. Bending of a lap joint may also occur under heavy tensile loads. A combination of a lap and butt joint is obtained in the *strap-butt* joint (*c*).

A scarf joint (*d*) eliminates many problems related to thickness and bending found in lap joints. However, the scarf joint is difficult to make and equally difficult to fit properly during joining.

Figure **11-5** Typical soldered joints

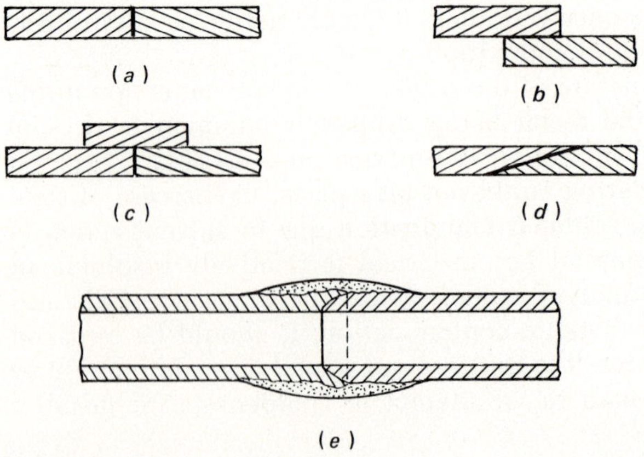
(a)
(b)
(c)
(d)
(e)

In the *wiped joint*, the parts are fitted together and the wiping solder molded, while pasty, into a smooth mass about the joint.

The soldered *sleeve joint* (e) has widely supplanted the wiped or threaded joint in plumbing installations containing copper and brass tubing. Sleeve joints, properly designed and made, will withstand loads up to the point of failure of the tubing itself.

In some solder applications, the principal role of the solder is that of a low-resistance electrical joint or a smooth attractive surface. For such applications, mechanical strength may be provided prior to soldering by bolting, twisting, or interlocking the parts, or by other suitable joining methods.

11-7 PROPERTIES OF SOLDERED JOINTS

Galvanic corrosion occurs at a soldered joint due to the presence of dissimilar metals. Galvanic corrosion is not severe in joints soldered with lead-base alloys, with silver solders, or with low-melting brazing alloys. Unprotected aluminum alloys will corrode rapidly in moist atmospheres. Solder joints in aluminum exposed to moist or marine conditions should be either of the long scarf or lap types. Soldered joints in aluminum or its alloys exposed to salt water or salt-water spray can be cathodically protected by attaching strips of zinc or cadmium close to the joint. It is considered good practice to use a moisture-proof lacquer or paint coating over solder joints in aluminum exposed to normal atmospheres.

Joints made with lead-tin alloys should not be used at temperatures higher than about 350°F (177°C); those soldered with tin-antimony solder would melt apart at about 450°F (232°C); and those made with cadmium-silver solder at about 640°F (338°C).

REVIEW QUESTIONS

1. Why are the properties of solder in a solder joint different from those of the original solder?

2. In what forms are solders available?

3. What is the function of flux used in soldering?

4. Why is it essential that the metal being soldered be free of nonmetallic impurities?

5. What is meant by the term *tinning* as used in soldering?

6. What is the average thickness (in inches) of a film of solder?

7. Why is it important that the thermal capacity of the soldering iron be adequate for the job intended?

8. Which solder joint will withstand tensile stresses corresponding to several times the strength of the solder?

9. What should be used to protect a solder joint exposed to ordinary humid atmospheres?

10. Differentiate between: (a) fusion welding and brazing; (b) brazing and soldering.

11. Why can soldering not be done without a flux?

12. Name the type of flux required to solder: manganese bronze; magnesium; aluminum; steel; copper.

13. What is the function of the soldering iron?

14. Sketch, and discuss the advantages and disadvantages of, three types of soldering joints.

CHAPTER 12
ADHESIVE BONDING

EPOXY

While the use of glues and cements is almost as old as humanity itself, adhesive bonding may not yet be considered a full-fledged science, because the data in this field are largely empirical and the explanations are still, to some degree, speculative.

Large-scale practical advances, such as the development of high-strength adhesives (epoxy and phenolic resins), has made it possible to bond virtually all known solids, often achieving higher efficiency and lower costs than are possible with other joining methods.

12-1 SYNTHETIC THERMOPLASTICS

Synthetic thermoplastics have excellent properties of adhesion to most metals and thus are widely employed in metal adhesive formulations. Since these resins are heat fusible, they are not usually employed in an unmodified state for the structural bonding of metals. Combinations of thermoplastic and thermosetting resins are employed for structural bonding.

Vinyl-acetate adhesives are available as milky-appearing, non-flammable water emulsions and as colorless organic-solvent solutions. Most vinyl-acetate adhesives set by water loss. As the last traces of water disappear, the surface becomes tacky; pressure should then be exerted to achieve bonding. The presence of some green strength at this time means that effective bonding of thin materials, such as foil, can be achieved with the momentary contact afforded by a nip roll. Clamping may be necessary where more rigid materials, such as sheet metal, are to be joined. A clamped joint is usually strong enough to handle in about 30 minutes. Bonds may be obtained by application of adhesive to one or both surfaces, depending on the method of lamination and the materials to be bonded. Adhesive application to both surfaces is preferred when a nip-roll operation is desired. If the materials to be combined are under such tension that adhesive application on one surface will not generate sufficient green strength to maintain an intimate bond after passing through the nip roll, then two-surface coating must be employed. For applications where clamping is practical and one or more surfaces are porous, a high-viscosity version of the vinyl-acetate adhesive can minimize the risk of producing an adhesive-starved joint when adhesive is applied to only one surface. When bonding two nonporous surfaces, the solvent is completely removed and bonds are obtained by heat sealing or by organic-solvent reactivation of the dried films.

Maximum service temperatures of about 120°F (unless greatly modified with higher-temperature resins) restrict the use of these adhesives in many exterior applications exposed to sunlight. Creep occurs in these adhesives at low loads, thus making them unsuited for structural applications. Vinyl-acetate adhesives have only moderate resistance to water and poor resistance to organic solvents. They do, however, have good resistance to oils and bacterial contamination.

12-2 POLYVINYL-CHLORIDE ADHESIVES

Polyvinyl-chloride adhesives are used mainly to bond vinyl plastics to metal or glass. In some instances, polyvinyl chloride has been copolymerized with neoprene or nitrile rubbers for bonding the same surfaces. If water solutions are employed and long-time service is desired, the compatibility of the adhesive should be checked against a metal surface in the presence of humidity.

12-3 ACRYLIC ADHESIVES

Acrylic adhesives have been used for bonding plastics, glass, leather, cloth, and metals. Bond strengths vary, depending on the kind of acrylate employed, as well as on the method of application. Adhesives of this type are recommended mainly for bonding metals to nonmetallic (leather, ceramic, and other) porous materials, where there is need for only medium water resistance. To obtain sufficient bond strength for structural applications, these adhesives must usually be heat cured; such applications would still be limited by the low maximum service temperature.

Acrylic adhesives are available as solutions, emulsions, or polymer-monomer mixtures, which may be cured by ultraviolet, heat, or chemical catalysts. Bonding from a solution consists of the usual procedure—drying to the point of tackiness and then joining two tacky surfaces. If one or both surfaces are porous, it is possible to join the surfaces by clamping or pressing while the solvent dissipates through the material. Some acrylics may be applied as a solution, but the solvent is then removed by thorough drying. The tack-free surface is then held under pressure against a second surface, and the bond is made by application of external heat.

The maximum service temperature of the simplest acrylic adhesive is about 125°F, but blends containing higher homologous acrylics can significantly raise this service temperature. The acrylics have medium water resistance, and good resistance to oil and thermal shock, but poor organic-solvent resistance.

12-4 HOT-MELT ADHESIVES

Hot-melt adhesives are particularly adaptable to high-speed assembly operations because they set up and reach their maximum strength as soon as they cool. Other advantages are their indefinite shelf life, reusability, and freedom from waste. They are offered in a wide range of polymers, with softening points from 100 to 500°F, and are employed widely in the packaging field, sometimes to bond various materials to aluminum foil. Hot melts are also employed as side-seam cements in fabricating metal, paper, and combination metal-paper containers.

Another large class of hot-melt materials is the microcrystalline waxes. These are for making temporary bonds to nonporous materials, such as metals and glasses. They have good resistance to water and bacterial growth, but low solvent resistance.

12-5 SYNTHETIC THERMOSETTING ADHESIVES

These adhesives, commonly known as *epoxy adhesives,* are most commonly used in joints that require high strength and structural integrity. There are many available modifications of epoxy adhesives. This abundance of variations results partly from the wide range of problems a designer encounters in adapting adhesive bonding to actual applications. Many adhesive manufacturers formulate custom-made adhesives to fit particular applications.

Epoxy adhesives are thermosetting; that is, they undergo a chemical change when cured, and there is no means by which they can be returned to their original state. Curing is accomplished by the addition of hardeners to the basic resin. The hardeners (or activators) present in one-component systems are activated by exposure to elevated temperatures—usually from 250 to 500°F—for a specified period of time. In epoxy adhesives, where curing is accomplished at room temperature, the hardener component is added just prior to application. It should be mentioned, however, that some multiple-component epoxy adhesives require elevated-temperature curing.

In order to examine the various attributes of epoxy adhesives, they will be classified for subsequent discussion as follows:

1. One-component:
   a. High-strength, flexible
   b. High-strength, semigrid, better elevated-temperature service and chemical resistance

2. Multiple-component

12-6 ONE-COMPONENT, HIGH-STRENGTH, FLEXIBLE EPOXIES

A compromise has existed in epoxy adhesives with respect to strength and flexibility. That is, those adhesives that exhibited high strength when subjected to shear or tensile stresses were rather poor in peel or cleavage because of the rigidity of the cured product. Additional modification to gain more flexibility usually resulted in a decrease of tensile or shear properties.

Several innovations have made it possible to obtain even higher strengths than those previously available with epoxy adhesives, while providing relatively high flexibility and peel or cleavage resistance. One of these was accomplished through nylon modification. The use of nylon-modified epoxies is often limited, since they are normally in film form and are relatively expensive.

Another innovation concerns a special process of polymerization. While this epoxy system is about as strong as nylon-modified systems, it is only half as flexible.

These two types of epoxy formulations may be used in applications where both high strength and flexibility are required. They do not retain their properties much above 200°F, which limits their use in elevated-temperature service. These adhesives are serviceable down to −67°F, however, and some of the nylon-epoxy films can be used in cryogenic applications with fairly good strength maintained to −400°F.

Most of the adhesives in this class are cured for about one hour at 350°F, with contact pressure to obtain optimum properties. Much shorter curing times at 400°F will still provide properties higher than other epoxy adhesives.

Typical applications for this class of epoxies are pressure vessels, architectural structure components, structural joints in the transportation field, and aircraft honeycomb panels.

12-7 ONE-COMPONENT, HIGH-STRENGTH, SEMIRIGID EPOXIES

Applications for early epoxy formulations were limited because of poor peel or cleavage strength. Despite the fact that stronger and more flexible epoxy adhesives are available today, there is still need for these rigid adhesives because of their good elevated-temperature properties.

For most of these materials, there is a wide latitude in curing cycles from a few hours at 300°F to a few minutes at 500°F. Most of the formulations are in paste form, but some are available by immersing the heated part into the powder or into a fluidized bed consisting of air-suspended powder. The rod is applied to a heated part in a manner similar to that employed in soldering. Again, the powder or rod can be melted at a temperature below that which would thermoset the material. The prepared parts are then coated by dipping them into the molten adhesive.

Most of these adhesives are serviceable from −30 to 400°F or above. Epoxies modified with a phenolic resin have been used in cryogenic applications down to −400°F. It should be reiterated that most epoxy adhesives require only contact pressure in curing. One exception to the rule is the family of phenolic-modified epoxy adhesives. These require pressure during the curing cycle. In severe exposures, such as boiling-water immersion or halogenated hydrocarbon propellants, epoxy adhesives of the semirigid type are the only materials known to be durable. The durability is, of course, relative (some percentage of strength is lost), but most other adhesives would fail completely in these types of exposures.

Typical applications for these epoxy materials may be found in joining refrigerator or evaporator tubing, automotive engine components, and aircraft honeycombs. One of the adhesives used in the USAF B-58 Hustler bomber is a phenolic-modified epoxy film.

12-8 MULTIPLE-COMPONENT EPOXIES

Essentially, most multiple-component systems sold as packaged adhesives contain two parts: the epoxy resin with fillers and/or modifiers already added, and the activator or hardener. Most of these adhesives can be cured at room temperature and will take from a few to as many as 48 hours.

Like the one-component epoxies, two-part formulations are available in various combinations of flexibility and strength. This is normally a matter of compromise. The stronger the adhesive in shear, the less flexible it is likely to be.

These materials are usually in liquid or paste form. The parts must be proportioned accurately and mixed thoroughly just prior to their use. Most of these formulations have a pot life of one-half hour to two hours, after which they will be too rigid to make application possible. The activator, which represents the force for cross-linking the resin polymers, must be added in the prescribed amount for all possible reactions to occur, and no excess of activator should remain in the mixture. For the same reasons, the combined material must be adequately mixed to make sure all reaction does take place.

For high rates of production, equipment is available to store the two parts separately; mixing and proportioning of components are accomplished in the head of a pressure gun. In some of the available two-component epoxies, flexibility is controlled by adding varying amounts of the hardener. It is obvious that in these instances the mixing ratio is less critical, since there are several ratios available.

Two-component adhesives that cure under ambient conditions are employed in applications where heat curing is either impossible or uneconomical. The performance of two-part epoxies can be improved by heating in the 150 to 250°F range; but if higher temperatures (above 250°F) are permissible, one-component epoxies might as well be employed.

Generally, the chemical, elevated-temperature, and water resistance of two-component formulations are not as good as those of one-part epoxies. Presumably, this may to some extent be caused by the sensitivity of these materials to surface preparation, particularly when they are cured at room temperature.

In many cases, heat-curing adhesives have been employed with only solvent cleaning or vapor degreasing as a surface preparation. Adequate adhesion to the surface is possible because of the increased wetting ability of these materials under the stimulus of heat.

12-9 ELASTOMERIC-BASED (NATURAL-RUBBER) ADHESIVES

Elastomeric-based adhesives find their widest applications in bonding nonmetallic materials, such as leather, fabrics, and paper and rubber products. However, they may also be used for attaching these nonmetallics to metal surfaces. The elasticity of the natural-rubber bond is advantageous where the metal surface is relatively unyielding to dimensional change in humidity, or where it moves more than the nonmetallic material under the stimulus of heat.

Methods of bonding with natural-rubber adhesives are, for the most part, similar to the procedures for the entire class of elastomers. They are generally available in organic solvent or water solutions of a wide viscosity range. These may be sprayed, roll coated, curtain coated, brushed, knifed, or troweled on one or both surfaces.

When making the bond at ambient temperature, it is often most convenient—and most conducive to bond reliability—to apply the adhesive to both surfaces, let the solvents evaporate until the surfaces are tacky, and then press them together (minimum pressures suffice to make the bond). If the adhesive is applied to only one surface, higher pressures are usually necessary to make satisfactory bonds. This increased pressure may be particulary effective when one surface is relatively porous and fingers of adhesive are literally forced into its surface pores. Higher strength and heat resistance can be achieved either by accelerating the adhesive-curing reaction with a chemical additive or by applying external heat. External heating must always be accompanied by the simultaneous application of pressure, so that the surfaces are held in intimate contact while the adhesive is achieving maximum strength. The choice can also be made as to whether heating is to be performed before the adhesive has lost all of its surface tack (some solvent remains in the adhesive) or after it has achieved a tack-free state (all solvent has been removed).

Resins of different softening points may be added to natural rubbers to achieve varying degrees of tackiness and ultimate strength. The recommendations of adhesive suppliers will vary accordingly as to drying time for optimum tack or temperature for maximum bond strength. An important advantage of these adhesives is that they manifest high initial tack and immediate bond strength. Hence, they are generally classified as contact-type adhesives.

Natural rubber has good resistance to deterioration by water, but poor resistance to oils, organic solvents, and chemical oxidizing agents. The natural-rubber molecule crystallizes (stiffens) at about −30°F, and water-latex, natural-rubber emulsions must be protected from freezing, which can readily damage their adhesive properties.

12-10 CHLORINATED NATURAL-RUBBER ADHESIVES

These adhesives are used mainly to bond natural and synthetic rubbers to metals. Such adhesives have actually been used to prime metal surfaces prior to bonding with neoprene- or nitrile-based adhesives.

One procedure for bonding a rubber product to a metal is to apply a chlorinated-rubber adhesive to the metal surface, to evaporate the solvent and obtain a tacky surface, and then press the rubber-product section on this surface. The assembly may finally be heat cured to produce the ultimate bond.

Chlorinated rubber itself has good resistance to water, some oils, salt spray, and bacterial growths. Similar behavior is found in chlorinated-rubber adhesives. In general, resistance to organic solvents is poor, and the practical temperature service limit is around 280°F.

12-11 CYCLICIZED- OR ISOMERIZED-RUBBER ADHESIVES

These adhesives have been successfully used in lining storage tanks, since they exhibit good adhesion to metals. Solutions of cyclicized rubber are used, therefore, to prime metal surfaces prior to bonding with other rubber-based adhesives. Adhesives formulated with cyclicized rubber form a tough bond to metal surfaces.

The adhesive is applied to both surfaces to be bonded and is dried to a nontack condition. The surfaces are joined and held in intimate contact while the assembly is heated to achieve vulcanization of the rubber interface.

Adhesives containing cyclicized rubber have good resistance to water and bacterial contamination, but are generally poor in the presence of oils and aromatic or chlorinated solvents. The maximum service temperature is about 140°F.

12-12 RUBBER-HYDROCHLORIDE ADHESIVES

These adhesives have been used principally in the form of thin, transparent films for general utility in the packaging industry. Some combinations have been made into adhesives, because of their inherent property of good adhesion to metal surfaces. The hydrochloride-containing adhesive acts as a primer for bonding a variety of rubbers to metal.

The methods of using cyclicized-rubber adhesives and hydrochloride-containing adhesives are similar. Application of adhesive to both surfaces is followed by drying to a nontack condition and then by heat-vulcanizing the adhesive interface.

Good resistance to water is common to all the natural rubber–base adhesives. Rubber hydrochloride is essentially insoluble in ethers, alcohols, and esters, but is soluble in aromatic hydrocarbons and chlorinated hydrocarbons. Its maximum service temperature is about 230°F.

12-13 NEOPRENE (POLYCHLOROPANE) RUBBER ADHESIVES

These adhesives are particularly useful in bonding metals to nonmetallic plastic insulating materials, or of metal to wood products for architectural applications. High-speed production methods may be employed, since these adhesives have immediate green strength. Also, no dwell time at elevated temperature is necessary to secure an adequate and durable bond. Varying degrees of strength and flexibility are available in neoprene adhesives. Accommodation can thus be obtained in the bond between metallic and nonmetallic laminated structures whose component parts may have different coefficients of heat expansion or undergo dimensional change under varying humidity.

Neoprene adhesives are generally suitable for applications involving continuous exposure at temperatures up to 200°F. Further, they can withstand short exposures to considerably higher temperatures. The combination of flexibility, good heat resistance, and durability makes neoprene adhesives ideal for bonding aluminum-faced sandwich panels in curtain walls and residential building construction. The neoprenes are versatile enough to find general application in bonding plastics, leather, rubbers, fabrics, and plywood to metal. Neoprene adhesives may also bond a wide variety of nonmetallic insulation and commercial building materials, such as hardboard, cement-asbestos board, gypsum board, glass fiber–reinforced polyester, and vegetable-fiber boards to metals.

Most of the neoprene adhesives set by a solvent-release mechanism (organic solvent or water). Four different procedures have been employed in bonding metals with neoprene adhesives:

1. The adhesive is applied (by spray, roller-coater, curtain-coater, brush, or mechanical spreader) to one or both surfaces and allowed to air-dry until tacky. The surfaces to be joined are pressed together (preferably in a rotary or platen press). Considerable immediate green strength is present, and the bond will develop about 85% of final strength within a few hours. Within a period of about 24 hours, sufficient strength is developed for most applications.

2. After applying the adhesive, the surfaces are heated by hot-air convection or infrared rays until the solvent is volatilized out of the adhesive. The surfaces are then combined immediately and pressed firmly together in a rotary or platen press. If the other surface to be bonded to the metal is nonmetallic, its different coefficient of thermal expansion may present a problem in trying to secure a flat laminate. It may be necessary to provide some cooling while maintaining flatness by pressing. The lower the temperature of satisfactory bonding, the better the opportunity to secure a flat laminate.

3. After applying the adhesive, the surfaces are heated or allowed to air-dry until a no-tack surface is obtained. In this condition, many neoprene adhesive–coated surfaces can be stacked and stored for periods ranging from a few days to a few weeks and held for subsequent reactivation. When the final metal laminate is to be made, the surface is mist sprayed with an appropriate organic solvent, and the adhesive develops a tacky surface condition. Pressure applied in the usual fashion should then furnish a satisfactory bond. More reliable bonds may be made by adhesive coating both surfaces, although it is necessary to restore tack to only one surface.

4. The same steps are employed to arrive at a tack-free condition; but after tack is restored by heating the surfaces, they are combined in a rotary or platen press. Higher reactivating temperatures are necessary for some neoprene adhesives that contain other resins blended in to secure maximum service temperatures. The high combining temperatures used may produce bowed laminates on cooling. This problem must be anticipated.

Neoprene adhesives have good resistance to water, some oils, aliphatic hydrocarbons, weak acids and alkalies, and bacterial contamination. They are generally unsuitable in contact with aromatic hydrocarbons, ketonic solvents, and strong oxidizing agents. Neoprene adhesives have adequate room-temperature shear strengths (200 to 800 psi) for many applications involving the joining of metals to nonmetallic materials, but they are unsuited for many metal-to-metal applications where the adhesive might creep under certain conditions of loading and temperature.

The tensile and stress-rupture values for many neoprene adhesives have also been determined at ambient temperature and at 180°F. These test results are considered necessary for predicting the performance of neoprene adhesives in outdoor applications.

12-14 NITRILE-RUBBER ADHESIVES

Nitrile-rubber adhesives will bond well to a wide variety of surfaces, such as plastics, wood, rubber, paper products, fabrics, glass, and metals. This adhesive is preferred for bonding a plastic-like polyvinyl chloride for bonding metals, because the plasticizer will generally not soften the adhesive if plasticizer migration occurs. The overall cohesive strength of the nitrile adhesive is maintained in water exposures, as well as its strength of adhesion to a metal surface. The greater range of adhesive properties available in neoprene types has meant that neoprenes continue to have the widest application for general bonding.

One of the outstanding advantages of nitrile rubber itself is its excellent resistance to oils; hence, metal bonding is often open to question in high-humidity environments, because of the alkaline nature of the latex. A nitrile-rubber, water-latex adhesive is available that is compatible with metal even when exposed to continuous humidity. This adhesive, referred to a *carboxylated nitrile,* may be of particular interest where the material bonded to metal is sensitive to organic solvents and a cold-bonding process is dictated.

REVIEW QUESTIONS

1. Name the types of adhesives used to bond: (a) vinyl plastics to metal; (b) ceramic to metal.

2. What kinds of adhesives would be used to bond material that must withstand service temperatures ranging from 250 to 350°F?

3. Explain how vinyl-acetate adhesives set.

4. Describe the general procedure for joining wood to metal.

5. What are some of the advantages of hot-melt adhesives? Name some applications of these adhesives.

6. Name an acceptable adhesive for joining magnesium to aluminum. Describe the general procedure that must be followed to make the bond.

7. Identify the family of single-component thermosetting adhesives that require pressure during the curing cycle.

8. Identify the components of multiple-component adhesives.

9. How can the performance of two-part epoxies be improved?

10. In general terms, describe the procedure to be followed for bonding two rubber parts and identify the type of adhesive used.

11. Name the type of adhesive that is most commonly used to bond rubber to metal.

12. What is the maximum service temperature of rubber-hydrochloride adhesives? In what industry are these adhesives commonly used?

13. What type of adhesive is commonly used for bonding aluminum to gypsum board? How is the adhesive applied?

14. Which adhesive is preferred for bonding plastics to aluminum? Why?

15. What is the maximum service temperature of the following adhesives?

a. Vinyl acetate.
b. Epoxy-thermosetting.
c. Chlorinated rubber.
d. Isomerized rubber.

16. Which of the synthetic thermoplastic adhesives are available in the form of a milky-appearing water emulsion and also in the form of a colorless organic-solvent solution?

PART 3
WELDING PRACTICE

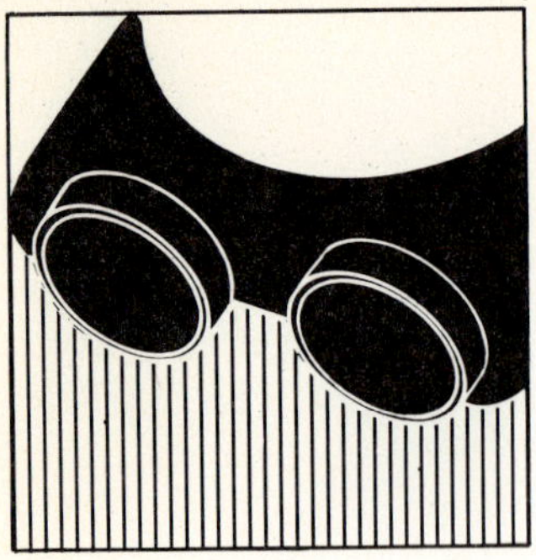

CHAPTER 13
SAFE PRACTICES IN WELDING AND CUTTING

The safe practices discussed in this chapter are based on the specifications and standards issued by the American Welding Society (AWS), the United States of America Standards Institute (USASI), the Compressed Gas Association (CGA), and the National Fire Protection Association (NFPA). These make specific reference to welding and are part of the Williams-Steiger Occupational Safety and Health Act of 1971, more commonly known as OSHA.

Specific aspects of the selection and installation of welding equipment were discussed in detail with each of the processes in the previous part of this book. Therefore, a general statement about the safety factors involved in the selection of equipment will suffice here.

Too often the welding equipment for a job is selected because of availability, cost, or other factors, with safety considered last. An inexpensive piece of equipment can be very costly if it cannot be operated safely. It cannot be overemphasized that a major, if not the major, consideration in equipment selection must be its safe operation under the conditions required for the job.

SAFE HANDLING OF EQUIPMENT

13-1 GENERAL SAFETY PRECAUTIONS

Although cooperation between the welding industry and fire-prevention and accident-prevention organizations has resulted in apparatus with built-in safeguards insofar as practicable, the use of common sense and the precautions listed below will further reduce the likelihood of accidents.

To minimize the risk of fires and explosions:

1. Never use compressed-gas cylinders to support the work that is being welded or cut, and never use cylinders as rollers.

2. Never perform welding or cutting in the presence of flammable gases or vapors (such as gasoline).

3. Always use an inert or nonflammable gas, such as argon, helium, carbon dioxide and nitrogen, or steam, to purge any drum, container, or hollow structure suspected of having contained a flammable or explosive substance before welding, heating, cutting, or brazing is begun.

4. Always vent by drilling or puncturing (as appropriate) any structural voids, jacketed containers, or castings suspected of being hollow before welding, cutting, heating, or brazing is begun. Gases expand on being heated. Expanded gases produce increased pressure if the space in which they are contained is not larger than the one in which they were contained prior to being heated. The increased pressure may result in the sudden bursting of the part being welded.

5. Never place work that is to be heated or welded on a concrete floor because, when sufficiently heated, concrete may fragment and fly, with possible injury to the welder or others.

6. Always leave oxygen and acetylene cylinders outside of tanks and other confined areas.

7. Never, under any circumstances, allow even a small part of an extra charge of carbide to be run into one charge of water in an acetylene generator.

8. Never attempt to transfer gas from one cylinder to another.

9. Never mix gases in one cylinder.

10. Never use a cylinder that is leaking gas.

To minimize the possibility of maimed or broken limbs:

1. Always place adequate (conforming to OSHA standards) guards on mechanical power transmission equipment, such as gears, shafting, or clutches, with which the hands or fingers of the welder may come into contact.

2. Never work on scaffolds, platforms, or runways unless provided with adequate railings, safety belts, safety lines, or some equally effective safeguard (conforming to OSHA standards).

3. Never discard stub ends of electrodes or welding rods where they can get underfoot and cause workers to fall.

4. Never work in a manhole or other confined space unless provided with safety belt (conforming to OSHA standards) or life lines, and a helper to quickly remove the welder from the confined space and render assistance if needed.

5. Always make sure that the wheels of heavy portable equipment are securely blocked to prevent accidental movement. Also block any internal moving parts.

6. To prevent flashback (flame going from welding tip back into cylinder) always use reverse- or back-pressure check valves (Figure 13-1) at cylinders, generators, and whenever possible at the torch butt.

7. To minimize the chances of spontaneous combustion caused by mixing oxygen with oil or grease or copper with acetylene:

a. Never handle oxygen cylinders, valves, regulators, hoses, or fittings with oily hands, gloves, or greasy equipment.

b. Never bring acetylene into contact with unalloyed copper, except in a torch tip or nozzle.

8. To enable the quick closing of the acetylene cylinder in an emergency:

a. Never open an acetylene cylinder valve more than 1½ turns.

b. Always leave the T wrench or key in position on the valve stem while acetylene is in use.

9. To prevent damage to the cylinder or confusion for you and other users:

Figure **13-1** Design of torch and regulator back-pressure check valves meeting OSHA requirements

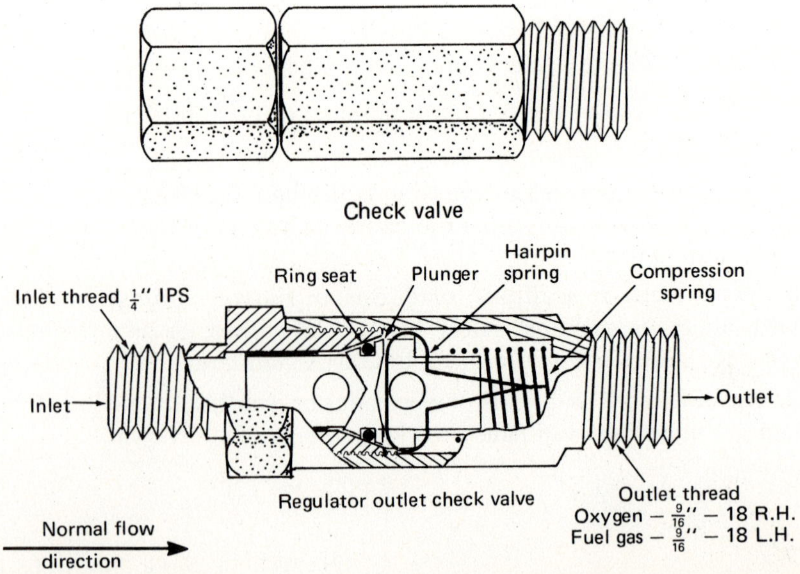
Check valve
Inlet thread $\frac{1}{4}$" IPS
Ring seat
Plunger
Hairpin spring
Compression spring
Inlet
Outlet
Regulator outlet check valve
Outlet thread
Oxygen – $\frac{9}{16}$" – 18 R.H.
Fuel gas – $\frac{9}{16}$" – 18 L.H.
Normal flow direction

a. Always (except when cylinders are in use) keep valve protecting caps in place.
b. Never use valve protection caps to lift cylinders from one vertical position to another (use specially designed slings).
c. Always mark empty cylinders plainly with the word *Empty* or the letters *MT*.
d. Always secure cylinders in a vertical position with straps, clamps, chains, or similar devices during use.

13-2 ARC WELDING SAFETY PRACTICES

The following safety rules are to be observed:

1. Never operate electric generators powered by internal combustion engines inside buildings or confined areas unless adequate provisions have been made to exhaust the carbon monoxide gas.

2. Never allow the power-supply cables of portable welding machines to become entangled with the welding cables or to be near enough to the welding operation to sustain possible damage to the insulation from sparks or hot metal.

3. Always keep welding leads and primary power-supply cables clear of ladders, passageways, or doors.

4. Always repair or replace defective cables immediately. Disconnect power before splicing cables. Only use insulated cable connectors of the locking-pin type (Figure 13-2) having a capacity not less than the capacity of the cable (see Table 13-1).

Figure **13-2** Locking pin–type connector on a screw-on terminal

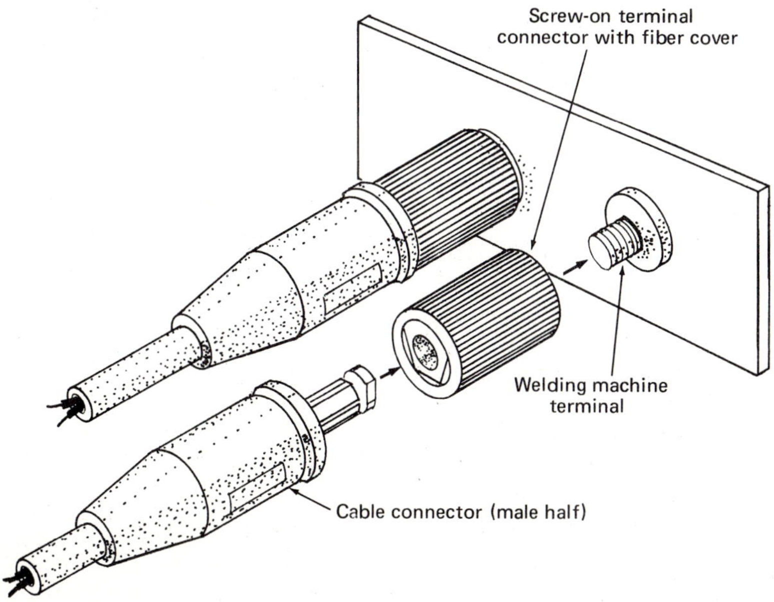
Screw-on terminal
connector with fiber cover
Welding machine
terminal
Cable connector (male half)

Table **13-1** Recommended cable sizes for manual welding[a]

MACHINE SIZE IN AMPERES)	DUTY CYCLE (%)	COPPER CABLE SIZES FOR COMBINED LENGTHS OF ELECTRODE PLUS GROUND CABLE				
		UP TO 50 FT	50–100 FT	100–150 FT	150–200 FT	200–250 FT
100	20	#8	#4	#3	#2	#1
180	20	#5	#4	#3	#2	#1
180	30	#4	#4	#3	#2	#1
200	50	#3	#3	#2	#1	#1/0
200	60	#2	#2	#2	#1	#1/0
225	20	#4	#3	#2	#1	#1/0
250[d]	30	#3	#3	#2	#1	#1/0
300	60	#1/0	#1/0	#1/0	#2/0	#3/0
400	60	#2/0	#2/0	#2/0	#3/0	#4/0
500	60	#2/0	#2/0	#3/0	#3/0	#4/0
600	60	#3/0	#3/0	#3/0	#4/0	c
650	60	#3/0	#3/0	#4/0	b	c

[a]For fully automatic welding, use two 4/0 cables for less than 1200 amperes or three 4/0 cables for up to 1500 amperes.
[b]Use double strand of #2/0.
[c]Use double strand of #3/0.
[d]For 225-amp 40% duty-cycle machines, use same cable size as 250-amp 30% duty cycle machines.

5. Always turn the welding machine off whenever leaving it for an extended period of time (rest room, lunch, end of work day, and so on).

6. Never dip hot electrode holders in water.

7. Always keep welding cables free of grease and oil.

8. Never allow welding cables to lie in water, soil, ditches, or tank bottoms.

9. Always install welding machines in accordance with the provisions of the National Electric Code.

10. Never repair welding equipment unless the power to the machine is shut off.

11. Never change the polarity switch while the machine is under a load. Wait until the machine idles and the circuit is open. Otherwise the contact surface of the switch may be burned and the resulting arcing could cause injury to the welder.

12. Never overload a welding cable (see Table 13-1).

13. Never operate a machine with poor connections. Avoid damp areas and keep the hands and clothing dry at all times.

14. Never strike an arc on a compressed gas cylinder.

15. Do not strike an arc if someone is nearby without proper eye protection, face shield, or screen (see Figure 13-3).

Figure **13-3** Typical eye protection devices conforming to OSHA standards. (Courtesy of Jackson Products, A Division of Airco, Inc., and Singer Safety Products, Inc.)

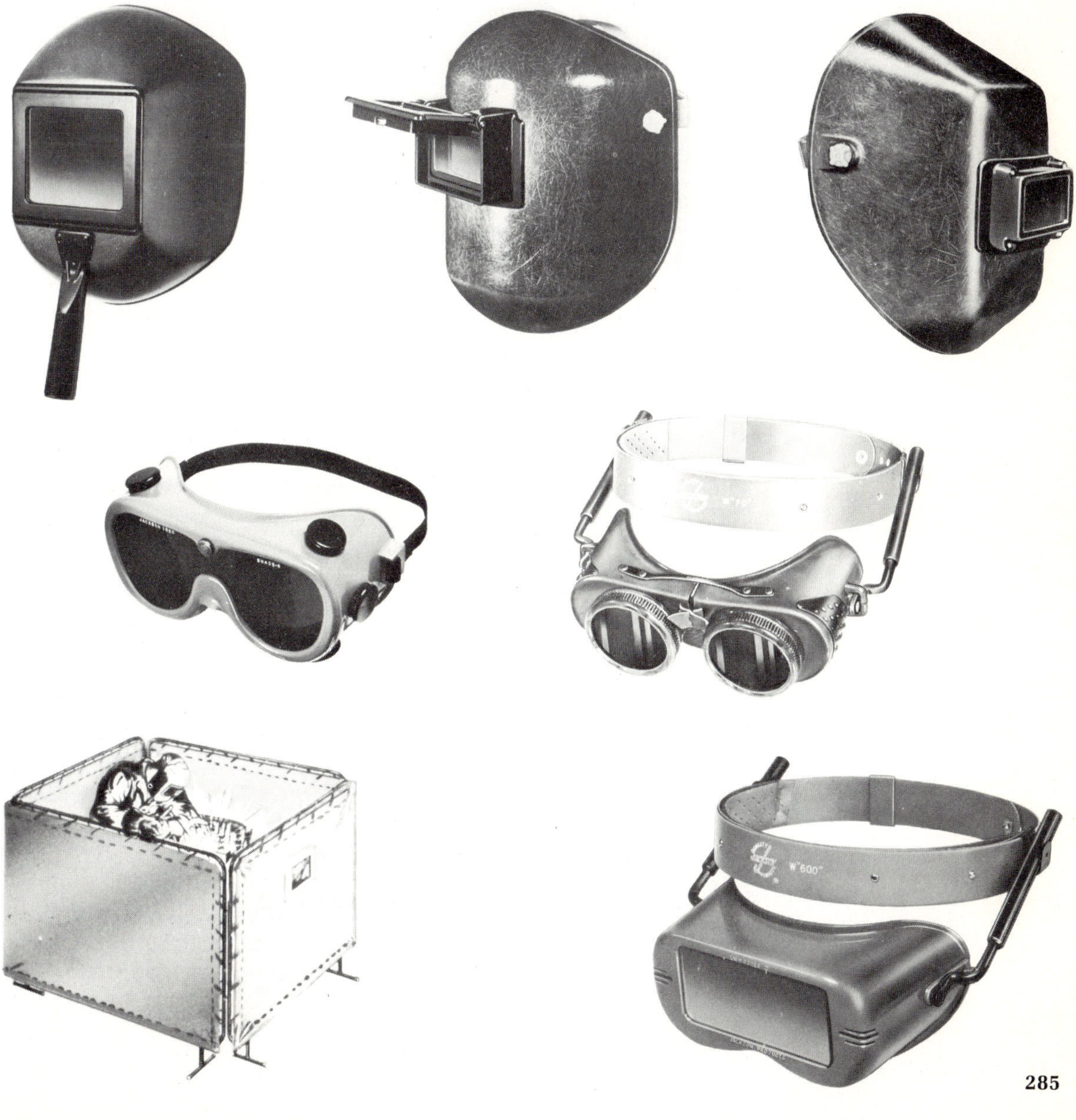

16. Always keep the uninsulated portion of the electrode holder from touching the welding ground when the current is on.

17. Never carry welding cables coiled around the shoulder when they are carrying power.

13-3 PROTECTIVE EQUIPMENT FOR WELDERS (ARC AND FUEL-GAS WORK)

Always wear goggles with suitable filter lenses when using a torch. Also, wear a heat shield or helmet with suitable filter lenses when arc welding (Figure 13-3). Wear flash goggles having side shields and a suitable lens at all times, even when adjusting controls. Goggles and helmets protect the eyes from sparks and flying slag and also from the strong light and injurious rays of the flame or arc. They also help to see the work better.

Wear leather gloves and aprons, and suitable shoes and other protective clothing (Figure 13-4).

Keep protective equipment dry and free of oil, and take care that clothing is not oily, that pockets do not contain matches or cigarette lighters, and that cuffs are not open and ready to receive sparks or hot slag.

13-4 SAFETY PRACTICES FOR OTHER WELDING PROCESSES

For information on safety practices for welding processes for which no practice exercises are given in this text, the reader should refer to the latest edition of the applicable safety instructions for the particular welding process. The publications are available from the American Welding Society, 2501 Northwest 7th Street, Miami, Florida 33125.

FIRE PREVENTION

13-5 RULES FOR FIRE PREVENTION

Certain definite rules for the prevention of fires during cutting and welding have been formulated by the NFPA in their bulletin No. 51B *(Rules for the Prevention of Fires During Cutting and Welding)*. However, in the absence of this bulletin, the following general safe practices should be followed.

Welding or cutting should not be done where an open flame or arc would be dangerous, as in the presence of explosive atmospheres (mixtures of flammable gases, vapors, liquids, or dusts with air) or near the storage of large quantities of readily ignitable materials.

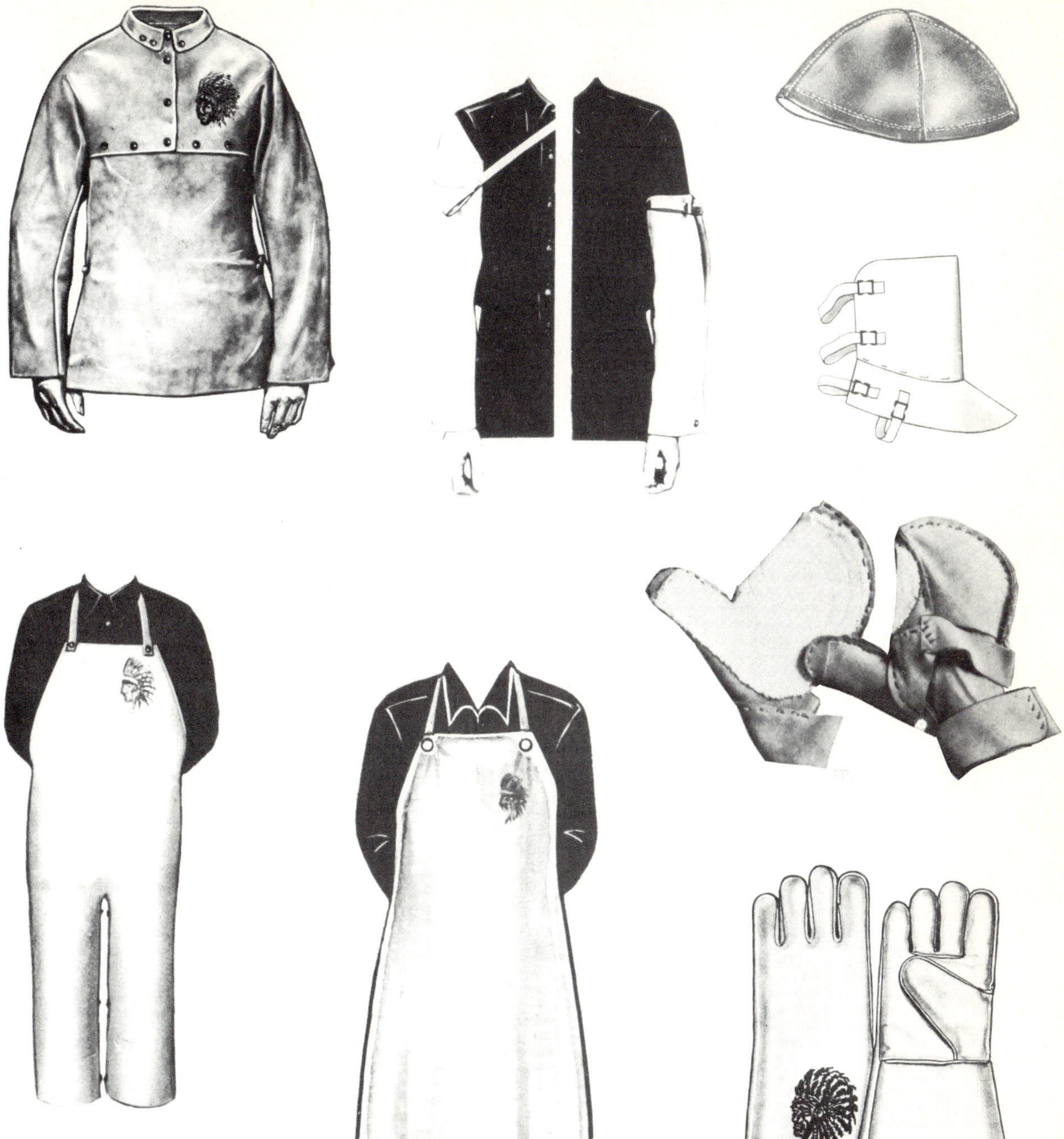

Figure **13-4** Protective clothing for welders. (Courtesy of Jackson Products, A Division of Airco, Inc.)

When welding or cutting near combustible materials, special precautions should be taken to make certain that the sparks or hot slag from cutting operations particularly do not come in contact with combustible material and thus start a fire. Combustible material should be moved a safe distance away, at least 35 feet, if the work to be welded or cut cannot be moved. Where relocation is impractical, protect combustibles with flameproofed covers or otherwise shield them with metal or asbestos guards or curtains. Edges of covers on the floor should be tightened to prevent sparks from going under them.

Floors should be swept clean for a radius of 35 feet. Combustible floors should also be wet down; in which case, personnel using electric equipment should be protected from shock. It is preferable to cover wooden floors, where sparks or hot metal are likely to fall, with metal or some other suitable noncombustible material. Whenever there are openings or cracks in the flooring, cover with metal or some other suitable noncombustible material. Make certain that there are no highly combustible materials on the floor below, where hot metal or slag might drop through the floor. Precautions should be taken to prevent hot slag or sparks from falling into machine tool pits.

Observe the same precautions outlined in the preceding paragraph with regard to cracks or holes in the walls, open doorways, and open or broken windows. Use sheet-metal guards or asbestos curtains where needed to protect against arc flashes and to assist in screening against sparks and slag (recognizing that protection at floor level may not be provided).

When it is necessary to do welding or cutting close to wooden construction or in locations where combustible materials cannot be removed, fire-extinguishing equipment conforming to OSHA requirements should be provided, suitable for the type of fire that may be encountered.

Whenever combustible material has been exposed to molten metal or hot slag from cutting or welding operations, keep a person at the source of the work for at least a half hour after completion to make sure that smoldering fires have not been started.

A welder or cutter should check with his or her supervisor before starting to weld in an area other than a production area. Issuance of written permits signed by the area supervisor or plant protection department has been effective in reducing fires in many plants.

13-6 FIRE-EXTINGUISHING EQUIPMENT

Suitable fire-extinguishing equipment conforming to OSHA standards must be maintained near all welding and cutting operations. The suitability of the equipment is determined by an analysis of the conditions at the scene of operations. If, for instance, the only combustible material within range of the welding or cutting operations or sparks therefrom is on asphalt-type waterproofing, a CO_2 extinguisher may be adequate. However, in a small space with a very small access opening, the operator may not be able to get out quickly in case of fire, and the use of CO_2 might be injurious. Under such conditions, the use of water from a 1½-inch water line or water-pump tank would be preferable.

13-7 COMBATING ELECTRICAL FIRES

If the insulation of some electrical equipment that cannot be removed or adequately protected is the only combustible material present, then a water spray may be more perilous than the fire itself. For combating electrical fires, CO_2 extinguishers should be provided.

13-8 CARBON TETRACHLORIDE

Carbon tetrachloride extinguishers must never be used. Carbon tetrachloride decomposes on hot metal to form phosgene, a very deadly gas.

EYE AND FACE PROTECTION

13-9 GOGGLES AND FACE SHIELDS

The eyes and faces of not only welding and cutting operators but also of other personnel, such as helpers, chippers, and inspectors, who remain in the vicinity of the welding and cutting operation, must be protected from stray flashes, glare, and flying particles by suitable helmets, hand-held shields, and goggles (see Figure 13-3).

13-10 TYPES OF GOGGLES

There are two general types of goggles:

1. *Spectacle.* Spectacle-type goggles are made both with and without metal side shields. They may have either a rigid nonadjustable or adjustable metallic bridge.

2. *Eyecup.* Eyecup goggles have flexible, connected lens containers shaped to conform to the configuration of the face. A cover-type eyecup is designed to be worn over corrective spectacles, whereas eyecup goggles are worn alone. The choice of proper eye and face protection equipment should be made by a responsible person who fully understands what protection is needed. Further, she or he should be guided by the specifications as detailed in *USA Standards* Z87, covering eye protection; Z88, covering respiratory protection; and Z89, covering industrial head protection.

13-11 OVERHEAD WELDING

Only the eyecup or cover-type goggles should be used when welding or cutting near or above eye level.

13-12 GOGGLES FOR GAS WELDING

Spectacle-type (side-shielded), eyecup, or cover-type goggles should be used during all gas-welding or cutting operations. Spectacle-type goggles without side shields and with suitable filter lenses are permitted for inspection or for use with gas-welding operations on light work.

13-13 PROTECTION FOR ELECTRIC-ARC WELDING

Helmets or hand-held shields should be used during all arc-welding or cutting operations. Spectacle-type (side-shielded) goggles should also be worn on these operations to provide protection from injurious rays from adjacent work and from flying objects.

13-14 SHADE OF LENSES

The object of tinted filter lenses is not only to diminish the intensity of visible light to a point where glare is reduced to a minimum, so that the welding zone can be readily seen, but also to protect the welder from harmful infrared and ultraviolet radiation from the arc of flame. Table 13-2 can be used for guidance in selecting goggles.

13-15 TRANSFER OF INTERPERSONAL EQUIPMENT

Helmets and goggles should not be transferred from one person to another without antiseptic cleaning. Most common household antiseptic sprays are suitable for this purpose.

13-16 WELDING BAYS PAINTED BLACK

Where welding with the electric arc is regularly done, the walls of the welding bay should be painted flat black or some other nonreflecting color to prevent flickering reflections. Otherwise, the work should be enclosed in a booth.

Table **13-2** Recommended lens shades for various welding processes

TYPE OF WELDING	SHADE OF LENS
Stray light from welding or cutting	1.7–4
Metal pouring and furnace work	1.7–4
Light gas cutting and welding; light electric spot welding	5
Gas cutting, medium gas welding, and arc welding and cutting up to 30 amperes	6–7
Heavy gas welding and arc welding and cutting up to 75 amperes	8–9
Arc welding and cutting up to 200 amperes	10–11
Arc welding and cutting up to 400 amperes	12–13
Arc welding and cutting above 400 amperes	14

13-17 PORTABLE BOOTHS

Where the work permits, workers or other personnel adjacent to the welding areas should be protected from the rays by enclosing the work area with flameproof screens or with individual booths that have been painted with a nonreflecting color, such as zinc oxide and lamp black (see Figure 13-3).

RESPIRATORY PROTECTION

13-18 RESPIRATORY HEALTH HAZARDS

The respiratory health hazards associated with welding operations are due largely to the inhalation of gases, dusts, and metal fumes. With only a few, relatively simple precautions, the chance of respiratory damage can be eliminated.

The amount of fumes or gases that the welder is liable to inhale is governed by factors such as the dimensions of the welding area, the number of welders, the arc time, the ventilation afforded, the type of

Table **13-3** Abbreviated listing of toxic materials with their 1966 threshold limits[a]

SUBSTANCE	PPM	mg/m^2
Acetylene tetrabromide	1	14
Acetone	1000	2400
Antimony and compounds (as Sb)	—	0.5
Arsenic and compounds (as As)	—	0.5
Boron oxide	1	15
Boron tribromide	1	10
Boron trifluoride	1	3
Butyl acetate	150	710
Butyl alcohol	100	300
Carbon black	—	5
Chromious salts (as Cr)	—	0.5
Chromium metal and insol. salts	—	1
Cobalt metal fume and dust	—	0.1
Copper fume	—	0.1
Copper dusts and mists	—	1
Ethyl alcohol	1000	1900
Hafnium	—	0.5
Ketene	0.5	0.9
LPG. (liquid petroleum gas)	1000	1800
Manganese and compound (as Mn)	—	5
Magnesium oxide fume	—	15
Methyl acetylene propadiene (MAPP)	1000	1800

[a]For a complete and up-to-date listing contact the American National Standards Institute, 1430 Broadway, New York, New York, 10018.

welding materials involved, and the workpiece size. Probably the single most important factor, though, is that governed by the welder—the position of his or her head with respect to the plume of the fumes. Two welders doing the same job may have an exposure ratio of 1:10 or more, all depending on the positions of their heads.

The nature of any toxic materials to which the welder may be exposed will depend on the type of welding, the filler and base metals, the presence of contamination in the base metal, and the presence of volatile solvents in the air. The degrees of toxicity of these materials can differ greatly. This is best illustrated by Table 13-3, which lists several of the more common materials that may be

SUBSTANCE	PPM	mg/m²
Methyl acetylene (propyne)	1000	1650
Methyl alcohol (methanol)	200	260
Molybdenum		
Soluble compounds	—	5
Insoluble compounds	—	15
Nickel carbonyl	0.001	0.007
Nickel metal and soluble compounds (as Ni)	—	1
Nitric acid	2	5
Sulfuric acid	—	1
Tantalum	—	5
Tellurium	—	0.1
Thallium (soluble compounds)	—	0.1
Tin (inorganic compound, except SnH_4 and SnO_2)	—	2
Tin (organic compounds)	—	0.1
Tungsten and compounds (as W)		
Soluble	—	1
Insoluble	—	5
Turpentine	100	560
Uranium (natural) soluble and insoluble compounds (as U)	—	0.2
Vanadium		
V_2O_5 dust	—	0.5
V_2O_5 fume	—	0.1
Yttrium	—	1
Zinc oxide fume	—	5
Zirconium compounds (as Zr)	—	5

encountered in welding, with their 1966 threshold-limit values. It should be emphasized that this represents only the degree of toxicity. Actual hazard of body damage cannot occur unless an individual inhales these materials in substantially greater amounts for long periods of time.

Inhalation of zinc and magnesium fumes in amounts exceeding the threshold limit will produce a condition of chills, fever, and nausea, occurring 4 to 8 hours after exposure and disappearing, almost invariably, within 24 hours. Copper produces a similar condition, although apparently more severe and longer lasting, and precipitated by less exposure. Nickel, cobalt, and mercury are strongly suspected of causing a chemical pneumonitis that may have severe, even fatal, results. Cadmium produces a severe fume pneumonitis, and beryllium is extremely dangerous in this respect. Iron and aluminum apparently do not cause a metal-fume fever, although inhalation of very large amounts of iron fumes for a period of years may produce the seemingly harmless condition known as *siderosis*, a deposit of iron in the lungs. Very little is known regarding the effects, if any, of the less common metals on the human respiratory system.

13-19 FUMES PRODUCED BY SHIELDED METAL–ARC AND SUBMERGED-ARC ELECTRODES

Under certain conditions, the fumes produced by some electrodes may be a respiratory irritant to some welders and, under extreme conditions, may cause permanent damage to the respiratory system.

The plain-carbon or low alloy–steel electrodes with EXXX0, EXXX1, EXXX2, and EXXX3 types of coverings, when used on uncoated steels, produce fumes that consist mostly of iron oxide and varying amounts of flux materials. The threshold limit for iron oxide is 10 milligrams per cubic meter of air, a concentration high enough to markedly reduce visibility but not sufficient to produce physical impairment, even after many years of exposure. However, when welding is done inside tanks or other confined areas where the fumes cannot be removed readily or diluted by convection currents, the hazard increases. Oxides of nitrogen, always caused by arc welding, may accumulate to the point where they can cause damage to the lungs.

The flux coverings of low-hydrogen electrodes EXXX15, EXX16, EXX18, EXX20, and EXX28, and of those electrodes used for arc welding stainless steel, nickel, aluminum, and other alloys contain fluoride compounds in the order of possibly 5 to 10%. The fumes produced also contain fluoride salts, usually in amounts somewhat higher than the coverings. Under some conditions, the fumes may also contain hydrogen fluoride, but usually in extremely small amounts. The threshold limit for fluoride as a salt is 2.5 mg/m^3; for hydrogen fluoride, 2.0 mg/m^3. In addition to the fluoride fumes, stainless steel electrodes produce fumes that may contain up to 6% chromates, apparently produced by oxidation of the chromium. Fume concentration in the breathing zone of the welder can exceed the threshold limit of 0.1 mg/m^3.

The amount of fluorides or chromates inhaled by the welder depends on the composition of the fumes; the size, shape, and position of the work; and the individual work habits of the welder. An increased hazard may exist, for example, in a well-ventilated area if the position of the work and the work habits of the welder are such that he or she works with his or her head in the path of the fumes. A welder may complain of nose and throat irritation after using these electrodes for the first time. Although the irritation may stop after a few days, it does not necessarily mean that the situation has improved, only that the affected body tissues have become desensitized.

Because it is more difficult to control the arc and the flux in submerged welding than in manual welding, the hazards from fumes produced by this process are greater.

13-20 RESPIRATORY HAZARDS ASSOCIATED WITH FUEL-GAS WELDING, CUTTING, AND BRAZING

The most commonly used fuel gas (acetylene) in these processes is classified as a simple asphyxiant. It does not contain an appreciable amount of toxic impurities, but in very high concentrations it may cause suffocation by decreasing the amount of oxygen available. In these concentrations it would also be in the explosive range, where a spark or flame might create an explosion.

Welding or cutting steel does not produce harmful fumes under normal conditions. In an enclosed area, where a large amount of cutting, welding, or heating of steel is being done, oxides of nitrogen may be formed in such quantity as to be harmful.

The amount of carbon monoxide formed by a well-operated torch is generally too small to be significant except in a very confined space.

Fluxes containing fluorides are frequently used in silver brazing and for brazing and welding aluminum, magnesium, and their alloys. The hydrogen fluoride released by the heat has a threshold limit of 2.0 mg/m^3. At this concentration it is just detectable by odor. Concentrations produced will depend upon type and amount of flux used and on the heat applied. Ordinarily, concentrations are quite high in the sometimes invisible plume of fumes, but low elsewhere in the room. In small, confined areas, the general atmospheric concentration may exceed safe limits. Skin contact with these fluxes should be avoided. If they penetrate the skin through nicks, cracks, or around the fingernails, they produce a severe irritation.

Some silver-brazing alloys contain cadmium, which may be volatilized if heated appreciably above its melting point. The yellow-brown fumes are extremely dangerous, producing severe and even fatal lung damage on short exposure. The amount of cadmium fumes produced depends largely on the temperature reached. Low concentrations have been reported when using a gas-air torch; higher concentrations with oxyacetylene. Cadmium-bearing brazing filler metals are now subject to the Federal Hazardous Substances Labelling Act and are required to carry the standard warning notice.

No health hazard is involved in the gas welding of lead where temperatures are not high enough to volatilize the lead. Oxygen cutting of stainless steel, using a chemical flux or iron powder, produces a very large amount of fumes, reportedly containing chromates. Ventilation is usually required.

13-21 OTHER RESPIRATORY HAZARDS ASSOCIATED WITH WELDING

Welding of parts coated (even accidentally) with toxic materials, such as lead, cadmium, zinc, mercury, or paint-containing toxic materials, produce what is probably the greatest health hazard in welding and cutting. Any such coatings should be removed prior to welding.

The welding of materials containing beryllium or welding on tanks, pipelines, or containers containing halogenated materials or plating solutions may produce an extremely dangerous condition. No such welding should be undertaken until complete control of fumes has been provided.

13-22 INDIVIDUAL VENTILATION DEVICES

Individual ventilation devices in the form of hoods or compressed-air ejectors should be used whenever fume removal, as opposed to fume dilution, is required. The hood must consist of at least a top and at least two sides surrounding the welding or cutting operation. The hood (Figure 13-5) should provide an air flow of not less than

Figure **13-5** A single-welder welding station with exhaust hood. (Courtesy of Teledyne McKay, Manufacturers of Electrodes and Welding Wire.)

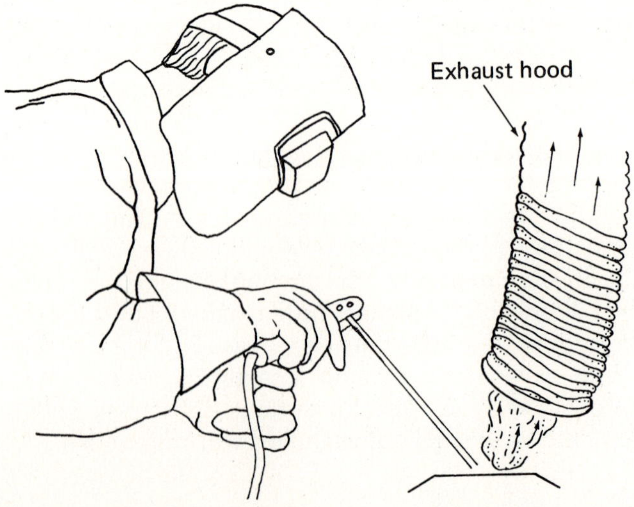
Exhaust hood

Table **13-4** Air capacities per cubic feet of space

SPACE	MINIMUM AIR FLOW (FT^3/MIN PER OPERATOR)	OR COMPLETE AIR CHANGE[a] EACH (MINUTES)
50,000 ft^3 or over	350	20
5000 to 50,000 ft^3	350	15
Under[b] 5000 ft^3		
Space per operator		
4000–5000 ft^3		4
3000–4000 ft^3		3
2000–3000 ft^3		2
Less than 2000 ft^3		1

[a]The volume of the space divided by the ft^3/min capacity of the blower equals the theoretical time to change the air once.
[b]At least 1000 ft^3 as well as the indicated space ventilation should be assured each operator.

100 linear feet per minute in a direction away from the welder and should (for welding uncoated ferrous metals) have the capacities shown in Table 13-4.

13-23 POSITIVE VENTILATION DEVICES FOR INDOOR WELDING

A compressed-air ejector may be used to either exhaust the fumes from the welding operation or dilute them. When used as an exhauster, the fumes are usually discharged into the room air. Compressed-air ejectors are frequently used in semiconfined areas where only dilution of fumes is required. They are too cumbersome to be of much use in exhausting the fumes.

13-24 RESPIRATORY PROTECTIVE EQUIPMENT

Respiratory protective equipment (nose masks) may be used where the use of local exhaust ventilation is not practicable or, in the case of very toxic materials, to supplement local exhaust ventilation. Airline respirators or nose masks will give adequate respiratory protection for all types of contaminants and are generally the preferred equipment. Air-supplied welding helmets are available commercially but have found little acceptance among welders. Filter-type respirators, approved by the U.S. Bureau of Mines for metal fumes, will give adequate protection against metal fumes not more toxic than lead, provided that they are selected, used, and maintained correctly. Their general use is not recommended because of the difficulty in determining that they were properly selected and maintained. No filter or cartridge respirator will protect against carbon monoxide, nitrogen dioxide, or mercury vapor—an airline respirator hose, mask, or gas mask is required when welding under these conditions. More detailed information on ventilation requirements will be found in *USA Standard* Z49.1-68, *Safety in Welding and Cutting*.

13-25 PROTECTIVE CLOTHING

The requirement for protective clothing (*USA Standard* L18, *Specification for Protective Occupational Clothing*) will vary with the size, nature, and location of the work; but in all cases it should be sufficient to protect the welder from burns, spatter, and, in the case of arc welding or cutting, from the radiant energy of the arc. If leather clothing is not available, woolen clothing is preferable to cotton because it is not so readily ignited. Cotton clothing, if used, should be chemically treated to reduce its flammability.

PROTECTION FROM ELECTRIC SHOCK

13-26 ELECTRIC SHOCK HAZARDS

Although the voltages required for most electric welding operations are low, they are enough to be a potential source of serious shock under unfavorable conditions. To minimize the exposure to and consequences of electric shock, the following precautions should be observed:

1. Never work out of sight of other persons.

2. Always handle any electric circuit as if it were live.

3. Always keep the body (of the welder) insulated from both the work, and the metal electrode and holder.

4. When practicable, stand on dry wooden mats or similar insulating material rather than on a grounded metal structure.

REVIEW QUESTIONS

1. Name some of the national organizations that issue specifications and standards regarding safe practices in welding and cutting in the United States.

2. What is the name of the federal act covering occupational safety and health?

3. If welding equipment is to be used commercially, it should be approved and listed by which laboratories?

4. List ten safe practices that should be followed to minimize the risk of fires and/or explosions when welding.

5. What markings does a welder place on a gas cylinder to indicate that it is empty?

6. What precautions should be taken to operate generators powered by internal combustion engines when operated inside buildings or confined areas?

7. Name the code that governs the installation of electric welding machines.

8. Why should the polarity never be changed while the welding machine is under load?

9. What types of goggles should be worn at all times by welders?

10. What is the title of the NFPA publication with whose contents welders doing mostly brazing or soldering should become thoroughly familiar prior to starting the job?

11. At what minimum distance (in feet) from combustible material is the relative risk of fire resulting from welding minimized?

12. List the precautions to be observed when the floor of the welding area is made of combustible material.

13. What precaution should be observed whenever combustible material is exposed to molten metal or hot slag from cutting or welding operations?

14. When welding overhead or above eye level, only one type of goggle should be worn. Name it.

15. What shade of lens should be worn when arc welding and/or fuel-gas cutting material of a thickness requiring between 75 and 200 amperes for welding?

16. What biological condition occurs in a welder after inhalation of copper fumes in excess of the generally accepted threshold limit?

17. Describe the commonly used ventilation device used at each welding station. What volume of air flow must this type of device produce?

18. What clothing material is also used as the preferred substitute for leather protective clothing?

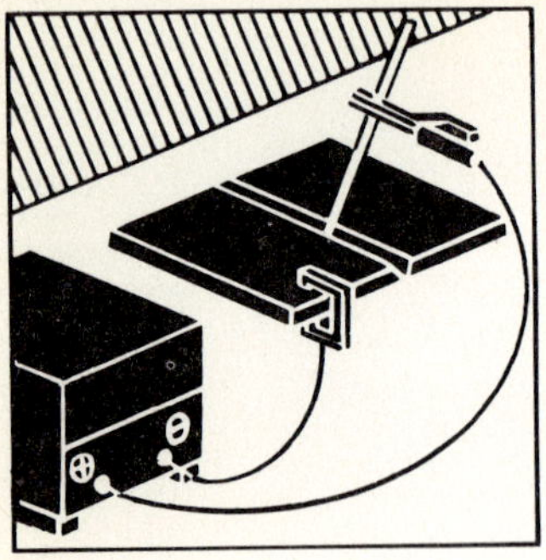

CHAPTER 14
SHIELDED METAL–ARC WELDING PRACTICE

The following is a series of exercises to familiarize the reader with shielded metal–arc welding. It is extremely important that the safe practices discussed in Chapter 13 be observed. In addition, the following merit reexamination:

1. Make sure that the equipment is in proper operating order.

2. Use only electrode holders specifically designed for shielded metal–arc welding and of a capacity capable of handling the maximum rated current required by the electrodes to be used on the job.

3. Make sure the welding machine and its accessories are clean, free from grease and oil, and free of other potential hazards, such as liquid and metal shavings and chips which could cause short circuits in the machine.

4. Use the welding equipment at a safe distance from flammable and explosive materials and gas cylinders. *Note:* If this is not possible, use fire-resistant covers or screens to prevent sparks from contacting combustibles.

5. Wear a helmet or carry a hand shield. Wear goggles beneath the helmet to provide protection from injurious rays from adjacent work or from sparks and flying particles.

6. Protect nearby persons from arc flashes by screening off work area or by providing them with flash goggles.

7. Wear the protective clothing specified in Chapter 13, especially when welding in close-quarters, vertical, or overhead positions.

8. Never work in damp areas that are not insulated against electrical shock; keep hands and feet dry.

9. Never cool hot electrode holders by dipping them in water—it may expose you to electric shock. *Warning:* Never touch two electrode holders with individual power sources at the same time—severe electric shock or burns may result.

14-1 ASSEMBLING THE SHIELDED METAL–ARC WELDING EQUIPMENT

OBJECTIVE To learn to set up the shielded metal–arc welding equipment.

HOURS REQUIRED ¼.

EQUIPMENT, TOOLS, AND MATERIALS REQUIRED The equipment, tools, and materials listed below are standard items for use in shielded metal–arc welding operations and will be required in all subsequent welding exercises.

1. Protective clothing and equipment

2. Arc welding machine, including:
   a. Ground and electrode holder cables
   b. Ground clamps
   c. Electrode holder

3. Nonelectrical equipment:
   a. Welding table (for small parts) with swivel vise, clamps, position holes and slots, ground cable holder, and electrode containers
   b. Chipping hammer to remove slag and spatter
   c. Wire brush to clean weld area
   d. Backing material to prevent drip of molten metal through a joint, to lend support, and to disperse heat
   e. Miscellaneous equipment to align, measure, mark, position, and clean the welding material

PROCEDURE *Warning:* Never turn the welding machine on until you have made sure that the grounding cable or work lead is connected from the machine to the work or to the work table.

1. Connect one end of the ground cable or work lead to the welding machine and the other end to the work table.

2. Connect the electrode cable to the welding machine.

3. Hang the electrode holder on the cable rack, where it cannot come in contact with the grounded cable.

4. Check with instructor for mistakes.

5. Reverse steps 1 through 4 for tearing down the set-up.

14-2 STRIKING AND HOLDING AN ARC

OBJECTIVE To learn how to strike and hold an arc.

HOURS REQUIRED ¾.

EQUIPMENT, TOOLS, AND MATERIALS REQUIRED The equipment, tools, and materials required for this exercise are the same as in exercise 14-1, plus a $1/8 \times 2 \times 6$-inch mild-steel plate and a supply of appropriate E-6010 electrodes.

PROCEDURE Read and become thoroughly familiar with the following instructions before proceeding.

1. Review the safe practices for arc welding described in Chapter 13.

2. Place a steel plate flat on the bench; brush it free of dirt and scale.

3. Attach the ground lead securely to the plate.

4. Set the welding machine amperage at 120 to 140.

5. Fit the electrode into the electrode holder.

6. Turn on the welding machine.

7. Strike an arc by brushing or tapping the base metal (workpiece) with the electrode (Figure 14-1). The distance between the electrode and the base metal should be about equal to the diameter of the electrode. *Note:* If the electrode sticks to the base metal, a quick sideways twist will usually free it. If not, remove the electrode from the holder and stop the machine. Then tap the electrode lightly with a chisel.

8. Strike an arc and, without oscillating (moving the electrode from side to side), lay several beads.

a. Keep the electrode in front of the puddle.

b. Hold the electrode perpendicular to the work (laterally), but with the welding end pointing slightly backwards toward the crater. (This will cause the slag to wash back and float on top of the bead and help overcome any tendency toward undercutting along the edge of the bead.)

c. Move the electrode forward just fast enough to deposit the weld metal uniformly.

d. The width of the bead should be about 1½ times the diameter of the electrode.

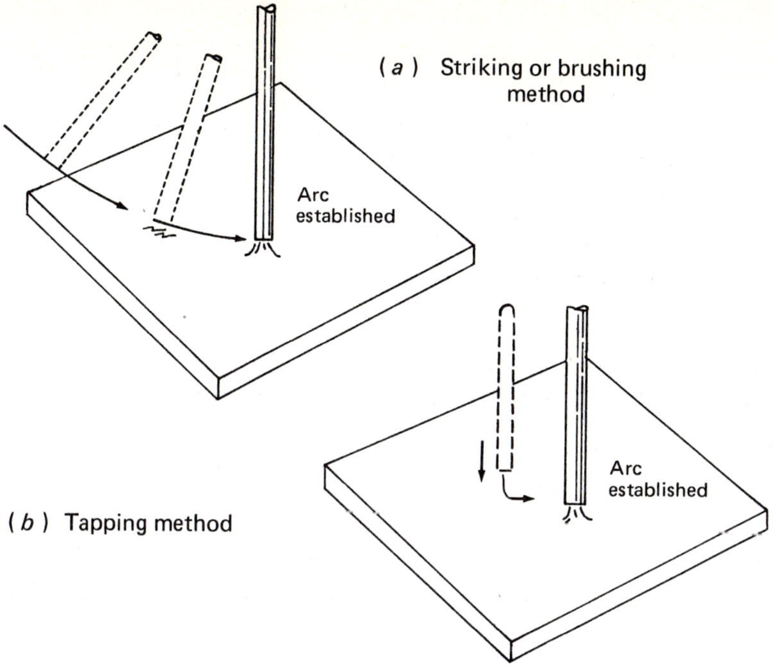
(a) Striking or brushing method
Arc established
(b) Tapping method
Arc established

Figure **14-1**

e. If the arc is broken while welding, do not restart it in the molten puddle, but rather just ahead of the bead on the work.
f. Do not change the amperage while the arc is sustained. If it becomes necessary to adjust the amperage, turn the machine off.

9. Chip the weld beads thoroughly and then wire-brush.

10. Turn the welding machine off and ask your instructor to check your work. *Note:* Turning the welding machine off before the job is finished, only to have to turn it on again to continue the job, is a waste of electricity. Unnecessary turning on and off of a welding machine should be avoided.

14-3 BUILDING A PAD OF WEAVE BEADS

OBJECTIVE To learn how to lay beads using one or more of the three standard weave patterns.

HOURS REQUIRED 2.

EQUIPMENT, TOOLS, AND MATERIALS REQUIRED The equipment, tools, and materials required for this exercise are the same as in exercise 14-1, plus appropriate $\frac{5}{32}$-inch E-6010 electrodes.

PROCEDURE Read and become thoroughly familiar with the following instructions before proceeding.

1. Review the safe practices for arc welding in Chapter 13.

2. Place a scrap metal plate flat on the bench; brush it clean of dirt and scale.

3. Attach the ground lead securely to the plate.

4. Set the welding amperage at 120 to 140.

5. Fit the electrode into the electrode holder.

6. Turn the welding machine on.

7. Strike an arc and form a puddle. (Remember to keep the electrode on the front edge of the puddle and to keep the puddle in a good molten condition.)

8. Run a wide bead by weaving the rod from side to side. Select a weaving motion (Figure 14-2) that is easiest for you to use.

a. Weaving is accomplished by oscillating the electrode back and forth, crossways in the direction of travel, and at the same time moving the electrode forward to advance the bead.

b. Weaving is used to float out slag, deposit a wider bead, secure good penetration, allow gas to escape, and avoid porosity.

c. Limit the weaving to 2½ times the electrode diameter to prevent unequal heating (Figure 14-3).

9. Pause at the end of each turn of the weave to avoid undercutting.

10. Run the second bead so that its edge overlaps the first bead (Figure 14-3).

11. Keep repeating the bead until the complete surface of the plate is filled.

12. Clean the slag between the layers.

13. Deposit each additional layer of beads crossways to the layer below, cleaning the slag between each bead and each layer (Figure 14-4).

Figure **14-2**

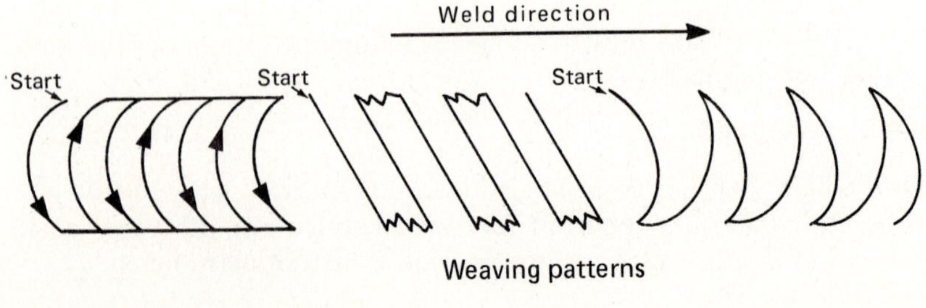
Weld direction
Start
Start
Start
Weaving patterns

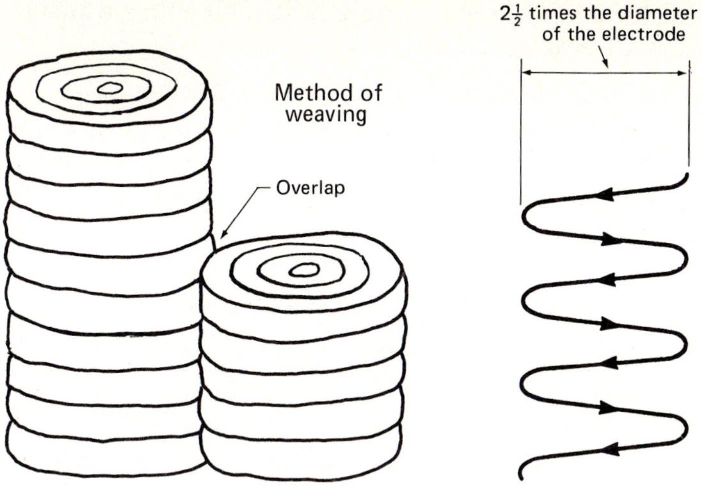
2½ times the diameter of the electrode
Method of weaving
Overlap

Figure **14-3**

14. Turn off the welding machine.

15. Check with your instructor for suggestions on how to improve your skill. (As part of this critique, saw through or break the plate to expose any entrapped oxide or porous spots.)

14-4 WELDING A SQUARE-GROOVE BUTT JOINT IN THE FLAT POSITION

OBJECTIVE To learn to make an acceptable square-groove butt joint in the flat position on mild-steel plate less than ¼-inch thick.

HOURS REQUIRED 1.

EQUIPMENT, TOOLS, AND MATERIALS REQUIRED The equipment, tools, and materials required are the same as for exercise 14-1, plus sufficient mild-steel plates of appropriate size.

Figure **14-4**

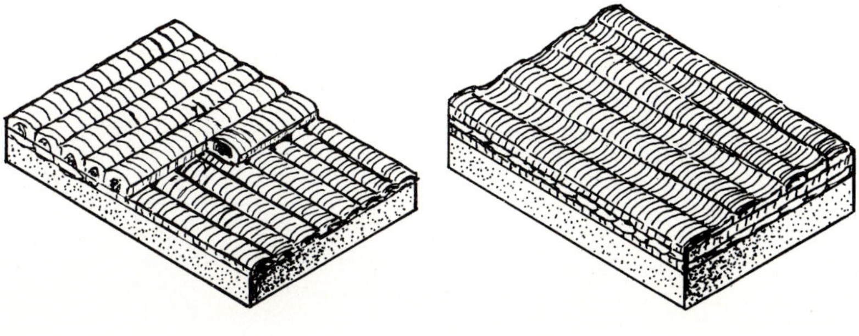

PROCEDURE Read and become thoroughly familiar with the following instructions before proceeding.

1. Review the safe practices for arc welding in Chapter 13.

2. Brush the steel plates free of dirt and scale and lay them on the bench, their edges parallel and about ⅛ inch apart. *Note:* Plates less than ¼-inch thick do not require any edge preparation and can be welded in one pass.

3. Attach the ground lead securely to one of the plates.

4. Set the welding machine amperage at 120 to 140.

5. Fit the electrode into the electrode holder.

6. Turn on the welding machine.

7. Tack weld the plates together at one end to keep them aligned (Figure 14-5); then strike an arc at the opposite end of the space and begin to weld the butt joint.

8. Run one bead holding the electrode perpendicular to the plates, using a slightly oscillating movement (Figure 14-6). *Note:* The correct width of this weld should be about 1½ times the diameter of the electrode, and there should be penetration to the bottom of the base metal.

9. Build up the bead with reinforcement (Figure 14-5) and penetrate clear through the bottom of the groove.

10. Chip and wire-brush.

11. Turn off the welding machine.

12. Check with your instructor for suggestions on how to improve your skill. (Inspect the butt weld for uniform density without holes or porosity, and for thorough penetration in the joint with good fusion in both plates.)

Figure **14-5**

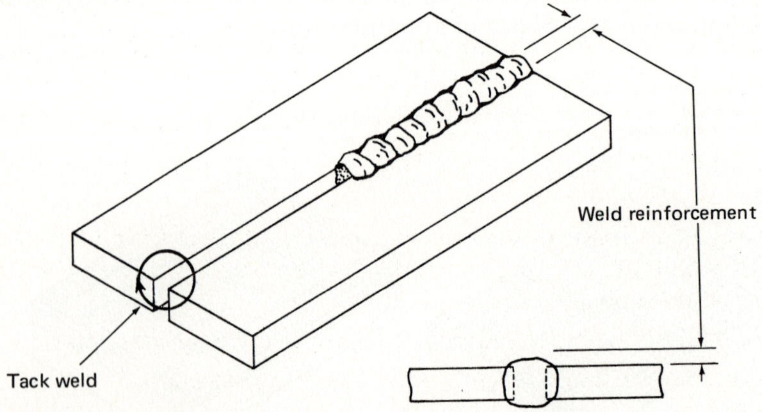
Weld reinforcement
Tack weld

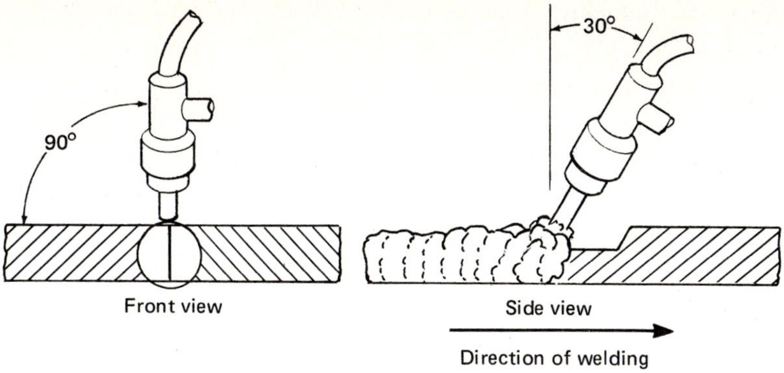

90°
Front view
30°
Side view
Direction of welding

Figure **14-6**

14-5 WELDING A SINGLE V-GROOVE BUTT JOINT IN THE FLAT POSITION

OBJECTIVE To learn to prepare and weld an acceptable single V-groove butt joint in ¼-inch-thick mild steel plate in the flat position.

HOURS REQUIRED 2.

EQUIPMENT, TOOLS, AND MATERIALS REQUIRED The equipment, tools, and materials required to perform this exercise are:

1. Two 2 × 6 × ¼-inch pieces of mild steel.

2. Several 5/32-inch-diameter electrodes (E-6010 for dc, E-6011 for ac).

3. One 1 × 6 × 3/16-inch piece of mild steel to be used as a backing strip.

4. One pedestal grinder or similar.

5. One welding power supply and accessories, as in the previous exercise.

PROCEDURE Read and become thoroughly familiar with the instructions below before beginning to weld.

1. Review the safe practices for arc welding in Chapter 13.

2. Grind a 30-degree bevel in both pieces of steel along the 6-inch side.

3. Lay the pieces of steel on the backing strip with the beveled edges parallel and facing each other, with a space of about ⅛ inch between them.

4. Attach the ground lead securely to one of the pieces of steel.

5. Set the welding machine amperage at 120 to 140.

6. Fit the electrode into the electrode holder.

7. Turn on the welding machine.

8. Tack weld the metal to the backing strip at both ends of the joint to keep the metal aligned.

9. Strike an arc and run a single bead (Figure 14-7). The first bead is deposited to seal the space between the two pieces of the joint and to weld the two pieces. Strive for equal fusion between both plates.

10. Thoroughly clean all slag from the first bead.

11. The second and subsequent layers can be deposited using a weave motion. (If weaving motion is used, pause at the end of each turn of the weave in order to avoid undercutting.)

12. Remove the backing strip with a cutting torch (Chapter 16) and, if necessary, add a seal bead at the back of the weld.

13. Clean all slag from the weld.

14. Turn off the welding machine.

15. Break or saw the weld and inspect it for uniform density and thorough penetration. (Ask your instructor for suggestions on how to improve your skill in welding this type of joint.)

14-6 WELDING A SINGLE V-GROOVE BUTT JOINT IN THE FLAT POSITION

OBJECTIVE To learn to prepare and weld an acceptable single V-groove butt joint in the flat position on ½-inch-thick mild-steel plate.

HOURS REQUIRED 2.

EQUIPMENT, TOOLS, AND MATERIALS REQUIRED The equipment, tools, and materials required to perform this exercise are:

1. Two 2 × 6 × ½-inch pieces of mild steel.

Figure **14-7**

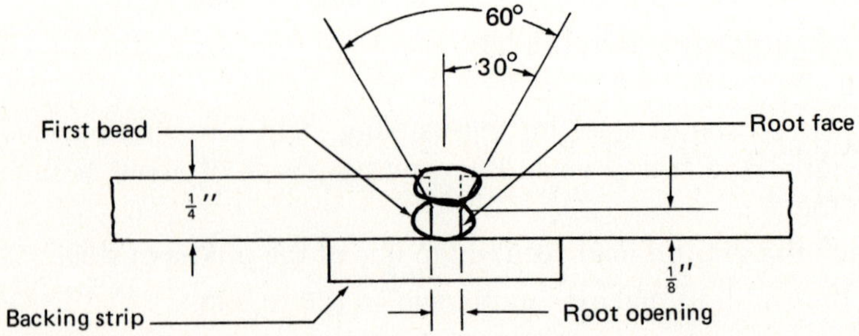
60°
30°
First bead
Root face
$\frac{1}{4}''$
$\frac{1}{8}''$
Backing strip
Root opening

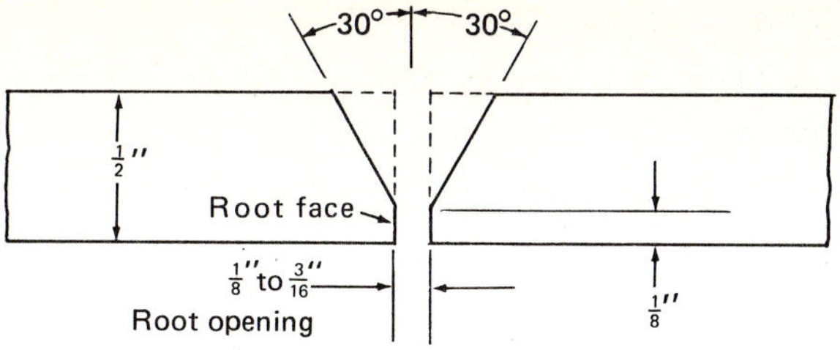
30°
30°
1/2"
Root face
1/8" to 3/16"
Root opening
1/8"

Figure **14-8**

2. Several 5/32-inch-diameter electrodes (E-6010 for dc, E-6011 for ac).

3. One 1 × 6 × 3/16-inch piece of mild steel to be used as a backing strip.

4. One pedestal grinder or similar.

5. One welding power supply and accessories, as in the previous exercise.

PROCEDURE Read and become thoroughly familiar with the instructions below before beginning to weld.

1. Review the safe practices for arc welding in Chapter 13.

2. Bevel one edge of each of the two steel plates with a cutting torch to an angle of 30 degrees. Leave a 1/8-inch root face (Figure 14-8).

3. Lay the pieces of steel on the backing strip with beveled edges parallel and facing each other, with a space of about 1/8 to 3/16 inch between them.

4. Attach the ground lead securely to one of the pieces of steel.

5. Set the welding machine amperage at 120 to 140.

6. Fit the electrode into the electrode holder.

7. Turn on the welding machine.

8. Tack weld the plates to the backing strip at both ends of the joint (Figure 14-9).

9. Strike an arc and run a single bead down the root of the joint ensuring equal fusion in both plates and into the backing strip. (Carefully remove all slag from this and all other beads before depositing additional beads.)

10. Run the subsequent beads as shown in Figure 14-10.

11. The *finish bead,* that is, the top bead, should be a wide bead made by weaving. (This should be flush with the upper surface of the plates or slightly convex.)

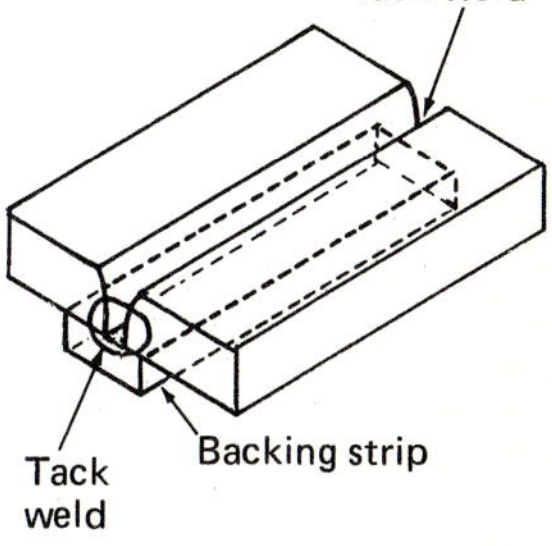
Tack weld
Tack weld
Backing strip

Figure **14-9**

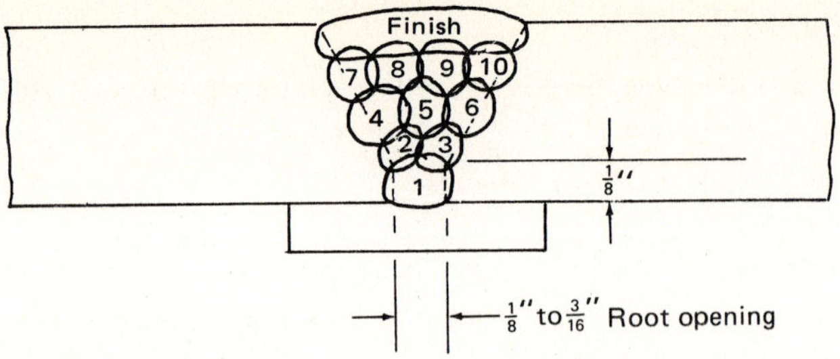
Finish
7
8
9
10
4
5
6
2
3
1
1/8″
1/8″ to 3/16″ Root opening

Figure **14-10**

12. Remove the backing strip with a cutting torch and, if necessary, add a seal bead at the back of the weld.

13. Clean all slag from the weld.

14. Turn off the welding machine.

15. Break or saw through the weld and inspect it. (Ask your instructor for suggestions on how to improve your skill in welding this type of joint.)

14-7 MAKING A SINGLE-BEAD FILLET WELD IN THE FLAT POSITION

OBJECTIVE To learn to make an acceptable single-bead fillet weld in the flat position.

HOURS REQUIRED 2.

EQUIPMENT, TOOLS, AND MATERIALS REQUIRED The equipment, tools, and materials required to perform this exercise are:

1. Two 8 × 3 × ¼-inch pieces of mild steel.

2. Several 5/32-inch-diameter electrodes (E-6010 for dc, E-6011 for ac).

3. One welding power supply and accessories, as in the last exercise.

PROCEDURE Read and become thoroughly familiar with the instructions below before beginning to weld.

1. Review the safe practices for arc welding in Chapter 13.
2. Lay one plate flat on the bench.
3. Attach the ground lead securely to the plate.
4. Set the welding machine amperage at 120 to 140.
5. Fit the electrode into the electrode holder.

6. Turn on the welding machine.

7. Set the second plate perpendicular to the first plate so that the two plates together form an L (Figure 14-11).

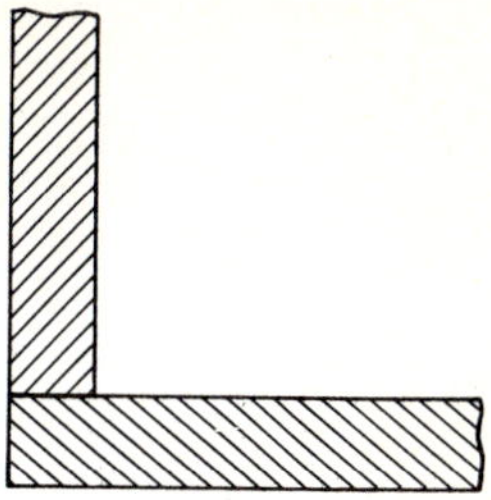

Figure **14-11**

8. Tack weld the plates in this L position.
a. Flat-position fillet welds are made on plates positioned at an angle of 45 degrees to the horizontal (Figure 14-12).
b. A good fillet weld must have complete penetration at the root or heel and be free of undercut and overlap at the toe.
c. The size of a fillet weld is measured by the length of the weld leg. There are two kinds of fillets: those with legs of equal length and those with legs of unequal length, or those that, although equal in length, may have concave or convex surfaces.

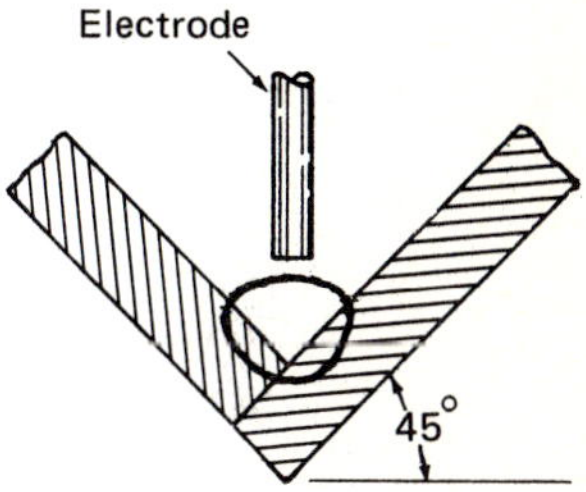
Electrode
45°

Figure **14-12**

9. Strike an arc and run a bead penetrating both plates down into the root of their intersection.

10. Clean all slag from the bead.

11. Turn off the welding machine.

12. Break the weld by means of a hammer blow. The broken surface should be sound and show complete penetration.

13. Ask your instructor to evaluate the weld.

14-8 LAYING HORIZONTAL BEADS ON A VERTICAL PLATE

OBJECTIVE To learn to lay horizontal beads on a vertical plate.

HOURS REQUIRED 1.

EQUIPMENT, TOOLS, AND MATERIALS REQUIRED The equipment, tools, and materials required to perform this exercise are the same as in exercise 14-7.

PROCEDURE Read and become thoroughly familiar with the instructions below before beginning to weld.

1. Review the safe practices for arc welding in Chapter 13.

2. Brush both plates free of dirt and scale.

3. Lay one of the scrap-steel plates on the bench.

4. Attach the ground lead securely to the plate on the bench.

5. Set the welding machine amperage at 120 to 140.

6. Fit the electrode into the electrode holder.

7. Turn on the welding machine.

8. Stand the second scrap-steel plate on its edge and along one edge of the first plate; tack weld the plates (Figure 14-13) at their intersection.

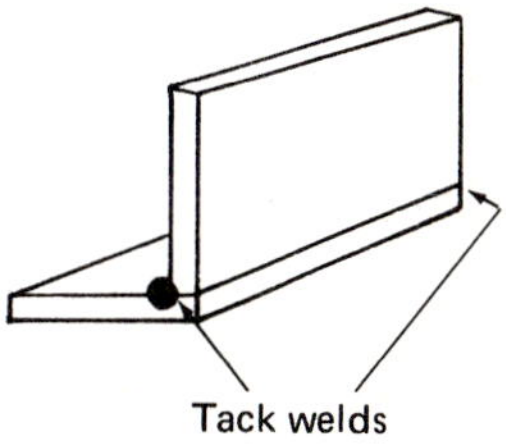
Tack welds

Figure **14-13**

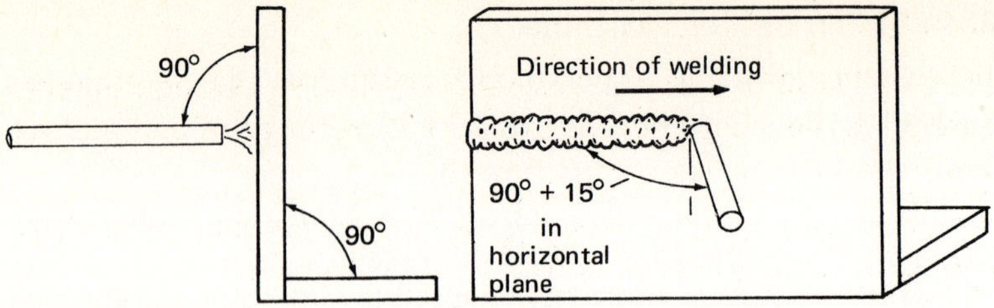
90°
90°
Direction of welding
90° + 15°
in
horizontal
plane

Figure **14-14**

9. Strike an arc and lay horizontal beads across the upright plate, starting at the bottom and alternating from left to right and from right to left (Figure 14-14).

10. Clean all slag from the beads.

11. Turn the plate on its side and deposit another layer, with beads running crosswise to those of the first layer (Figure 14-15). Build up a pad several layers thick.

12. Turn the welding machine off.

13. Check with your instructor for suggestions as to how to improve your skill. (As part of this critique, saw through or break the plate to expose any entrapped oxide or porous spots.)

14-9 MAKING A MULTIPLE-BEAD FILLET WELD IN THE HORIZONTAL POSITION

OBJECTIVE To learn to make an acceptable multiple-bead fillet weld in the horizontal position on a vertical plate (T joint).

Figure **14-15**

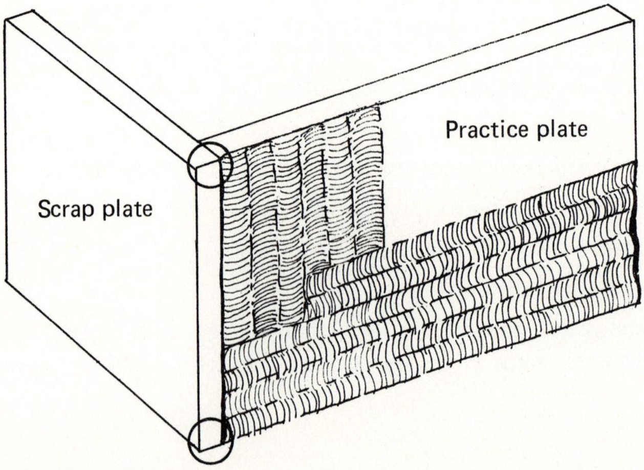
Practice plate
Scrap plate

HOURS REQUIRED 2.

EQUIPMENT, TOOLS, AND MATERIALS REQUIRED The equipment, tools, and materials required to perform this exercise are the same as in exercise 14-7.

PROCEDURE Read and become thoroughly familiar with the instructions below before beginning to weld.

1. Review the safe practices for arc welding in Chapter 13.

2. Lay one plate flat on the bench.

3. Attach the ground lead securely to the plate.

4. Set the welding machine amperage at 120 to 140.

5. Fit the electrode into the electrode holder.

6. Turn on the welding machine.

7. Position the second plate perpendicular to the first, making an inverted T (Figure 14-16). Tack weld the plates at their intersection.

8. Strike an arc and run a bead penetrating both plates down into the root of their intersection (Figure 14-17).

a. Remember to hold the electrode perpendicular to the line of the weld.

b. The electrode may be slightly inclined forward (though not more than 5 degrees) so that it points back toward the puddle.

c. Do not allow the slag to run to the front of the puddle.

9. Clean all slag from the bead.

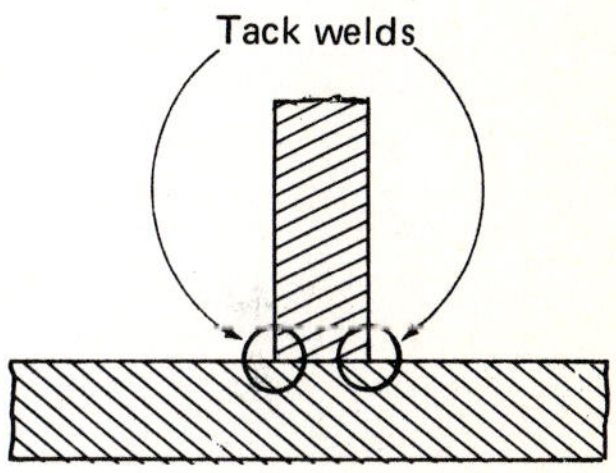
Tack welds

Figure **14-16**

Figure **14-17**

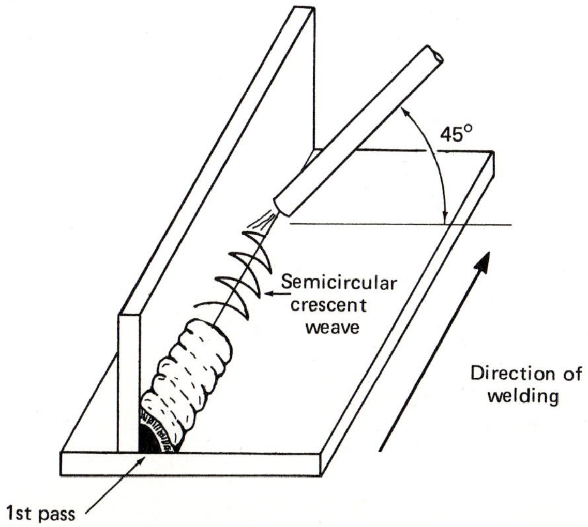
45°
Semicircular
crescent
weave
Direction of
welding
1st pass

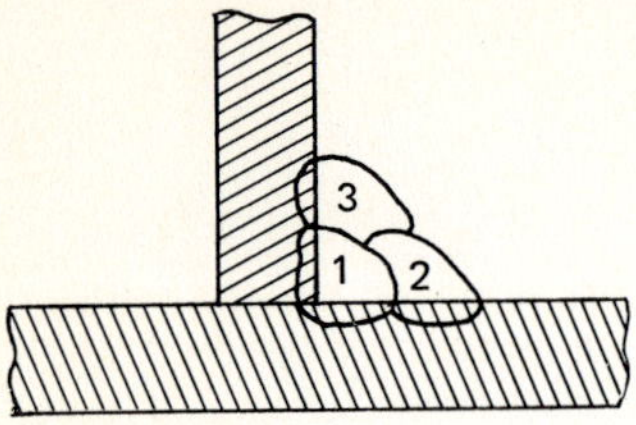

3
1
2

Figure **14-18**

10. Strike an arc and run a second bead joining the base metal of the bottom plate and the first bead (Figure 14-18).

11. Clean the slag from this bead, then run a third bead joining the base metal of the vertical plate and the first two beads.

12. Clean all slag from the final bead.

13. Turn the welding machine off.

14. Inspect the weld. Ask your instructor for suggestions on how you might improve your skill in welding this kind of joint.

14-10 WELDING A SINGLE V-GROOVE BUTT JOINT IN THE HORIZONTAL POSITION

OBJECTIVE To learn to prepare and weld an acceptable single V-groove butt joint in mild steel in the horizontal position.

HOURS REQUIRED 2.

EQUIPMENT, TOOLS, AND MATERIALS REQUIRED The equipment, tools, and materials required to perform this exercise are the same as for exercise 14-7.

PROCEDURE Read and become thoroughly familiar with the instructions below before beginning to weld.

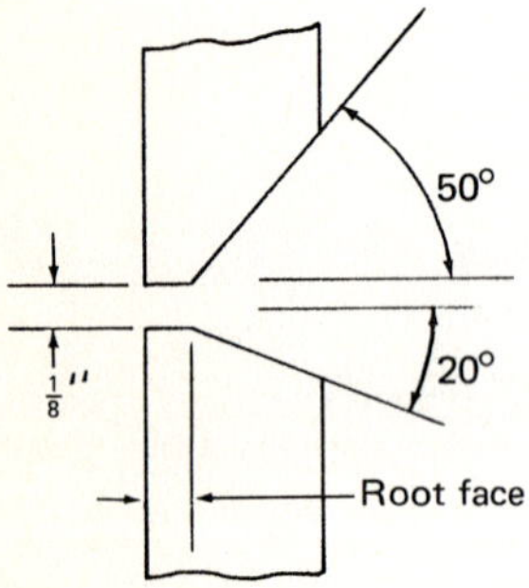

50°
20°
1/8"
Root face

Figure **14-19**

1. Review the safe practices for arc welding in Chapter 13.

2. Bevel the edge of the lower plate at an angle of 20 degrees (Figure 14-19).

3. Bevel the edge of the upper plate at an angle of 50 degrees. (Leave a ⅛-inch root face on both plates.)

4. Lay the plates flat on the bench.

5. Attach the ground lead securely to one of the plates.

6. Set the welding machine amperage at 120 to 140.

7. Fit the electrode into the electrode holder.

8. Turn on the welding machine.

9. Tack weld the plates in position, leaving a ⅛-inch root opening.

10. Weld or clamp the assembly in an upright position, perpendicular to the bench.

11. Strike an arc and run the first bead in the root of the joint (Figure 14-20). Be sure to get penetration through the root with equal fusion in both plates.

12. Clean all slag from this bead.

13. Run subsequent beads in the sequence shown in Figure 14-21. (Clean slag from each bead after it is run. The first bead in each layer

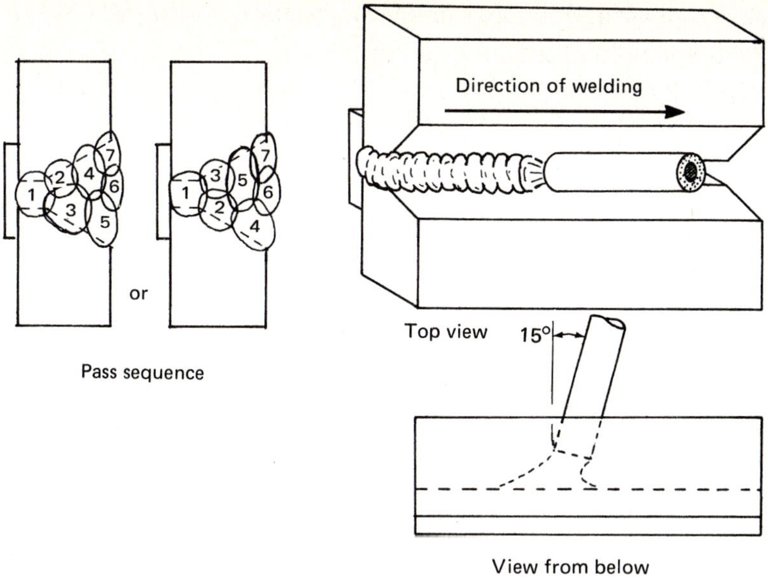
Direction of welding
1
2
3
4
5
6
7
or
Pass sequence
Top view
15°
View from below

Figure **14-20**

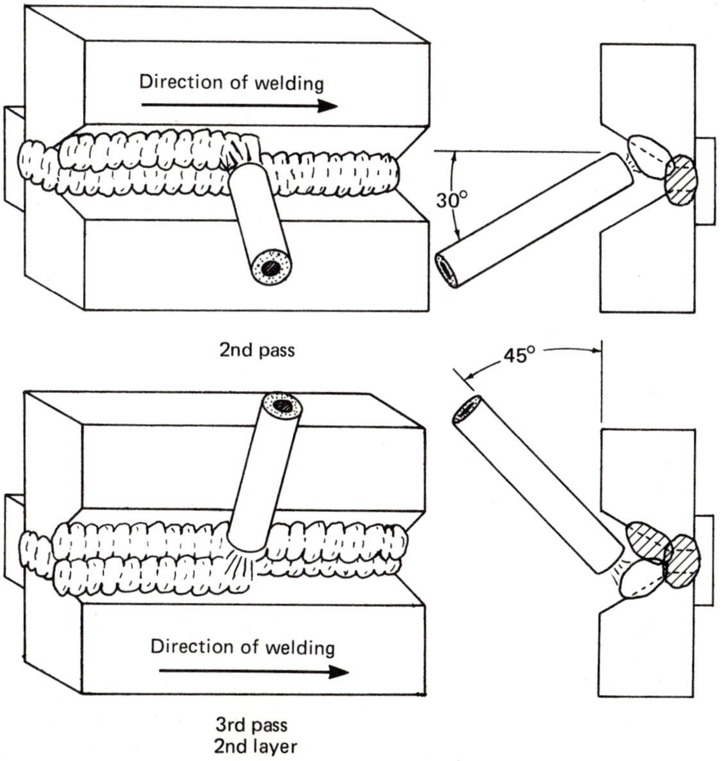
Direction of welding
30°
2nd pass
45°
Direction of welding
3rd pass
2nd layer

See Figure 14-20 for choice of laying down beads

Figure **14-21**

is deposited on the lower plate, so that gravity assists in getting penetration and holding the metal in place.)

14. Turn off the welding machine.

15. Ask your instructor to check the fillet weld.

14-11 PRACTICE IN LAYING VERTICAL BEADS AND VERTICAL WEAVING

OBJECTIVE To learn to lay beads and weave the electrode in the vertical-up and vertical-down positions.

HOURS REQUIRED 2.

EQUIPMENT, TOOLS, AND MATERIALS REQUIRED The equipment, tools, and materials required to perform this exercise are the same as in exercise 14-7.

PROCEDURE Read and become thoroughly familiar with the instructions below before beginning to weld.

1. Review the safe practices for arc welding in Chapter 13.

2. Brush both plates free of dirt and scale.

3. Lay one of the scrap metal plates on the bench.

4. Attach the ground lead securely to the plate on the bench.

5. Set the welding machine amperage at 120 to 140.

6. Fit the electrode into the electrode holder.

7. Turn on the welding machine.

8. Stand the second scrap metal plate on its edge and along one edge of the first plate; tack weld the plates at their intersection.

9. Strike an arc and lay vertical beads along the plate from top to bottom and from bottom to top (Figure 14-22).

10. Clean all slag from the beads.

11. Next, lay vertical weave beads both top to bottom and bottom to top.

12. Turn off the welding machine.

13. Check with your instructor for suggestions on how to improve your skill. (As part of this critique, saw through or break through the plate to expose any entrapped oxide or porous spots).

14-12 WELDING A SINGLE V-GROOVE BUTT JOINT IN THE VERTICAL POSITION

OBJECTIVE To learn to prepare and weld an acceptable single-V butt joint in the vertical position.

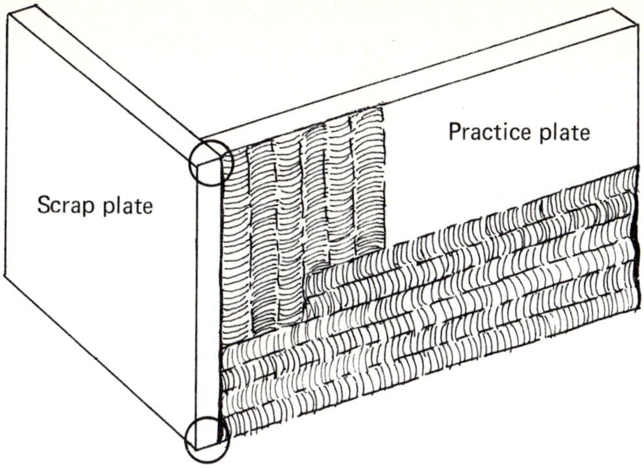

Practice plate
Scrap plate

Figure **14-22** When the practice plate is filled with beads, turn the plate on its side and continue making beads over the original layer, always in the vertical-up or vertical-down position.

HOURS REQUIRED 2.

EQUIPMENT, TOOLS, AND MATERIALS REQUIRED The equipment, tools, and materials required to perform this exercise are the same as in exercise 14-7.

PROCEDURE Read, and become thoroughly familiar with, the instructions below before beginning to weld.

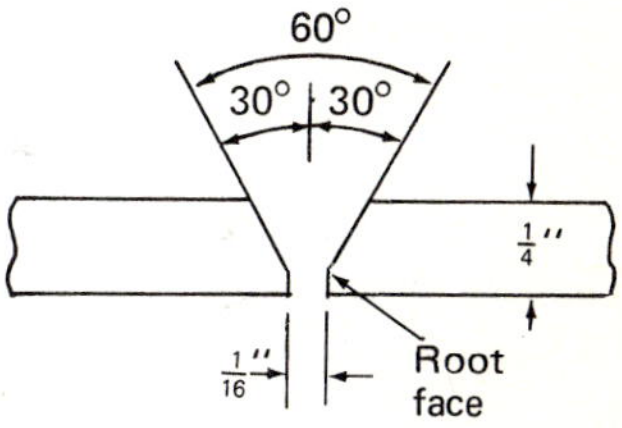

60°
30°
30°
1/4"
1/16"
Root face

Figure **14-23**

1. Review the safe practices for arc welding in Chapter 13.

2. Bevel the two ¼-inch-thick mild-steel plates to form a 60-degree V. Leave a root face of about $\frac{1}{16}$ inch (Figure 14-23).

3. Lay the plates flat on the bench.

4. Attach the ground lead securely to one of the plates.

5. Set the welding machine amperage at 120 to 140.

6. Fit the electrode into the electrode holder.

7. Turn on the welding machine.

8. Lay the second plate in position alongside the first plate and tack weld them together. Leave a root opening of about $\frac{1}{16}$ inch.

9. Tack weld or clamp the assembly into an upright position. Form a single-V butt joint in the vertical position (Figure 14-24).

10. Strike an arc and lay three straight beads from bottom to top, using a rocking motion (see Figure 14-25).

11. Clean all slag from between the beads.

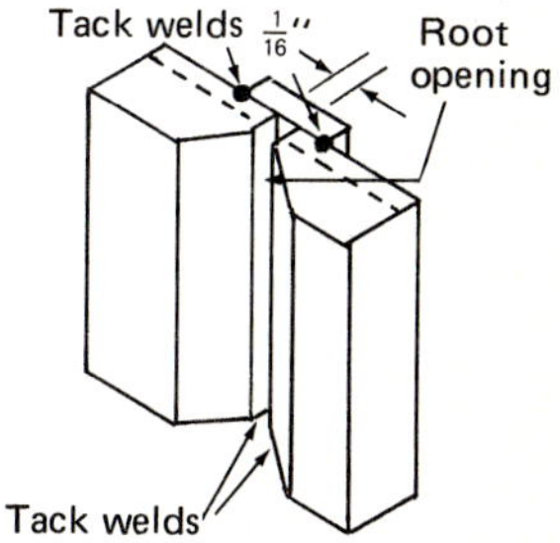

Tack welds
1/16"
Root opening
Tack welds

Figure **14-24**

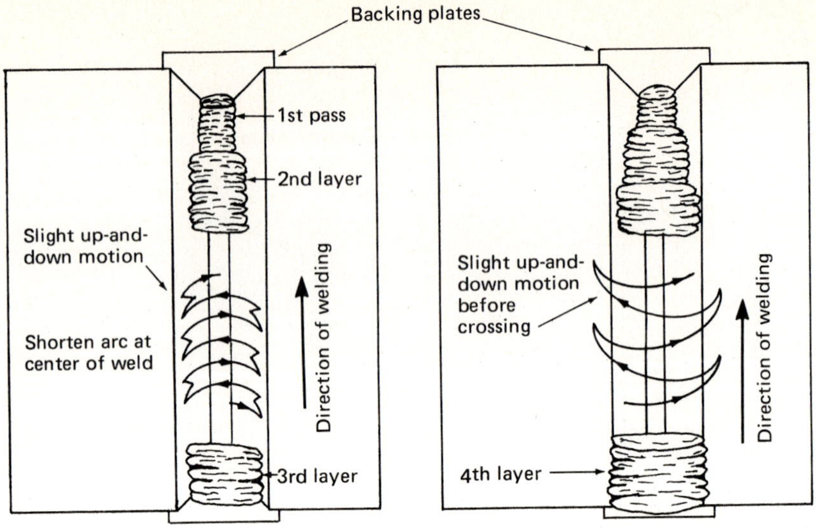
Backing plates
1st pass
2nd layer
Slight up-and-down motion
Shorten arc at center of weld
Direction of welding
3rd layer
Slight up-and-down motion before crossing
4th layer
Direction of welding

Figure **14-25**

12. Turn off the welding machine.

13. Inspect the weld and ask your instructor to check your work.

14-13 WELDING A VERTICAL CORNER JOINT

OBJECTIVE To learn to prepare and weld an acceptable corner joint in the vertical position.

HOURS REQUIRED 2.

EQUIPMENT, TOOLS, AND MATERIALS REQUIRED The equipment, tools, and materials required to perform this exercise are the same as for exercise 14-7.

PROCEDURE Read and become thoroughly familiar with the instructions below before beginning to weld.

1. Review the safe practices for arc welding in Chapter 13.

2. Lay one plate flat on the bench.

3. Attach the ground lead securely to the plate.

4. Set the welding machine amperage at 120 to 140.

5. Fit the electrode into the electrode holder.

6. Turn on the welding machine.

7. Tack weld the second plate perpendicular to the first plate, forming an outside corner joint, as shown in Figure 14-26.

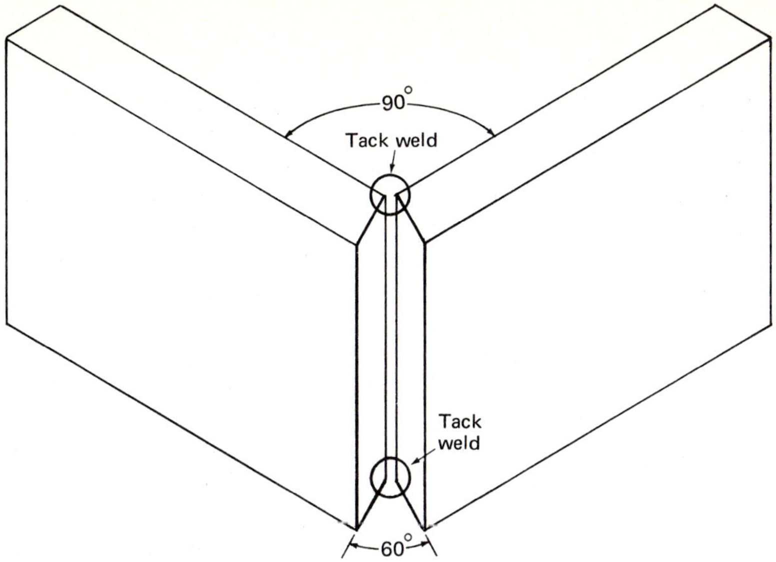
90°
Tack weld
Tack
weld
60°

Figure **14-26**

8. Stand the assembly in position on the bench to weld the outside corner joint.

9. Make the first pass using a straight bead laid from bottom to top with the standard rocking motion (Figure 14-27).

10. Clean all slag from the first bead.

Figure **14-27**

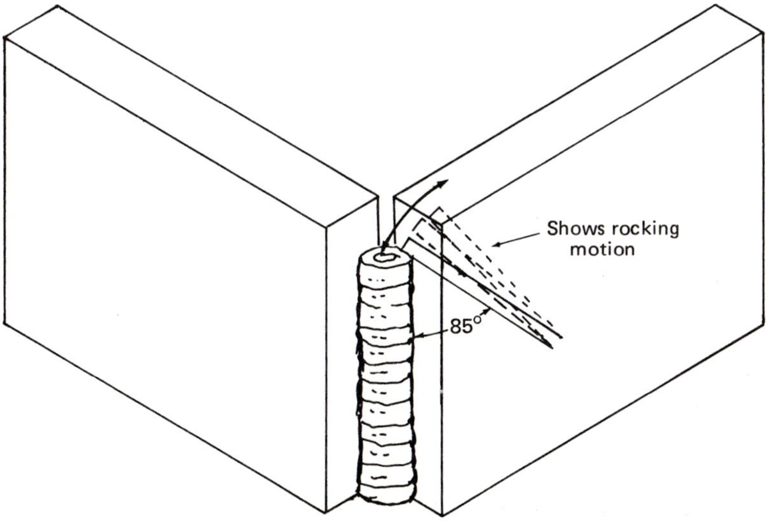
Shows rocking
motion
85°

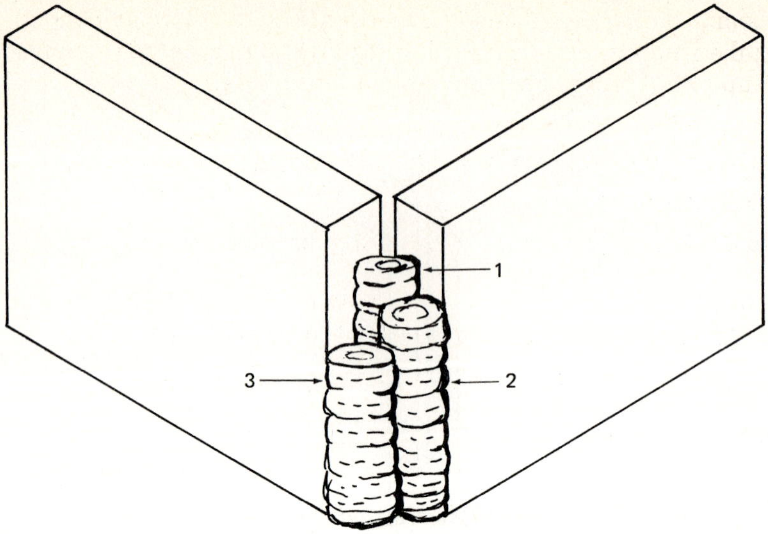
1
2
3

Figure **14-28**

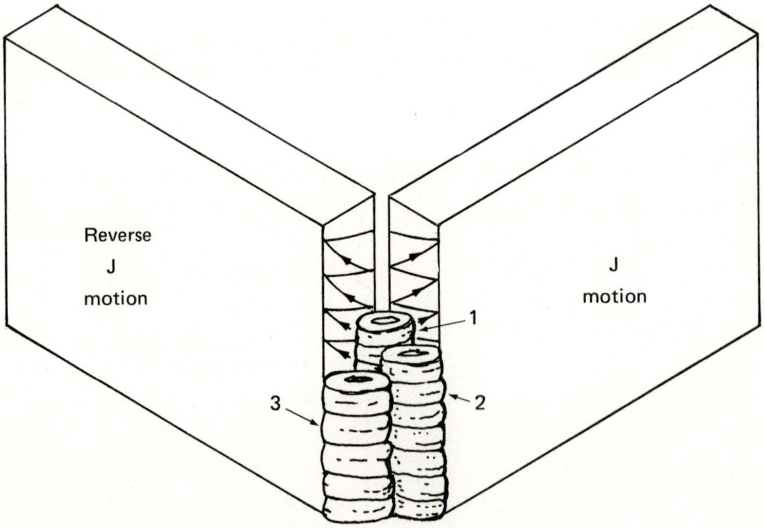
Reverse
J
motion
J
motion
1
2
3

Figure **14-29**

11. Complete the second and third passes (Figures 14-28 and 14-29) using the J and reverse-J motions, respectively, in order to bring the weld up to full height and provide a uniform face.

12. Turn the assembly around and weld the inside corner. (Lay as many beads as required in the sequence shown in Figure 14-30.)

13. Turn off the welding machine.

14. Inspect the weld and ask your instructor to check your work.

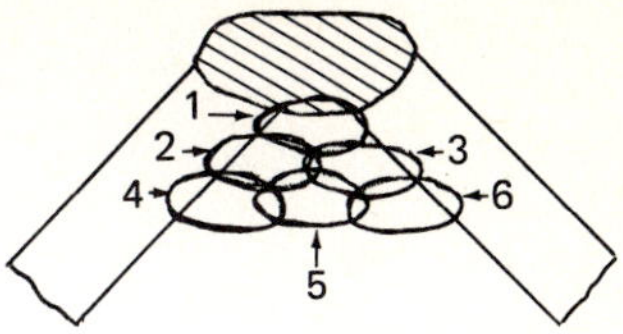
1
2
3
4
5
6
Figure **14-30**

14-14 LAYING STRAIGHT BEADS IN THE OVERHEAD POSITION

OBJECTIVE To learn to lay straight beads in the overhead position.

HOURS REQUIRED 2.

EQUIPMENT, TOOLS, AND MATERIALS REQUIRED The equipment, tools, and materials required to perform this exercise are the same as in exercise 14-7, plus a welding jig (fixture) for overhead position welding, a vest, and a skull cap.

PROCEDURE Read and become thoroughly familiar with the instructions below before beginning to weld.

1. Review the safe practices for arc welding in Chapter 13.

2. Secure the steel plate in the jig so that it is in a position parallel to the floor and high enough to permit welding comfortably from the underside.

3. Attach the ground lead securely to the plate.

4. Set the welding machine amperage at 120 to 140.

5. Fit the electrode into the electrode holder.

6. Turn on the welding machine.
   a. Overhead welding requires a short arc and a quick wrist pivot or rocking motion to counteract the tendency of the molten metal to fall out of the weld.
   b. Assume the most comfortable position possible. Your grip on the electrode holder should permit free wrist action. The welding cable should not drag on the hand.
   c. Hold the electrode perpendicular to the plate laterally and slightly inclined (5 to 15 degrees) so as to point away from the crater (Figure 14-31).

7. Strike an arc and lay straight beads using a wrist pivot or rocking motion. (The line of the weld may be in any direction. Remember to clean the slag from each bead before depositing subsequent beads.)

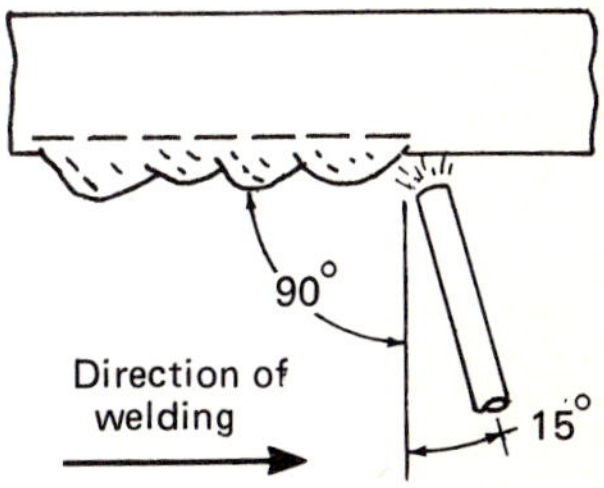
90°
Direction of welding
15°
Figure **14-31**

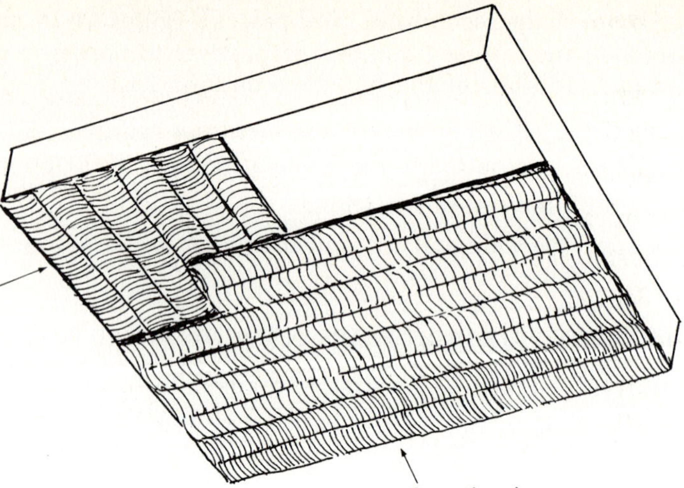
First pad
of beads
Second pad
of beads

Figure **14-32**

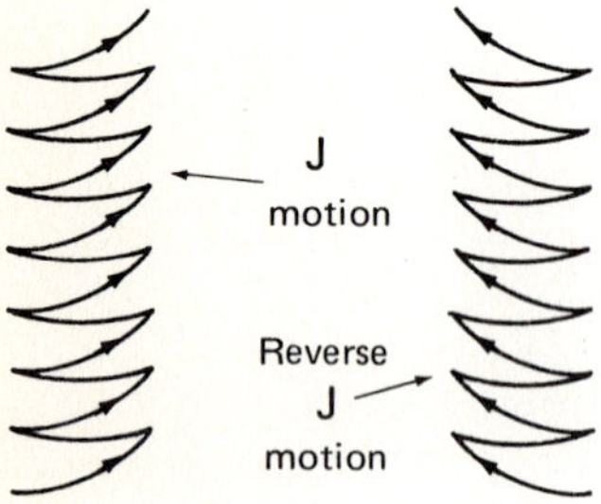
J
motion
Reverse
J
motion

Figure **14-33**

8. Lay a second pad of beads crossways on the original pad. Strive for uniform thickness (Figure 14-32).

9. Practice the J and reverse-J motions in a third pad of beads (Figure 14-33).

10. Turn off the welding machine.

11. Inspect the weld and ask your instructor to check your work.

14-15 MAKING MULTIPLE-PASS FILLET WELDS IN THE OVERHEAD POSITION

OBJECTIVE To learn to make acceptable multiple-pass fillet welds in the overhead position on mild-steel plate.

HOURS REQUIRED 3.

EQUIPMENT, TOOLS, AND MATERIALS REQUIRED The equipment, tools, and materials required to perform this exercise consist of a welding power supply as used in exercise 14-7 and the following:

1. An overhead jig.

2. Two 3 × 8 × ¼-inch mild-steel plates.

3. Two 3 × 8 × ½-inch mild-steel plates.

4. Several 5/32-inch-diameter electrodes (E-6010 for dc, E-6011 for ac).

5. Several 3/16-inch-diameter electrodes (E-6010 for dc, E-6011 for ac).

PROCEDURE Read and become thoroughly familiar with the instructions below before beginning to weld.

1. Review the safe practices for arc welding in Chapter 13.

2. Lay one 1/4-inch-thick mild-steel plate flat on the bench.

3. Attach the ground lead securely to the plate.

4. Set the welding machine amperage at 120 to 140.

5. Fit a 5/32-inch-diameter electrode into the electrode holder.

6. Turn on the welding machine.

7. Lap the second 1/4-inch-thick mild-steel plate over the first one and tack weld the plates together at the ends of the joint at both sides (Figure 14-34).

8. Clamp the assembly in the overhead jig so that the narrow side of the V is uppermost (Figure 14-35).

9. Deposit the first bead deep in the root of the V using a wrist-pivot motion.

10. Remove all slag from this bead.

11. Deposit the second and third beads in the sequence indicated in Figure 14-35.

Figure **14-34**

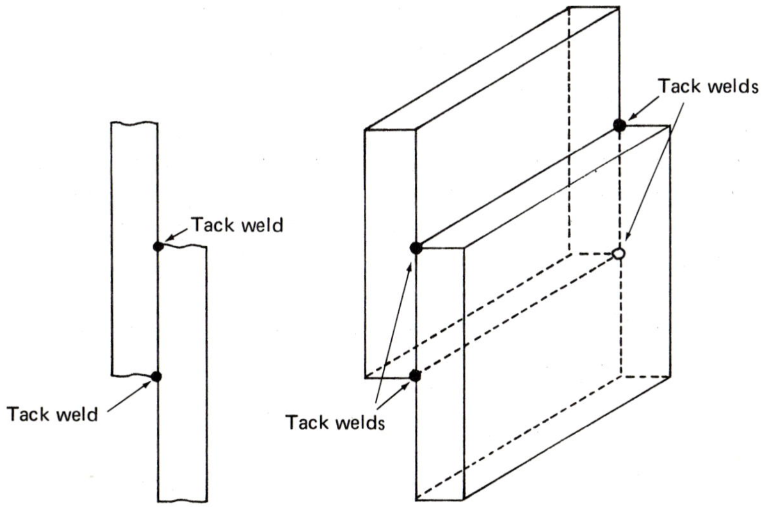
Tack weld
Tack weld
Tack welds
Tack welds

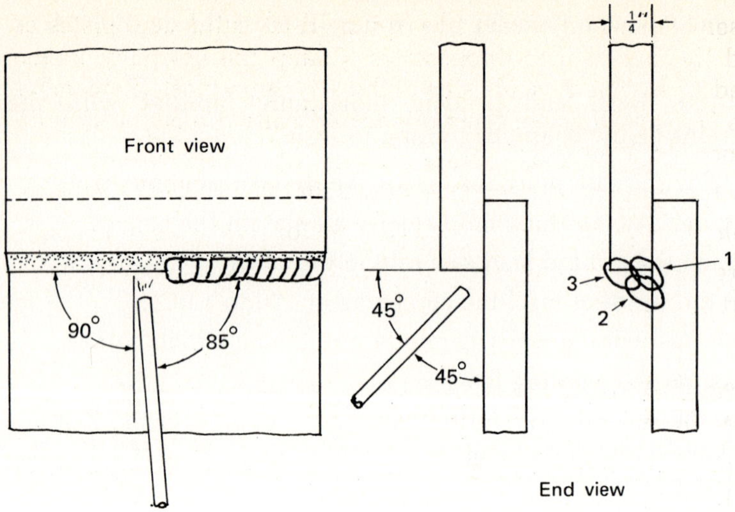

Front view
90°
85°
45°
45°
1/4"
1
3
2
End view

Figure **14-35**

12. Turn the assembly over and weld the opposite side (Figure 14-36).

13. Turn off the welding machine and ask your instructor for a critique of your work.

14. Adjust the welding machine amperage to 170 to 190 and turn the machine on.

Figure **14-36**

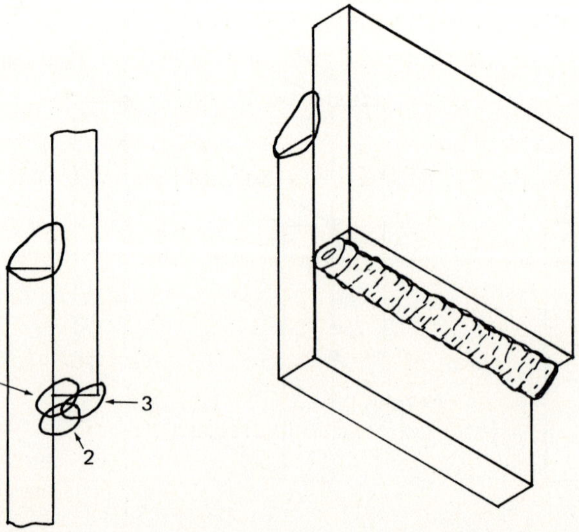

1
3
2

15. Assemble and tack weld two ½-inch-thick mild-steel plates as you did the two ¼-inch-thick plates. Clamp the assembly in the overhead jig so that the wide side of the V is uppermost. (Remember to use a 3/16-inch-diameter electrode for all parts of your work with the ½-inch-thick plates.)

16. Lay the beads in the sequence shown for the ¼-inch-thick plates.

17. Weld the opposite side of the assembly for additional practice.

18. Turn off the welding machine and ask your instructor to check this part of the job.

14-16 PREPARING AND WELDING A SINGLE V-GROOVE BUTT JOINT IN THE OVERHEAD POSITION

OBJECTIVE To learn to prepare and weld an acceptable single V-groove butt joint in the overhead position.

HOURS REQUIRED 2.

EQUIPMENT, TOOLS, AND MATERIALS REQUIRED The equipment, tools, and materials required to perform this exercise are the same as in exercise 14-15, except that only the ½-inch-thick mild-steel plates need be used.

PROCEDURE Read and become thoroughly familiar with the instructions below before beginning to weld.

1. Review the safe practices for arc welding in Chapter 13.

2. Bevel one edge of each of the plates at a 30-degree angle. Leave a root face of about ⅛ inch.

3. Lay one plate flat on the bench.

4. Attach the ground lead securely to the plate.

5. Set the welding machine amperage at 120 to 140.

6. Fit the electrode into the electrode holder.

7. Turn on the welding machine.

8. Tack weld the plates together at the ends of the V with a 1/16-inch root opening (Figure 14-37).

Figure **14-37**

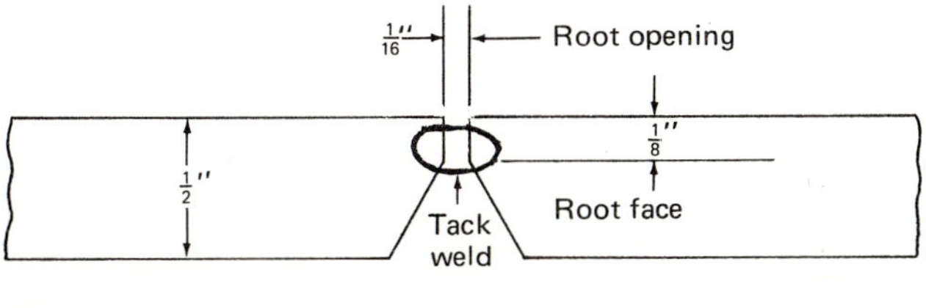
1/16"
Root opening
1/8"
1/2"
Root face
Tack weld

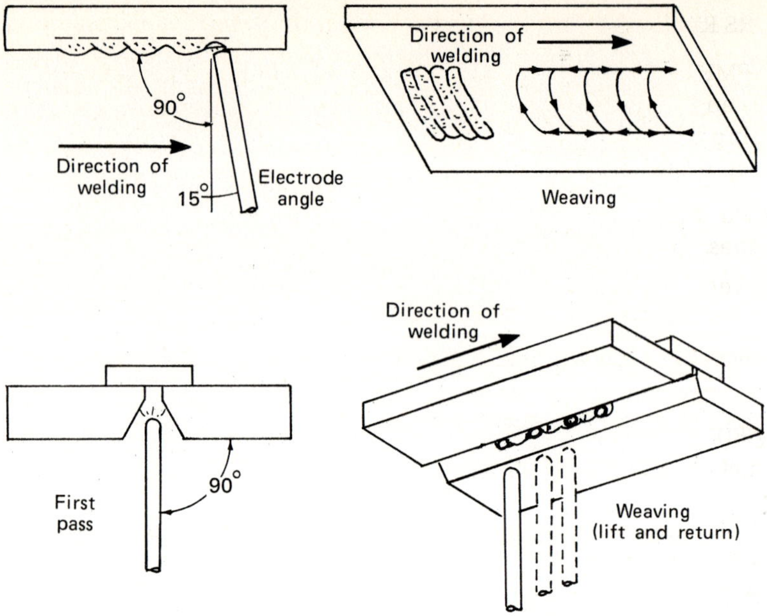
90°
Direction of welding
15°
Electrode angle
Direction of welding
Weaving
Direction of welding
90°
First pass
Weaving (lift and return)

Figure **14-38**

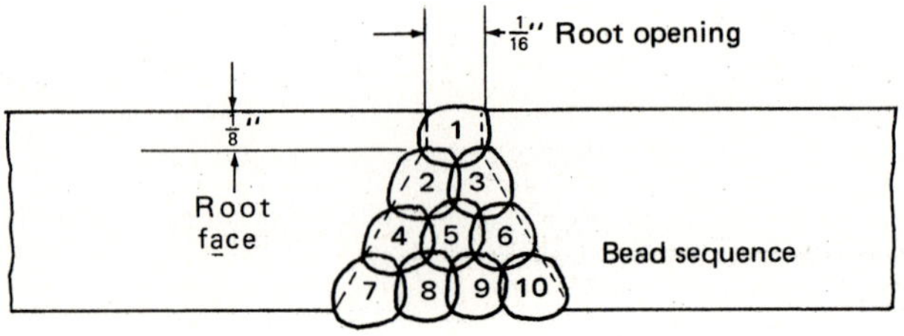
1/16" Root opening
1/8"
Root face
1
2
3
4
5
6
7
8
9
10
Bead sequence

Figure **14-39**

9. Clamp the assembly in the overhead jig with the V on the underside, as shown in Figure 14-38.

10. Lay beads in the sequence shown in Figure 14-39.

11. Turn off the welding machine and ask your instructor to check your work.

14-17 WELDING A SQUARE-GROOVE BUTT JOINT IN A STEEL PIPE IN THE 5G POSITION

OBJECTIVE To practice welding in all welding positions on a single square-groove pipe joint where the conditions are such that the pipe cannot be turned [in the fixed horizontal (5 G) position].

HOURS REQUIRED 2.

EQUIPMENT, TOOLS, AND MATERIALS REQUIRED The equipment, tools, and materials required to perform this exercise consist of a welding power supply, accessories, and an overhead welding jig, as used in exercise 14-16, plus:

1. Two 3-inch unbeveled pieces of steel pipe with a ¼-inch wall thickness.

2. Several 5/32-inch-diameter electrodes (E-6010 for dc, E-6011 for ac).

3. Several 3/16-inch-diameter electrodes (E-6010 for dc, E-6011 for ac).

PROCEDURE Read and become thoroughly familiar with the instructions below before beginning to weld.

1. Review the safe practices for arc welding in Chapter 13.
2. Lay the two pieces of scrap pipe on the bench.
3. Attach the ground lead securely to one of the pieces.
4. Set the welding machine amperage at 120 to 140.
5. Fit a 5/32-inch-diameter electrode into the electrode holder.
6. Turn on the welding machine.
7. Tack weld the pipes together, end to end, without beveling, as shown in Figure 14-40.
8. Clamp the assembly in the overhead jig in a horizontal position, as shown in Figure 14-41. *Note:* Locate the pipe about waist high, so

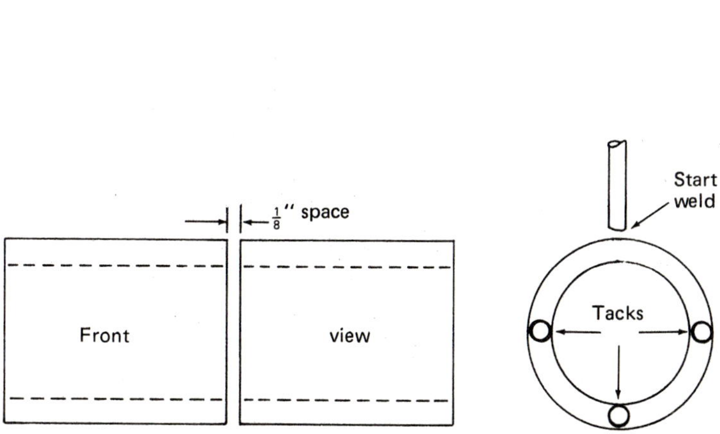

$\frac{1}{8}$" space
Front
view
Start weld
Tacks
End view

Figure **14-40**

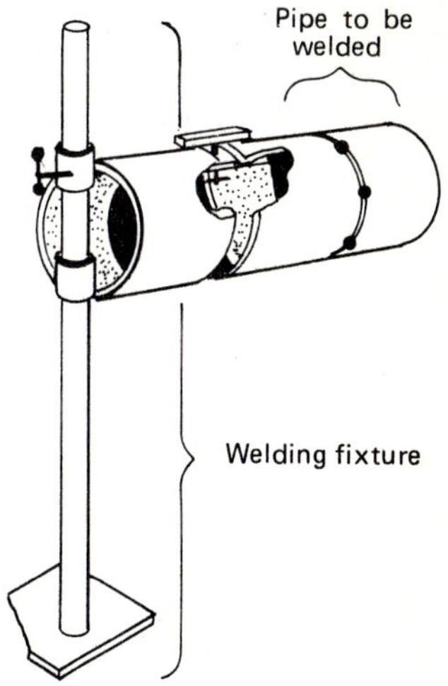

Pipe to be welded
Welding fixture

Figure **14-41**

that you can weld on top while standing, but can get down on your knees or even sit on the floor when welding the side or underneath.

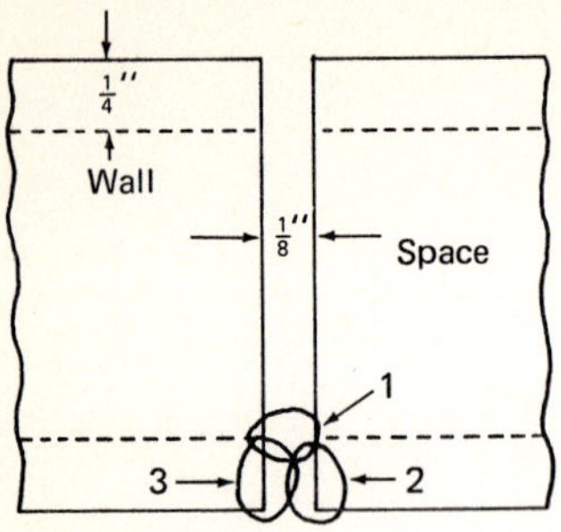

1/4"
Wall
1/8"
Space
1
2
3

Figure **14-42**

9. Strike an arc at the top and lay a straight bead completely around the pipe, getting penetration clear through the joint.

10. Clean all slag from this weld.

11. Finish the weld with two more passes (Figure 14-42), using the 3/16-inch-diameter electrode. (Adjust the welding machine amperage to between 170 and 190, and use a weaving motion.) *Note:* Make the weld more convex than welds on flat plates, since many piping jobs require extra reinforcement. Remember to clean the weld of all slag after each pass.

12. Turn off the welding machine.

13. Inspect the weld and ask your instructor to check your work.

14-18 WELDING A SQUARE-GROOVE BUTT JOINT IN A STEEL PIPE IN THE 2G POSITION

OBJECTIVE To learn to weld an acceptable square-groove butt joint in pipe in the vertical-fixed (2G) position.

HOURS REQUIRED 1.

EQUIPMENT, TOOLS, AND MATERIALS REQUIRED The equipment, tools, and materials required to perform this exercise consist of an overhead welding jig, a welding power supply and accessories, and pieces of pipe as described in the last exercise, plus several 1/8-inch-diameter and 5/32-inch-diameter electrodes (E-6010 for dc, E-6011 for ac).

PROCEDURE Read, and become thoroughly familiar with, the instructions below before beginning to weld.

1. Review the safe practices for arc welding in Chapter 13. *Note:* To weld pipe in the fixed-vertical position, use the same techniques as for welding butt welds on flat plates in the horizontal position.

2. Lay the two pieces of scrap pipe on the bench.

3. Attach the ground lead securely to one of the pieces.

4. Set the welding machine amperage at 100 to 120.

5. Fit a 1/8-inch-diameter electrode into the electrode holder.

6. Turn on the welding machine.

7. Tack weld the pipes together, end to end, without beveling, as shown in Figure 14-40.

8. Clamp the assembly in the overhead jig in the vertical position, so that the joint is in a horizontal position, as shown in Figure 14-43.

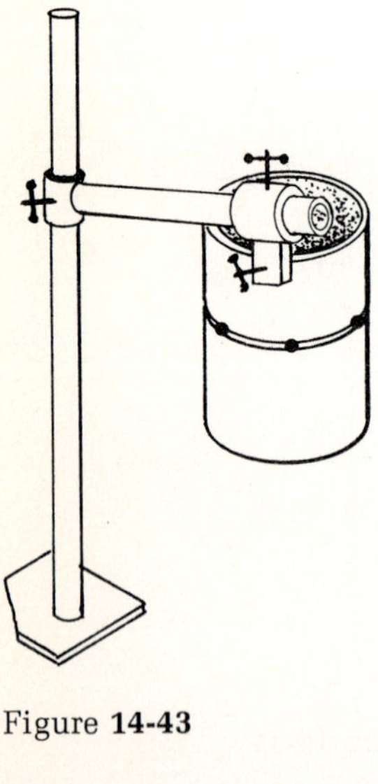
Figure **14-43**

9. Strike an arc and lay the first bead so that it penetrates clear through to the inside of the pipe wall.

10. Clean the slag from this weld.

11. Adjust the welding machine amperage to 120 to 140.

12. Fit a 5/32-inch-diameter electrode into the electrode holder.

13. Lay two more beads, in the sequence shown in Figure 14-42. (Be sure to clean the slag from each bead.) *Note:* All three beads should be straight beads without weaving, and the surface of the beads should be made more convex than the surface of joints in flat plates.

14. Turn off the welding machine.

15. Inspect the weld and ask your instructor to check your work.

14-19 ROLL WELDING A SINGLE-V BUTT JOINT IN STEEL PIPE

OBJECTIVE To learn to control the arc while the pipe is being turned at a constant rate under it.

HOURS REQUIRED 1.

EQUIPMENT, TOOLS, AND MATERIALS REQUIRED The equipment, tools, and materials required to perform this exercise consist of a power supply and accessories, plus the following:

1. A simple rolling device.

2. Two 8- or 10-inch-diameter 6-inch-long pieces of steel pipe with a 1/4-inch-thick wall.

3. Several 5/32-inch-diameter electrodes (E-6010 for dc, E-6011 for ac).

4. Several 3/16-inch-diameter electrodes (E-6010 for dc, E-6011 for ac).

PROCEDURE Read and become thoroughly familiar with the instructions below before beginning to weld.

1. Review the safe practices for arc welding in Chapter 13.

2. Bevel one end of each piece of pipe at an angle of 30 degrees. Leave a root face of 3/32 inch (Figure 14-44).

Figure **14-44**

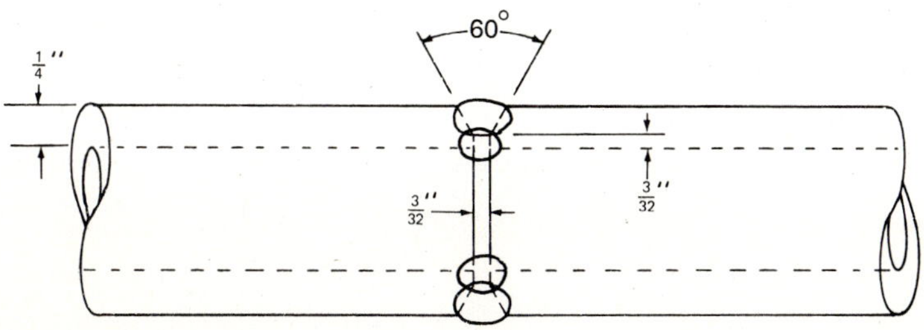

60°
1/4''
3/32''
3/32''

3. Place the two pieces together on the rollers with a 3⁄32-inch root opening, as shown in Figure 14-45.

4. Attach the ground cable securely to one of the pipes.

5. Set the welding machine amperage at 120 to 140.

6. Fit a 5⁄32-inch-diameter electrode into the electrode holder.

7. Turn on the welding machine.

8. Tack weld the pipes on top. Give the pipe a quarter-turn and tack weld again on top.

9. After another quarter-turn, tack weld the pipes on top a third time.

10. After a third quarter-turn, start welding deep in the root of the V with a straight bead. *Note:* Be sure to keep the electrode in the perpendicular position, simultaneously turning the pipe at the proper speed so that one inch of bead is produced for each inch of electrode melted.

11. Clean all slag from the joint.

12. Set the welding machine amperage at 170 to 190.

13. Fit a 3⁄16-inch-diameter electrode into the electrode holder.

14. Deposit the second bead with a slight side-to-side weaving motion, getting penetration into the first bead and into both sides of the V.

15. Clean the slag from this bead.

16. Turn off the welding machine.

17. Inspect the joint and ask your instructor to check your work.

Figure **14-45**

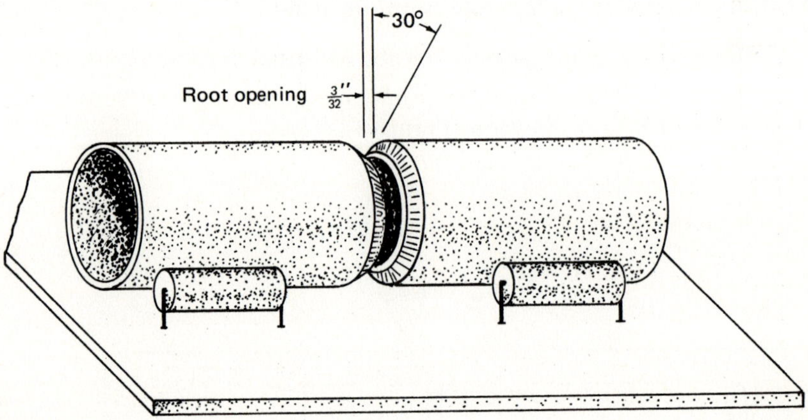
30°
Root opening
3/32"

14-20 WELDING A PIPE FITTING

OBJECTIVE To learn to use a standard pattern, cut the pattern, and weld a pipe T fitting.

HOURS REQUIRED 3.

EQUIPMENT, TOOLS, AND MATERIALS REQUIRED The equipment, tools, and materials required to perform this exercise are:

1. Two pieces of mild-steel pipe 6 inches long, 2 inches in diameter, with a ¼-inch-thick wall.

2. Set of 2-inch-diameter pipe cutting curves (Figure 14-46).

3. Oxyacetylene welding outfit with an appropriate cutting torch.

Figure **14-46**

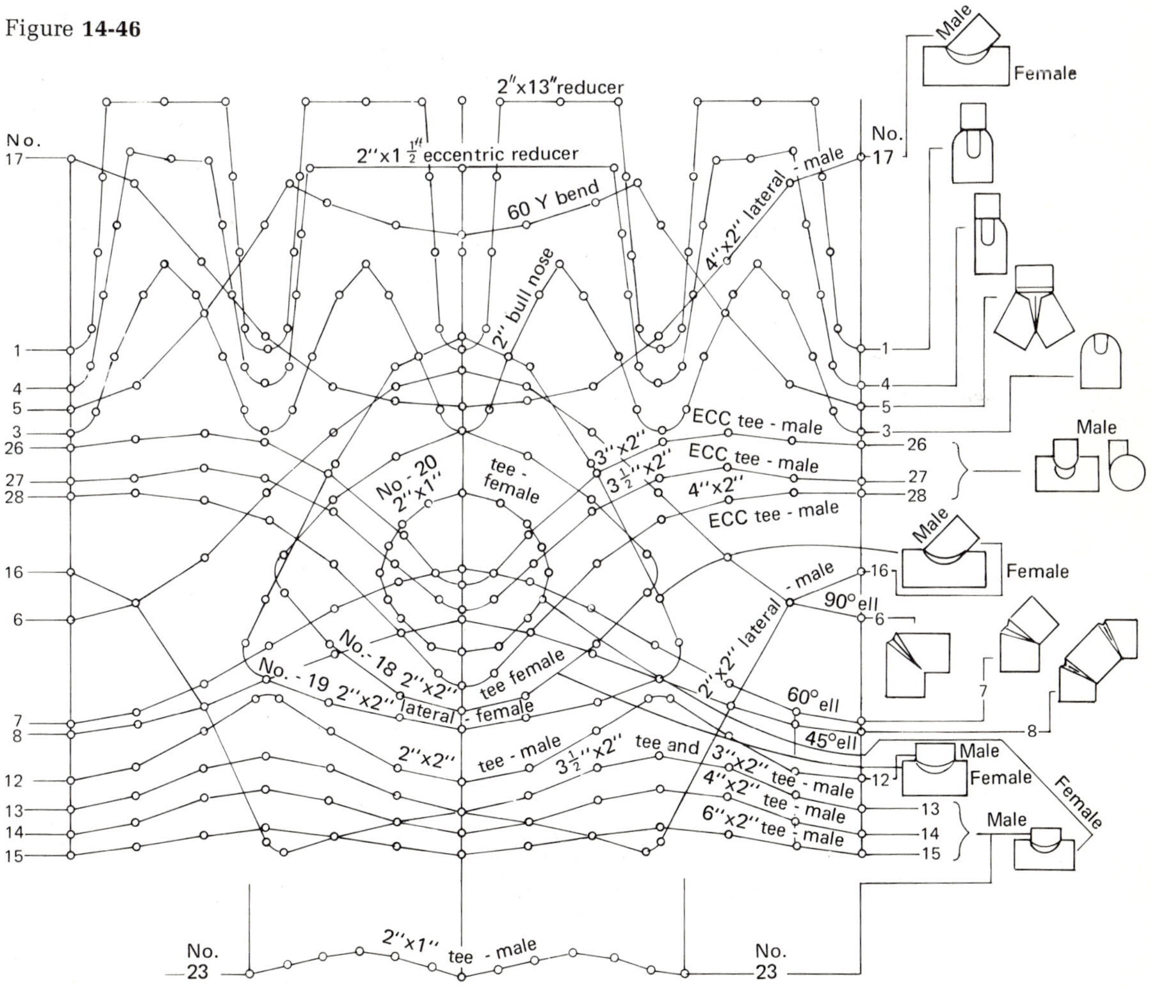

2"x13"reducer
2"x1½" eccentric reducer
60 Y bend
2" bull nose
4"x2" lateral - male
No. 17
ECC tee - male
3"x2"
3½"x2"
4"x2"
No - 20 2"x1"
tee - female
No.- 18 2"x2" tee female
No. - 19 2"x2" lateral - female
2"x2" lateral - male
90° ell
60° ell
45° ell
2"x2" tee - male
3½"x2" tee and 3"x2" tee - male
4"x2" tee - male
6"x2" tee - male
2"x1" tee - male
No. 23
Male
Female
1
3
4
5
6
7
8
12
13
14
15
16
26
27
28

4. Marking chalk and prick punch.

5. 12-ounce ball-peen hammer.

6. Arc welding power supply and accessories.

7. 11 × 7-inch (B size) tracing paper.

8. 3H pencil.

9. One 11 × 17-inch (B size) piece of paste board, wrapping paper, sheet metal, or similar thin, flat material.

10. Tinners snips or similar.

PROCEDURE Read and become thoroughly familiar with the instructions below before beginning to weld.

1. Review the safe practices for arc welding in Chapter 13.
a. Since it is necessary to know how to use the oxyacetylene cutting torch to perform this exercise, it is recommended that this exercise not be performed until after the student has learned to use the oxyacetylene equipment.
b. Templates for standard fittings in sizes of 2, 2½, 3, 3½, 4, 6, 8, 10, 12, and 16 inches can be obtained from many of the large welding equipment distributors, such as AIRCO and Linde Division of Union Carbide.
c. Once the type of fitting to be made has been decided upon, trace the fitting on the tracing paper.

2. Using chalk or the prick punch and hammer, mark the centerline of the pipe(s) on which the template is to be used.

3. Accurately cut the template traced to the lines (see Figure 14-46, lines 12 and 18).

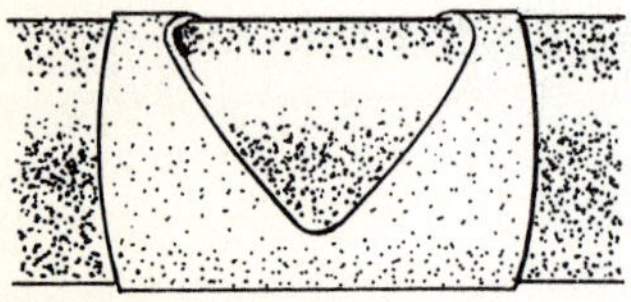

Figure **14-47**

4. Wrap the template around the pipe (Figure 14-47). *Note:* Align centerline of template with centerline of pipe.

5. Using chalk or prick punch and hammer, outline the template on the pipe.

6. Use the oxyacetylene torch to cut out the template (see exercise 16-15).

7. Lay the header pipe on the bench.

8. Attach the ground lead securely to the pipe.

9. Set the welding machine amperage at 120 to 140.

10. Fit the electrode into the electrode holder.

11. Turn on the welding machine.

12. Place the branch line in position, as shown in Figure 14-48, and tack weld at the points shown.

13. Weld from top to bottom (Figure 14-49) on the four sides of the points in the end of the branch-line pipe.

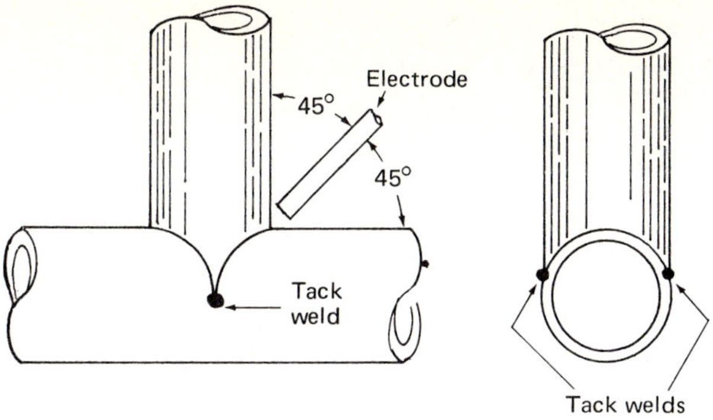

Electrode
45°
45°
Tack
weld
Tack welds

Figure **14-48**

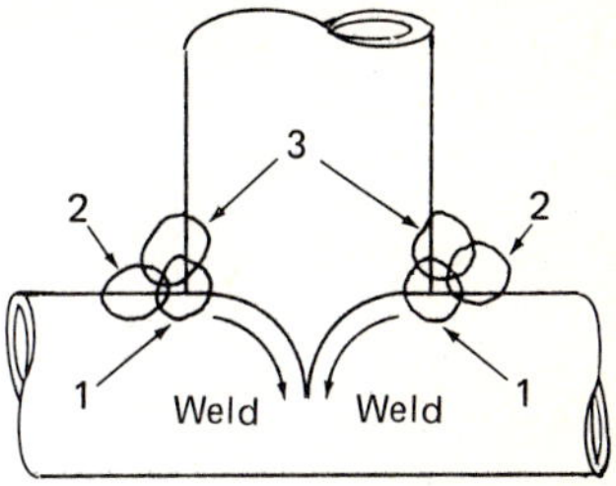

3
2
2
1
Weld
Weld
1

Figure **14-49**

14. Deposit the first bead in all four joints and remove the slag.

15. Deposit the second and third beads as indicated, removing the slag after each is laid.

16. Turn off the welding machine and ask your instructor to check your work.

REVIEW QUESTIONS

1. Describe or sketch what the cross-section of a good fillet weld should look like.

2. Name and sketch the shape of three edge preparations used for butt welds.

3. Name and describe the two methods used for vertical welding.

4. Why is the whipping motion used in vertical welding?

5. Name two reasons for using different weave patterns in the vertical-up position.

6. What electrode angle is used for overhead welding?

7. What can be done to keep the puddle from dropping or sagging during overhead welding?

8. Name the AWS electrode type, diameter, and amperage setting required to weld ⅜-inch-thick mild steel in the flat position.

9. Why is the weaving motion used?

10. What is the thickness of the thinnest plate that requires preparation of the edges for welding?

11. Why are backing strips sometimes used when welding mild-steel plates?

12. List the items of safety clothing that should be worn by welders welding in the overhead position.

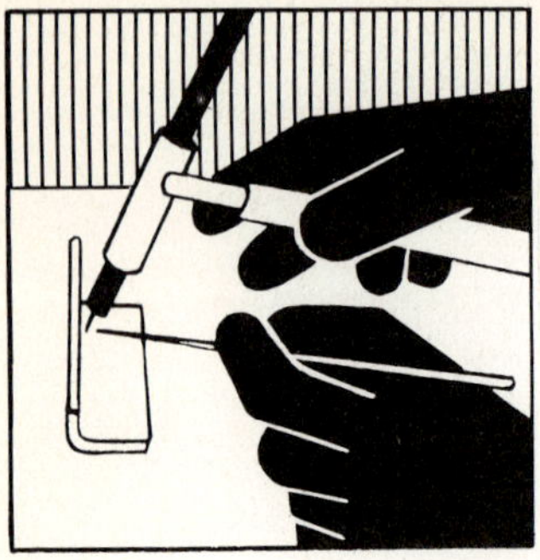

CHAPTER 15
GAS SHIELDED–ARC WELDING PRACTICE

GAS METAL–ARC WELDING (GMAW) PRACTICE

The gas metal–arc welding process is a dc, reverse-polarity process in which solid, bare, continuously fed consumable electrodes are shielded from the atmosphere by an inert gas. MIG welding can be performed with any one of three types of voltage characteristics: drooping, constant arc, or rising arc. The power supplies are designed for complete control of slope and voltage. With adjustable slope, voltage, and stabilizer, it is possible to select and maintain the best possible arc characteristics.

As discussed in Chapter 5, metal transfer in the GMAW process is either by the spray-arc or short-circuiting method. In the spray-arc method, the electrodes are larger than in the short-circuit method and the arc is always on. This produces a heavy deposition of filler metal and is well suited for single- or multiple-pass heavy weldments in the flat position. The short-circuit method accomplishes transfer by allowing molten metal droplets to "short out" the arc. The *slope*, or amount of voltage drop per 100 amperes, controls the amount of current available from the power supply (Figure 15-1). This current acts in a manner to exert a "pinch force" on the wire, which causes the wire to "neck down" and finally separate from the workpiece (Figure 15-2). The amount of current at this point is important because the manner in which the wire parts from the work can be either violent or smooth. If the current is too high, a violent parting occurs. This causes spatter and a pronounced solidification pattern in the bead. If the current is at the correct value, the parting is smooth, with little or no spatter, and a tightly knit freeze pattern

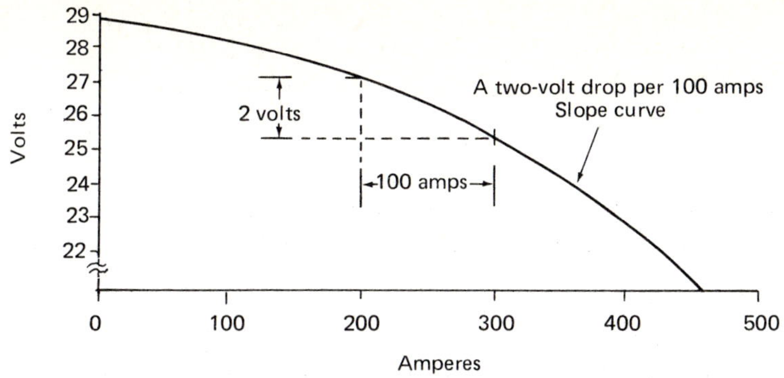

Volts
29
28
27
26
25
24
23
22
2 volts
100 amps
A two-volt drop per 100 amps
Slope curve
0
100
200
300
400
500
Amperes

Figure **15-1** The meaning of slope in MIG welding. (Redrawn by permission from Miller Electric Manufacturing Company, *Fundamentals of Gas Metal-Arc(MIG) Welding,* Form: MTS-6A 3/71.)

appears in the bead. The amount of current available on short circuit is controlled by means of slope selection.

The short-circuiting method of transfer, also called the *short-arc method,* is exceptionally well suited to the welding of thin sections in any welding position. Carbon dioxide, argon–carbon dioxide mixtures, and helium-base gases can be used as shielding gases. The short-arc method is the one we will use most here. However, Tables 15-1 to 15-9 (pages 336–341) list recommended gas mixtures, types of joints, slope characteristics, and other selected information for good welds in aluminum, stainless steel, copper, and carbon steel with the short-arc method and the spray method.

(*text continues page* 342)

Figure **15-2** The relationship between current, spatter, quantity, and rate of pinch. (Redrawn by permission from Union Carbide Corporation, Linde Division, *How to Do Manual Metal Inert Gas (Heliarc) Welding,* F-51-110A 5/71.)

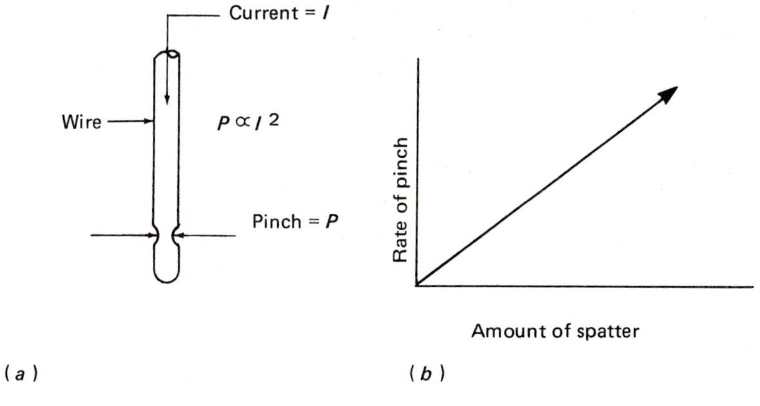

Current = *I*
Wire
$P \propto I^2$
Pinch = *P*
(*a*)
Rate of pinch
Amount of spatter
(*b*)

Table **15-1** General welding conditions* for carbon steel

PLATE THICKNESS (IN.)	JOINT AND EDGE PREPARATION	WIRE DIAMETER (IN.)	GAS FLOW (CFH)	AMPC (DCRP)	VOLTAGE	WIRE FEED SPEED (IPM)	WELDING SPEED (IPM)	PASSES
0.035	↑	↑	↑	55	16*	117	13	1
0.047	↕	↕	↕	65	17*	140	15	1
0.063	Non-positioned fillet or lap	0.030	10–15	85	17*	170	15	1
0.078	↕	↕	↕	105	18*	225	15	1
0.100	↕	↕	↕	110	18*	225	15	1
⅛	↓	↓	↓	130	19*	300	16	1
⅛	Butt (square edge)	1/16	↑	280	—	165	—	1
3/16	Butt (square edge)	1/16	↕	375	—	260	—	2
3/16	Fillet or lap	1/16	↕	350	—	230	—	1
¼	Double-V butt (60° included angle, no nose)	↑	↕	375 (1st pass) 430 (2nd pass)	27	83 (1st) 95 (2nd)	24	2
5/16	Double-V butt (60° included angle, no nose)	↕	40–50	400 (1st pass) 420 (2nd pass)	28	87 (1st) 92 (2nd)	20	2
5/16	Non-positioned fillet	↕	↕	400	27	87	16	1
½	Double-V butt (60° included angle, no nose)	3/32	↕	400 (1st pass) 450 (2nd pass)	28	87 (1st) 100 (2nd)	14	2
½	Non-positioned fillet	↕	↕	450	28	100	12	1
¾	Double-V butt (90° included angle, no nose)	↕	↕	450 (all 4 passes)	29	100	12	4
¾	Positioned fillet	↕	↕	475	30	110	9	1
1	Fillet	↓	↓	450 (all 4 passes)	28	100	7	4

Sigma 0–5 argon mixture (95% argon + 5% O_2) C-25 mixture (75% A + 25% CO_2)

*Short arc

Source: Union Carbide Corporation, Linde Division, *How To Do Manual Metal Inert Gas Welding*, F-51-110-A 5/71.

Table **15-2** Manual short-arc welding techniques for aluminum

WELDING POSITION	JOINT TYPE	TECHNIQUE	REMARKS
Flat and horizontal	Butt	Forehand	Best visibility
Flat and horizontal	Fillet	Backhand or forehand	—
Vertical	Butt	Vertical down	Flattest weld bead
Vertical	Fillet	Vertical down	Flattest weld bead
Overhead	Butt	Forehand	Flattest weld bead
Overhead	Fillet	Backhand or forehand	—

Source: Union Carbide Corporation, Linde Division, *How To Do Manual Metal Inert Gas Welding,* F-51-110-A 5/71.

Table **15-3** Shielding gas preference for inert-gas welding of aluminum

PLATE THICKNESS (INCHES)	GAS MIXTURE
0 to 1	Argon
1 to 2	Argon
	50% Argon–50% Helium
2 to 3	50% Argon–50% Helium
	25% Argon–75% Helium
3+	25% Argon–75% Helium

Source: Union Carbide Corporation, Linde Division, *How To Do Manual Metal Inert Gas Welding,* F-51-110-A 5/71.

Table **15-4** Manual short-arc welding conditions for aluminum

PLATE THICKNESS (INCHES)	TYPE OF JOINT	WIRE DIAM. (INCHES)	ARGON FLOW (CFH)	AMPERES (DCRP)	VOLTAGE (VOLTS)	APPROXIMATE WIRE FEED SPEED (IPM)	APPROXIMATE WELDING SPEED (IPM)
0.040	Fillet or tight butt	0.030	30	40	15	240	20
0.050	Fillet or tight butt	0.030	15	50	15	290	15
0.063	Fillet or tight butt	0.030	15	60	15	340	15
0.093 to 0.125	Fillet or tight butt	0.030	15	90	15	410	15

Source: Union Carbide Corporation, Linde Division, *How To Do Manual Metal Inert Gas Welding,* F-51-110-A 5/71.

Table **15-5** Conditions for manual spray–arc welding of butt joints in aluminum in the downhand position

PLATE THICKNESS	PREPARATION	WIRE DIAMETER (IN.)	ARGON FLOW (CFH)	AMPS (DCRP)	VOLTAGE	WELDING SPEED (IPM)	NO. OF PASSES
0.250	Single-V butt (60° included angle) sharp nose, backup strip used	3/64	35	180	24	15	1
	Square butt with backup strip	3/64	40	250	26	16	1
	Square butt with no backup strip	3/64	35	220	24	24	2
0.375	Single-V butt (60° included angle) sharp nose, backup strip used	1/16	40	280	27	24	2
	Double-V butt (75° included angle, 1/16-in. nose). No backup, back chip after root pass	1/16	40	260	26	18	2
	Square butt with no backup strip	1/16	50	270	26	22	2
0.500	Single-V butt (60° included angle) sharp nose, backup strip used	1/16	50	310	27	18	2
	Double-V butt (75° included angle 1/16-in. nose). No backup, back chip after root pass	1/16	50	300	27	18	3

Source: Union Carbide Corporation, Linde Division, *How To Do Manual Metal Inert Gas Welding*, F-51-110-A 5/71.

Table **15-6** Welding conditions for manual out-of-position butt welds in aluminum (spray-arc welding)

PLATE THICKNESS (IN.)	POSITION	EDGE PREPARATION[1]	BACKING	WIRE DIA. (IN.)	SHIELDING GAS ARGON (CFH)	NO. OF PASSES	VOLTS	CURRENT (AMPS) DCRP	WELDING SPEED (IPM)
1/4	Vertical	Single-V 60° included angle, 1/16-in. nose back chip after root pass	None	3/64 or 1/16	40	2	23	180	20
	Overhead	Single-V 60° included angle, 1/16-in. nose back chip after root pass	None	3/64 or 1/16	40	2	23–24	200	22
3/8	Vertical	Single-V 60° included angle, 1/16-in. nose back chip after root pass	None	3/64 or 1/16	40	3	23	210	18
	Overhead	Single-V 60° included angle, 1/16-in. nose back chip after root pass	None	1/16	45	3	23–24	220	20
1/2	Vertical	Single-V 60° included angle, 1/16-in. nose back chip after root pass	None	1/16	45	3	22–23	215	12
	Overhead	Single-V 60° included angle, 1/16-in. nose back chip after root pass	None	1/16	50	4	23–24	225	16

[1]Abutting surfaces and groove faces should be draw filed, and adjacent surfaces wire brushed.

Table **15-6** *(cont.)*

PLATE THICKNESS (IN.)	POSITION	EDGE PREPARATION[1]	BACKING	WIRE DIA. (IN.)	SHIELDING GAS ARGON (CFH)	NO. OF PASSES	VOLTS	CURRENT (AMPS) DCRP	WELDING SPEED (IPM)
3/4	Vertical	Single-V 75° included angle, 1/16-in. nose back chip after root pass	None	1/16	50	4	23–24	225	10
	Overhead	Single-V 75° included angle, 1/16-in. nose back chip after root pass	None	1/16	50	6	24	240	14

Source: Union Carbide Corporation, Linde Division, *How To Do Manual Metal Inert Gas Welding,* F-51-110-A 5/71.

Table **15-7** General welding conditions for manual spray-arc welding of stainless steel

PLATE THICKNESS (IN.)	JOINT & EDGE PREPARATION	WIRE DIA.	GAS FLOW (CFH)	CURRENT (AMPS) DCRP	WIRE FEED (IPM)	WELDING SPEED	PASSES
0.125	Square butt with backing	1/16	35	200–250	110–150	20	1
0.250	Single-V butt 60° inc. angle No nose	1/16	35	250–300	150–200	15	2
0.375	Single-V butt 60° inc. angle 1/16-in. nose	1/16	(O_2–1)	275–325	225–250	20	2
0.500	Single-V butt 60° inc. angle 1/16-in. nose	3/32	(O_2–1)	300–350	75–85	5	3–4
0.750	Single-V butt 90° inc. angle 1/16-in. nose	3/32	(O_2–1)	350–375	85–95	4	5–6
1.000	Single-V butt 90° welded angle 1/16-in. nose	3/32	(O_2–1)	350–375	85–95	2	7–8

Source: Union Carbide Corporation, Linde Division, *How To Do Manual Metal Inert Gas Welding,* F-51-110-A 5/71.

Table **15-8** General welding conditions for manual short-arc welding of stainless steel

PLATE THICKNESS (IN.)	JOINT AND EDGE PREPARATION	WIRE DIA. (IN.)	GAS FLOW (CFH)	CURRENT (AMPS) DCRP	VOLTAGE*	WIRE FEED SPEED (IPM)	WELDING SPEED (IPM)	PASSES
0.063	Non-positioned fillet or lap	0.030	15–20	85	15	184	18	1
0.063	Butt (square edge)	0.030	O_2–2	85	15	184	20	1
0.078	Non-positioned fillet or lap	0.030	O_2–2	90	15	192	14	1
0.078	Butt (square edge)	0.030	O_2–2	90	15	192	12	1
0.093	Non-positioned fillet or lap	0.030	O_2–2	105	17	232	15	1
0.125	Non-positioned fillet or lap	0.030	O_2–2	125	17	280	16	1

*Voltage values are for C-25 gas or O_2-2 gas. For 90% HE−10% C-25 voltage will be 6 to 7 volts higher.

Source: Union Carbide Corporation, Linde Division, *How To Do Manual Metal Inert Gas Welding,* F-51-110-A 5/71.

Table **15-9** Nominal conditions for manual GMAW of commercial coppers[a]

THICKNESS (IN.)	AMPS (DCRP)	VOLTS	TRAVEL (IPM)	WIRE DIA. (IN.)	WIRE FEED SPEED (IPM)	JOINT DESIGN
1/8	310	27	30	1/16	200	Square butt, steel backup strip required
1/4 (1)[b]	460	26	20	3/32	135	Square butt
1/4 (2)	500				150	
3/8 (1)	500	27	14	3/32	150	Double bevel, 90° included angle,
3/8 (2)	550				170	3/16-in. nose
1/2 (1)	540	27	12	3/32	165	Double bevel, 90° included angle,
1/2 (2)	600		10		180	1/4-in. nose

[a]See also Table 21-11. [b]Pass number.

Source: Union Carbide Corporation, Linde Division, *How To Do Manual Metal Inert Gas Welding,* F-51-110-A 5/71.

15-1 SAFE PRACTICES

In addition to the general safety practices and the specific safe practices for shielded metal–arc welding detailed in Chapter 13, welders should use lenses at least one shade deeper than those used for shielded metal–arc welding.

The following sections are a series of exercises in gas metal–arc welding.

15-2 BECOMING FAMILIAR WITH EQUIPMENT (SHORT-ARC METHOD)

OBJECTIVES To learn to:

1. Visually identify and correctly name all component parts of a GMAW outfit.

2. Correctly set up and make operational a GMAW outfit.

3. Correctly start a GMAW outfit.

HOURS REQUIRED ½ to ¾.

EQUIPMENT, TOOLS, AND MATERIALS REQUIRED One GMAW welding outfit and accessories per student.

PREVIOUS PREPARATION OF EQUIPMENT BY INSTRUCTOR It is assumed that the instructor has made the following preparations prior to turning the equipment over to the student:

1. Set up the equipment according to the instructions supplied by the equipment manufacturer.

2. Hooked up the welding machine to the power supply.

3. Checked to see that:

a. Wire feeder and wire-feeder controls have been properly installed and adjusted.

b. Gun-cable assembly has been connected to the wire feeder and aligned.

c. Hoses have been connected from the gas supply system to the wire feeder, and from the wire feeder to the gun-cable assembly.

d. Wire and shielding gas are supplied as required to perform the exercise.

e. Contactor connections and power connections between the welding machine and the wire feeder have been made.

PREWELDING PROCEDURE BY STUDENT

1. Check the power cable connections. *Note:* Remember that dc reverse-polarity current is the most widely used.

2. Start the welding machine. *Note:* If the machine is different from the one shown in Figure 15-3, refer to the instructions supplied by the welding manufacturer.

3. Start the wire feeder. *Note:* If the machine is different from the one shown in Figure 15-3, refer to the instructions supplied by the manufacturer.

Figure **15-3** Commonly used MIG power supply

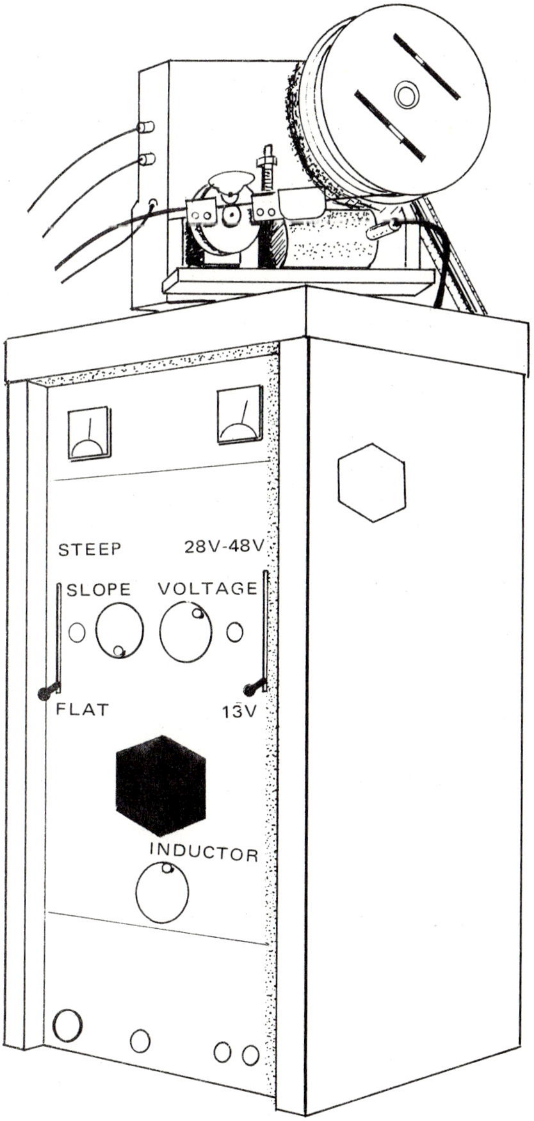
STEEP
28V-48V
SLOPE
VOLTAGE
FLAT
13V
INDUCTOR

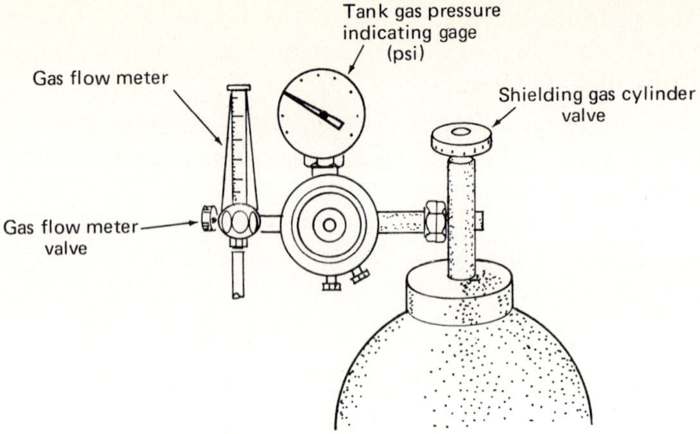
Tank gas pressure
indicating gage
(psi)
Gas flow meter
Shielding gas cylinder
valve
Gas flow meter
valve

Figure **15-4** Shielding-gas outlet valves

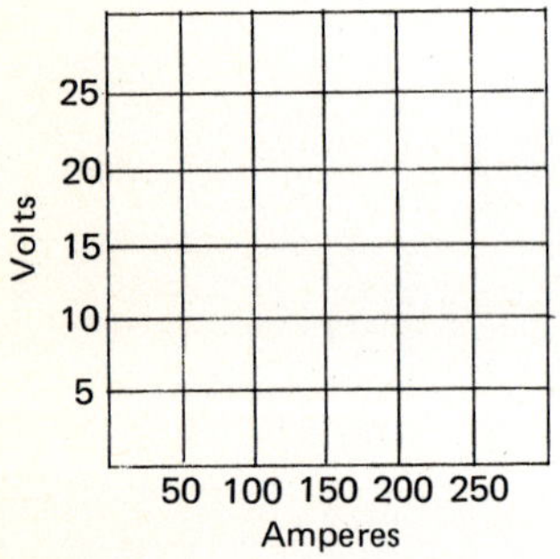
Volts
25
20
15
10
5
50 100 150 200 250
Amperes

Figure **15-5** Basic graph for constructing slope for a MIG power supply

4. Set the wire-feed speed control on the zero position. (This prevents the wire from feeding while you make any necessary adjustments.)

5. Open the shielding-gas cylinder outlet valve (Figure 15-4).

6. Slowly open the flowmeter valve, simultaneously squeezing and holding down the gun trigger.

Note: The following steps are for determining and setting the slope. Usually this information is contained in the instructions supplied by the equipment manufacturer, but if it isn't, starting with a graph like the one in Figure 15-5 and continuing with the following steps will suffice, except when using silicon-rectifyer types of machines, which will be damaged by this procedure.

7. Set the slope control crank (Figure 15-3) at ½ the maximum number of turns. If the machine has a 14-turn maximum, set it at 8.

8. Crank the (voltage) open circuit (Figure 15-3) so that the open-circuit voltage meter reads 24 volts.

9. Check that the voltage is at 24 by pulling the trigger on the gun. This closes the main contactor. *Note:* If the voltage is not set at 24, repeat steps 8 and 9 until it is.

10. Begin to plot the graph (Figure 15-6) by placing a dot at 24 on the vertical (voltage) line.

11. Remove the shielding cup from the torch and place the contact tip in good contact with the grounded work table.

12. Momentarily depress the trigger and simultaneously read the amperage.

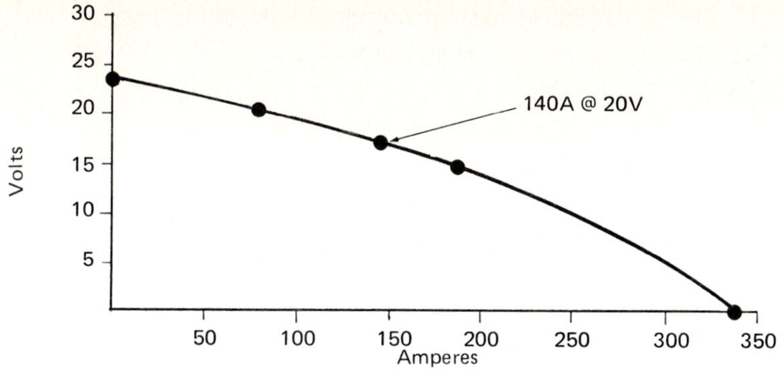
30
25
20
15
10
5
Volts
140A @ 20V
50
100
150
200
250
300
350
Amperes

Figure **15-6** Sample plot of an MIG power supply slope. (Redrawn by permission from Miller Electric Manufacturing Company, *Fundamentals of Gas Metal-Arc (MIG) Welding,* Form: MTS-6A 3/71.)

13. Place a dot on the horizontal axis (Figure 15-6), indicating the amperage you have just read.

14. Replace the shielding cup on the torch.

15. Try to strike an arc.

a. If no arc is struck, vary the wire speed (Figure 15-7) until you get an arc.

b. When there is a good arc, have someone read the current and voltage of the arc and plot them on the graph.

Figure **15-7** Commonly used MIG wire-feed control. (Courtesy of the Miller Electric Manufacturing Company.)

10-E

16. Continue welding and slowly increase wire-feed speed to just the point where the wire begins to stub. Read the voltage and amperage and plot them.

17. Connect the plotted points with a swaying curve (Figure 15-6).

18. Continue welding and slowly decrease the wire feed to the point where large drops are going through the arc. Read the voltage and amperage at this point and plot them.

19. Connect the plotted points with a swaying curve (Figure 15-6).

The swaying curve is the slope of the power supply for the settings that were made. If the short-circuit current is not sufficient for the diameter of wire used, occasional stubbing will be seen and the response time will seem too slow. On the other hand, if the short-circuit current is too high, the arc will be harsh, spatter will be evident, and the arc will feel unstable, as if the response time were too quick. The bead with too much short-circuit current will have pronounced freeze lines, more digging or penetration, and a higher convex bead.

Once the proper slope setting (volt-ampere curve) for a particular size wire is obtained, there is no need to change it for that size wire.

While constructing the graph, it will probably have been noticed that when only the wire speed was changed, the current (amperage) also changed. This means that the wire feed is the current control. By changing the slope values, the shape of the curve is changed (Figure 15-8). Changing the voltage *does not* change the shape of the curve (Figure 15-9). This is what permits welding different material thicknesses without changing the slope.

Figure **15-8** Sample plot of slopes obtained by varying turns of the slope crank. (Redrawn by permission from Miller Electric Manufacturing Company, *Fundamentals of Gas Metal-Arc (MIG) Welding*, Form: MTS-6A 3/71.)

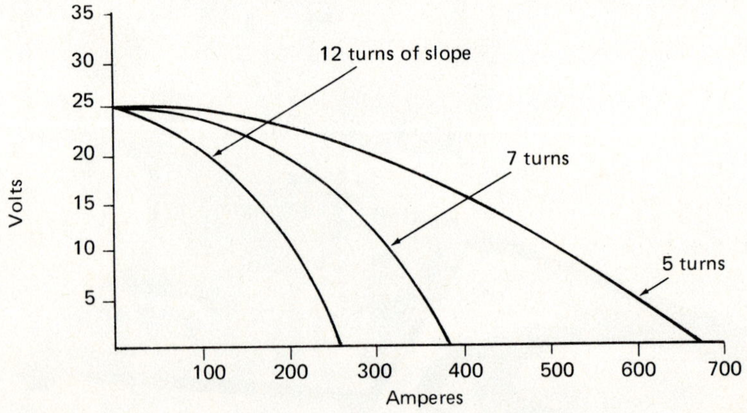
35
30
25
20
15
10
5
Volts
12 turns of slope
7 turns
5 turns
100
200
300
400
500
600
700
Amperes

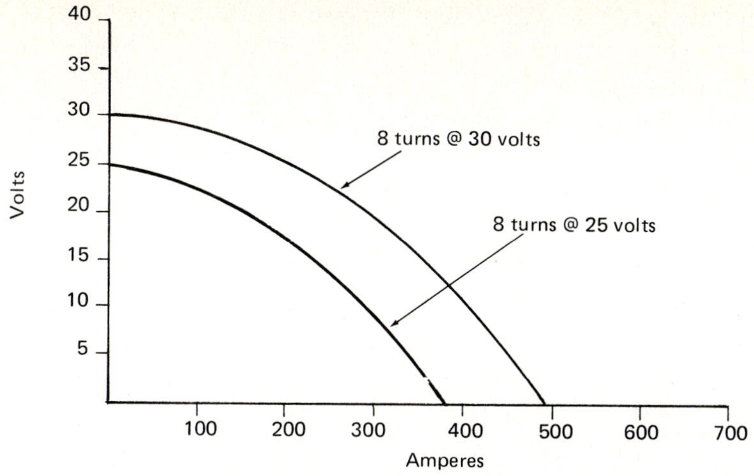

Volts
40
35
30
25
20
15
10
5
8 turns @ 30 volts
8 turns @ 25 volts
100
200
300
400
500
600
700
Amperes

Figure **15-9** Sample plot showing that a change in voltage does not change slope. (Redrawn by permission from Miller Electric Manufacturing Company, *Fundamentals of Gas Metal-Arc (MIG) Welding,* Form: MTS-6A 3/71.)

15-3 WELDING SQUARE-GROOVE BUTT JOINTS IN CARBON STEEL

OBJECTIVES To learn to:

1. Correctly set the tip-to-work (stick-out) distance of the electrode (filler metal).

2. Make square-groove welds in mild or similar steel in the flat position.

3. Correctly shut off and tear down the welding outfit.

HOURS REQUIRED 2.

EQUIPMENT, TOOLS, AND MATERIALS REQUIRED The equipment, tools, and materials required to perform this exercise are:

1. One GMAW welding outfit, as shown in Figure 15-10.

2. Several 2 × 4 × 1/16-inch pieces of mild steel.

3. Several 2 × 4 × 3/16-inch pieces of mild steel.

4. Sufficient 0.030-inch-diameter filler metal.

5. Sufficient 1/16-inch-diameter filler metal.

6. 75% argon + 25% CO_2 shielding gas, or 95% argon + 5% CO_2 shielding gas.

PROCEDURE

1. Set voltage at 19 volts.

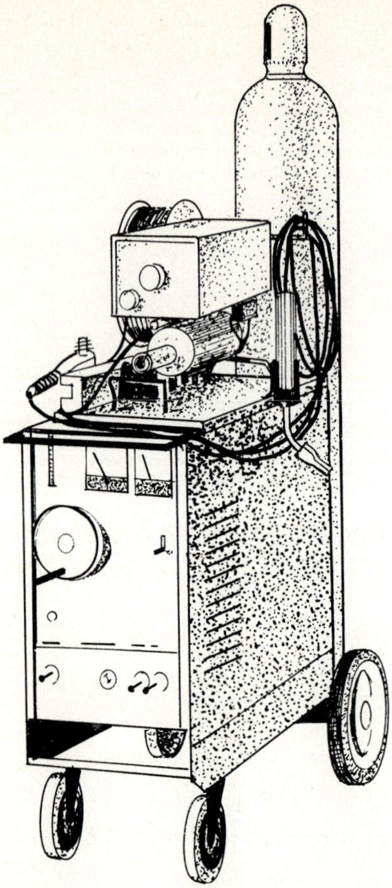

Figure **15-10**

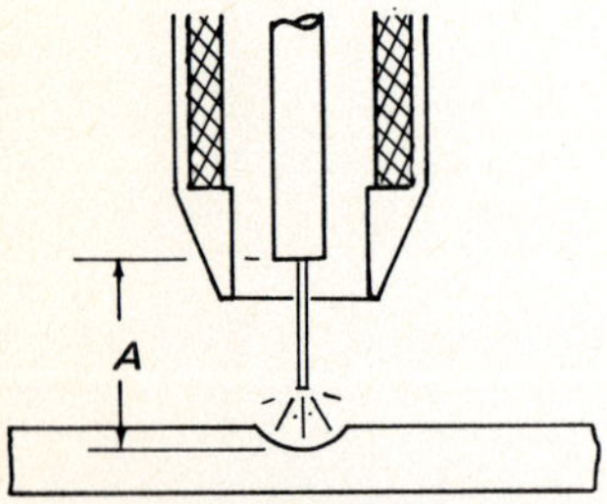
A
A = tip to work distance
(stickout)

Figure **15-11**

2. Set wire-feed speed control to produce a current of approximately 130 amperes, or a wire-feed speed of 300 inches per minute (in./min).

3. Adjust the gas flow to approximately 15 cubic feet per hour (ft^3/h).

4. Adjust stick-out to ¼ inch (Figure 15-11).

5. Arrange two pieces of steel to form a butt joint.

6. Lower your face helmet, squeeze the gun trigger activating the controls, and establish an arc.

7. Tack weld the two pieces of steel, leaving a ¹⁄₁₆-inch root opening.

8. Hold your gun at the 90-degree position, lean 5 degrees with the gun pointing in the direction of travel, and weld the butt joint traveling from right to left, using a leading gun angle. *Note:* If you do not get a smooth arc, adjust the welding current. Remember the ammeter on the welding machine registers only while welding.

9. Stop welding and release the gun trigger.

10. Repeat steps 5 through 9 using 3/16-inch-thick plate.

Note: Refer to Table 15-1 for suggested welding conditions.

15-4 WELDING LAP JOINTS IN CARBON STEEL

OBJECTIVE To learn to make acceptable fillet welds in a lap joint in 1/16- and 3/16-inch mild or similar steel in the horizontal position.

HOURS REQUIRED 2.

EQUIPMENT, TOOLS, AND MATERIALS REQUIRED The equipment, tools, and materials required to perform this exercise are described in exercise 15-2.

PROCEDURE:

1. Refer to Table 15-1 to determine the welding conditions suggested for making a fillet weld in 1/16-inch-thick carbon steel.

2. Set the welding machine to meet the appropriate conditions.

3. Arrange two pieces of 1/16-inch-thick carbon steel to make the lap joint.

4. Lower your face helmet.

5. Angle your gun nozzle about 60 degrees from the lower plate. Squeeze the trigger to establish an arc and weld the lap joint using a trailing gun angle (Figure 15-12), making a single horizontal-fillet weld.

Figure **15-12**

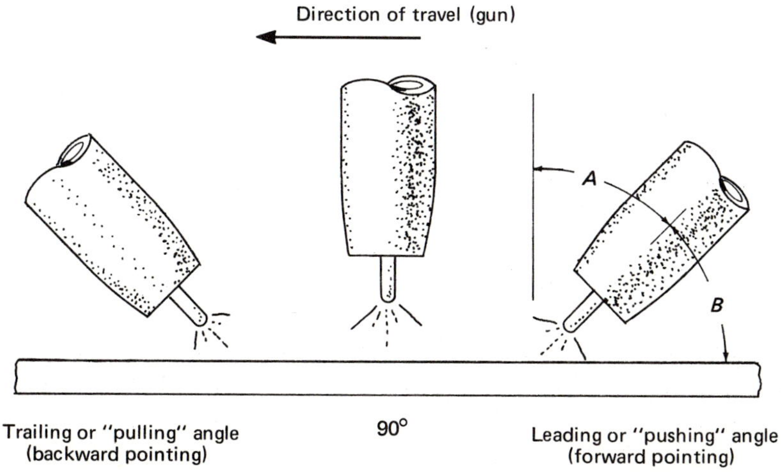

Direction of travel (gun)
A
B
Trailing or "pulling" angle (backward pointing)
90°
Leading or "pushing" angle (forward pointing)

6. Repeat this procedure on various joints until you are proficient in making the fillet weld in this joint.

7. Repeat steps 1 through 6 and practice making fillet welds in $\frac{3}{16}$-inch-thick carbon steel.

15-5 WELDING T JOINTS IN CARBON STEEL

OBJECTIVE To learn to make acceptable fillet welds in a T joint in $\frac{1}{16}$- and $\frac{3}{16}$-inch mild or similar steel in the horizontal position.

HOURS REQUIRED 2.

EQUIPMENT, TOOLS, AND MATERIALS REQUIRED The equipment, tools, and materials required to perform this exercise are the same as in exercise 15-2.

PROCEDURE

1. Refer to Table 15-1 to determine the welding conditions suggested for making a fillet weld in a T joint in $\frac{1}{16}$-inch-thick carbon steel.

2. Set the welding machine to meet the appropriate conditions.

3. Arrange two pieces of $\frac{1}{16}$-inch-thick carbon steel to make the T joint.

4. Lower your face helmet, squeeze the trigger, and tack weld both ends of the steel sheets.

5. Hold your gun so that the gun nozzle bisects the joint. Lean 5 to 10 degrees with the gun pointing away from the direction of travel (Figure 15-13) and weld the T joint using a horizontal fillet on each side.

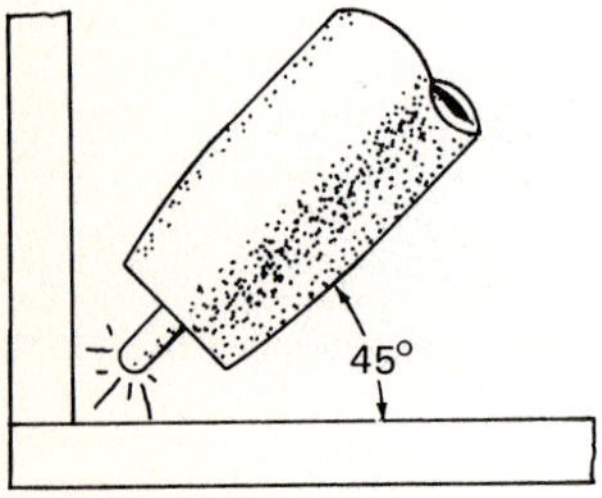
45°

Nozzle angle (end view)

Figure **15-13**

6. Repeat this procedure on various joints until you are proficient in making the fillet weld in this joint.

7. Repeat steps 1 through 6 using $\frac{3}{16}$-inch-thick carbon steel.

15-6 PROPERLY ADJUSTING THE EQUIPMENT

OBJECTIVE To learn to give reasons for and perform appropriate corrective measures when there is insufficient or too much penetration or burn-through.

HOURS REQUIRED 1.

EQUIPMENT, TOOLS, AND MATERIALS REQUIRED The equipment, tools, and materials required to perform this exercise are the same as in exercise 15-2.

PROCEDURE

1. Set up to weld butt joints.

2. Turn the welding machine on and make the same settings as in exercise 15-2.

3. Try to make the weld using a stick-out of approximately ¼ inch. Notice that the penetration is deep and may even burn through.

4. Increase stick-out to approximately ½ inch and try to weld the butt joint. Notice that with the longer stick-out there is less penetration and the electrode (filler metal) may possibly have welded itself to the gap without melting or fusing.

5. Set the plate in an upright position with the edges touching.

6. With ¼-inch stick-out, try to make the weld using the vertical-up technique. Notice that with no spacing you are getting insufficient penetration.

7. Set another set of properly spaced plates in an upright position; increase the stick-out to ⅜ inch and, using the up technique with a slight weaving motion and letting the electrode stack back in the puddle instead of on the leading edge of the puddle, complete the first pass. The bead should have turned out to be acceptable.

8. Set up another pair of correctly spaced plates to make a butt joint in the flat position.

9. Set the filler-metal (electrode) feed control to the 9 or 10 (90% or 100%) position. This will deposit more metal; in fact, it will probably deposit too much metal, and the wire will most probably also jab into the weld. This results from the faster wire-feed speed, which produces a higher amperage.

10. Set the wire-feed speed control at the 2 or 3 (20% or 30%) position. This setting will produce a very low amperage that will result in one or a combination of the following conditions:

a. Unstable, brief arc with irregular metal deposits.
b. Electrode burnoff, with little or no metal deposit.
c. Electrode sticking to the contact tip.

15-7 WELDING ALUMINUM

OBJECTIVE To learn to produce acceptable fillet welds in the horizontal and vertical positions on aluminum.

HOURS REQUIRED 1.

EQUIPMENT, TOOLS, AND MATERIALS REQUIRED

1. One GMAW welding outfit as described in exercise 15-2. Set up for aluminum welding (check with instructor).

2. Several 2 × 4 × ⅛-inch pieces of aluminum.

3. Sufficient 3/64-inch-diameter 4043 aluminum filler metal.

4. Sufficient argon shielding gas.

PROCEDURE

1. Install aluminum filler-metal wire feeder.

2. Install argon shielding-gas tank on welding machine.

3. Install grip-type water-cooled welding gun.

4. Refer to Table 15-4 for recommended welding machine conditions and set welding outfit accordingly.

5. Arrange two pieces of ⅛-inch-thick aluminum plate to form a T joint. Make sure your fit-up is good. Clean the area of the joint thoroughly.

6. Turn on the welding machine, lower your helmet, and tack weld both ends of the joint.

7. Place the tack-welded plates so that the joint is in the horizontal position.

8. Holding the gun nozzle about ⅝ inch away from the work and at approximately 15 degrees in the direction of travel, make a single horizontal fillet weld on each side of the joint.
   a. If the speed of welding (travel) is too fast, the weld bead will be too small and high crowned and you will not get enough penetration.
   b. If the speed of welding (travel) is too slow, the weld bead will be too large and you will get excessive penetration.

9. Repeat steps 5 through 8 until you can produce acceptable welds.

10. Repeat step 5, lower your helmet, tack weld the plates, and place them so that the joint is in a vertical position.

11. Hold the gun nozzle about ⅝ inch away from the work and at approximately 15 degrees to the direction of travel.

12. Weld the joint by the vertical-up method (from the bottom of the joint to the top of the joint) and make a single fillet weld on each side of the joint.

13. Repeat steps 5 through 12 until you can produce an acceptable weld.

15-8 WELDING STAINLESS STEEL

OBJECTIVE To learn to produce acceptable-quality fillet square-groove butt welds in the horizontal and flat positions on stainless steel.

HOURS REQUIRED 2.

EQUIPMENT, TOOLS, AND MATERIALS REQUIRED

1. One GMAW outfit as described in exercise 15-2.

2. Several 2 × 4 × 0.125-inch pieces of stainless steel.

3. Sufficient triple-mixture shielding gas [90% helium and 10% C-25 (75% percent argon + 25% carbon dioxide)].

4. Sufficient O-2 (argon with 2% oxygen) gas.

PROCEDURE

1. Determine the type of joint to be made and the welding position to be used.

2. Refer to Table 15-7 to determine welding conditions, welding position, and type of joints to be made.

3. Set up welding machine according to the conditions determined in step 2.

4. Determine the appropriate arrangement of the plates to be welded.

5. If you are going to make a fillet on a T joint, angle the welding nozzle about 45 degrees from the lower plate.

6. Lower your helmet and start the welding machine.

7. Holding the gun nozzle about ¼ inch from the work, squeeze the trigger and weld the joint in the horizontal position.

8. Repeat steps 5 through 7 until you can produce an acceptable weld. Once you can produce an acceptable weld, go on to step 9.

9. Refer to Table 15-7 to determine the conditions for making a square-groove butt joint in the flat position.

10. Set up the welding machine according to the welding conditions determined in step 9.

11. Determine the appropriate arrangement of the plates to be welded.

12. Lower your helmet and, holding the nozzle so that it is about ¼ inch from the work and bisects the work, depress the trigger, and weld the joint.

13. Repeat step 12 until you can produce acceptable square-groove butt joints in the flat position.

15-9 WELDING STEEL PIPE IN THE HORIZONTAL-FIXED (5G) POSITION

OBJECTIVE To learn to make an acceptable root pass in the horizontal-fixed (5G) position, using the downhill technique in 0.322-inch-wall, Schedule 40, 8-inch-diameter mild-steel pipe.

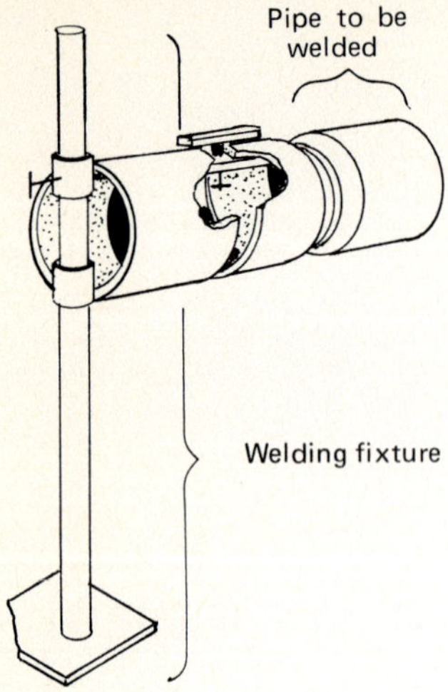

Pipe to be welded
Welding fixture

Figure **15-14**

EQUIPMENT, TOOLS, AND MATERIALS REQUIRED

1. Welding fixture (Figure 15-14).
2. Sufficient 3½-inch lengths of mild-steel pipes prepared as shown in Figure 15-15.
3. Sufficient 0.035-inch-diameter carbon-steel filler metal.
4. Sufficient shielding gas (75% Ar and 25% CO_2).
5. One GMAW outfit as described in exercise 15-2.

PROCEDURE

1. Refer to Table 15-1 to determine the welding machine settings.
2. Set the welding machine to the appropriate setting.
3. Place a piece of scrap pipe on the welding table.
4. Adjust the electrode wire stick-out to approximately ¼ inch.
5. Lower your helmet, depress the trigger of the welding gun, and make a test weld. Inspect the weld and make whatever adjustments are necessary.
6. Place one pipe nipple in a vertical position with the groove upward on the work table (Figure 15-15).
7. Take a sufficiently long piece of 3⁄32-inch-diameter mild-steel electrode and bend it to form a V (Figure 15-15).

Figure **15-15**

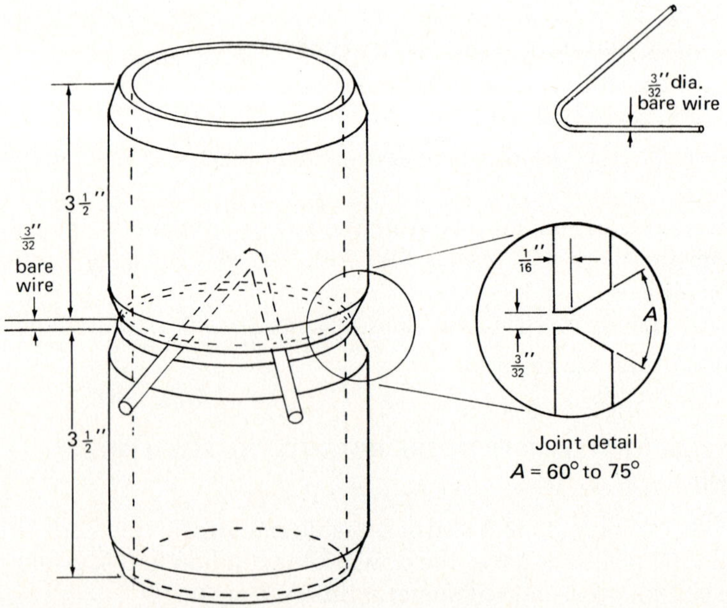

3/32" dia. bare wire
3 1/2"
3/32" bare wire
3 1/2"
1/16"
A
3/32"
Joint detail
A = 60° to 75°

8. Lay the bent mild-steel electrode across the pipe nipple so that the wire touches the pipe nipple at four places.

9. Place the second pipe nipple with the groove downward on top of the first pipe nipple (Figure 15-15).

10. Lower your helmet and tack weld the two pipe nipples together at four equally spaced places. *Note:* Tack welds should be between ¾ and 1 inch long.

11. Place the tack-welded pipe nipples in the horizontal-fixed (5G) position (Figure 15-14) and tighten the fixture clamps.

12. Start welding the root pass in the vertical-down position and continue until you reach the 6 o'clock position.

a. If you must stop before you reach the 6 o'clock position of the root pass, stop at one of the tack welds (Figure 15-16). A stop at any other location in welding the root pass can cause shrinkage, cracks, cavities, or cratering.

b. If you have to stop in the open groove while welding the root pass, carry the weld puddle halfway up the pipe bevel (Figure 15-17).

c. Prior to restarting the weld, chisel or grind the bead and surrounding area until all slag and inclusions are removed.

Figure **15-16**

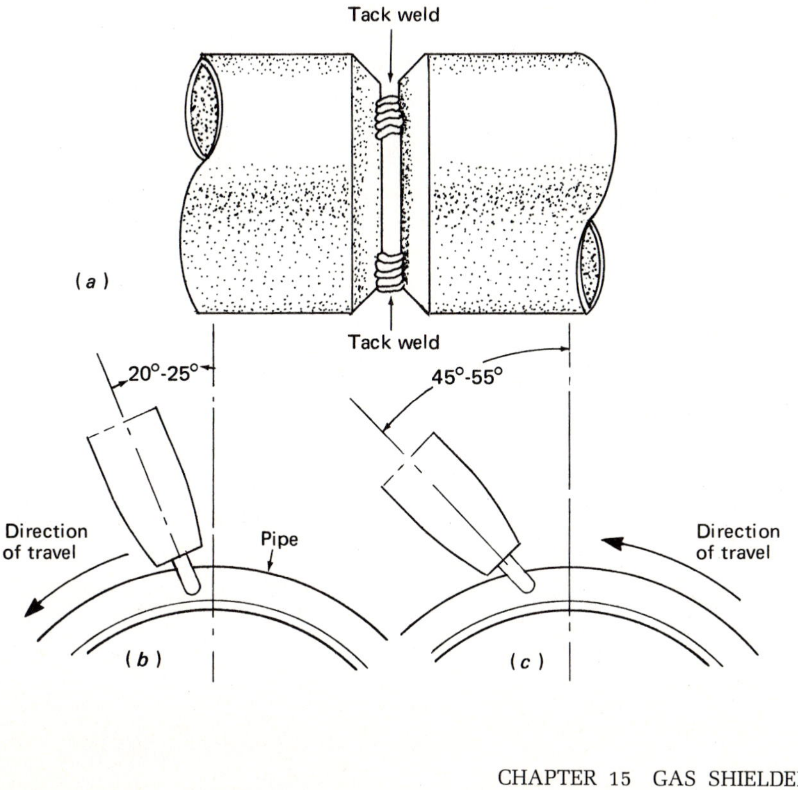

Tack weld
(a)
Tack weld
20°-25°
45°-55°
Direction
of travel
Pipe
Direction
of travel
(b)
(c)

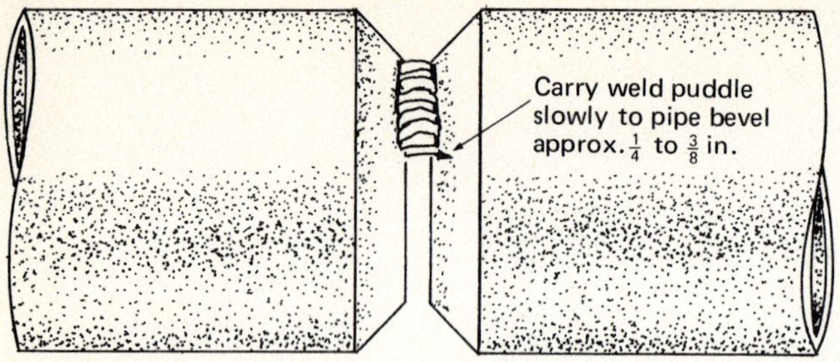
Carry weld puddle
slowly to pipe bevel
approx. $\frac{1}{4}$ to $\frac{3}{8}$ in.

Figure **15-17**

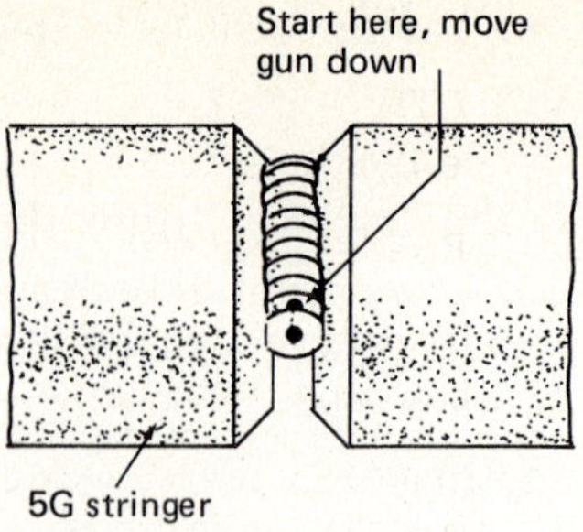
Start here, move
gun down
5G stringer

Figure **15-18**

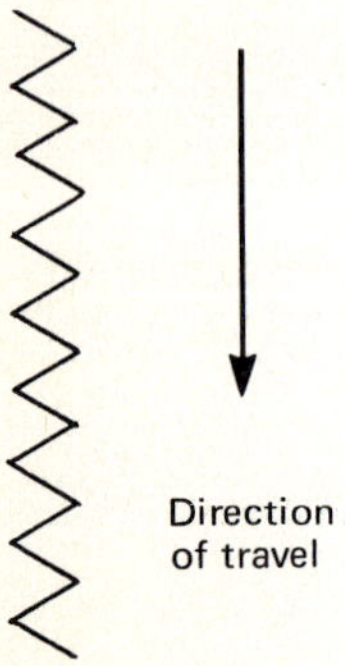
Direction
of travel

Figure **15-19**

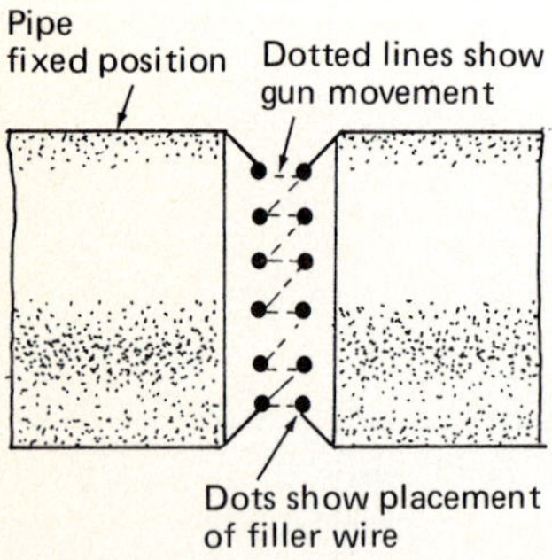
Pipe
fixed position
Dotted lines show
gun movement
Dots show placement
of filler wire

Figure **15-20**

d. To restart the weld, use the "tie-in" technique (Figure 15-18).

e. When you reach the overhead position you may need to increase the trailing angle of the gun (torch) and resume the narrow weaving motion (Figure 15-19).

f. If the travel speed is too slow, the base metal will melt too much, resulting in lack of penetration.

13. Rotate the pipe 180 degrees, return to the 12 o'clock position, and repeat steps 12 and 13 until you have completed the circumferential root bead.

14. Wearing safety glasses and using a brush, scraper, chisel, and/or grinder remove all slag from the surface of the root pass.

15. Refer to Table 15-1 for recommended welding conditions and, if necessary, reset the welding machine.

16. Lower your helmet and, using a slight backhand torch angle of 15 to 20 degrees, start the fill pass.

a. To ensure a good bead contour and tie-in at the edge, manipulate the torch from side-to-side to the exact width of the root pass (Figure 15-20).

b. An important rule of fill and cover-pass welding is to keep the arc ahead of the puddle. Cold tapping is a common defect and is almost certain to occur if the welding speed is reduced in an attempt to deposit more metal on any given pass. *Keep it thin.* Failure to observe this rule is the reason why some welders cannot meet the ASME code quality with downhill technique.

c. To avoid excessive build-up while making the fill pass, stop the fill pass at 5 o'clock on one side of the pipe and at 7 o'clock on the other side.

17. Study Figures 15-18 and 15-20 and then repeat steps 15 through 16 for the cover pass.

18. Repeat steps 6 through 17 until you can produce acceptable welds.

15-10 WELDING PIPE IN THE HORIZONTAL (2G) POSITION

OBJECTIVE To learn to produce acceptable welds in mild-steel pipe in the (2G) horizontal position.

HOURS REQUIRED 2.

EQUIPMENT, TOOLS, AND MATERIALS REQUIRED The equipment, tools, and materials required are the same as in exercise 15-9.

PROCEDURE

1. Refer to Table 15-1 to determine the machine settings.

2. Set the welding machine to the appropriate settings.

3. Place a piece of scrap pipe on the welding table.

4. Adjust the electrode wire stick-out to approximately ¼ inch.

5. Lower your helmet, depress the trigger of the welding gun, and make a test weld. Inspect the weld and make whatever adjustments are necessary.

6. Place one pipe nipple in a vertical position with the groove upward on the work table (Figure 15-15).

7. Take a sufficiently long piece of 3/32-inch-diameter mild-steel electrode and bend it to form a V (Figure 15-15).

8. Lay the bent mild-steel electrode across the pipe nipple at four places.

9. Place the second pipe nipple, with the groove downward, on top of the first pipe nipple (Figure 15-15).

10. Lower your helmet and tack weld the two pipe nipples together at four equally spaced places (Figure 15-16).

11. Place the tack-welded pipe nipples in the horizontal (2G) position (Figure 15-21) and tighten fixture clamps.

12. Hold the welding gun as shown in Figure 15-22 and start welding the root pass at a tack point.

13. Refer to Table 15-1 and, if necessary, change the settings of the welding machine as required to make the second pass.

14. Make the second pass.

a. Hold the gun nozzle as in the root pass.

b. Move the gun in close, short ovals, as shown in Figure 15-23.

c. If the pipe groove is too wide to fill by using the oval gun movement, use as many stringer beads as required to fill the joints.

d. If more than two passes are required, refer to Figure 15-24 for pass position sequence.

15. Repeat steps 1–15 until you can produce acceptable welds.

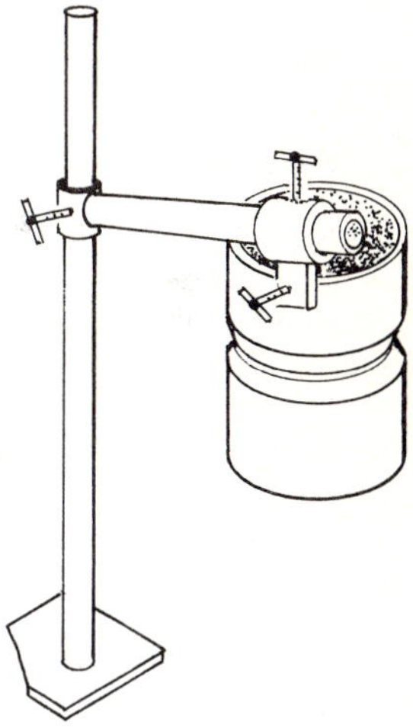

Figure **15-21**

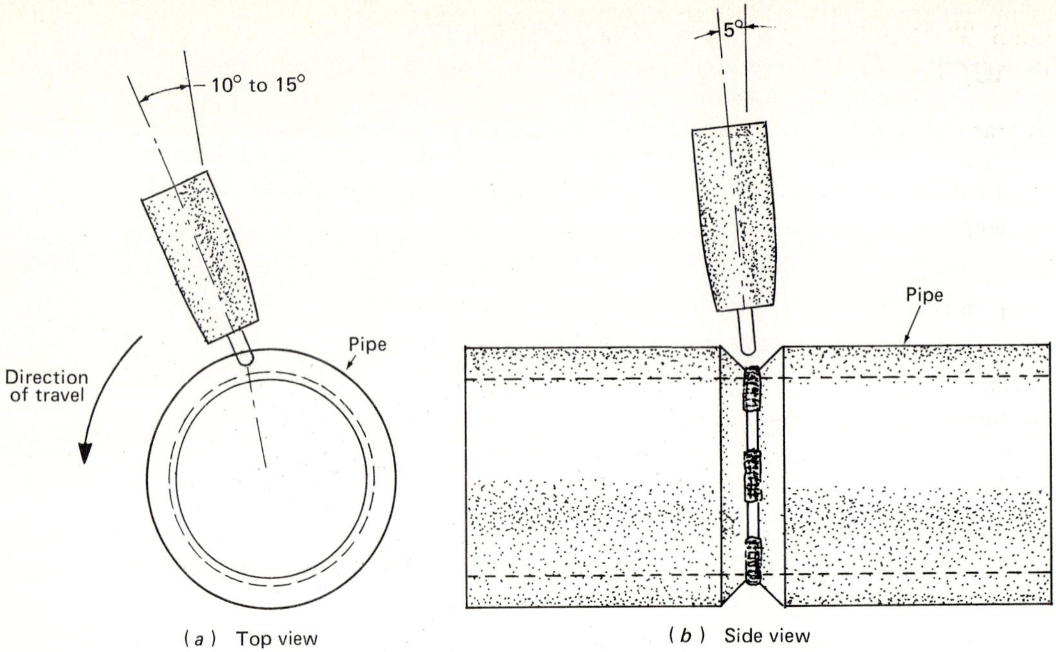

10° to 15°
Pipe
Direction
of travel
(a) Top view
5°
Pipe
(b) Side view

Figure **15-22**

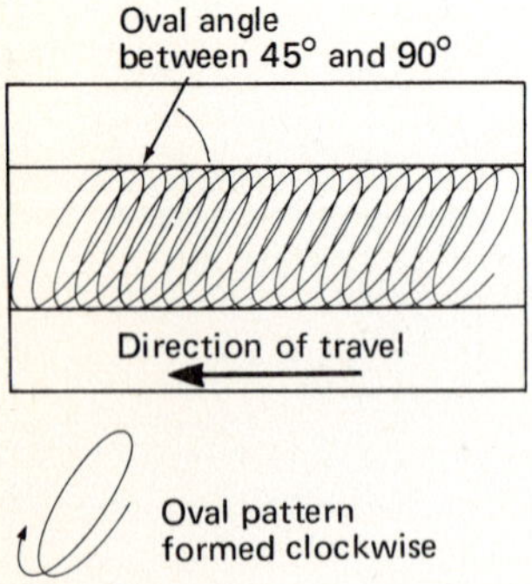

Oval angle
between 45° and 90°
Direction of travel
Oval pattern
formed clockwise

Figure **15-23**

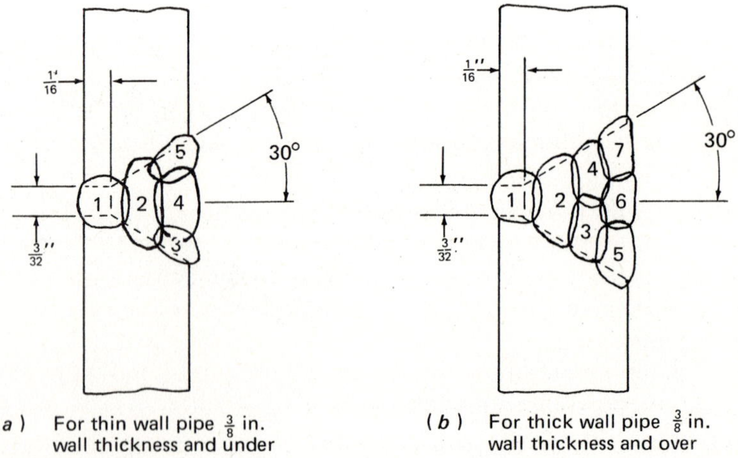

1/16'
3/32''
30°
1
2
3
4
5
(a) For thin wall pipe 3/8 in. wall thickness and under
1/16''
3/32''
30°
1
2
3
4
5
6
7
(b) For thick wall pipe 3/8 in. wall thickness and over

Figure **15-24**

GAS TUNGSTEN–ARC WELDING (GTAW) PRACTICE

As was discussed in Chapter 5, the GTAW process is an arc process that uses a nonconsumable tungsten electrode and externally supplied shielding gases (Figure 15-25). This process is used primarily for deep-penetration, high-quality welds.

15-11 POWER SOURCES

The power sources for TIG welding may be either ac or dc. Alternating current is normally used for manual welding of aluminum and magnesium. Direct current is usually preferred for the ferrous materials and other nonferrous metals. Some automatic welding applications on aluminum are done with direct current. Amperage requirements may range from a few to several hundred amperes. The power

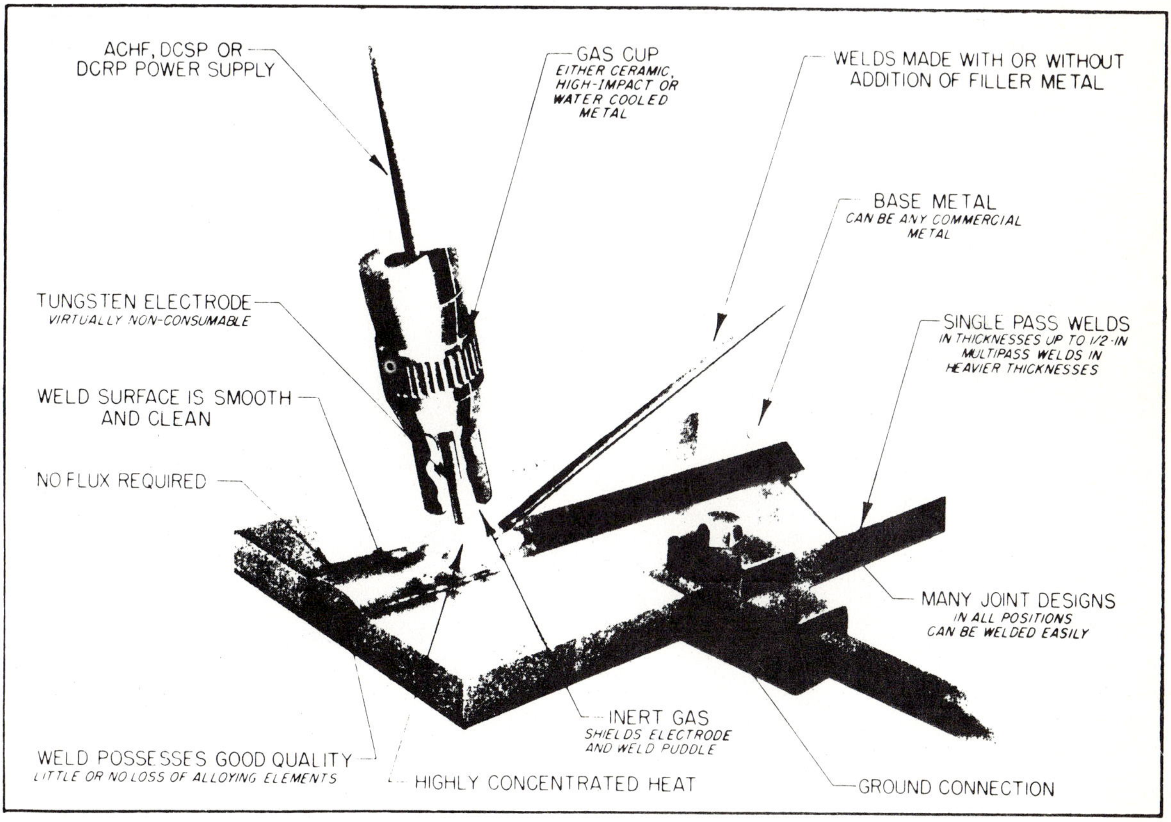

ACHF, DCSP OR DCRP POWER SUPPLY
GAS CUP
EITHER CERAMIC, HIGH-IMPACT OR WATER COOLED METAL
WELDS MADE WITH OR WITHOUT ADDITION OF FILLER METAL
BASE METAL
CAN BE ANY COMMERCIAL METAL
TUNGSTEN ELECTRODE
VIRTUALLY NON-CONSUMABLE
SINGLE PASS WELDS
IN THICKNESSES UP TO 1/2-IN
MULTIPASS WELDS IN HEAVIER THICKNESSES
WELD SURFACE IS SMOOTH AND CLEAN
NO FLUX REQUIRED
MANY JOINT DESIGNS
IN ALL POSITIONS CAN BE WELDED EASILY
INERT GAS
SHIELDS ELECTRODE AND WELD PUDDLE
WELD POSSESSES GOOD QUALITY
LITTLE OR NO LOSS OF ALLOYING ELEMENTS
HIGHLY CONCENTRATED HEAT
GROUND CONNECTION

Figure **15-25** Essentials of the TIG welding process. (From Union Carbide Corporation, Linde Division, *How to do TIG (Heliarc) Welding*, F-1607-D 5/70, p. 3.)

sources (Figure 15-26) used may be the same ac or ac/dc machine used for stick-electrode welding or, depending on job requirements, may be a sophisticated power supply capable of being programmed to complete welds automatically. When applications call for high amperages, or when welding continuously, a water-cooled torch and cable assembly (Figure 15-27) may be used. The water is brought to the torch through a hose, circulates through the torch, and is returned through another hose that contains the power cable. This is possible because the cooling water returning from the torch cools the cable and allows it to carry a much higher amperage. This wouldn't be possible if it were not cooled. Cooling water may be supplied from municipal mains, from a central cooling system, or from individual self-contained cooling systems (Figure 15-28).

miller
Dialarc 250-AC DC
miller
miller
miller

(*a*) (*b*)

Figure **15-26** Power supplies used for TIG welding: (*a*) ac/dc machine used for stick-electrode welding which can also be used for TIG applications; (*b*) ac/dc machine designed for TIG welding. (Courtesy of the Miller Electric Manufacturing Company.)

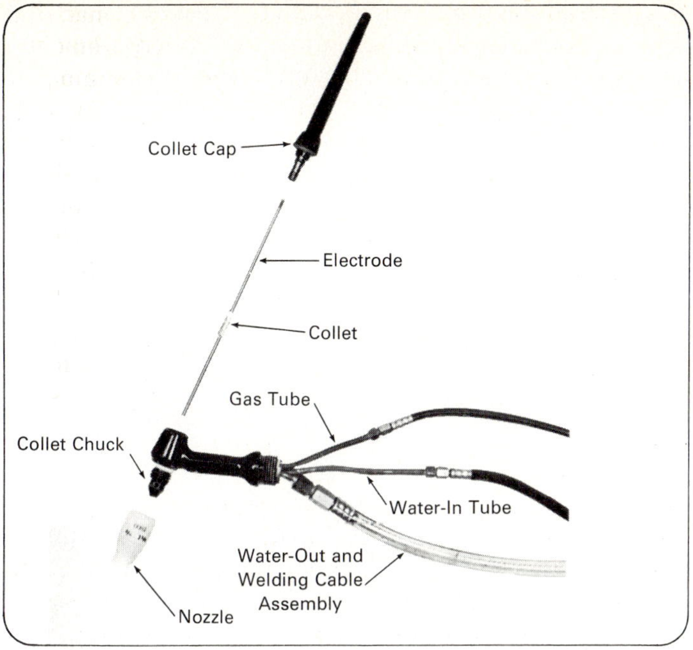
Collet Cap
Electrode
Collet
Gas Tube
Collet Chuck
Water-In Tube
Water-Out and
Welding Cable
Assembly
Nozzle

Figure **15-27** Exploded view of a water-cooled TIG torch and cable assembly. (Courtesy of the Miller Electric Manufacturing Company.)

MOTOR 115 VOLTS A.C.
USE SOLUBLE OIL IN TANK
miller
WELDERS

Figure **15-28** Commonly used individual, self-contained, TIG power-supply cooling system. (Courtesy of the Miller Electric Manufacturing Company.)

When water is taken from a municipal main it is passed through a solenoid valve. While welding is in progress, the valve is open and water is circulating. The water runs into a drain after it passes out of the torch. When welding is interrupted, the valve will close, and the water flow stops. This system is not desirable where city water is too expensive, contains too high a mineral content, or may be scarce during certain times of the year.

A central cooling system, consisting of a large holding tank and high-volume pump, may circulate water to a large number of machines. Some major difficulties of a central cooling system include:

1. Inflexibility.
2. Any failure of the cooling system, which shuts down all units.
3. Very high original cost.

The third type of cooling system, *individual self-contained*, consists of a tank, an electric motor, and a high-pressure pump to circulate water. The pump runs all the time the machine is on and recirculates the water through the system. It is sometimes desirable to use the solenoid valve even when the water-circulating system is used. Excessive cooling of the welding torch during idle periods can cause condensation of ambient air moisture within the torch, which can result in weld porosity in subsequent welding. In cold environments, care must be taken to prevent the water from freezing. Water-soluble oil is normally added to the water for pump lubrication purposes. Where freezing conditions may be encountered, permanent-type antifreeze may be substituted for the water soluble oil. Permanent antifreeze has sufficient lubricating qualities for adequate pump lubrication.

15-12 WELDING POWER

The operator has three choices of welding current. They are: dc straight polarity, dc reverse polarity, and ac with high-frequency stabilization. Each of these current types has its applications (Table 15-10), its advantages, and its disadvantages. The type of current used will have a great effect on the penetration pattern, as well as the bead configuration. Figure 15-29 shows details of the arc area for each current type.

15-13 SAFE PRACTICES

Prior to beginning the weld, it is recommended that the student review the safety practices for arc welding detailed in Chapter 13.

Table **15-10** Suitability of types of current for gas tungsten–arc welding of various metals

METAL WELDED	ALTERNATING CURRENT[a]	DIRECT CURRENT *STRAIGHT POLARITY*	DIRECT CURRENT *REVERSE POLARITY*
Low-carbon steel:			
0.015 to 0.030 in.[a]	G[b]	E	NR
0.030 to 0.125 in.	NR	E	NR
High-carbon steel	G[b]	E	NR
Cast iron	G[b]	E	NR
Stainless steel	G[b]	E	NR
Heat-resisting alloys	G[b]	E	NR
Refractory metals	NR	E	NR
Aluminum alloys:			
Up to 0.025 in.	E	NR[c]	G
Over 0.025 in.	E	NR[c]	NR
Castings	E	NR[c]	NR
Beryllium	G[b]	E	NR
Copper and alloys:			
Brass	G[b]	E	NR
Deoxidized copper	NR	E	NR
Silicon bronze	NR	E	NR
Magnesium alloys:			
Up to 1/8 in.	E	NR[c]	G
Over 3/16 in.	E	NR[c]	NR
Castings	E	NR[c]	NR
Silver	G[b]	E	NR
Titanium alloys	NR	E	NR

E = Excellent; G = Good; NR = Not recommended

[a]Stabilized. Do not use alternating current on tightly jigged assemblies. [b]Amperage should be about 25% higher than when straight-polarity direct current is used. [c]Unless work is mechanically or chemically cleaned in the areas to be welded.

Source: By permission, from *Metals Handbook* Volume 6, Copyright American Society for Metals, 1971.

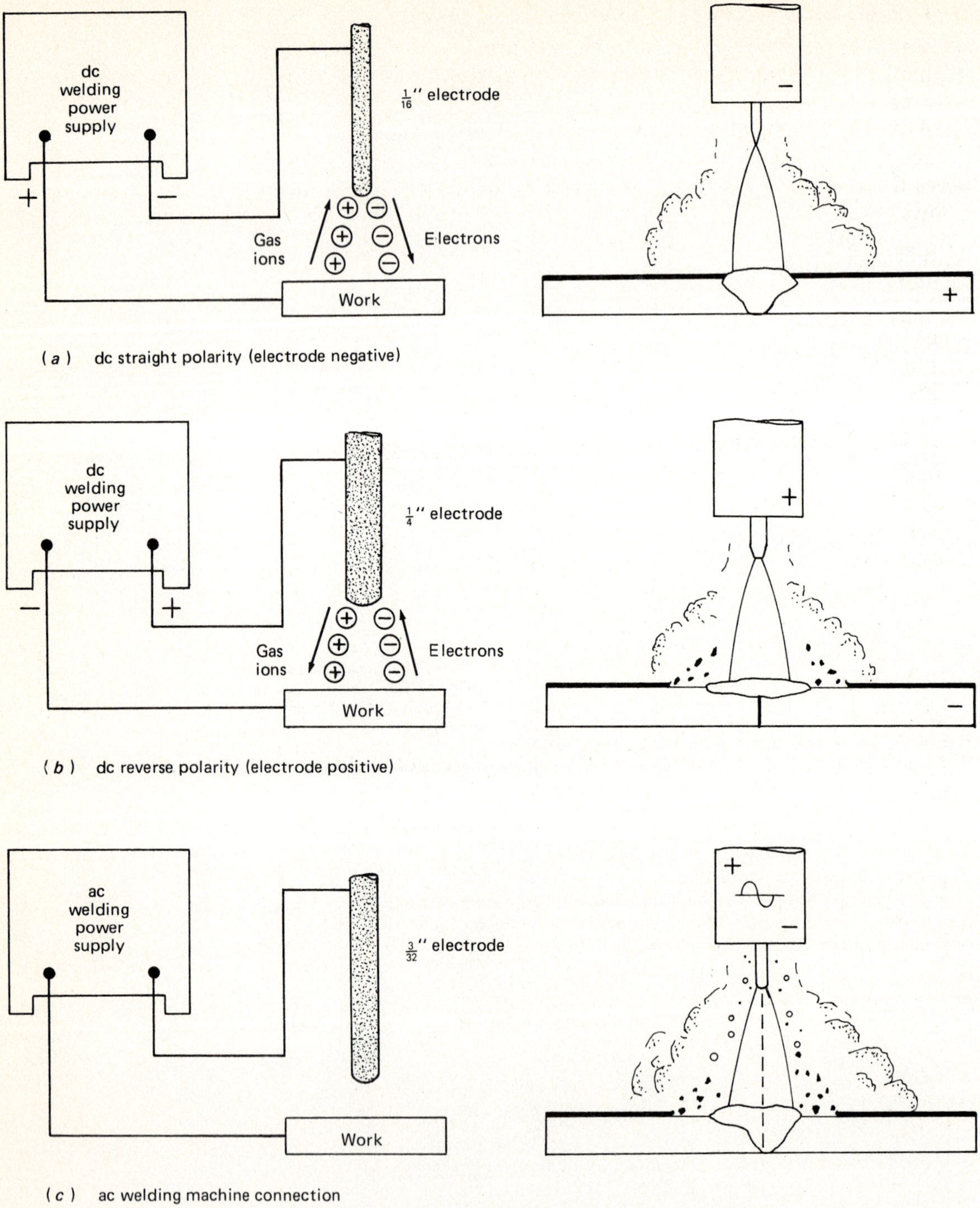
dc
welding
power
supply
+
−
1/16'' electrode
Gas
ions
Electrons
Work
−
+
(a) dc straight polarity (electrode negative)
dc
welding
power
supply
−
+
1/4'' electrode
Gas
ions
Electrons
Work
+
−
(b) dc reverse polarity (electrode positive)
ac
welding
power
supply
3/32'' electrode
Work
+
−
(c) ac welding machine connection

Figure **15-29** Details of the arc area produced with various types of currents. (Courtesy of the Miller Electric Manufacturing Company.)

15-14 PREPARATION FOR WELDING

Certain basic preparations should be made prior to establishing an arc. Included in these will be base metal preparation and setup of the machine along with its controls. Figure 15-30 illustrates the front panel of a typical machine designed for GTAW welding. Each of the various controls has a specific function, and the operator changes or varies them as the application changes. The function of each is as follows.

HIGH-FREQUENCY SWITCH The high-frequency switch has three positions: "Start," "Off," and "Continuous." When it is desired to use high frequency only to start the arc, the switch is placed in the "Start" position. The high frequency remains on only until an arc is established and then is automatically removed from the circuit.

This allows for starting the arc without touching the electrode to the work. This is advantageous on materials that might contaminate the electrode. The "Off" position is used when high frequency is not desired, such as when using the machine for stick-electrode welding. The switch is placed in the "Continuous" position when welding with alternating current on aluminum or magnesium. This provides continuous high frequency for arc stabilization.

Figure **15-30** Control panel of a typical ac/dc power supply specifically designed for TIG welding. (Courtesy of the Miller Electric Manufacturing Company.)

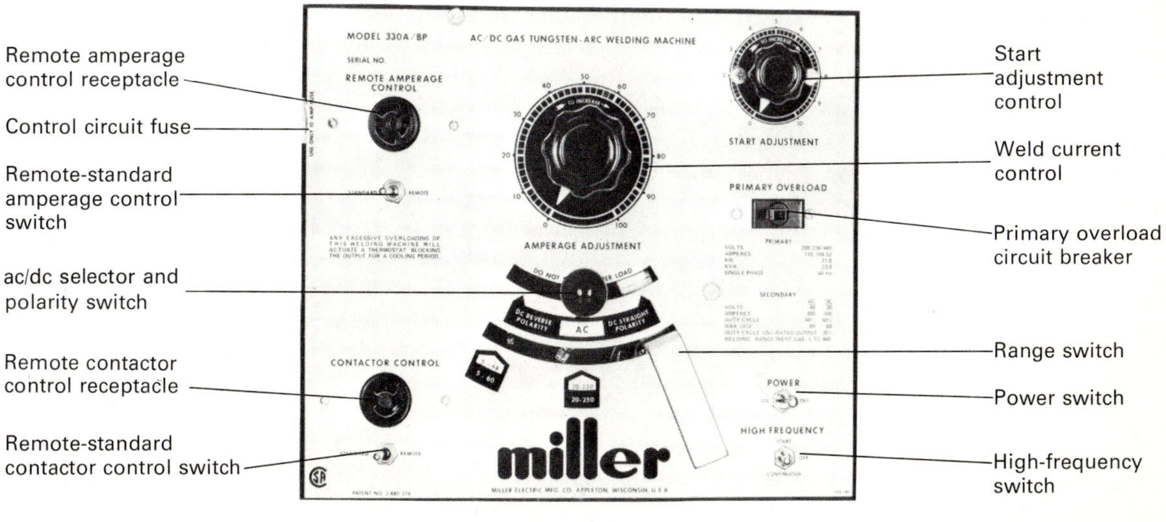

Remote amperage control receptacle
Control circuit fuse
Remote-standard amperage control switch
ac/dc selector and polarity switch
Remote contactor control receptacle
Remote-standard contactor control switch
Start adjustment control
Weld current control
Primary overload circuit breaker
Range switch
Power switch
High-frequency switch
MODEL 330A/BP
AC/DC GAS TUNGSTEN-ARC WELDING MACHINE
REMOTE AMPERAGE CONTROL
START ADJUSTMENT
PRIMARY OVERLOAD
AMPERAGE ADJUSTMENT
CONTACTOR CONTROL
POWER
HIGH FREQUENCY
AC
miller

POWER SWITCH This switch controls the primary line power to the transformer. When the switch is in the "On" position, the fan will operate and voltage is applied to the control circuit.

RANGE SWITCH The range switch has three positions. Adjust the range switch to select the desired amperage range. Control from minimum to maximum within the range is available with the weld-current control.

PRIMARY-OVERLOAD CIRCUIT BREAKER The circuit breaker provides protection against overloading of the welding machine's main components. The circuit breaker must be "On" before the primary contactor of the machine can be energized.

WELD-CURRENT CONTROL This current control can be adjusted to provide a percentage of current between minimum and maximum of the range selected.

START-ADJUSTMENT CONTROL The start-adjustment control is used to control the starting current for approximately ⅔ of a second. After this time delay, the current automatically changes to the current setting of the weld-current control. The initial amperage may be either higher or lower than the welding current. A low-amperage "soft" start may be desirable on thin metals where burn-through at the start may be a problem.

The higher-amperage "hot" start is primarily used on thicker materials or on metals having good thermal conductivity. As an example of how the start control functions, Figure 15-31 illustrates the positions of the start- and current-control rheostats for a low-amperage start and a high-amperage start.

The main current-control rheostat is set at 50% of the welding range. The start adjustment is set at 20% for the low-amperage start. The starting amperage, therefore, will be approximately 20% of the range for ⅔ of a second and then go to the 50% setting on the main rheostat. Again, the main rheostat is set at 50%. The start adjustment is set for 80%. The starting amperage will now be approximately 80% of the range and then drop off to the 50% setting, providing a "hot" start.

REMOTE AMPERAGE CONTROL RECEPTACLE This receptacle is provided for connecting a remote hand control or a remote foot control. This allows the operator to have amperage control while welding at the work station, which may be a considerable distance from the power source. With the foot control, the operator can vary the amperage with progression along a joint. This is particularly helpful when starting on a cold workpiece. Amperage may be increased to establish a weld puddle quickly, and as the material heats up, the operator can decrease the amperage. When coming to the end of a

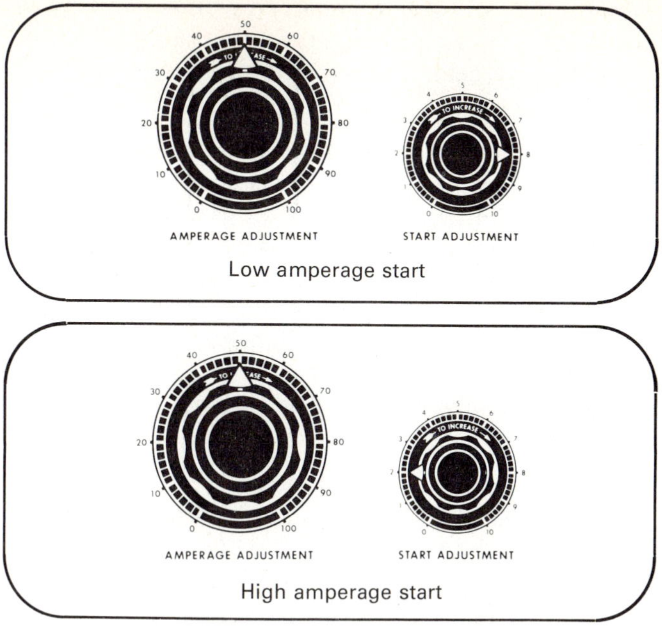
AMPERAGE ADJUSTMENT
START ADJUSTMENT
Low amperage start
AMPERAGE ADJUSTMENT
START ADJUSTMENT
High amperage start

Figure **15-31** Settings for low- and high-amperage starts. (Courtesy of the Miller Electric Manufacturing Company.)

joint, the amperage can be further decreased to taper off and "crater out."

CONTROL-CIRCUIT FUSE This fuse protects the control circuit of the machine. It also protects the remote-control circuit and the control rectifier. If this fuse is defective, the output of the machine will be only the minimum of a range.

REMOTE-STANDARD AMPERAGE CONTROL SWITCH When a remote control device is being used, the switch must be in the "remote" position. When amperage control is to be at the machine panel, the switch must be in the "standard" position.

AC/DC SELECTOR AND POLARITY SWITCH This three-position switch permits the operator to select ac, dc straight polarity, or dc reverse polarity.

REMOTE CONTACTOR CONTROL RECEPTACLE A remote control switch can be used to open and close the contactor. This is normally used with the remote amperage control. The arc cannot be established until the contactor is closed.

STANDARD-REMOTE CONTACTOR CONTROL SWITCH When a remote contactor switch is being used, it must be in the "Remote" position. When the switch is in the "Standard" position, the contactor remains closed as long as primary power is applied to the machine. Whenever the contactor is closed, the gas and water valves open to allow flow to the torch.

POSTFLOW TIMER In addition to the controls shown on the front panel in Figure 15-30, a postflow timer is located behind the access door on the machine (see Figure 15-32).

The postflow gas and water-valve timer controls the length of time the gas and water will continue to flow after the arc is broken. This timer should be set so that the gas flow continues for a sufficient length of time to allow the tungsten to cool to a point where it will no longer be contaminated when exposed to air. This time is usually 10 seconds for each 100 amperes of welding current. The tungsten should cool bright and shiny. Any blueing or blackening indicates a lack of postflow.

HIGH-FREQUENCY INTENSITY CONTROL This control allows the operator to choose the proper intensity for the high-frequency output. As the setting of this control is increased, the current in the

Figure **15-32** Rear view of control panel shown in Figure 15-30. (Courtesy of the Miller Electric Manufacturing Company.)

Spark Gap Assembly
Balancing Resistor Switch
Water Valve
Post-Flow Gas & Water Valve Timer
Work Terminal
High Frequency Intensity Control
Weld Timer (Spot Panel Model Only)
Start Circuit Switch
Shielding Gas Valve
Electrode Terminal

high-frequency circuit is increased. This should be set for the required intensity to start the arc. It is recommended that this control be kept at a minimum setting that will provide satisfactory weld starts. The higher the setting, the greater the amount of radiation, which will cause interference with communications equipment.

SPARK-GAP ASSEMBLY The gap points are preset at the factory. The device (assembly) provides the proper voltage in the circuit. A feeler gauge can be used to check the spacing or make adjustments. *Warning:* This inspection adjustment must be made only with the power turned off. Consult the manufacturer's manual for appropriate gap size.

BALANCING-RESISTER KNIFE SWITCH This switch has two positions. The SHIELDED METAL–ARC position is for stick-electrode welding. The GAS TUNGSTEN–ARC position is for TIG welding.

15-15 RECOMMENDED WELDING PARAMETERS

Tables 15-11 through 15-18 (pages 370–379) list the recommended shielding gas, welding positions, gas flow, and selected other information necessary to produce good welds in aluminum, stainless steel, copper, bronze, magnesium, and carbon steel by the manual GTAW (TIG) process.

The following sections contain exercises in gas tungsten–arc welding.

15-16 BECOMING FAMILIAR WITH THE GTAW EQUIPMENT

OBJECTIVES To learn to:

1. Visually identify and correctly name all major GTAW equipment components.

2. Correctly set up and make operational all component parts of the GTAW outfit.

3. Correctly start a GTAW outfit.

HOURS REQUIRED ½ to ¾.

EQUIPMENT, TOOLS, AND MATERIALS REQUIRED One GTAW welding outfit (Figures 15-25 and 15-26) and safety accessories such as apron, gloves, and face shield.

PREVIOUS PREPARATION OF EQUIPMENT BY INSTRUCTOR It is assumed that the instructor has made the following preparations prior to turning over the equipment to the student:

(*text continues on page* 380)

Table **15-11** TIG hand-welding of aluminum[a]

			WELDING CURRENT				
			AMPERES			ELECTRODE DIAMETER* (IN.)	WELDING SPEED** (IPM)
THICK-NESS (IN.)	TYPE OF WELD	*TYPE*	FLAT	VERTICAL	OVERHEAD		
1/16	Butt	Alternating current—high-frequency stabilized	60–80	60–80 Down	60–80	1/16	12
	Lap		70–90	55–75 Up	60–80	1/16	10
	Corner		60–80	60–80 Up	60–80	1/16	12
	Fillets		70–90	70–90 Down	70–90	1/16	10
1/8	Butt		125–145	115–135 Down	120–140	3/32	12
	Lap		140–160	125–145 Up	130–160	3/32	10
	Corner		125–145	115–135 Up	130–150	3/32	12
	Fillets		140–160	115–135 Down	140–160	3/32	10
3/16	Butt		190–220	190–220 Up or Down	180–210	1/8	11
	Lap		210–240	190–220 Up	180–210	1/8	9
	Corner		190–220	180–210 Up	180–210	1/8	11
	Fillet		210–240	190–220 Up	180–210	1/8	9
1/4	Butt		260–300	220–260 Up	210–250	3/16	10
	Lap		290–340	220–260 Up	210–250	3/16	8
	Corner		280–320	220–260 Up	210–250	3/16	10
	Fillet		280–320	220–260 Up	210–250	3/16	8
3/8	Butt		330–380	250–300	250–300	3/16, 1/4	5
	Lap		330–380	250–300	250–300	3/16, 1/4	5
	Tee Fillets		350–400	250–300	250–300	3/16, 1/4	5
	Corner		330–380	250–300	250–300	3/16, 1/4	5
1/2	Butt		400–450	290–350 Up	250–300	3/16, 1/4	3
	Lap		400–450	300–350 Up	275–325	3/16, 1/4	3
	Tee Fillets		420–470	300–350 Up	275–325	3/16, 1/4	3
	Corner		400–450	300–350 Up	275–325	3/16, 1/4	3

Note: For aluminum heavier than 1/2 in., argon-helium mixtures provide greatly improved performance.

*If two sizes are listed, the smaller is for vertical and overhead welding. Use a larger electrode or slightly lower welding current when balanced wave transformer is used.

**Welding speed for flat position.

[a]See also Table 21-7.

Source: Union Carbide Corporation, Linde Division, *How To Do TIG (Heliarc) Welding,* F-1607-D 5/70.

WELDING ROD*** SIZE (IN.)	GAS CUP OR NOZZLE SIZE**** CERAMIC (LAVA) CUP (250 amp. max.)	HIGH IMPACT (ALUMINA) CUP (300 amp. max.)	METAL NOZZLE (WHERE AVAILABLE)	GAS FLOW ARGON (CFH)	REMARKS
None or 1/16	4,5,6	4,5,6	6	15	
None or 1/16	4,5,6	4,5,6	6	15	
None or 1/16	4,5,6	4,5,6	6	15	
None or 1/16	4,5,6	4,5,6	6	15	
3/32 or 1/8	6,7	6,7	6,7	20	3/32″ Rod—Overhead
None or 3/32	6,7	6,7	6,7	20	3/32″ Rod—Overhead
None or 3/32	6,7	6,7	6,7	20	3/32″ Rod—Vertical
1/16 or 3/32	6,7	6,7	6,7	20	
1/8	7,8	7,8	6,8	20	
1/8	7,8	7,8	6,8	20	
1/8	7,8	7,8	6,8	20	
1/8	7,8	7,8	6,8	20	
1/8 or 3/16		8,10,12	8,10	25	1/8″ Rod—Vertical and
1/8 or 3/16		8,10,12	8,10	25	overhead two passes
1/8 or 3/16		8,10,12	8,10	25	
1/8 or 3/16		8,10,12	8,10	25	1/8″ Rod—Vertical
3/16 or 1/4			10	30	Two passes
3/16 or 1/4			10	30	Two passes
3/16 or 1/4			10	30	Two passes
3/16 or 1/4			10	30	Two passes
3/16 or 1/4			10	30	Two or three passes
3/16 or 1/4			10	30	Three passes
3/16 or 1/4			10	30	Three passes
3/16 or 1/4			10	30	Three passes

***Refer to Chapter 4.
****Consult the torch instruction booklet for the maximum current ratings of the torch and for sizes and part numbers of available cups or nozzles.

Table **15-12** TIG hand-welding of stainless steel*

THICKNESS (in.)	TYPE OF WELD	*CHARACTERISTICS*	WELDING CURRENT *AMPERES* FLAT	VERTICAL	OVERHEAD	ELECTRODE DIAMETER (in.)	WELDING SPEED** (I.P.M.)
1/16	Butt	Straight polarity—direct current	80–100	70– 90 up	70–90	1/16	12
	Lap		100–120	80–100 up	80–100	1/16	10
	Corner		80–100	70– 90 up	70–90	1/16	12
	Fillet		90–110	80–100 up	80–100	1/16	10
3/32	Butt		100–120	90–110 up	90–110	1/16	12
	Lap		110–130	100–120 up	100–120	1/16	10
	Corner		100–120	90–110 up	90–110	1/16	12
	Fillet		110–130	100–120 up	100–120	1/16	10
1/8	Butt		120–140	110–130 up	105–125	1/16	12
	Lap		130–150	120–140 up	120–140	1/16	10
	Corner		120–140	110–130 up	115–135	1/16	12
	Fillet		130–150	115–135 up	120–140	1/16	10
3/16	Butt		200–250	150–200 up	150–200	3/32	10
	Lap		225–275	175–225 up	175–225	3/32, 1/8	8
	Corner		200–250	150–200 up	150–200	3/32	10
	Fillet		225–275	175–225 up	175–225	3/32, 1/8	8
1/4	Butt		275–350	200–250 up	200–250	1/8	—
	Lap		300–375	225–275 up	225–275	1/8	—
	Corner		275–350	200–250 up	200–250	1/8	—
	Fillet		300–375	225–275 up	225–275	1/8	—
1/2	Butt		350–450	225–275 up	225–275	1/8, 3/16	—
	Lap		375–475	230–280 up	230–280	1/8, 3/16	—
	Fillet		375–475	230–280 up	230–280	1/8, 3/16	—

*Conditions very similar for HASTELLOY B & C and other similar alloys.
**Welding speed for flat position.

Source: Union Carbide Corporation, Linde Division, *How To Do TIG (Heliarc) Welding*, F-1607-D 5/70.

WELDING ROD: FILLER METAL TYPE NO.	WELDING ROD: SIZE (in.)	GAS CUP OR NOZZLE SIZE***: CERAMIC (LAVA) CUP (250 amp. max.)	GAS CUP OR NOZZLE SIZE***: HIGH IMPACT (ALUMINA) CUP (300 amp. max.)	GAS CUP OR NOZZLE SIZE***: METAL NOZZLE (WHERE AVAILABLE)	GAS FLOW ARGON (CFH)	REMARKS
For welding rod to use, refer to "Principal Weldable Types of Stainless Steel."	1/16	4,5,6	4,5,6	6	10	
	1/16	4,5,6	4,5,6	6	10	
	1/16	4,5,6	4,5,6	6	10	
	1/16	4,5,6	4,5,6	6	10	
	1/16 or 3/32	4,5,6	4,5,6	6	10	
	1/16 or 3/32	4,5,6	4,5,6	6	10	
	1/16 or 3/32	4,5,6	4,5,6	6	10	
	1/16 or 3/32	4,5,6	4,5,6	6	10	
	3/32	4,5,6	4,5,6	6	10	
	3/32	4,5,6	4,5,6	6	10	
	3/32	4,5,6	4,5,6	6	10	
	3/32	4,5,6	4,5,6	6	10	
	1/8	6,7,8	6,7,8	6,8	15	
	1/8	6,7,8	6,7,8	6,8	15	
	1/8	6,7,8	6,7,8	6,8	15	
	1/8	6,7,8	6,7,8	6,8	15	
	3/16			8	15	One or two passes
	3/16			8	15	One or two passes
	3/16			8	15	One pass
	3/16			8	15	
	1/4			8	15	Two or three passes
	1/4			8	15	Three passes
	1/4			8	15	Three passes

***Consult the torch instruction booklet for the maximum current ratings of the torch and for sizes and part numbers of available cups or nozzles.

Table **15-13** TIG hand-welding of deoxidized copper

THICKNESS (in.)	TYPE OF WELD	WELDING CURRENT AMPS		ELECTRODE DIAMETER (in.)	WELDING SPEED (I.P.M.)	WELDING ROD* SIZE (in.)
		TYPE	*FLAT*			
1/16	Butt	Direct current straight polarity	110–140	1/16	12	1/16
	Lap		130–150	1/16	10	1/16
	Corner		110–140	1/16	12	1/16
	Fillet		130–150	1/16	10	1/16
1/8	Butt		175–225	3/32	11	3/32 or 1/8
	Lap		200–250	3/32	9	3/32 or 1/8
	Corner		175–225	3/32	11	3/32 or 1/8
	Fillet		200–250	3/32	9	3/32 or 1/8
3/16	Butt		190–225	1/8	10	1/8
	Lap		205–250	1/8	8	1/8
	Corner		190–225	1/8	10	1/8
	Fillet		205–250	1/8	8	1/8
1/4	Butt		225–260	1/8	9	1/8
	Lap		250–280	1/8	7	1/8
	Corner		225–260	1/8	9	1/8
	Fillet		250–280	1/8	7	1/8
3/8	Butt		280–320	3/16		3/16
	Lap		300–340	3/16		3/16
	Corner		280–320	3/16		3/16
	Fillet		300–340	3/16		3/16
1/2	Butt		375–525	3/16, 1/4		1/4

*OXWELD 63. (copper)

Source: Union Carbide Corporation, Linde Division, *How To Do TIG (Heliarc) Welding*, F-1607-D 5/70.

GAS CUP OR NOZZLE SIZE**			GAS FLOW		
CERAMIC (LAVA) CUP (250 amp. max.)	*HIGH IMPACT (ALUMINA) CUP (300 amp. max.)*	*METAL NOZZLE (WHERE AVAILABLE)*	*ARGON (CFH)*	*HELIUM (CFH)*	REMARKS
4,5,6	4,5,6	6	15		One pass
4,5,6	4,5,6	6	15		One pass
4,5,6	4,5,6	6	15		One pass
4,5,6	4,5,6	6	15		One pass
6,7,8	6,7,8	6,8	15		One pass
6,7,8	6,7,8	6,8	15		One pass
6,7,8	6,7,8	6,8	15		One pass
6,7,8	6,7,8	6,8	15		One pass
	8,10	8		30	One pass; preheat to 200°F.
	8,10	8		30	One pass; preheat to 200°F.
	8,10	8		30	One pass; preheat to 200°F.
	8,10	8		30	One pass; preheat to 200°F.
	8,10	8		30	One pass; preheat to 300°F.
	8,10	8		30	One pass; preheat to 300°F.
	8,10	8		30	One pass; preheat to 300°F.
	8,10	8		30	One pass; preheat to 300°F.
		8		40	Two passes; preheat to 500°F.
		8		40	Three passes; preheat to 500°F.
		8		40	Two passes; preheat to 500°F.
		8		40	Three passes; preheat to 500°F.
		8,10		40	Three passes; preheat to 500°F.

**Consult the torch instruction booklet for the maximum current ratings of the torch and for sizes and part numbers of available cups or nozzles.

Table **15-14** TIG hand-welding of silicon bronze

THICKNESS (in.)	TYPE OF WELD	WELDING CURRENT TYPE	AMPERES FLAT	AMPERES VERTICAL	AMPERES OVERHEAD	ELECTRODE DIAMETER (IN.)	WELDING SPEED* (IPM)
1/16	Butt	Direct current straight polarity	100–120	90–110 Up	90–110	1/16	12
	Lap		110–130	100–120 Up	100–120	1/16	10
	Corner		100–130	90–110 Up	90–110	1/16	12
	Fillet		110–130	100–120 Up	100–120	1/16	10
1/8	Butt		130–150	120–140 Up	120–140	1/16	12
	Lap		140–160	130–150 Up	130–150	1/16, 3/32	10
	Corner		130–150	120–140 Up	120–140	1/16	12
	Fillet		140–160	130–150 Up	130–150	1/16, 3/32	10
3/16	Butt		150–200	—	—	3/32	—
	Lap		175–225	—	—	3/32	—
	Corner		150–200	—	—	3/32	—
	Fillet		175–225	—	—	3/32	—
1/4	Butt		150–200	—	—	3/32	—
	Butt		250–300	—	—	1/8	—
	Lap		175–225	—	—	3/32	—
	Fillet		175–225	—	—	3/32	—
3/8	Butt		230–280	—	—	1/8	—
	Lap		250–300	—	—	1/8	—
	Fillet		230–280	—	—	1/8	—
1/2	Butt		250–300	—	—	1/8	—
	Lap		275–325	—	—	1/8	—
	Fillet		275–325	—	—	1/8	—
3/4	Butt		300–350	—	—	1/8	—
	Lap		300–350	—	—	1/8	—
	Fillet		300–350	—	—	1/8	—
1	Butt		300–350	—	—	1/8	—
	Lap		325–350	—	—	1/8	—
	Fillet		325–350	—	—	1/8	—

*For welding in flat position.

Source: Union Carbide Corporation, Linde Division, *How To Do TIG (Heliarc) Welding*, F-1607-D 5/70.

WELDING ROD		GAS CUP OR NOZZLE SIZE***				
MATERIAL NO.**	SIZE (in.)	CERAMIC (LAVA) CUP (250 amp. max.)	HIGH IMPACT (ALUMINA) CUP (300 amp. max.)	METAL NOZZLE (WHERE AVAILABLE)	GAS FLOW ARGON (CFH)	REMARKS
Everdur	1/16	4,5,6	4,5,6	6	15	
Everdur	1/16	4,5,6	4,5,6	6	15	
Everdur	1/16	4,5,6	4,5,6	6	15	
Everdur	1/16	4,5,6	4,5,6	6	15	
Everdur	3/32	6,7,8	6,7,8	6,8	15	
Everdur	3/32	6,7,8	6,7,8	6,8	15	
Everdur	3/32	6,7,8	6,7,8	6,8	15	
Everdur	3/32	6,7,8	6,7,8	6,8	15	
Everdur	1/8	6,7,8	6,7,8	6,8	20	
Everdur	1/8	6,7,8	6,7,8	6,8	20	
Everdur	1/8	6,7,8	6,7,8	6,8	20	
Everdur	1/8	6,7,8	6,7,8	6,8	20	
Everdur	1/8 or 3/16	7,8,10	7,8,10	6,8	20	Three passes
Everdur	1/8 or 3/16	7,8,10	7,8,10	6,8	20	One pass—square butt
Everdur	1/8 or 3/16	7,8,10	7,8,10	6,8	20	Three passes
Everdur	1/8 or 3/16	7,8,10	7,8,10	6,8	20	Three passes
Everdur	1/8, 3/16		8,10,12	8	20	Three or four passes
Everdur	1/8, 3/16		8,10,12	8	20	Three passes
Everdur	1/8, 3/16		8,10,12	8	20	Three passes
Everdur	1/8, 3/16		8,10,12	8	20	Four or five passes
Everdur	1/8, 3/16		8,10,12	8	20	Six passes
Everdur	1/8, 3/16		8,10,12	8	20	Seven passes
Everdur	3/16			8	20	Nine or ten passes
Everdur	3/16			8	20	Twelve passes
Everdur	3/16			8	20	Fourteen passes
Everdur	3/16, 1/4			8	20	Thirteen passes
Everdur	3/16, 1/4			8	20	Sixteen passes
Everdur	3/16, 1/4			8	20	Twenty passes

**OXWELD No. 26 Welding Rod for "Everdur" alloy.

***Consult the torch instruction booklet for the maximum current ratings of the torch and for sizes and part numbers of available cups or nozzles.

Table **15-15** Manual welding of magnesium (AZ31B & HK31A alloys) using A.C.H.F.

THICKNESS (in.)	TYPE OF WELD (1)	WELDING CURRENT AMPERES* FLAT POSITION		WELDING ROD (2) SIZE (in.)	SHIELDING GAS FLOW		REMARKS
		*HK31A***	*AZ31B****		*ARGON*	*HELIUM*	
0.040	Butt	40	35	3/32, 1/8	10	25	Backup
0.040	Butt	30	25	3/32, 1/8	10	25	No backing
0.040	Fillet	40	35	3/32, 1/8	10	25	
0.064	Butt	55	50	3/32, 1/8	10	25	Backup
0.064	Butt & Corner	35	30	3/32, 1/8	10	25	No backing
0.064	Fillet	55	50	3/32, 1/8	10	25	
0.081	Butt	75	65	1/8	10	25	Backup
0.081	Butt, corner & edge	45	40	1/8	10	25	No backing
0.081	Fillet	75	65	1/8	10	25	
0.102	Butt	95	85	1/8	20	30	Backup
0.102	Butt, corner & edge	70	60	1/8	20	30	No backing
0.102	Fillet	95	85	1/8	20	30	
0.128	Butt	110	100	1/8, 5/32	20	35	Backup
0.128	Butt, corner & edge	80	70	1/8, 5/32	20	35	No backup
0.128	Fillet	110	100	1/8, 5/32	20	35	
3/16	Butt	155	140	1/8, 5/32	20	35	1 pass
3/16	Butt	110	100	1/8, 5/32	20	35	2 passes
1/4	Butt	200	180	5/32, 3/16	25	50	1 pass
1/4	Butt	125	115	5/32	20	35	2 passes
3/8	Butt	270	250	5/32, 3/16	25	50	1 pass
3/8	Butt	160	140	5/32, 3/16	25	50	2 passes
1/2	Butt	330	310	3/16	25	50	2 passes
3/4	Butt	450	420	3/16, 1/4	35	75	2 passes

*Current values given for butt joint welding are with backing plate. Slightly lower values used for welding without a backing plate.

**Welding Rod: EZ33A

***Welding Rod: AZ92A or AZ61A

Source: Union Carbide Corporation, Linde Division, *How To Do TIG (Heliarc) Welding*, F-1607-D 5/70.

Table **15-16** TIG hand welding of other copper alloys

	TYPE OF WELDING CURRENT	FLUX[a]	WELDING TECHNIQUE	ROD[a]
Red brass	DCSP on thick-			
Zinc bronze	nesses greater			
Low-zinc brass	than 0.050-in.		Forehand	
Common brass	ACHF on thinner			
Muntz metal	material.			
Phosphor bronze	DCSP	None	Forehand	
Leaded bronzes	DCSP	None	Forehand	
Beryllium-copper	ACHF	None	Forehand	Be-Cu

[a]Refer to Chapter 4

Source: Union Carbide Corporation, Linde Division, *How To Do TIG (Heliarc) Welding*, F-1607-D 5/70.

Table **15-17** TIG hand welding of aluminum bronze

PLATE THICKNESS (in.)	EDGE PREP.	WELDING CURRENT AMPERES (ACHF)	ARGON FLOW AT RECOMMENDED 20 (PSI) (CFH)	ELECTRODE DIAMETER (in.)	NUMBER OF PASSES
1/4	90° V, sharp nose	200	15–20	1/8	2
3/8	60° V, sharp nose	250	15–20	5/32	3
1/2	60° V, sharp nose	260	15–20	5/32	4

Source: Union Carbide Corporation, Linde Division, *How To Do TIG (Heliarc) Welding*, F-1607-D 5/70.

Table **15-18** TIG hand welding of carbon steel

THICKNESS (in.)	AMPERES (DCSP)	SUGGESTED ROD SIZE (in.)	AVERAGE WELDING SPEED (in.)	ARGON FLOW (CFH)
0.035	100	1/16	12–15	8–10
0.049	100–125	1/16	12–18	8–10
0.060	100–140	1/16	12–18	8–10
0.089	140–170	3/32	12–18	8–10
0.125	150–200	1/8	10–12	8–10

Source: Union Carbide Corporation, Linde Division, *How To Do TIG (Heliarc) Welding*, F-1607-D 5/70.

1. Set up the equipment according to instructions furnished by manufacturer.

2. Hooked up the welding machine to the power supply.

3. Hooked up the welding machine to the cooling-water supply.

4. Provided and hooked up the correct shielding gas.

PREWELDING PROCEDURE BY STUDENT

1. See that welding table is clean.

2. Ascertain that necessary protective equipment has been provided.

a. Helmet (with correct shade lens).
b. Leather apron.
c. Leather gloves.
d. Welder's cap.
e. Spats (if instructor requires them).

The following steps assume that you are going to weld aluminum.

3. Obtain 2 × 6 × ⅛-inch pieces of aluminum.

4. Sufficient 4043 aluminum filler rod.

5. Check welding torch (Figure 15-27) for the following:

a. *Nozzle.* Must be ⅜ inch (I.D.) and have an unburnt, uncracked, even end.
b. *Collet chuck.* Must be 3/32-inch capacity.
c. *Collet.* Must be 3/32-inch capacity.
d. *Electrode.* Must be pure tungsten 3/32-inch diameter in good condition [Figure 15-33(B)].

Figure **15-33** Shapes of commonly used tungsten electrodes. (Courtesy of the Miller Electric Manufacturing Company.)

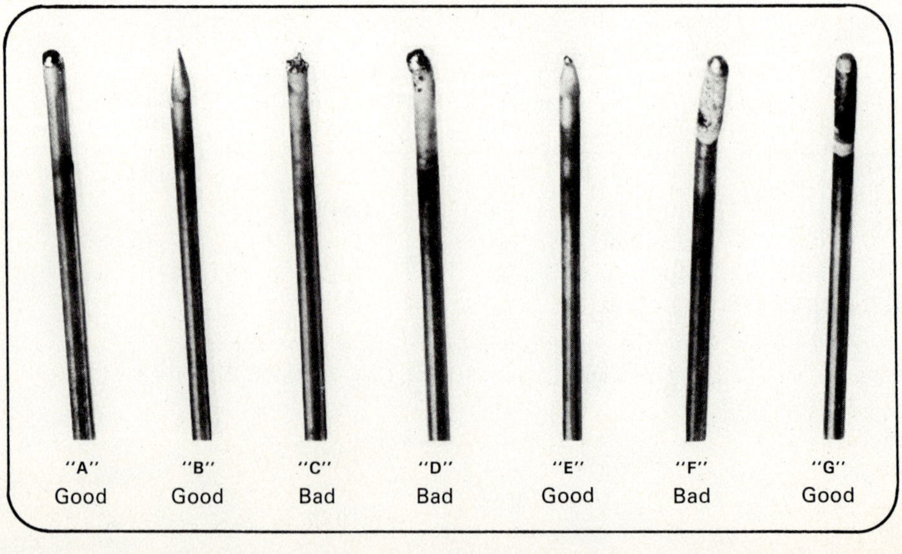
"A"
Good
"B"
Good
"C"
Bad
"D"
Bad
"E"
Good
"F"
Bad
"G"
Good

6. Adjust electrode as follows:

a. Using the appropriate wrench furnished with the torch, loosen the cap.

b. Remove the cap from the torch.

c. Insert the collet into the torch body.

d. Insert the electrode into the collet and push, so that it sticks out approximately ½ inch beyond the end of the nozzle (Figure 15-11).

e. Tighten the cap lightly.

f. Tap the electrode against the edge of the welding table until it extends only a length equal to 1½ to 2 times its diameter.

g. Finger tighten cap.

7. Clamp work lead (ground wire) to a location on welding table where it will not interfere with welding.

8. Refer to Table 15-11 to determine the recommended welding-machine settings and flow of shielding gas.

9. Set the welding machine to the appropriate setting.

10. Turn on the welding machine.

11. Slowly open the shielding-gas–tank valve as far as it will open. *Warning:* Do not stand directly in front of the gauge dial while opening the valve—the dial could blow out and injure you.

12. Press the foot pedal *once* and release *immediately*. This opens the gas valve long enough to adjust the gas-flow valve.

13. Adjust the gas flow to ft^3/h determined from Table 15-11.

15-17 MAKING A STRINGER BEAD IN THE FLAT POSITION

OBJECTIVE To learn how to properly manipulate the GTAW (TIG) outfit described in exercise 15-16.

1. Sufficient supply of argon shielding gas.

2. One or two 3/32-inch-diameter pure-tungsten electrodes.

3. Sufficient ⅛-inch-diameter 4043 aluminum filler metal.

4. Ample supply of 6 × 2 × ⅛-inch aluminum strips.

PROCEDURE

1. Remove oxides from aluminum.

2. Place a piece of aluminum on the welding table, as shown in Figure 15-34.

3. Refer to Table 15-11 to determine the welding conditions.

4. Set the welding machine to the appropriate conditions.

5. Start the welding machine.

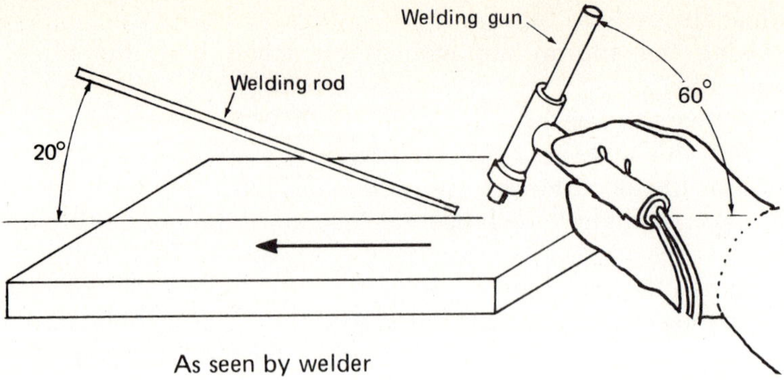
Welding gun
Welding rod
60°
20°
As seen by welder

Figure **15-34**

6. Refer to Figure 15-34 for position of the torch and filler metal relative to the base metal.

7. Lower the welding hood.

8. Start the arc by pressing the foot pedal enough to melt the base metal and form a ¼-inch puddle.

9. Start laying down the bead.

a. Arc length should be equal to diameter of electrode.

b. Width of bead face should be equal to twice the diameters of the electrode.

c. After traveling the entire length of the aluminum strip, inspect the welds.

10. Repeat steps 6–9 until you can produce acceptable beads. If you fill the strip and do not want to build it up or do not want to lay down beads on both sides of the strip, repeat steps 1, 2, and 6–9.

11. Shut down the welding machine as follows:

a. Close gas-tank valve snug-tight by hand.

b. Press foot pedal once and release immediately to bleed gas line.

c. Close gas-flow valve when ball rests on bottom.

d. Shut off power at welding machine.

15-18 MAKING A SQUARE-GROOVE BUTT JOINT ON ALUMINUM IN THE FLAT POSITION

OBJECTIVE To learn to make acceptable square-groove butt joints in the flat position with the GTAW process.

HOURS REQUIRED 1.

EQUIPMENT, TOOLS, AND MATERIALS REQUIRED Same as for exercise 15-16.

PROCEDURE

1. Remove oxides from the aluminum.

2. Position the aluminum on the welding table as shown in Figure 15-34.

3. Refer to Table 15-11 to determine the welding conditions.

4. Set the welding machine to the appropriate settings.

5. Clamp the work lead (ground clamp) to the welding table so that it will not interfere with the welding.

6. Place the torch on the welding table so that it will not arc.

7. Start the welding machine.

8. Lower your helmet.

9. Tack weld one end of the aluminum strip as follows:
   a. Rest nozzle on welding table 1/16 inch from end of groove.
   b. Start arc and melt metal 1/16 inch into each side of groove.
   c. Feed filler rod in and out of melted area to complete first tack weld.

10. Turn the work around and tack weld the other end.

11. Lay down the torch and lift your helmet.

12. Refer to Figure 15-34 for the angle of the torch and filler metal relative to the work piece.

13. Lower your helmet and, holding the torch as shown in Figure 15-34, start welding utilizing same manipulative process as in exercise 15-17.

14. Complete the weld. Lay down the torch and lift your helmet.

15. If necessary, pick up another piece of aluminum and, using the welded piece as the other half, repeat steps 1 through 14 until you can produce an acceptable weld.

16. Shut down the welding outfit.

15-19 MAKING A FILLET-WELD LAP JOINT IN THE HORIZONTAL POSITION ON ALUMINUM WITHOUT USING FILLER METAL

OBJECTIVE To learn to make an acceptable fillet weld in the horizontal position on aluminum with the GTAW process without using filler metal.

HOURS REQUIRED 1.

EQUIPMENT, TOOLS, AND MATERIALS REQUIRED Same as in exercise 15-16.

PROCEDURE Read and become thoroughly familiar with the procedure below prior to welding.

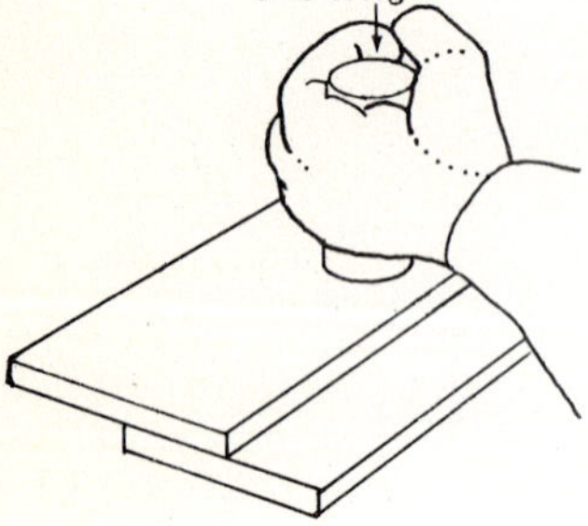

5-lb weight

Figure **15-35**

1. Remove oxides from the aluminum.

2. Refer to Table 15-11 to determine the welding conditions.

3. Set the welding machine to the appropriate settings.

4. Place aluminum pieces on the welding table with a heavy (5-pound) weight on them as shown in Figure 15-35.

5. Turn on the welding machine.

6. Tack weld both ends of the upper joint (Figure 15-36) without using filler metal.

7. Remove the heavy weight from the aluminum pieces and place it in an out-of-the-way location.

8. Turn the tack-welded pieces over and tack weld other joint.

9. Refer to Figure 15-37 for the position of the torch. Take care to keep the arc length equal to the diameter of the electrode and proceed to weld the joint at a speed that will produce complete penetration (Figure 15-38) into the corner of joint and a bead face equal to twice the diameter of the electrode.

10. Complete the welds, lay down the torch, lift your helmet, and refer to Figure 15-39 for weld quality.

11. If necessary, repeat this exercise; otherwise shut down the machine.

Figure **15-36**

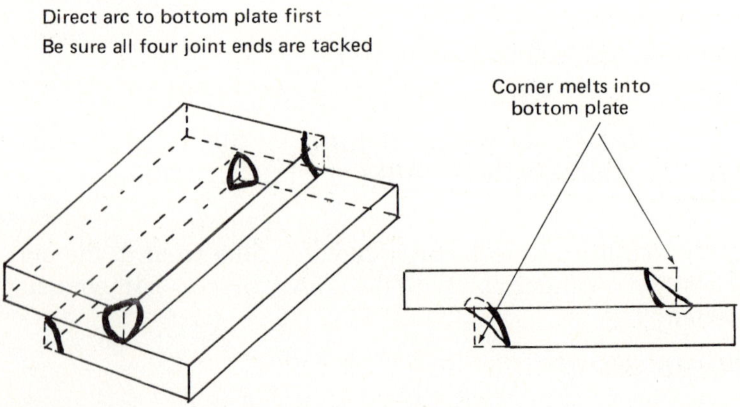

Direct arc to bottom plate first
Be sure all four joint ends are tacked
Corner melts into bottom plate

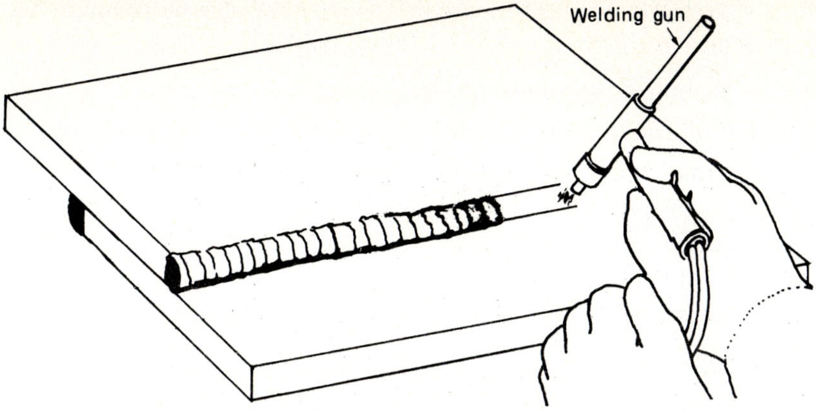

Welding gun

Figure **15-37**

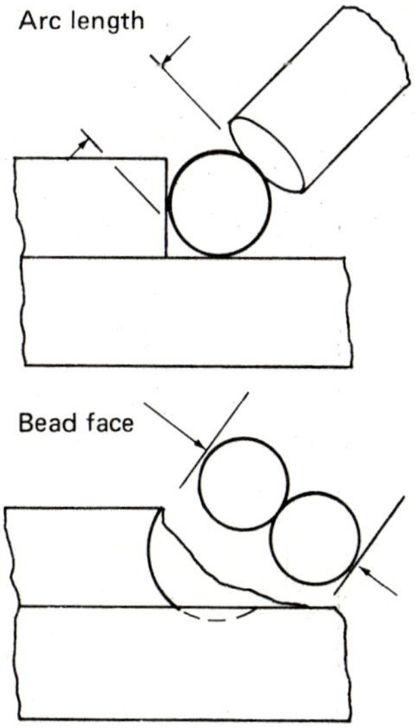

Arc length
Bead face

Figure **15-38**

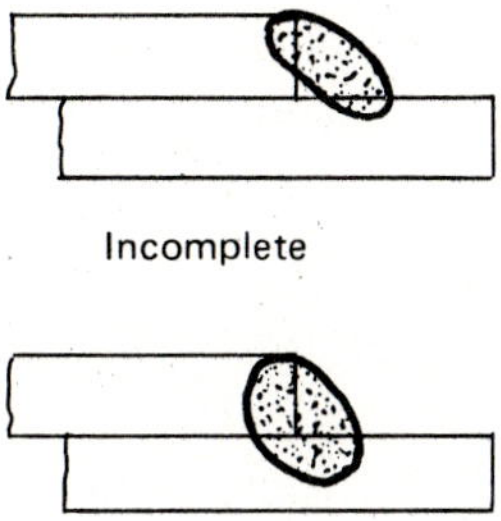

Penetration
Incomplete
Complete

Figure **15-39**

15-20 MAKING A FILLET-WELD LAP JOINT IN THE HORIZONTAL POSITION ON ALUMINUM USING FILLER METAL

OBJECTIVE To learn to make an acceptable fillet weld on aluminum in the horizontal position using filler metal.

HOURS REQUIRED 1.

EQUIPMENT, TOOLS, AND MATERIALS REQUIRED Same as in exercise 15-16, plus filler metal as recommended in Table 15-11.

PROCEDURE Read and become thoroughly familiar with the procedure below prior to welding.

1. Repeat steps 1 through 9 of exercise 15-19.

2. Position the torch and filler metal as shown in Figure 15-37.

3. Start the arc and, using filler metal, proceed as described in step 10 of exercise 15-19.

4. Complete the welds, lay down the torch, lift your helmet, and refer to Figure 15-39 for weld quality.

5. If necessary, repeat this exercise; otherwise shut down the machine.

15-21 MAKING A FILLET WELD TO FORM AN OUTSIDE CORNER IN THE FLAT POSITION ON ALUMINUM

OBJECTIVE To learn to make an acceptable fillet weld to form an outside corner in the flat position on aluminum.

HOURS REQUIRED 1.

EQUIPMENT, TOOLS, AND MATERIALS REQUIRED Same as in exercise 15-20 plus an angle iron of appropriate size for assembling the two pieces of aluminum to be joined and a C clamp.

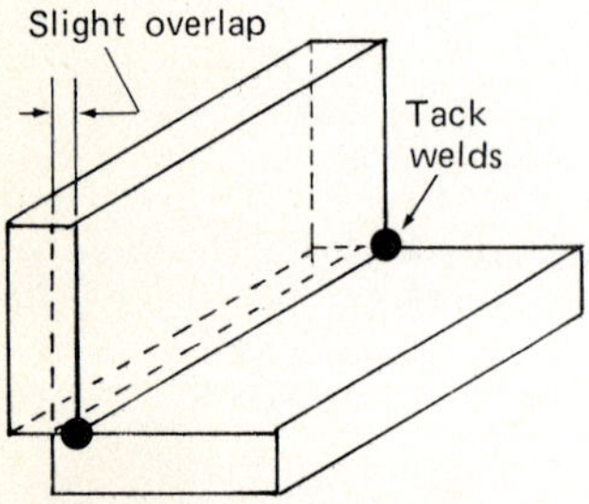
Slight overlap
Tack
welds

Figure **15-40**

PROCEDURE Before beginning to weld, read and become thoroughly familiar with the procedure below.

1. Position the two pieces on the welding table as shown in Figure 15-40.

2. Set the welding machine to the conditions determined from Table 15-11.

3. Turn on the welding machine.

4. Tack weld both ends of the joint.

5. Remove the C clamp from the tack-welded assembly and the tack-welded assembly from the angle iron.

6. Place the C clamp and angle iron in an out-of-way location.

7. Position the workpiece on the welding table.

8. Position the nozzle near the far end of the joint and, holding the torch at a 60-degree angle *with no side angle,* start the arc and make the weld [Figures 15-41(*a*) and (*b*)].

a. Position the arc and hold the arc length as shown in Figure 15-42.

b. Adjust your travel speed so that the arc produces rounded edges in the joint, as shown in Figure 15-43.

9. Inspect the weld. A quality joint, as shown in Figure 15-44, should appear.

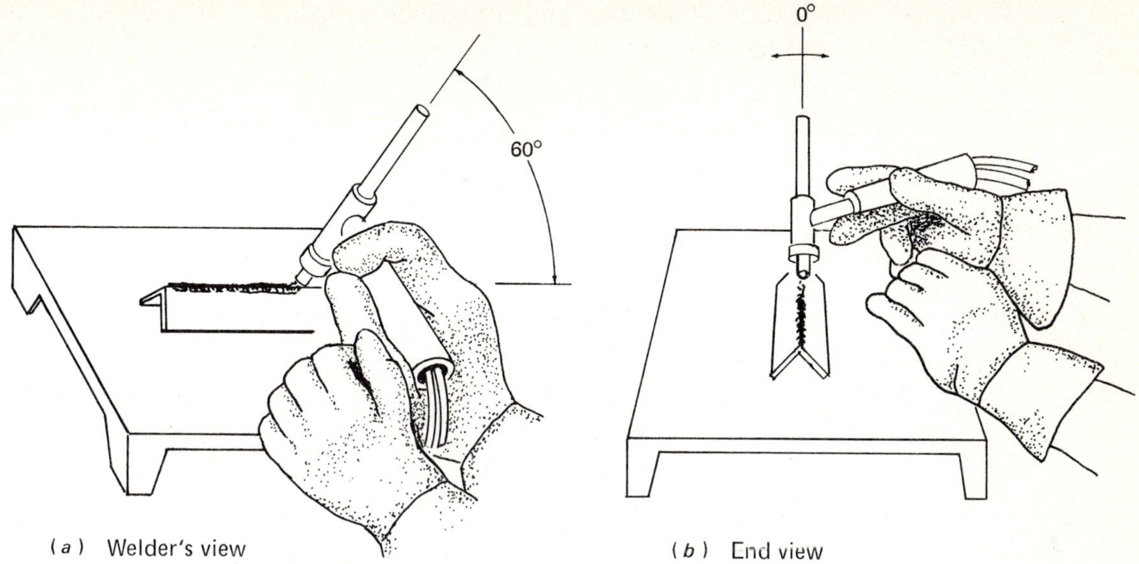
60°
0°
(a) Welder's view
(b) End view

Figure **15-41** (Filler metal not shown)

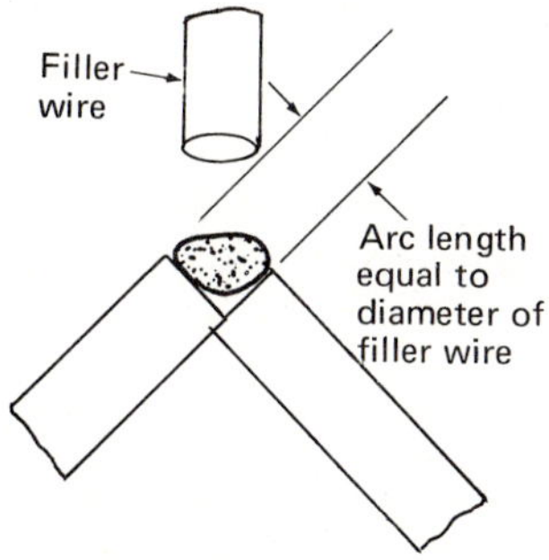
Filler wire
Arc length equal to diameter of filler wire

Figure **15-42**

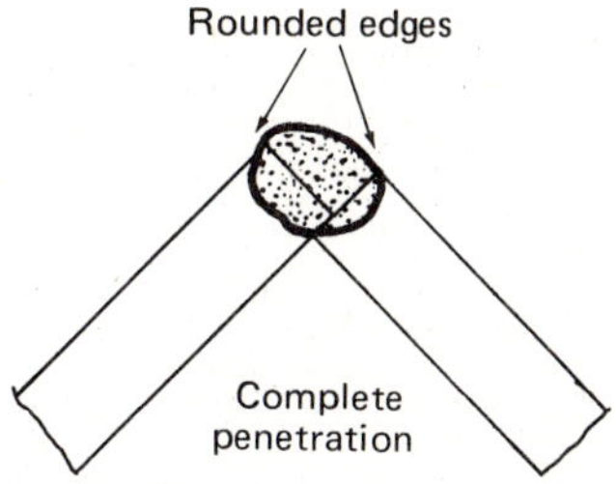
Rounded edges
Complete penetration

Figure **15-43**

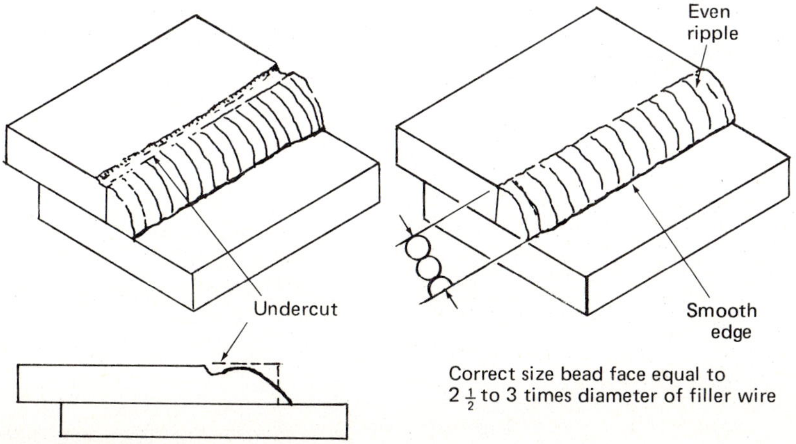
Undercut
Even ripple
Smooth edge
Correct size bead face equal to 2 ½ to 3 times diameter of filler wire

Figure **15-44**

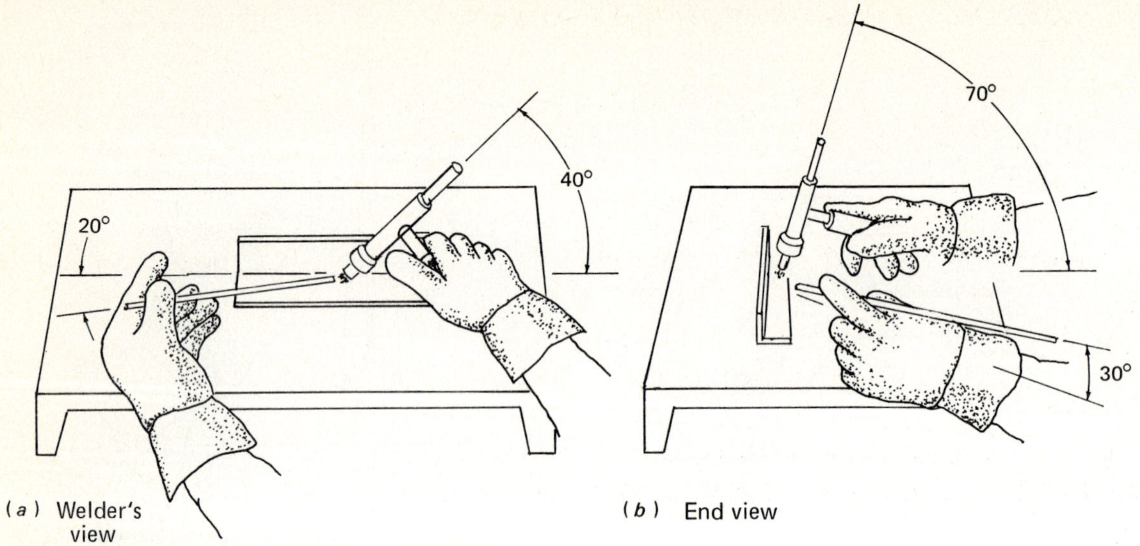
20°
40°
(a) Welder's view
70°
30°
(b) End view

Figure **15-45**

15-22 WELDING AN INSIDE CORNER JOINT IN THE HORIZONTAL POSITION ON ALUMINUM

OBJECTIVE To learn to weld an acceptable inside corner joint in the horizontal position on aluminum.

HOURS REQUIRED 1.

EQUIPMENT, TOOLS, AND MATERIALS REQUIRED The equipment, tools, and materials required for this exercise are the same as those in exercise 15-21.

PROCEDURE Read and become thoroughly familiar with the procedure below before beginning to weld.

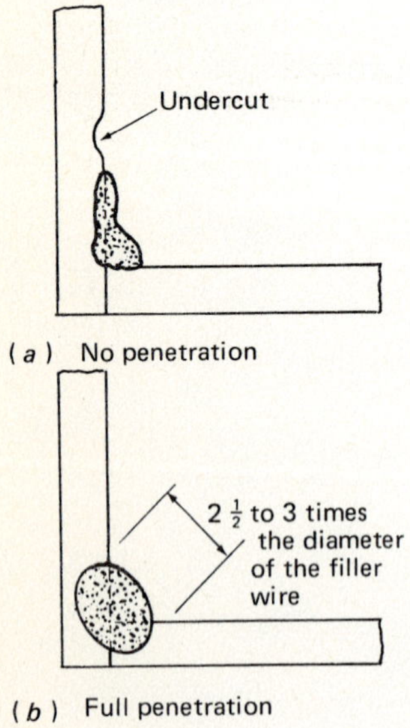
Undercut
(a) No penetration
2 ½ to 3 times the diameter of the filler wire
(b) Full penetration

Figure **15-46**

1. Perform steps 1 through 7 of exercise 15-21.
2. Position the tack-welded pieces on the work table.
3. Holding the torch and rod as shown in Figures 15-45(*a*) and 15-45(*b*), and using same welding movements as used for lap joints, weld toward the far end.
   a. Adjust your travel speed so that the width of the deposited bead is 2½ to 3 times the diameter of the electrode (Figure 15-46).
   b. Adjust the arc length so that full penetration is obtained with no undercut.

15-23 MAKING BEAD WELDS IN THE VERTICAL POSITION ON ALUMINUM

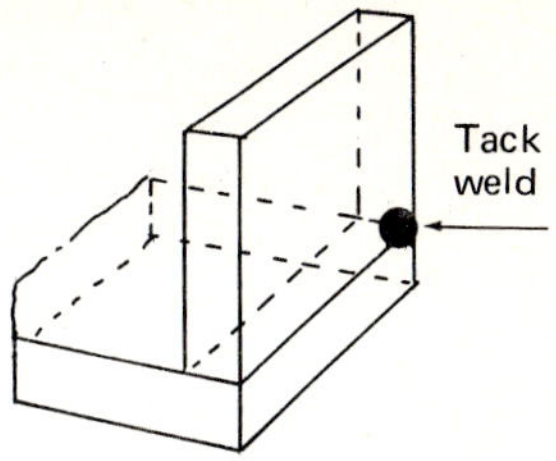
Tack
weld

Figure **15-47**

OBJECTIVE To learn to make acceptable welds in the vertical position on aluminum.

HOURS REQUIRED 1.

EQUIPMENT, TOOLS, AND MATERIALS REQUIRED The equipment, tools, and materials required to perform this exercise are the same as those in exercise 15-21, plus the weight used in exercise 15-20.

PROCEDURE Read and become thoroughly familiar with the procedure below before beginning to weld.

1. Perform steps 1 through 3 of exercise 15-21.

2. Tack the pieces together at one outside corner only.

3. Remove the C clamp from the tack-welded assembly and the tack-welded assembly from the angle iron (welding fixture).

4. Place the C clamp and angle iron in an out-of-way location.

5. Place the tack-welded assembly on the work table with one leg in the vertical-up position (Figure 15-47) and place the weight on the leg resting on the table.

6. Holding the rod and torch as shown in Figure 15-48, start welding at the bottom of the plate and, using the manipulative techniques described in exercise 15-17 and welding upwards, complete the weld.

7. Turn off the welding machine.

8. Inspect the weld.

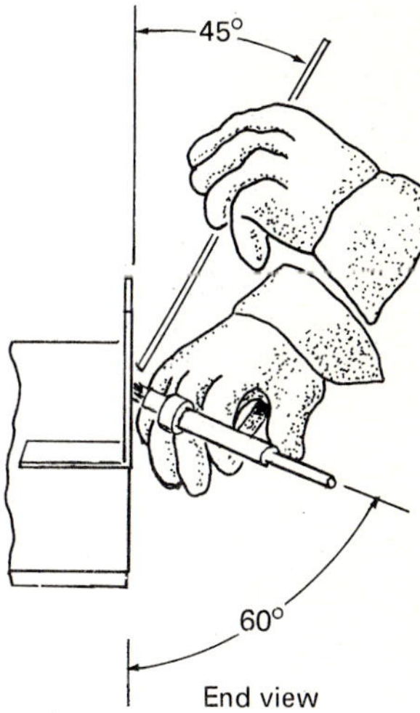
45°
60°
End view

Figure **15-48**

REVIEW QUESTIONS

1. Name the two methods by which metal is transferred in the GMAW process.

2. Compare the shade of the lens worn by MIG welders to that worn by shielded metal–arc welders.

3. What is meant by *slope* in the GMAW process? What is the function of slope?

4. Why should the use of the spray-arc method of metal transfer be restricted to heavy weldments in the flat position?

5. Name all the components and accessories of a MIG welding outfit.

6. What is meant by *electrode stick-out?*

7. What type of electric current is most frequently used for the GMAW process?

8. In addition to feeding wire (filler metal), what other function does the wire-feed control perform?

9. What conditions should be met to weld the metals listed below in the indicated thicknesses in the flat, horizontal, and vertical positions with the MIG welding process?

a. *Aluminum:* 0.050 inch; 0.375 inch
b. *Carbon steel:* 1/8 inch; 5/16 inch
c. *Copper:* 3/8 inch

10. Why is it advisable to learn the manipulative techniques necessary for welding with a fuel-gas process before attempting to weld with the GTAW process?

11. Can a power supply used or designed for shielded metal–arc welding be used for TIG welding? If your answer is yes, will the duty cycle stay the same? Explain.

12. Name the essential components of the GTAW process.

13. Design a joint and list the materials (in the thicknesses indicated) for welding with the TIG process in the flat and vertical positions.

a. *Aluminum:* 1/8-inch thick
b. *Stainless steel:* 1/4-inch thick
c. *Silicon bronze:* 3/8-inch thick

14. List the four steps in the TIG preweld checking procedure that must definitely be observed to avoid damaging the equipment and/or electrodes.

15. What are the proper filler rods and torch angles for TIG welding aluminum in the flat position?

16. What precaution must be taken when installing a flow meter on a gas cylinder?

17. How does the diameter of the water-cooled cable to the torch compare with the diameter of the ground or work cable used in the TIG welding process?

18. What effect does a loose gas cap or ceramic nozzle have on the electrode and work zone?

19. Why is argon recommended as the shielding gas for most TIG welding?

CHAPTER 16
FUEL-GAS (OXYACETYLENE) WELDING AND CUTTING PRACTICE

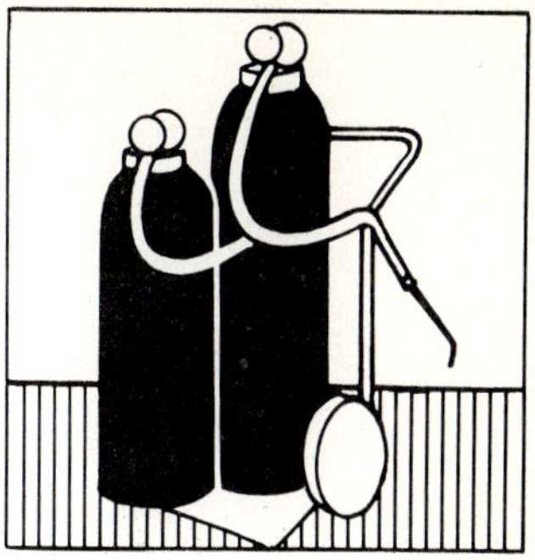

The following is a series of exercises in oxyacetylene welding. Before beginning these exercises, be sure to review the safe practices discussed in Chapter 13, and specifically:

1. Always wear goggles with suitable filter lenses when using a lighted torch.

2. Wear gauntlet-style gloves of heat-resistant leather to protect your hands and wrists.

3. Take care that your clothing is not oily, and that pockets and cuffs are not open and ready to receive sparks or hot slag.

4. Wear a heat-resistant helmet or skull cap.

5. Do not use equipment which you suspect is defective.

6. Never use a match or hot metal to light or relight a torch.

7. Never use acetylene at pressures above 15 psi gage.

8. Always open oxygen-cylinder valves fully.

9. Never open acetylene cylinder valves more than 1½ turns.

10. Use only the wrench supplied with the cylinder to open a cylinder valve.

11. Always keep the acetylene cylinder–valve wrench on the cylinder valve until the job is done and the hose is drained.

12. Keep appropriate fire extinguishers handy at all times.

16-1 ASSEMBLING THE EQUIPMENT

OBJECTIVE To learn to assemble the oxyacetylene welding equipment in accordance with standard procedures.

HOURS REQUIRED 1/4.

EQUIPMENT, TOOLS, AND MATERIALS REQUIRED The equipment, tools, and materials required for this exercise consist of a welding outfit, as shown in Figure 16-1, and a bucket of soapy water.

PROCEDURE Read the instructions below thoroughly before proceeding.

1. Assemble the equipment for lighting:
   a. Fasten the cylinder in a vertical position.
   b. Turn the acetylene cylinder so that the valve outlet will point away from the oxygen cylinder.

Figure **16-1**

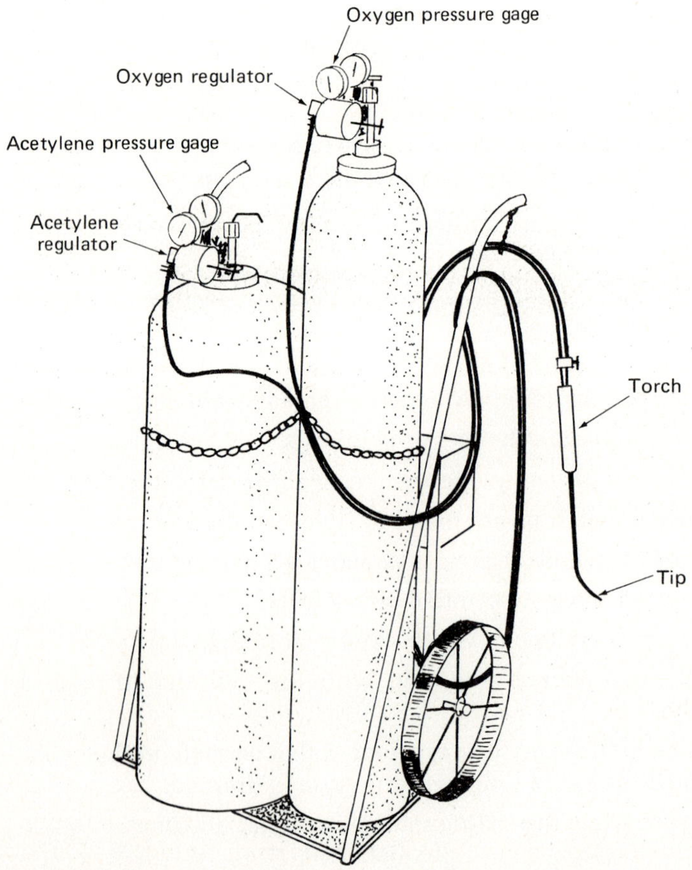
Oxygen pressure gage
Oxygen regulator
Acetylene pressure gage
Acetylene
regulator
Torch
Tip

c. Remove the protective bottle caps from the cylinders.
d. Crack the valves on each cylinder to remove dirt from the nozzles.
e. Connect the oxygen regulator to the oxygen cylinder.
f. Connect the acetylene regulator to the acetylene cylinder.
g. Connect the acetylene hose to the acetylene regulator.
h. Connect the oxygen hose to the oxygen regulator.
i. Connect the torch to the oxygen and acetylene hoses.
j. Screw the tip into the torch body.
k. Loosen the adjusting screw by turning counterclockwise on both the oxygen and acetylene regulators.
l. Open the oxygen-cylinder valve fully.
m. Open the acetylene-cylinder valve one-half turn.
n. Turn the adjusting screw on the gas-pressure regulators clockwise until the correct pressure is indicated on the pressure-gage scales.
o. Test all the fitting connections for leaks with soapy water.

2. Secure equipment:
a. Close the acetylene-cylinder valve.
b. Close the oxygen-cylinder valve.
c. Open the acetylene valve on the torch to release the pressure.
d. Release the regulating screw on the acetylene regulator by turning counterclockwise.
e. Close the acetylene valve on the torch.
f. Open the oxygen valve on the torch to release the pressure.
g. Release the regulating screw on the oxygen regulator by turning counterclockwise.
h. Close the oxygen valve on the torch.
i. Put the torch and hose on a stowage rack, not on the regulators.

16-2 LIGHTING AND SHUTTING OFF THE CUTTING AND WELDING TORCH

OBJECTIVE To learn how to light and shut off the cutting and welding torch in accordance with standard procedures.

HOURS REQUIRED ¼.

EQUIPMENT, TOOLS, AND MATERIALS The equipment, tools, and materials required to perform this exercise consist of the welding outfit shown in Figure 16-1.

PROCEDURE Study the instructions below until you are thoroughly familiar with them.

1. Preparation:
a. Open the acetylene valve slowly, approximately one and one-half turns. (It has been found that maximum acetylene flow is

reached when the cylinder valve has been opened one and one-half turns). *Note:* Do not remove the wrench from the cylinder valve.

b. Open the oxygen-cylinder valve slowly.

2. Fuel gas:

a. Open the fuel-gas (acetylene) inlet-valve on the torch (*see* Figure 16-2).

b. Turn the pressure-regulating screw on the pressure regulator [Figure 16-3(*a*)] clockwise until the needle on the delivery (working) pressure gage shows the desired pressure.

c. Close the acetylene valve on the torch.

3. Oxygen:

a. Open the oxygen-inlet valve on the torch (Figure 16-2).

b. Turn the pressure-regulating screw on the pressure regulator until the desired pressure is reached [Figure 16-3(*a*)].

c. Close the oxygen valve on the torch.

4. Lighting the torch:

a. Open the acetylene-inlet valve on the torch.

b. Light the torch (Figure 16-4).

c. Open the oxygen valve on the torch.

5. Adjusting the flame (Figure 16-5, page 396):

a. Adjust to a neutral flame.

b. Adjust to a carburizing flame.

c. Adjust to an oxidizing flame.

d. Adjust to a neutral flame.

6. Shutting off the torch:

a. Close the acetylene-inlet valve on the torch.

b. Close the oxygen-inlet valve on the torch.

c. Close the acetylene-cylinder valve.

d. Close the oxygen-cylinder valve.

e. Open the acetylene valve on the torch to release the pressure.

f. Release (back off) regulating screw on the acetylene regulator by turning counterclockwise.

g. Close the acetylene valve on the torch.

h. Open the oxygen valve on the torch to release pressure.

i. Release (back off) regulating screw on the oxygen regulator by turning counterclockwise.

j. Close the oxygen valve on the torch.

k. Put the torch and hose on a stowage rack.

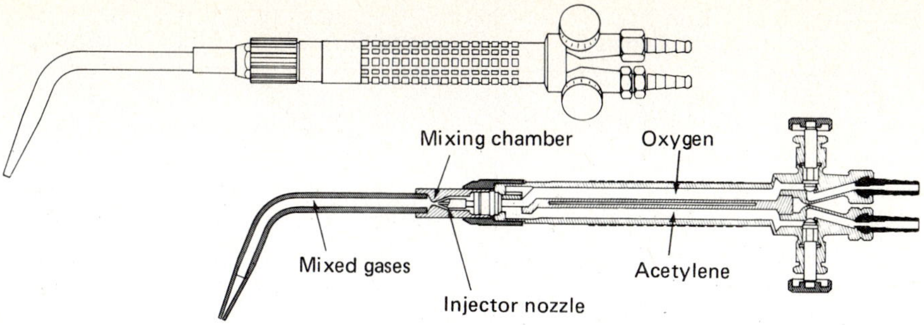
Mixing chamber
Oxygen
Mixed gases
Acetylene
Injector nozzle

Figure **16-2**

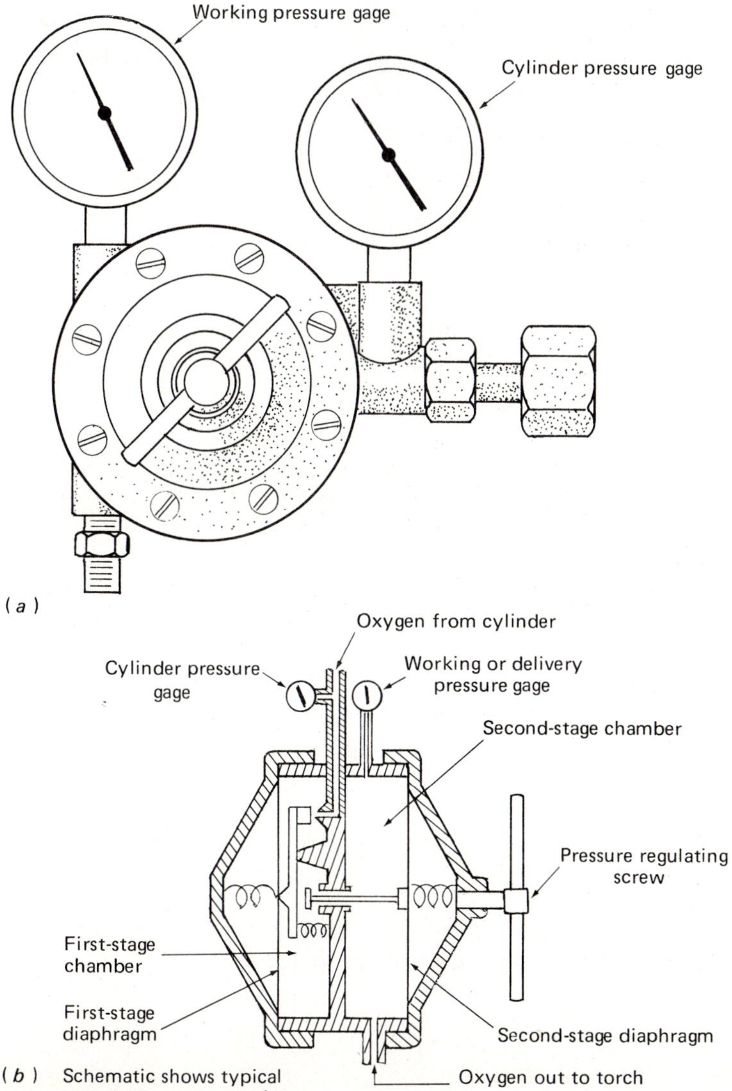
Working pressure gage
Cylinder pressure gage
(a)
Oxygen from cylinder
Cylinder pressure gage
Working or delivery pressure gage
Second-stage chamber
Pressure regulating screw
First-stage chamber
First-stage diaphragm
Second-stage diaphragm
Oxygen out to torch

(*b*) Schematic shows typical two-stage oxygen regulator

Figure **16-3**

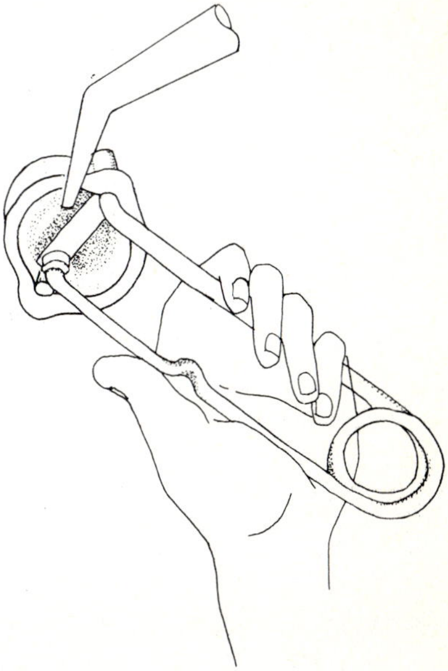

Figure **16-4**

Torch flames	Ratio O/A	Effect on metal
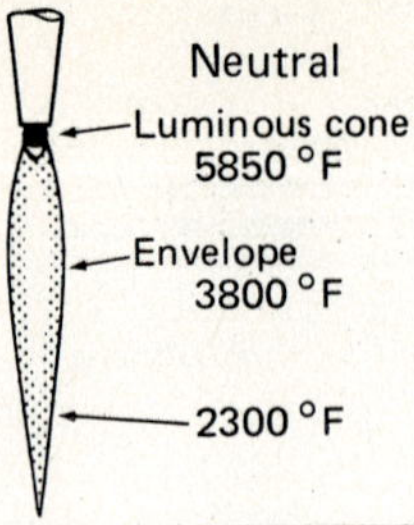Neutral Luminous cone 5850 °F Envelope 3800 °F 2300 °F	$\frac{1.04 - 1.14}{1}$	Metal is clean and clear, flowing easily
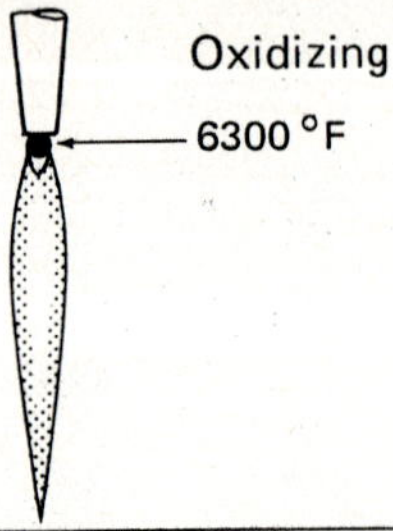Oxidizing 6300 °F	$\frac{1.15 - 1.70}{1}$	Excessive foaming and sparking of metal
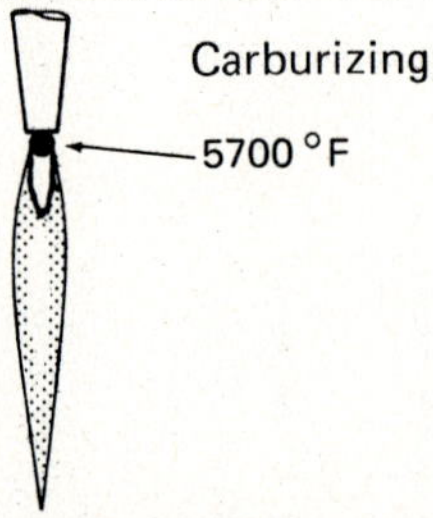Carburizing 5700 °F	$\frac{0.85 - 0.95}{1}$	Metal boils and is not clear

For complete combustion of acetylene - One molecular volume (380 cubic feet at 60° F) of acetylene plus 2 molecular volumes of oxygen burns to form 2 molecular volumes of carbon dioxide plus 1 molecular volume of water vapor liberating 542,700 Btus of heat.

Figure **16-5**

16-3 LAYING BEADS ON FLAT PLATE WITHOUT USING FILLER METAL

OBJECTIVE To learn to carry a bead without using filler metal.

HOURS REQUIRED 4.

EQUIPMENT, TOOLS, AND MATERIALS REQUIRED The equipment, tools, and materials required to perform this exercise are:

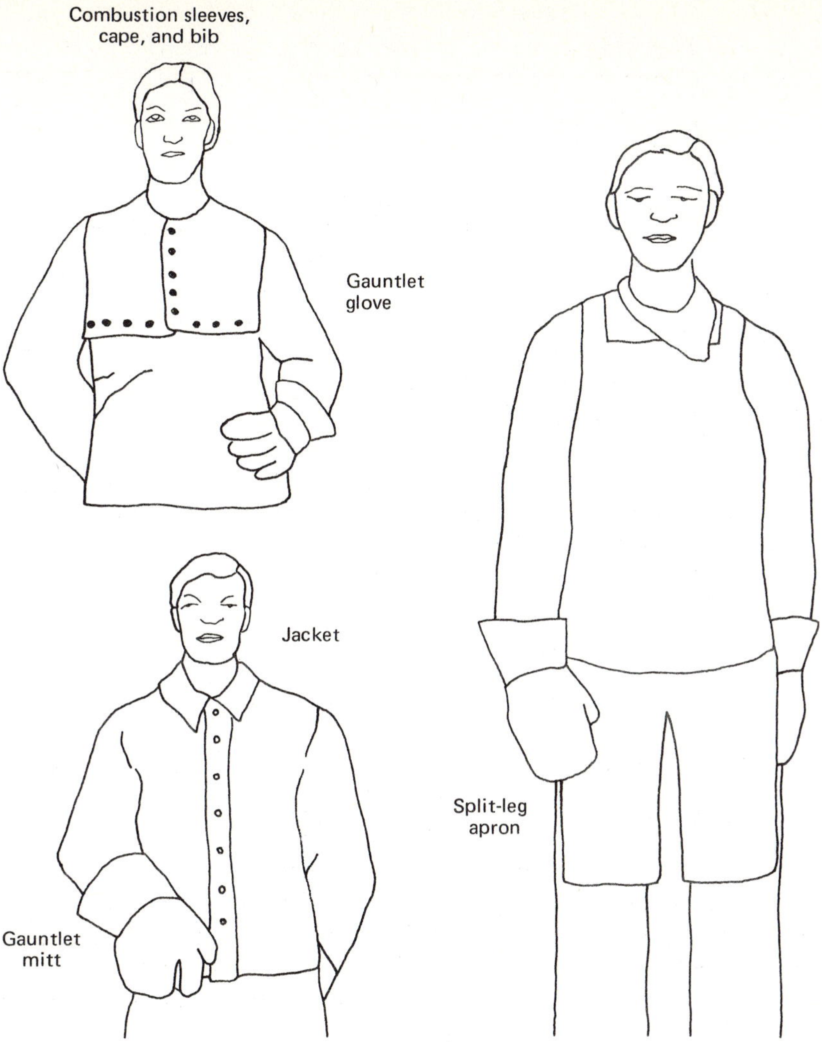
Combustion sleeves, cape, and bib
Gauntlet glove
Jacket
Gauntlet mitt
Split-leg apron

Figure **16-6**

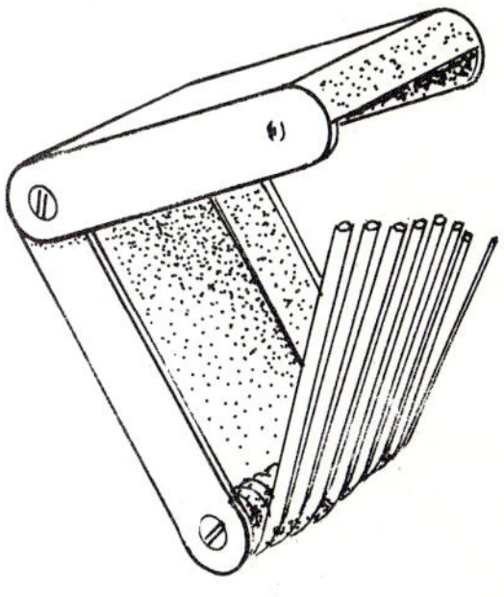

Figure **16-7**

1. Protective clothing (Figure 16-6).
2. Oxyacetylene welding outfit.
3. Welding torch with a proper-size tip.
4. Firebrick.
5. Tip cleaner (Figure 16-7).
6. Several $2 \times 4 \times \frac{1}{8}$-inch pieces of hot-rolled mild steel.

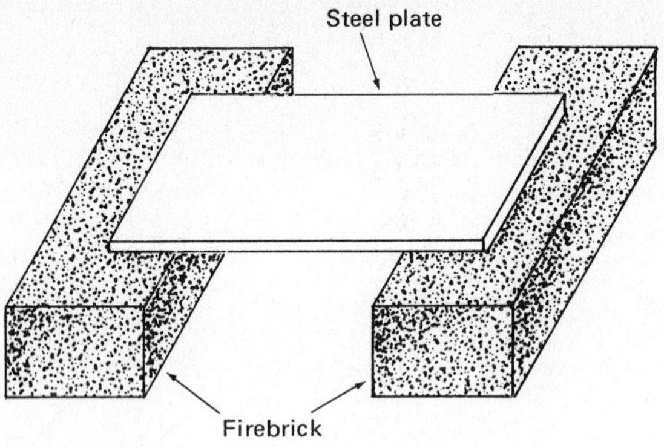
Steel plate
Firebrick

Figure **16-8**

PROCEDURE Read the instructions below until you are thoroughly familiar with them.

1. Select the proper-size torch tip.

2. Place the steel plate on the firebrick (Figure 16-8).

3. Light the torch and adjust to a neutral flame.

4. Work from right to left if right handed, and melt the base metal to form a molten puddle; reverse the procedure if left handed [Figure 16-9(*a*)].

5. Ensure that the inner cone does not touch the molten puddle. [Keep it about $\frac{1}{16}$ to $\frac{1}{8}$ inch away from the puddle, as in Figure 16-9(*b*) and (*c*).]

Figure **16-9**

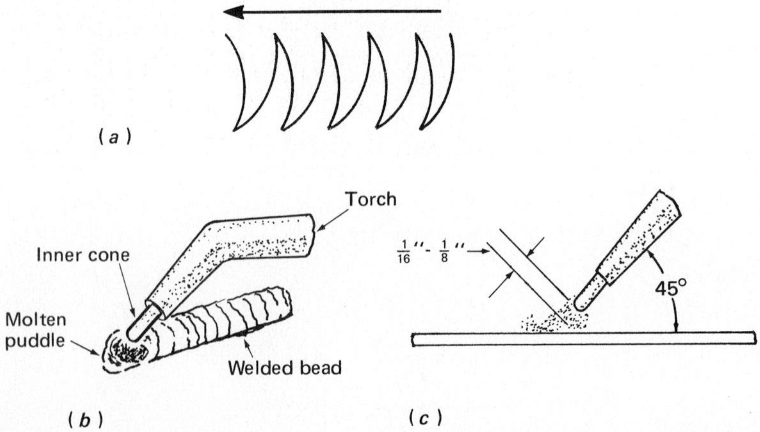
(a)
Torch
Inner cone
Molten puddle
Welded bead
(b)
1/16"- 1/8"
45°
(c)

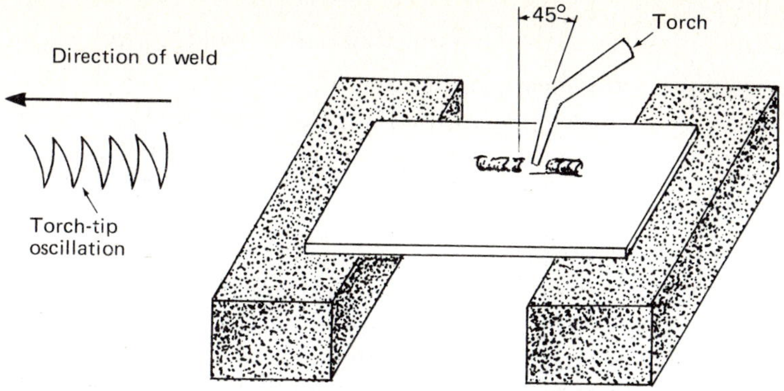
45°
Torch
Direction of weld
Torch-tip
oscillation

Figure **16-10**

6. When the puddle is formed, move the torch forward with an oscillating motion (Figure 16-10).

7. Ensure that both the forward motion and the oscillating motion are uniform.

8. Obtain a good steady speed in the movement of the torch and continue the weld the full length of the metal plate.

a. A large puddle will be obtained in places where the speed was slow. When the speed is too slow, there is a danger of overheating the part and burning a hole in it.

b. A small puddle will be obtained if the speed is too fast. When the speed is too fast, the result will be little or no fusion.

c. The results of incorrect speed are shown in Figure 16-11.

d. A correct, satisfactory bead must be uniform in width and slightly depressed below the surface. The underside of the metal must be free from icicles, and a thin film of oxide will cover the top and bottom surfaces.

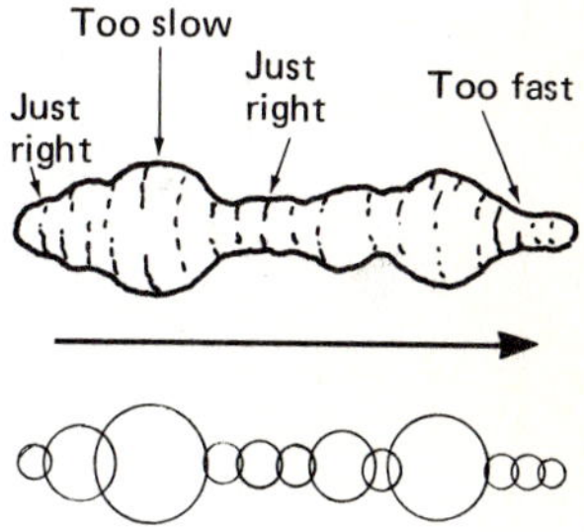
Too slow
Just
right
Just
right
Too fast

Figure **16-11**

9. When through with the first bead, make three or four more beads and check each one against the first bead. (It takes practice to be able to make a good welded bead.)

10. Shut off the torch and clean torch tip using the correct-size tip cleaner.

11. Present the welds to your instructor for approval.

16-4 LAYING BEADS ON FLAT PLATE USING FILLER METAL

OBJECTIVE To learn to carry a bead in the flat position using a welding rod.

HOURS REQUIRED 4.

EQUIPMENT, TOOLS, AND MATERIALS REQUIRED

1. A ⅛-inch-diameter AWS Classification GA50 welding rod.

2. The welding outfit shown in Figure 16-1.

3. Several 2 × 4 × ⅛-inch hot-rolled mild-steel plates.

PROCEDURE Read the instructions below until you are thoroughly familiar with them.

1. The position of the metal on the firebricks and the manipulative skills required to perform this exercise are the same as those for exercise 16-3, except that in this exercise the bead will be built up above the surface of the sheet and efforts must be made to keep the height, width, and spacing of the ripples uniform.

2. The position of the welding rod is similar to that of the blowpipe except that the welding rod is held with the opposite hand (Figure 16-12).

3. To bring both the base metal and welding rod to melting temperature at the same time, hold the welding rod just inside the outer envelope of the flame while concentrating the flame on a spot of the base metal at the beginning of the bead.

4. As soon as the metal begins to melt, feed the welding rod into the molten puddle, simultaneously moving the torch tip across the base metal.

Figure **16-12**

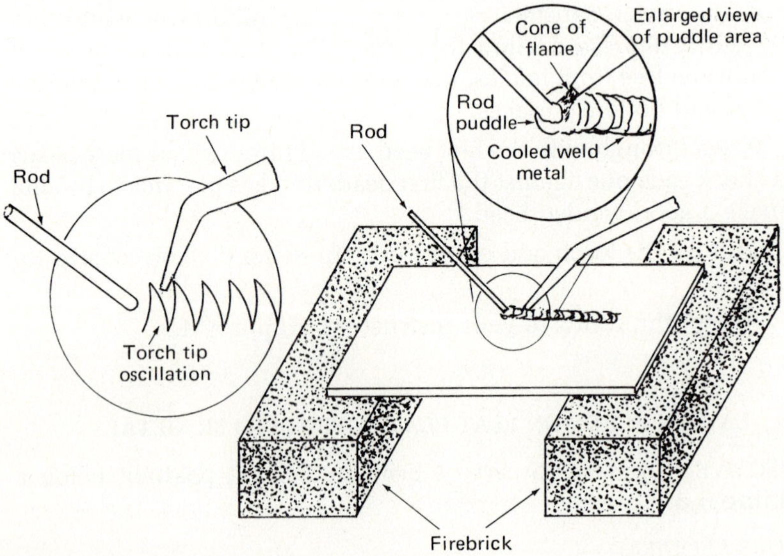

Cone of flame
Enlarged view of puddle area
Rod
puddle
Cooled weld metal
Torch tip
Rod
Rod
Torch tip oscillation
Firebrick

5. If the flame is not aimed directly at the welding rod, the rod may stick (freeze or weld) to the base metal. If this occurs, do not jerk the welding rod loose, simply aim the tip of the torch at it and the rod will immediately loosen itself from the base metal.

6. If the flame is held too long at one spot, the molten puddle will become too large and may even melt through. If the puddle becomes too large, withdraw the welding rod from the puddle and raise the flame until only the outer envelope of the flame is in contact with the puddle. Removing the flame entirely from the puddle will allow air to strike the base metal and oxidize it. If the puddle melts through, try filling in the hole before continuing.

7. Before proceeding to the next step, notice how the welded sheets have curled and buckled after the weld was completed. If you pick up the sheets with a pair of pliers and examine their undersides you will notice that there is a certain amount of oxidation and scaling directly under the weld. You will also notice that where the welding action was slow and the puddle large, and consequently deep, there is more oxidation than in places where the puddle was shallow.

16-5 WELDING A SQUARE-GROOVE BUTT JOINT IN THE FLAT POSITION

OBJECTIVE To learn to weld square-groove butt joints in the flat position.

HOURS REQUIRED 2.

EQUIPMENT, TOOLS, AND MATERIALS REQUIRED The equipment, tools, and materials required to perform this exercise are the same as for exercise 16-4, plus a ball-peen hammer and an anvil.

PROCEDURE Read the instructions below until you are thoroughly familiar with them.

1. Select the proper size of torch tip for welding ⅛-inch-thick steel plate (see Table 16-1).

2. Place two pieces of steel plate flat on the firebrick (Figure 16-13, page 403), leaving a slight gap for the root opening (approximately 1/16 inch).

3. Light the torch and adjust it to a neutral flame.

4. Heat both pieces of plate, starting at the left and working towards the right if right handed; the reverse if left handed. (This is called *backhand welding.*)

5. Keep the cone of the flame pointed toward the leading edge of the molten puddle, making sure that the inner cone does not touch the puddle.

Table **16-1** Torch-tip sizes for various metal thicknesses

THICKNESS OF METAL*		OXWELD WELDING BLOWPIPES WELDING HEAD SIZES*					OTHER WELDING BLOWPIPES
IN.	GA.	W-15	W-29	W-47	W-17 W-22	W-26 HEATING	
	32	1(6–7)	1(5–6)				
	28	1(16–24)	2(7–8)	2(5)			
	25	2(16–24)	2(7–10)	2(5)	4(7–12)		
			4(7–12)				
1/32	22	2(16–24)	4(7–18)	4(5)	4(7–18)		
		3(16–24)					
1/16	16	3(16–24)	6(8–20)	6(5)	6(8–20)		
		4(16–24)					
3/32	13	4(16–24)	6(15–20)	6(5)	6(15–20)		
		5(16–24)	9(10–18)	9(5)	9(10–18)		
1/8	11	5(16–24)	9(12–24)	9(5)	9(12–24)		
		6(16–24)	12(11–19)	12(5)	12(11–19)		
		7(16–24)					
3/16		6(16–24)	12(16–25)	12(5)	12(16–25)		
		7(16–24)	15(11–19)	15(5)	15(11–18)		
1/4			15(16–25)	15(5)	15(17–26)		
			20(20–29)	20(5)	20(12–23)		
3/8			20(29–34)	20(5)	30(18–29)		
			30(30–40)	30(5)			
1/2				40(5)	40(24–33)		
5/8				55(5)	55(25–32)		
3/4				70(5)	55(30–39)		
					70(22–30)		
1				85(6)	90(30–42)		
1½				100(7)		100(50–60)	
2				125(10)		125(50–60)	
3				150(12)		150(50–60)	
Heating				200(14)		175(55–65)	
				250(22)		250(60–70)	
				300(28)			

*Oxygen pressure range in parentheses. Acetylene pressures 5 lb. per sq. in. except when supplied from low-pressure generators.

Source: Union Carbide Corporation, Linde Division, *The Oxyacetylene Handbook*.

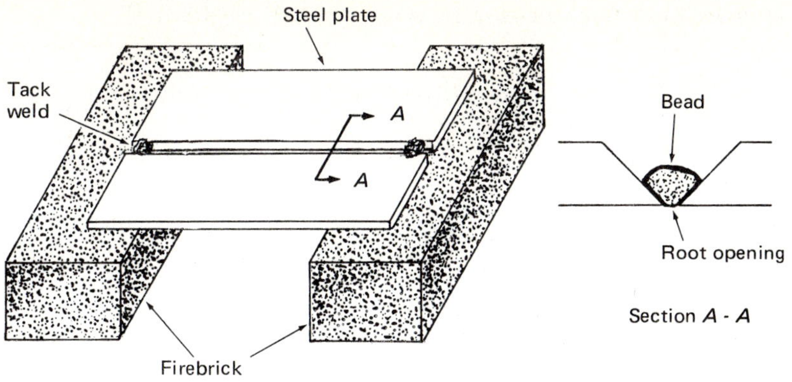
Steel plate
Tack weld
A
A
Bead
Root opening
Section A - A
Firebrick

Figure **16-13**

6. Add the filler rod, keeping the cone of the flame between the rod and the molten puddle, and tack weld at each end of the two pieces of steel (Figure 16-13).

7. Let the tack welds cool.

8. Begin at the right and heat both pieces of steel, forming a molten puddle. Watch for the keyhole (see step 9 in section 16-10, page 409).

9. Add the filler rod, keeping the cone of the flame between the rod and the molten puddle.

10. Oscillate the torch and tip in a slow motion, rolling the filler rod from side to side, and moving the weld from right to left (forehand) along the root opening.

11. Travel slowly enough to allow the base metal to thoroughly fuse throughout the cross-section of the groove and build it up with a 25% reinforcement. *Note:* As the end of the weld is approached, you will observe that the accumulation of heat causes the weld puddle to become enlarged. When this happens, stop welding and allow the weld to cool to a black color before continuing to weld. If you try to finish the weld without allowing it to cool, the molten metal will run away and it will be impossible to build up the required reinforcement.

12. Continue welding until both pieces of steel are completely fused.

13. Shut off the torch and clean the tip using the correct-size tip cleaner.

14. Test the weld by placing it on the raised portion of the anvil and hitting it with a hammer.

15. Present it to instructor for approval.

16-6 WELDING CORNER JOINTS WITH AND WITHOUT THE USE OF FILLER METAL

OBJECTIVE To learn to weld acceptable corner joints with and without the use of filler metal.

HOURS REQUIRED 2.

EQUIPMENT, TOOLS, AND MATERIALS REQUIRED The equipment, tools, and materials required to perform this exercise are the same as in exercise 16-4.

PROCEDURE *Note:* When two pieces of sheet metal come together at a corner, either of two common methods of designing the joint may be used to produce a weld with the desired characteristics of strength and tightness (Figure 16-14). The joint shown in Figure 16-14(*b*) can be made without filler metal by melting down the edges, although it is better for the student operator to use a welding rod and build up the weld to fill the corner.

1. Take two of the steel plates and set them up on the firebrick at right angles, as shown in Figure 16-14(*a*), or, if the instructor directs, as shown in Figure 16-14(*b*).

a. If filler metal is to be used, separate the edges to be welded by a space of 1/16 inch.

b. In any event, make sure that the plates are at an angle of 90 degrees to one another. If they are less than 90 degrees to one another, the hammer test will impose severe stresses likely to cause failure. If the plates are welded at more than 90 degrees, the weld will not be stressed enough when tested.

2. Light the torch and adjust it to a neutral flame.

Figure **16-14**

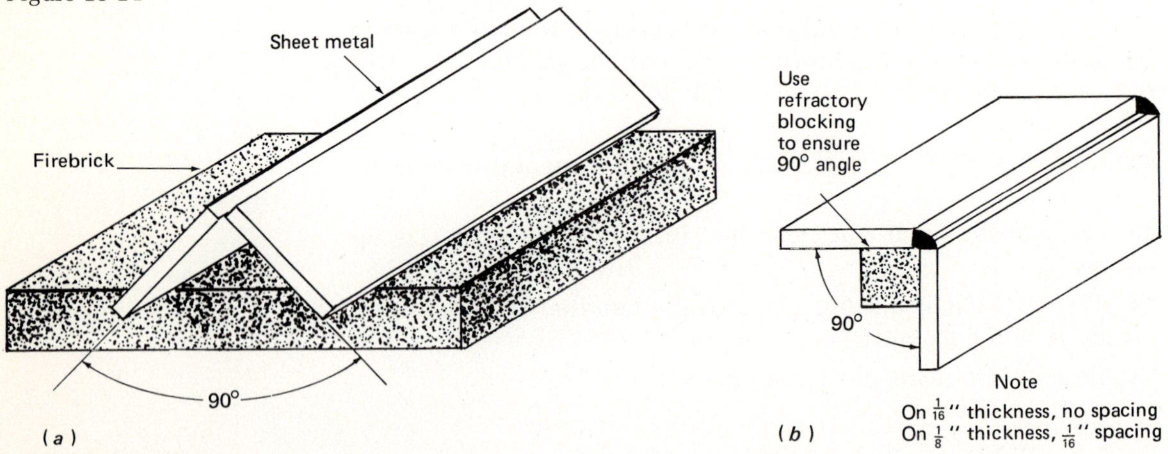

Sheet metal
Firebrick
90°
(a)
Use refractory blocking to ensure 90° angle
90°
(b)
Note
On 1/16" thickness, no spacing
On 1/8" thickness, 1/16" spacing

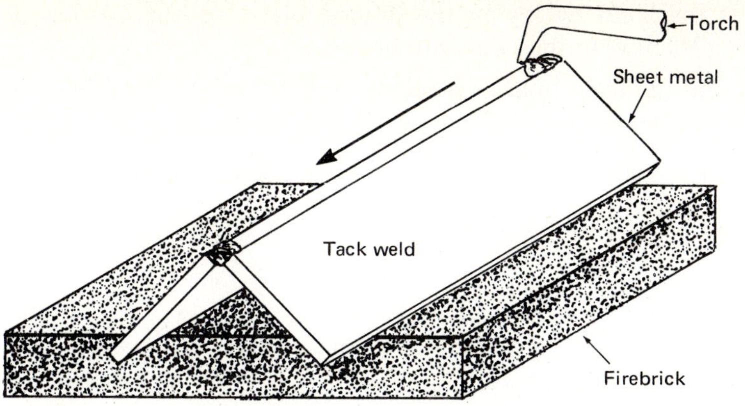
Torch
Sheet metal
Tack weld
Firebrick

Figure **16-15**

3. Working from righ to left (reverse, if left handed), heat the corner edges of the sheet metal to form a molten puddle.

4. Be sure that the inner cone of the flame does not touch the molten puddle.

5. Run the puddle approximately ¼ inch to the left, forming a tack weld.

6. Remove the torch and tack weld the opposite end.

7. Let the tack welds cool and place the assembly on the welding table, as shown in Figure 16-15.

8. Working from right to left, form a molten puddle along the corner edge towards the left. *Note:* Too fast a travel will result in fusing only across the upper surfaces of the V. (Figure 16-16 shows the results of right and wrong speeds of travel.)

9. Make one complete weld along the corner edges, ensuring good penetration. Do not overheat the metal.

10. Shut off the torch and clean the tip using the correct-size tip cleaner.

Figure **16-16**

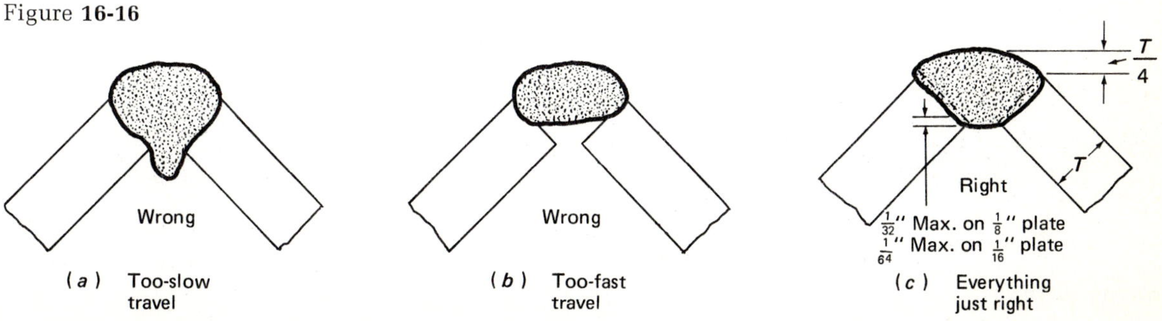
Wrong
(a) Too-slow travel
Wrong
(b) Too-fast travel
T
4
T
Right
1/32" Max. on 1/8" plate
1/64" Max. on 1/16" plate
(c) Everything just right

11. Test the weld by laying the jointed plates on the anvil and flattening them with the ball-peen hammer.

12. Present the weld to the instructor for approval.

16-7 MAKING A LAP JOINT IN THE FLAT POSITION

OBJECTIVE To learn to make an acceptable lap joint in the flat position.

HOURS REQUIRED 2.

EQUIPMENT, TOOLS, AND MATERIALS REQUIRED The equipment, tools, and materials required to perform this exercise are the same as in exercise 16-4.

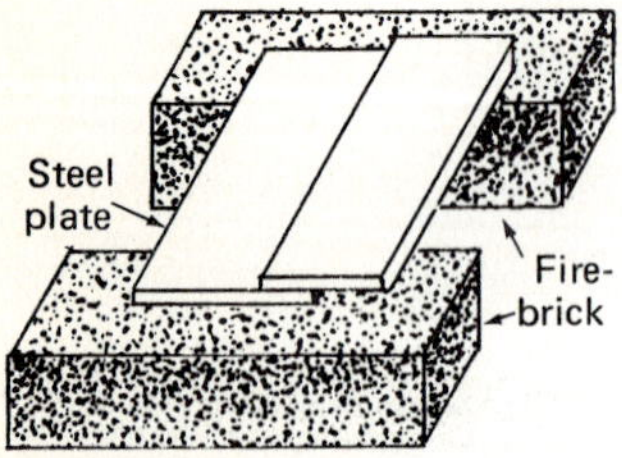

Steel plate
Fire-brick

Figure **16-17**

PROCEDURE

1. Set up two sheets so that the top one overlaps the lower one by half its width and tack weld the ends at the weld face (Figure 16-17).

2. Light and adjust the torch to a neutral or slightly carburizing flame.

3. Melt down the edge of the upper sheet for a distance inward of about half the puddle and make the fillet using forehand welding and an oscillating motion (Figure 16-18).

   a. To obtain complete penetration at the root, carry fusion forward in this area for a distance about equal to the puddle diameter before building the weld up to the required dimensions (Figure 16-19).

   b. If the torch is not held at the correct angle, the face of the weld opposite the side toward which the weld is inclined will most probably not be fused.

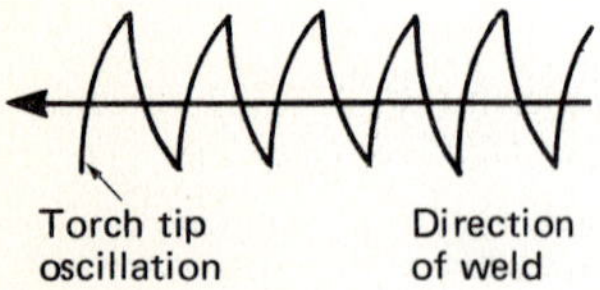

Torch tip oscillation
Direction of weld

Figure **16-18**

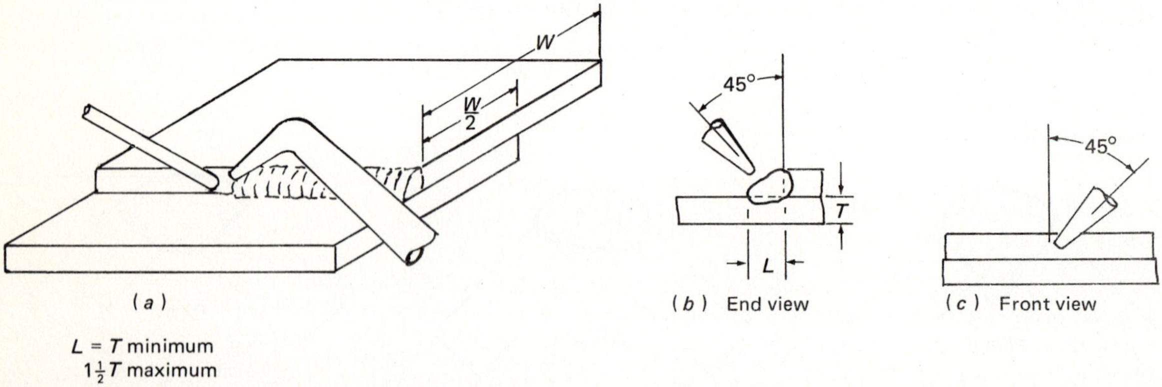

W
W/2
(a)
45°
T
L
(b) End view
45°
(c) Front view
L = T minimum
1½T maximum

Figure **16-19**

4. Continue to add filler rod and move the puddle and weld to the left until both pieces of metal are welded together.

5. Turn the welded pieces over on the firebrick and, using the same procedures as above, weld the reverse side.

6. Shut the torch off and clean the tip using the correct-size tip cleaner.

7. Test the weld by placing it on the raised portion of the anvil and hitting it with the hammer.

8. Present the weld to the instructor for approval.

16-8 MAKING FLANGE WELDS WITHOUT USING FILLER METAL

OBJECTIVE To learn to prepare metal for, and make flange welds in, sheet metal without the use of filler metal.

HOURS REQUIRED 2.

EQUIPMENT, TOOLS, AND MATERIALS REQUIRED The equipment, tools, and materials required to perform this exercise consist of:

1. Several 2 × 4 × 16-gage (0.062-inch-thick) pieces of hot-rolled mild-steel sheet.
2. Welding tip of appropriate size for the job.
3. Bench-mounted machinist vise.
4. 16-ounce ball-peen hammer.
5. Welding outfit.
6. Appropriate protective clothing.

PROCEDURE

1. Insert the metal in the vise, allowing the edge to be welded to protrude over the vise jaws a distance approximately equal to the thickness of the metal (1/16 inch in this case).

2. By hammering, bend the protruding part of the metal about 90 degrees over the vise jaws (Figure 16-20).

3. Place the two pieces of sheet metal on the welding table with the flanged-up edges just holding.

4. Light the blowpipe and obtain a neutral flame.

5. Hold the blowpipe so that the flame points the same way that you are going to weld, and make an angle of about 45 degrees with the horizon.

6. Use the welding rod and tack weld the sheets at both ends of the joint. *Note:* The rod and sheet should be brought to melting temperature at the same time. This is best done by holding the rod just inside the outer envelope of the flame while the flame is being concentrated on the starting spot of the weld.

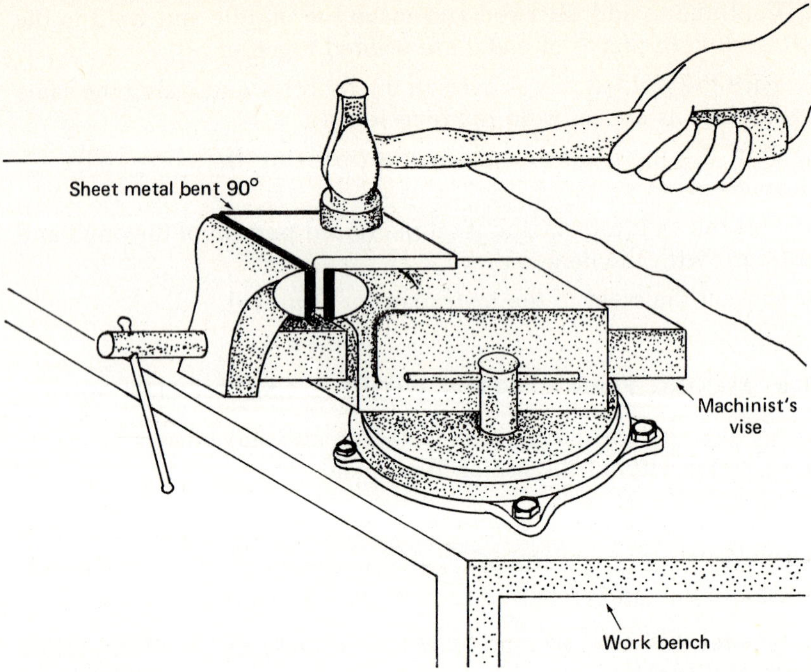
Sheet metal bent 90°
Machinist's vise
Work bench

Figure **16-20**

7. Return with the blowpipe to the first tack weld and hold it until a puddle of the proper size is formed (about twice the diameter of the hole in the tip of the blowpipe). Then carry the puddle as described in exercise 16-3.

8. Shut off the torch.

9. Present the weld to instructor for approval.

16-9 WELDING A T JOINT IN THE HORIZONTAL POSITION

OBJECTIVE To learn to weld an acceptable T joint.

HOURS REQUIRED 2.

EQUIPMENT, TOOLS, AND MATERIALS REQUIRED The equipment, tools, and materials required to perform this exercise are the same as in exercise 16-4.

PROCEDURE

1. Place one metal plate flat on a firebrick.

2. Place the second piece of metal plate on edge in a vertical position, lengthwise down the center of the horizontal plate (Figure 16-21).

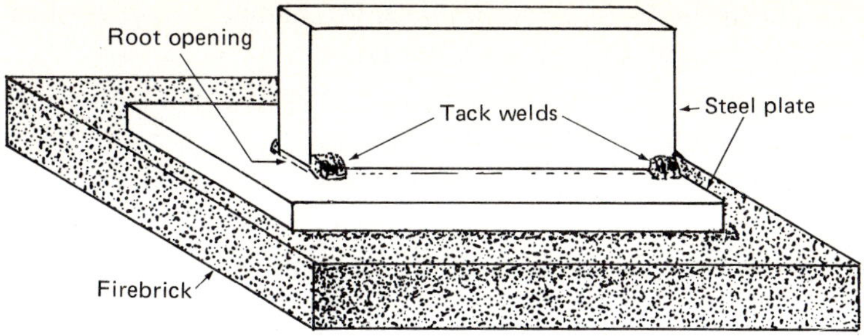
Root opening
Tack welds
Steel plate
Firebrick

Figure **16-21**

3. Shim the vertical plate up off the horizontal plate about 1/32 inch for the root opening.

4. Light the torch and adjust to a neutral flame.

5. Heat both pieces of steel plate at the bottom of the vertical plate and the base of the horizontal plate until a molten puddle is formed.

6. Keep the cone of the flame pointed toward the molten puddle at a 45-degree angle. Ensuring that the inner cone of the flame does not touch the puddle, tack weld the parts, moving from left to right (if right handed).

7. Add filler rod, keeping the rod between the puddle and the cone of the flame, and tack weld the two pieces of metal plate at each end.

8. Let the tack welds cool.

9. Begin welding at the left and heat both pieces of metal, forming a molten puddle (Figure 16-22).

10. Add filler rod between the puddle and the cone of the flame, leaving the rod in the puddle until the weld is complete.

11. Oscillate the torch and tip slowly, roll the filler rod from side to

Figure **16-22**

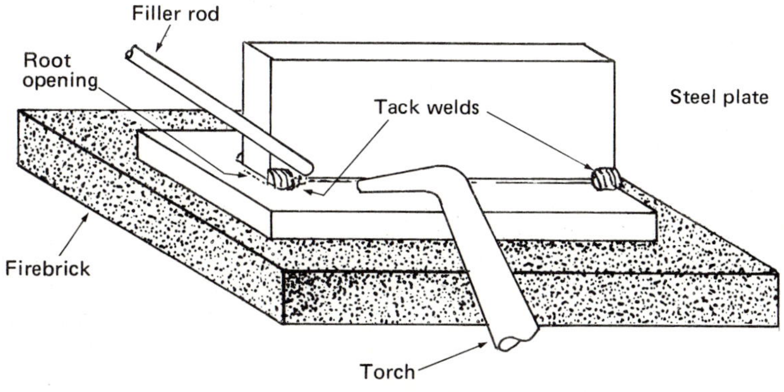
Filler rod
Root opening
Tack welds
Steel plate
Firebrick
Torch

side in the opposite direction of the torch oscillation, and move the weld from left to right along the root opening.

12. Travel slowly enough to allow the base metal to melt before the filler rod builds up the root opening.

13. Continue welding until both pieces of steel plate are completely fused.

14. Shut off the torch and clean the tip using the correct-size tip cleaner.

15. Test the weld by placing it on the raised portion of the anvil and hitting it with a ball-peen hammer.

16. Present the weld to the instructor for approval.

16-10 BUTT WELDING A SQUARE-GROOVE JOINT IN THE VERTICAL POSITION

OBJECTIVE To learn to weld an acceptable square-groove joint in the vertical position.

HOURS REQUIRED 2.

EQUIPMENT, TOOLS, AND MATERIALS REQUIRED The equipment, tools, and materials required to perform this exercise consist of the same equipment, tools, and materials as in exercise 16-4, plus a vertical-position welding jig.

PROCEDURE

1. Place the metal plates to be welded on the welding table, leaving $\frac{1}{32}$ inch between them for root opening.

2. Light the torch and adjust it to a neutral flame.

3. Tack weld the two pieces at both ends.

4. Shut off the torch.

5. Bend a welding rod to a 90-degree angle, about two inches from the end.

6. Mount the tack-welded assembly on the jig so that the first tack weld or, if root opening is uneven, the end of the assembly with the narrower spacing, will be at the bottom.

7. Light the torch.

8. Start the weld by building up a shelf at the bottom of the groove that will support the molten puddle.

9. Hold the torch and welding rod so as to form the angles shown in Figure 16-23 with the metal to be welded, and using the same oscillating motions for torch and rod as described in exercise 16-7, proceed upward along the joint. *Note:* If a constant upward speed is maintained, a keyhole-shaped opening will be established ahead of the puddle (Figure 16-24). Maintenance of this opening ahead of the

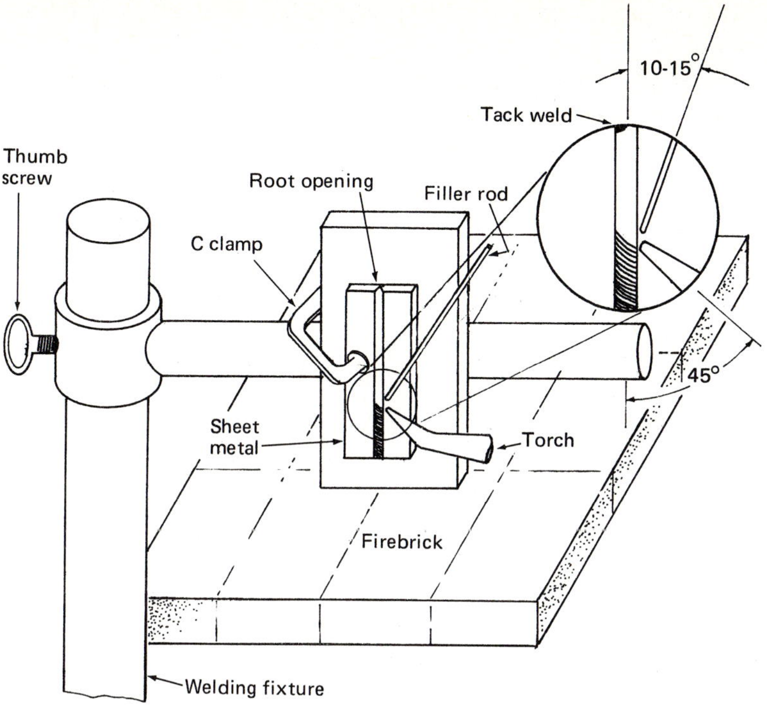
10-15°
Tack weld
Thumb screw
Root opening
Filler rod
C clamp
45°
Sheet metal
Torch
Firebrick
Welding fixture

Figure **16-23**

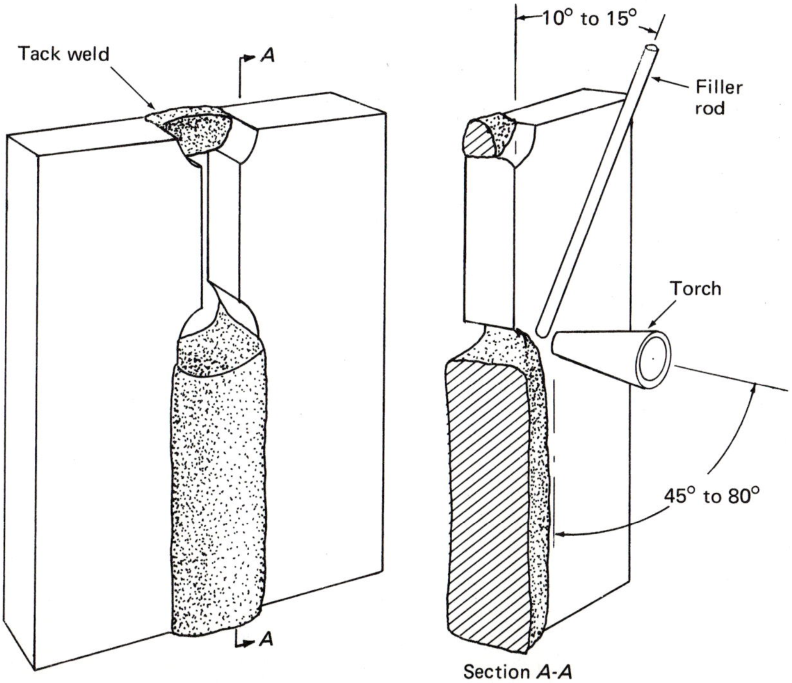
Tack weld
A
A
10° to 15°
Filler rod
Torch
45° to 80°
Section A-A

Figure **16-24** (By permission, from *Metals Handbook* Volume 6, Copyright American Society for Metals, 1971.)

puddle is assurance that proper penetration at the root is being obtained.

10. Continue welding until the groove is filled and the joint made.

11. Shut the torch off and clean the tip using the correct-size tip cleaner.

12. Remove the welded pieces of sheet metal from the welding jig and check the back side of weld to ensure good penetration.

13. Place the welded sheet metal on the raised portion of the anvil and hit it with a ball-peen hammer to test the strength of the weld.

14. Present the weld to the instructor for approval.

16-11 WELDING A V-GROOVE BUTT JOINT IN THE HORIZONTAL POSITION

OBJECTIVE To learn to prepare and weld an acceptable V-groove butt joint in the horizontal position.

HOURS REQUIRED 2.

EQUIPMENT, TOOLS, AND MATERIALS REQUIRED The equipment, tools, and materials required to perform this exercise are:

1. Several pieces of 2 × 4 × ¼-inch mild-steel plate.
2. The appropriate protective clothing.
3. A welding outfit.
4. A horizontal-position welding jig.
5. A 16-ounce ball-peen hammer.
6. An anvil or similar platform.
7. A torch-tip cleaner.
8. A firebrick.

PROCEDURE

1. Select the proper-size torch tip.

2. Prepare the V groove as shown in Figure 16-25.

3. Place the two pieces of steel plate with the beveled sides end to end on a firebrick, ensuring that there is a 1/16-inch gap for the root opening.

4. Tack weld each end of the metal plates together (Figure 16-25).

5. Let the tack welds cool, then place the tacked pieces on the welding jig so that the weld will be in a horizontal position (Figure 16-26).

6. Light the torch and adjust it to a neutral flame.

7. Working from right to left (the reverse if left handed), position

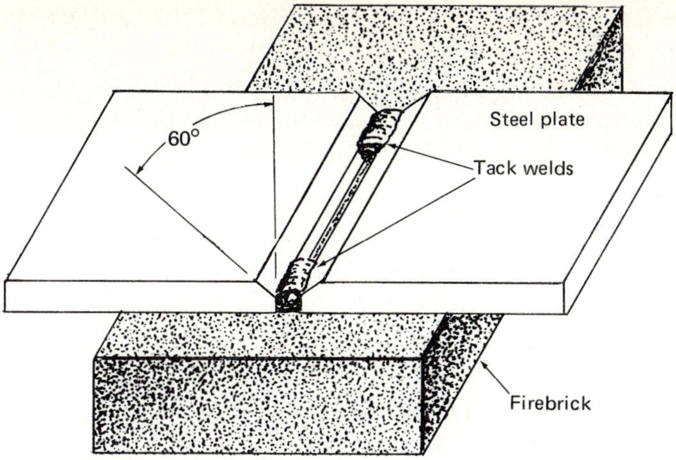
60°
Steel plate
Tack welds
Firebrick

Figure **16-25**

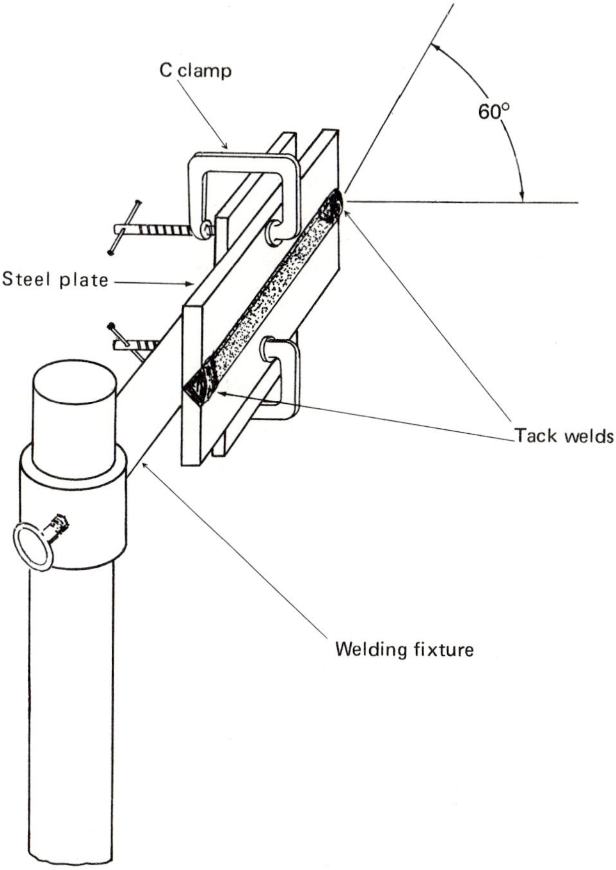
C clamp
60°
Steel plate
Tack welds
Welding fixture

Figure **16-26**

the torch so that the cone of the flame is pointed 5 to 10 degrees toward the direction of travel and 5 to 10 degrees down from the horizontal. Then form a molten puddle in the root opening.

8. Add filler with the rod pointed toward the molten puddle opposite the torch and 45 degrees up from the horizontal.

9. Ensure that the inner cone does not touch the molten puddle.

10. Oscillate both the filler rod and the torch in opposite directions and move the weld toward the left, keeping the weld small and even without undercutting.

11. If the molten puddle begins to run down, remove the torch momentarily to let the molten puddle cool.

12. Continue to weld the full length of the metal plates.

13. Shut off the torch and clean the tip using the correct-size tip cleaner.

14. Remove the weld from the welding jig and place it on the raised portion of the anvil.

15. Hit the weld with the ball-peen hammer to test its strength.

16. Present the weld to the instructor for approval.

16-12 WELDING A V-GROOVE BUTT JOINT IN THE OVERHEAD POSITION

OBJECTIVE To learn to prepare and weld an acceptable V-groove butt joint in the overhead position.

HOURS REQUIRED 2.

EQUIPMENT, TOOLS, AND MATERIALS REQUIRED The equipment, tools, and materials required to perform this exercise are the same equipment, tools, and materials as used in exercise 16-11, plus an overhead-position welding jig. *Warning:* Failure to wear a nonflammable vest and an appropriate headcover will most likely result in burns to the welder's chest and head.

PROCEDURE

1. Prepare the V groove as shown in Figure 16-25.

2. Place the plates on the firebrick, leaving an appropriate root opening.

3. Light the torch and adjust it to a neutral flame.

4. Tack weld both ends of the V groove.

5. Shut off the torch.

6. Place the tack-welded assembly on the welding jig in an overhead position (Figure 16-27).

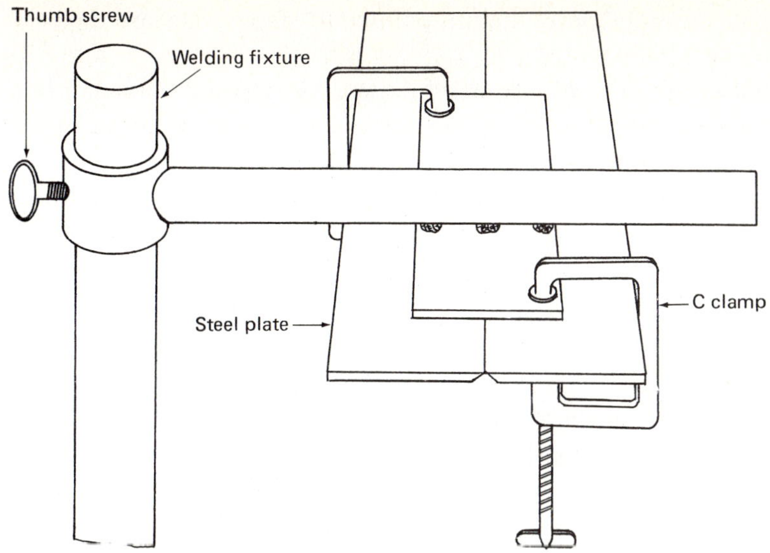
Thumb screw
Welding fixture
Steel plate
C clamp

Figure **16-27**

7. Bend the welding rod to an angle of 45 degrees approximately 2 inches from the end.

8. Light the torch and adjust it to a neutral flame.

9. Working from right to left if right handed (the reverse if left handed), with the torch pointed 20 degrees to the base of the steel plate, heat the plate, forming a small molten puddle (Figure 16-28).

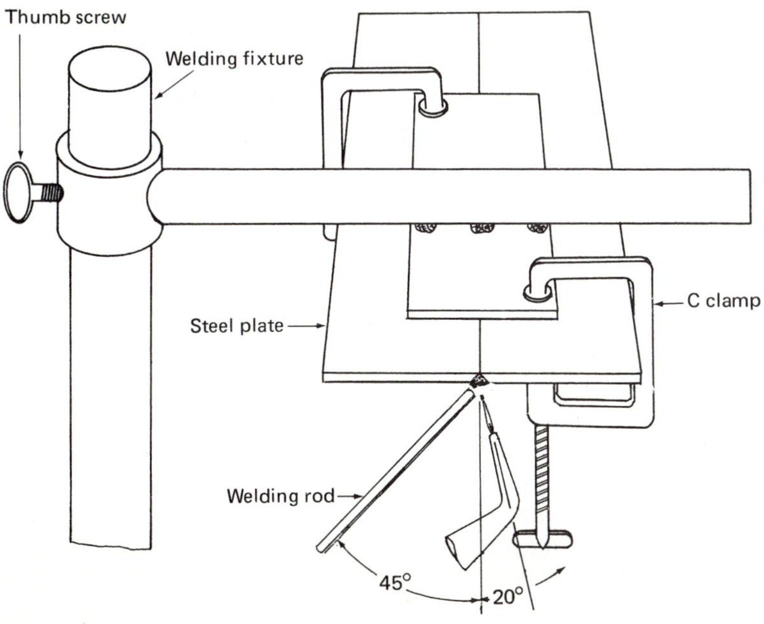
Thumb screw
Welding fixture
Steel plate
C clamp
Welding rod
45°
20°

Figure **16-28**

10. Add filler rod, with the rod pointed 45 degrees toward the base of the steel plate and the body of the torch.

11. Keeping the puddle as small as possible, move it to the left in a slow, even manner.

12. Oscillate both torch and rod in opposite directions.

13. If a molten puddle begins to fall downward, remove the heat momentarily to let the molten puddle cool.

14. Continue the bead for the full length of the steel plate.

15. Shut off the torch and clean the tip using the correct-size tip cleaner.

16. Remove the welded plates from the welding jig and place them on the raised portion of anvil.

17. Hit the weld with the ball-peen hammer to test it.

18. Present the weld to the instructor for approval.

16-13 NOMENCLATURE OF THE OXYACETYLENE CUTTING TORCH

OBJECTIVE To learn to recognize and name all parts of the standard oxyacetylene cutting torch.

HOURS REQUIRED ¼.

EQUIPMENT, TOOLS, AND MATERIALS REQUIRED The equipment, tools, and materials required to perform this exercise are:

1. A standard cutting torch.

2. Different-sized cutting tips.

3. Different-angled cutting heads.

4. Welding-tip cleaners.

5. An instruction manual supplied by the welding-equipment manufacturer.

PROCEDURE Look at the various items illustrated in Figure 16-29, read the instruction booklet supplied by the welding-equipment manufacturer, and learn the correct nomenclature of all parts of a standard cutting torch.

16-14 FREEHAND CUTTING OF ¼-INCH-THICK STEEL PLATE

OBJECTIVE

1. To learn to prepare metal for freehand cutting with the standard oxyacetylene torch

2. To select the proper cutting tip size and gas pressures (Table 16-2) necessary for cutting through a given thickness of metal

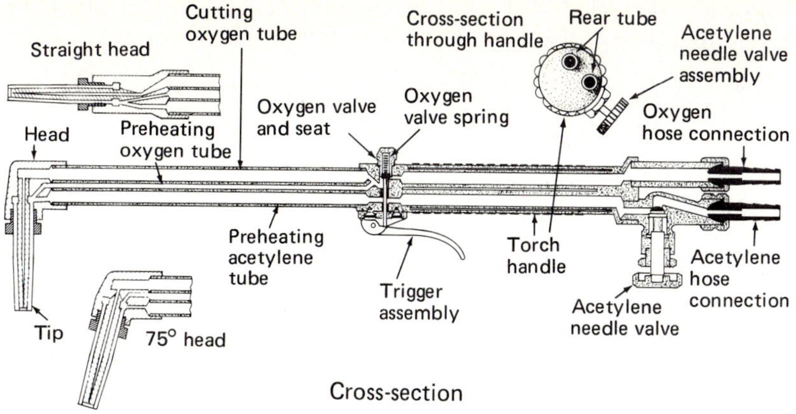
Straight head
Cutting oxygen tube
Cross-section through handle
Rear tube
Acetylene needle valve assembly
Oxygen valve and seat
Oxygen valve spring
Oxygen hose connection
Head
Preheating oxygen tube
Preheating acetylene tube
Trigger assembly
Torch handle
Acetylene hose connection
Acetylene needle valve
Tip
75° head
Cross-section

Figure **16-29**

Table **16-2** Torch-tip sizes for various gas pressures*

PUROX WELDING BLOWPIPES WELDING HEAD SIZES*			PREST-O-LITE WELDING BLOWPIPES WELDING HEAD SIZES*
W-200	W-201	W-202	No. 420
2(5–7)	2(5–7)	2(5–7)	2(5)
2(5–7)	2(5–7)	2(5–7)	2(5)
4(5–7)	4(5–7)	4(5–7)	2(5)
4(5–7)	4(5–7)	4(5–7)	6(5)
6(5–7)	6(5–7)	6(5–7)	
6(5–7)	6(5–7)	6(5–7)	6(5)
9(5–7)	9(5–7)	9(5–7)	
9(5–7)	9(5–7)	9(5–7)	6(5)
12(5–7)	12(5–7)	12(5–7)	
12(5–7)	12(5–7)	12(5–7)	15(5)
15(5–7)	15(5–7)	15(5–7)	
15(5–7)	15(5–7)	15(5–7)	15(5)
20(5–7)	20(5–7)	20(5–7)	
20(5–7)	20(5–7)	20(5–7)	20(6)
30(5–7)	30(5–7)	30(5–7)	
30(5–7)	30(5–7)	30(5–7)	20(6)
	40(5–7)	40(5–7)	30(9)
	40(5–7)	40(5–7)	
	55(6–8)	55(6–8)	
	55(6–8)	55(6–8)	
		70(6–8)	
		85(6–8)	
		100(6–8)	

*Oxygen pressure range in parentheses. Acetylene pressure approximately equal to oxygen pressure.

Source: Union Carbide Corporation, Linde Division, *The Oxyacetylene Handbook*.

3. To learn to make an acceptable freehand cut using the standard oxyacetylene cutting torch

EQUIPMENT, TOOLS, AND MATERIALS REQUIRED The equipment, tools, and materials required to perform this exercise are:

1. Protective clothing.

2. An oxyacetylene welding outfit.

3. An oxyacetylene cutting torch with an appropriately sized tip.

4. A bending slab.

5. A 16-ounce ball-peen hammer.

6. A prick-punch.

7. Welding-tip cleaners.

8. One piece of 8 × 12 × ¼-inch-thick mild-steel plate.

PROCEDURE

1. Select the proper-size torch tip for cutting ¼-inch-thick steel plate.

2. Place the plate to be cut on the bending slab and, using the prick-punch and ball-peen hammer, mark off the metal in sections approximately 2 × 6 inches (Figure 16-30).

3. Light the torch and adjust it to a neutral flame using preheating gas valves.

4. Point the cutting torch so that the tip is at an angle of 45 degrees to the edge of the steel plate, and approximately 1/16 inch above the

Figure **16-30**

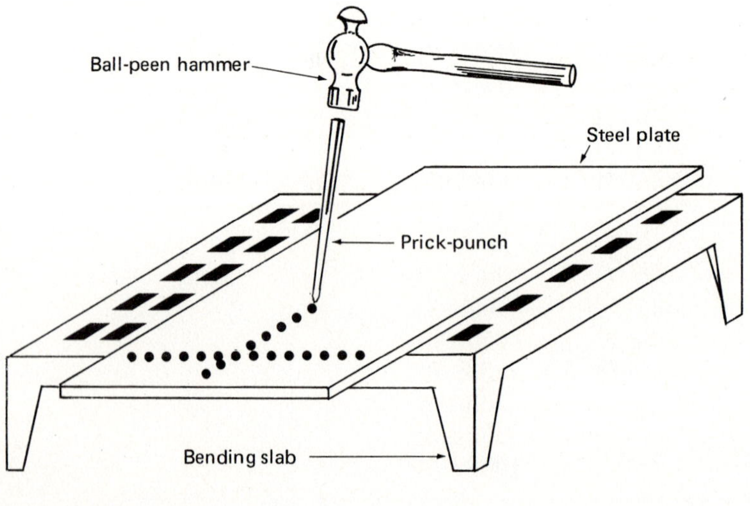
Ball-peen hammer
Steel plate
Prick-punch
Bending slab

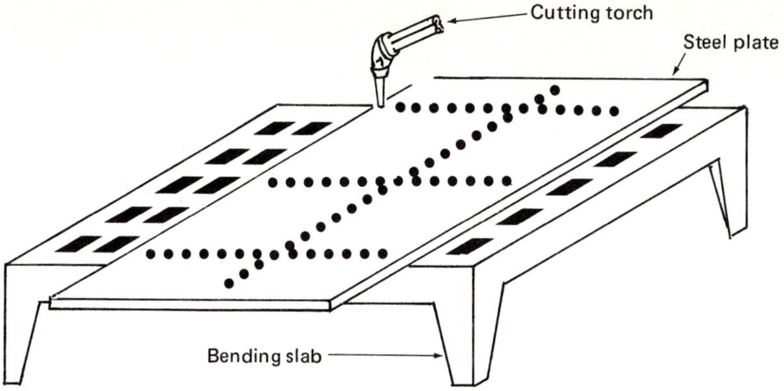
Cutting torch
Steel plate
Bending slab

Figure **16-31**

surface of the plate (Figure 16-31). *Note:* Hold the torch in this position until the metal is a bright-red color.

5. Keeping the preheating flame 1/16 inch above the surface of the metal, slowly depress cutting oxygen trigger and simultaneously begin to position the torch so that the cutting tip is perpendicular to the surface of the plate to be cut.

6. Maintaining a steady hand, pull the torch toward you slowly enough to cut the metal, yet fast enough for a clean cut.

7. Continue to cut the metal until all eight pieces are cut.

8. Shut off the torch and clean the tip using the correct-size tip cleaner.

9. Allow all pieces of metal to cool, then present the weld to the instructor for approval.

16-15 FREEHAND CUTTING A BEVEL ON 1/4-INCH-THICK STEEL PLATE

OBJECTIVE To learn to cut acceptable freehand bevels on 1/4-inch-thick steel plate using the standard oxyacetylene cutting torch.

HOURS REQUIRED 1.

EQUIPMENT, TOOLS, AND MATERIALS REQUIRED

1. Protective clothing.

2. The related welding equipment.

3. A cutting torch with a #0 tip.

4. One piece of 4 × 6 × 1/4-inch-thick steel plate.

5. A bending slab.

6. 12 inches of 1¼-inch angle iron.

7. Two four-inch C clamps.

8. A tip cleaner.

PROCEDURE

1. Select the proper-size torch tip for cutting ¼-inch-thick steel plate on a bending slab with one edge protruding, and clamp a piece of 1¼-inch angle iron down the center using C clamps (Figure 16-32).

2. Light the torch and set the preheating flame to neutral.

3. Work from the outward side with the cutting torch in front of you and heat the steel plate, keeping the preheating flame 1/16 inch above the metal (Figure 16-33).

4. Heat the steel plate until the metal turns bright red in color.

5. Open the cutting oxygen valve slowly but steadily.

6. Pull the torch toward you slowly, but fast enough to cut the metal cleanly.

7. Keep the torch against the angle iron the full length of the cut, making a beveled cut.

8. Shut off the torch and clean the tip using the correct-size tip cleaner.

9. Allow both pieces of metal to cool and present them to the instructor for approval.

16-16 FILLING IN A ½-INCH-DIAMETER HOLE IN A 1-INCH-DIAMETER, ⅛-INCH-THICK WALL STEEL PIPE IN THE FLAT POSITION

OBJECTIVE To learn to fill in a hole in a steel pipe in the flat position.

Figure **16-32**

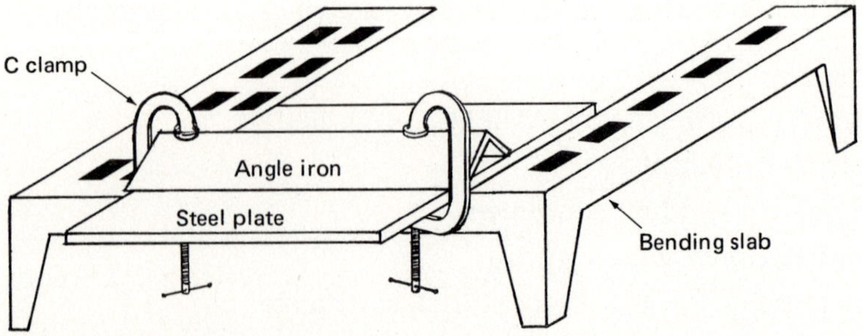
C clamp
Angle iron
Steel plate
Bending slab

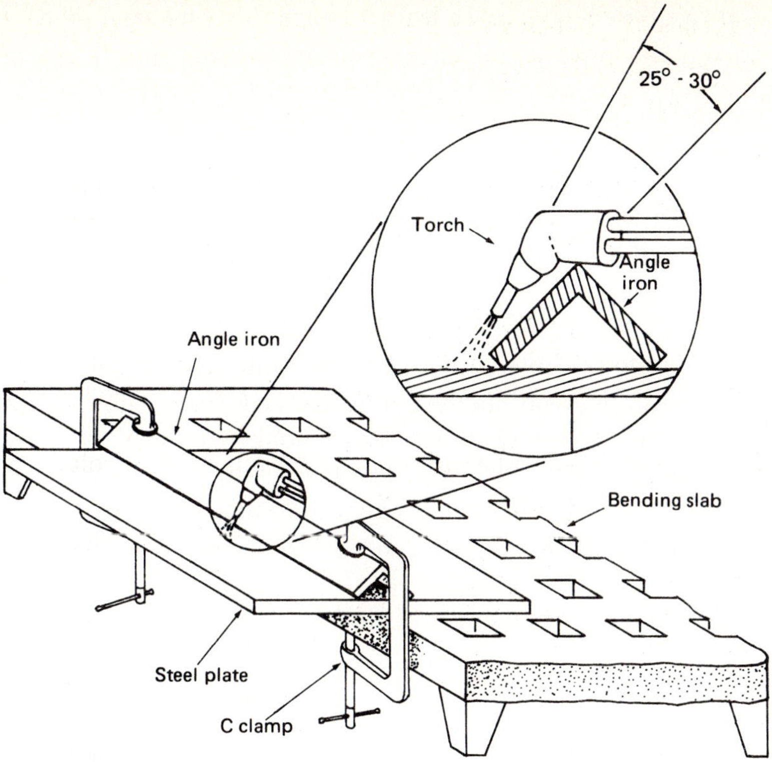
25° - 30°
Torch
Angle iron
Angle iron
Bending slab
Steel plate
C clamp

Figure **16-33**

HOURS REQUIRED 1.

EQUIPMENT, TOOLS, AND MATERIALS REQUIRED

1. Protective clothing.

2. An oxyacetylene welding outfit.

3. A ⅛-inch-diameter, AWS Classification AG50 welding rod.

4. One 6-inch-long piece of 1-inch-diameter, ⅛-inch-thick wall steel pipe.

5. A firebrick.

6. A hacksaw frame and blade with 18 teeth per inch.

7. A drill press and drill-press vise.

8. A ½-inch-diameter twist drill.

9. Welding-tip cleaners.

PROCEDURE

1. Select the proper-size tip for welding ⅛-inch-diameter thick steel.

2. Cut one 6-inch-long piece from a 1-inch-diameter steel pipe.

3. Clamp the pipe in the drill vise and set the drill press at approximately 800 rpm. Then drill four ½-inch-diameter holes facing upward (Figure 16-34).

4. Place the pipe on the firebrick with holes facing upward (Figure 16-34).

5. Light the torch and adjust it to a neutral flame. Then move the torch to one of the end holes.

6. Point the torch so that the cone of the flame is pointed toward the hole in the pipe at a 45-degree angle and heat the metal immediately adjacent to the hole, until it forms a molten puddle.

7. Place a welding rod into the puddle and, keeping the inner cone of the flame out of the molten puddle, oscillate both the rod and the torch in opposite directions until the hole is filled.

8. Continue to weld the remaining holes in the same manner.

9. Shut off the torch and clean the tip using the correct-size tip cleaner.

10. Present the weld to the instructor for approval.

Figure **16-34**

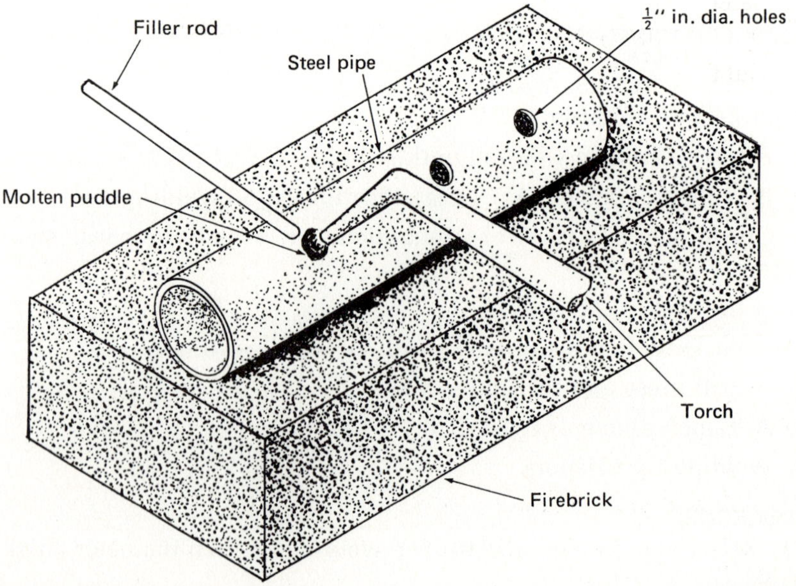
Filler rod
Steel pipe
½" in. dia. holes
Molten puddle
Torch
Firebrick

16-17 WELDING PIPE IN THE HORIZONTAL-FIXED POSITION

OBJECTIVE

1. To learn to make a 30-degree bevel on pipe, freehand, with the oxyacetylene cutting torch.

2. To learn to weld an acceptable butt joint in pipe with the pipe in the horizontal-fixed position.

HOURS REQUIRED 2.

EQUIPMENT, TOOLS, AND MATERIALS REQUIRED

1. Protective clothing.

2. An oxyacetylene welding outfit.

3. A cutting torch with a proper-size tip.

4. A welding torch with a proper-size tip.

5. Two 3-inch-long pieces of 2-inch-diameter, 1/4-inch-thick wall steel pipe.

6. A 5/32-inch-diameter, AWS Classification GA50 welding rod.

7. A firebrick.

8. A bench-mounted or similar grinder.

9. Vise-grip pliers of suitable size.

10. A wire brush.

11. A welding-tip cleaner.

PROCEDURE

1. Select the proper-size torch tip for cutting 1/4-inch-thick steel plate.

2. Cut two pieces of 2-inch-thick steel pipe approximately 3 inches long.

3. Using the cutting torch (Figure 16-35), cut a 30-degree bevel on one end of each piece of pipe.

4. Grind the beveled ends clean with a bench grinder, keeping a 30-degree angle on the pipe.

5. Replace the cutting torch with a welding torch.

6. Select the proper-size torch tip for welding on 1/4-inch-thick steel plate.

7. Place two pieces of 2-inch steel pipe with the beveled ends butted together, leaving a 1/16-inch root opening between two firebricks (Figure 16-36).

8. Light the torch and adjust it to a neutral flame.

9. Tack weld the two pieces of 2-inch steel pipe at four equal places (Figure 16-36).

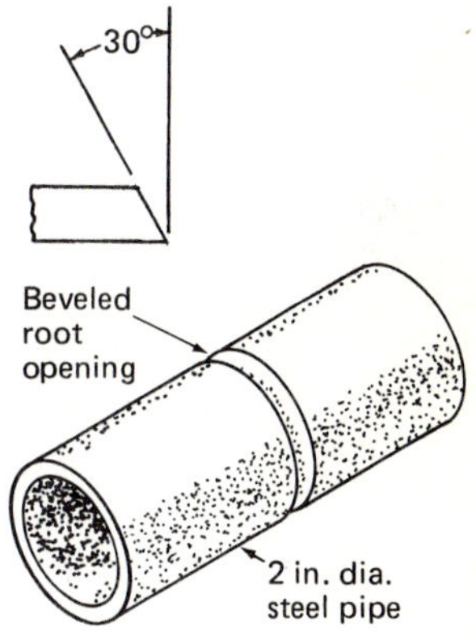
30°
Beveled
root
opening
2 in. dia.
steel pipe

Figure **16-35**

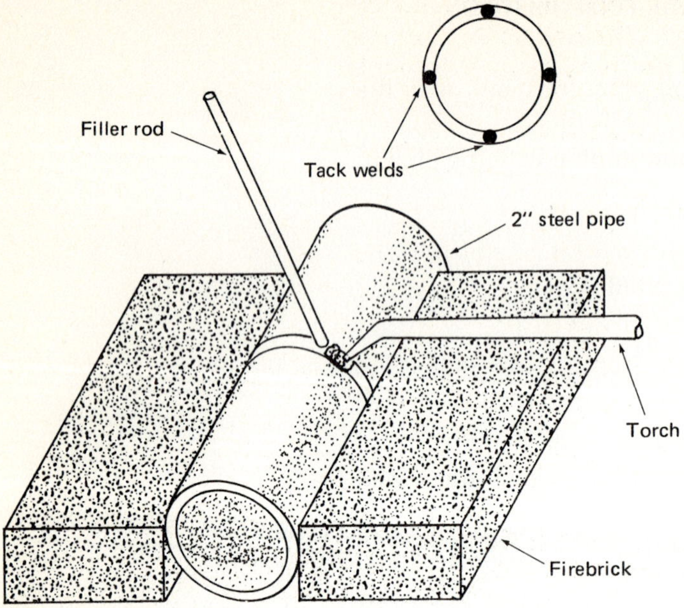
Filler rod
Tack welds
2" steel pipe
Torch
Firebrick

Figure **16-36**

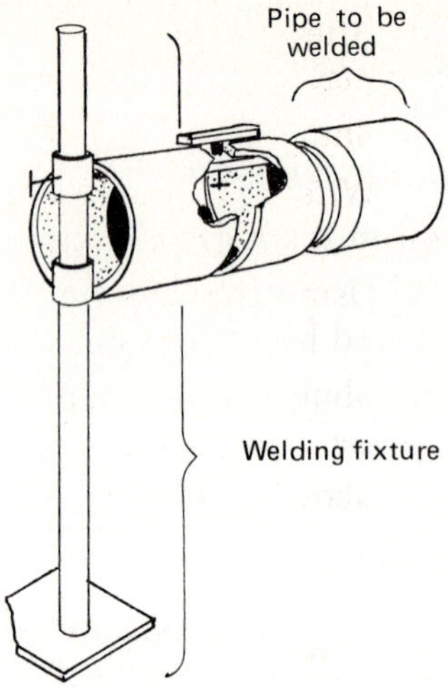
Pipe to be welded
Welding fixture

Figure **16-37**

10. Shut off the torch and let the tack welds cool.

11. Clamp the pipe on the welding jig in a horizontal position. Leave it in this position for the entire weld (Figure 16-37).

12. Starting at the bottom of the pipe with the torch pointed so that the cone of the flame is at a 5-degree angle from the vertical, heat the metal at the root opening and form a molten puddle (Figure 16-38).

Figure **16-38**

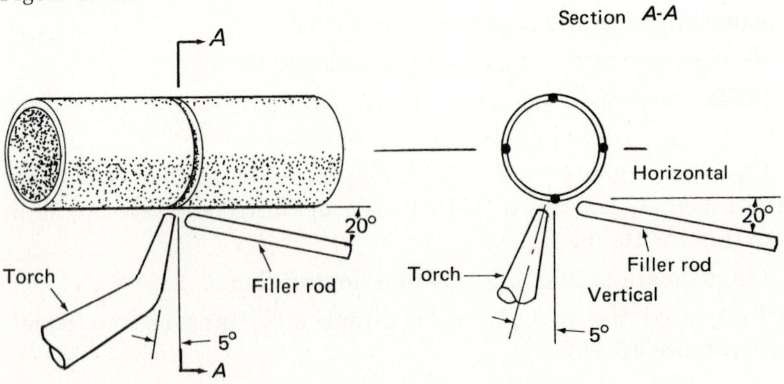
Section A-A
A
A
20°
Torch
Filler rod
5°
Horizontal
20°
Filler rod
Torch
Vertical
5°

13. Add filler with the rod pointed toward the molten puddle 20 degrees from the horizontal, working from the bottom around and up toward the top in a counterclockwise position, keeping the torch and the rod pointed at the appropriate angle.

14. After completing the right side, weld the left side in the same manner, keeping the bead small and even.

15. Shut off the torch and let the weld cool.

16. Wire brush the first bead clean. Then light the torch again and adjust it back to a neutral flame.

17. Using the same procedure as with the first bead, make the second bead around the pipe (Figure 16-39).

18. Shut off the torch and clean the tip using the correct-size tip cleaner.

19. Show the weld to the instructor for approval.

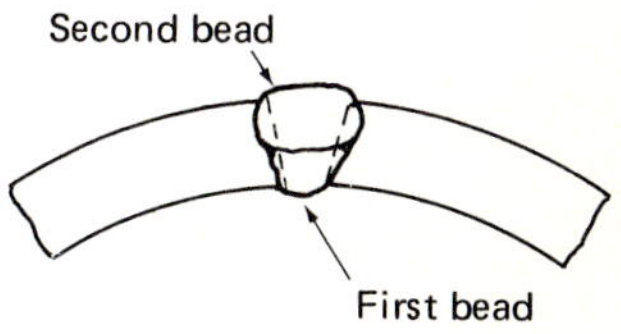

Second bead
First bead

Figure **16-39**

16-18 WELDING PIPE IN THE VERTICAL-FIXED POSITION

OBJECTIVES

1. To learn to make a 45-degree bevel on pipe, freehand, with the oxyacetylene torch.

2. To learn to weld an acceptable butt joint in pipe with the pipe in the vertical-fixed position.

HOURS REQUIRED 2.

EQUIPMENT, TOOLS, AND MATERIALS REQUIRED The equipment, tools, and materials required to perform this exercise are the same as those required for exercise 16-17.

PROCEDURE

1. Select the proper-size torch tip for cutting ¼-inch-thick steel plate.

2. Cut two pieces of 2-inch-diameter steel pipe approximately 3 inches long.

3. Using the cutting torch (Figure 16-40), cut a 45-degree bevel on one end of each piece of pipe.

4. Grind the beveled ends clean with a bench grinder, keeping a 45-degree angle on the pipe.

5. Replace the cutting torch with a welding torch.

6. Select the proper-size torch tip for welding on ¼-inch steel plate.

7. Place two pieces of 2-inch-diameter steel pipe with the beveled

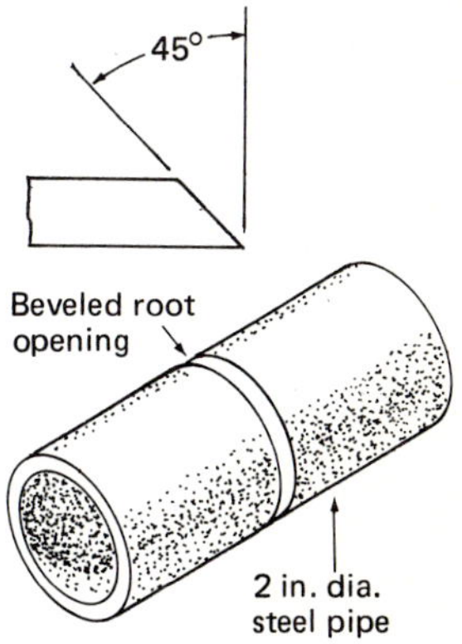

45°
Beveled root opening
2 in. dia. steel pipe

Figure **16-40**

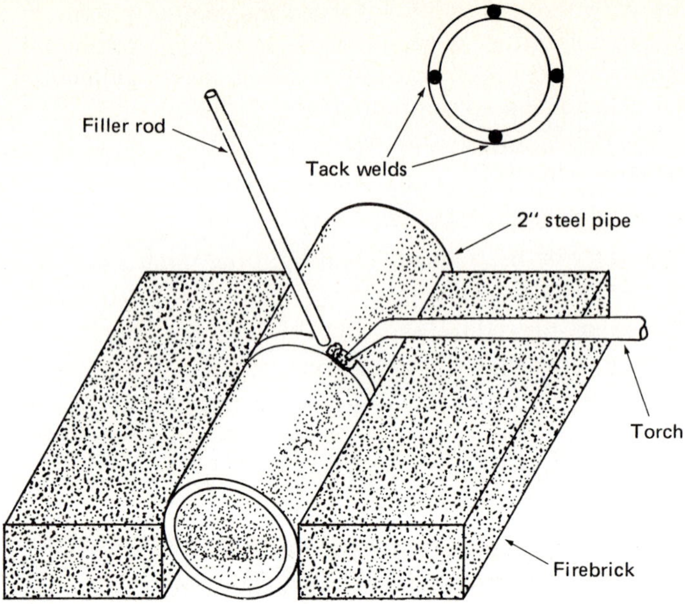
Filler rod
Tack welds
2" steel pipe
Torch
Firebrick

Figure **16-41**

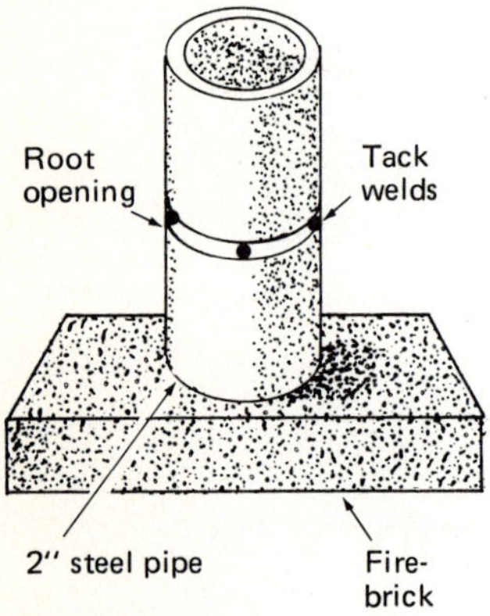
Root
opening
Tack
welds
2" steel pipe
Fire-
brick

Figure **16-42**

ends butted together, leaving a 1/16-inch root opening between two firebricks (Figure 16-41).

8. Light the torch and adjust it to a neutral flame.

9. Tack weld the two pieces of 2-inch-diameter steel pipe at four equal places (Figure 16-41).

10. Shut off the torch and let the tack welds cool.

11. Place the tack-welded pipe on a firebrick in the vertical position (Figure 16-42).

12. Light the torch and adjust it to a neutral flame.

13. Starting at the front, tack weld and, moving toward the left if right handed, position the torch so that the cone of the flame is pointed in the direction of the weld at 5 to 10 degrees down from the vertical, with a 5- to 10-degree side angle. Form a molten puddle. (Reverse the procedure if left handed.)

14. Add filler with the rod pointed toward the molten puddle at 45 degrees up from the horizontal, with a 45-degree side angle opposite the cone of the flame (Figure 16-43).

15. Oscillate both the torch and the filler rod in opposite directions and keep the first bead as small as possible (Figure 16-44).

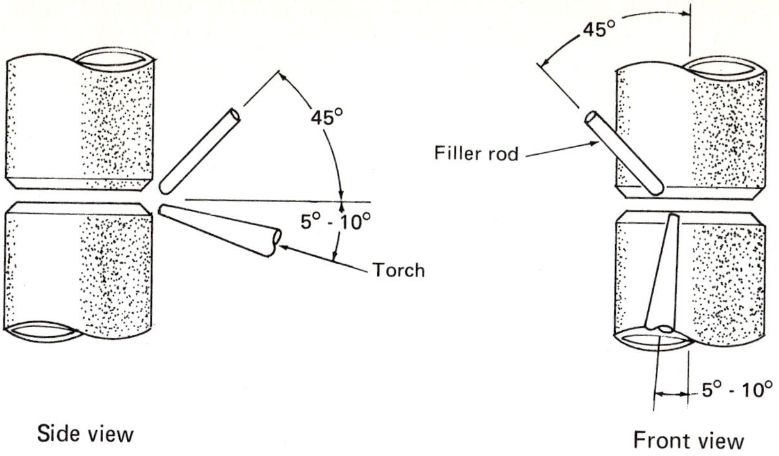
45°
5° - 10°
Torch
Side view
45°
Filler rod
5° - 10°
Front view

Figure **16-43**

16. Continue the first bead completely around the pipe.

17. Shut off the torch and let the pipe cool.

18. Wire-brush the first bead clean and place the steel pipe back on the firebrick in the same position as before.

19. Light the torch and adjust it to a neutral flame.

20. Keeping the torch and rod in the same position as before, lay the

Figure **16-44**

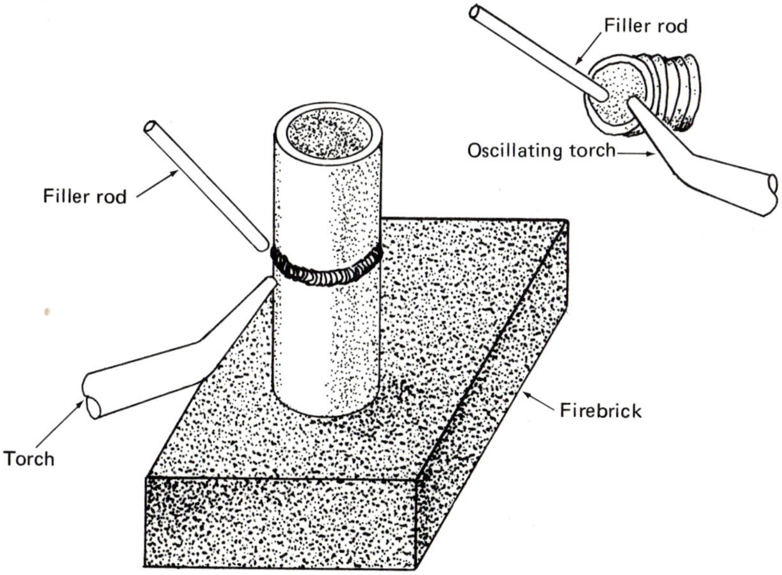
Filler rod
Oscillating torch
Filler rod
Firebrick
Torch

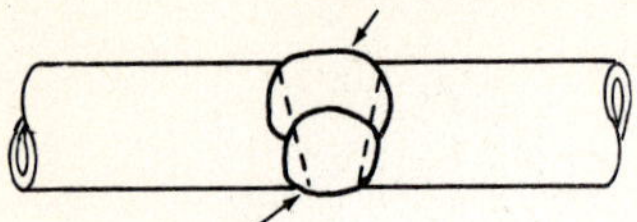
Second bead
First bead
Figure **16-45**

second bead around the steel pipe, starting in front of you and working to the right (Figure 16-45).

21. Complete the second bead and shut off the torch.

22. Clean the torch tip using the correct-size tip cleaner.

23. Present the weld to the instructor for approval.

REVIEW QUESTIONS

1. After both the oxygen and acetylene gases are in the line hose as far as the torch, which valve on the torch is opened first in order to light?

2. When lighting the torch, how is it possible to determine, without looking at the gage, whether enough pressure is on the acetylene line?

3. What is a neutral flame? How does it look? Sketch a neutral flame.

4. What are the main features in the construction of a cutting torch?

5. How does the strength of a good weld compare with the strength of the original part?

6. In applying the torch to metals to be welded, what kind of motion should be used with the torch? Explain.

7. Why should one use particular care to see that acetylene cylinders are tightly closed?

8. Why is it dangerous to use oil on oxygen cylinders, regulators, and fittings?

9. Why should copper tubing never be used to pipe acetylene?

10. In cutting by the oxyacetylene process, which does the cutting, the oxygen jet or the neutral flame?

11. What part does the neutral flame play in cutting?

12. Why is it important not to heat one section more than another when welding two pieces together?

13. What is the use of oxygen in oxyacetylene welding?

14. What is the temperature of the oxyacetylene flame?

15. How is the flame adjusted for steel welding?

CHAPTER 17
HARD-SURFACING PRACTICE

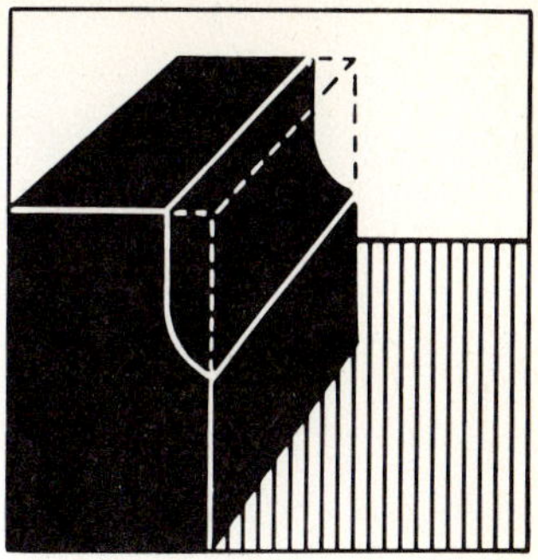

In the AWS *Welding Handbook*, *hard surfacing*, sometimes called *hard facing*, is defined as "the deposition of filler metal on a metal surface to obtain the desired properties and/or dimensions." The desired properties are those that will resist abrasion, heat, and/or corrosion. Another way to say the same thing is: Hard surfacing is the deposition of some kind of special alloy material on a metal part by any one of various welding practices, to form a surface that resists abrasion, heat, impact, and corrosion, or any combination of these.

17-1 SELECTION OF HARD-SURFACING ALLOYS

The selection of a hard-surfacing alloy for any application is based on the resultant savings and advantages from the application of the alloy. These savings and advantages come from increased production, fewer replacement parts, and less downtime. In virtually all hard-surfacing applications, the surfacing materials represent the least significant element in the total cost. Wages, lost production during downtime, and overhead rates are far more important. Accordingly, the alloys listed in Chapter 4 are those that have proved hundreds of times to provide maximum savings. In some instances, however, it may be desirable to use an alloy other than the one listed. For example, on larger equipment, the amount of coverage may dictate a lower-cost electrode; out-of-position welding may be required, or wear conditions in any given area may indicate that a hard-surfacing alloy, of lower or higher wear resistance or impact than that which is recommended, should be used.

17-2 MAJOR WEAR FACTORS: ABRASION, HEAT, AND CORROSION

Before discussing each of the three major wear factors in detail, it is necessary to clear up the misconceptions and fallacies regarding the hardness of the hard-surfacing deposit and its relationship to wear. The hardness of the deposit, as shown in Figure 17-1 (and Table 17-2), is no true criterion for wear.

Abrasion of metals has been classified into three general types: scratching, grinding, and gouging. Most abrasive situations involve combinations of the three.

Scratching abrasion or *low-stress abrasion* is the least severe of the three types of abrasion. It depends on hard and usually sharp particles for metal removal. In this type of abrasion, the original hardness and sharpness of the abrasive are important. Greater abra-

Figure **17-1** Graph showing relative abrasion resistance and impact strength of various hard-surfacing deposits. (Courtesy of the Stoody Company.)

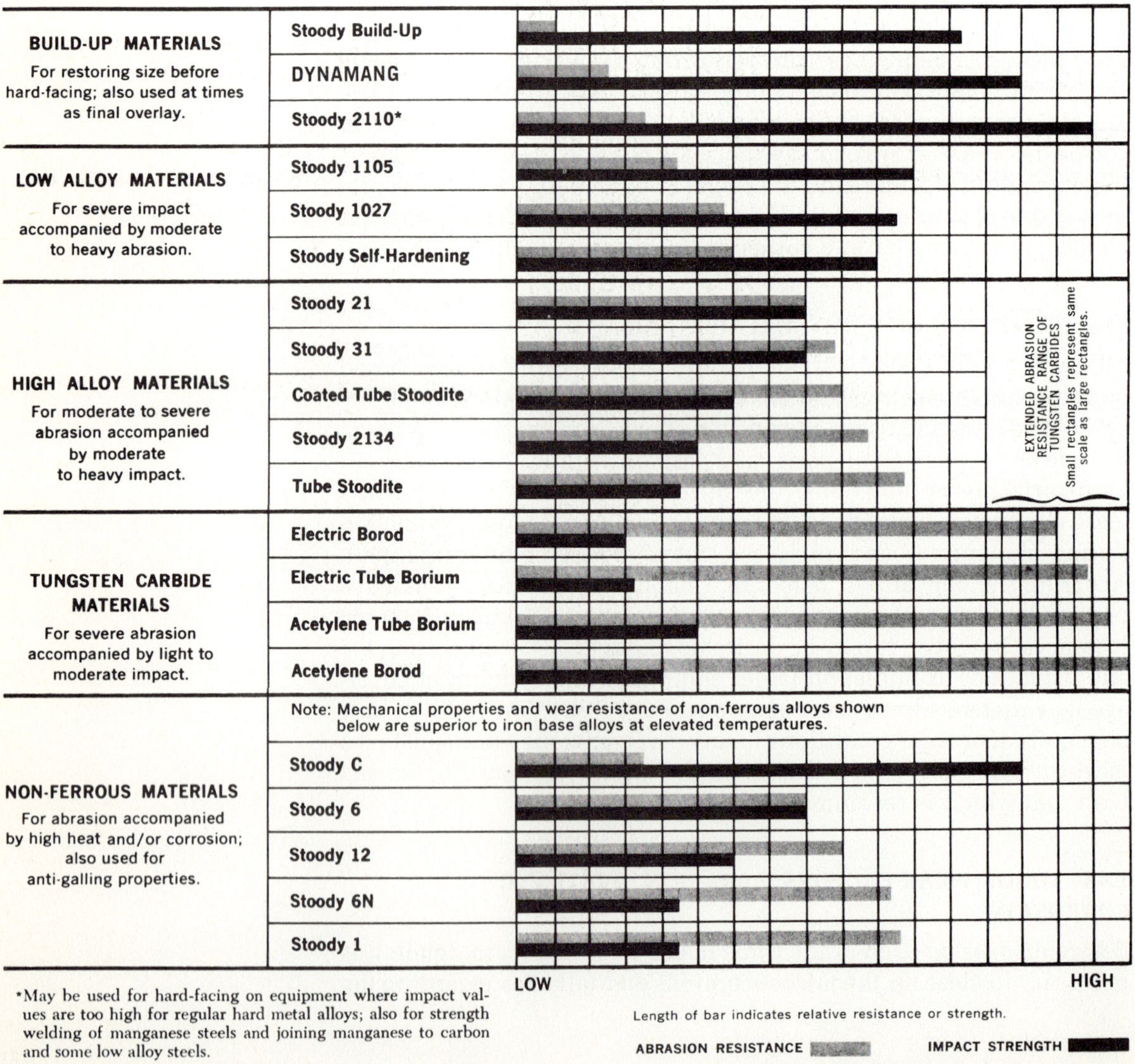

BUILD-UP MATERIALS
For restoring size before hard-facing; also used at times as final overlay.
Stoody Build-Up
DYNAMANG
Stoody 2110*
LOW ALLOY MATERIALS
For severe impact accompanied by moderate to heavy abrasion.
Stoody 1105
Stoody 1027
Stoody Self-Hardening
HIGH ALLOY MATERIALS
For moderate to severe abrasion accompanied by moderate to heavy impact.
Stoody 21
Stoody 31
Coated Tube Stoodite
Stoody 2134
Tube Stoodite
EXTENDED ABRASION RESISTANCE RANGE OF TUNGSTEN CARBIDES
Small rectangles represent same scale as large rectangles.
TUNGSTEN CARBIDE MATERIALS
For severe abrasion accompanied by light to moderate impact.
Electric Borod
Electric Tube Borium
Acetylene Tube Borium
Acetylene Borod
Note: Mechanical properties and wear resistance of non-ferrous alloys shown below are superior to iron base alloys at elevated temperatures.
NON-FERROUS MATERIALS
For abrasion accompanied by high heat and/or corrosion; also used for anti-galling properties.
Stoody C
Stoody 6
Stoody 12
Stoody 6N
Stoody 1
LOW
HIGH
Length of bar indicates relative resistance or strength.
ABRASION RESISTANCE
IMPACT STRENGTH

*May be used for hard-facing on equipment where impact values are too high for regular hard metal alloys; also for strength welding of manganese steels and joining manganese to carbon and some low alloy steels.

sion hardness and sharper cutting edges increase the severity of the scouring. Increased velocity of the abrasive particle, as in an air blast or the turbulent flow of a liquid containing the abrasive, rapidly increases the abrasive action. Impact forces are usually negligible. Sand slingers and dredge-pump impellers, fans, blowers, plowshares, coke chutes, cement chutes, and brake drums are examples of equipment in which scratching abrasion takes place.

Grinding abrasion or *high-stress abrasion* is caused by the fragmentation of small, hard, abrasive grains, usually between surfaces. The broken abrasive grains are sharp and can cause deep furrowing. Impact forces are usually negligible. Augers, concrete-mixer blades, scraper blades, screw conveyors, cement die rings, muller tires, ball milling, and other machinery parts that rub together in a gritty environment suffer grinding abrasion.

Gouging abrasion implies abrasion on a gross scale and is usually associated with impact. Sometimes the forces are applied at a relatively low velocity, as in the case of a power shovel digging in rock; in other cases, they may be applied at high velocity, as in the case of hammer or breaker bars in an impact-type pulverizer. The mechanism of metal removal is similar to that produced by machining with a cutting tool or high-speed abrasive grinding wheel. Prominent gouges are cut in or torn from the wearing surface in gouging abrasion. Power shovels digging in rock and rock-crusher operations are prime examples of equipment subjected to gouging abrasion.

The last two factors, corrosion and heat, or a combination of the two, or simultaneous exposure to corrosion, heat, and grinding abrasion usually occur in hot-work tools and dies, hot-shear blades, forging-hammer dies, rolling-mill mill guides, and high-speed steel-cutting tools.

17-3 HARD-SURFACING PROCESSES

The welding procedures that are used for hard surfacing are, in order of their frequency of use:

1. Manual: by oxyacetylene or electric arc.
2. Semiautomatic: with open arc, submerged arc, or shielded metal–arc.
3. Automatic: with open or submerged arc.
4. Spray powder.

17-4 SELECTION OF HARD-SURFACING PROCEDURE

For a given hard-surfacing application, the selection of the most suitable welding process may be as important as the selection of the alloy. Along with service requirements, the physical characteristics of the workpiece, the metallurgical properties of the base metal, the form and composition of the hard-surfacing alloy, the property and quality requirements of the weld deposit, the skill of the welder, and the cost of the operation must be considered in selecting a welding process.

At least three selection factors must be coordinated—the base metal, the composition and form of the hard-surfacing alloy, and the welding process. The following discussion will clarify the above.

Very large, heavy parts that are difficult or impossible to transport usually require selection of a welding process that can be moved to the site of the workpiece. In such applications, welding is most often done by manual or semiautomatic means, particularly when hard surfacing of difficult-to-reach areas is involved. In contrast, parts that can be easily moved to the welding equipment and that are to be processed in large quantities can be processed most efficiently by automatic or semiautomatic methods.

The properties of the base metal determine the preheating, in-process heating, and postweld heating rates (see Chapters 18 and 19). In a nutshell, preheating of the base metal to be hard surfaced is sometimes necessary to minimize distortion, to prevent spalling or cracking, and to avoid thermal shock. In order to determine the correct preheating temperature, the composition of the metal to be hard surfaced must be known or determined (see Chapter 19).

In general, according to *Welding Engineer Data Sheets,* 3rd edition, the need for preheating increases as the following factors change:

1. The larger the mass being welded
2. The lower the temperature of the pieces being welded
3. The lower the atmospheric (ambient) temperature
4. The smaller the diameter of the welding rod
5. The greater the speed of welding
6. The higher the carbon content of the steel
7. The higher the manganese content in plain carbon or low-alloy steels (*Note:* this does not apply to austenitic Hadfield manganese steel.)
8. The greater the alloy content in air-hardening steels
9. The greater the air-hardening capacity of the steel
10. The more complicated the shape or section of the parts

Some surfacing alloys, especially the harder, more brittle ones, are produced as a powder mixture. To make some of these powders available in a solid-wire–type shape, they are inserted into a tubular carbon-steel wire. Powders can be applied with standard oxyacetylene spraying equipment and fused after spraying with a standard oxyacetylene torch.

In addition to the composition of the hard-surfacing alloy itself, the extent of base metal dilution of the hard-surfaced part must be considered. *Dilution* is the interalloying of the hard-surfacing metal and metal of the workpiece, usually expressed as a percentage of base metal in the hard-surfacing deposit. For example, a dilution of 10% means that the deposit contains 10% base metal and 90% hard-surfacing alloy. As dilution increases, the hardness, wear resistance, and other desirable properties of the alloy deposit are reduced. Sometimes, in order to control composition and to counteract the adverse effects of differences of thermal expansion—contraction between the workpiece and the hard-surfacing alloy—a buffer layer of weld metal is deposited between the workpiece and the hard-surfacing alloy. Table 17-1 relates the welding process, mode of application, form of hard-surfacing alloy, and other factors to weld-metal dilution.

It is also important to relate the requirements of the hard-surfaced part to the skill of the welder doing the hard surfacing. It is not necessary, for example, to use highly skilled welders to hard surface mining and earth-moving equipment. The hard surfacing of engine valves, on the other hand, requires highly skilled welders and precise control of the welding operation.

Table **17-1** Characteristics of welding processes used in hard surfacing

WELDING PROCESS	MODE OF APPLICATION	FORM OF HARD FACING ALLOY	WELD-METAL DILUTION, %	DEPOSITION, (lb per hr)	MINIMUM THICKNESS, in.[a]	APPLICABLE HARD FACING ALLOYS
Oxyacetylene	Manual	Bare cast rod; tubular rod	1 to 10	1 to 6	1/32	All[b]
	Manual	Powder	1 to 10	1 to 15	1/32	All[b]
	Automatic	Extra-long bare cast rod; tubular wire	1 to 10	1 to 6	1/32	All[b]
Shielded metal–arc	Manual	Flux-covered cast rod; flux-covered tubular rod	15 to 25	1 to 6	1/8	All[b]
Open-arc	Semiautomatic	Alloy-cored tubular wire	15 to 25	5 to 25	1/8	Iron-base
	Automatic	Alloy-cored tubular wire	15 to 25	5 to 25	1/8	Iron-base
Gas tungsten–arc	Manual	Bare cast rod; tubular rod	10 to 20	1 to 8	3/32	All[b]
	Automatic	Various forms[c]	10 to 20	1 to 8	3/32	All[b]
Submerged–arc	Semiautomatic	Bare tubular wire	20 to 60	10 to 20	1/8	Iron-base
	Automatic, single wire	Bare tubular wire	30 to 60	10 to 25	1/8	Iron-base
	Automatic, multiwire	Bare tubular wire	15 to 25	25 to 60	3/16	Iron-base
	Automatic, series arc	Bare tubular wire	10 to 25	25 to 35	3/16	Iron-base
Plasma–arc	Automatic	Powder[d]	5 to 30	1 to 15	1/32	All[b]

[a]Recommended minimum thickness of deposit. [b]Iron-base, nickel-base, and cobalt-base alloys; tungsten carbide composites. [c]Bare tubular wire; extra-long (8-ft) bare cast rod; tungsten carbide powder with cast rod or bare tubular wire. [d]With or without tungsten carbide granules.

Source: By permission, from *Metals Handbook* Volume 6, Copyright American Society for Metals, 1971.

HARD SURFACING WITH THE OXYACETYLENE TORCH

17-5 SAFE PRACTICES

In addition to reviewing the safe practices applicable to oxyfuel-gas welding detailed in Chapter 13, make certain that:

1. The equipment you are going to use has been examined and tested for safety.

2. The gas cylinders are fastened in a vertical position and located away from the heating device.

3. All hose and regulator connections are clean and free of dirt.

17-6 PREPARING THE WORKPIECE FOR HARD SURFACING

To minimize the chances for porosity and/or spalling, the surface of the workpiece that is to receive the hardened layer should be thoroughly cleaned of all dirt, scale, grease, or other foreign material. Cleaning may be accomplished by grinding, filing, or power brushing.

Parts such as shear blades, punches, or dies that are subjected to severe shock while in use should be machined as shown in Figure 17-2(*a*). Parts such as cement-mill gudgeons, thrust collars, sprocket teeth, and expeller screws should be machined as shown in Figure 17-2(*b*). Parts subject to heat abrasion and impact, such as hot-work tools and dies, hot-shear blades, forging-hammer dies, rolling-mill mill guides, and shafts, should be prepared as shown in Figure 17-2(*c*).

Figure **17-2** Preparing surfaces for hard surfacing. (Redrawn by permission from Union Carbide Corporation, Linde Division, *The Oxy-Acetylene Handbook,* 2d ed., 1960, p. 360.)

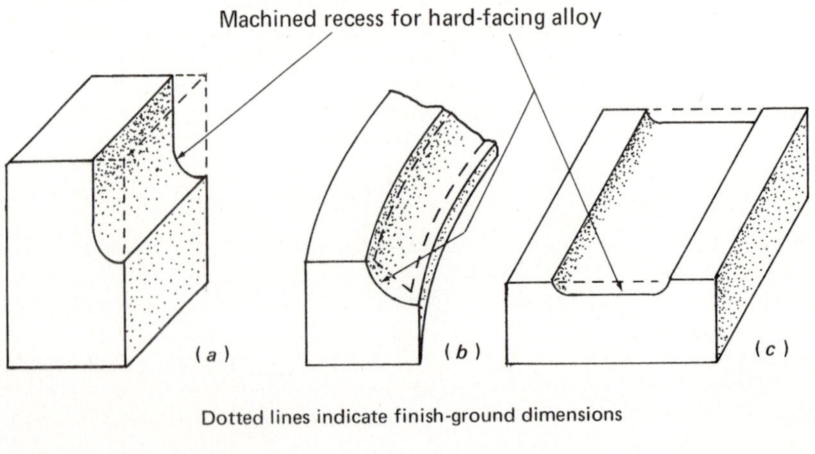
Machined recess for hard-facing alloy
(a)
(b)
(c)
Dotted lines indicate finish-ground dimensions

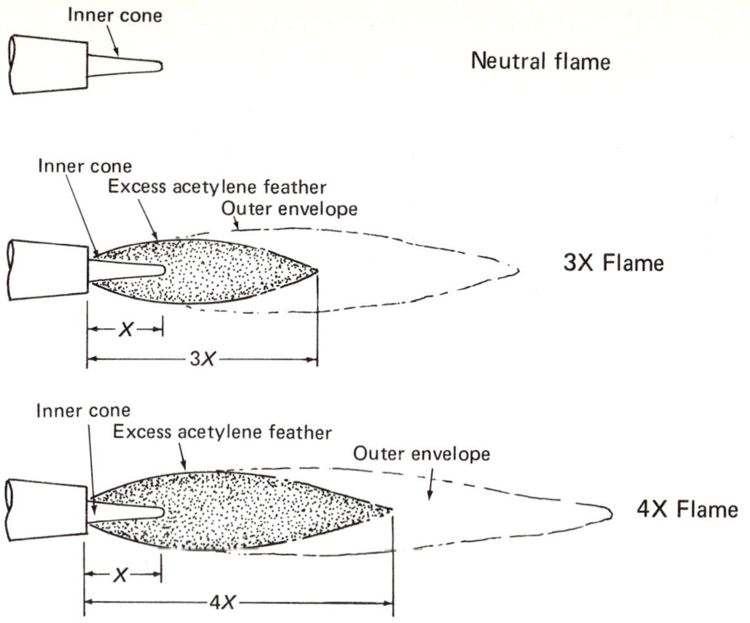
Inner cone
Neutral flame
Inner cone
Excess acetylene feather
Outer envelope
3X Flame
X
3X
Inner cone
Excess acetylene feather
Outer envelope
4X Flame
X
4X

Figure **17-3** Various oxyacetylene flames used for hard surfacing

17-7 TYPE AND LENGTH OF OXYACETYLENE FLAME USED FOR HARD SURFACING

An excess acetylene or carbonizing flame should always be used for hard surfacing. The excess acetylene flames are used (1) to spread the heat to minimize possible burn-through on thin edges, and (2) to add excess carbon to the surface of the part being hard faced. The additional carbon lowers the melting point and "sweating" temperature, facilitating the deposit of the hard-surfacing alloy.

With very few exceptions, the acetylene feather is three times longer than the inner cone (Figure 17-3). Because of these few exceptions, it is recommended that the manufacturer of the particular alloy being used be consulted as to the precise shape of the oxyacetylene flame to be used.

17-8 HARD SURFACING STEEL WITH THE OXYACETYLENE TORCH

OBJECTIVE To learn to rebuild the surface of carbon steel.

HOURS REQUIRED 2.

EQUIPMENT, TOOLS, AND MATERIALS REQUIRED

1. One oxyacetylene welding outfit and accessories. (*Note:* The torch tip should be approximately three times larger than would be used to weld steel of the same thickness).

2. Several 3 × 4 × ¼-inch low-carbon steel pieces.

3. Sufficient ⅛-inch-diameter AWS Group I welding rods.

4. Equipment and/or material for annealing.

PROCEDURE: Read the entire procedure below until you are thoroughly familiar with it.

1. Assemble the equipment for lighting.

2. Place the steel plate to be surfaced on the welding table.

3. Light the torch.

4. Adjust the flame to suit the hard-surfacing alloy to be used.

5. Put your goggles over your eyes.

6. Hold the pipe flame on a small area of the part to be surfaced until the carbon from the excess acetylene flame disappears and the surface of the steel at the point being heated assumes the watery, glazed appearance known as *sweating* [Figure 17-4(*a*)].

7. Bring the end of the welding rod into the flame and allow it to melt and spread over the sweating area [Figure 17-4(*b*)].

a. If the procedure is done correctly, the welding rod will immediately begin to flow in a manner similar to a brazing rod on a properly heated surface. If this does not happen, the metal is not hot enough. *Stop,* and repeat steps 6 and 7.

b. Penetration and puddling should be avoided because they will cause dilution of the surfacing material with the iron from the base metal, as well as excess carbon pick-up from the acetylene flame.

c. An excess of carbon in the deposit, especially in deposits of the cobalt-based alloys, will cause them to become magnetic and lower their corrosion resistance.

8. As the alloy spreads, bring the rod into the flame again and, with the end of the rod touching the puddle, melt more alloy into the spreading puddle.

9. If any particle of dirt, rust, or scale appears in the puddle, float it to the surface. If it will not float, dislodge it with the welding rod.

10. Repeat step 9 until the area to be surfaced is fully coated.

11. Move the puddle in the desired direction by means of flame pressure, not by stirring with the rod. Usually the work is done by use of the forehand method; however, the backhand method can be used if the sections are less than ⅛ inch thick or scale badly.

12. Inspect the completed surface for pinholes. If any pinholes are found, heat the area around each pinhole to a dull red, then melt

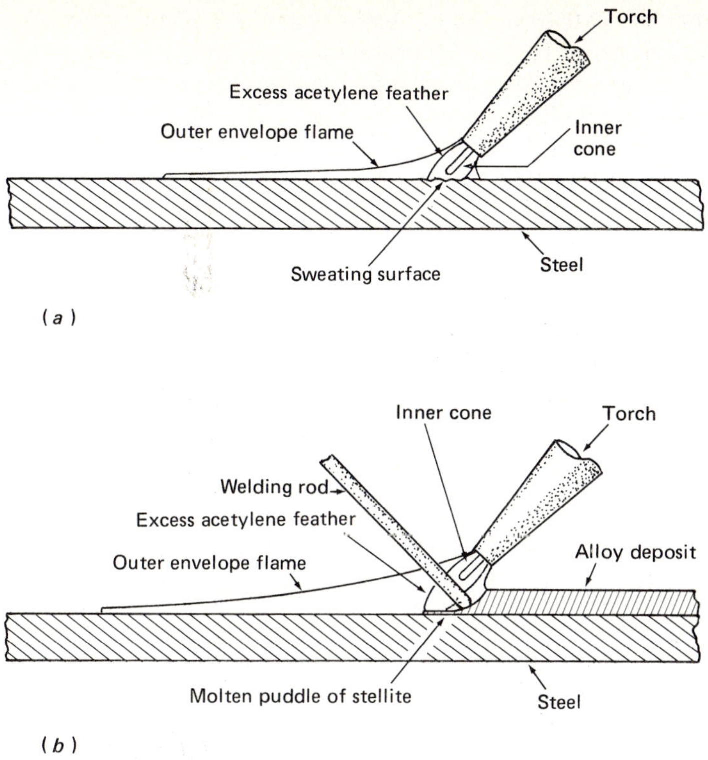
Torch
Excess acetylene feather
Outer envelope flame
Inner cone
Sweating surface
Steel
(a)
Inner cone
Torch
Welding rod
Excess acetylene feather
Outer envelope flame
Alloy deposit
Molten puddle of stellite
Steel
(b)

Figure **17-4** Sweating surface of metal to be hard surfaced and applying metal manually with the oxyacetylene torch. (Redrawn by permission from Union Carbide Corporation, Linde Division, *The Oxy-Acetylene Handbook*, 2d ed., 1960, Figs. 265, 266.)

down into the hole to float off impurities. Allow the molten metal to close in slowly and add a little filler metal. Slowly draw the heat away from the hot spot into the body of the workpiece. This will prevent quick cooling and shrinkage.

13. If a subsequent pass is to be made, care must be taken to melt only the surfaces or the hard-surfacing materials of the previous pass. Melting down into the base metal will cause dilution.

14. After the hard-surfacing job has been completed, the entire part should be warmed to bring it to approximately the same temperature, then covered with a ½–¾-inch layer of powdered lime or warm construction sand, or placed in an annealing furnace that has been preheated to the preheating temperature of the material, and allowed to cool slowly. Of course, any other method that will produce the same annealing effect may be used instead of the methods mentioned above.

15. Remove the part from the annealing envelope (lime, sand, furnace, or other). *Note:* The workpiece may be brought to an exact size or have high spots removed from it by hand grinding on a machine equipped with a grinding wheel of the A-60-L-5-V or similar specification. The wheel should rotate at a velocity of somewhere between 2800 and 4200 surface feet per minute.

17-9 SOME TIPS FOR HARD SURFACING CAST IRON

Cast iron is extremely crack sensitive, so rebuilding and hard facing generally are not recommended. However, some cast-iron parts subject primarily to straight abrasion are in use. They wear out and are hard faced. Although the preparative and manipulative skills required are the same as those required for steel, there are some procedural differences:

1. Because cast irons do not "sweat" like steel, the oxyacetylene feather should be 1½ times longer than the inner cone.

2. Apply a cast-iron welding flux to the area to be welded.

3. Test the workpiece for the correct welding temperature by breaking the surface crust with the end of the rod. This procedure should be followed until the entire area to be hard surfaced is covered.

4. Peen the entire surface deposit—this will help relieve stresses that build up during cooling.

5. If necessary, repeat steps 2–4 until the desired thickness has been reached.

a. Peening must always be accomplished after every deposit, including the last one.

b. To restrict the melting to the metal that is very close to the surface, cast-iron parts ¼ inch or less in thickness should be backed up with carbon paste or any other suitable heat sink.

17-10 HARD SURFACING WITH TUNGSTEN CARBIDE

The acetylene feather used for this purpose should be approximately 1½ times longer than the inner cone, and the surfacing alloy should be laid down without penetrating as deep into the workpiece, the same as for mild steel.

17-11 HARD-SURFACING MATERIALS OTHER THAN CARBON STEEL

Unless it is absolutely necessary, the oxyacetylene or GTAW process should not be used for hard surfacing such heat-sensitive materials as the 321- and 400-series stainless steels, or all alloy steels containing more than 0.75% manganese silicon. If it is necessary to use the oxyacetylene flame, the steels must be carefully preheated and postheated (Chapter 20).

High-speed steels should be fully annealed before surfacing and should be kept heated as evenly as possible during the surfacing operation. As a general rule, however, the hard surfacing of high-speed steel is not recommended because often cracks will be formed in the base metal under the hard surface.

Because of the difficulty in controlling heating and cooling and, therefore, in preventing melting, beginning welders should not attempt to hard surface copper, brass, or other alloys having similar low melting points.

HARD SURFACING BY ELECTRIC-ARC WELDING

17-12 SAFE PRACTICES

The safe practices detailed in Chapters 13–15 are, depending on the process used, also applicable to hard surfacing.

17-13 PREPARING THE WORKPIECE FOR HARD SURFACING

The preparatory procedures listed in section 17-6 are also applicable for hard surfacing by arc welding.

17-14 HARD-SURFACING MANIPULATIVE SKILLS AND WORKPIECE POSITIONING

When possible, the part to be hard surfaced should be positioned for welding from left to right (downhand welding). When downhand welding is not practical, an all-position electrode should be used. The manipulative skills needed for hard surfacing by manual electric-arc welding are also described in Chapter 14.

17-15 BUILD-UP MATERIALS AND BASE METALS FOR MANUAL ELECTRIC-ARC WELDING

Considerable difference exists between welding materials used to build up worn equipment and those used for hard surfacing. Because most hard-surfacing materials should be limited to two layers, badly worn parts must be restored within 3/16 to 3/8 inch of finished size, with an appropriate build-up material prior to hard surfacing. The Stoody Company of Whittier, California, recommends the conditions detailed in Table 17-2, beginning on page 439, as suitable for manual welding with the electric arc.

(*Text continues on page* 454)

Table **17-2** Conditions recommended for manual electric-arc welding

BUILD-UP AND JOINING MATERIALS FOR MANUAL APPLICATION			
NOMENCLATURE	NOMINAL COMPOSITION	GENERAL DESCRIPTION	MECHANICAL PROPERTIES (All hardness readings based on Rockwell C scale and designated as Rc)
Stoody Build-up TWIN-COTE® (coated only)	Alloy content—6% Chromium Manganese Silicon Carbon Iron base	Carbon steel core wire with alloys in extruded iron-powder coating; for AC or DC electric application to carbon and low alloy steels (not manganese steel or cast iron) as a build-up material or an underbase for hard-facing.	Hardness: 2 passes (weave beads)—med. carbon steel 24–28 Rc 2 passes (weave beads)—med. carbon steel with 500°F interpass temp. 20–22 Rc 2 passes (stringer beads)—med. carbon steel 31–35 Rc 2 passes (stringer beads)—mild steel 29–33 Rc 5 passes (stringer beads) 32–36 Rc *Hardness can be increased by water quenching from 1600°F. Anneal by slow cooling from 1600°F to 1300°F and air or furnace cool from 1300°F.* Tensile strength 118,000 p.s.i. Yield strength 113,000 p.s.i. Elongation in 2 in. 6%
Stoody Dynamang Coated only	Alloy content—21% Manganese Nickel Chromium Silicon Carbon Iron base	Carbon steel core wire with alloys in extruded coating.	Hardness: 1 pass (weave bead)—mang. steel 14–16 Rc; work-hardens to 45–50 Rc 2 passes (weave beads)—mang. steel 17–20 Rc; work-hardens to 45–50 Rc All weld metal 9–12 Rc; work-hardens to 45–50 Rc Tensile strength DC reverse polarity 125,000 psi Yield strength DC reverse polarity 81,000 psi Elongation in 2 in. DC reverse polarity 46.0%
Stoody 2110 Coated only	Alloy content—37% Chromium Manganese Nickel Silicon Carbon Iron base	AC-DC chromium tubular electrode with extruded titania-type coating. This alloy is a modified high chromium-high manganese stainless steel that combines toughness and wear-resistance.	Hardness: 2 passes—1020 steel 15–17 Rc As work-hardened 40–45 Rc 2 passes—mang. steel 19–23 Rc As work-hardened 42–47 Rc 5 passes—mang. steel 21–24 Rc As work-hardened 46–48 Rc Tensile strength 129,000 p.s.i. Yield strength 85,000 p.s.i. Elongation in 2 in. 33½%

Stoody Build-up TWIN-COTE

WELDING PROCEDURE	DIA.	LENGTH	RECOMMENDED AMPERAGE		NO. RODS PER POUND	AREA COVERED PER POUND	DEPOSIT CHARACTERISTICS	RECOMMENDED USES
			DC	*AC*				
Can be applied AC or DC, either polarity in stringer or weave beads. Maximum deposition with DC straight polarity or AC. Operates well with drag arc or normal arc; short arc reduces spatter. Can be welded vertically, starting at the bottom and welding upward. Note: Use wide weave beads (1″ to 2″) when depositing high alloy hard-facing materials over build-up. Avoid severe quench.	1/8″ 5/32″ 3/16″ 1/4″	14″ 14″ 14″ 18″	80–160 140–230 180–280 200–330	90–185 155–250 200–310 220–360	14 8.8 6.3 2.9	1/8″ depth 23–25 sq. in.	Tough, dense, flawless deposits of high compressive strength provide ample support for hard-facing overlays. Forgeable and resistant to cold flow. Doesn't work-harden. Preheat and slow cool prior to machining; carbide tools are recommended. Strongly magnetic on carbon and low alloy steel.	Build-up of carbon and low alloy steels; final overlay on parts that must be machined; underbase for hard-facing alloys. Should not be used for joining.

Table **17-2** *(cont.)*

	WELDING PROCEDURE	DIA.	LENGTH	RECOMMENDED AMPERAGE	NO. RODS PER POUND	AREA COVERED PER POUND	DEPOSIT CHARACTERISTICS	RECOMMENDED USES
	BUILD-UP AND JOINING MATERIALS FOR MANUAL APPLICATION							
Stoody Dynamang	Coated electrode runs well on both polarities; for fastest deposition use straight polarity. AC stability is excellent. Use crescent weave. Use DC reverse polarity for out-of-position and fusion welding; use DC straight polarity for rebuilding manganese parts.	*Coated* 5/32″ 3/16″ 1/4″ 5/16″	*Coated* 14″ 14″ 18″ 18″	*Coated—AC–DC* 140–185 175–230 200–275 225–325	*Coated* 8 5.5 2.3 1	1/8″ depth 21–23 sq. in. 3/16″ depth 16–17 sq. in.	Deposits are tough, free of cracks and porosity; work harden rapidly and are non-magnetic.	For joining manganese wear plates or castings to earth-moving equipment, repair of large, difficult to position manganese parts, and a variety of other applications. Can be used for joining dissimilar metals, when proper welding technique is maintained to insure low dilution of the base metal.
Stoody 2110	Operates well on DC, either polarity, and AC; DC reverse is generally preferred. Weldability is excellent with low spatter. Build-up is superior to other electrodes of this type. Use stringer or weave beads.	5/32″ 3/16″ 1/4″	14″ 14″ 14″	*DC* 125–190 150–260 240–325	7 5 3	1/8″ depth 27–29 sq. in.	Deposits are dense, porosity-free, extremely tough and work-harden rapidly. Cannot be flame cut; machinable with carbide tools; nonmagnetic.	For rebuilding manganese and carbon steel parts (shovel pads, rail frogs and switch points, etc.); for a wear-resistant overlay on very high impact applications (roll crushers, hammers, etc.); and joining manganese, manganese and carbon steels and some low alloy steels. Not recommended for cast iron.

LOW ALLOY HARD-FACING MATERIALS FOR MANUAL APPLICATION

NOMENCLATURE	NOMINAL COMPOSITION	GENERAL DESCRIPTION	MECHANICAL PROPERTIES (All hardness readings based on Rockwell C scale and designated as Rc)
Stoody 1105 Coated only	Alloy content—8% Manganese Chromium Silicon Molybdenum Vanadium Carbon Iron base	Iron-powder coating extruded on carbon steel core wire provides superior weldability and extremely high deposition rate.	Hardness: (½" weave beads—air cooled) 2 passes—mild steel (reverse polarity) 36–39 Rc 2 passes—1045 steel (reverse polarity) 38–42 Rc 2 passes—mild steel (straight polarity) 37–40 Rc 2 passes—1045 steel (straight polarity) 42–44 Rc 2 passes—mild steel (AC) 39–42 Rc 2 passes—1045 steel (AC) 42–45 Rc 4 passes—1045 steel (straight polarity) 44–47 Rc
Stoody 1027 Coated only	Alloy content—10% Chromium Manganese Silicon Carbon Iron base	Extruded graphitic-coated electrode for AC-DC electric application. Excellent for all-position work	Hardness: 2 passes (weave beads)—med. carbon steel 45–49 Rc 2 passes (weave beads)—mang. steel 15–19 Rc (work-hardens to 45–50 Rc) 2 passes (weave beads)—cast iron 48–50 Rc (500°F interpass temp.) Melting point 2600°F Specific gravity 7.8
Stoody Self-hardening Coated only	Alloy content—13% Chromium Manganese Silicon Carbon Iron base	Available as carbon steel core wire with alloys in extruded iron powder coating for AC-DC electric application. Best of low alloy group for hard-facing parts subject to high impact and moderate abrasion.	Hardness: All weld metal 54–58 Rc 2 passes (weave beads)—mild steel 52–56 Rc Water-quenched from 1700°F 56–59 Rc Furnace-cooled from 1700°F 19–23 Rc 2 passes (weave beads)—1045 steel 54–58 Rc Water-quenched from 1700°F 56–60 Rc Furnace-cooled from 1700°F 19–23 Rc Melting point 2525°F Specific gravity 7.8

Table **17-2** (*cont.*)

LOW ALLOY HARD-FACING MATERIALS FOR MANUAL APPLICATION

	WELDING PROCEDURE	DIA.	LENGTH	RECOMMENDED AMPERAGE DC	RECOMMENDED AMPERAGE AC	NO. RODS PER POUND	AREA COVERED PER POUND	DEPOSIT CHARACTERISTICS	RECOMMENDED USES
Stoody 1105	Can be applied in stringer or weave beads; operates well with drag or normal arc. Hold electrode at 30° to 45° in direction of welding. Runs well on AC and DC, either polarity; smoothest operation and highest deposition rate with DC straight polarity or AC. Limit layers to four.	5/32″ 3/16″ 1/4″	14″ 14″ 18″	140–210 165–250 230–320	150–230 180–275 250–350	9 6 2.8	1/8″ depth 23–25 sq. in.	Bonds readily to carbon and low alloy steels; not recommended for cast iron. Deposit properties are same as those of Stoody 105. Forgeable and machinable with Grade 883 Carboloy or equivalent. Magnetic on carbon and low alloy steels; nonmagnetic on manganese.	Recommended for hard-facing tractor rollers and idlers, arch wheels, shovel rollers and idlers, sprockets, drive tumblers, churn drills, charging car wheels and similar parts involving high impact, abrasion and metal-to-metal wear.
Stoody 1027	Operates well on AC or DC; DC reverse polarity generally preferred. Multiple passes of any type of bead can be used. Can be operated over wide amperage range depending on part size and type of bead.	1/8″ 5/32″ 3/16″ 1/4″	14″ 14″ 14″ 14″	90–130 120–160 140–220 175–250		16 11 7.5 4	1/8″ depth 26–28 sq. in.	Deposits are sound and smooth with minimum spatter loss; graphitic coating eliminates slag removal problems. Magnetic on carbon, low alloy steels and cast iron; nonmagnetic on manganese. Forgeable at red heat; subject to heat treatment.	For surfacing new or worn parts to resist impact and moderate abrasion. Also for applications involving metal-to-metal wear. Best choice for cast iron parts.

Stoody Self-hardening	Apply AC or DC, straight or reverse polarity; use straight polarity for maximum density—drag, normal or long arc. Can be applied vertically or out-of-position. Limit weave beads to ¾″ or use stringers.	⅛″ 5/32″ 3/16″ ¼″	14″ 14″ 14″ 18″	100–150 150–200 175–275 240–400	12 7 5 2	⅛″ depth 22–24 sq. in.	Can be applied to plain or alloy steels; magnetic on carbon or low alloy steels. Can be forged at red heat; not readily machinable.	Slightly higher alloy content gives this material more wear resistance than Stoody 1027, although both materials are generally recommended for similar applications.

HIGH ALLOY HARD-FACING MATERIALS FOR MANUAL APPLICATION

NOMENCLATURE	NOMINAL COMPOSITION	GENERAL DESCRIPTION	MECHANICAL PROPERTIES (All hardness readings based on Rockwell C scale and designated as Rc)
Stoody 21 Coated only	Alloy content—25% Chromium Manganese Silicon Molybdenum Zirconium Carbon Iron base	Furnished as fabricated tubular electrode with extruded graphitic coating.	Hardness: 2 passes—mang. steel 46–50 Rc 2 passes—med. carbon 52–56 Rc 2 passes—cast iron 55–59 Rc (500°F interpass temp.) Melting point 2450°F Specific gravity 7.75
Stoody 31 Coated only	Alloy content—35% Chromium Manganese Silicon Molybdenum Carbon Iron base	Extruded coated fabricated tubular electrode for AC-DC application. For abrasion resistance coupled with moderate impact where a minimum of cross checking and sound deposits are desirable. Good anti-galling properties. An all position electrode.	Hardness: 2 passes (weave beads) med. carbon steel 47–49 Rc 2 passes (weave beads) mang. steel 45–48 Rc Melting point 2400°F Specific gravity 7.80 For hot wear applications up to 950°F
Coated tube Stoodite Coated only	Alloy content—39% Chromium Manganese Silicon Molybdenum Zirconium Carbon Iron base	Graphitic-coated fabricated tubular electrode for AC-DC application. Generally has higher abrasion resistance than Stoody 21, although both alloys are often recommended for like applications.	Hardness: 2 passes (weave beads)—med. carbon steel 57–61 Rc 2 passes (weave beads)—mang. steel 47–51 Rc Melting point 2350°F Specific gravity 7.90 For hot wear applications up to 950°F

Table **17-2** *(cont.)*

HIGH ALLOY HARD-FACING MATERIALS FOR MANUAL APPLICATION

NOMENCLATURE	NOMINAL COMPOSITION	GENERAL DESCRIPTION	MECHANICAL PROPERTIES (All hardness readings based on Rockwell C scale and designated as Rc)
Tube Stoodite Bare only	Alloy Content—36% Chromium Manganese Boron Silicon Carbon Iron base	Bare fabricated tubular rod for oxy-acetylene application	As deposited hardness 56–58 Rc Melting point 2275°F Specific gravity 7.45
Stoody 2134 Coated only	Alloy content—44% Chromium Molybdenum Manganese Silicon Nickel Vanadium Carbon Iron base	Graphitic-coated fabricated tubular electrode for AC-DC application.	Hardness: 2 passes (weave beads)—1045 plate As welded 56–60 Rc Lime-cooled from 1750°F 48–51 Rc Water-quenched from 1750°F 63–65 Rc 2 passes (weave beads)—mang. steel As welded 45–50 Rc Deposits may work-harden 5 to 6 points. For hot wear applications up to 950°F

	WELDING PROCEDURE	DIA.	LENGTH	RECOMMENDED AMPERAGE	NO. RODS PER POUND	AREA COVERED PER POUND	DEPOSIT CHARACTERISTICS	RECOMMENDED USES
Stoody 21	Electrode can be applied AC or DC; DC reverse polarity is recommended. No slag interference. Avoid severe quench of work. Limit layers to two. Apply in weave beads.	 ⅛″ 5⁄32″ 3⁄16″ ¼″	 14″ 14″ 14″ 14″	*AC–DC* 90–130 120–160 140–220 175–300	 18 11 8.2 5	 ⅛″ depth 29–31 sq. in.	Bonds well with carbon or alloy steel, including manganese and cast iron. Surface checks relieve stresses and help prevent warpage. Slightly magnetic on carbon and low alloy steels; non-magnetic on manganese. Not machinable or forgeable.	Recommended as overlay on new or worn parts to resist severe abrasion and impact. Used primarily in construction, rock products, brick and clay, mining, agriculture.

Stoody 31	Can be applied AC or DC either polarity. Use DC reverse polarity for out of position welding. Can be applied with stringer or weave beads.	1/8" 5/32" 3/16" 1/4"	14" 14" 14" 14"	*AC–DC* 100–120 120–165 170–240 250–325	13.5 8.2 6 4	1/8" depth 28–30 sq. in.	Bonds well with carbon or alloy steels and manganese. Deposits polish to mirror finish. Slightly magnetic on carbon steels; nonmagnetic on manganese steels. Not machinable or forgeable. Multi-layer deposits are extremely sound. Can be heat-treated. Has minimum tendency to develop cross-checks.	An outstanding high alloy electrode with exceptional welding characteristics. Recommended where sound deposits with low coefficient of friction are desired in construction, cement and steel industries. Deposits provide excellent bearing surfaces on friction type guides, cement mill gudgeons, etc. Also recommended for dredge parts, runners, and pumps.
Coated tube Stoodite	Can be applied AC or DC; best results generally obtained with DC reverse polarity. Use minimum amperage for out-of-position work. Weave passes are recommended over stringer beads.	1/8" 5/32" 3/16" 1/4"	14" 14" 14" 18"	90–120 100–150 120–175 175–250	22 14 9 4	1/8" depth 28–30 sq. in.	Bonds with carbon or low alloy steels, manganese and cast iron. Surface checks relieve stresses and reduce warpage. Slightly magnetic on carbon and low alloy steels; nonmagnetic on manganese steels. Not machinable or forgeable.	An outstanding high alloy material used chiefly in construction, brick and clay, mining, rock products, and cement industries.

Table **17-2** (*cont.*)

HIGH ALLOY HARD-FACING MATERIALS FOR MANUAL APPLICATION

	WELDING PROCEDURE	DIA.	LENGTH	RECOMMENDED AMPERAGE	NO. RODS PER POUND	AREA COVERED PER POUND	DEPOSIT CHARACTERISTICS	RECOMMENDED USES
Tube Stoodite	Use 20 to 40 drill-size tip; use excess acetylene feather four times length of inner cone. Clean rust, scale and dirt from part. Heat part to "sweating" temperature and apply Tube Stoodite with minimum of penetration. 1/16″ minimum deposit thickness is recommended.		14″,28″ 14″,28″ 28″	— — —	13,6.5 8,4.3 2.6	1/16″ depth 56–60 sq. in.	Deposits polish to a mirror finish under earth abrasion; low coefficient of friction. Not forgeable or machinable; magnetic on carbon and low alloy steels.	Ideal for metal-to-metal wear and earth abrasion. Particularly suited where thin deposits are required. Recommended for all types of farm implements.
Stoody 2134	Can be applied AC or DC, straight or reverse polarity. Use DC straight polarity for maximum deposition rate. Can be applied in stringer beads, but weave passes are recommended. Limit deposits to two layers.	5/32″ 3/16″ 1/4″	14″ 14″ 18″	*DC* *AC* 100–140 110–155 150–190 165–210 180–275 200–300	12.3 8 4	1/8″ depth 28–30 sq. in.	Bonds readily to carbon, low alloy and manganese steels. Magnetic on carbon and low alloy steels; non-magnetic on manganese steel. Deposits are sound and dense. Surface checks relieve stresses and reduce warpage. Not machinable or forgeable.	Developed primarily for earth-working equipment, mill hammers, crushing and similar tools subject to severe abrasion with moderate to heavy impact; compressive strength is very high.

NON-FERROUS HARD-FACING MATERIALS FOR MANUAL APPLICATION

NOMENCLATURE	NOMINAL COMPOSITION	GENERAL DESCRIPTION	MECHANICAL PROPERTIES (All hardness readings based on Rockwell C scale and designated as Rc)
Stoody 6 Coated and bare	Chromium Tungsten Carbon Silicon Manganese Cobalt base	Cast rod bare for oxy-acetylene and coated for DC electric application. Non-ferrous, cobalt-chromium-tungsten alloy. Has greater impact strength than Stoody 1, but is slightly less wear-resistant. Stoody 6 is manufactured to conform to the chemistry and usability requirements of the following specifications: MIL-R-17131A*, Type MIL-RCoCr-A; AWS A5. 13-70, Types RCoCr-A (Bare) and ECoCr-A (Coated); AMS 5788.	Hardness: 2 passes (weave beads) Reverse polarity 38–40 Rc Straight polarity 43–45 Rc 1 pass (oxy-acetylene) 41–46 Rc Melting point 2325°F Specific gravity 8.57 For hot wear applications up to 1300°F
Stoody 12 Bare only	Chromium Tungsten Carbon Silicon Manganese Cobalt base	Cast rod bare for oxy-acetylene application. Non-ferrous, cobalt-chromium-tungsten alloy. Mechanical properties are intermediate to Stoody 1 and 6. Stoody 12 is manufactured to conform to the chemistry and usability requirements of AWS A5. 13-70, Type RCoCr-B.	Hardness: 1 pass (oxy-acetylene) 47–52 Rc Melting point 2300°F Specific gravity 8.71 For hot wear applications up to 1300°F
Stoody 1 Coated and bare	Chromium Tungsten Carbon Silicon Manganese Cobalt base	Cast rod bare for oxy-acetylene and coated for DC electric application. Non-ferrous, cobalt-chromium-tungsten alloy. Stoody 1 is manufactured to conform to the chemistry and usability requirements of the following specifications: MIL-R-17131A*, Type MIL-RCoCr-C; AWS A5. 13-70, Types RCoCr-A (Bare) and ECoCr-C (Coated).	Hardness: 2 passes (weave beads) Reverse polarity 51–53 Rc Straight polarity 56–58 Rc 1 pass (oxy-acetylene) 53–55 Rc Melting point 2275°F Specific gravity 8.85 For hot wear applications up to 1300°F

Table **17-2** *(cont.)*

NON-FERROUS HARD-FACING MATERIALS FOR MANUAL APPLICATION

NOMENCLATURE	NOMINAL COMPOSITION	GENERAL DESCRIPTION	MECHANICAL PROPERTIES (All hardness readings based on Rockwell C scale and designated as Rc)
Stoody C Coated only	Alloy content 43% Chromium Manganese Silicon Molybdenum Carbon Vanadium Tungsten Iron Nickel base	Cast nickel-base coated electrode for DC reverse electric arc application. Resistant to high heat and corrosion accompanied by abrasion and/or impact.	Hardness: 2 passes—mild steel 13–15 Rc 2 passes—stainless steel 19–21 Rc Deposits will work-harden and age-harden in service to 40–45 Rc with minimum deformation.
Stoody 6N Bare only	Alloy content 27% Chromium Carbon Silicon Iron Boron Nickel base	Cast nickel base hard-facing rod for oxy-acetylene application.	Hardness: As deposited 56 Rc Typical hot hardness values 1200°F 53 Rc 1400°F 49 Rc 1600°F 38 Rc

	WELDING PROCEDURE	DIA.	LENGTH	RECOMMENDED AMPERAGE	NO. RODS PER POUND	AREA COVERED PER POUND	DEPOSIT CHARACTERISTICS	RECOMMENDED USES
Stoody 6	Weld DC reverse polarity. Use minimum amperage; apply weave bead ¾″ to 1½″ wide. For check-free deposits, preheat and slow cool. In applying bare rod, use a larger torch tip than is generally used for same diameter mild steel. Use excess acetylene feather three times length of inner cone. Generally limited to two layers.	⅛″ 5/32″ 3/16″ ¼″ 5/16″ ⅜″	*Coated* *Bare** 9″–10″ 8″–14″ 11″–14″ 8″–14″ 12″–14″ 8″–14″ 12″–14″ 8″–14″ — 8″–14″ — 8″–14″	*Coated* 75–100 100–150 125–185 175–250 — —	*Coated* *Bare* 22 22 10 13 7.4 9 4.3 4.8 — 3.3 — 2.2	Bare—⅛″ depth—24–26 sq. in. Coated—⅛″ depth—21–23 sq. in.	Withstands many common corrosive media. Deposits are smooth and normally acquire mirror-like finish in use. Deposits retain wear resistance at high temperatures. Nonmagnetic; not forgeable; machinable with carbide tools. Bonds well with weldable alloy steels, including stainless.	Recommended for metal-to-metal abrasion and high impact applications involving high temperatures and/or corrosive media. Typical applications are valves of all kinds, shear blades, hot punches, saw guides, etc.
Stoody 12	Use a larger torch tip than is generally recommended for same diameter mild steel. Use excess acetylene feather three times length of inner cone. Generally limited to two layers.	 ⅛″ 5/32″ 3/16″ ¼″ 5/16″	*Bare** 8″–14″ 8″–14″ 8″–14″ 8″–14″ 8″–14″	 — — — — —	*Bare* 22 13 9 4.8 3.3	Bare—⅛″ depth—21–23 sq. in.	Deposits are smooth and acquire high polish in use. Mechanical properties retained at high temperatures. Nonmagnetic; not forgeable. Machined with difficulty using carbide tools. Bonds well with weldable alloy steels, including stainless	Recommended for metal-to-metal abrasion involving high temperature and/or corrosive media with moderate impact.

Table **17-2** *(cont.)*

NON-FERROUS HARD-FACING MATERIALS FOR MANUAL APPLICATION

	WELDING PROCEDURE	DIA.	LENGTH		RECOMMENDED AMPERAGE	NO. RODS PER POUND		AREA COVERED PER POUND	DEPOSIT CHARACTERISTICS	RECOMMENDED USES
			Coated	*Bare**	*Coated*	*Coated*	*Bare*			
Stoody 1	Weld DC reverse polarity. Use minimum amperage; apply weave bead ¾″ to 1½″ wide. For check-free deposits, preheat and slow cool. In applying bare rod, use a larger torch tip than is generally used for same diameter mild steel. Use excess acetylene feather three times length of inner cone. Generally limited to two layers.	⅛″	9″–10″	8″–14″	75–100	22	22	Coated—⅛″ depth—21–23 sq. in. Bare—⅛″ depth—21–23 sq. in.	Deposits are smooth; acquire mirror-like finish in use and retain wear resistance at high temperatures. Non-magnetic; not forgeable. Machined with difficulty using carbide tools. Bonds well with weldable alloy steels, including stainless.	For applications involving severe abrasion accompanied by heat and/or corrosion with moderate impact.
		5/32″	11″–14″	8″–14″	100–150	10	13			
		3/16″	12″–14″	8″–14″	125–185	7.4	9			
		¼″	12″–14″	8″–14″	175–250	4.3	4.8			
		5/16″	—	8″–14″	—	—	3.3			
		⅜″	—	8″–14″	—	—	2.2			
			Coated		*DC*	*Coated*				
Stoody C	DC reverse polarity is recommended; AC and DC straight polarity are not advisable. Use a reasonably low current, consistent with good weldability, to minimize penetration and dilution. Hold a very short arc. Two or three-layer deposits are recommended for maximum effectiveness.	5/32″	14″		70–150	8		⅛″ depth 23 sq. in. ¼″ depth 16 sq. in.	Deposits are dense, smooth and resistant to high temperature deformation and corrosive agents. Although extremely tough, deposits are readily machinable with high speed or cemented carbide tools.	For applications involving heat accompanied by abrasion and/or impact. Also corrosion resistant. Typical applications are hot work tools and dies such as hot forge dies, hot shear blades, forging hammer dies, mill guides, shafts, etc.
		3/16″	14″		80–180	6				
		¼″	14″		110–200	3.7				

	Normal preheating and post heating requirements should be observed, depending on type of base metal.							
Stoody 6N	Can be deposited to cast iron (with flux), nickel-base alloys, low and medium carbon steels and austenitic stainless steels. Thoroughly clean work piece prior to welding. Observe preheat and postheat requirements where necessary. Contrary to suggested procedure for other hard-facing rods, 6N should be deposited with a neutral oxy-acetylene flame. Can also be applied with TIG process.		*Bare*		*Bare*		Deposits have exceptional hot hardness properties and will not yield to most corrosive media. Resistance to pitting, galling, heat and abrasion is excellent. Developed primarily for oxy-acetylene application, 6 N wets and flows easily in thin deposits or, where necessary, can be stacked. Air-cooled deposits of 6 N are best finished by grinding; machining is not recommended. This alloy can be hot-formed while in the plastic stage.	Deposits have a low-coefficient of friction and are outstanding in metal-to-metal wear. 6N should be used where corrosion is a problem and/or high hot hardness values are necessary. Especially effective on machined pump parts, thrust collars, guides, de-watering equipment, sprocket teeth, expeller screws, etc.
		5/32″	8″–14″	—	13	1/8″depth		
		3/16″	8″–14″	—	9	24–26 sq. in.		
		1/4″	8″–14″	—	4.8			
		5/16″	8″–14″	—	3.3			
		3/8″	8″–14″	—	2.2			

*For certification to MIL-R-17131A, this requirement must be specified at the time order is placed with Stoody Company.
*Length of all bare rods are in conformance to MIL-R-17131A (8″ to 14″)
"TWIN-COTE" is a registered Trade Mark of Stoody Co., Whittier, Calif.

Source: The Stoody Company.

17-16 SINGLE SUBMERGED-ARC HARD-SURFACING PROCEDURE

Most submerged-arc hard-surfacing fluxes are designed to be used with mild-steel electrodes. The alloying elements are in the flux; when the flux is melted, the alloys blend with the molten electrode and base metal to produce an alloy deposit.

The recommended power source is dc variable voltage, although dc constant voltage or ac variable voltage can be used. Direct-current negative (straight polarity) is recommended for the best deposition rate and less penetration into the base metal; ac or dc positive (reverse polarity) can be used, but the deposition rate and build-up are lower, and the alloy content of the deposit tends to be slightly higher. The electrode stick-out should be about 1 inch, with a flux depth just enough to prevent flash-through of the arc.

Almost any automatic or semiautomatic wire feed can be used, but it should be equipped with good meters. Select a travel speed that produces weld beads that are sound, well shaped, and of the desired width and thickness. With a single arc and stringer-bead technique, the travel speed should be within the range listed in Table 17-3.

The content of the alloy in the deposit can be varied by changing the welding procedure (arc voltage and welding current). Once a satisfactory weld deposit has been obtained, it is necessary to strictly adhere to that procedure to obtain consistent results.

The arc voltage and current settings shown in Figure 17-5 were determined by the Lincoln Electric Company and, if used properly, will provide for a constant percentage of alloy deposition.

Table **17-3** Recommended travel speeds for single-arc hard-surfacing techniques

FLUX	ELECTRODE	TRAVEL SPEED (ipm)
Low to medium-high alloy	5/64–3/16	15–40
High alloy	3/32–5/32	25–35

17-17 TWIN-ELECTRODE AND OSCILLATING SUBMERGED-ARC HARD-SURFACING PROCEDURE

Where equipment is available, this procedure will result in more economical beads than would be possible with the single submerged-arc procedure.

Automatic heads can be equipped to feed two electrodes (5/64, 3/32, and 1/8 inch) through one set of drive rolls into one weld puddle. Twin electrodes used on an oscillating welding head produce smooth beads up to 4 inches wide. Typical procedures are given in Table 17-4.

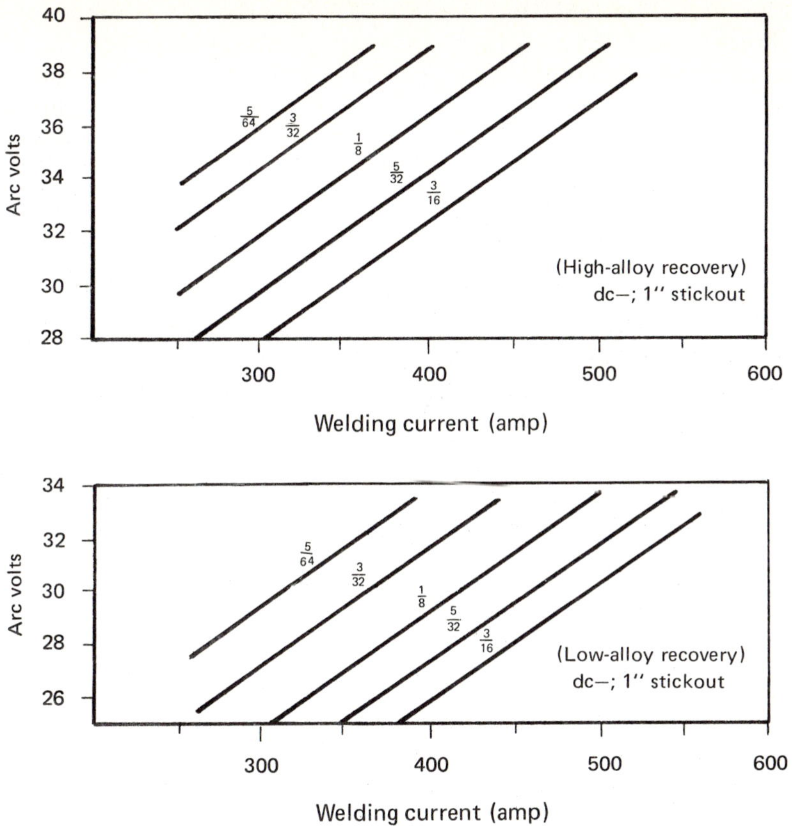

Arc volts
Welding current (amp)
5/64
3/32
1/8
5/32
3/16
(High-alloy recovery)
dc–; 1'' stickout
(Low-alloy recovery)
dc–; 1'' stickout

Figure **17-5** Submerged-arc hard-surfacing procedure developed by the Lincoln Electric Company for high- and low-alloy recovery. (Redrawn by permission from *The Procedure Handbook of Arc Welding*, 11th ed., Lincoln Electric Company, Cleveland, Ohio, pp. 13.7–13.17.)

Table **17-4** Twin-electrode hard-surfacing procedures

FLUX	BEAD SIZE WIDTH (in.)	BEAD SIZE THICKNESS† (in.)	WIRE SPACING (in.)	OSCILLATIONS* PER MIN.	CURRENT DC—AMP	VOLTS	TRAVEL SPEED (ipm)	MIN. PLATE THICKNESS (in.)
Low-alloy	2	1/8–5/32	5/8	30	700	32	8–9	5/16
Medium-high-alloy	4	1/8–3/16	5/8	15	800	35	5	3/8
High-alloy	3	3/16–1/4	3/8	20	700–750	33	4.5	3/8

*One oscillation is a complete cycle over and back.

†Thickness is for one layer. Use one layer only for high-alloy flux. Two layers can be used for medium-high-alloy flux. More than two layers can be used for low-alloy flux. DC positive may also be used but the buildup will be less. Electrode used is 1/8-inch mild steel submerged-arc classification EL12

Source: *The Procedure Handbook of Arc Welding* (Cleveland, Ohio: Lincoln Electric Company), pp 13.7–13.18.

Approximate displacement (*d*) of the electrode

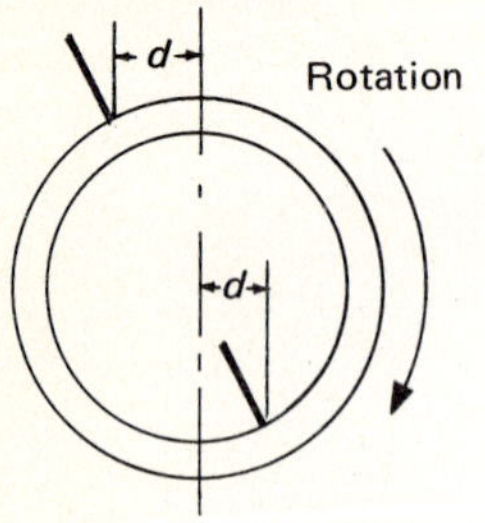
d
Rotation
d

Girth dia. (in.)	Electrode position "*d*" (in. ahead of vertical center)
3-18	$\frac{3}{4}$-1
18-36	$1\frac{1}{4}$-$1\frac{1}{2}$
36-48	$1\frac{1}{2}$-2
48-73	2-$2\frac{1}{2}$
over 18	3

Figure **17-6** Guidelines for circumferential hard surfacing with the submerged arc. (Redrawn by permission from *The Procedure Handbook of Arc Welding,* 11th ed., Lincoln Electric Company, Cleveland, Ohio, pp. 13.7–13.18.)

The oscillation speed can be decreased or increased as necessary to improve bead appearance, increase build-up at the center of the bead, or eliminate scalloped edges.

17-18 CIRCUMFERENTIAL SUBMERGED-ARC HARD-SURFACING PROCEDURES

The Lincoln Electric Company has developed guidelines for the circumferential surfacing of cylindrical objects of various diameters (Figure 17-6).

Recommended circumferential hard-surfacing voltage, current, and off-center distance ranges are given in the Figure 17-7. The following procedures should specifically be noted when attempting to hard-surface cylindrical objects.

THREADING The temperature of the work should be kept below 700°F for easy slag removal and control of spilling. In addition to depositing small beads and using air jets or water cooling (when practical), temperature can be controlled by depositing a stringer bead like a screw thread along the work. Fill the space between the beads with successive threads along the work. Fill the space between the beads with successive welds (see Figures 17-7 and 17-8).

Figure **17-7** Correct method of threading hard surfacing with the submerged arc. (Redrawn by permission from *Hardsurfacing,* Lincoln Electric Company, Cleveland, Ohio, Bulletin No. 3000.2.)

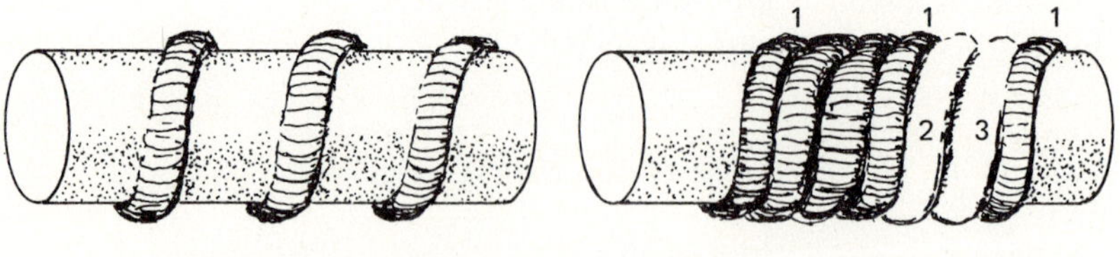
1
1
1
2
3
First threaded bead
Space will be filled with 2nd or 2nd and 3rd beads

Welding current and voltage

Diameter (in.)	Current (amp)
3-6	250-350*
6-12	300-400
12-18	350-500
Over 18	Standard hard-surfacing procedures (single electrode or twin electrodes)

* "Threading" may be necessary.
Normal voltage range is 28-32 volts.

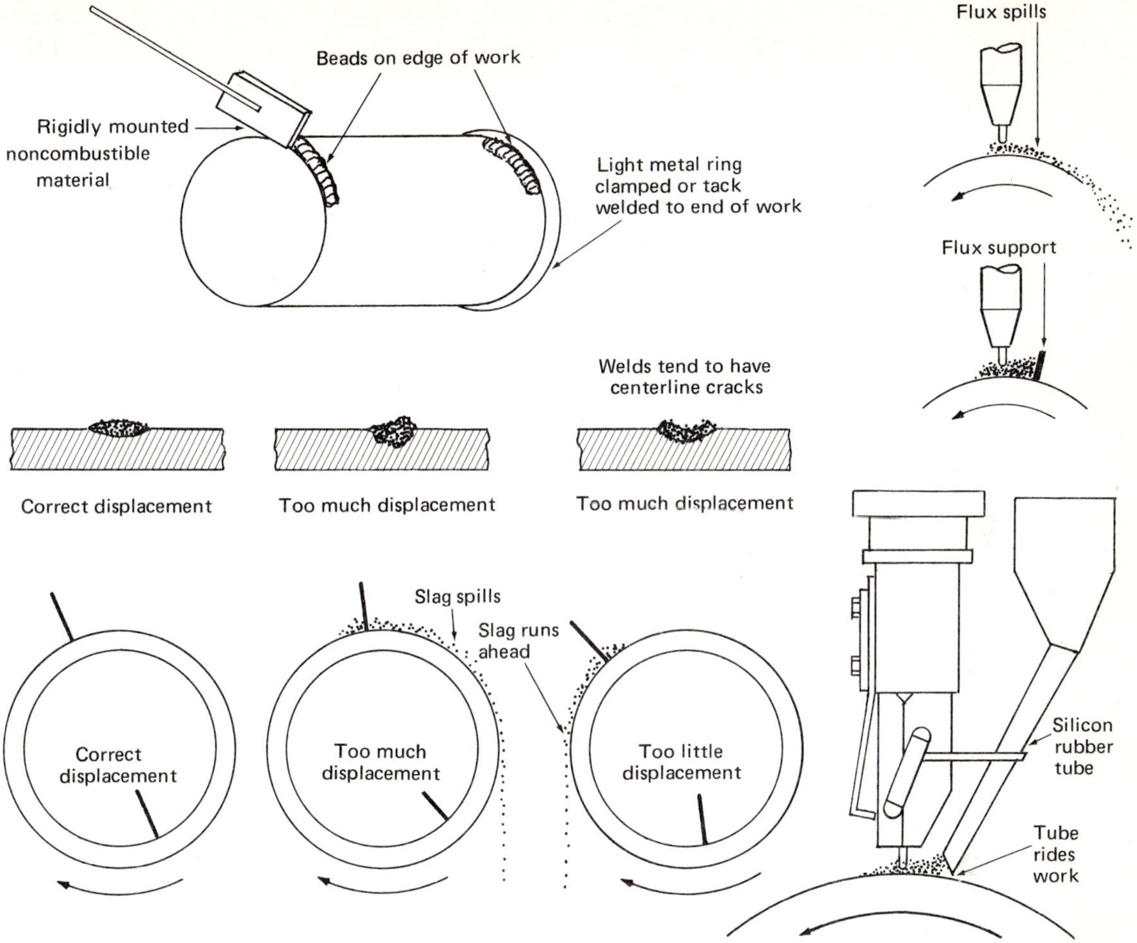
Beads on edge of work
Rigidly mounted noncombustible material
Light metal ring clamped or tack welded to end of work
Flux spills
Flux support
Welds tend to have centerline cracks
Correct displacement
Too much displacement
Too much displacement
Slag spills
Slag runs ahead
Correct displacement
Too much displacement
Too little displacement
Silicon rubber tube
Tube rides work

Figure **17-8** Correct and incorrect methods of applying hard-surfacing materials with the submerged arc. (Redrawn by permission from *Hardsurfacing,* Lincoln Electric Company, Cleveland, Ohio, Bulletin No. 3000.2.)

OVERLAP The amount that one bead overlaps the adjacent bead affects both the admixture of base metal into the weld metal (greater overlap reduces admixture) and the appearance of the finished weld. Control the overlap by adjusting the amount of longitudinal travel with each revolution.

Longitudinal travel is accomplished either by spiraling the bead or indexing the welding head across the work after each complete revolution (step-over). Unless a lathe with a slow screw-feed mechanism is available, the step-over method is recommended because of ease of operation.

For automatic step-over, mount a limit switch that is operated by a cam-type trip on the rotating fixture. Connect the limit switch into the travel carriage–motor circuit so that the motor runs when the switch is operated. The distance moved is controlled by the size of the cam and speed of the travel motor. A time delay can be used in place of the cam.

Slag must be removed before each bead makes a complete revolution. If slag removal is a problem, the threading technique should be considered.

17-19 TEST FOR ABRASION AND IMPACT RESISTANCE

There are no standardized methods for testing abrasion or impact resistance because of the many different types of wear and service environments. There are, however, various tests that are commonly applied to give relative measures of these properties.

One type of test drags the specimen over a copper slab in a mixture of quartz sand and water; the abrasion factor is the loss of weight of the specimen compared to the loss of weight of a standard annealed SAE 1020 steel specimen. A material that loses the same weight as the SAE 1020 steel will have an abrasion factor of 1.00. The ideal material would not lose any weight and would have a factor of 0.00.

Another test presses the specimen against a notched rubber wheel, which is rotated in a slurry of sand and water; the loss of weight is taken as the abrasion value.

A preferred method is the use of a grinding wheel with aluminum oxide as the abradant. The amount of wheel that is ground away to remove a standard weight of the hard-surfacing material is taken as the abrasion index. This test is excellent for evaluating the cutting ability of hard-surfacing materials used to dig earth. The loss of weight of the specimen when run against the wheel at standard conditions is taken as the abrasion value for the material when it is to be used for wear-resistant purposes. It is important to have both values because the material that shows the least loss of weight may not be the one that cuts the best on for example, a rotary well-drilling bit.

The grinding-wheel test is preferred to the wet-sand test because the wet-sand results do not show as much difference in hard-surfacing materials as is usually found in the field. The grinding-wheel method gives a very large difference in weight loss: 0.02-grams loss for a hard grade of sintered tungsten carbide, and about 10.0-grams loss for mild steel—a ratio of 1:500.

Abrasion values are not directly applicable to field service. Some of the variables that can change the order of abrasion resistance of hard-surfacing are temperature, size, properties of the abradant, and the amount of impact.

Impact tests are commonly made on commercial hard-surfacing materials by pounding weld deposits against each other and against standards. The materials are then arranged according to their resistance to cracking, chipping, and spalling.

FLAME-SPRAYED COATINGS

17-20 FLAME-SPRAYED COATINGS

Flame spraying, or *particle impacting* as it is sometimes called, consists of spraying molten material onto a previously prepared surface, to form a coating. The coating material is melted in a flame and then atomized into a fine spray. The impacting particles flatten, interlock, and overlap one another so that they are securely bonded together to form a dense, coherent coating. Because the molten material is accompanied by a blast of air, the part being sprayed is not excessively heated.

The four flame-sprayed methods in common use are:

1. Oxyacetylene (wire, powder, and rod).
2. Oxyhydrogen (powder).
3. Detonation (powder).
4. Plasma (powder).

Of the four methods, the oxyacetylene-flame method is the one most commonly used.

All common metals and alloys can be sprayed by the oxyacetylene wire and powder processes. Metals that can be sprayed by the wire process are limited, of course, to those materials that can be formed into wire. As previously mentioned, the powder process permits the use of an almost infinite variety of metals. With each metal powder, consideration must be given to mesh size and range, grain type, flow properties, thermal properties, and the effects of minor deviations in composition.

Typical of the metals that can be flame sprayed are aluminum, zinc, lead, tin, copper-base alloys, nickel, molybdenum, and mild, high-carbon, and stainless steels, as well as nickel-chromium self-fluxing alloys for hard surfacing. Tungsten and other high-melting-point metals are often applied by the plasma-arc process.

17-21 FLAME-SPRAYING PROCESSES

In the oxyacetylene method, shown in Figure 17-9, metallic and nonmetallic wires, powders, or rods are fed into a chamber where they are melted in an oxyacetylene flame at a temperature above 5000°F. The molten material is then atomized by a compressed-air blast that carries the particles to a previously prepared surface.

The design of the gun used for spraying with powder differs from that used for spraying with wires and rods (Figures 17-10 and 17-11).

In powder spraying, the powder is fed from a reservoir attached to the gun. The feed rate (which controls the quantity deposited) can be controlled by changing the powder orifice. The powder method is preferred to the wire and rod methods because it allows the use of an almost unlimited variety of metal alloys, ceramics, and cermet powders to be used for surfacing.

The oxyacetylene wire gun can use wires ranging from ⅛ to 3/16 inch in diameter. The gun is relatively easy to operate, in that it has only three controls: the main valve, which shuts gases and air on and off, the speed control for wire feed, and a lever that starts and stops the wire.

In this method, the distance of the gun from the deposit depends on the wire size, type of metal, and spraying speed. For example, with 3/16-inch-diameter wire, spraying may be nearly twice as fast as with ⅛-inch-diameter wire. At this speed, however, it is advisable to spray from a distance of about 8 to 10 inches.

The oxyacetylene rod method is used to spray ceramics, and is basically the same as the wire method except that a ¼-inch or larger-diameter ceramic rod is used instead of a metal wire.

The *oxyhydrogen method* is considerably more expensive to use than the oxyacetylene method and is used mainly to apply certain grades of self-fluxing nickel-chromium alloys. Its advantage over the

Figure **17-9** Cross-section of an oxyacetylene flame and spray gun

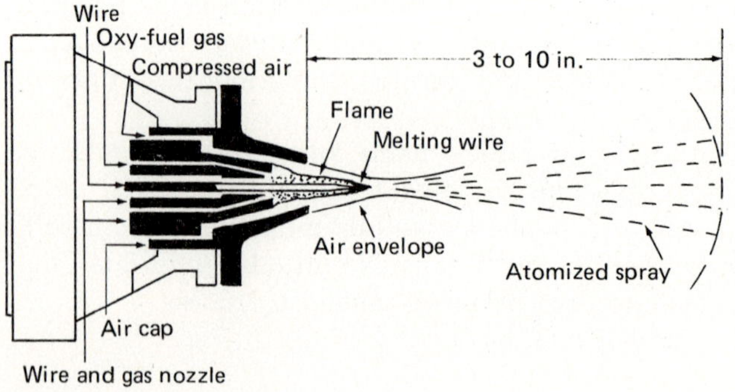

Wire
Oxy-fuel gas
Compressed air
3 to 10 in.
Flame
Melting wire
Air envelope
Atomized spray
Air cap
Wire and gas nozzle

oxyacetylene method is that its flame is not as bright, and therefore allows closer observation of the details of the work during the spraying operation. The only technical difference between the oxy-acetylene and the oxyhydrogen methods is that in the oxyhydrogen method hydrogen is substituted for acetylene.

The *detonation method* uses a rifle-size gun. It is essentially an oxyacetylene-powder method and differs from it in that precisely measured quantities of oxygen and acetylene gases are pressure-fed into the detonation chamber of the gun (Figure 17-10). A timed spark plug ignites the charge and creates a detonation (explosion) that hurls the coating particles, now heated to a somewhat plastic state (6000°F), out of the gun barrel at a speed of 2500 ft/s toward the part to be coated. The part to be coated is 2 to 4 inches from the end of the barrel.

Because the controlled detonations (explosions) occur at a rate of four times per second and produce a noise in excess of 150 decibels (120 decibels is painful), the gun is housed in a double-walled, sound-insulated concrete-block cubicle. The coating operation is completely automatic and is operated from a remote-control panel located outside the concrete-block enclosure.

The torch used with the plasma-type spray gun (Figure 17-11) is capable of producing and maintaining a high-temperature (over 20,000°F), high-velocity inert gas (usually nitrogen) for periods of more than one hour. The mechanism or heat source of the gun is referred to as *thermal plasma,* hence the term *plasma spraying.* The guns usually consume gas at the rate of 100 to 300 ft^3/hour and operate at 20 to 40 kW.

As in the other flame-spraying processes, the molten particles of the coating materials—which were introduced into the plasma arc in powder from—upon striking and impacting the workpiece form a dense, high-purity coating.

Figure **17-10** Cross-section of a detonation-type hard-surfacing device

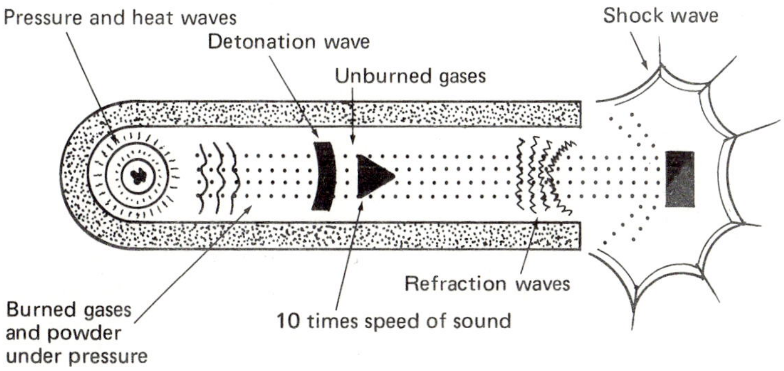

Pressure and heat waves
Detonation wave
Unburned gases
Shock wave
Refraction waves
10 times speed of sound
Burned gases and powder under pressure

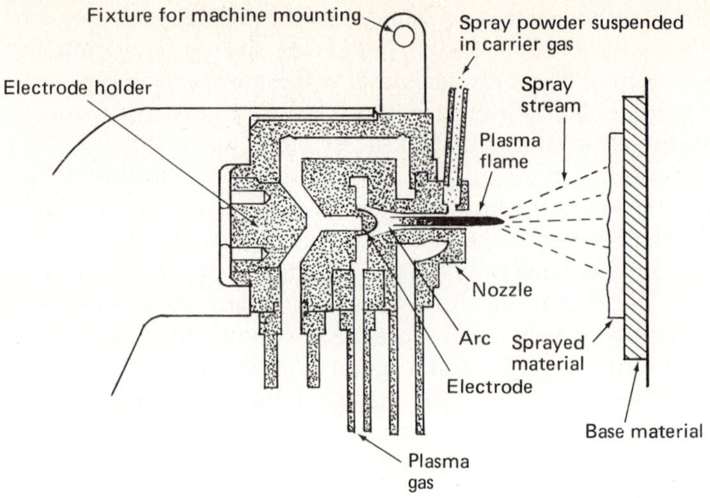
Fixture for machine mounting
Spray powder suspended in carrier gas
Electrode holder
Spray stream
Plasma flame
Nozzle
Arc
Sprayed material
Electrode
Base material
Plasma gas

Figure **17-11** Cross-section of a plasma-type spray gun used in hard surfacing

Plasma-sprayed coatings are generally denser and better bonded, contain less oxides, and have higher tensile strengths than metal and ceramic coatings flame sprayed by the oxyacetylene and oxyhydrogen powder methods.

The parameters (limitations) of the various flame-spraying processes are shown in Table 17-5. Table 17-6 lists the properties of several commonly used surface materials.

17-22 SURFACE PREPARATION AND FINISHING

To obtain a good bond with flame-sprayed coatings, it is necessary to apply the coatings to a clean, roughened surface. Oil, grease, paint, and other foreign matter should be removed not only from the surface to be coated, but from adjacent surfaces. Otherwise, the heat of the process may cause grease or oil to run over onto the coated area. Castings, particularly oil- or grease-soaked porous castings are troublesome. Thus it is good practice to solvent wash or degrease castings and then heat them to 500 to 600°F to char and drive out oil from the pores.

Before flame spraying shafts or similar objects, it is almost always necessary to undercut the surface which is to be built up. The depth to which a shaft should be undercut is determined by many factors, such as the diameter of the shaft, the severity of service, and the amount of wear to be expected in service. In general, if the maximum wear allowance is 0.020 inch on the shaft radius, the part should be undercut 0.025 to 0.030 inch on the radius in order to leave a

Table **17-5** Application parameters of the various flame-spraying methods

	METHOD			
		OXYACETYLENE SPRAYING		
	PLASMA ARC	UNFUSED	FUSED[a]	DETONATION[b]
Choice of coating material	Any powder or mixture that melts; limited by particle bond in coating	Nonreactive metals; refractories with melting point <5000 F	"Self-fluxing" alloys	Tungsten carbide with selected matrices; selected oxides
Choice of base material	Almost all metals and ceramics; some organic materials	Almost all metals and ceramics	Metals with melting point >2000 F	Almost all metals and ceramics
Normal processing temp. at the base material, F	Usually less than 250 F; up to 400 F for a few coatings	200–400	1850–2150	<400
Type of bond	Mechanical, sometimes quasi-metallurgical	Mechanical	Metallurgical	Intimate mechanical
Dimensional limits on base[c]	0.025 in. dia min, no max limit	0.004 in. dia min, no max limit	0.06 in. dia min, no max limit	0.2 in. dia min, to 60 in. max dia
Coating thickness & tolerances, in.[d]	0.002–0.1, ±0.001	0.005–0.2, ±0.003	0.005–0.2, ±0.005	0.001–0.012, ±0.001
Surface fin, rms μin. as applied after grinding	75–125 10 typically, as low as 2–4	150–300 1–2	150–300, <5	125 As low as 1

[a]A fused coating is heated to its melting point to consolidate the coating and to produce metallurgical alloying of the coating with the base metal. [b]Flame-Plating. [c]These are practical limits: in some cases special techniques permit coating application to sections as thin as 0.002 in. [d]These are common values; limits often vary with base and coating materials.

Source: *Materials Selector, Materials Engineering*, a Penton/IPC Reinhold Publication.

Table **17-6** Properties and applications of selected materials commonly applied by flame spraying

	TYPE[a]						
	ALUMINUM	BABBITT A[b]	BRASS(65:35)	BRONZE AA[c]	COMMERCIAL BRONZE	MANGANESE BRONZE	PHOSPHOR BRONZE
Specific gravity	2.41	6.67	7.45	7.06	7.57	7.26	7.68
Ult ten str, 1000 psi	19.5	—	12	29	11.5	12	18
Strain at ult str, %	0.23	—	0.45	0.46	0.42	0.46	0.35
Rockwell hardness	H72	H58	B22	B78	B18	B27	B20
Shrinkage, in./in.	0.0068	—	0.009	0.0055	0.011	0.009	0.010
Spraying speed, lb/hr	18	95	32	24	24	36	31
Spraying efficiency, %[d]	89	69	81	77	82	79	85
Major characteristics and uses[e]	Good corrosion and heat resistance	Good bearing properties	Sprays fast; fair machine finish	Hard, very wear resistant; easily machined	Softest bronze; fair machine finish	Excellent machine finish; special uses only	Fair machine finish; special uses only
	TOBIN BRONZE	COPPER	LEAD	MOLYBDENUM	MONEL	NICKEL	18-8 STAINLESS
Specific gravity	7.46	7.54	10.21	8.86	7.67	7.55	6.93
Ult ten str, 1000 psi	13	—	—	7.5	21	17.5	30
Strain at ult str, %	0.51	—	—	0.30	0.26	0.30	0.27
Rockwell hardness	B27	B32	—	C38	B39	B49	B78
Shrinkage, in./in.	0.0104	—	—	0.003	0.009	0.008	0.012
Spraying speed, lb/hr	36	29	80	8	17	18	21
Spraying efficiency, %[d]	80	80	65	87	85	79	81
Major characteristics and uses[e]	General purpose; fair machine finish	Electrical uses; brazing	Good corrosion resistance; X-ray shielding	Used as bonding coating; excellent bearing properties	Good corrosion resistance; good machine finish	Good corrosion resistance; fair machine finish	High corrosion resistance; good wearing properties

	HIGH Cr STAINLESS	STEEL (LS)[f]	1010 STEEL (0.10%C)	1025 STEEL (0.25%C)	1080 STEEL (0.80%C)	TUNGSTEN[g]	ZINC
Specific gravity	6.74	6.78	6.67	6.78	6.36	16.5	6.36
Ult ten str, 1000 psi	40	33.5	30	34.7	27.5	7.0	13
Strain at ult str, %	0.50	0.45	0.30	0.46	0.42	—	1.43
Rockwell hardness	C29	C25	B89	B90	C36	A50	H46
Shrinkage, in./in.	0.0018	0.002	0.008	0.006	0.0014	—	0.010
Spraying speed, lb/hr	19	18	19	19	19	5.5	61
Spraying efficiency, %[d]	81	87	87	87	87	—	66
Major characteristics and uses[e]	High hardness and wear resistance; grind finish	Good mechanical and finishing properties	Simple bearing surfaces and press fits; excellent machine finish	Harder and lower shrinkage than 0.10C; excellent machine finish	Very hard and wear resistant; good bearing properties; grind finish	High heat resistance	Good all-around corrosion resistance

[a]All metals, except tungsten, applied by oxyacetylene wire metallizing. Data supplied by Metco Inc. [b]Lead-free, high tin alloy. [c]Aluminum-iron-bronze. [d]Percent of metal deposited. [e]All metals have about the same shiny surface after spraying, but surfaces of various metals differ after machining. [f]Low shrinkage. [g]Plasma sprayed powder.

Source: *Materials Selector, Materials Engineering*, a Penton/IPC Reinhold Publication.

continuous coating after maximum wear has taken place. Similarly, flat surfaces should be undercut to obtain a satisfactory thickness of flame-spray deposit. As with shafts, the depth of undercutting for flat surfaces is determined by factors such as strength requirements and the amount of wear expected.

Prior to flame spraying inside diameters, it is usually necessary to machine the bore oversize in order to obtain a satisfactory thickness of spray deposit on the finished part. For the restoration of press fits, the minimum allowance for coating should be 0.002 inch on the radius. For all other types of service, the minimum allowance for coating depends on the method of preparation, as follows:

1. Where the surface is prepared by blasting followed by a bond coat of molybdenum or nickel aluminide, the minimum bore oversize should be 0.020 inch on the radius.

2. Where other means of surface preparation are used, the minimum bore oversize should be 0.020 inch on the radius.

When spraying metals inside bores or diameters, spraying is usually done from one end only, with the spray striking the surface at an angle of 45 degrees. The leading edge, therefore, should not be dovetailed, as it would on a shaft. Where practicable, the back end of the oversize bore should be dovetailed to an angle of 15 to 20 degrees, with a radius at the bottom if the spraying is to be performed from one end of the bore only.

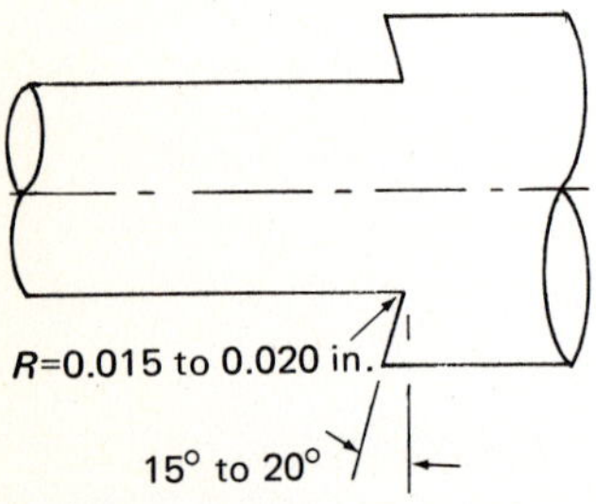

R=0.015 to 0.020 in.
15° to 20°

Figure **17-12** Standard preparation of cylindrical base material for flame spraying

On shafts, sprayed metals should be mechanically anchored into the base metal at the ends of the undercut section. This is done by providing suitable shoulders or other positive anchorages at each end of the undercut section. The dovetail shoulder, at an angle of 15 to 20 degrees and with a radius at the bottom, as shown in Figure 17-12, provides a satisfactory anchorage. An equally effective method, shown in Figure 17-13, is to machine a straight cut at right angles to the shaft with a radius at the bottom, in which case it is necessary to roughen or otherwise prepare the sidewalls of the undercut.

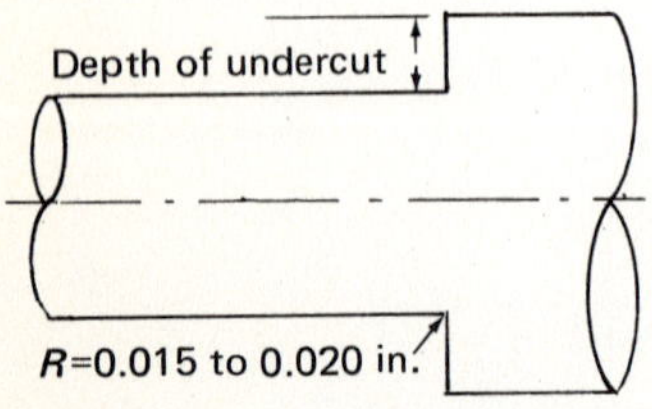

Depth of undercut
R=0.015 to 0.020 in.

Figure **17-13** Alternate preparation of cylindrical base material for flame spraying

After undercutting, the next operation is the preparation of the undercut section that is to receive the flame-sprayed material. Abrasive blasting is a common and versatile method of surface roughening. It depends on several variables, including type and mesh size of the abrasive, type of blast equipment, air pressure, and hardness of the surface. Angular chilled iron produces the greatest roughness on surfaces having a Rockwell hardness of up to C-40. Aluminum abrasives do a better job on surfaces with a Rockwell hardness over C-40.

Another surface-preparation method frequently used is molybdenum and nickel-aluminide bonding, which consists of spraying molybdenum or nickel aluminide directly onto a clean surface.

Since molybdenum oxidizes rapidly at temperatures above 900°F, it should not be used on parts that will be subject to temperatures much over 600°F.

The groove and rotary-roughening method of surface preparation gives one of the strongest bonds possible on all machinable metals, such as low-carbon steel, stainless steel, Monel, nickel, brass, and bronze. It requires two separate operations, consisting of:

1. Cutting spiral or annular grooves in the undercut section.

2. Spreading and roughening the tops of the ridges produced in the first operation.

The rough-threading method is a fast, one-step operation; however, it should only be used on low-carbon steels where the surface can be torn and roughened. The objective of rough threading is to obtain roughened threads with the sides torn and jagged and with a radius of not more than 0.015 inch at the bottom. The number of threads normally required is from 16 to 24 per inch, depending on the diameter of the workpiece, the thickness of the deposit, and the intended service.

The threading and knurling method also requires two separate operations and, like the groove and rotary-roughening method, gives one of the strongest bonds possible on machinable metals. The threading operation consists of cutting V threads; the knurling operation consists of upsetting or cold forming a uniform pattern of hooks or keys in the tops or ridges of the V threads.

Preheating before spraying usually is necessary to produce a satisfactory bond. One of the two end products of an oxyacetylene flame is water vapor, and the entire product of an oxyhydrogen flame is water vapor. If the flame strikes a cold surface, the water vapor condenses and the surface becomes momentarily wet, and when molten metal or ceramic particles strike this wet surface, the water film instantly vaporizes. This amounts to a minute but violent explosion beneath the particle, and it prevents the particle from bonding properly. Thus condensation on the surface must be avoided, and preheating to 200°F or higher should immediately precede the flame-spray operation.

Cracks or lifting at corners or edges may occasionally appear in a sprayed deposit. These problems can be minimized by selecting a metal or ceramic that gives a coating with the lowest possible shrinkage. Another way to avoid cracking is to keep the workpiece at or below 300°F. If necessary, spraying should be periodically stopped to allow cooling, or a diffused blast of clean air should be introduced near the area being sprayed. The part should be allowed to cool normally to room temperature after spraying is completed. Quenching or spraying with water or any other liquid should be avoided.

Flame-sprayed metals and ceramics can be finished either by machining, grinding, or both. There are several types of grinding operations, including surface, internal, and centerless. Any of these may be done wet or dry, although wet is preferable to dry grinding and should be used whenever possible. In general, medium-hard and medium-dense vitrified-bonded silicon-carbide wheels are sufficient for finishing metal and some ceramic coatings. However, diamond wheels are used most often for finishing hard ceramics and cermets.

Rough machining should be started near the center of the sprayed area and worked out to the edges. Rough cuts should be taken to within 0.020 inch of the finishing size, followed by finishing. Wheel surface speed should be 6000 ft/min for dry grinding and 6500 ft/min for wet grinding. Very light passes should be used for finish grinding.

Sprayed metals can be polished and buffed to a high luster. For example, tin, babbitt, zinc, and aluminum are easy to polish with ordinary polishing equipment; copper, brass, and bronze are more difficult. Sprayed steels, including carbon and stainless, are extremely difficult to polish and should not be considered for any application that calls for a lustrous finish.

17-23 GENERAL INSTRUCTIONS FOR FLAME SPRAYING

The instructions that follow are generally applicable to oxyacetylene and oxyhydrogen flame-spraying processes using wire, rod, and powder. Instructions for the plasma-arc and detonation methods are not given because they are not used as frequently as the oxyhydrogen and oxyacetylene processes, and require that the operator undergo a short training course that is usually offered by the equipment manufacturer or the distributor. In any event, it is always advisable that the instruction booklet supplied with the equipment be consulted prior to beginning a flame-spraying operation. *Note:* Because the metal fumes and dust that occur during flame spraying are extremely hazardous to health, all metal spraying should be conducted in a well-ventilated area (see Chapter 13.)

1. The wire speed, amount of spray, and gas and oxygen pressures must be regulated per recommendations for the equipment used and the type of metallizing. As a general rule, air pressure is set for 60 psi. The use of a flowmeter will ensure more accurate control of the gases. A slight increase in air pressure provides a finer coating, while a decrease of air pressure produces a coarser coating.

2. The tip of the melting wire should project beyond the end of the air cap. This length depends to a large extent on the material being used. A recommended practice is to speed up the wire until chunks are ejected, then reduce the wire feed until the ejection stops.

3. Keep each coating as light as possible—about 0.003 to 0.005 inch thick. Too heavy a coat will produce an irregular and stratified surface. The actual movement of the gun is very similar to paint spraying. Keep the nozzle 4 to 10 inches away from the surface and move it uniformly. If the gun is held too close to the work, minute cracks will produce a soft, spongy deposit. The rate of gun travel is also important. When the travel is too rapid, the coating develops a high oxide content.

4. When spraying a flat surface, move the gun back and forth to allow a full, uniform deposit. Spraying should begin beyond the edge of the area to be covered and continued beyond the end of the area. After the first layer, the work, or the gun, is often rotated 90 degrees. This technique is repeated for each subsequent coating until the required thickness is built up. On cylindrical pieces, the work is generally fastened in a lathe with the gun mounted on the traveling carriage.

The easiest, although not very popular, flame-spraying process, is the one which applies powder through a special hopper and spray control that can be attached to most oxyacetylene torches. In this method, metal powder is simply placed in the hopper, the torch lighted, and the torch moved over the area that is to receive an overlay.

REVIEW QUESTIONS

1. How is hard surfacing defined in the AWS *Welding Handbook?*

2. Differentiate between *scratching abrasion* and *gouging abrasion.*

3. Name the welding procedure that is used most frequently for hard surfacing.

4. What determines whether the welding process to be used for hard surfacing is manual or semiautomatic?

5. What is the range of thickness (in inches) of hard-surfacing material applied by means of the fused oxyacetylene spray method?

6. What should be done to a metal surface to make it ready for hard surfacing with the oxyacetylene torch?

7. What type of oxyacetylene flame should always be used for hard surfacing?

8. Why is it good practice to consult the manufacturer of a hard-surfacing alloy being used for the first time as to the shape of the oxyacetylene flame with which it is to be applied?

9. Why is it unusual to hard surface cast-iron parts?

10. How long should the acetylene feather be for hard surfacing with tungsten carbide?

11. a. Refer to Table 17-2 and determine the hard-surfacing material to be used with the SMAW and GTAW processes for hard-surfacing the following machinery parts: (1) forging die block; (2) dredge parts; (3) horseshoe; and (4) earth-moving shovel latch bar.
 b. Determine the number of passes recommended, the type of current to be used, and the recommended amperage setting.

12. Why have standardized test methods not been developed for testing abrasions and/or impact resistance of hard-surface deposits?

13. Which of the four particle-impacting methods is most commonly used?

14. The oxyacetylene process is used primarily to apply what composition of hard-surfacing alloys?

15. What temperature (°F) is produced by the plasma-type spray gun?

16. How should: (a) inside diameters be prepared for flame spraying? (b) outside surfaces to be surfaced with the plasma-spray method be readied for spraying?

17. Why is it necessary to preheat parts that are to be flame sprayed?

PART 4
WELDING METALLURGY

CHAPTER 18
PROPERTIES OF METALS OF IMPORTANCE TO THE WELDER

Metals can be broken, bent, twisted, dented, scratched, and otherwise damaged. Some metals cannot be pulled apart by the force of an automobile, while others of the same size can be bent by a child. Some metals can be scratched by the fingernail, while others will withstand hours of pounding against solid rock. With such widely varying properties, it becomes a problem to express in a few words exactly what type of service a piece of metal will withstand without failure.

In this chapter, we will examine some of the physical properties of metals and some of the tests that have been developed for the purpose of accurately measuring metallic properties. We will also develop and examine some basic rules for dealing with the properties of metals in welding.

18-1 STRENGTH OF METALS

Strength is the capacity of a metal to withstand destruction under the action of external loads (Figure 18-1). The strength value indicates the force required to overcome the bond holding together the molecules that make up the crystal lattices. Two of these loads (tension and compression) are determined by means of a universal testing machine (Figure 18-2), that is, a machine capable of producing both tensile (pulling) and compressive (pushing) forces. The most frequently used standard tensile test specimen is a round bar

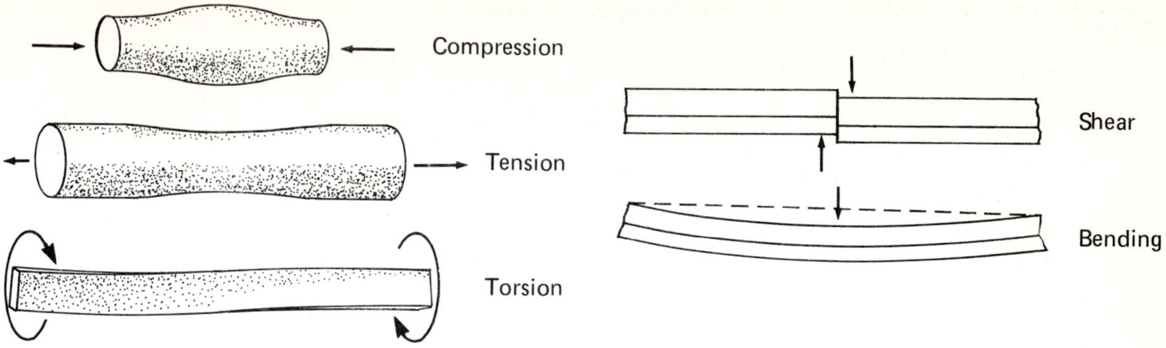
Compression
Tension
Torsion
Shear
Bending

Figure **18-1** Results of external forces applied to metals. (By permission from N. Makiyenko, *Benchwork*, MIR Publishers, Moscow, U.S.S.R.)

EHLE

Figure **18-2** A commonly used type of universal (tension-compression) testing machine. (Courtesy of Teledyne McKay, Manufacturers of Electrodes and Welding Wires.)

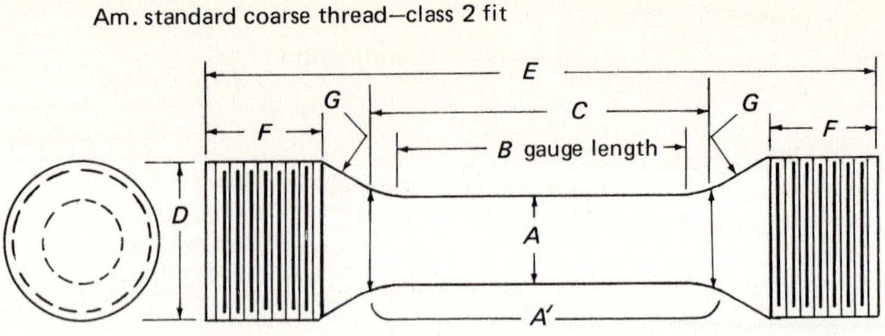

Am. standard coarse thread—class 2 fit
E
G
C
G
F
F
B gauge length
D
A
A'

Test specimen dimensions								
Speci-men	$A^{(2)}$	Area$^{(1)}$	B	$C^{(3)}$	D	$E^{(4)}$	$F^{(4)}$	$G^{(4)}$
	in.	sq in.		in.	in.	in.	in.	in.
C-1	0.505	0.200	2	$2\frac{1}{4}$	$\frac{3}{4}$	$4\frac{1}{4}$	$\frac{3}{4}$	$\frac{5}{8}$

$A' = A$ min.
(1) Cross-sectional area =0.785 x A^2
(2) Tolerance, $\pm 1\%$
(3) Approximate
(4) Minimum
Note: Dimensions ***A, B, C,*** and ***G*** shall be as shown, but the ends may be of any shape to fit the holders of the testing machine in such a way that the load shall be axial.

Figure **18-3** Standard tensile test specimen 505, or C-1. (Redrawn by permission from James F. Lincoln Arc Welding Foundation, *Metals and How to Weld Them,* 2d ed., 1954, 1962, p. 16.)

(Figure 18-3) having a diameter of 0.505 inch over a distance of at least 2 inches. To calculate the load (in pounds) that the cross-sectional area of a given part can support, the material's tensile strength [in pounds/in^2 (psi)] is multiplied by the cross-sectional area (in square inches; in the case of the 0.505-inch specimen, 0.200 inch). There are times when we don't know the tensile strength of the metal, so it is convenient to test an actual part made from the metal. The tensile strength of the metal (in psi) can be calculated by dividing the maximum load (in pounds) by the cross-sectional area (in square inches). This can be written as a formula:

$$\text{Tensile strength (psi)} = \frac{\text{maximum load (lb)}}{\text{cross-sectional area in}^2}$$

Sometimes it is desirable to test for the tensile strength of a welded steel plate (Figure 18-4). The sample cut from the plate will be larger than the standard 0.505-inch specimen (Figure 18-3) and will be of full plate thickness. Calculations will be as follows: A maximum load of 50,000 pounds is required to break a welded-steel sample

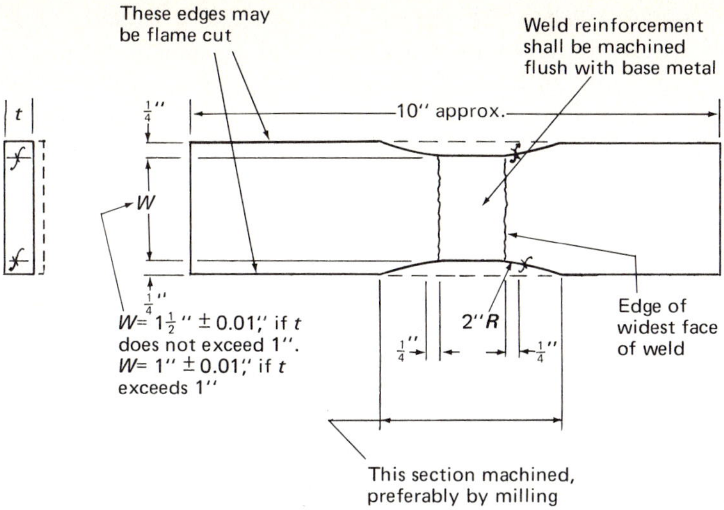
These edges may be flame cut
Weld reinforcement shall be machined flush with base metal
10" approx.
t
W
1/4"
1/4"
W= 1 1/2" ± 0.01" if t does not exceed 1". W= 1" ± 0.01" if t exceeds 1"
2"R
1/4"
1/4"
Edge of widest face of weld
This section machined, preferably by milling

Figure **18-4** Standard tensile test specimen for testing welds made in plate. (Redrawn by permission from James F. Lincoln Arc Welding Foundation, *Metals and How to Weld Them.* 2d ed., 1954, 1962, p. 17.)

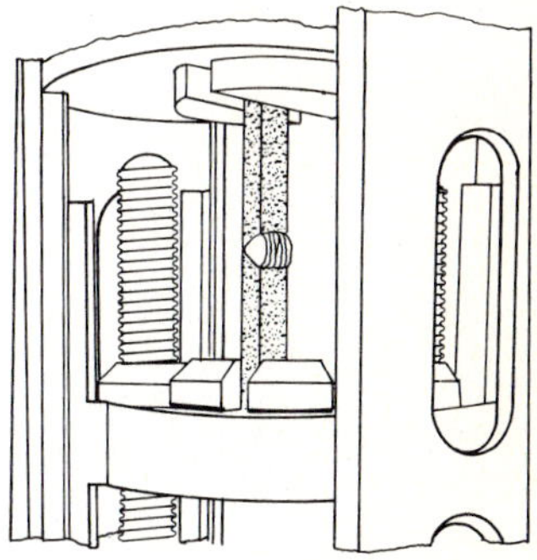

Figure **18-5** Plate-weld tensile test specimen in position in a universal testing machine. (Redrawn by permission from Union Carbide Corporation, Linde Division, *The Oxy-Acetylene Handbook,* 2d ed., 1960, p. 105.)

(Figure 18-5) that is ½ × 1½ inch at the weld. The tensile strength of the steel is calculated by using the formula. The cross-sectional area is ½ × 1½ = ¾ sq in.; therefore

$$\text{Tensile strength} = \frac{50{,}000}{3/4} = 66{,}666.6 \text{ psi}$$

18-2 ELASTICITY

Most metals do not break off all at once. As the load is gradually applied in a tension-test machine (Figure 18-6), for instance, the specimen will be seen to stretch for some time. Then a noticeable "necking-in" will occur at some point, and later, as the load increases, the part will break. The stretch that is observed in such a test is not uniform. At first it is an elastic stretch, and later a permanent stretch.

If you were to take a piece of rubber and stretch it, it would return to its original size just about as soon as you let go. If the rubber were pulled harder, it would snap. No matter how hard it was pulled, twisted, or squeezed, as long as it did not break it would go back to its original size when the force was removed.

Up to a certain point, metals act like rubber. They are elastic. They stretch or bend or twist under force, and then return to their original size when the force is removed, in just the same way, but not to as great an extent, as a piece of rubber.

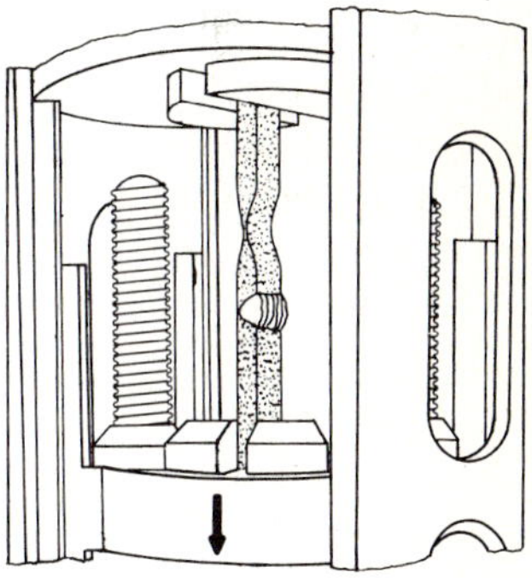

Figure **18-6** Plate-weld test specimen showing necking-in and failure. (Redrawn by permission from Union Carbide Corporation, Linde Division, *The Oxy-Acetylene Handbook,* 2d ed., 1960, p. 105.)

When rubber reaches the end of its elasticity, it breaks. Most metals, however, do not break when they reach the elastic limit, but keep on stretching for some time before they break. Beyond the elastic limit, therefore, metals act more like taffy—they change shape with the force. However, when the force is removed, they do not return to the original shape but are permanently deformed.

In short, when a metal part is acted upon by a force—pulled, twisted, bent, or squeezed (compressed)—it changes shape elastically for a while, or until the force reaches the elastic limit of the metal, then it permanently deforms. The *elastic limit*, therefore, is the point at which permanent deformation begins. In industrial applications, this point is approximately determined and expressed as the *yield point* or *yield strength* of the metal.

18-3 DUCTILITY

Figure **18-7** Ductility: permanent deformation without failure. (Redrawn by permission from Union Carbide Corporation, Linde Division, *The Oxy-Acetylene Handbook*, 2d ed., 1960.)

A *ductile* material is one that can be permanently deformed without failure (Figure 18-7). Ductility is a frequently misused term, however. The fact that a metal bends easily does not necessarily mean that it is ductile, unless such bending is a permanent deformation. For instance, a spring may be flexible and bend easily, yet it may withstand little permanent deformation, or set, without breaking. Such a spring is said to have low ductility. As pointed out previously, rubber is elastic, but molasses candy or chewing gum is ductile. All metals are both elastic and ductile to some extent.

Four methods of measuring ductility are in common use. One is a method whereby the ductility may be stated as the amount of permanent stretch (over a distance of 2 inches) a part will undergo in a tension test. The second method for measuring ductility uses the difference between the original cross-sectional area and the smallest area at the point of rupture in a tensile test. The ductility is expressed as a percentage of the original cross-section (the difference in area is divided by the original area). This method of measuring ductility is called the *reduction-in-area test.* The third method employs a *free-bend test* (Figure 18-8) for a comparable determination. The fourth method is referred to as the *guided-bend test.*

The guided-bend test fixture is shown in Figure 18-9. In this test, the specimen is laid horizontally across the supports of a female die with the weld at midspan. The male punch is forced down onto the weld specimen until the specimen is bent to a U shape; and until it becomes impossible to insert a 1/32-in.-diameter wire between the specimen and the punch. If, after bending, the convex surface of the specimen does not exhibit cracks or other open defects exceeding 1/8 inch in length, the specimen (weld) is considered to have passed the test. Cracks on the corners of the specimen are disregarded. Two

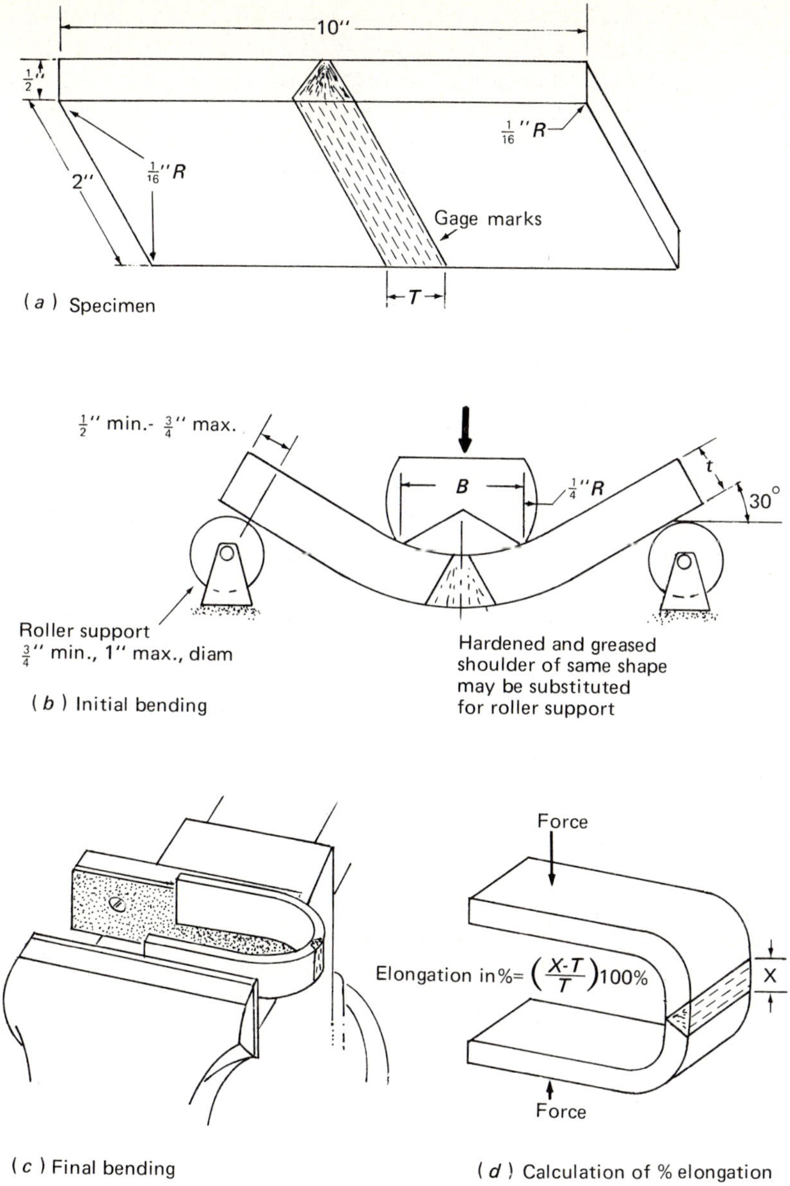

10″
½″
1/16″R
2″
1/16″R
Gage marks
T
(a) Specimen
½″ min.- ¾″ max.
B
¼″R
t
30°
Roller support
¾″ min., 1″ max., diam
Hardened and greased
shoulder of same shape
may be substituted
for roller support
(b) Initial bending
Force
Elongation in%= ((X-T)/T)100%
X
Force
(c) Final bending
(d) Calculation of % elongation

Figure **18-8** Specimen and procedure for free-bend testing of a weld. (Redrawn by premission from James F. Lincoln Arc Welding Foundation, *Metals and How to Weld Them*, 2d ed., 1954, 1962.)

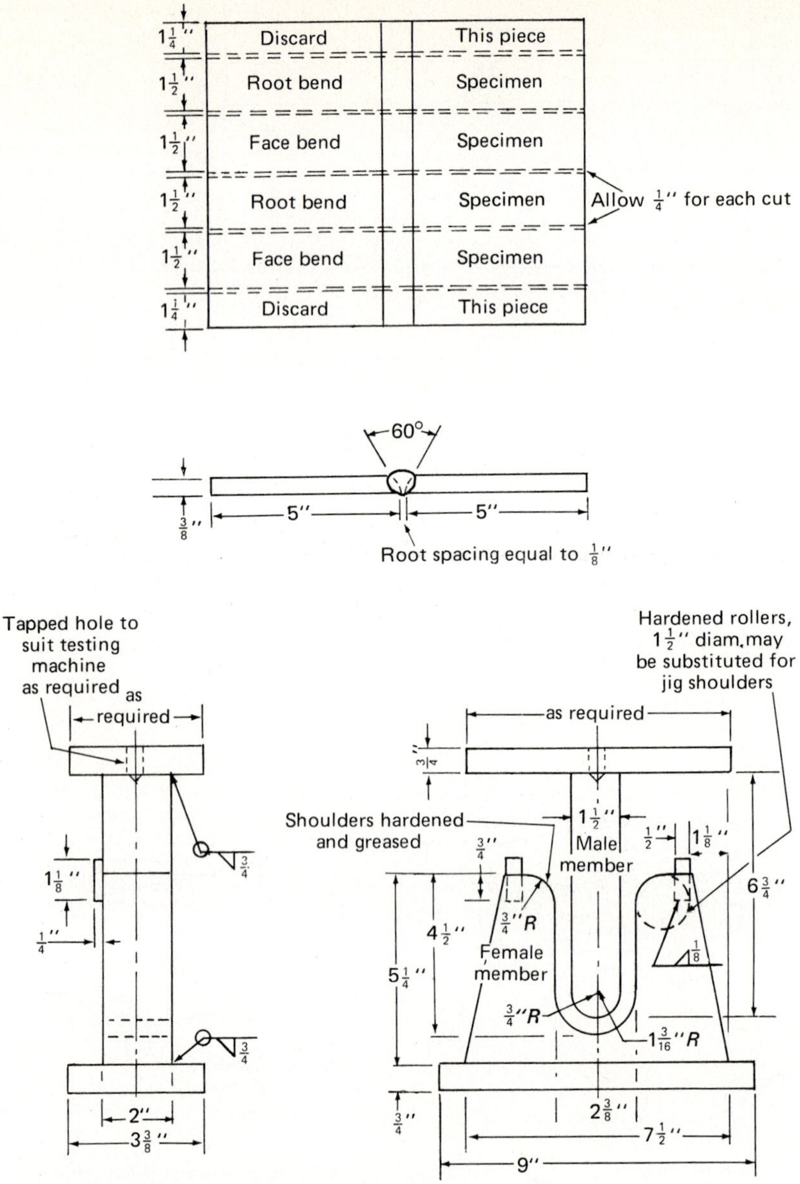
Discard
This piece
Root bend
Specimen
Face bend
Specimen
Root bend
Specimen
Face bend
Specimen
Discard
This piece
Allow 1/4" for each cut
1 1/4"
1 1/2"
1 1/2"
1 1/2"
1 1/2"
1 1/4"
60°
3/8"
5"
5"
Root spacing equal to 1/8"
Tapped hole to suit testing machine as required
as required
Hardened rollers, 1 1/2" diam. may be substituted for jig shoulders
as required
Shoulders hardened and greased
Male member
Female member
3/4"R
3/4"R
1 3/16"R
1 1/8"
1/4"
3/4
3/4
2"
3 3/8"
3/4"
1 1/2"
1/2"
1 1/8"
6 3/4"
4 1/2"
5 1/4"
1/8
2 3/8"
7 1/2"
9"

Figure **18-9** Standard guided-bend testing machine. (Redrawn by permission from James F. Lincoln Arc Welding Foundation, *Metals and How to Weld Them,* 2d ed., 1954, 1962.)

tests are made in this fashion: a face test and a root test. The only difference between the two tests is that in a *face test* the face of the weld is placed toward the punch, whereas in the *root test* the root of the weld is placed toward the punch.

18-4 BRITTLENESS

Brittleness is the opposite of ductility. *Brittle* materials are substances that fail (Figure 18-10) without appreciable permanent deformation. A brittle substance also has a low resistance to shocks or loads applied rapidly. An example of a brittle metal is ordinary white cast iron.

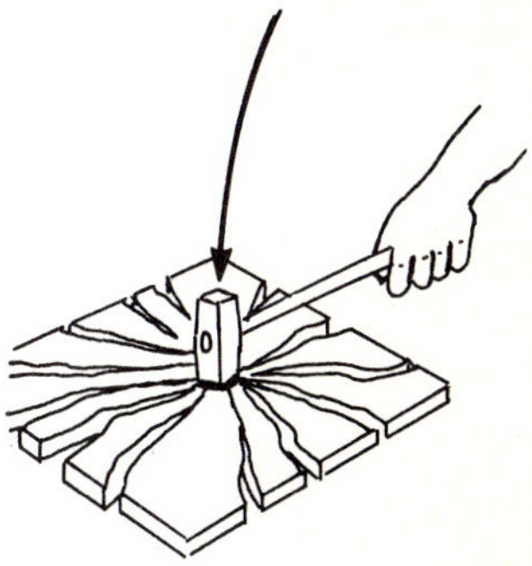

Figure **18-10** Brittleness: failure of a material without appreciable deformation. (Redrawn by permission from Union Carbide Corporation, Linde Division, *The Oxy-Acetylene Handbook*, 2d ed., 1960, p. 111.)

18-5 TOUGHNESS

Toughness is that property of a metal that enables the metal to withstand considerable stress, slowly or suddenly applied, continuously or often applied, and to deform before failure. The test most commonly used to determine the toughness in metals is the *impact test*. In this test, a rectangular specimen is cut from the part to be tested, prepared as shown in Figure 18-11, and tested in a machine

Figure **18-11** Test principles and standard dimensions of test samples in the IZOD (left) and Charpy (right) impact tests. (Redrawn by permission from James F. Lincoln Arc Welding Foundation, *Metals and How to Weld Them*, 2d ed., 1954, 1962, p. 35.)

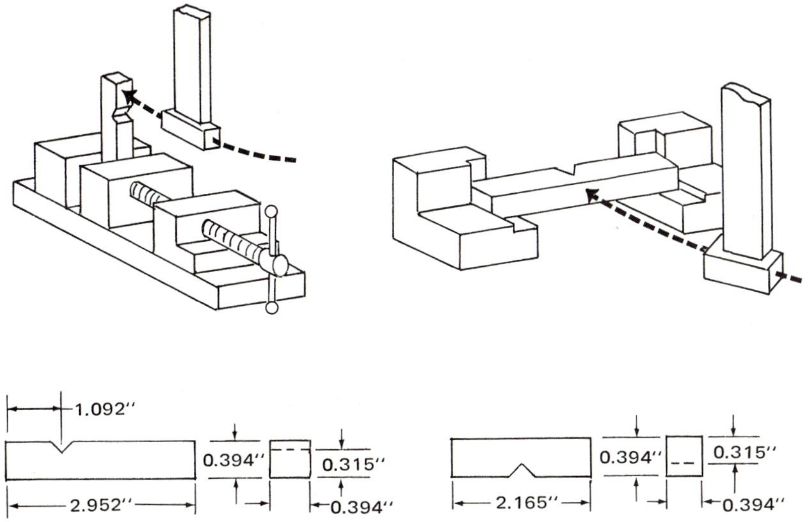
1.092″
0.394″
0.315″
2.952″
0.394″
0.394″
0.315″
2.165″
0.394″

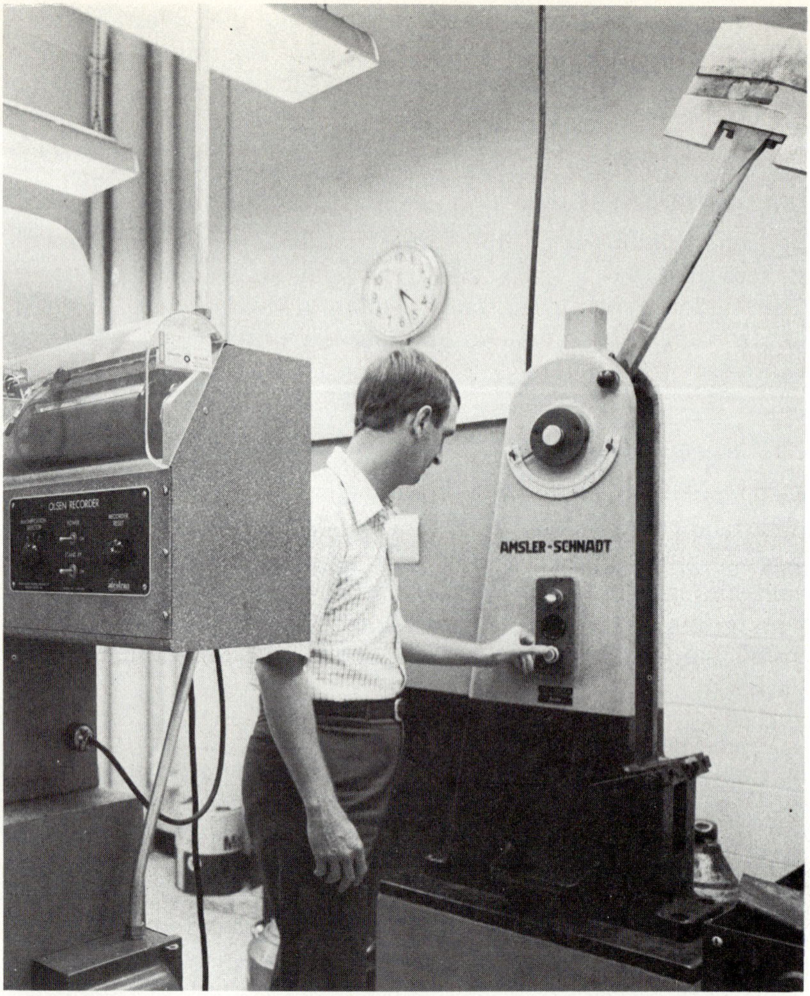
OLSEN RECORDER
AMSLER-SCHNADT

Figure **18-12** A commonly used type of impact machine. (Courtesy of Teledyne McKay, Manufacturers of Electrodes and Welding Wires.)

such as the one shown in Figure 18-12. Impact strength is measured in foot-pounds. The specimen is placed in the machine and the operator releases a heavy pendulum that swings from a standard height to strike the specimen (Figure 18-13).

The strength of the specimen is determined by the amount of energy needed to break or bend it. The energy of the falling pendulum is known. The distance through which the pendulum swings after breaking the specimen indicates how much of the total energy (foot-pounds) was used in breaking it. With no specimen in the

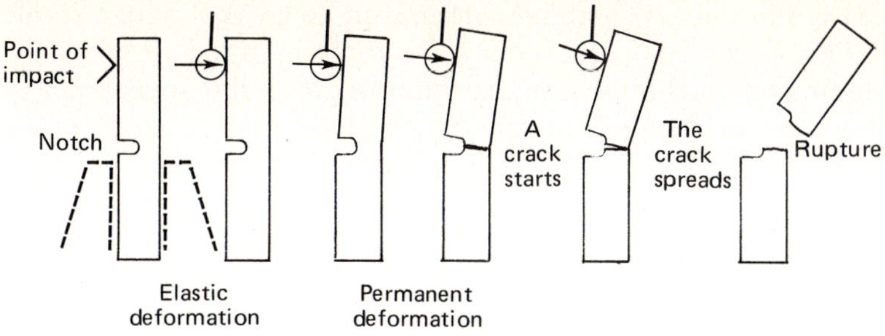
Point of impact
Notch
Elastic deformation
Permanent deformation
A crack starts
The crack spreads
Rupture

Figure **18-13** Toughness test specimen being deformed on impact by the pendulum of the impact tester. (Courtesy of Teledyne McKay, Manufacturers of Electrodes and Welding Wire.)

machine, the pendulum swings to the zero reading on the scale. The tougher the metal specimen broken by the swing, the shorter the distance the pendulum travels beyond the point of impact. The shorter the distance, the higher the reading on the scale.

Steels or welds are commonly tested for notch toughness or impact strength without such elaborate testing equipment, right in the shop. The welder or inspector (see Chapter 23) first saws two notches in a small specimen of the metal and secures it in a bench vise (Figure 18-14). He then hits the piece on the notch side with a

Figure **18-14** Standard specimen and procedure for performing the nick-break test

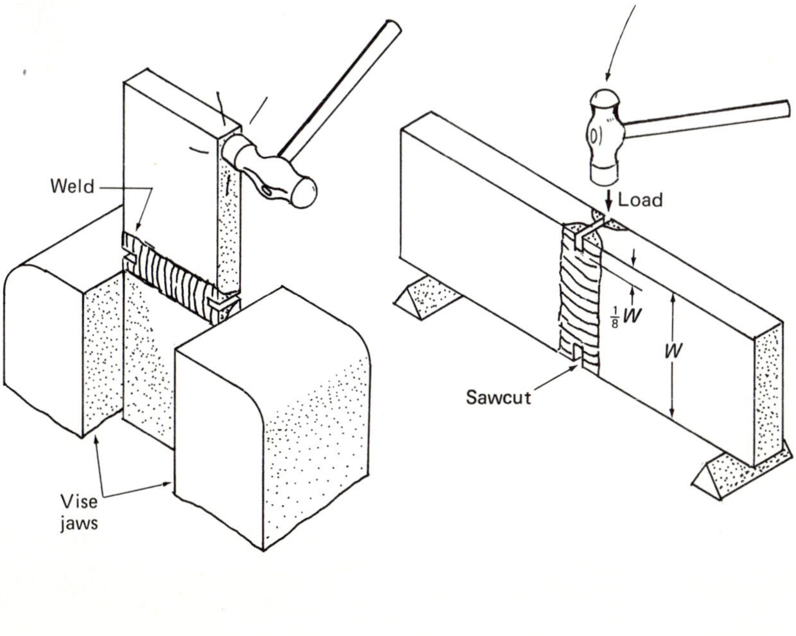
Weld
Vise jaws
Load
$\frac{1}{8}W$
W
Sawcut

sledge (or other heavy) hammer, attempting to break it with a single blow. This is the *nick-break test.* The force required to break the sample, the amount of bending at the break, and the appearance of the fractured surfaces all indicate the relative impact strength of the metals being tested.

18-6 HARDNESS

Hardness is a property with which the welder must become thoroughly familiar. The heat of welding may change the hardness of the metals being welded (Figure 18-15), or the end result may be a difference in hardness between the deposited weld metal and the parent metal. A difference in hardness usually indicates a difference in strength and some other property or properties. Through the study of welding metallurgy, one can learn more about the causes of, and how to control, hardness. However, in many cases no change or difference in hardness accompanies the welding process, in which case there is no concern over hardness control.

Hardness means different things to different people. The metallurgist thinks of hardness as the ability of a material to resist indentation or penetration. The mineralogist thinks of hardness as the ability of a material to resist abrasion or scratching. A machinist, on the other hand, considers hardness as an index of machinability.

From a metallurgical point of view, the main reason for making a hardness test is for what it will tell about other properties. For example, the tensile strength of a material is directly related (Table 18-1) to the hardness.

Hardness tests are both inexpensive to make and nondestructive. A hardness test, therefore, may be substituted for the more difficult and destructive tensile test.

In general, the harder of two metals of similar composition has a higher tensile strength, lower ductility, and more resistance to abrasive wear. High hardness also indicates low impact strength, although some steels, when properly treated, have both high hardness and good impact strength.

Hardness tests are widely used to check the uniformity of the material of metal parts during production. Any lack of uniformity in the material is quickly revealed by its being either too hard or too soft.

The Brinell test is the oldest commercial method of determining indentation hardness. A steel ball of 10-millimeters (about ⅜-inch) diameter is pressed into the surface of the metal with a load of 3000 kilograms (6614 pounds). Then the diameter of the impression is measured with a special microscope and the reading is converted to the Brinell hardness number by consulting a table. Soft iron is about 100 Brinell, file-hard steel about 600. A Brinell test machine is

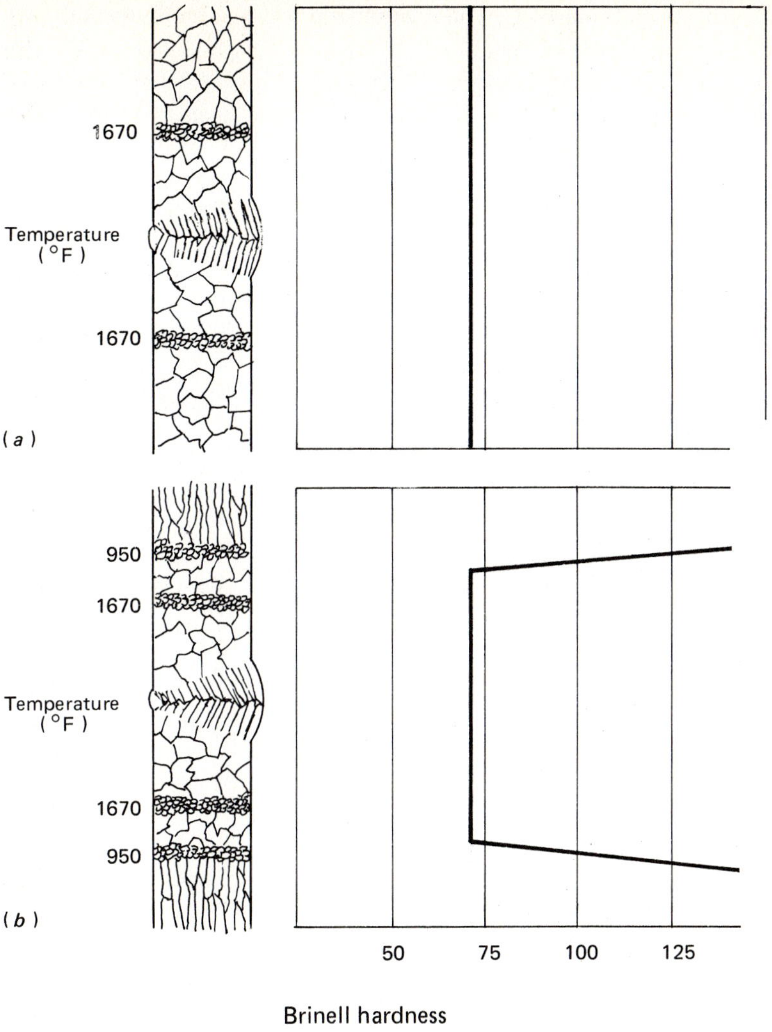
1670
Temperature
(°F)
1670
(a)
950
1670
Temperature
(°F)
1670
950
(b)
50
75
100
125
Brinell hardness

Figure **18-15** (*a*) Iron annealed before welding. Considerable grain refinement occurred when the metal reached 1670°F during welding but, as the chart indicates, the welding did not affect hardness. (*b*) In the cold-worked iron, grain refinement occurred in both the 950° and 1670°F zones and, as the chart indicates, the welding considerably softened the iron. (From O. H. Henry and G. E. Claussen, as revised by G. E. Linnert, *Welding Metallurgy (Iron and Steel)*, p. 190. Copyright 1967 by the American Welding Society. Used with permission of the American Welding Society.)

Table **18-1** The relationship between hardness and tensile strength

BRINELL HARDNESS NO.	ROCKWELL C	SCLEROSCOPE NO.	TENSILE STRENGTH 1000 psi
898	—	—	440
857	—	—	420
817	—	—	401
780	70	106	384
745	68	100	368
712	66	95	352
682	64	91	337
653	62	87	324
627	60	84	311
601	58	81	298
578	57	78	287
555	55	75	276
534	53	72	266
514	52	70	256
495	50	67	247
477	49	65	238
461	47	63	229
444	46	61	220
429	45	59	212
415	44	57	204
401	42	55	196
388	41	54	189
375	40	52	182
363	38	51	176
352	37	49	170
341	36	48	165
331	35	46	160
321	34	45	155
311	33	44	150
302	32	43	146
293	31	42	142
285	30	40	138
277	29	39	134
269	28	38	131
262	26	37	128
255	25	37	125
248	24	36	122
241	23	35	119
235	22	34	116
229	21	33	113

BRINELL HARDNESS NO.	ROCKWELL C	SCLEROSCOPE NO.	TENSILE STRENGTH 1000 psi
223	20	32	110
217	18	31	107
212	17	31	104
207	16	30	101
202	15	30	99
197	13	29	97
192	12	28	95
187	10	28	93
183	9	27	91
179	8	27	89
174	7	26	87
170	6	26	85
166	4	25	83
163	3	25	82
159	2	24	80
156	1	24	78
153	—	23	76
149	—	23	75
146	—	22	74
143	—	22	72
140	—	21	71
137	—	21	70
134	—	21	68
131	—	20	66
128	—	20	65
126	—	—	64
124	—	—	63
121	—	—	62
118	—	—	61
116	—	—	60
114	—	—	59
112	—	—	58
109	—	—	56
107	—	—	56
105	—	—	54
103	—	—	53
101	—	—	52
99	—	—	51
97	—	—	50
95	—	—	49

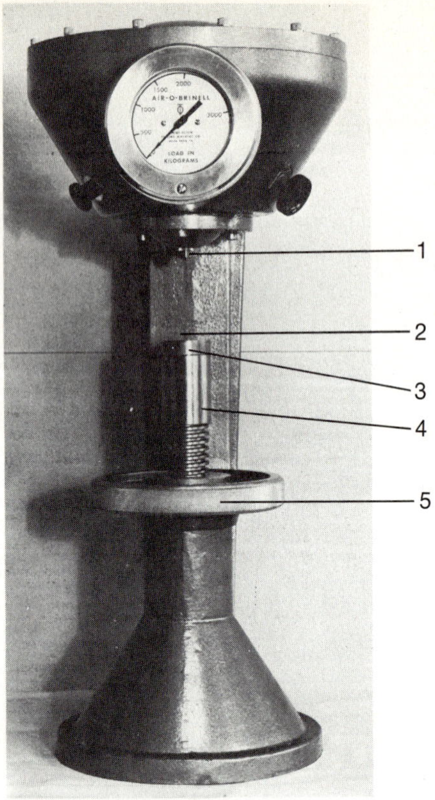
AIR-O-BRINELL
LOAD IN KILOGRAMS
1
2
3
4
5

(a)

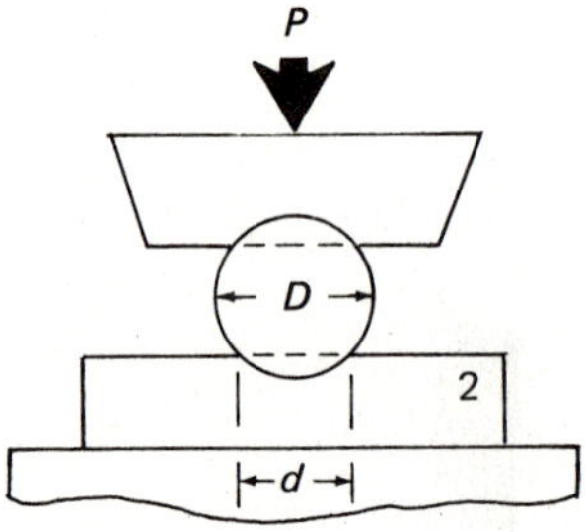
P
D
2
d

(b)

Figure **18-16** (*a*) Brinell hardness-testing machine: (1) spindle, (2) specimen goes here, (3) table, (4) sleeve, (5) hand wheel. (*b*) Penetration of the ball and boundary of the impression: P = load, D = diameter of the ball, 2 = specimen, and d = diameter of indentation. (Courtesy of Teledyne McKay, Manufacturers of Electrodes and Welding Wires.)

pictured in Figure 18-16. Physical-property tables in catalogs and handbooks often use the abbreviation BHN or Bhn for Brinell hardness number.

The Rockwell hardness test is widely used in production inspection. This test is also of the indentation type, but the penetrator is smaller and the loads are lighter than those used for the Brinell test. With a sample of the work positioned on the anvil of the machine [Figure 18-17(*a*)], a diamond cone is pressed into the metal with a major load of 150 kilograms (about 330 pounds) and a minor load, which depending on the hardness may be either 10 or 15 kilograms (Figure 18-17). The depth of the impression is indicated on a dial, and the reading is called the *Rockwell C hardness*. A typical reading might appear on specifications as 52 Rockwell C.

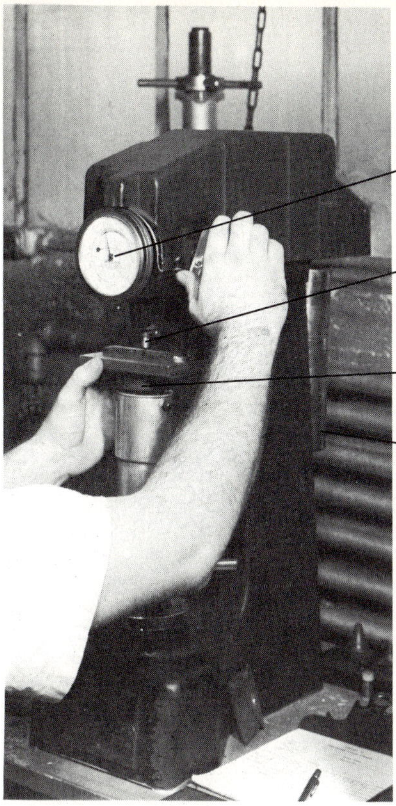
1
2
3
4
(a)

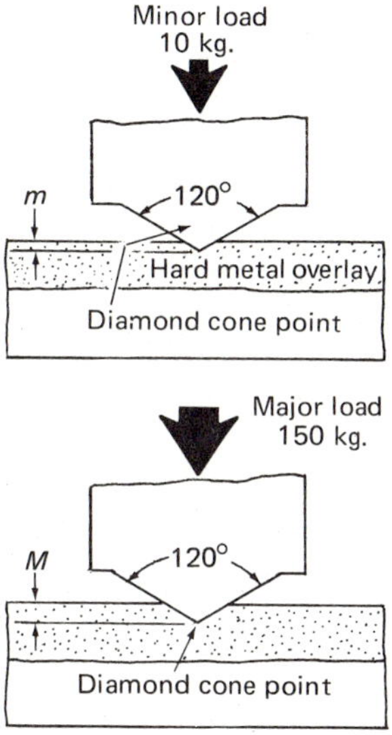
Minor load
10 kg.
m
120°
Hard metal overlay
Diamond cone point
Major load
150 kg.
M
120°
Diamond cone point
(b)

Figure **18-17** (*a*) Rockwell hardness-testing machine: (1) hardness scale, (2) penetrator, (3) anvil in use, (4) major load. (*b*) M-m is recorded on C scale in Rockwell hardness numbers. (Courtesy of Teledyne McKay, Manufacturers of Electrodes and Welding Wires.)

To determine the hardness of the softer metals, the hardest metals, and very thin metals, other major loads are used on the diamond cone. For soft metals, the diamond is replaced by a steel ball of 1/16-inch diameter. A load of 100 kilograms is used and the reading is obtained from the B scale.

The Shore scleroscope hardness test measures the height of rebound of a diamond-pointed hammer (Figure 18-18). Hard metals cause a higher rebound than soft metals.

The file-and-scratch test is a quick procedure for checking hardness. It consists of simply trying to scratch or cut the surface of a metal with a file or pointed object of known hardness. All scratch tests reveal only a superficial outer-skin hardness. They tell nothing about the hardness 1/8 inch below the surface.

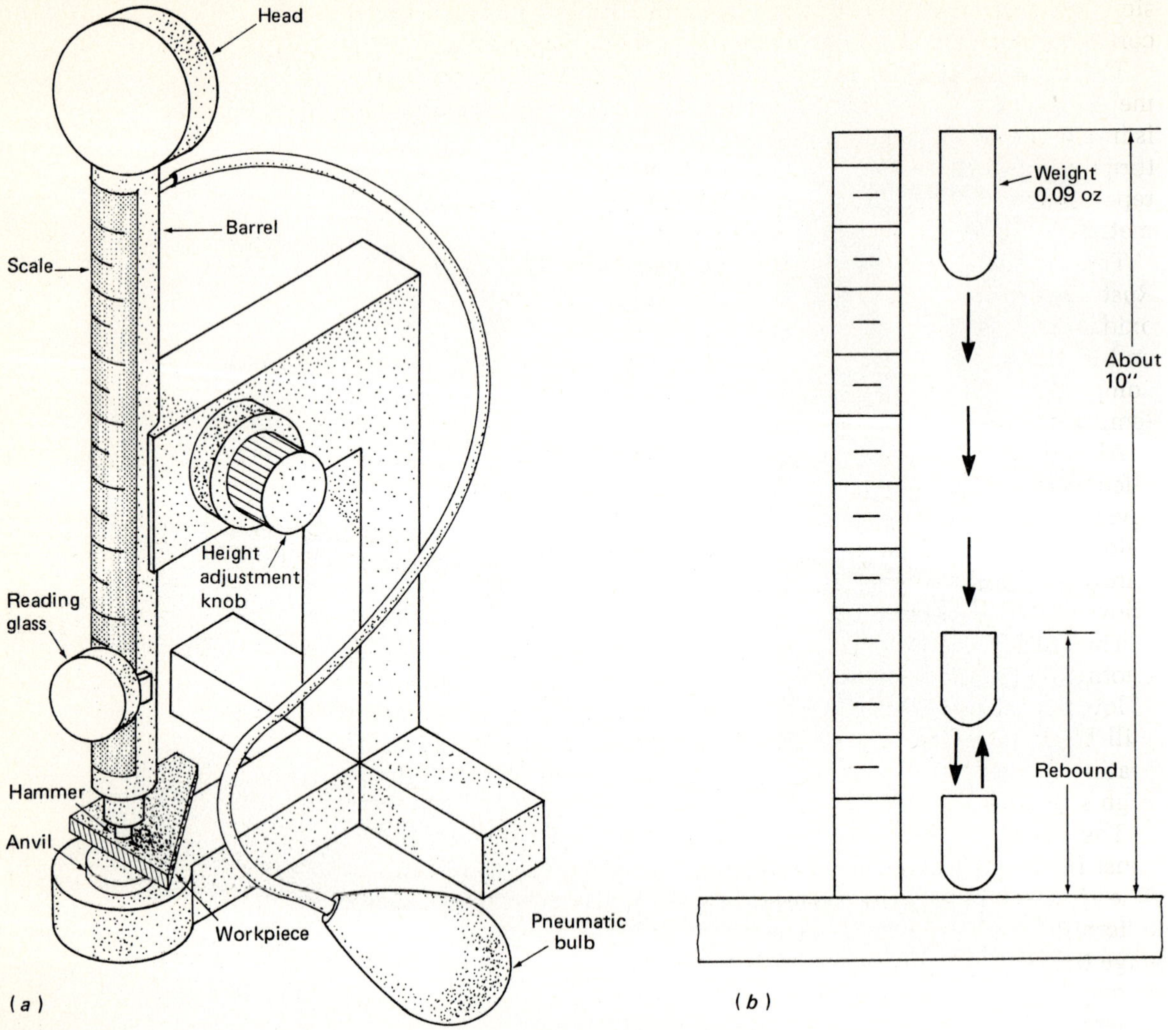
Head
Barrel
Scale
Height
adjustment
knob
Reading
glass
Hammer
Anvil
Workpiece
Pneumatic
bulb
(a)
Weight
0.09 oz
About
10"
Rebound
(b)

Figure **18-18** The Shore scleroscope and its working principle

18-7 CHEMICAL PROPERTIES

Of the many chemical properties of a metal, the one that is of importance to the welder is the ability of a metal to withstand corrosion. *Corrosion* is the wasting away of metals by slow gradual combination with other elements and chemical compounds. Corrosion resistance is the ability of a metal to resist such attack. The chemical attack may be made by a gas or a liquid, either hot or cold. A common gas, or a combination of gases such as air, or a common liquid like water may cause metals to corrode. The effects of corrosion are generally enhanced by heat, though for most purposes corrosion is considered to be an attack at room temperature.

The most familiar form of corrosion is that which occurs when metals react with oxygen in the air to form oxides. Rust, for example, is iron oxide. If the attack continues, eventually all of the iron will be turned into iron oxide. Bridges and other structures must be repainted periodically to protect them against rusting, since loss of metal may seriously weaken the structure.

This chemical attack by oxygen on metals is called *oxidation*. Rusting is a form of oxidation. In the case of some metals, the rate of oxidation is so fast that heat is given off by the chemical reaction, and the heat is sufficient to maintain the reaction until the metal is completely burned up. This effect is especially noticeable at high temperatures.

Aluminum oxidizes very quickly at room temperature, but the effects are not the same as those of rust. Aluminum oxide forms an invisible film over the surface of the metal that protects the metal below against further reaction. Because of the use of nickel and chromium as alloying elements, stainless steels and other stainless alloys do not react with oxygen in the air, even at high temperatures.

The corrosion rate of a metal may be changed in the presence of another metal. If one metal is brought into contact with another that is lower on the list (Table 18-2), or more cathodic, the higher metal will be protected from corrosion. The technique in galvanizing or plating is sometimes used in protection of a metal of low cost but high strength.

The danger of loss of metal through corrosion is ever present and must be guarded against in the manufacture of any metal product. The rates at which particular metals corrode, and the tests used to determine their rates of corrosion, do not constitute useful knowledge to the welder and so will not be described. The welder should know, however, which metals are more resistant to corrosion than others, and what effects corrosion has on welding procedures.

18-8 ELECTRICAL PROPERTIES

The electrical properties of a metal, of interest to the welder, are the electrical resistivity of the metal and, therefore, its electrical conductivity.

Electrical resistance is the "friction" that an electric current encounters when it flows through a material. As the resistance offered by a material increases, a higher voltage is required to force a given current (amperes) through the metal.

The voltage required to force a given current through the metal can be calculated by means of the following equation:

Table **18-2** The galvanic series

Anodic	Magnesium	
↑	Magnesium alloys	
	Zinc	
	Aluminum, 2S	
	Cadmium	
	Aluminum alloy 17S-T	
	Carbon steel	
	Copper steel	
	Cast iron	
	4 to 6% Cr steel	
	12 to 14% Cr steel	Active
	16 to 18% Cr steel	Active
	23 to 30% Cr steel	Active
	Ni-resist	
	7% Ni, 17% Cr steel	Active
	8% Ni, 18% Cr steel	Active
	14% Ni, 23% Cr steel	Active
	20% Ni, 25% Cr steel	Active
	12% Ni, 18% Cr, 3% Mo steel	Active
	Lead-tin solder	
	Lead	
	Tin	
	Nickel	Active
	60% Ni, 15% Cr	Active
	Inconel	Active
	80% Ni, 20% Cr	Active
	Brasses	
	Copper	
	Bronzes	
	Nickel-silver	
	Copper-nickel	
	Monel metal	
	Nickel	Passive
	60% Ni, 15% Cr	Passive
	Inconel	Passive
	80% Ni, 20% Cr	Passive
	12 to 14% Cr steel	Passive
	16 to 18% Cr steel	Passive
	7% Ni, 17% Cr steel	Passive
	8% Ni, 18% Cr steel	Passive
	14% Ni, 23% Cr steel	Passive
	23 to 30% Cr steel	Passive
	20% Ni, 25% Cr steel	Passive
	12% Ni, 18% Cr, 3% Mo steel	Passive
↓	Silver	
Cathodic	Graphite	

Source: *Stainless Steels*, C. A. Zapffe, American Society for Metals.

Resistance (ohms) × current (amperes) = volts

Some of the resistance of the metal is converted into heat (watts) and can be calculated by means of the following equation:

Watt-hours = (amperes)^2 × ohms × hours

A good conductor heats up less than a poor conductor when the same current is passed through each. In spot welding, therefore, the poorer conductor requires less current than the good conductor, other factors being equal. In arc welding, electrical resistivity is only of slight importance in heating the electrode. Like thermal conductivity, electrical resistivity depends upon the content of alloying elements. A solid solution (alloy) of two or more metals always has greater electrical resistivity than the pure metal (Table 18-3, pages 492–493). Resistivity further increases as the temperature is raised.

18-9 THERMAL PROPERTIES

The thermal properties of importance to the welder are thermal conductivity, the coefficient of thermal expansion, fusibility, and heat of fusion.

As will be noted during the discussion below, thermal conductivity and the coefficient of thermal expansion are of great importance to the welder, because these properties determine the kind and quantity of fixturing required to minimize distortion of the workpiece (base metal) during welding.

Thermal conductivity is a measure of the rate at which heat will flow through a material. The difference in thermal conductivity between iron and copper is easily demonstrated (Figure 18-19). The copper conducts heat much faster than the iron.

Figure **18-19** The difference in thermal conductivity between iron and copper. Copper, the metal that conducts heat faster, lights the match first. (Redrawn by permission from James F. Lincoln Arc Welding Foundation, *Metals and How to Weld Them,* 2d ed., 1954, 1962, p. 43.)

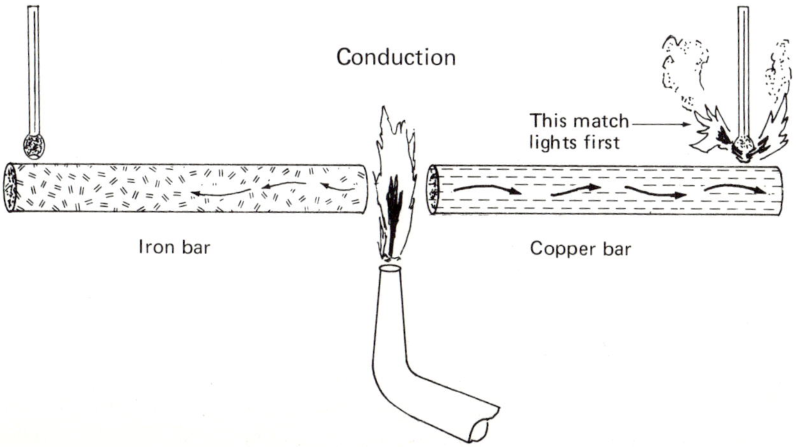
Conduction
This match lights first
Iron bar
Copper bar

Table **18-3** Mechanical properties and physical constants (including electrical resistivity of selected metals and alloys)

MATERIAL	NOMINAL COMPOSITION (ESSENTIAL ELEMENTS), %	FORM AND CONDITION	TYPICAL MECHANICAL PROPERTIES: Yield Strength (0.2% offset) 1,000 psi	Tensile Strength, 1,000 psi	Elongation in 2 in., %	Hardness, Brinell	TYPICAL PHYSICAL CONSTANTS: Density, lb/in³	Specific Gravity	Melting Point, °F	Specific Heat (32–212°F) Btu/(lb)(°F)	Thermal Expansion Coefficient (32–212°F) × 10⁻⁶ in./(in.)(°F)	Thermal Conductivity (32–212°F) Btu/(ft²)(hr)(°F)(in.)	Electrical Resistivity (68°F), ohms/cir mil ft	Tensile Modulus of Elasticity × 10⁶ psi
Aluminum Alloy No. 1100	Al 99 plus	Annealed-0	5	13	45	23	—	—	—	—	—	1540		
QQ-A-411C; ASTM B211 wire, rod, and bar		Cold-rolled-H14	17	18	20	32	.098	2.71	1190–1215	.23	13.1 (68°F)	—	18	10
		Cold-rolled-H18	22	24	15	44						1510 (68°F)	19	
3003	Al bal., Mn 1.2	Annealed-0	6	16	40	28	—	—	—	—	—	1340	21	
QQ-A-357; ASTM B221 bar, rod, and shapes extruded		Cold-rolled-H14	21	22	16	40	.099	2.73	1190–1210	.23	12.9 (68°F)	1100	25	10
		Cold-rolled-H18	27	29	10	55						1075 (68°F)	26	
2014	Al bal., Cu 4.4, Si 0.8, Mn 0.8, Mg 0.4	Annealed-0	14	27	18	45	—	—	—	—	—	1335	21	
ASTM B211 QQ-A-266 rod, bar, extrusions, forgings		Heat-treated T4	42	62	20	105	.101	2.8	950–1215	.23	12.5	840	34	10.6
		Heat-treated and artificially aged-T6	60	70	13	135	—	—	—	—	—	1075	26.5	
Carbon steel AISI-SAE 1020	Fe bal., Mn 0.45, Si 0.25, C 0.20	Annealed	38	65	30	130	0.284	7.86	2760	0.107	6.7	360	60	30
		Hot-rolled	42	68	32	135								
		Hardened (water quench 1000°F temper)	62	90	25	179								
Cast gray iron (ASTM A48-48, Class 25)	C 3.4, Si 1.8, Mn 0.5, Fe bal.	Cast (as cast)	—	25 min.	0.5 max.	180	.260	7.20	2150	—	6.7	310	400	13 ±1.5
Ductile iron (Mg-containing)	C 3.4, Si 2.5, Mn 0.40, P 0.1 max., Ni 0–1, Mg 0.06, Fe bal.	Cast	53	70	18	170								
		Cast (as cast)	68	90	7	235	.26	7.2	2100	—	7.5	228	360	25
		Cast (quench, temper)	108	135	5	310								
Low-carbon nickel Rods, bars, forgings	Ni(+Co) 99.50 C 0.02, Mn 0.20,	Annealed	15	60	50	90	.321	8.89	2615–2635	.11	7.2	420	50	30
		Hot-rolled	25	60	45	105								

ASME SB-160 ASTM B160 Plate, sheet, strip ASME SB-162 ASTM B162 Seamless pipe, tubing ASME SB-161 ASTM B161	Fe 0.15, S 0.005, Si 0.05, Cu 0.05	Cold-drawn Cold-rolled	65	95	15	150								
Malleable iron	C 2.5, Si 1, Mn 0.55 max.	Cast (annealed)	33	52	12	130	.264	7.32	2250	.122	6.6	—	180	25
Red brass (wrought) Sheet, strip, and plate, ASTM B36 Wire ASTM B134 Tubes ASTM B135	Cu 85, ZN 15	Annealed Cold-drawn Cold-rolled	15 55 60	40 70 75	50 15 7	50 120 135	.316	8.75	1875	.09	9.8	1100	28	17
Red brass (cast)	Cu 85, Zn 5, Pb 5, Sn 5	Cast (as cast)	17	35	25	60	.317	8.75	1810–1840	—	10.2	500	63	13
Stainless steel type 201	C 0.15 max., Mn 5.5-7.5, Cr 16.0-18.0, Ni 3.5-5.5, N 0.25 max.	Mill-annealed strip	50	115	60	194	.28	7.7	2550–2650	.12	—	113	414	28.6
Stainless steel type 202	C 0.15 max., Mn 7.5-10.0, Cr 17.0-19.0, Ni 4.0-6.0, N 0.25 max.	Mill-annealed strip	50	100	60	184	.28	7.7	2550–2650	.12	—	113	414	28.6
Stainless steel type 301	Fe bal., Cr 17, Ni 7, C 0.08-0.20	Annealed Cold-rolled	30 up to 165	100 up to 200	72 15	160 385	.29	8.02	2550–2590	.12	9.4	112.8	435	28

If one end of a copper bar is kept in boiling water (212°F) and the other end in chipped ice, heat will flow into the bar from the water, then through the bar to the ice, causing it to melt. The rate at which the ice melts indicates the rate at which heat is flowing through the bar.

The amount of ice that melts depends on:

1. *Time.* The longer the time, the more ice will melt.

2. *Size of the bar.* The larger the cross-sectional area of the bar, the more heat will flow.

3. *Length of the bar.* The shorter the bar, the faster the ice will melt.

4. *Temperature difference.* The higher the temperature of the hot end of the bar, the faster the ice will melt.

5. *Thermal conductivity.* The higher the thermal conductivity of the bar, the more heat will flow.

The amount of heat flow, therefore, depends upon time, area, length, temperature difference, and thermal conductivity.

The thermal conductivity, or the heat-conducting capacity, of a material is frequently expressed in Btu (British thermal units) per square foot of area per inch of length (or thickness) per hour per degree Fahrenheit. The thermal conductivities of a few metals are given in Table 18-4.

Thermal expansion is the increase in the dimensions of a body due to a change in its temperature. (See Figure 18-20 for the linear thermal expansion of selected materials.)

The coefficient of linear expansion is the ratio of the change in length of a material, caused by heating it one unit of temperature divided by the original length. The coefficient of linear expansion of iron at room temperature is 0.0000065/°F (6.5×10^{-6}/°F). The total increase in length of an iron bar 100 feet long, heated from 10 to 110°F, will be:

$0.0000065 \times (110-10) \times 100 = 0.065$ ft, or 0.78 in.

The coefficient of cubical expansion approximately equals three times the coefficient of linear expansion.

Fusibility is a measure of the ease of melting. Mercury (the metal with the lowest melting point) melts at 38°F, while tungsten, which has the highest melting point, melts at 6100°F.

A pure metal has a definite melting point, which is the same temperature as its freezing point. Alloys and mixtures of metals, however, have a temperature at which melting starts (solidus) and a higher temperature at which the melting is complete (liquidus).

Table **18-4** Thermal conductivity of selected metals

METAL	CHEMICAL SYMBOL	Btu/sq ft/in./hr/°F	RELATIVE CONDUCTIVITY BASED ON SILVER AS 100%
Aluminum	Al	1428	49.7%
Copper	Cu	2664	92.7
Gold	Au	2037	70.9
Iron-pure	Fe	467	16
Iron-steel		313	10.9
Iron-cast		316	11
Lead	Pb	241	8
Mercury	Hg	476	17
Molybdenum	Mo	1004	34.9
Nickel	Ni	412	14.3
Platinum	Pt	483	16.8
Silver	Ag	2873	100
Tin	Sn	450	15.6
Tungsten	W	1381	48
Zinc	Zn	770	27

Source: *Metals and How to Weld Them* (Cleveland, Ohio: James F. Lincoln Arc Welding Foundation), p. 44.

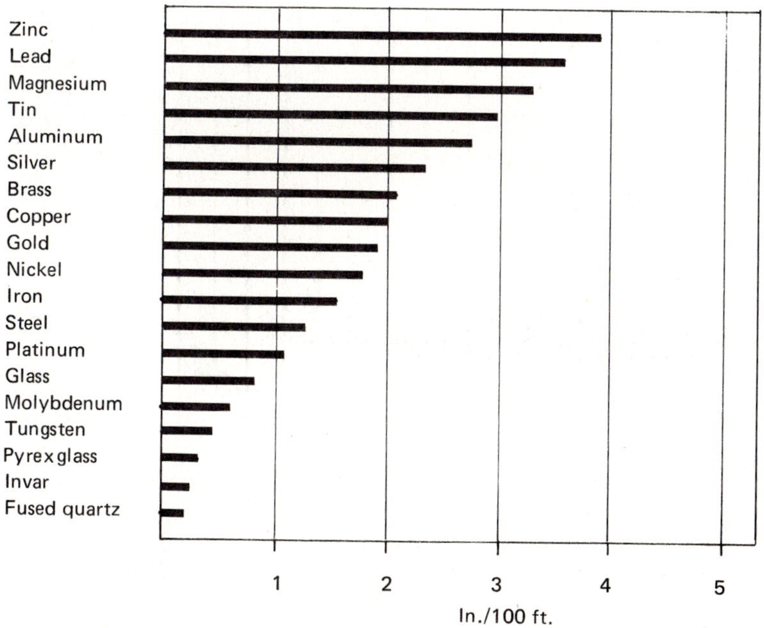
Zinc
Lead
Magnesium
Tin
Aluminum
Silver
Brass
Copper
Gold
Nickel
Iron
Steel
Platinum
Glass
Molybdenum
Tungsten
Pyrexglass
Invar
Fused quartz
1
2
3
4
5
In./100 ft.

Figure **18-20** Differences in linear thermal expansion of selected materials when heated from 32° to 212°F. (Redrawn by permission from James F. Lincoln Arc Welding Foundation, *Metals and How to Weld Them*, 2d ed., 1954, 1962, p. 45.)

The melting points of a few metals are given in Table 18-5. A chart for rapid conversion of temperature readings from centigrade to Fahrenheit or from Fahrenheit to centigrade is located in the Reference Tables at the end of the book.

The *heat of fusion* is the quantity of energy necessary to change a solid material to a liquid. Heat is the usual energy source, and the

Table **18-5** Melting points of selected metals and alloys

METAL OR ALLOY	MELTING POINT, °F
Allegheny metal	2640
Aluminum, cast, 8% copper	1175
Aluminum, pure	1218
Aluminum, 5% silicon	1117
Ambrac, A	2100
Antimony	1166
Arsenic	1497
Bismuth	520
Boron	3992
Brass, commercial high	1660
Bronze, tobin	1625
Bronze, muntz metal	1625
Bronze, manganese	1598
Bronze, phosphor	1922
Cadmium	610
Calcium	1490
Carbon (more than)	6332
Chromium	2740
Cobalt	2700
Copper, deoxidized	1981
Copper, electrolytic	1981
Duriron	2310
Everdur	1866
Gold	1945
Inconel	2540
Iridium	311
Iridium	4262
Iron, cast	2300
Iron, malleable	2300
Iron, pure	2786
Iron, wrought	2900
Lead, pure	620
Lead, chemical	620
Magnesium	1240
Manganal	2450
Manganese	2246
Mercury	−38

heat-of-fusion number is generally the amount of heat necessary to change one pound of the solid to a liquid. The British thermal unit (Btu) is used to measure the quantity of heat. For all practical purposes, a Btu is the amount of heat required to raise the temperature of 1 pound of water 1°F. The heat of fusion of ice, therefore, is 144 Btu/pound. Table 18-6 gives the heats of fusion of a few metals.

METAL OR ALLOY	MELTING POINT, °F
Molybdenum	4532
Monel metal	2400
Nichrome	2460
Nickel	2646
Nickel silver, 18%	1955
Osmium	4900
Palladium	2831
Phosphorus	114
Platinum	3218
Rhodium	3550
Ruthenium	4440
Silicon	2588
Sil-phos	1300
Silver, pure	1762
Steel, hard (0.40% to 0.70% carbon)	2500
Steel, low-carbon (less than 0.15%)	2700
Steel, medium (0.15% to 0.40% carbon)	2600
Steel manganese	2450
Steel, amsco nickel manganese	2450
Steel, nickel, 3½%	2600
Steel, cast	2600
Stainless steel, 18% chromium, 8% nickel	2550
Stainless steel, 18-8, low carbon	2640
Stellite (average)	2336
Stoodite	2420
Sulphur	240
Tantalum	5160
Thorium	3353
Titanium	3270
Tin	450
Timang	2450
Tungsten	6152
Uranium	3360
Vanadium	3182
Zinc, cast or rolled	786
Zirconium	3090

Table **18-6** The heats of fusion of selected metals

METAL	HEAT OF FUSION	METAL	HEAT OF FUSION
Aluminum	170 Btu per lb	Tungsten	79 Btu per lb
Magnesium	160	Silver	45
Chromium	136	Zinc	43
Nickel	133	Gold	29
Molybdenum	126	Tin	26
Iron	117	Lead	11
Manganese	115	Mercury	5
Copper	91		

Source: *Metals and How to Weld Them* (Cleveland, Ohio: James F. Lincoln Arc Welding Foundation), p. 43.

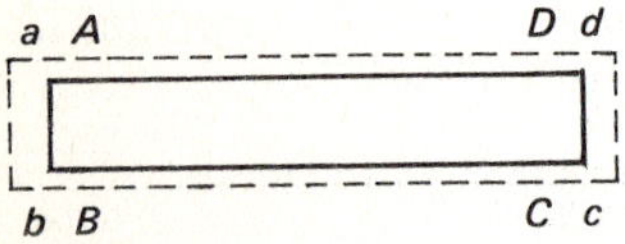
a A
D d
b B
C c

Figure **18-21** Unrestricted thermal expansion. (Redrawn by permission from *The Procedure Handbook of Arc Welding*, 11th ed., Lincoln Electric Company, Cleveland, Ohio, Fig. 2-86.)

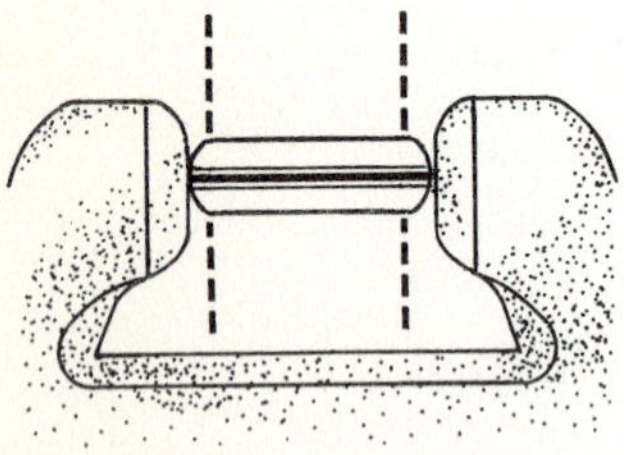

Figure **18-22** Cubical expansion resulting from restrained thermal expansion. (Redrawn by permission from *The Procedure Handbook of Arc Welding*, 11th ed., Lincoln Electric Company, Cleveland, Ohio, Fig. 2-88.)

18-10 INTERRELATIONSHIP BETWEEN THERMAL CONDUCTIVITY AND THERMAL EXPANSION

Distortion of welds resulting from a combination of thermal conductivity and thermal expansion can be minimized if the welder understands the causes, and knows how to calculate the deformation expected as a result of thermal expansion.

Consider the effect of heat on an ordinary 2-inch-diameter × 6-inch-long mild-steel bar. If the bar is heated thoroughly and uniformly to a temperature of, say, 1600°F (or any other temperature above 68°F), its entire volume will expand in all directions, and if this natural expansion is entirely free and unrestrained, the bar will increase by the amount indicated in Figure 18-21.

The coefficient of thermal expansion for steel is 0.00000645 in./in./°F (inches per inch per degree Fahrenheit). Therefore, the amount of expansion in our steel bar can be calculated as follows:

$$(1600°F - 68°F) \times 0.00000645 \text{ in./in./°F} = 0.00988$$

Therefore, length will expand $0.00988 \times 6 = 0.05928$ in., and diameter will expand $0.00988 \times 2 = 0.01976$ in.

However, if the two ends of the bar are restrained (placed in a vise or between immovable fixtures) and the bar is heated uniformly as before, expansion towards the ends will be prevented and the expansion will occur only laterally. The result of displacement of metal within the bar will be noticed when the bar cools and contracts. The bar will now be shorter and thicker, as shown in Figure 18-22. The amount of lengthwise contraction and lateral expansion is referred to as the *cubical expansion*. In this case, the quantity to which the metal would have expanded lengthwise is distributed in proportional amounts over the width and thickness.

Now instead of heating the bar uniformly, suppose the heat is applied only to one side. In this case, expansion is localized and uneven. The surrounding cool metal prevents or hinders expansion in all directions except on the surface, so that the displacement occurs there (Figure 18-23). When that area starts to cool and contract, a certain amount of this original displacement becomes permanent, causing an uneven contraction throughout the entire area. The result is distortion and shrinkage (Figure 18-24).

Several such bars treated in this way, laid side by side (Figure 18-25), may be compared to a steel plate of the same thickness. Thus the principles of the effect on the steel bar apply to plates.

Applying the discussed principles to a simple welding job, assume that two plates are to be jointed with a butt weld. As the weld is applied, the molten weld metal and the arc transmit heat out into the surrounding areas, causing considerable uneven expansion. As the weld progresses, the molten weld metal begins to cool and contract immediately, but, at the same time, the heat of the arc causes considerable expansion ahead of this contraction. It should be understood that while the weld metal is cooling, and therefore contracting, the temperature of the surrounding plates is rising, and therefore the plates are expanding (see Figure 18-26). As the plates themselves cool, they also contract.

If expansion and contraction on this particular welding operation are allowed to occur without any control, distortion results (Figure 18-27). However, three simple rules can be followed that will aid materially in the prevention and control of distortion. In many cases, the application of a single rule will be sufficient. In others, a combination of the rules may be required.

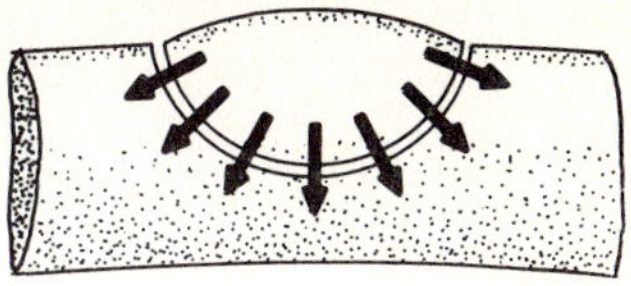

Figure **18-23** Surface displacement resulting from localized heating of a metal bar. (Redrawn by permission from *The Procedure Handbook of Arc Welding,* 11th ed., Lincoln Electric Company, Cleveland, Ohio, Fig. 2-89.)

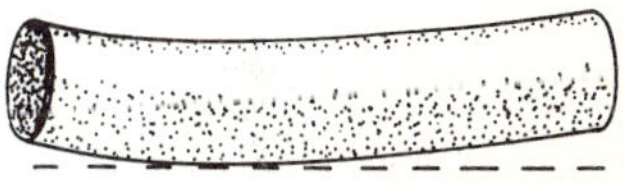

Figure **18-24** Permanent deformation of a metal bar resulting from localized heating. (Redrawn by permission from *The Procedure Handbook of Arc Welding,* 11th ed., Lincoln Electric Company, Cleveland, Ohio, Fig. 2-90.)

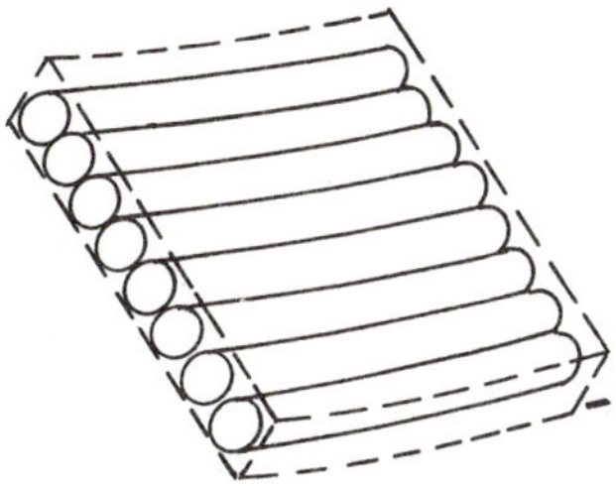

Figure **18-25** Permanent deformation of a plate resulting from localized heating. (Redrawn by permission from *The Procedure Handbook of Arc Welding,* 11th ed., Lincoln Electric Company, Cleveland, Ohio, Fig. 2-91.)

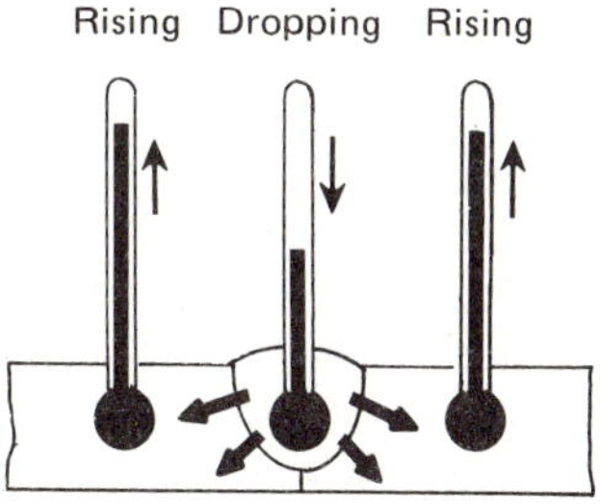
Rising
Dropping
Rising

Figure **18-26** The uneven heating and cooling of two plates during butt welding. (Redrawn by permission from *The Procedure Handbook of Arc Welding,* 11th ed., Lincoln Electric Company, Cleveland, Ohio, Fig. 2-92.)

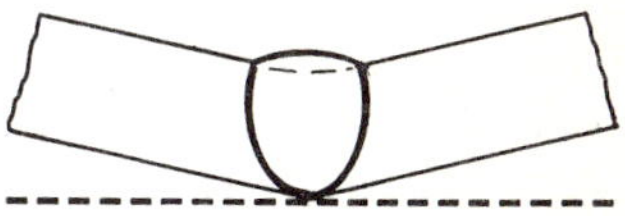

Figure **18-27** Distortion of a butt weld resulting from uncontrolled thermal expansion and contraction. (Redrawn by permission from *The Procedure Handbook of Arc Welding,* 11th ed., Lincoln Electric Company, Cleveland, Ohio, Fig. 2-93.)

- *Rule 1.* Reduce the effective shrinkage force.
- *Rule 2.* Make shrinkage forces work to minimize distortion.
- *Rule 3.* Balance shrinkage forces with other forces.

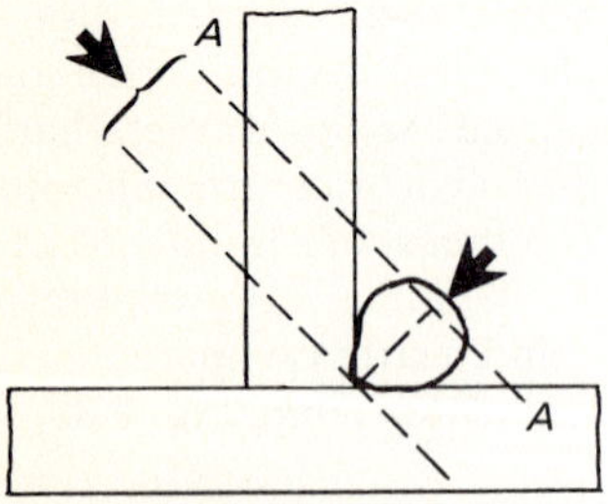
A
A

Figure **18-28** Excess weld metal above line *A-A* does not increase strength but does increase shrinkage forces. (Redrawn by permission from *The Procedure Handbook of Arc Welding*, 11th ed., Lincoln Electric Company, Cleveland, Ohio, Fig. 2-94.)

1. *Reduce the effective shrinkage force.* One way to reduce shrinkage forces is to avoid overwelding. The addition of excess weld metal not needed to meet the service requirement of the joint is known as *overwelding*. This causes the kind of distortion shown in Figure 18-27 and contributes nothing to the strength and performance of the joint. Actually, it is a waste of time and money. Weld metal should be kept at a minimum, consistent with the service requirements of the joint.

Another way of stating this principle is: Use as little weld as possible and make intelligent use of the needed weld metal. It is known that the strength of a conventional fillet weld for a T joint is determined by the effective throat (see Figure 18-28). Any excess of weld metal above the line A-A does not increase strength, but obviously increases the effective shrinkage force. Less shrinkage force may be obtained with no loss of strength by making a flat or concave weld. Less weld metal means less distortion.

It also is possible to reduce the effective shrinkage force through proper edge preparation. To obtain the proper fusion at the root of the weld with a minimum of weld metal, the bevel for steel should not exceed 30 degrees. Proper fit-up also is important, so that a minimum amount of weld metal will be needed to produce a strong joint.

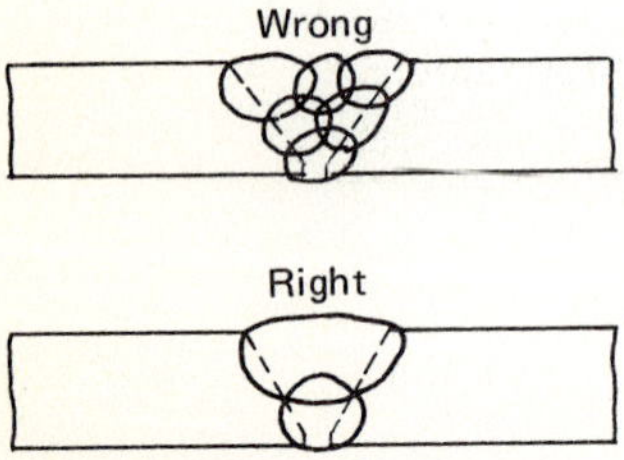
Wrong
Right

Figure **18-29** Lateral distortion can be reduced by using larger electrodes and fewer passes. (Redrawn by permission from *The Procedure Handbook of Arc Welding*, 11th ed., Lincoln Electric Company, Cleveland, Ohio, Fig. 2-95.)

Another way to make an intelligent use of weld metal is to use fewer passes (Figure 18-29). Distortion in the lateral direction is always a major problem. The use of one or two passes with large electrodes reduces distortion in this direction.

In some cases, however, distortion in the longitudinal direction is a problem and then, owing to the greater ability of a small bead to stretch longitudinally (compared to a large bead), the number of passes should be increased rather than decreased. This apparently paradoxical relationship is a function of the thickness of the plate and its natural resistance to distortion. There is inherent rigidity against the longitudinal bending of a plate, providing the plate is thick enough. Light-gage sheets have little rigidity in this direction and therefore will buckle easily. Unless the two plates to be welded are restrained, there is no lateral rigidity whatsoever, since each of the two plates is free to move angularly with relation to one another; lateral distortion, therefore, is more common.

Another means of reducing the effective shrinkage force is to place welds as close as possible to the neutral axis so that there is not sufficient leverage to pull the plates out of alignment (Figure 18-30).

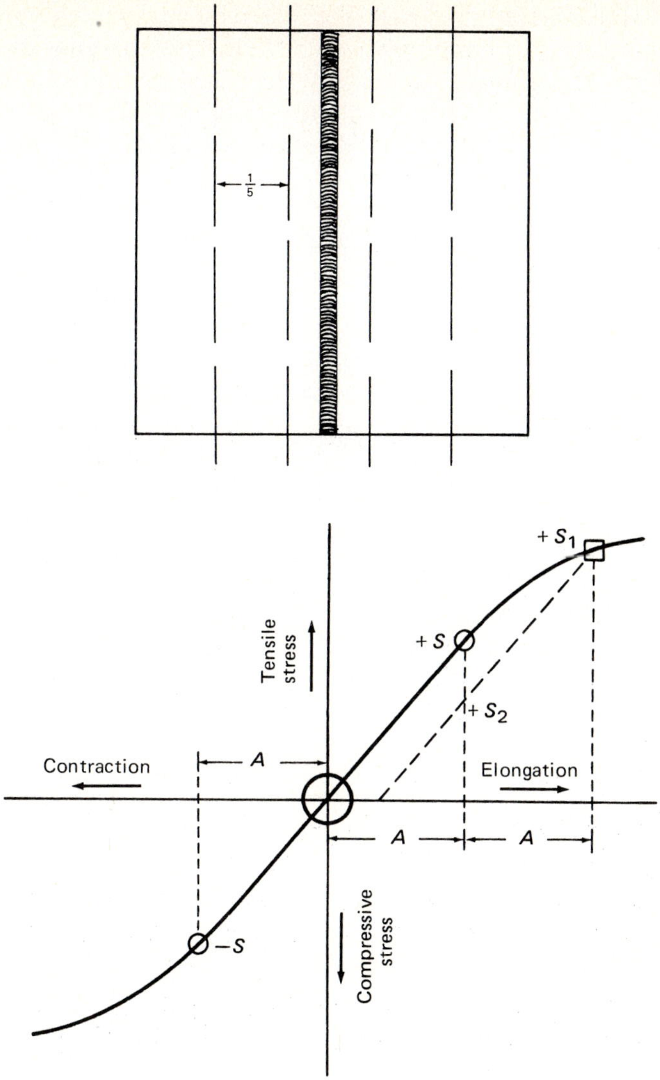

1/5
+ S1
Tensile stress
+ S
+ S2
Contraction
A
Elongation
A
A
–S
Compressive stress

Figure **18-30** Placing welds close to the neutral axis reduces leverage and therefore shrinkage forces. (From O. H. Henry and G. E. Claussen, as revised by G. E. Linnert, *Welding Metallurgy (Iron and Steel)*, p. 127. Copyright 1967 by the American Welding Society. Used with permission of the American Welding Society.)

To further reduce the effective shrinkage force by minimizing the amount of weld metal, intermittent welds (Figure 18-31) may be used in many cases instead of continuous welds. Often, it is possible to use up to two-thirds less weld metal and still obtain the strength required. The use of intermittent welds also distributes the heat more widely throughout the structure.

If the job requires a continuous weld, it still is possible to reduce the effective shrinkage force by the backstep technique. With this technique, shown in Figure 18-32, the general direction of welding progression is, say, from left to right, but each bead is deposited from right to left. As each bead is applied, the heat from the weld along the edges causes expansion that temporarily separates the plates at end B, but as the heat moves out across to C, the expansion along the outer edges C and D brings the plate back together. This occurs when the first bead is laid. The same will be true with each successive bead as it is laid, the plates expanding to a less and less degree with each bead because of the locking effect of each weld.

2. *Make shrinkage forces work to minimize distortion.* A simple way to use the shrinkage force of weld metal to advantage is to locate the parts out of position before welding. Figure 18-33 shows a T weld being made with the vertical plates out of alignment before the weld is deposited. When the weld shrinks it will pull the vertical plate to its correct 90-degree position.

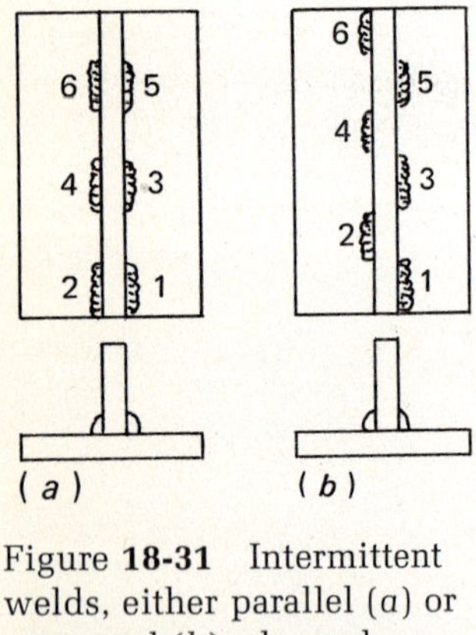
6
5
4
3
2
1
(a)
6
5
4
3
2
1
(b)

Figure **18-31** Intermittent welds, either parallel (*a*) or staggered (*b*), also reduce shrinkage forces without weakening the weld strength. (Redrawn by permission from *The Procedure Handbook of Arc Welding*, 11th ed., Lincoln Electric Company, Cleveland, Ohio, Fig. 2-101.)

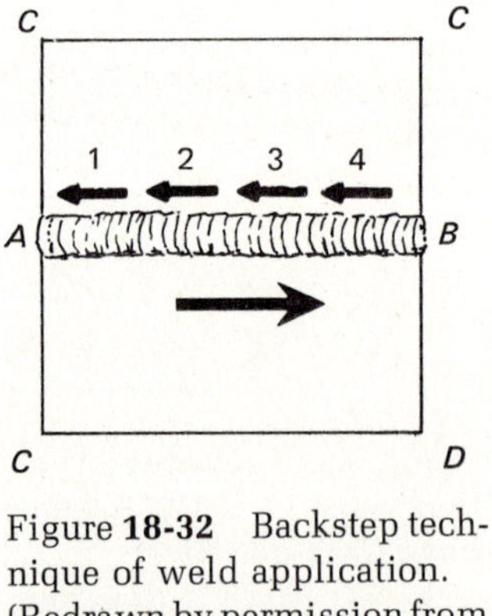
C
C
1
2
3
4
A
B
C
D

Figure **18-32** Backstep technique of weld application. (Redrawn by permission from *The Procedure Handbook of Arc Welding*, 11th ed., Lincoln Electric Company, Cleveland, Ohio, Fig. 2-96.)

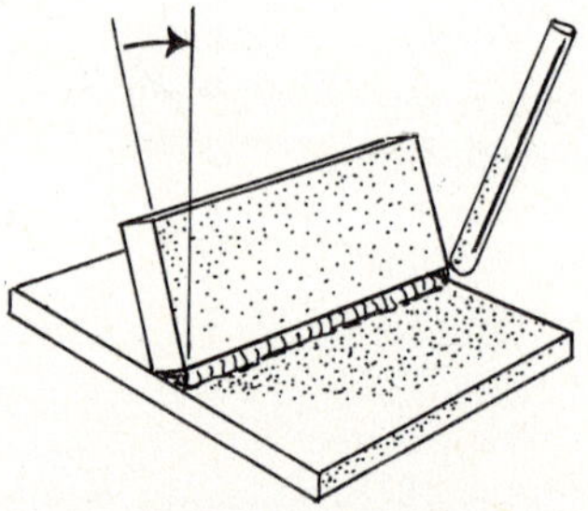

Figure **18-33** Making a T weld with the vertical plate out of alignment allows the shrinkage forces to work to straighten the joint. (Redrawn by permission from *The Procedure Handbook of Arc Welding*, 11th ed., Lincoln Electric Company, Cleveland, Ohio, Fig. 2-97.)

Another method is to space parts before welding. Experience indicates just how much space should be allowed for any given job, so that the parts will be in correct alignment after welding is completed. For example, the distance between trunnion arms of a large searchlight, shown in Figure 18-34, had to be accurately controlled. Correct spacing of the parts prior to welding allowed the arms to be pulled into the correct position by the shrinkage forces of the welding.

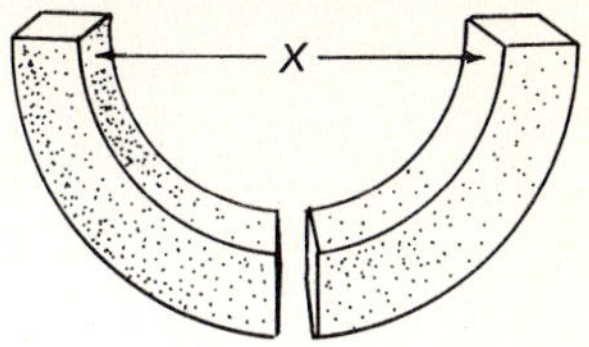
X

Figure **18-34** Correct spacing of the two parts before welding allows the shrinkage forces to pull the arms into the correct position. (Redrawn by permission from *The Procedure Handbook of Arc Welding*, 11th ed., Lincoln Electric Company, Cleveland, Ohio, Fig. 2-98.)

In many cases, shrinkage force can be put to work by prebending or prespringing the parts to be welded. For example, when the plates in Figure 18-35 are sprung away from the weld side, the counterforce exerted by the clamps overcomes most of the shrinkage tendency of the weld metal, causing it to yield. But when the clamps are removed, there still is a slight tendency for the weld to contract, and this contraction, or shrinkage force, pulls the plates into exact alignment.

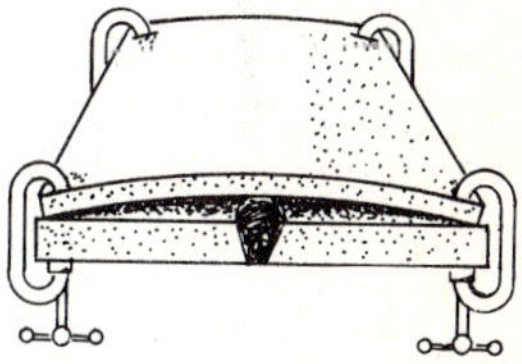

Figure **18-35** When the clamps are removed after welding, the shrinkage forces allow for enough contraction to pull the plates into alignment. (Redrawn by permission from *The Procedure Handbook of Arc Welding*, 11th ed., Lincoln Electric Company, Cleveland, Ohio, Fig. 2-99.)

3. *Balance shrinkage forces with other forces.* Often the structural nature of parts to be welded is such as to provide sufficient rigid balancing forces to offset welding shrinkage forces. This is particularly true in heavy sections where there is inherent rigidity because of the arrangement of the parts. If, however, these natural balancing forces are not present, it is necessary to balance the shrinkage forces in the weld metal in order to prevent distortion.

This can be accomplished by the use of a welding sequence that places weld metal at different points about the structure, so that as one section of weld metal shrinks it counteracts the shrinkage forces of previous welds. A simple example of this is the alternative welding on both sides of the neutral axis of a simple butt weld, as shown in Figure 18-36.

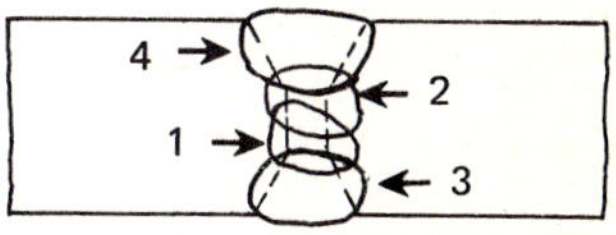
4
2
1
3

Figure **18-36** Welding both sides of the neutral axis sets up counteractive and neutralizing shrinkage forces. (Redrawn by permission from *The Procedure Handbook of Arc Welding*, 11th ed., Lincoln Electric Company, Cleveland, Ohio, Fig. 2-83.)

Another application of this principle is the staggering of intermittent welds applied in a sequence, such as shown in Figure 18-31(*a*) and (*b*). Here the shrinkage force of weld number 1 is balanced by that of weld number 2; the shrinkage force of weld number 2 is balanced by that of weld number 3, and so on.

Peening a bead, another application of this rule, actually stretches the weld, counteracting its tendency to contract and shrink as it cools. Peening should be used with great care, for too much peening may damage the weld metal.

The most important method of avoiding distortion, and one that is the most frequent application of Rule 3, is the use of clamps, jigs, or fixtures to hold the work in a rigid position during welding. In this way, the shrinkage forces of the weld are balanced with sufficient counterforces to minimize distortion. What actually happens is that the balancing forces of the jig or fixture cause the weld metal to stretch, thus preventing most of the distortion.

REVIEW QUESTIONS

1. What is meant by the *strength* of a metal?

2. Calculate the following:

a. The tensile strength of a ½-inch-diameter bar that failed when a load of 6250 pounds was applied to it.

b. The percent elongation of a ⅜-inch-gage-length weld that measured 7/16 inch after the load was removed.

c. The energy expended by the hammer of an impact tester after swinging through an arc of 60 degrees. The hammer weighed 150 pounds and was attached to a 4-foot-long arm.

3. What is meant by *brittleness*?

4. For what purpose is the nick-break test used?

5. What is the Rockwell C number corresponding to a Bhn of 578?

6. What is the principle of Shore scleroscope hardness test? Is it widely used?

7. A 50-hp engine drives a dc generator. If the generator has an efficiency of 80%, how many kW and hp does it deliver? (1 hp = 0.746 kW)

8. Calculate the thermal expansion of the length and diameter of a ¾-inch-diameter, 2-foot-long bar of aluminum when it is heated from 68 to 300°F.

9. What is the melting point of:

a. SAE 1050 steel?

b. SAE 1018 steel?

c. Molybdenum?

CHAPTER 19
FUNDAMENTALS OF WELDING METALLURGY

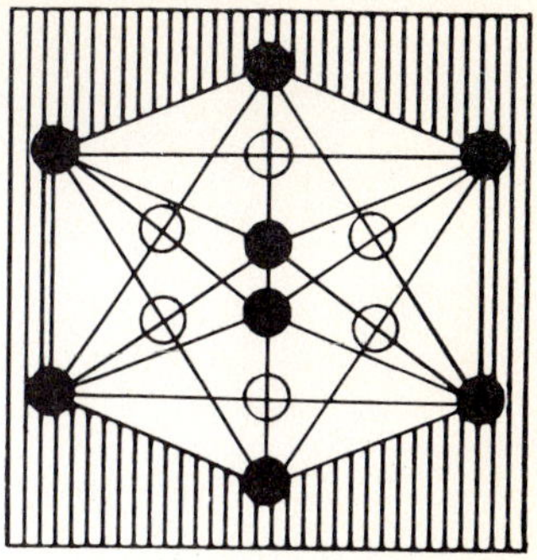

Metals are undoubtedly one of the most familiar construction materials. Space ships, the framework of industrial and commercial buildings, bridges, computers, and household appliances are all composed to a large extent of one or more kinds of metal. Because metals are frequently joined by one or more of the welding processes, it is important for those engaged in welding to know what a metal is, what properties it has, and the effect of heat on some of its properties.

A great deal was learned in the last chapter about the general properties of metals and about some basic welding rules that accommodate the properties of metals. In this chapter we will examine the structure of various metals and alloys: the metallurgy of welding.

19-1 MICROSTRUCTURE OF METALS AND ALLOYS

Although the logical starting point for a discussion of metal and alloy properties is a consideration of the atomic structure, it is beyond the scope of this book to attempt any correlation between the characteristics of individual atoms and the weld. For the purposes of this book, the crystal lattice will be the smallest unit of metals and alloys discussed.

Each metal has a definite crystalline structure. Iron in the cold state, for example, consists of unit cells (Figure 19-1) having a body-centered cubic structure (BCC). Copper has a face-centered cubic structure (FCC). And zinc has a close-packed hexagonal crystal lattice (CPH). The shapes of the crystal lattices are determined by an x-ray method referred to as *x-ray diffraction*. In one of those methods (Figure 19-2), a small particle of the metal or alloy to be studied is placed in a container in the center of a wheel containing a specially designed strip of film and rotated about its horizontal axis. When the x rays being shot at the rotating particle hit an atom (solid), they cannot pass through, and the film is therefore not exposed at those places (Figure 19-3). These unexposed areas indicate the location of solids (atoms). By applying an equation known as *Bragg's equation*, the shape of the crystal can be calculated.

The properties of metals depend on the shape of the crystals, the number of atoms comprising each crystal lattice, the distance between the atoms in the crystal lattice, and the interrelationship of the crystal lattices.

Other important factors upon which the properties depend are the features of the solidification process, or the phenomena associated with the transformation of metals from the liquid to the solid state. This process is shown schematically in Figure 19-4 and graphically in Figure 19-5. The diagram in Figure 19-5 is referred to interchangeably as an *equilibrium diagram*, a *constitutional diagram*, or a *phase diagram*. A phase diagram is essentially a plot of the temperature over which the phases of a metal are stable. When the word *equilibrium* is used, it implies that any change in a pure metal or an alloy will be a reversible change. That is, any change resulting from a temperature rise, for example, will be reversed by a corresponding lowering of the temperature. From these diagrams, something of the constitution or microstructure of the solid alloys can be predicted as well as what phases can exist. A *phase* refers to the state of a pure metal or alloy, such as a metal in gaseous state, a metal in liquid state, or a metal in one of its different crystal structures.

At this point, another concept should be introduced, that of *solid solution*. A solid solution exists when a pure parent metal, or solvent, is able to retain within it atoms of a second metal, or solute, to such a degree that there is no positive evidence to the unaided eye, or through the microscope, that the second metal is present. X-ray evidence would also indicate that the second metal is atomically dispersed within or upon the structure (lattice) of the solvent metal. In other words, if the solution were liquid, upon freezing it would not separate into different types of solid metal but remain as one (crystalline) phase. Except for possible color differences, its appearance would be that of a pure metal. It is customary to use Greek letters alpha (α), beta (β), and so on, in connection with studies of

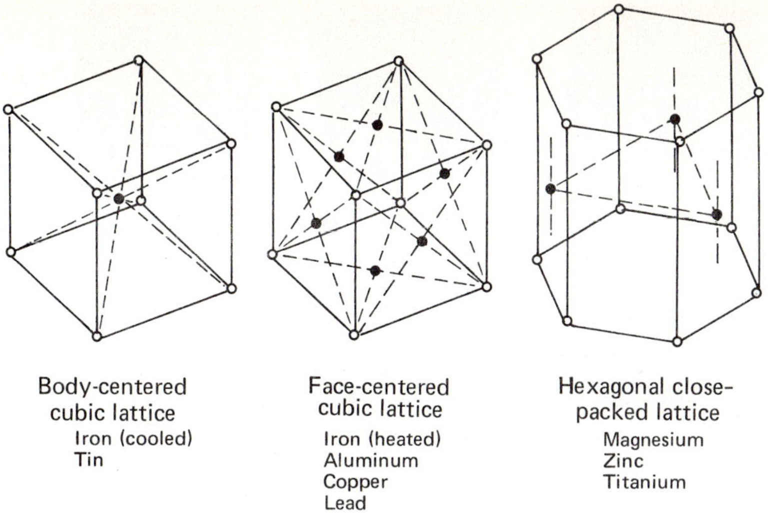
Body-centered cubic lattice
Iron (cooled)
Tin
Face-centered cubic lattice
Iron (heated)
Aluminum
Copper
Lead
Hexagonal close-packed lattice
Magnesium
Zinc
Titanium

Figure **19-1** Crystal lattice structure of some commonly welded materials. (Redrawn by permission from *Fundamentals of Service Welding*, John Deere Service Publications, 1971.)

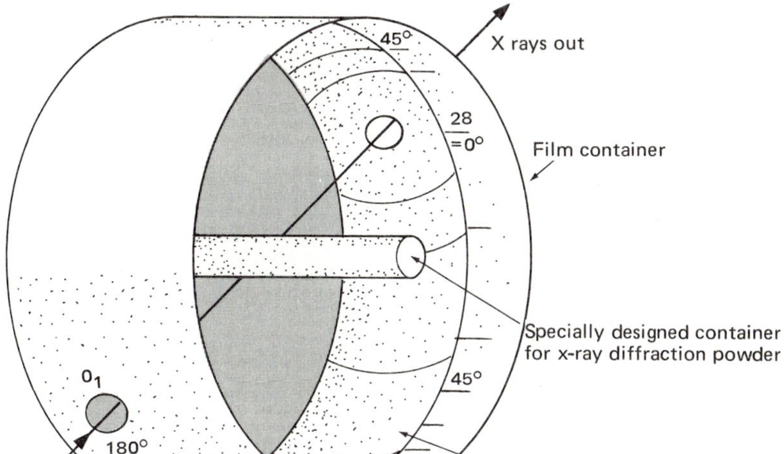
45°
X rays out
28
=0°
Film container
Specially designed container for x-ray diffraction powder
45°
0_1
180°
90°
Film being exposed to x rays
X rays in

Figure **19-2** Schematic diagram of an x-ray diffraction method

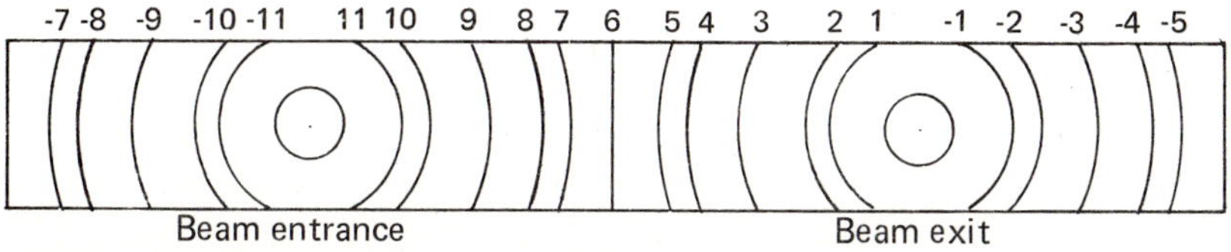
-7 -8 -9 -10 -11 11 10 9 8 7 6 5 4 3 2 1 -1 -2 -3 -4 -5
Beam entrance
Beam exit

Figure **19-3** Schematic diagram of a diffraction pattern of a crystal lattice

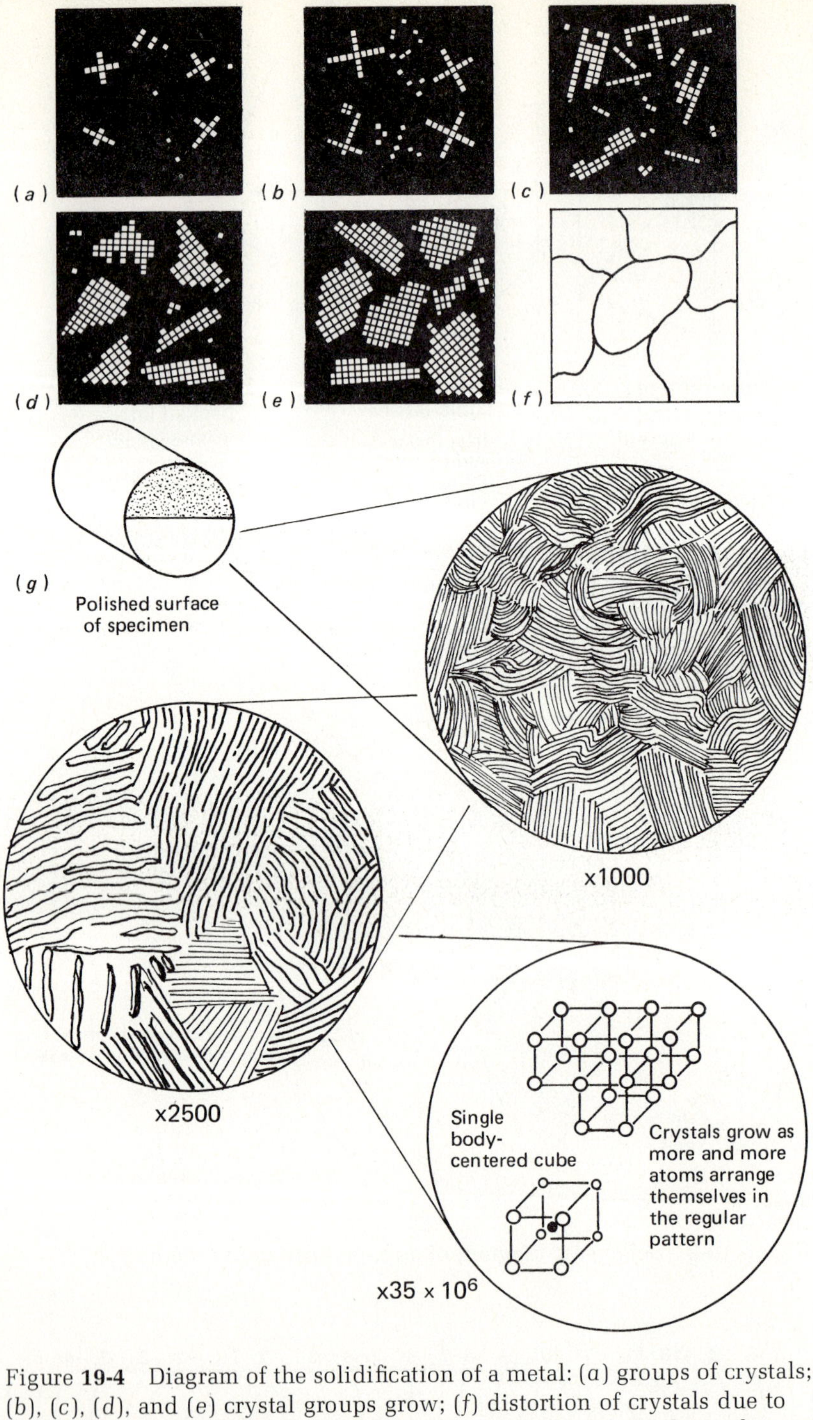
(a)
(b)
(c)
(d)
(e)
(f)
(g)
Polished surface of specimen
x1000
x2500
Single body-centered cube
Crystals grow as more and more atoms arrange themselves in the regular pattern
x35 × 10^6

Figure **19-4** Diagram of the solidification of a metal: (*a*) groups of crystals; (*b*), (*c*), (*d*), and (*e*) crystal groups grow; (*f*) distortion of crystals due to obstruction; (g) magnification of the surface of the metal. (Redrawn by permission from James F. Lincoln Arc Welding Foundation, *Metals and How to Weld Them*, 1954, 1962, p. 114.)

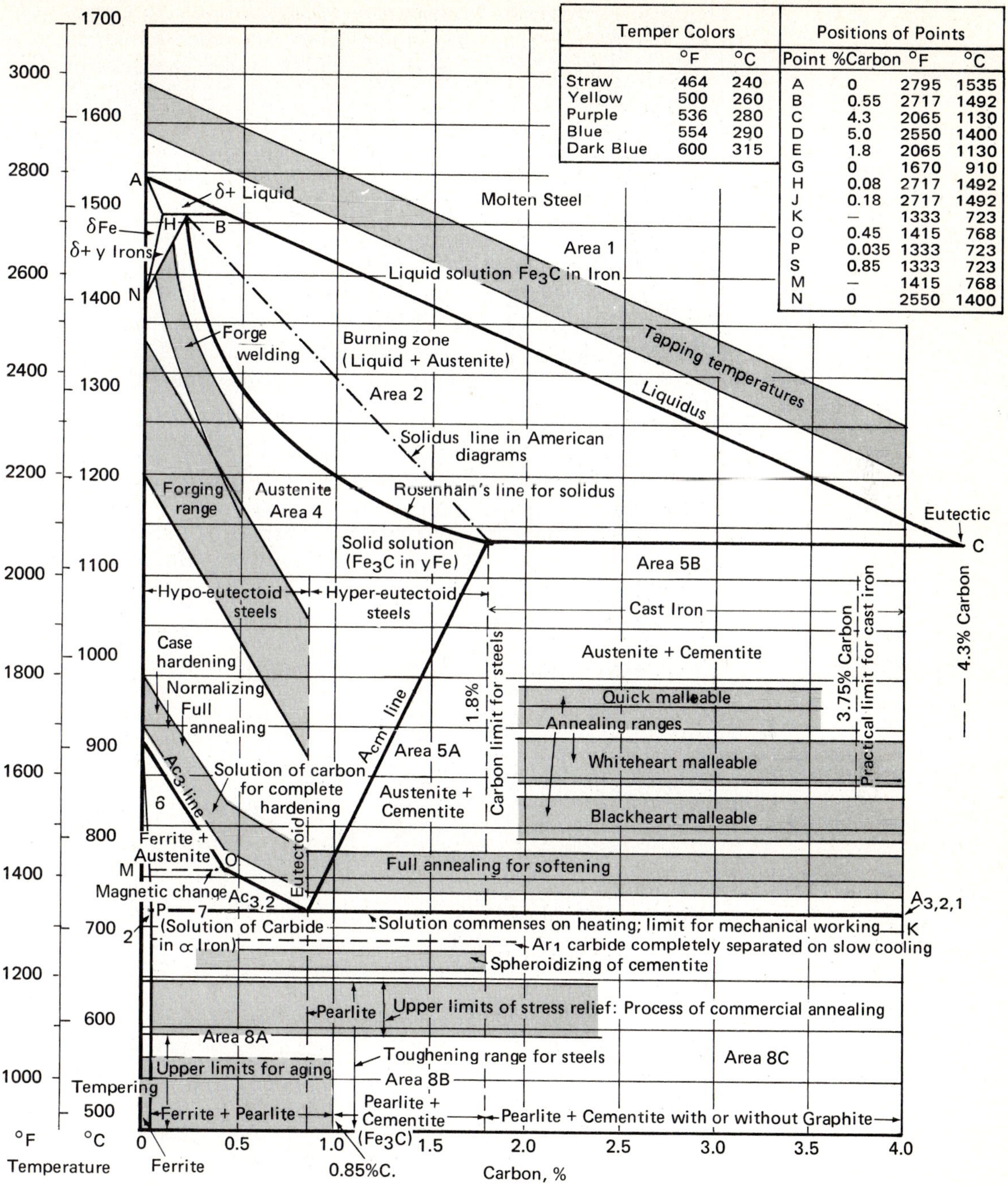

Temper Colors	°F	°C
Straw	464	240
Yellow	500	260
Purple	536	280
Blue	554	290
Dark Blue	600	315

Point	%Carbon	°F	°C
A	0	2795	1535
B	0.55	2717	1492
C	4.3	2065	1130
D	5.0	2550	1400
E	1.8	2065	1130
G	0	1670	910
H	0.08	2717	1492
J	0.18	2717	1492
K	—	1333	723
O	0.45	1415	768
P	0.035	1333	723
S	0.85	1333	723
M	—	1415	768
N	0	2550	1400

Temper Colors
Positions of Points
Molten Steel
Area 1
δ+ Liquid
δFe
δ+ γ Irons
Liquid solution Fe3C in Iron
Tapping temperatures
Liquidus
Forge welding
Burning zone (Liquid + Austenite)
Area 2
Solidus line in American diagrams
Rosenhain's line for solidus
Forging range
Austenite Area 4
Solid solution (Fe3C in γFe)
Area 5B
Eutectic
Hypo-eutectoid steels
Hyper-eutectoid steels
Cast Iron
Case hardening
Normalizing
Full annealing
Austenite + Cementite
Quick malleable
Annealing ranges
Whiteheart malleable
Blackheart malleable
1.8%
Carbon limit for steels
3.75% Carbon
Practical limit for cast iron
4.3% Carbon
Acm line
Area 5A
Ac3 line
Solution of carbon for complete hardening
Austenite + Cementite
6
Ferrite + Austenite
Eutectoid
Full annealing for softening
Magnetic change
Ac3,2
A3,2,1
Solution commenses on heating; limit for mechanical working
(Solution of Carbide in α Iron)
Ar1 carbide completely separated on slow cooling
Spheroidizing of cementite
Pearlite
Upper limits of stress relief: Process of commercial annealing
Area 8A
Toughening range for steels
Area 8C
Upper limits for aging
Area 8B
Tempering
Ferrite + Pearlite
Pearlite + Cementite (Fe3C)
Pearlite + Cementite with or without Graphite
°F
°C
Temperature
Ferrite
0.85%C.
Carbon, %
A
B
C
H
K
M
N
O
P
2
7
0
0.5
1.0
1.5
2.0
2.5
3.0
3.5
4.0
1700
1600
1500
1400
1300
1200
1100
1000
900
800
700
600
500
3000
2800
2600
2400
2200
2000
1800
1600
1400
1200
1000

Figure **19-5** Iron-carbon equilibrium diagram. (Courtesy *Metal Progress*.)

equilibrium diagrams to designate solid solutions. In considering such diagrams, it may also be helpful to remember that the coordinates of any point are expressed in terms of composition and temperature.

19-2 EQUILIBRIUM DIAGRAMS

Equilibrium diagrams enable the following predictions to be made for specific alloys: (1) temperatures at which the solid alloy will start melting, the *solidus*, and finish melting, the *liquidus;* and (2) possible phase changes that will occur as a result of altering composition or temperature. Because of the great interest in heat-treatment changes in the solid state of an alloy a summary of the solubility possibilities for any pair of metals in the solid state follows:

1. The two metals may be completely insoluble in each other in the solid state at all temperatures. Such alloys are known as *eutectic-type alloys.* A eutectic-type alloy may be defined as an alloy whose components are in such proportions as to allow the lowest possible melting point for that alloy. This is theoretically very improbable. However, numerous alloys approximate this condition. For example, cadmium does not retain any appreciable amount of bismuth in solid solution (see Figure 19-6).

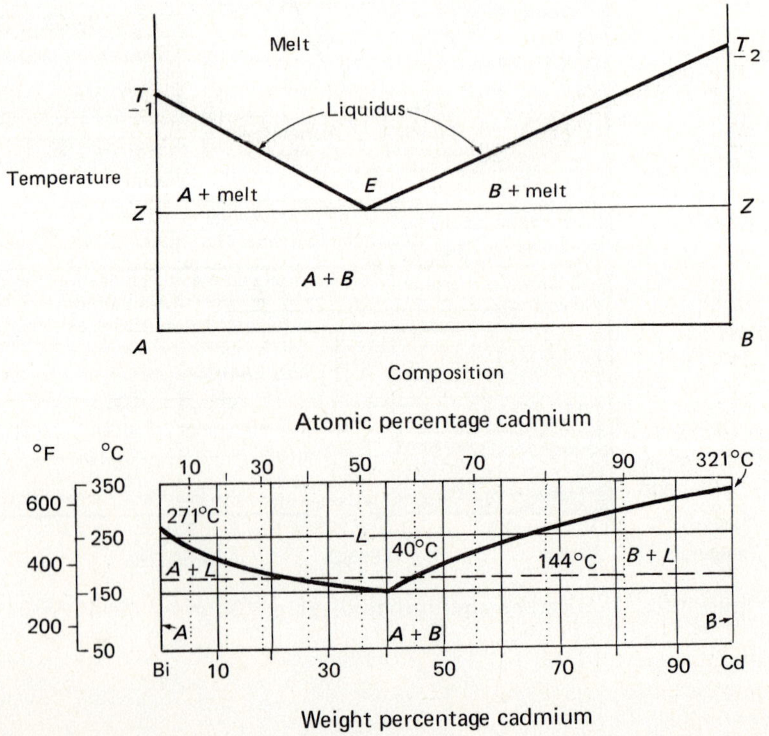
Melt
T_1
T_2
Liquidus
Temperature
A + melt
E
B + melt
Z
Z
A + B
A
B
Composition
Atomic percentage cadmium
°F
°C
10
30
50
70
90
321°C
350
600
271°C
250
L
400
40°C
144°C
B + L
A + L
150
200
A
A + B
B
50
Bi
10
30
50
70
90
Cd
Weight percentage cadmium

Figure **19-6** Bismuth-cadmium equilibrium diagram. E is the eutectic point and Z-Z is the eutectic temperature.

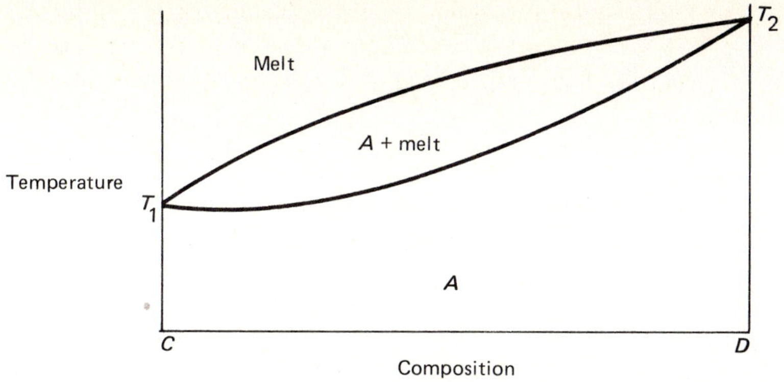

Melt
A + melt
Temperature
T_1
T_2
A
C
D
Composition

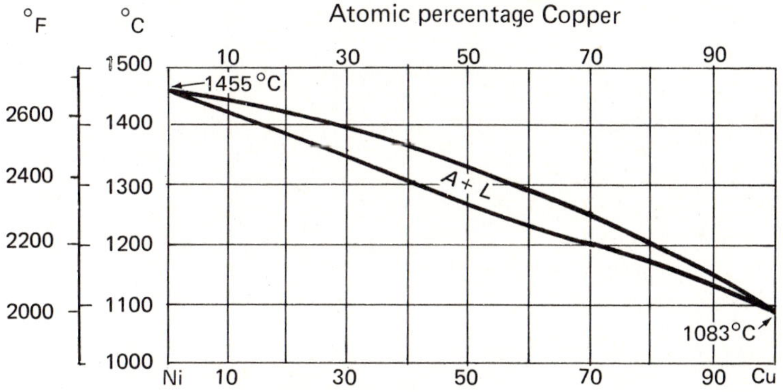

°F
°C
Atomic percentage Copper
1500
1400
1300
1200
1100
1000
2600
2400
2200
2000
10
30
50
70
90
1455 °C
A + L
1083°C
Ni
Cu

Figure **19-7** Copper-nickel equilibrium diagram. The two metals are mutually soluble in the solid state in all proportions at all temperatures.

2. The two metals may be mutually soluble in each other in all proportions at all temperatures in the solid state (see Figure 19-7). The single solid solution would commonly be designated as α (alpha). An example would be alloys of copper and nickel.

3. The two metals may be mutually soluble in each other in the solid state (see Figure 19-8). As would be expected, the metals may be partly soluble in each other; that is, the degree of solubility may vary with temperature. Usually, although not necessarily, the solubilities decrease at lower temperatures. If more than one solid solution exists, additional Greek letters are used to identify additional solid solutions.

4. If an intermetallic compound is formed, it may be considered, for the purpose of considering solubility relationships, as though it were a new element (see Figure 19-9). Metals can form chemical compounds with each other. Such compounds are known as *intermetallic compounds,* and although they do not necessarily follow

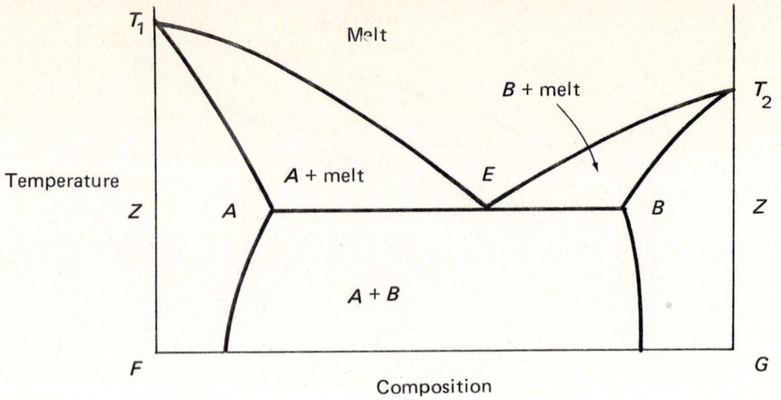

T_1
Melt
B + melt
T_2
Temperature
A + melt
E
Z
A
B
Z
A + B
F
G
Composition

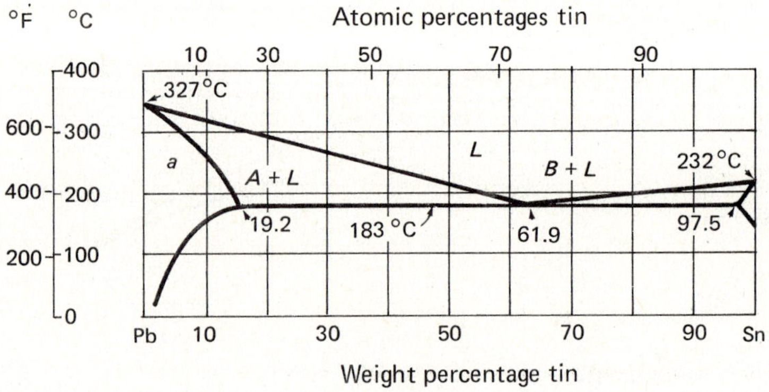

°F
°C
Atomic percentages tin
10
30
50
70
90
400
327°C
600
300
a
A + L
L
B + L
232°C
400
200
19.2
183 °C
61.9
97.5
200
100
0
Pb
10
30
50
70
90
Sn
Weight percentage tin

Figure **19-8** Lead-tin equilibrium diagram. The two metals are mutually soluble in the solid state, but the solubility varies with temperature.

chemical rules, they are, nevertheless, definite compounds, for example, Fe_3C and $CuAl_2$. If we regard a compound as a new metal, then with respect to either of the original metals the compound may be either insoluble, completely soluble, or partially soluble.

So far in this chapter we have assumed that the desired metals have been obtained in a form and of sufficient purity for our purposes. The story of how each metal is obtained from its ore is fascinating but, again, far beyond the scope and purpose of this book. As a reminder of the numerous operations involved in producing steel, however, Figure 19-10 shows the process of making steel, from iron ore to steel bars.

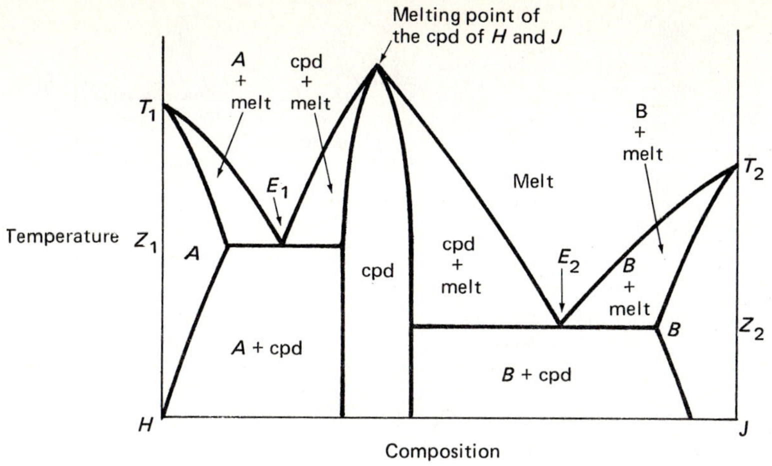
Melting point of the cpd of H and J
A + melt
cpd + melt
B + melt
T1
T2
E1
E2
Melt
Temperature
Z1
Z2
A
cpd
cpd + melt
B + melt
B
A + cpd
B + cpd
H
J
Composition

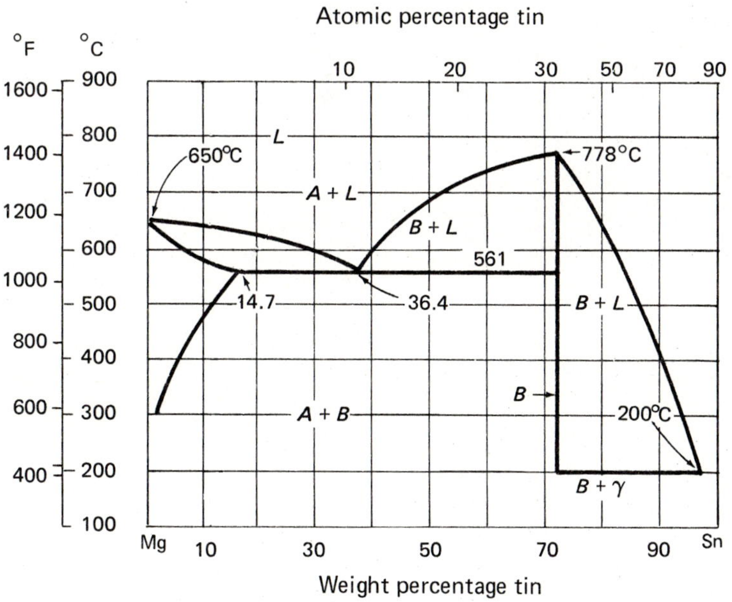
Atomic percentage tin
°F
°C
10 20 30 50 70 90
1600 1400 1200 1000 800 600 400
900 800 700 600 500 400 300 200 100
L
650°C
A + L
B + L
778°C
561
14.7
36.4
B + L
B
A + B
200°C
B + γ
Mg 10 30 50 70 90 Sn
Weight percentage tin

Figure **19-9** Magnesium-tin equilibrium diagram. An intermetallic compound (cpd) forms between the two metals.

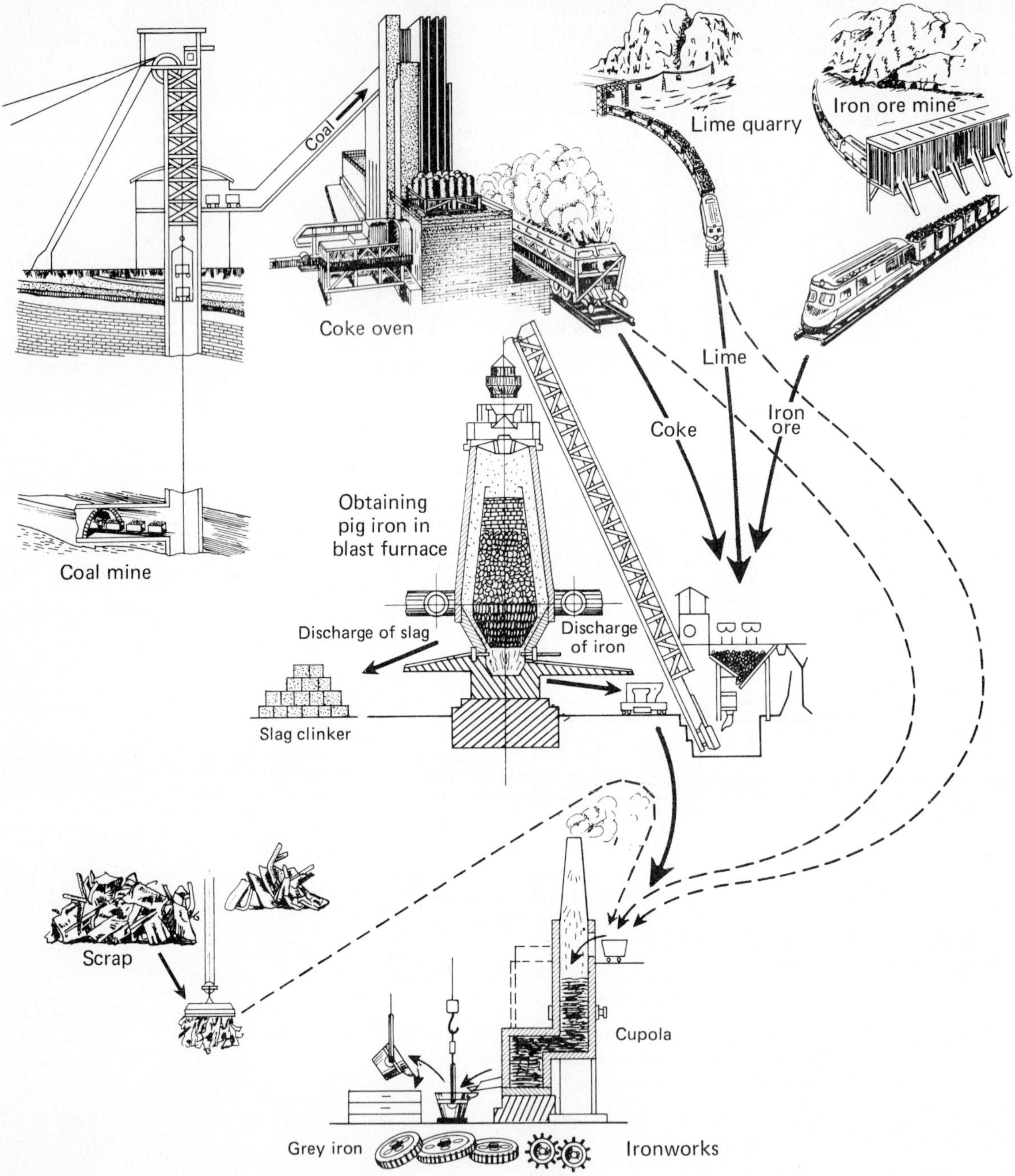
Coal
Lime quarry
Iron ore mine
Coke oven
Lime
Iron
ore
Coke
Obtaining
pig iron in
blast furnace
Coal mine
Discharge of slag
Discharge
of iron
Slag clinker
Scrap
Cupola
Grey iron
Ironworks

Figure **19-10** The process of steel making. (By permission from N. Makiyenko, *Benchwork*, MIR Publishers, Moscow, U.S.S.R.)

19-3 METALLURGICAL ASPECTS OF WELDING

The metallurgical aspects of what takes place in the weld zone during cooling differ somewhat from those observed during the cooling of a casting.

In welding, the molten metal solidifies in a matter of a few seconds. The amount of metal rarely exceeds one cubic inch. The source of heat and the metal pool have a temperature that is considerably higher than in melting furnaces. As a result of the quick cooling of the weld pool, the chemical reactions initiated in the molten metal and slag have no time to be completed.

The solidification of the molten metal in the weld pool is shown diagrammatically in Figure 19-11. As the bead progresses, the temperature of the weld pool drops due to heat abstraction into the base metal and radiation into the ambient atmosphere, and the metal solidifies.

Figure 19-11 Progressive solidification of molten metal in a weld pool: (*a*) cooling curve, calling out different structures; (*b*) top view of weld pool (*W*) and isotherm lines around pool. The isotherms are numbered according to the curve in (*a*); all points on an isotherm are the same temperature. (By permission from V. Tsegelsky, *The Electric Welder (A Manual)*, Foreign Language Publishing House, Moscow, U.S.S.R.)

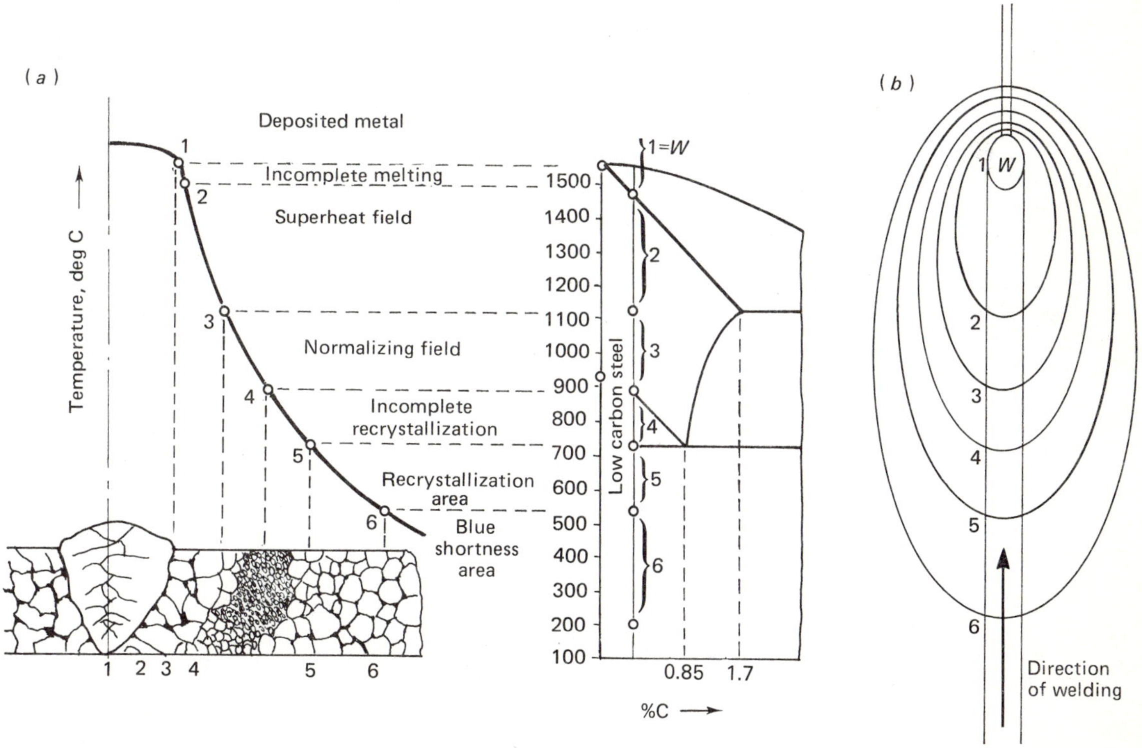
(a)
Deposited metal
Incomplete melting
Superheat field
Normalizing field
Incomplete recrystallization
Recrystallization area
Blue shortness area
Temperature, deg C
1500
1400
1300
1200
1100
1000
900
800
700
600
500
400
300
200
100
Low carbon steel
1=W
0.85
1.7
%C
(b)
W
Direction of welding

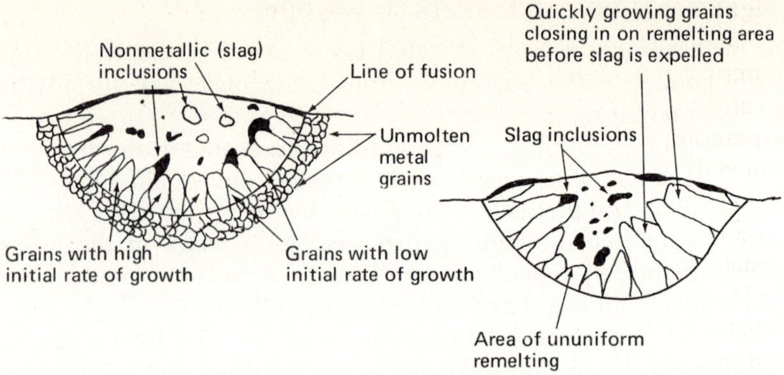
Quickly growing grains
closing in on remelting area
before slag is expelled
Nonmetallic (slag)
inclusions
Line of fusion
Unmolten
metal
grains
Slag inclusions
Grains with high
initial rate of growth
Grains with low
initial rate of growth
Area of ununiform
remelting

Figure **19-12** Cross-section of two weld beads showing slag inclusions. (By permission from V. Tsegelsky, *The Electric Welder (A Manual)*, Foreign Language Publishing House, Moscow, U.S.S.R.)

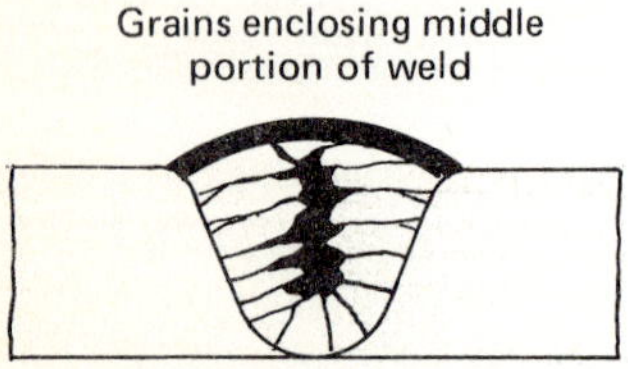
Grains enclosing middle
portion of weld

Figure **19-13** Cross-section of a weld with entrapped inclusions resulting from an aspect ratio of less than unity. (By permission from V. Tsegelsky, *The Electric Welder (A Manual)*, Foreign Language Publishing House, Mowcow, U.S.S.R.)

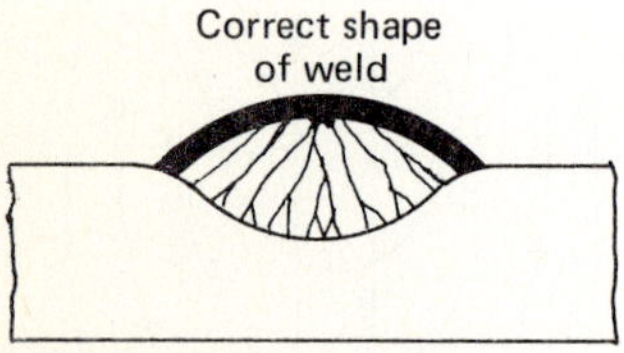
Correct shape
of weld

Figure **19-14** Cross-section of a weld bead in which aspect is greater than unity and inclusions are expelled. (By permission from V. Tsegelsky, *The Electric Welder (A Manual)*, Foreign Language Publishing House, Moscow, U.S.S.R.)

Grains first appear at the line of fusion, where the temperature is relatively low. Grains grow at different rates because as the crystals grow and press against each other, each acts on the other according to the conditions of its growth. However, the growing grains may push nonmetallic inclusions out to the surface of the weld. This is why, in overhead welding, slag appears on the weld surface rather than floating up to the weld root. It is not a matter of floating at all, but rather one of nonmetallic material being forced out of the liquid metal as the crystals begin to form and press against one another.

Steady solidification of the metal pool is upset by remelting when subsequent passes are applied. This may result in pockets of molten metal where the grain growth is retarded. These spots are therefore likely to have slag inclusions (Figure 19-12). The structure of the weld metal and the proper solidification of the metal pool depend to some extent on the *aspect ratio* of the weld, that is, the ratio of the weld width to its depth of penetration. If the aspect ratio of the weld groove is less than unity (deeper than it is wide), the pockets of molten metal that form last will be found in the center of the weld's cross-section (Figure 19-13), and the accumulations of slag, gas, and the like are possible there. In a groove that is wider than it is deep (aspect ratio greater than unity), molten metal pockets that form last (Figure 19-14) will be in the middle of the weld surface, and all impurities will be removed from the weld metal.

The portion of the parent metal that is immediately adjacent to the weld is called the *near-weld zone*. As the structure of this zone is altered by the heat of welding, this zone is also termed the *heat-affected area* or *heat-affected zone* (HAZ).

Figure 19-15 shows alterations in the structure of the heat-affected area in a low-carbon steel. Adjacent to the weld is an area of incomplete melting (1) where the metal is heated to a high temperature and coarse grains form. As we move away from the weld (2), the temperature and amount of overheating decrease, and so also the grain size. In the normalizing field (3) the grain is fine, as the time of heating is not long enough for austenitic grains to intergrow, and the subsequent cooling rejects fine grains of pearlite and ferrite. The normalizing field is followed by an area of incomplete recrystallization (4), where grains of pearlite break up into still finer grains. The recrystallization area (5) is characterized by the recovery of grains deformed by rolling. Structural alterations in the heat-affected area usually vary with the content of carbon and alloying elements in a steel.

As the temperature of the near-weld zone varies from place to place, the metal of the heat-affected area also varies in structure and mechanical properties. In the normalizing area, the weld metal may be superior to the parent metal. In the overheated area where the grain is coarse, the metal loses some of its ductility, especially impact strength. The heat-affected area also shows changes in hardness, especially in the case of steels sensitive to heat treatment. An

Figure **19-15** Diagram of the heat-affected zone (HAZ) of low-carbon steel (see Figure 19-11). (By permission from V. Tsegelsky, *The Electric Welder (A Manual)*, Foreign Language Publishing House, Moscow, U.S.S.R.)

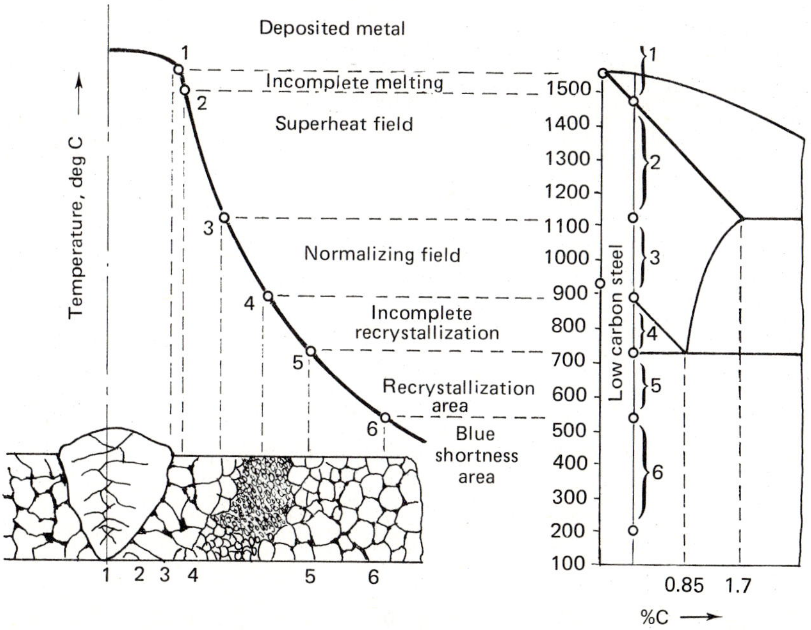

Deposited metal
Incomplete melting
Superheat field
Normalizing field
Incomplete recrystallization
Recrystallization area
Blue shortness area
Temperature, deg C
1500
1400
1300
1200
1100
1000
900
800
700
600
500
400
300
200
100
Low carbon steel
0.85
1.7
%C
1 2 3 4 5 6

increase in hardness is usually accompanied by increased brittleness and reduced ductility.

It should be noted that in welding plain low-carbon steels, structural changes in the near-weld zone do not appreciably affect the strength of weldments.

19-4 MINIMIZING WELD CRACKING IN FERROUS METALS

The various methods employed to minimize the cracking of welds in ferrous metals include (1) the use of the correct filler metal, (2) preheating, and (3) postheating. The types of filler metals for the various methods and the effects of each were discussed earlier. The important thing is that the filler selected be proper for both the welding process and the parent metal.

Preheating essentially is the raising of the temperature of the entire part, or the area of the weld, to a temperature above that of the surroundings. Preheating is primarily used with ferrous alloys that contain more than 0.25% carbon prior to welding, and its purpose is to produce a slower and therefore more even cooling.

Sometimes an entire part is placed in a permanent oven or in a temporary brick furnace built around the part. When local preheating is necessary, say, when the part can't be moved or is too large to move, an oxyacetylene or some other preheating torch could be used; or an electric induction preheater, as shown in Figure 19-16, could be used.

The preheat temperature will vary with the analysis, size, and shape of the work. Where mild-steel electrodes (see Chapter 4) are used, the preheat temperature usually ranges between 350 and 700°F. If the low-hydrogen-type electrodes are used, it is usually sufficient to use 300°F, less preheat than is recommended for mild-steel electrodes. The temperature to which a part being preheated has reached can be determined by use of such commercially available devices as temperature-indicating crayons and inks. A device such as the Lincoln Electric Company preheat calculator (Figure 19-17) can be used to calculate the preheat temperature, and the approximate preheat temperature could also be calculated by means of some simple equations if the calculator is not available. However, in both cases, the thickness and the chemical analysis of the steel to be welded must be known.

When calculating preheat temperature by equations, step 1 is to determine the chemical carbon equivalent of the steel:

$$[C]_c = C + \frac{Mn}{20} + \frac{Ni}{15} + \frac{(Cr + Mo + V)}{10}$$

Step 2 is to determine the carbon equivalent for plate thickness:

(a)

(b)

Figure **19-16** (a) Attaching the EXO-LEC electrical preheating system. (b) Completed installation of EXO-LEC preheating system with a pyrometer. (Courtesy of Exomet, Inc., a subsidiary of Air Products and Chemicals, Inc.)

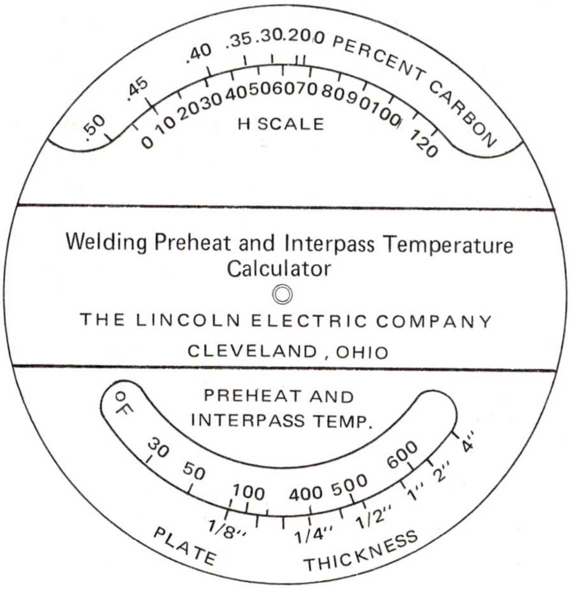
.50 .45 .40 .35 .30 .20 0 PERCENT CARBON
0 10 20 30 40 50 60 70 80 90 100 120
H SCALE
Welding Preheat and Interpass Temperature Calculator
THE LINCOLN ELECTRIC COMPANY
CLEVELAND, OHIO
PREHEAT AND INTERPASS TEMP.
°F
30 50 100 400 500 600
1/8″ 1/4″ 1/2″ 1″ 2″ 4″
PLATE THICKNESS

Figure **19-17** The Lincoln Electric preheat calculator. (Courtesy of the Lincoln Electric Company, Cleveland, Ohio.)

$[C]_t = [1 + 0.005 \text{ (thickness in millimeters)}]$

Step 3 is to determine the total carbon equivalent:

$[C]_T = [C]_c[C]_t$

Step 4 gives the preheat temperature:

$$T_{°F} = 630\,(\sqrt{[C]_T} - 0.25)^{1/2} + 32$$

The postheat treatment of a welded joint or part may consist of stress-relief heat treatment, annealing, normalizing, hardening, hardening and tempering, cross-tempering, and/or mar-tempering. The use of these terms has been quite loose, so misunderstandings have arisen. Nevertheless, each of these terms has a fairly definite meaning, as we shall see.

The American Welding Society defines *stress-relief heat treatment* as "the uniform heating of a structure to a suitable temperature, below the critical range of the base metal, followed by uniform cooling." Heat treatment within the critical range is usually undesirable because it often changes grain structure and dimension and is therefore injurious to the part. Consequently, stress-relief heat treatment is performed below the critical range in most cases.

The temperature at which stress relief is performed for welded power boilers and unfired pressure vessels is prescribed by the Boiler Construction Code of the American Society of Mechanical Engineers. The rules specify stress relieving as follows:

> Where stress relieving is required, it shall be done by heating uniformly to at least 1100°F, and up to 1200°F, or higher, if this can be done without distortion. Different temperatures may be used to obtain proper stress relieving when required by the characteristics of the material. The structure or parts of the structure shall be brought slowly up to the specified temperature and held at that temperature for a period of time proportioned on the basis of at least one hour per inch of thickness and shall be allowed to cool slowly in a still atmosphere to a temperature not exceeding 600°F.

The ASME code permits stress relieving by any of the following methods:

1. Heating the complete vessel as a unit.
2. Heating a complete section of the vessel (head or course) containing the part or parts to be stress relieved before attachment to other sections of the vessel.
3. [To] stress relieve in sections, stress relieving the final girth joints by heating uniformly a circumferential band having a minimum width of 6 times the plate thickness on each side of the welded seam in such a manner that the entire band shall be brought up to the temperature and held for the time specified. . . .

4. Nozzles or other welded attachments . . . may be locally stress relieved by heating a circumferential band around the entire vessel with the connection at the middle of the band, the band width to be at least 12 times the shell thickness wider than the attachment, . . . the entire band shall be brought up to temperature and held for the time prescribed. . . .

5. [For] welded joints in piping and tubing, the width of the . . . band shall be at least 3 times the width of the widest part of the welding groove but [not] less than twice the width of the weld reinforcement.

The temperatures commonly used for stress-relief heat treatment can be obtained by referring to the tabulated temperatures in Chapters 20 and 21, using a preheat calculator, or calculating them by means of the equation previously given.

Although the stress-relief treatment is expected only to relieve stresses and not necessarily to produce any changes in the microstructure of the steel, the general effects of a stress-relief heat treatment are as follows:

1. Recovery.
2. Relaxation.
3. Tempering or drawing (removal of hard zones).
4. Recrystallization.
5. Spheroidizing.

The first effect is universal; the second effect is achieved when the stress-relief heat treatment is conducted at a sufficiently high temperature for an adequate length of time; the third effect is exerted only if hard zones have formed as a result of welding; and the last two effects have minor significance for welding.

Unless the shrinkage stresses are reduced to nearly zero, the part is subject to brittle failure due to tensile shrinkage stresses and tensile service stresses acting in two or three directions at right angles. Metal can stand very little deformation under multiaxial tensile stresses, particularly shock, and so will fracture in a brittle manner with very little deformation. In addition, if the welded structure—for example, an engine stand—is machined before stress-relief heat treatment, the machining operation removes metal under shrinkage stress, thus creating redistribution of stress and causing distortion. The machinist cannot be certain that the correct dimensions are being machined unless the part has been stress relieved beforehand.

Besides having shrinkage stresses, the welded part may have undergone a little permanent deformation, not only at high temperatures but at low temperatures as well. Permanent deformation below 850 or 1000°F is accompanied by slip lines in the microstructure. *Slip planes* represent planes in the grains of steel that can move over each other. Therefore, a slippage of one or several atomic distances

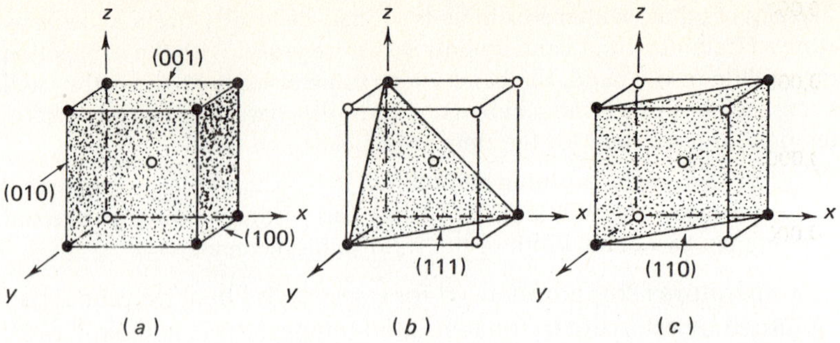
z
(001)
(010)
x
(100)
y
(a)
z
x
(111)
y
(b)
z
x
(110)
y
(c)

Figure **19-18** Atomic planes in body-centered cubic lattices. Slippage occurs on any of the indicated planes. (By permission from Y. Lakhtin, *Engineering Physical Metallurgy*, MIR Publishers, Moscow, U.S.S.R.)

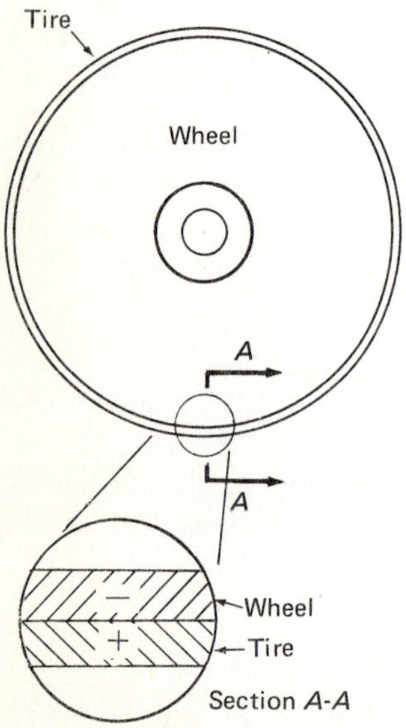
Tire
Wheel
A
A
−
Wheel
+
Tire
Section A-A

Figure **19-19** The shrunk-on tire illustrates the kind of stresses working on a welded structure that is to be stress relieved. (From O. H. Henry and G. E. Claussen, as revised by G. E. Linnert, *Welding Metallurgy (Iron and Steel)*, p. 162. Copyright 1967 by the American Welding Society. Used with permission of the American Welding Society.)

could occur along any of the planes indicated in Figure 19-18. Slip due to permanent deformation at higher temperatures occurs in the same way, but the slipped planes instantaneously straighten out and lose all traces of curvature involved in the slip.

The welded structure that is to be stress relieved thus contains tensile and compressive shrinkage stresses of all magnitudes up to the yield strength, and also has some cold-worked crystals with slip lines. The welded structure is in a condition similar to the wheel with the shrunk-on tire in Figure 19-19, which, in turn, resembles a girth weld joining a head to a drum. The tangential shrinkage stresses are tensile at the circumference, compressive toward the center. Hard zones will be considered absent for the moment.

RECOVERY The first effect to be found as the temperature is raised for a stress-relieving treatment is *recovery*. The temperature is uniformly raised in order to keep all parts of the structure at as nearly the same temperature as possible at all times, thus preventing thermal stresses. As the temperature is raised through the first 400°F or so, there is no observable change in grain structure, yet the shrinkage stresses decrease a little. The decrease is due to recovery. It is a general rule that internal stresses in a material decrease when the temperature is raised. In many metals, such as tin and lead, internal stresses decrease to nearly zero at room temperature, and there is evidence that shrinkage stresses in steel may decrease slightly at room temperature. Recovery also causes changes in magnetic and electrical properties, suggesting that its effect on internal stresses is related to some obscure movements among the electrons and atoms of steel as its temperature is raised.

However, the release of shrinkage stresses at the lower temperatures (recovery) is not always adequate; therefore the temperature must be raised still more. At all temperatures, an increase in time

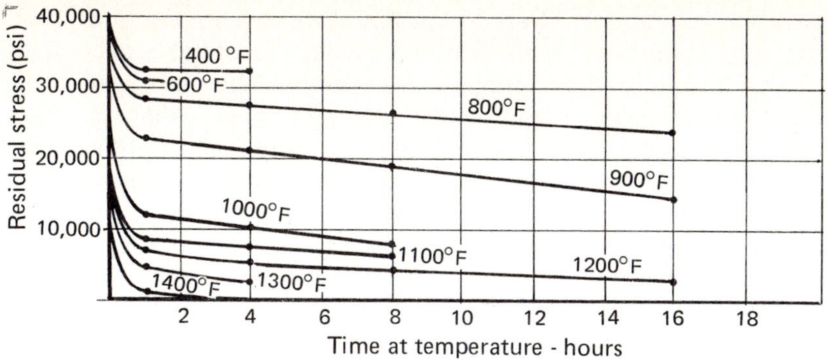
Residual stress (psi)
40,000
30,000
20,000
10,000
400 °F
600°F
800°F
900°F
1000°F
1100°F
1200°F
1300°F
1400°F
2
4
6
8
10
12
14
16
18
Time at temperature - hours

Figure **19-20** Influence of temperature and time on relieving stress in steel. (From O. H. Henry and G. E. Claussen, as revised by G. E. Linnert, *Welding Metallurgy (Iron and Steel)*, p. 163. Copyright 1967 by the American Welding Society. Used with permission of the American Welding Society.)

reduces the stress still further, but the major release of stress occurs during the first hour.

RELAXATION Upon raising the temperature to 1000°F or higher, we find that relaxation occurs and shrinkage stresses are relieved swiftly and completely, as indicated in Figure 19-20. The change during relaxation is shown in Figure 19-21. A bar of steel is

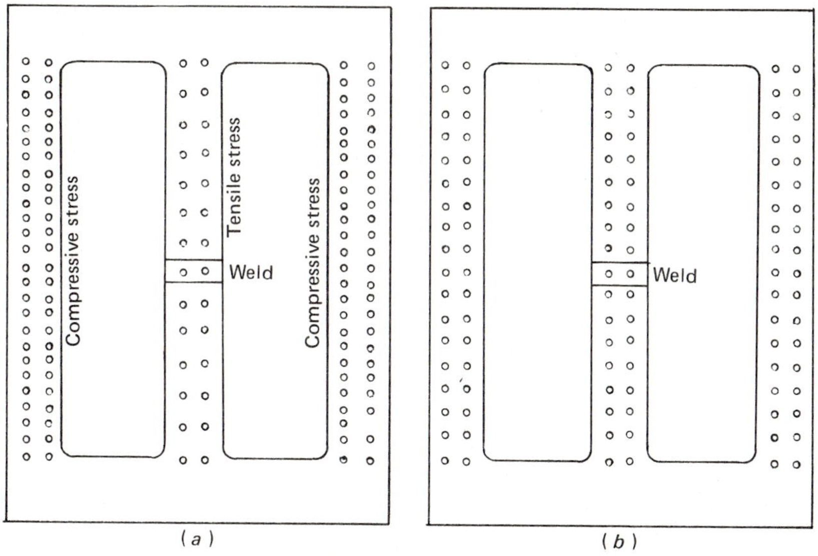
Compressive stress
Tensile stress
Weld
Compressive stress
(a)
Weld
(b)

Figure **19-21** (*a*) In the as-welded state, the atoms in the outer arms are compressed together vertically; the inner arm atoms are drawn apart. (*b*) At 1000°F the stresses are relieved as the atoms move to equalize stress. (From O. H. Henry and G. E. Claussen, as revised by G. E. Linnert, *Welding Metallurgy (Iron and Steel)*, p. 164. Copyright 1967 by the American Welding Society. Used with permission of the American Welding Society.)

stretched to the yield stress in tension at room temperature [Figure 19-21(a)] upon being welded into a frame. The atoms stretch apart parallel to the load and move close together perpendicular to the stress, which indeed is the basic meaning of *stress*—a movement of the atoms to positions capable of withstanding load. At 1000°F, the short-time yield strength is only a fraction of the yield strength at room temperature. That is to say, the atoms can no longer withstand the stress, and therefore return closer and closer to the equidistant positions occupied by atoms in an unstressed crystal. Since the temperature causes the yield strength to fall below the stress imposed, the metal flows plastically until the stress (whether compression or tension) is reduced to the yield strength of the metal at 1000°F. The stress decreases until the atoms no longer yield.

Steel can withstand stress up to the yield value virtually indefinitely at room temperature. At 1000°F, however, atoms can only withstand considerable stress for a short period of time without yielding. More precisely stated, atoms move continuously under all stresses; however, the motion, and therefore stress relief, is so slow and minute as to be undetectable after hundreds of years at room temperature (Figure 19-22). At 1000°F the motion of the atoms increases considerably and may involve undetermined motions at

Figure **19-22** Schematic diagram showing the decrease in shrinkage stress with time at different temperatures. (From O. H. Henry and G. E. Claussen, as revised by G. E. Linnert, *Welding Metallurgy (Iron and Steel)*, p. 165. Copyright 1967 by the American Welding Society. Used with permission of the American Welding Society.)

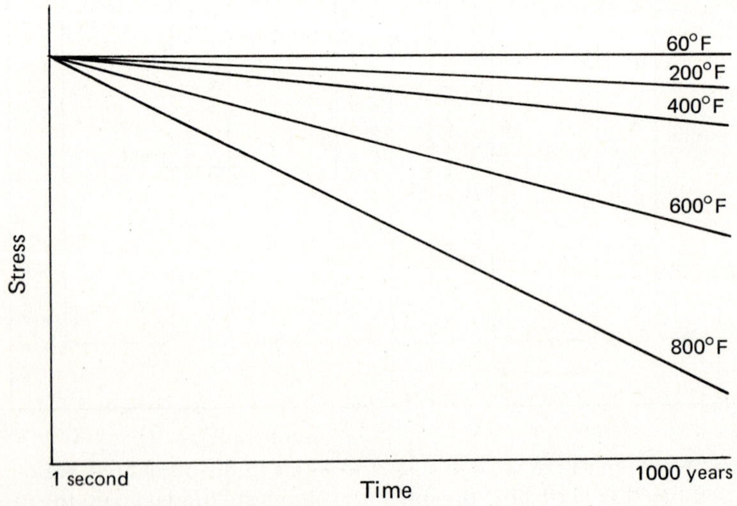
60°F
200°F
400°F
600°F
800°F
Stress
1 second
Time
1000 years

the grain boundaries. Both types of motions constitute flow or creep of the metal and greater stress relief.

The distortions created by relaxation (creep) or stress relief in a welded structure are difficult to compute, but generally are far smaller than the distortions created by machining nonstress-relieved (annealed or normalized) parts, because machining after welding generally removes the most highly stressed material.

Figure 19-20 indicates that the most effective method to obtain stress relief is to raise the temperature of the material. However, the extent to which the temperature can be raised is governed by the permissible distortion of the unsupported mass. Furthermore, it is seldom advisable to raise the temperature above the lower transformation point. Figure 19-20 also shows that increase in time favors stress relief. It is a general rule that coarse grain size exerts no detectable influence on stress relief at 1000 or 1200°F. The higher the initial shrinkage stress, the higher the stress after stress relief at a given temperature for a given time, owing to the progressive strengthening involved in the greater release of stress by strain. This strain may be looked upon as cold work during stress relief and strengthens the crystals. Since nearly all types of welds in a given steel have maximum shrinkage stresses near the yield strength, the time required for stress relief cannot be arbitrarily reduced (from the rule of thumb of a minimum of one hour per inch of cross-section thickness), simply because of an apparent absence of restraint during welding.

Structures composed of steels designed for exceptional resistance to creep at elevated temperatures will relax less rapidly than mild steel. The creep-resistant steels, for example, 0.20% and 0.5% Mo, require a higher temperature, or a longer time at constant temperature, for any given degree of stress relief, than unalloyed steel.

A following weld bead relieves the shrinkage stresses in the underlying bead, but creates new shrinkage stresses. Consequently, multipass welds have shrinkage stresses no lower than single-pass welds and require the same amount of stress relief.

TEMPERING OR DRAWING Raising the temperature of quenched steel to any point below the lower transformation point is called *tempering* or *drawing*. The heat-affected zone close to welds made without preheat or with insufficient preheat in medium-carbon steel and many other high-tensile steels cools very rapidly and therefore is very hard. The *hardness* of any steel is a function of the speed and method of its cooling. The various methods of cooling are known as *quenching*. If the weld zone was above its critical range during welding, it consisted of austenite containing ten times as much carbon in solid solution as is soluble at room temperature. During quenching (in this case, cooling in the air), the austenite changes largely to martensite—the tetragonal crystalline form of steel intermediate between face-centered and body-centered. The carbon in the martensite is believed to exist as either carbon atoms or tiny crystals of iron carbide (or carbides containing alloying elements in alloy steels), which are believed to account for the great hardness of martensite compared with any other state of a steel.

When the temperature is raised, the zone containing martensite undergoes three changes:

1. Martensite changes to ferrite (body-centered cubic crystals) with the precipitation of fine carbide crystals from the supersaturated tetragonal crystal lattice.

2. Any austenite that has not changed to martensite during quenching changes to ferrite and carbide.

3. The minute crystals of carbides in the martensite and the larger crystals of carbide in other constituents, such as fine pearlite, increase in size.

The temperatures at which changes 1 and 2 occur are not known precisely. In plain-carbon steels with about 0.7% carbon, change 1 occurs at 300°F, and change 2 at 450°F. Small amounts of austenite can be retained on quenching plain-carbon steels of at least 0.4% carbon, but only by carefully controlled quenching. Austenite is retained with lower carbon contents if alloying elements are present. The growth in size of carbide particles is continuous as the temperature is raised. Change 3 seems to account for the major drop in hardness on tempering. That is, the hardness of martensitic steel depends upon a fine dispersion of carbide particles on each crystal plane. This hinders slip and raises hardness while simultaneously reducing ductility.

The hardness of steel after tempering or drawing depends primarily on the tempering temperature. The time at that temperature is of secondary importance. The influence of the time factor is illustrated in Figure 19-23. After one minute at 1200°F the hardness at room temperature drops from 56 to 27 Rockwell C for the quenched 0.35%-carbon steel. After an hour the hardness has dropped to C-18. The reduction in hardness caused by stress-relief heat treatment of a welded medium-carbon steel containing hard zones depends largely on the time of treatment. The greatest reduction in hardness, as well as in shrinkage stresses, occurs in the first few minutes. The hardness after the first few seconds at 1200°F is nearly the same, whether the steel consisted initially of martensite, bainite, or fine pearlite.

Concurrent with the softening during tempering, there is a rise in ductility, as revealed in the static tensile test, and a decrease in tensile strength. Notch-impact values and other measures of toughness, however, are often at a minimum after tempering at 400 to

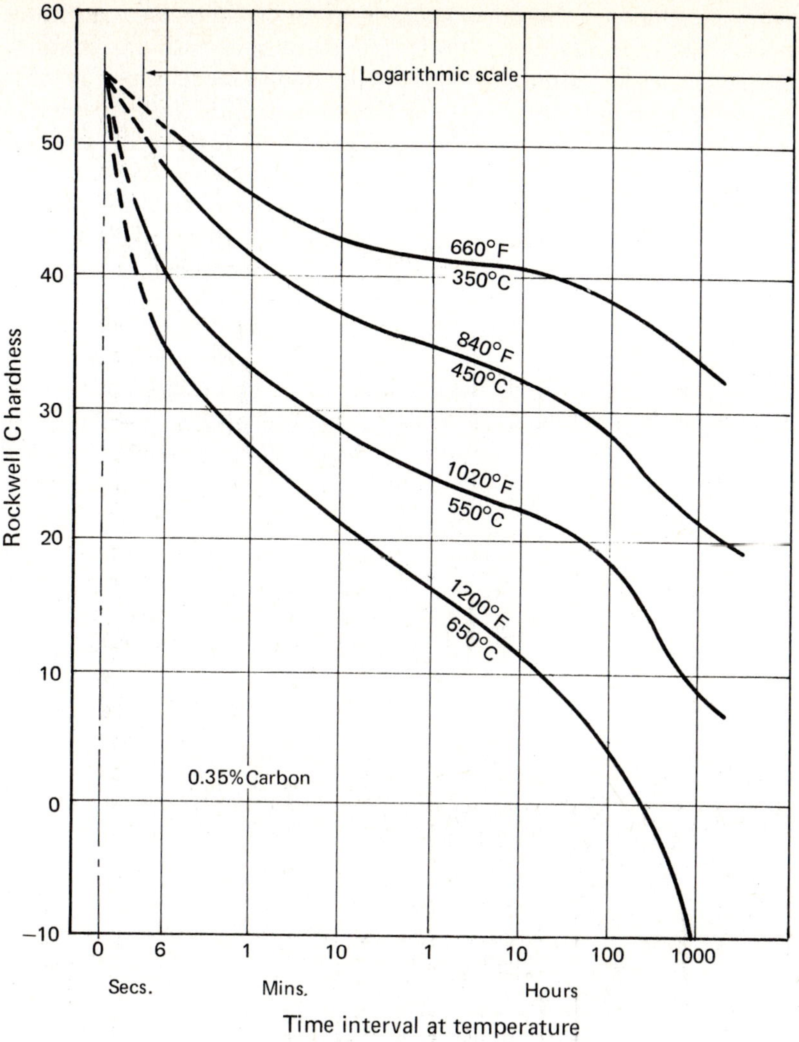

60
50
40
30
20
10
0
−10
Rockwell C hardness
Logarithmic scale
660°F
350°C
840°F
450°C
1020°F
550°C
1200°F
650°C
0.35% Carbon
0
6
1
10
1
10
100
1000
Secs.
Mins.
Hours
Time interval at temperature

Figure **19-23** Influence of temperature-time on Rockwell hardness of SAE 1035 steel. (From O. H. Henry and G. E. Claussen, as revised by G. E. Linnert, *Welding Metallurgy (Iron and Steel)*, p. 168. Copyright 1967 by the American Welding Society. Used with permission of the American Welding Society.)

700°F, rising to high values above 800°F. The reduction in notch toughness is sometimes attributed to the decomposition of internal stresses resulting from the temperature increase. The softening of plain-carbon steel is continuous as the tempering temperature is raised, because the main purpose is the agglomeration and growth of carbide particles. When carbide-forming elements are added to the steel, it may become harder instead of softer as tempering continues to 900°F after tempering at 700°F. The increase is called *secondary hardness,* and, as Bain discovered, is due to the formation of minute crystals of a second compound after the iron carbide has coalesced to a degree no longer very effective as a hardener. The minute crystals of alloy carbide form at elevated temperatures that provide the necessary opportunity for diffusion. Secondary hardness extends its influence through the alloying elements to the temperatures of stress-relief heat treatments, as shown in Figure 19-24. It is due to

Figure **19-24** Chart showing the effect known as *secondary hardness* in various chrome-molybdenum steels. (From O. H. Henry and G. E. Claussen, as revised by G. E. Linnert, *Welding Metallurgy (Iron and Steel),* p. 169. Copyright 1967 by the American Welding Society. Used with permission of the American Welding Society.)

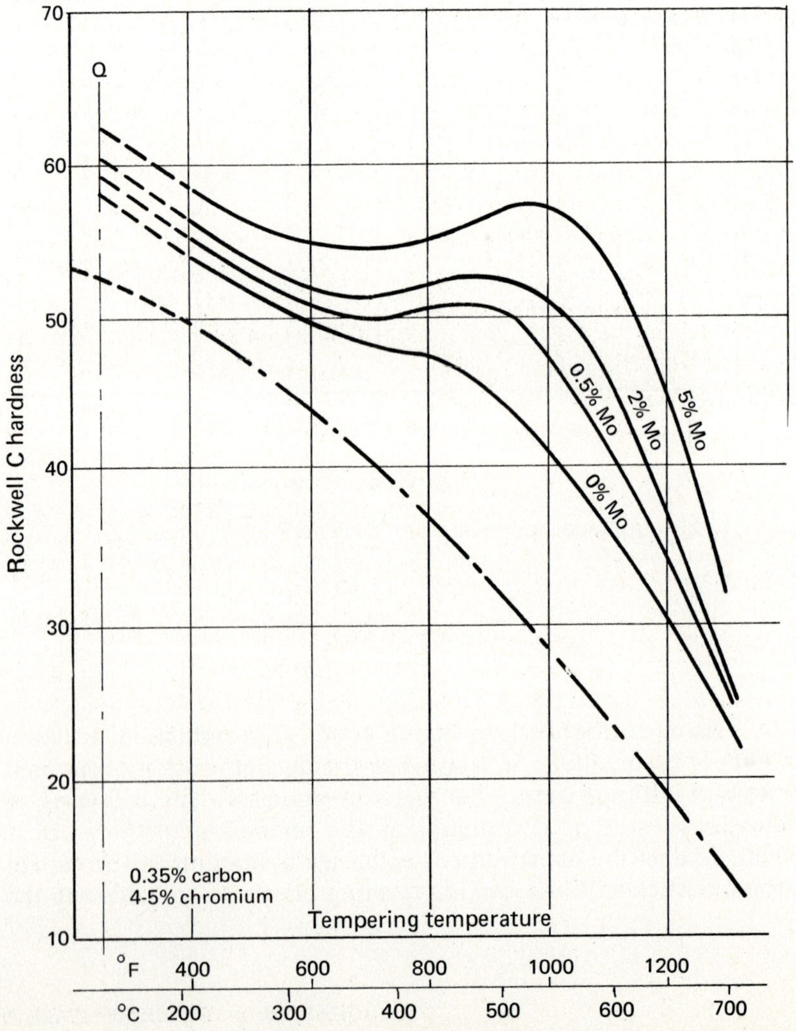

Rockwell C hardness
70
60
50
40
30
20
10
Q
0.5% Mo
2% Mo
5% Mo
0% Mo
0.35% carbon
4-5% chromium
Tempering temperature
°F 400 600 800 1000 1200
°C 200 300 400 500 600 700

their hardness as well as their creep resistance that the extra precautions are taken in stress relieving some of the high-alloy steels.

Tempering is important not only in the stress-relief heat treatment of welded parts containing hard zones, but also in multilayer welding of medium-carbon and other strongly hardening steels. In Figure 19-25(*a*), bead 1 has created a hard zone. Bead 2 has likewise created a hard zone by raising the temperature of the base metal close to the weld above the critical range and allowing it to cool rapidly. If the heat in bead 2 is sufficient to raise the entire hard zone created by bead 1 above 500°F (but not above the critical range, of course), the zone will be considerably softer and devoid of untempered martensite and austenite. Whether its notch impact strength will necessarily be raised is problematical. In general, however, it may be said that the tempered zone will be considerably tougher than the untempered zone. If bead 2 had been deposited directly after bead 1, it is possible that the hard zone of bead 1 would have been caught before it had been formed. That is, the zone would have been prevented from further cooling by bead 2 before martensite had begun to form. The hard zone might then have consisted of bainite or fine pearlite instead of martensite, depending on the precise temperature distributions. Furthermore, the heat from bead 1 might have retarded the cooling rates in bead 2, and so have prevented the formation of hard zones near it. Hard zones, it will be realized, are to some extent within the control of the operator, although he or she cannot change the composition of the base metal.

In Figure 19-25(*b*), bead 2 has been deposited after bead 1 has cooled, and is too small to bring the entire hard zone of bead 1 above, say, 200 to 300°F. Significant tempering has failed to occur and the hard zone is there to stay. It is too much to expect later beads to succeed where the second has failed. With proper multilayer welding, however, it is possible to temper the hard zones of every bead, as in Figure 19-25(*c*).

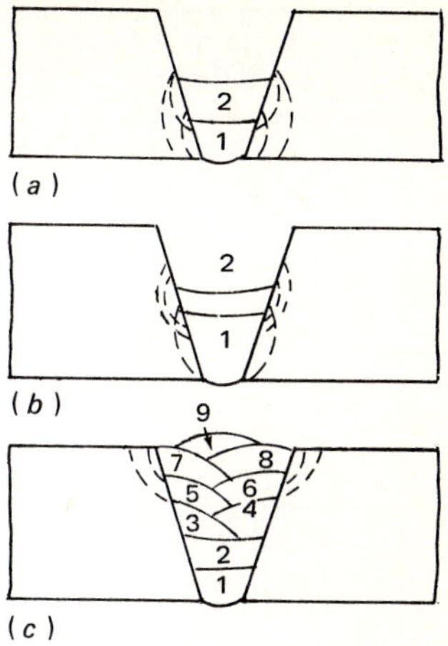

2
1
(a)
2
1
(b)
9
7
8
5
6
4
3
2
1
(c)

Figure **19-25** Tempering of hard zones by heat from later beads. (From O. H. Henry and G. E. Claussen, as revised by G. E. Linnert, *Welding Metallurgy (Iron and Steel)*, p. 170. Copyright 1967 by the American Welding Society. Used with permission of the American Welding Society.)

RECRYSTALLIZATION Cold work or permanent deformation below the recrystallization range breaks each perfect crystal into a large number of fragments due to slip. These fragments change into new grains with undistorted crystal lattices when the temperature is raised to the recrystallization range. When the new grains appear, first as minute crystals, usually at the grain boundaries of the cold-worked grains, recrystallization has begun. As the temperature is maintained or increased, the minute crystals grow. Eventually all the old crystals with slip lines have been absorbed by the new crystals, which are free from slip lines, and recrystallization is complete.

Recrystallization transforms the distorted fragments, which are the source of internal stresses, to new, undistorted crystals free from

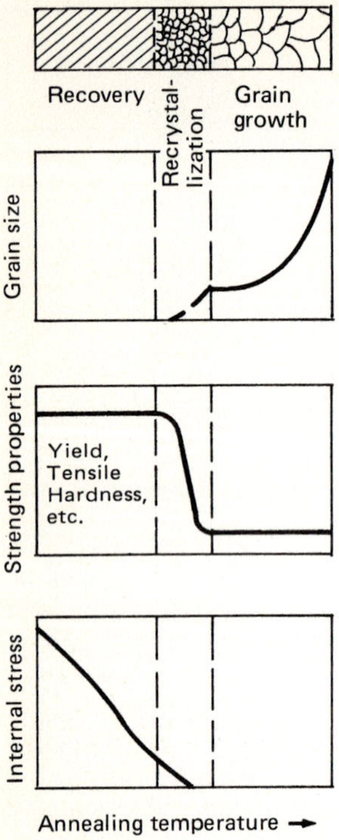
Recovery
Recrystal-lization
Grain growth
Grain size
Strength properties
Yield, Tensile Hardness, etc.
Internal stress
Annealing temperature →

Figure **19-26** Cross-section of heated low-carbon steel showing the recrystallization-recovery process. (From O. H. Henry and G. E. Claussen, as revised by G. E. Linnert, *Welding Metallurgy (Iron and Steel)*, p. 171. Copyright 1967 by the American Welding Society. Used with permission of the American Welding Society.)

internal stress. Recovery (Figure 19-26) partly removes shrinkage stresses; recrystallization completes their removal. The temperature at which recrystallization occurs in a given metal is lowered by increasing the degree of prior cold work. There is no single recrystallization temperature for a given metal. For iron and steel, the lowest temperature at which recrystallization has been observed is 840°F.

No recrystallization whatever occurs, however, if the dislocation, or cold work, is minor. For example, the cold work caused by the shrinkage of parts during postwelding cooling generally is too slight to cause recrystallization during stress-relief heat treatment. Stress relief occurs, therefore, primarily through recovery and relaxation, rather than through recrystallization.

SPHEROIDIZING The process of creating spherical carbides in the steel is called *spheroidizing*. If we start with pearlite, we spheroidize by converting the plates of carbides to spheres. If we start with martensite or bainite, on the other hand, in which the minute carbides are approximately spherical initially (although because of their extreme fineness their presence, let alone their size, is not certain), spheroidizing consists simply of coalescing the particles and rendering them visible under the microscope. Martensite needles do not spheroidize.

The carbide in the heat-affected zone of welded low-carbon steel is usually present as pearlite. During stress-relief heat treatment, we observe that the plates become less and less angular, the carbide particles becoming more nearly spherical [Figure 19-27(*a*)]. The longer the time at a given temperature and the higher that temperature (provided the critical range is not exceeded), the larger and more nearly spherical the carbides become [Figure 19-27(*b*)].

The carbide in the heat-affected zone of welded medium-carbon steel is extremely finely divided [Figure 19-27(*c*)]. The flecks within the martensite needles are small spheroids of carbide. As the temperature rises during stress-relief heat treatment, the carbides follow the general law of small particles in a metal. Carbon from the smaller particles enters the solution to precipitate again on the larger particles. In this way, a crowd of tiny spheroidal carbide particles are replaced by a few large spheroidal ones, as shown under magnification in Figure 19-27(*d*). Since the coarsely spheroidized condition generally is the softest and most ductile, the spheroidizing action of stress-relief heat treatment is desirable.

At 1200°F, steel has an appreciable solubility for several elements, particularly nitrogen. If the part is slowly cooled from the stress-relieving temperature, there is an opportunity for the complete precipitation of these elements. With rapid cooling, on the other hand, nitrogen is retained in supersaturated solution at room temperature, and the steel may have low ductility. Weld metal with a

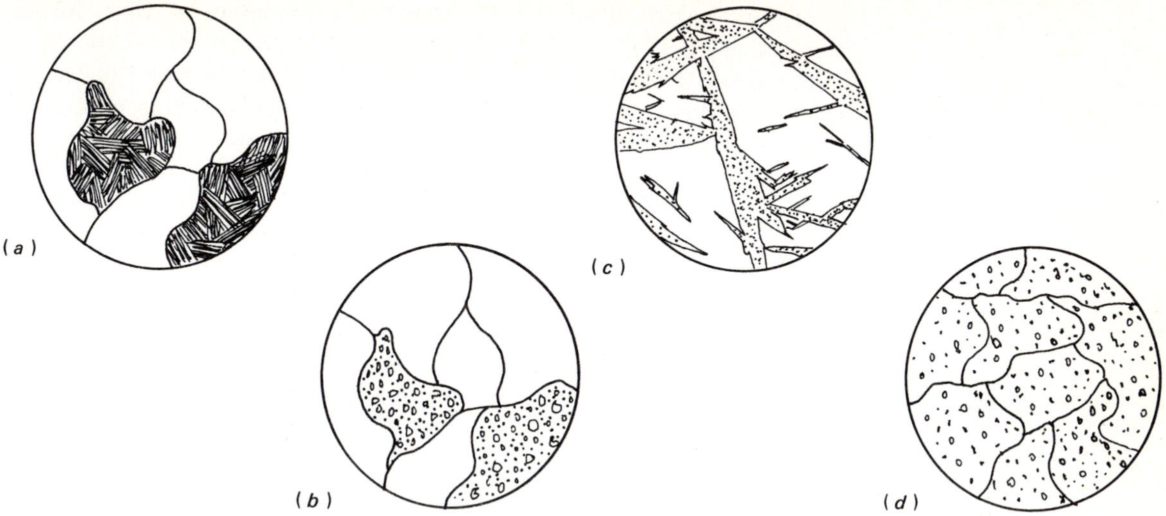
(a)
(b)
(c)
(d)

Figure **19-27** The spheroidization of carbide in pearlite (*a*, *b*) and martensite (*c*, *d*). (From O. H. Henry and G. E. Claussen, as revised by G. E. Linnert, *Welding Metallurgy (Iron and Steel)*, p. 172. Copyright 1967 by the American Welding Society. Used with permission of the American Welding Society.)

high-nitrogen content, such as bare electrode deposits, is particularly subject to embrittlement on rapid cooling from the temperature of stress-relief heat treatment. In any event, rapid cooling after stress relieving is undesirable to avoid shrinkage stresses due to the nonuniform distribution of temperature in any rapidly cooled object.

The spheroidizing of carbide near a welded joint is a process which begins when the weld is being made (Figure 19-28). The pearlite in the zone that is heated short of the critical ranges is more

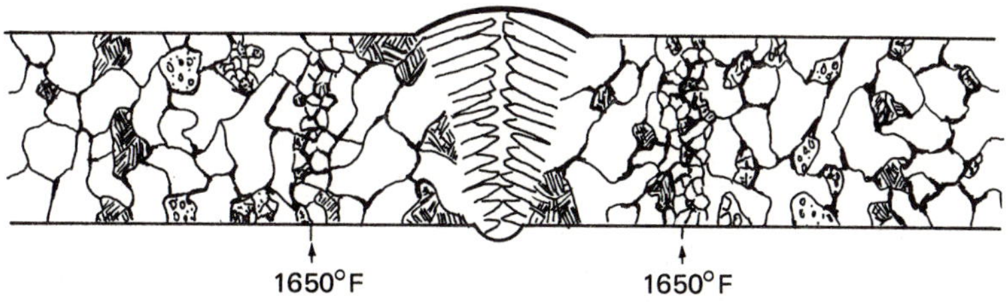
1650°F
1650°F

Figure **19-28** Sketch showing the zone in which pearlite tends to be spheroidized in the welding of mild steel (From O. H. Henry and G. E. Claussen, as revised by G. E. Linnert, *Welding Metallurgy (Iron and Steel)*, p. 173. Copyright 1967 by the American Welding Society. Used with permission of the American Welding Society.)

or less spheroidized, depending on the length of time it has remained close to the lower temperature of the critical range. Once the pearlite is heated into the critical range, the carbide in the pearlite dissolves in the austenite. In very rapid welding processes, such as spot welding, the carbide may not remain in the critical range long enough to dissolve completely in the austenite. Some spheroids may be observed, therefore, even in the zone that has been heated in the critical range. During stress-relief heat treatment, the spheroidization begun during welding is extended to the steel throughout the welded part. All pearlite is being spheroidized to some extent by stress-relief heat treatment. The outline of the heat-affected zone in the macrostructure of a stress-relieved weld is far less sharp than in the macrostructure before heat treatment. Unlike preheating, however, stress-relief heat treatment has no effect on the width of the heat-affected zone.

Although local stress relief has the same effect in softening hard zones as placing the entire structure in the furnace, it may not always reduce shrinkage stresses. Local stress relief is commonly used for high-pressure piping. Raising the temperature of the joint to 1200°F replaces the nonuniformly distributed shrinkage stresses in the immediate vicinity of the weld with more uniformly distributed reaction stresses. It is important that the deformation, necessitated by the wide heat-treated zone during cooling, be absorbed without causing cracks or endangering supports. In most structures, the severe shrinkage strains involved in local stress relief preclude its use. Instead of relieving stresses, the cycle of local heating and cooling often intensifies them. Local stress relief of local heat treatment of any kind is disastrous for austenitic steels of the unstabilized 18-8 type, because of a zone near the locally heated area that enters a temperature range in which a dangerous form of carbide precipitation occurs.

To soften a welded steel as much as possible and to create a uniformly fine-grained structure throughout, the steel is full annealed. *Annealing,* or *full annealing* as it is properly called, consists of heating the part 50 to 100°F above the critical temperature range; that is, the range in which ferrite changes to austenite. For 0.20%-carbon steel, the full-annealing temperature is 1560°F + 50°F = 1610°F. This temperature is maintained for 1 hour for each inch of section of the heaviest parts being treated. Ordinarily, the parts are cooled in the furnace or in some substance yielding a slow cooling rate, such as ashes. By using controlled rates of slow cooling it is possible to vary the coarseness of the lamellar pearlite or the degree of spheroidization formed.

Annealing from temperatures above the critical range is seldom applied to welded parts, because the high temperatures involved often cause excessive distortion in the welded part despite extensive means of support in the furnace. Where the shape of the weldment is self-supporting, annealing from temperatures above the critical range can be applied without distortion difficulties.

A second problem that arises in annealing, because of the high temperatures and long soaking and cooling cycles, is that of decarburization. The risk of decarburization can be minimized by controlling the oxygen content of the furnace atmosphere or by replacing the oxygen in the furnace with an inert gas. Decarburization is inappreciable at the lower temperatures employed for stress relieving.

Normalizing is a heat treatment somewhat similar to annealing, and is frequently employed for preweld and postweld heating of the steel. Normalizing involves heating the steel approximately 100°F above its critical range to transform the structure to austenite, followed by cooling in still air. Whereas annealing, with its very slow furnace cooling, produces a carbide structure of coarse lamellar pearlite or spheroidized carbide, normalizing with its air-cooling treatment creates a finer lamellar pearlite in most steels that, although slightly harder, is quite satisfactory for service. A normalizing treatment may be used to (1) reduce stresses from cold working or welding, (2) remove hardened zones adjacent to the weld, (3) create a more uniform and desirable microstructure in both the weld metal and the base metal, or (4) refine (by recrystallization) any coarse structure that may have been developed in the steel through high-temperature (perhaps above 1900°F) hot-working or forming operations.

Hardening is accomplished by heating the steel 50 to 100°F above the critical range and then cooling at a rate that exceeds the critical cooling rate of the steel. This may involve quenching in water, oil, or air, depending on the hardenability of the particular steel being treated. Above the critical range, the structure of the steel transforms to austenite (grains of gamma iron with carbon in solid solution). When cooled to room temperature at a rate exceeding the critical cooling rate, the structure of austenite transforms to martensite, which is the hard structural condition. Hardening frequently occurs in the heat-affected zone of the base metal, where it is often considered an undesirable condition. Nevertheless, the hardening treatment is commonly used in preparing high carbon–steel cutting and forming tools, treating surfaces to improve wear resistance, and so forth.

Hardening and tempering can generally produce the best combination of mechanical properties of which steel is capable. The steel is first hardened in the customary fashion, that is, heated and quenched to produce a martensitic structure. Then it is tempered or drawn, which involves reheating the steel to a particular temperature somewhere below the critical point (1335°F), holding it at that temperature for a specified length of time, and allowing it to cool to room temperature. The structure produced by reheating to temperatures ranging up to 600°F is often called *tempered martensite* because it is not too unlike the freshly quenched structure. The structure obtained at temperatures between 600 and 900°F is generally called *troostite,* or possibly *secondary troostite.* Troostite etches rapidly and appears very dark. The structure consists of a very fine aggregate of ferrite grains and cementite particles. Between 900 and 1300°F, the structure that forms is called *sorbite.* There is no sharp point of demarkation between troostite and sorbite. In the sorbitic structure, the carbide particles have grown so that the structure possesses a granular appearance. Examination under the microscope at high magnifications will reveal small globular carbides in a matrix of ferrite. As the tempering temperature is increased, the tensile strength, yield strength, and hardness of the steel will decrease, while the toughness and ductility generally increase.

Austempering is a heat treatment sometimes applied to relatively thin or small sections of hardenable steels for the purpose of securing a structure of bainite. By producing this structure, we obtain reasonable ductility at high hardness levels without resorting to a combination treatment of hardening and tempering. Further claims are made for better properties with the bainite structure than found with a tempered martensite or troostite. Thin sections of steel are treated more easily because all portions of the steel mass must be cooled at a rate exceeding the critical cooling rate during the treatment. The first step is to heat the steel 50 to 100°F above its critical range to form a structure of austenite. Then the steel is quenched rapidly (faster than the critical cooling rate) to a temperature somewhat above the M_s point (see Figure 19-29) and held at this temperature for a while. Thus the transformation of the austenite is allowed to proceed under isothermal conditions. The cooling and isothermal treatment is usually accomplished by quenching and holding in a molten-salt bath or lead pot.

Martempering is a heat treatment applied to steels of high hardenability that are prone to quench cracking. During the process, the piece of steel is first austenitized by heating it to 50 to 100°F above the critical range. The piece is then quenched in a molten-salt bath or lead pot held at a temperature just above the M_s point of the steel being treated. The piece is held here until the temperature throughout the mass becomes uniform; and before the transformation to bainite commences, the piece is withdrawn from the bath and slowly cooled through the martensite transformation range. Martensite forms uniformly throughout the section. The likelihood of cracking is markedly reduced because stresses created by temperature gradients are virtually absent.

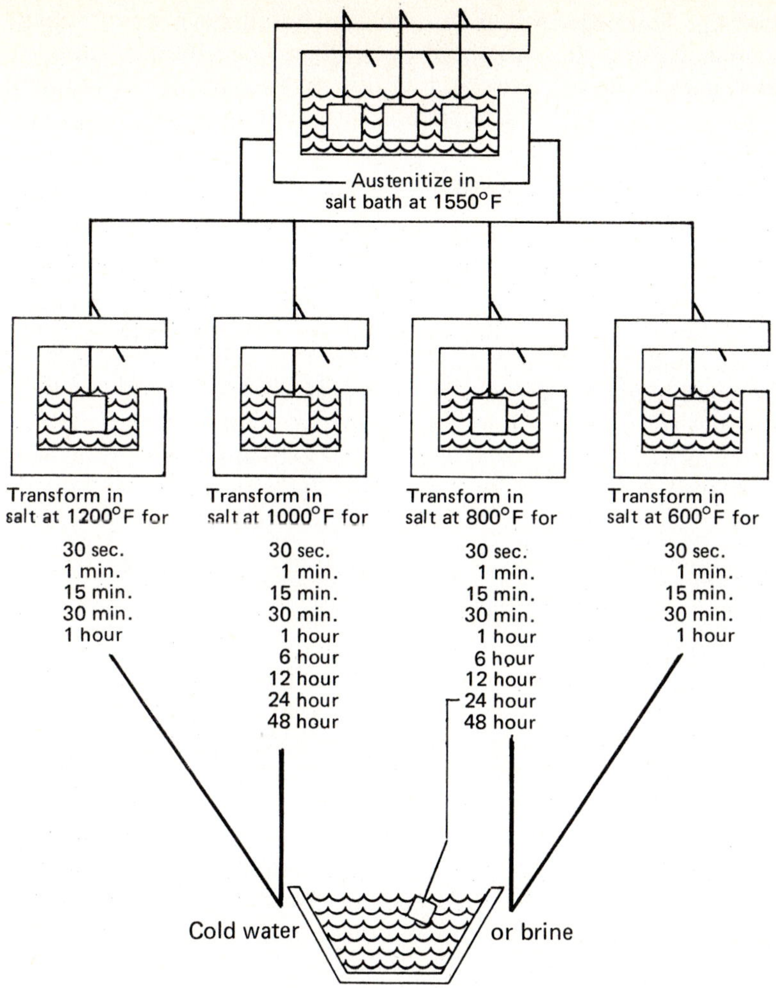
Austenitize in
salt bath at 1550°F
Transform in
salt at 1200°F for
30 sec.
1 min.
15 min.
30 min.
1 hour
Transform in
salt at 1000°F for
30 sec.
1 min.
15 min.
30 min.
1 hour
6 hour
12 hour
24 hour
48 hour
Transform in
salt at 800°F for
30 sec.
1 min.
15 min.
30 min.
1 hour
6 hour
12 hour
24 hour
48 hour
Transform in
salt at 600°F for
30 sec.
1 min.
15 min.
30 min.
1 hour
Cold water
or brine

Figure **19-29** The austempering process for a steel having an austenitizing temperature of 1450–1500°F

19-5 MINIMIZING WELD CRACKING IN NONFERROUS METALS

The methods used to minimize cracking of welds in nonferrous metals are similar to those for welding ferrous metals, and are described in detail in Chapter 21.

19-6 IDENTIFYING METALS

Often (especially when doing repair work), a welder must first identify the general type of metal from which a part is made before selecting a filler metal or deciding on the most appropriate welding procedure. This means that the welder should have one or more dependable, accurate, and rapid methods of identifying metals. A welder should know how to identify metals by (1) the surface appearance, (2) sound, (3) magnetic and acid spot tests, (4) spark testing, (5) the appearance of the surface of a fracture, (6) chisel and file tests, and (7) flame testing.

SURFACE APPEARANCE The surface appearance of the metal itself helps to classify it. Metal that has been cast in a sand mold has a rough surface appearance. This rough surface is due to the imprint made by the sand used in the casting molds. There is also a ridge left along the edge of the casting where the two halves of the casting mold come together in the casting process.

Castings have one or more enlarged areas along the side that are referred to as *gates*. On the surface, the gate appears as though it were broken or fractured. In some cases the surface area of the gate has been smoothed by grinding. Gates are formed by openings that are left in the sand mold for the metal to flow into the mold cavity during the casting process. The size of the gate varies with different metals. Cast steel, for example, has a comparatively large gate, while gray cast iron has a small gate.

Pieces that have been drop-forged have a rather rough, scaly surface appearance. Drop-forged pieces are usually simple in design. Pieces of steel heated to red heat are placed on the die, where they are forged under the pressure of a heavy press. The forging operation usually leaves ridges along the edge of the piece of metal where the two dies overlap each other in the forging operation. This ridge is always sheared and is sometimes ground to smooth up the edge. The part numbers, which are stamped on the part during the drop-forging operation, appear sharp and distinct as compared with those found on castings. *Forgings are usually made of medium-carbon or low-alloy steels.*

SOUND Metals can be identified by rapping them with a hammer and listening to the sound produced. One can tell by the ring of the metal the type of material used in the manufacture of the part.

It is difficult to explain a sound in words, so it is necessary for each individual to compare the sound from each different type of metal a number of times to establish a dependable personal standard. For example, when one strikes steel with a hammer, it resounds with a higher-pitched tone than that given off by cast metal. Gray cast iron and malleable cast iron can be compared by sound. Gray cast iron has a dull, chalky tone, while malleable cast iron has a higher, clearer-pitched tone.

MAGNETIC AND ACID SPOT TESTS Sometimes a welder or welding-machine operator has the problem of identifying a piece of metal that might be stainless iron (11 to 13% chromium type); stainless steel (18% chromium, 8% nickel type); monel (67% nickel, 33% copper); copper-nickel (70 to 80% copper, 30 to 20% nickel); or plain structural-grade low-carbon steel. A plain bar magnet can be used to separate these materials into three groups:

1. Those that are strongly magnetic—stainless iron and low-carbon steel.

2. Those that are slightly magnetic—monel (although some alloy variations are nonmagnetic), high-nickel alloys, and stainless steel (18Cr-8Ni type when cold worked).

3. Those that are nonmagnetic (stainless steel, 18Cr-nickel-type annealed); copper-base alloys; aluminum-base alloys; zinc-base alloys; some alloy modifications of monel.

Further identification can be made with a simple spot test on material whose surface is clean from oil, paint, dirt, rust, and scale. A drop of concentrated nitric acid will not attack the stainless steel or stainless iron; it will produce a green or blue-green color with monel and copper-nickel, and a brown color with carbon steel.

THE SPARK TEST Most ferrous metals may be roughly classified by observing the sparks given off when the surface or edge of the metal is touched against a grinding wheel. To classify the unknown metal, the sparks given off are compared to those of a known metal. Either a portable or stationary grinder may be used. A wheel of medium grit, such as 40 to 60 grain, gives the most satisfactory results. The wheel should operate at a speed of 7000 to 8000 surface feet per minute. An 8-inch-diameter grinding wheel turning at the rate of 3600 rpm provides a surface speed of 7500 feet per minute. The wheel should be dressed before testing a metal. This removes glazing, sharpens the wheel, and removes traces of metals ground previously.

Sparks can be observed more easily in diffused light. Avoid bright sunlight or a darkened room. If the sparks are given off against a dark background it is easier to distinguish the spark characteristics.

Different elements in the steel influence spark behavior. The presence of carbon causes a bursting stream of sparks; the higher the carbon content, the more plentiful the spark bursts (Figure 19-30). An exception occurs when the metal has alloys such as silicon, chromium, nickel, and tungsten, which tend to suppress the carbon bursts. As the manganese content is increased, the sparks tend to follow the surface of the wheel. Chromium gives off orange sparks, which make it difficult to get a spark stream. When nickel is added to a metal, forked tongues are produced on the end of the spark streams.

Heat treatment also changes spark patterns. The undersurface as well as the surface should be tested, since some metals may have a

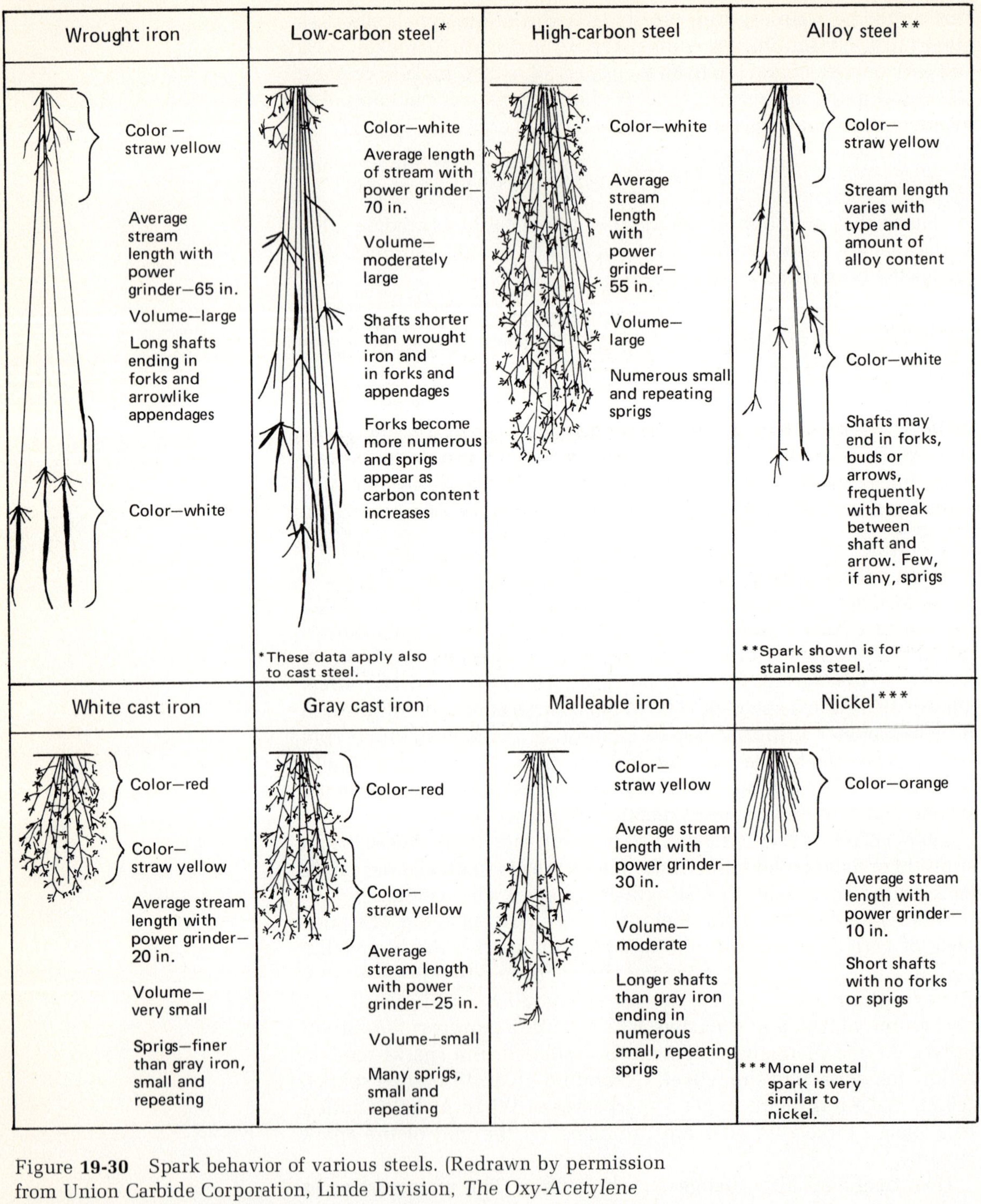
Wrought iron
Color — straw yellow
Average stream length with power grinder—65 in.
Volume—large
Long shafts ending in forks and arrowlike appendages
Color—white
Low-carbon steel*
Color—white
Average length of stream with power grinder—70 in.
Volume—moderately large
Shafts shorter than wrought iron and in forks and appendages
Forks become more numerous and sprigs appear as carbon content increases
*These data apply also to cast steel.
High-carbon steel
Color—white
Average stream length with power grinder—55 in.
Volume—large
Numerous small and repeating sprigs
Alloy steel**
Color—straw yellow
Stream length varies with type and amount of alloy content
Color—white
Shafts may end in forks, buds or arrows, frequently with break between shaft and arrow. Few, if any, sprigs
**Spark shown is for stainless steel.
White cast iron
Color—red
Color—straw yellow
Average stream length with power grinder—20 in.
Volume—very small
Sprigs—finer than gray iron, small and repeating
Gray cast iron
Color—red
Color—straw yellow
Average stream length with power grinder—25 in.
Volume—small
Many sprigs, small and repeating
Malleable iron
Color—straw yellow
Average stream length with power grinder—30 in.
Volume—moderate
Longer shafts than gray iron ending in numerous small, repeating sprigs
Nickel***
Color—orange
Average stream length with power grinder—10 in.
Short shafts with no forks or sprigs
***Monel metal spark is very similar to nickel.

Figure **19-30** Spark behavior of various steels. (Redrawn by permission from Union Carbide Corporation, Linde Division, *The Oxy-Acetylene Handbook*, 2d ed., 1960, p. 551.)

low or high carbon surface due to surface treatment, while the underneath composition is different.

It will take practice to classify metals by the spark test. Spend time observing sparks from common known metals, carefully noting the length, color, and end shape of the spark from the time it leaves the wheel until it disappears. It is advisable to keep samples of known metals by the grinder to use in making comparative tests. The color of the spark stream is important. It will vary from white to yellow, orange, and straw. Various terms are used to describe the pattern of the spark stream. Figure 19-31 illustrates the common patterns and terms.

THE SURFACE OF A FRACTURE Looking at the broken edge of the part is one of the first steps in metal identification. The fractured surface reveals such items as the nature of the break, type of grain, and color.

The surface of a fracture on a piece of gray cast iron is dark gray and will usually rub off black (graphite) on the fingertip. White cast iron has a silvery white appearance. Malleable cast iron shows a dark center with a light outer skin, due to its surface treatment.

The way in which the part breaks is of interest. Malleable cast-iron parts are ductile and usually bend before breaking. The metal along the edge of the break will indicate this characteristic by showing

Figure **19-31** Common terms and patterns in spark testing. (Redrawn by permission from Union Carbide Corporation, Linde Division, *The Oxy-Acetylene Handbook,* 2d ed., 1960, p. 550.)

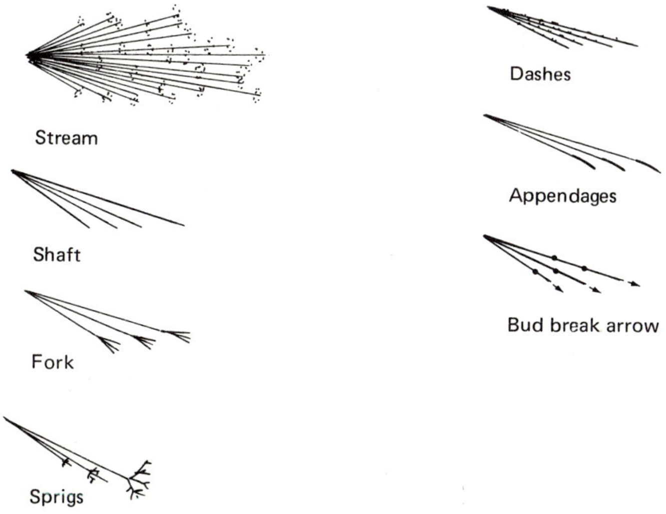
Stream
Shaft
Fork
Sprigs
Dashes
Appendages
Bud break arrow

some distortion. Gray and white cast irons are brittle and make a clean break.

The fractured surface of steel provides a definite grain pattern. Low-carbon steel is bright gray, while high-carbon is a little darker but still a light gray.

The fractured edge of a steel axle or rod has a distinctive grain pattern. The metal checks at a point of stress and, as the break continues across the piece, the friction of the two surfaces smooths the fractured grain pattern. When the part finally breaks, the portion of the metal that broke last has had no rubbing action and shows a crystalline surface. In the presence of oil or grease, this surface will often be free of grease, and the older portion of the break will be darkened by it.

Table 19-1 relates the appearance of the surface of a fracture to a specific type of metal. It should be noted that sometimes it is impossible to observe a clean fracture in the metal because it is oxidized or dirty, or the metal is still in one piece. When this is the case, the chisel test may be used.

THE CHISEL (CHIPPING) AND FILE TEST For the chisel part of the test, a cold chisel and hammer are used to make a shallow cut (not over 1/16 inch) along a flange, boss edge, or other nonoperating, nonconspicuous location. The kinds of chips produced by different types of metals are described in Table 19-2.

To differentiate between malleable iron and cast steel, a spot on the surface of the metal to be identified is cleaned of oil, grease, rust, paint, and dirt, and a few strokes are made over the spot with a bastard-cut file. The filing direction should be at a right angle to the length of the metal. Visual examination of the filed surface spot, or—if available—examination with a low-power magnifying glass will disclose numerous lines crossing the file marks if the material is malleable iron; no crossing marks are observed if the metal is cast steel.

THE FLAME TEST This test may be used in conjunction with the chipping test. For this test, it is necessary to have an oxyacetylene welding outfit and welding table. The chips are removed from the metal to be identified as described in the previous section and placed on top of a metal plate located on the welding table. The oxyacetylene flame is ignited and adjusted until it is neutral. The neutral flame is then brought to bear on the chips. Table 19-3 lists the various conditions obtained by heating different materials to the melting point.

Table **19-1** Fracture appearance versus other surfaces of various metals

	ALLOY** STEEL	COPPER	BRASS AND BRONZE	ALUMINUM AND ALLOYS†	MONEL METAL	NICKEL	LEAD††
Color of fracture	Medium gray	Red color	Red to yellow	White	Light gray	Almost white	White; crystalline
Unfinished surface	Dark gray; relatively rough; rolling or forging lines may be noticeable	Various degrees of reddish brown to green due to oxides; smooth	Various shades of green brown, or yellow due to oxides; smooth	Evidences of mold or rolls; very light gray	Smooth; dark gray	Smooth; dark gray	Smooth; velvety; white to gray
Newly machined	Very smooth; bright gray	Bright copper red color dulls with time	Red through to whitish yellow; very smooth	Smooth; very white	Very smooth; light gray	Very smooth; white	Very smooth; white

	WHITE CAST IRON°	GRAY CAST IRON	MALLEABLE* IRON	WROUGHT IRON	LOW-CARBON STEEL AND CAST STEEL	HIGH-CARBON STEEL
Color of fracture	Very fine silvery white silky crystalline formation	Dark gray	Dark gray	Bright gray	Bright gray	Very light gray
Unfinished surface	Evidence of sand mold; dull gray	Evidence of sand mold; very dull gray	Evidence of sand mold; dull gray	Light gray; smooth	Dark gray; forging marks may be noticeable; cast—evidences of mold	Dark gray; rolling or forging lines may be noticeable
Newly machined	Rarely machined	Fairly smooth; light gray	Smooth surface; light gray	Very smooth surface; light gray	Very smooth; bright gray	Very smooth; bright gray

°Very seldom used commercially. *Malleable iron should always be braze-welded. **Alloy steels vary so much in composition and consequently in results of tests that experience is the best solution to identification problems. †Due to white or light color and extremely light weight, aluminum is usually easily distinguishable from all other metals; aluminum alloys are usually harder and slightly darker in color than pure aluminum. ††Weight, softness, and great ductility are distinguishing characteristics of lead.

Source: Union Carbide Corporation, Linde Division, *The Oxyacetylene Handbook*, 2d., 1960, p. 546.

Table **19-2** Chip appearance of various metals

	ALLOY** STEEL	COPPER	BRASS AND BRONZE	ALUMINUM AND ALLOYS†	MONEL METAL	NICKEL	LEAD††
Appearance of chip	**	Smooth chips; saw edges where cut	Smooth chips; saw edges where cut	Smooth chips; saw edges where cut	Smooth edges	Smooth edges	Any shaped chip can be secured because of softness
Size of chip	**	Can be continuous if desired	Can be continuous if desired	Can be continuous if desired	Can be continuous if desired	Can be continuous if desired	Can be continuous if desired
Facility of chipping	**	Very easily cut	Easily cut; more brittle than copper	Very easily cut	Chips easily	Chips easily	Chips so easily it can be cut with penknife

	WHITE CAST IRON°	GRAY CAST IRON	MALLEABLE* IRON	WROUGHT IRON	LOW-CARBON STEEL AND CAST STEEL	HIGH-CARBON STEEL
Appearance of chip	Small broken fragments	Small partially broken chips but possible to chip a fairly smooth groove	Chips do not break short as in cast iron	Smooth edges where cut	Smooth edges where cut	Fine grain fracture; edges lighter in color than low-carbon steel
Size of chip		⅛ in.	¼–⅜ in.	Can be continuous if desired	Can be continuous if desired	Can be continuous if desired
Facility of chipping	Brittleness prevents chipping a path with smooth sides	Not easy to chip because chips break off from base metal	Very tough, therefore harder to chip than cast iron	Soft and easily cut or chipped	Easily cut or chipped	Metal is usually very hard, but can be chipped

°Very seldom used commercially. *Malleable iron should always be braze-welded. **Alloy steels vary so much in composition and consequently in results of tests that experience is the best solution to identification problems. †Due to white or light color and extremely light weight, aluminum is usually easily distinguishable from all other metals; aluminum alloys are usually harder and slightly darker in color than pure aluminum. ††Weight, softness, and great ductility are distinguishing characteristics of lead.

Source: Union Carbide Corporation, Linde Division, *The Oxyacetylene Handbook*, 2d ed., 1960, p. 547.

Table **19-3** Flame tests of various metals

	ALLOY** STEEL	COPPER	BRASS AND BRONZE	ALUMINUM AND ALLOYS†	MONEL METAL	NICKEL	LEAD††
Speed of melting (from cold state)	**	Slow	Moderate to fast	Faster than steel	Slower than steel	Slower than steel	Very fast
Color change while heating	**	May turn black and then red; copper color may become more intense	Becomes noticeably red before melting	No apparent change in color	Becomes red before melting	Becomes red before melting	No apparent change
Appearance of slag	**	So little slag that it is hardly noticeable	Various quantities of white fumes, though bronze may not have any	Stiff black scum	Gray scum; considerable amounts	Gray scum; less slag than Monel metal	Dull gray coating
Action of slag	**	Quiet	Appears as fumes	Quiet	Quiet; hard to break	Quiet; hard to break	Quiet
Appearance of molten puddle	**	Has mirrorlike surface directly under flame	Liquid	Same color as unheated metal; very fluid under slag	Fluid under slag	Fluid under slag film	White and fluid under slag
Action of molten puddle under blowpipe flame	**	Tendency to bubble; puddle solidifies slowly and may sink slightly	Like drops of water; with oxidizing flame will bubble	Quiet	Quiet	Quiet	Quiet; may boil if too hot

Table **19.3** (*cont.*)

	WHITE CAST IRON°	GRAY CAST IRON	MALLEABLE* IRON	WROUGHT IRON	LOW-CARBON STEEL AND CAST STEEL	HIGH-CARBON STEEL
Speed of melting (from cold state)	Moderate	Moderate	Moderate	Fast	Fast	Fast
Color change while heating	Becomes dull red before melting	Becomes dull red before melting	Becomes red before melting	Becomes bright red before melting	Becomes bright red before melting	Becomes bright red before melting
Appearance of slag	A medium film develops	A thick film develops	A medium film develops	Oily or greasy appearance with white lines	Similar to molten metal	Similar to molten metal
Action of slag	Quiet; tough, but can be broken up	Quiet; tough, but possible to break it up	Quiet; tough, but can be broken	Quiet; easily broken up	Quiet	Quiet
Appearance of molten puddle	Fluid and watery; reddish white	Fluid and watery; reddish white	Fluid and watery; straw color	Liquid; straw color	Liquid; straw color	Lighter than low-carbon steel; has a cellular appearance
Action of molten puddle under blowpipe flame	Quiet; no sparks; depression under flame disappears when flame is removed	Quiet; no sparks; depression under flame disappears when flame is removed	Boils and leaves blowholes; surface metal sparks; interior does not	Does not get viscous; generally quiet; may be slight tendency to spark	Molten metal sparks	Sparks more freely than low-carbon steel

°Very seldom used commercially. *Malleable iron should always be braze-welded. **Alloy steels vary so much in composition and consequently in results of tests that experience is the best solution to identification problems. †Due to white or light color and extremely light weight, aluminum is usually easily distinguishable from all other metals; aluminum alloys are usually harder and slightly darker in color than pure aluminum. ††Weight, softness, and great ductility are distinguishing characteristics of lead.

Source: Union Carbide Corporation, Linde Division, *The Oxyacetylene Handbook* 2d ed., 1960, p. 548.

REVIEW QUESTIONS

1. By what method are the shapes of the crystal lattices of metal determined?

2. What is the name of the diagram used to represent the transformation of metals from the liquid to the solid state?

3. Metals can form chemical compounds with each other. By what name are such compounds known?

4. Incomplete recrystallization occurs in which temperature range of the HAZ in low-carbon steel?

5. Name three methods that are commonly used for minimizing the cracking of welds in ferrous metals.

6. Calculate the nominal or average preheat temperature for the following one-inch-thick rectangular steel bars: (a) SAE 4140; (b) SAE 5150; (c) SAE 8640.

7. The *ASME Boiler and Pressure Vessel Code* requires that the area adjacent to welded joints in pipes and tubing be preheated. This area can be expressed as a ratio of the widest part of the welding groove. What is this ratio?

8. What are the general effects of stress-relief heat treatment?

9. What changes occur in the microstructure of a martensitic steel during the stress-relief heat treatment?

10. At what temperature do all iron-carbon alloys recrystallize?

11. What is meant by the term *spheroidizing*?

12. To harden SAE 1050 steel, from what temperature would it have to be quenched?

13. What is troostite? How is it formed?

14. What is martempering? How is it done?

15. What is the percentage of carbon in cast iron?

16. What shape does an iron crystal have below 1300°F? above 1650°F?

17. What simple method can be used to check the temperature of a piece of low-carbon steel at about 1415°F?

18. Why does a highly polished sample of steel fail to show grain structure when examined microscopically?

19. Describe the test method to be used for separating carbon steel from stainless steel. What method would you use to double check your first identification?

20. What should be the prevailing light condition in a room used to identify metals by means of the spark test?

21. Describe the difference in appearance between a newly fractured surface and the unfinished surface of gray cast iron.

22. When is the chisel test used?

23. What kind of oxyacetylene flame is used for the flame test?

24. A metal being flame tested changes color to red before melting. To make certain of the identity of the metal, what other test would have to be performed?

25. When flame testing cast iron, what would be a positive indication that the iron is gray cast iron and not malleable iron?

CHAPTER 20

FERROUS METALS: CLASSIFICATION, APPLICATION, AND WELDABILITY

Ferrous metals constitute the backbone of our industrial world. The industrial revolution was based on iron, the second (to aluminum) most common metal found on earth. Iron, in combination with other metals and elements, is made into several kinds of steels. In this chapter, we shall examine carbon, low-alloy, tool and die, cast, and stainless steels. We shall see how one element (iron) in combination with various other elements can result in myriad useful and necessary metals.

20-1 CARBON STEELS

Carbon steels (also referred to as *plain-carbon steels, ordinary steels,* and *straight-carbon steels*) consist of a broad range of steels (Table 20-1) containing up to:

Carbon	1.70% max.
Manganese	1.65% max.
Silicon	0.60% max.

They include iron-carbon alloys with a carbon level almost as low as wrought iron (which is virtually free from carbon) up to cast iron (which contains more than 1.7% carbon).

Carbon steels containing the alloying elements in the percentages indicated are readily weldable by all welding methods:

Carbon	Between 0.13 and 0.20%
Manganese	Between 0.40 and 0.60%

Table **20-1** AISI steel numbering system

SYSTEM OF IDENTIFICATION. A system of symbols is used to identify the grades of standard steels. In these symbols a capital letter prefix is used to indicate the steelmaking process. Numbers are used to indicate grades of steel by chemical composition as described below.

PREFIX LETTER DESIGNATIONS. The prefix letters *B* and *C* are used to designate the two principal steelmaking processes for carbon steel as follows:

B denotes acid bessemer carbon steel;
C denotes basic open hearth carbon steel;
(E denotes electric furnace steel).

NUMERICAL DESIGNATIONS OF GRADES. A four-numeral series is used to designate graduations of chemical composition of carbon steel, the last two numbers of which are intended to indicate the approximate middle of the carbon range. For example, in the grade designation 1035, 35 represents a carbon range of 0.32 to 0.38%.

It is necessary, however, to deviate from this rule and to interpolate numbers in the case of some carbon ranges; and for variations in manganese, phosphorus, or sulfur with the same carbon range.

The first two digits of the four-numeral series of the various grades of carbon steel and their meanings are as follows:

SERIES DESIGNATION	TYPES
10XX	Nonresulphurized basic open hearth and acid bessemer carbon steel grades (plain carbon steel)
11XX	Resulphurized basic open hearth and acid bessemer carbon steel grades (free-machining steel)
13XX	Manganese 1.75% (low-alloy carbon steel)
23XX	Nickel 3.50% (low-alloy carbon steel)
25XX	Nickel 5.00% (low-alloy carbon steel)
31XX	Nickel 1.25%—Chromium 0.65 to 0.80% (low-alloy carbon steel)
33XX	Nickel 3.50%—Chromium 1.55% (low-alloy carbon steel)
40XX	Molybdenum 0.25% (low-alloy carbon steel)
41XX	Chromium 0.95%—Molybdenum 0.20% (low-alloy carbon steel)
43XX	Nickel 1.80%—Chromium 0.50 or 0.80%—Molybdenum 0.25% (low-alloy carbon steel)
46XX	Nickel 1.80%—Molybdenum 0.25% (low-alloy carbon steel)

SERIES DESIGNATION	TYPES
48XX	Nickel 3.50%—Molybdenum 0.25% (low-alloy carbon steel)
50XX	Chromium 0.30 or 0.60% (low-alloy carbon steel)
51XX	Chromium 0.80, 0.95, or 1.05% (low-alloy carbon steel)
5XXXX	Carbon 1.00%—Chromium 0.50, 1.00 or 1.45% (low-alloy carbon steel)
61XX	Chromium 0.80 or 0.95%—Vanadium 0.10 or 0.15% min. (low-alloy carbon steel)
86XX	Nickel 0.55%—Chromium 0.50%—Molybdenum 0.20% (low-alloy carbon steel)
87XX	Nickel 0.55%—Chromium 0.50%—Molybdenum 0.25% (low-alloy carbon steel)
92XX	Manganese 0.85%—Silicon 2.00% (low-alloy carbon steel)
93XX	Nickel 3.25%—Chromium 1.20%—Molybdenum 0.12% (low-alloy carbon steel)
94XX	Manganese 1.00%—Nickel 0.45%—Chromium 0.40%—Molybdenum 0.12% (low-alloy carbon steel)
97XX	Nickel 0.55%—Chromium 0.17%—Molybdenum 0.20% (low-alloy carbon steel)
98XX	Nickel 1.00%—Chromium 0.80%—Molybdenum 0.25% (low-alloy carbon steel)

Phosphorus	Not more than 0.03%
Silicon	Not more than 0.10%
Sulfur	Not more than 0.035%

Changing the percentages of the five elements (C, Mn, P, S, and Si) or the addition of other elements has the following effects on weldability:

1. *Carbon* is the most potent hardening element. As the carbon content is increased, the steel will become more and more hardenable. If the carbon content is high enough (over 2.5%), sudden cooling below the welding temperature may result in a hard and sometimes brittle zone adjacent to the weld; and if considerable carbon is picked up through admixture, the weld deposit itself may be hard and brittle and tend to crack. Carbon steels with carbon content over 0.10% and up to 1.2% have a tensile strength greater than those whose carbon content is below 0.10%. However, these steels are not as weldable as the lower-carbon-content carbon steels. The addition of small quantities of other alloying elements to low-carbon-content carbon steels serves to increase their tensile strength without reducing their weldability.

2. *Silicon* is added mainly to produce homogeneity in the steel. When used in large percentages (in conjunction with lighter manganese) it increases the tensile strength. If the carbon content is fairly high, the addition of silicon aggravates cracking conditions.

3. *Manganese* also increases the hardenability and tensile strength. Over 0.60% minimum (with relatively high carbon content) will increase the cracking tendency. When below 0.30%, it may increase the susceptibility to internal porosity as well as cracking. Manganese between 0.20 and 0.30% and sulfur above 0.03% produce steels that crack very easily when welded.

4. *Sulfur* is added to carbon steel to improve its machinability. Carbon steels to which sulfur has been added for this purpose are referred to as *free-machining steels.* Free-machining steels have a tendency toward *hot-shortness.* In hot-shortness, the weld deposit tends to crack because it does not have sufficient strength to withstand the shrinkage stresses set up in the weld as it begins to solidify. In addition to cracking due to hot-shortness, carbon steels with sulfur content (in excess of 0.05%) have a tendency to become porous with any deep penetrating technique, especially when welding with AWS class E-6010 and E-6011 electrodes. This is due to the fact that the hydrogen gas produced by the combustion of the coating of these electrodes combines with the sulfur present in the steel being welded. Free-machining steels are, however, readily weldable with low-hydrogen electrodes.

Another common cause of poor welding quality, which is not apparent from a spectrographic analysis of a sample of the steels (taken before welding), are segregated layers of sulfur in the form of manganese or iron sulfide. These layers cause gas pockets or other defects to form at the fusion line when arc welding. They can be determined by microscopic examination, or more easily by deep etching a cross-section of the weld.

5. *Phosphorus* is generally classed as an impurity in welding and should be kept as low as possible. More than 0.40% tends to make welds brittle and reduce shock or fatigue values. At times it is effective in lowering the surface tension of the molten weld metal, making it more difficult to control. It increases the tendency for cracking.

6. As other elements are added, such as chrome, nickel, molybdenum, and vanadium, other welding characteristics are affected. The principal change is in the increased hardenability, which usually results as long as the carbon content is not reduced. This often requires preheat to prevent cracking and hard, sometimes brittle, weld zones. It is important to know that the addition of some of the above alloying elements can be of great benefit when used properly. For example, good weldability in high-strength steels can be retained by reducing the carbon content and getting the required yield strength by the use of the right alloying elements.

Based on carbon content, the AWS divides carbon steels into four groups:

Low-carbon steels	Up to 0.15% carbon
Mild-carbon steels	0.15 to 0.29% carbon
Medium-carbon steels	0.29 to 0.40% carbon
High-carbon steels	0.45 to 1.70% carbon

Most steels in these different categories are produced according to a national specification developed by the ASTM, ASME, SAE, and others to cover carbon steels for various applications.

LOW-CARBON STEELS *Low-carbon steels* are produced primarily in sheet or coils in both cold-rolled and hot-rolled conditions. They have low yield strength and are used in most applications requiring considerable cold forming, such as stampings, rolled or bent shapes in bar stock, and structural shapes. They can be joined by any welding process. On rare occasions, steels having a carbon content of less than 0.10% show a tendency to develop porous welds. Although porosity (unless the weld is very porous) does not present a serious problem from the standpoint of strength, surface holes in the weld are undesirable from the standpoint of appearance. To minimize this problem, one or more of these procedures should be followed:

1. Change the welding procedure.

2. Reduce the quantity of heat being used.

3. Make certain to thoroughly remove all slag and flux from each bead before depositing another.

4. Puddle the weld, keeping the metal molten sufficiently long to allow entrapped gases to escape.

MILD-CARBON STEELS The *mild steels* as rolled provide a yield strength in the 36,000 to 65,000 psi range and are defined in appropriate ASTM specifications (such as ASTM A36, A441, and A572). Steels in this carbon range are available in the hot-rolled, cold-rolled, and heat-treated conditions, providing yield strengths up to 100,000 psi. A large percentage of water tanks, welded structural shapes, machine bases, and machine parts are made from these steels.

Mild steels in thicknesses up to and including 5/16 inch are readily weldable without any precautions. Thicker sections of these steels, in which more than one of the elements is on the high side of the permissible limit of carbon content, show some tendency to crack. If cracking should occur, use one or more of the following techniques:

1. Use techniques that produce a flat to slightly convex bead shape.

2. Melt as little as possible of the base metal (workpiece) into the weld deposit.

3. Put in as large a weld as practical in the first pass by using a slower travel speed. This increases the cross-section of the weld. The use of a slow travel speed also increases the heat input for a given length of weld, causing the plate to heat more—thereby reducing the rate of cooling and hardening of the weld zone. Apply the second pass while the plate is still preheated by the first pass.

4. Leave a 1/32-inch gap between plates to allow for free movement while the weld contracts during cooling.

5. Weld towards the unrestrained end of a joint. Make tacks that do not excessively strain the joints.

If the above precautions do not wholly correct the difficulty with cracking, uniform preheating to the following temperatures is recommended.

Temperature (°F)	*Section Thickness (inches)*
70	Up to 1/4
150	1/4 to 1/2
300	1/2 to 1
400	1 or more

MEDIUM-CARBON STEELS Medium-carbon steels are easily hardened, and their hardness can range from dead soft [Figure 20-1(*e*)] in the annealed condition to a Rockwell reading of C 25. Cold- and hot-forged machine parts such as bolts, studs, connecting rods, and front axles are made from these steels.

Welding of these steels may require special procedures that may include preheat, postheat, and stress relieving. Without prior experience in welding these steels, it is advisable to check the steel for a tendency to crack by making a 12- or 14-inch-long fillet weld on the plate in question (at room temperature and without preheating), break the weld, and visually examine it for cracks. Cracks in broken-open fillet welds will usually show up as purple areas due to oxidation of the cracked surface. This indicates that the cracks are "hot-short" cracks; that is, they occurred at temperatures of 600°F or higher. Only rarely will a crack occur after a weld has cooled to room temperature. If the steel shows no tendency to crack, standard welding procedures can be used. If, however, a tendency to crack shows up, the procedures outlined under the welding of mild steel should be followed.

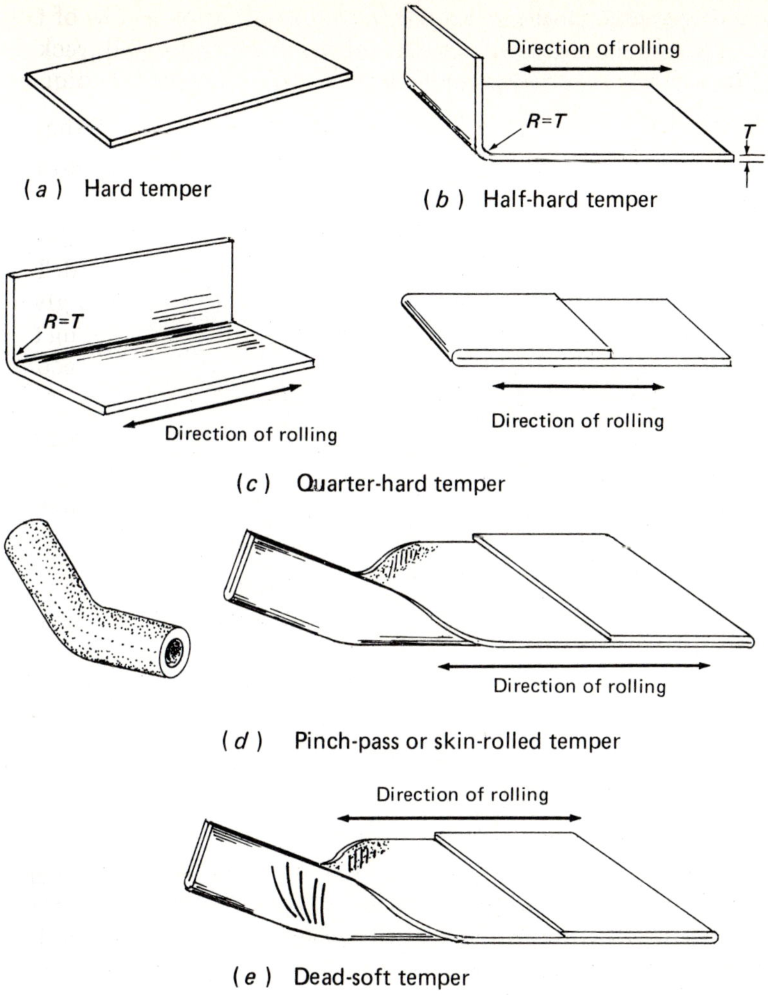

(a) Hard temper
Direction of rolling
R=T
T
(b) Half-hard temper
R=T
Direction of rolling
Direction of rolling
(c) Quarter-hard temper
Direction of rolling
(d) Pinch-pass or skin-rolled temper
Direction of rolling
(e) Dead-soft temper

Figure **20-1** Various tempers of carbon steel. (Redrawn from *New American Machinist's Handbook* by F. H. Calvin and F. A. Stanley, edited by R. LeGrand. Copyright 1950 McGraw-Hill Book Company. Used with permission of McGraw-Hill Book Company.)

If, because of cracking, preheating and postheating are required, the approximate preheating temperatures listed below may be used.

Steel	*¼-in. thick or less*	*½ in.*	*1 in.*	*2 in. or more*
SAE 1030	70°F	150°F	300°F	400°F
SAE 1035	70	200	400	500
SAE 1040	200	400	500	550

When required, postheating (stress relieving) after welding will also help reduce any hard zone resulting from the welding. The usual stress-relieving temperature is from 1100 to 1200°F. It is common practice to hold the steel for one hour at this temperature for each inch (or fraction thereof) of thickness up to a maximum of 8 hours.

HIGH-CARBON STEELS *High-carbon steels* include many steels that are generally used in a hardened condition, ranging between Rockwell C 40 and 63 full hard [Figure 20-1(*a*)]. Parts made from steels in this group include lathe tools, twist drills, drill bits, knives, scraper blades, plowshares, railroad car wheels, mill rolls, and other articles that do not require welding in manufacturing. However, these steels are frequently repair welded. Successful welding of these steels requires the development of specific welding procedures for each application. The welding procedure should be qualified (see Chapter 23) before it is adopted.

The problems that can arise during the welding of high-carbon steels are:

1. Cracking of the weld metal. The cracks may be across the bead (transverse) or they may be through the center of the bead (longitudinal). Longitudinal cracks are more prevalent. Sometimes they do not show on the surface of the bead, but the underportion of the bead will be cracked.

2. Porosity in the weld metal.

3. Excessive hardening of the parent metal.

4. Cracking of the parent metal: This includes both underbead cracking—cracks just under the fusion zone—and radial cracks that are at the fusion zone and extend into the parent metal.

5. Excessive softening of the parent metal.

To minimize the occurrence of these problems, follow the rules detailed below and make corrections as described under cracking of welds in mild- and medium-carbon steels.

• *Rule 1.* Prepare the weld joint carefully by shearing, machining, grinding, chipping, and so on. Remove all irregularities, such as nicks, cracks, and gouges, that could act as stress raisers, and be sure that all moisture and foreign material are removed from the joint and adjacent area.

• *Rule 2.* Avoid excessive penetration, and keep the weld metal as low in carbon content as practical. The deposit will then have the maximum ductility. Otherwise, the steel would be overstressed and fail (rupture or crack).

• *Rule 3.* Go slowly enough to deposit a substantial bead or layer of weld metal, but on wide welds weave rather than make parallel stringer beads. Avoid thin weld cross-sections. A thin, concave fillet weld between two rigid members is likely to crack. Crater cracks, also known as *hot cracks,* originate in the concave crater often left by the welder at the end of a weld bead. When welding in a groove or a fillet, it is the first or root bead that is most susceptible to cracking.

- *Rule 4.* Joints in steels containing 1.0% or more carbon should be surfaced with austenitic stainless steel.

After welding is completed, the usual procedure is to stress relieve the welded part as described under medium-carbon steels.

20-2 LOW-ALLOY STEELS

Low-alloy steels are carbon steels to which specific alloying elements have been added to obtain certain desirable properties (Table 20-1). Structural low-alloy steels are classified as shown in Table 20-2. However, for the purpose of describing their weldability they have been classified into four groups.

The steels in group 1 are commonly referred to as high-strength (45,000 to 70,000 psi) structural steels. They differ from carbon steel principally in that in addition to manganese, they are alloyed with varying amounts of chromium, vanadium, zirconium, nickel, and phosphorus. These elements strengthen the ferrite, promote hardenability, and control the grain size. Most of these steels are covered by ASTM specifications and are of proprietary compositions.

Steels from group 3 are used for welded structural fabrication and are covered by ASTM specifications A242, A441, A572, and A588.

Table **20-2** Low-alloy structural steels

1	2	3	4
A572 grades 42, 45, 50; A441; A242	Steels 13 and 14[a]	ASTM A36; A242 weldable grade; A375; A441; A529; A570 grades D & E; A572 grades 42, 45, 50; A588; Steels 13 and 14	ASTM A572 grades 55, 60, 65

[a]Cast steel composition:

13 0.24 C, 1.45 Mn, 0.49 Si, 0.028 P
14 0.23 C, 1.47 Mn, 0.44 Si, 0.016 P, 1.05 Cu

Those used for pressure-vessel fabrication are covered by ASTM specifications A203, A204, A225, A353, A553, and A537. Although these steels are as weldable as carbon steels when pressure vessels are welded according to the *ASME Boiler and Pressure Vessel Code,* the use of specified filler metals, as well as preheat and postheat treatment, is required.

Although acceptable preheat temperatures can be calculated, as described in Chapter 19, the AWS recommends the preheating temperatures listed in Table 20-3 and, in addition, the *ASME Boiler and*

Table **20-3** Recommended minimum preheat and interpass temperatures[1]

	WELDING PROCESS			
	SHIELDED METAL–ARC WELDING WITH OTHER THAN LOW-HYDROGEN ELECTRODES		*SHIELDED METAL–ARC WELDING WITH LOW-HYDROGEN ELECTRODES; SUBMERGED-ARC WELDING; GAS METAL–ARC WELDING; OR FLUX-CORED–ARC WELDING*	
THICKNESS OF THICKEST PART AT POINT OF WELDING—INCHES	A572 GR. 42, 45, 50 A441, A242	STEELS 13 AND 14	ASTM A36; A242 WELDABLE GRADE; A375; A441; A529; A570 GRADES D & E; A572 GRADES 42, 45, AND 50; A588 AND STEELS 13 AND 14	ASTM A572 GRADES 55, 60 AND 65
To 3/8″, incl.	None[2]	150	None[2]	None[2]
Over 3/8 to 3/4, incl.	70	250	None[2]	70°F
Over 3/4 to 1 1/2, incl.	Not recommended	300	70°F	150°F
Over 1 1/2 to 2 1/2, incl.	Not recommended	Not recommended	150°F	225°F
Over 2 1/2	Not recommended	Not recommended	225°F	300°F

[1]Welding shall not be done when the ambient temperature is lower than 0°F. When the base metal is below the temperature listed for the welding process being used and the thickness of material being welded, it shall be preheated (except as otherwise provided) in such manner that the surface of the parts on which weld metal is being deposited are at or above the specified minimum temperature for a distance equal to the thickness of the part being welded, but not less than 3 inches, both laterally and in advance of the welding. Preheat and interpass temperatures must be sufficient to prevent crack formation. Temperature above the minimum shown may be required for highly restrained welds.

[2]When the base metal temperature is below 32°F, preheat the base metal to at least 70°F and maintain this minimum temperature during welding.

Source: American Welding Society, *Welding Handbook,* 6th ed., section 4.

Pressure Vessel Code requires that all weldments made to code be stress relieved for one hour per inch of thickness, with a maximum of 8 hours for sections 8 inches thick or thicker, as follows:

Percent composition	*Temperature in °F (°C)*
2.25 Ni	1150 (621)
3.50 Ni	1100 (593)
9.00 Ni	1050 (566)

Steels in group 2 are used in the quenched and heat-treated conditions; have a carbon content that usually does not exceed 0.22%; and, depending on their chemical composition, thickness, and heat treatment, develop yield strengths ranging from 50,000 to 180,000 psi. These steels include:

ASTM A533, Grade B steel (Table 20-4) was developed for use in heavy sections in nuclear pressure-vessel construction. This steel has a higher strength level (Class 3) than the original grade, and is used as material for thin-walled and layered pressure vessels.

ASTM A537, Grade B steel (Table 20-4) is also intended for

Table **20-4** Heat treatment and microstructure for selected low-alloy structural steels

GRADE	TREATMENT	MICROSTRUCTURE
A533 Grade B	Water quenched from 1550°F and tempered at 1100°F	Tempered bainite and tempered martensite (thin plates); ferrite and tempered bainite (thick plates).
Steel A	Water quenched from 1625/1675°F and tempered at 1050°F	Ferrite and tempered bainite and/or tempered martensite.
A537 Grade B	Water quenched from 1625/1675°F and tempered at 1100°F	Ferrite and tempered bainite and/or tempered martensite.
A514/517	Water quenched from 1650°F and tempered at 1150°F	Tempered bainite and tempered martensite.
A543 Grade B	Water quenched from 1650°F and tempered at 1100°F	Tempered bainite and tempered martensite.
HY-130	Water quenched from 1475/1525°F and tempered at 1000°F	Tempered bainite and tempered martensite.
HP 9-4-20	Normalized from 1675°F, water quenched from 1550°F and tempered at 1000°F	Tempered martensite.
10 Ni-Cr-Mo-Co	Water quenched from 1700°F Water quenched from 1500°F and aged at 950°F for 10 hours	Tempered martensite.

Source: American Welding Society, *Welding Handbook*, 6th ed., section 4, p. 63.20.

pressure vessels. It is used where notch (impact) toughness (see Chapter 18) is required at a yield strength of 60,000 psi.

Steel A in Table 20-4 is very similar to the steel covered by ASTM A537 Grade B, except that its yield strength (see Chapter 19) has been increased to 80,000 psi.

ASTM A514/517 steels in Grades A through P are used principally for fabricating earth-moving equipment, pressure vessels, bridges, penstocks, scroll cases, buildings, steel-mill and mining equipment, TV antenna tower construction, fans, ships, and many other applications.

ASTM A543 steel is used to build nuclear-reactor vessels and for similar applications requiring premium-toughness, high-yield strength steel. A543 Grade-B Class 1 and 2 steels are known as US Navy Grades HY-80 and HY-100 steels, and are used in the construction of military, marine, and commercial hydrospace vehicles.

ASTM A553 steel is the quenched and tempered (see Chapter 19) counterpart of ASTM A353 steel (double-normalized and tempered 9% nickel steel). Both are used at temperatures as low as −320°F (−196°C) because of their toughness. Since the welding considerations are the same for these steels regardless of which heat treatment is used, both were included in the discussion of the as-rolled and normalized steels (group 1) and will not be further discussed here.

The HY-130, HP 9-4-20, and 10 Ni-Cr-Mo-Co steels are the newest of the premium-toughness, higher-yield-strength steels developed for use in critical hydrospace and aerospace applications.

All steels in this group are readily weldable by all standard welding processes. The electron-beam welding process can be used to weld any steel in this group having a yield strength of 150,000 psi or more.

Best results are obtained when groove-type joints are used and the steel is preheated to the temperature in the following tabulation.

Thickness (in.)	*Temperature (°F)*
To ⅝ inch	100
Over ⅝ to ⅞ inch	160
Over ⅞ to 1⅜ inch	230
Over 1⅜ inch	260

The steels of group 3 are used principally to make such parts as cams, roller bearings, heat-treated nuts, transmission shafts, and others. They can be readily welded by all standard welding procedures.

The low-alloy steels most frequently used in weldments that must be quenched and tempered after welding contain between 0.25 and 0.45% carbon. They are, by SAE (or other industrial) designation:

4027, 4037, 4130, 4135, 4140, 4320, 4340, 5130, 5140, 1630, 8640, 8740, 4335V, AMS6334, D-6, 300-M, H-11, and HP9-4-45.

Steels within this group containing less than 0.25% carbon are quenched and tempered before welding. However, regardless of carbon content, it is required that all steels within this group be preheated to the temperatures listed in Table 20-5.

The chromium-molybdenum steels of group 4 are air-hardenable, corrosion-resistant steels used to manufacture forgings, tubes, castings, and plate for use in the steam-power industry. They are readily weldable by most standard, commercially available welding processes.

Because the air-hardening characteristics of these steels make them prone to the development of cracks, the steels must be preheated and postheated to the temperatures indicated in Tables 20-6 and 20-7.

Table **20-5** Approximate preheat temperatures for group 3 steels

	PLATE THICKNESS (INCHES)			
AISI[a] NUMBER	*¼ OR LESS*	½	1	*2 OR MORE*
1330	95°F	200°F	400°F	450°F
1335	150	400	450	500
1340	300	450	500	550
1345	400	550	600	60
2315	Room temp.	Room temp.	250	400
2515	Room temp.	100	300	400
2517	Room temp.	250	450	500
3140	550	625	650	700
3310	200	425	500	550
3316	250	475	550	600
4027	Room temp.	150	400	450
4068	750	800	850	900
4130	100	350	450	525
4320	200	450	525	550
4335	400			
4620	Room temp.	150	300	450
5120	600	200	400	450
5150	Room temp.	750	750	750

[a]See Table 20-1 for an explanation.

Table **20-6** Recommended minimum preheat temperatures[1] for welding chromium-molybdenum steels

THICKNESS Cr–Mo STEEL	TO ½ IN.	½ TO 2¼ IN.	OVER 2¼ IN.
½Cr–½Mo	70	200	300
1Cr–½Mo	250	300	300
1¼Cr–½Mo	250	300	300
2Cr–½Mo	300	300	300
2¼Cr–1Mo	300	300	300
3Cr–1Mo	300	300	300
5Cr–½Mo	300	300	300
5Cr–½MoSi	300	300	300
5Cr–½MoTi	300	300	300
7Cr–½Mo	400	400	400
9Cr–1Mo	400	400	400

[1]Fahrenheit.

Source: American Welding Society, *Welding Handbook,* 6th ed., section 4, p. 63.39.

Table **20-7** Recommended postweld heat-treatment temperature ranges for chromium-molybdenum steels

ALLOY TYPE	TEMPERATURE RANGE, °F
½Cr–½Mo	1100–1300
1Cr–½Mo	1100–1350
1¼Cr–½Mo	1100–1375
2¼Cr–1Mo	1250–1400
3Cr–1Mo	1250–1400
5Cr–½Mo	1250–1400
5Cr–½MoSi	1250–1400
5Cr–½MoTi	1250–1400
7Cr–½Mo	1250–1400
9Cr–1Mo	1250–1400

Source: American Welding Society, *Welding Handbook,* 6th ed., section 4, p. 63.42.

20-3 TOOL AND DIE STEELS

Tool and die steels may be either carbon or alloy steels capable of being hardened and tempered. The American Iron and Steel Institute and the Society of Automotive Engineers group steels of similar properties as shown in Table 20-8. The specific uses for the various groups of tool steels are described below.

Water-hardening (W) carbon tool steels are only suitable for cold-work applications, since they have low *red-hardness* (ability to remain hard at temperatures above 1300°F). Also, because they develop a hard case with a soft core during heat treatment, they are widely used for percussion and impact tools, such as cold header dies, coining and restrike dies, shear blades, and engraving tools. The steels are readily weldable in the annealed and heat-treated condition. Preweld and postweld heat treatment is required of all tool and die steels, as indicated in Table 20-9 (p. 564).

Shock-resisting (S) tool steels have high impact strength and are used in riveting tools, impact hammers and chisels, rock tools, and similar applications. Although these steels are subject to considerable distortion during the preweld and postweld heat treatments, the distortion can be ignored because it is not very significant in the class of tools made from these steels.

Table **20-8** Classification and compositions of principal types of tool steels

DESIGNATION	C	Mn	Si OR Ni	Cr	V	W	Mo	Co
				WATER-HARDENING TOOL STEELS[a]				
W1*	0.60 to 1.40[a]	—	—	—	—	—	—	—
W2*	0.60 to 1.40[a]	—	—	—	0.25	—	—	—
W3	0.60 to 1.40[a]	—	—	—	0.50	—	—	—
W4	0.60 to 1.40[a]	—	—	0.25	—	—	—	—
W5	0.60 to 1.40[a]	—	—	0.50	—	—	—	—
W6	0.60 to 1.40[a]	—	—	0.25	0.25	—	—	—
W7	0.60 to 1.40[a]	—	—	0.50	0.20	—	—	—
				SHOCK-RESISTING TOOL STEELS				
S1*	0.50	—	—	1.50	—	2.50	—	—
S2	0.50	—	1.00 Si	—	—	—	0.50	—
S3	0.50	—	—	0.75	—	1.00	—	—
S4	0.50	0.80	2.00 Si	—	—	—	—	—
S5*	0.50	0.80	2.00 Si	—	—	—	0.40	—
				OIL-HARDENING COLD-WORK TOOL STEELS				
O1*	0.90	1.00	—	0.50	—	0.50	—	—
O2	0.90	1.60	—	—	—	—	—	—
O6	1.45	—	1.00 Si	—	—	—	0.25	—
O7	1.20	—	—	0.75	—	1.75	0.25 opt	—
				AIR-HARDENING MEDIUM-ALLOY COLD-WORK TOOL STEELS				
A2*[b]	1.00	—	—	5.00	—	—	1.00	—
A4	1.00	2.00	—	1.00	—	—	1.00	—
A5	1.00	3.00	—	1.00	—	—	1.00	—
A6	0.70	2.00	—	1.00	—	—	1.00	—
A7	2.25	—	—	5.25	4.50	—	1.00	—
				HIGH-CARBON HIGH-CHROMIUM COLD-WORK STEELS				
D1	1.00	—	—	12.00	—	—	1.00	—
D2*[b]	1.50	—	—	12.00	—	—	1.00	—
D3*[b]	2.25	—	—	12.00	—	—	—	—
D4*	2.25	—	—	12.00	—	—	1.00	—
D5[b]	1.50	—	—	12.00	—	—	1.00	3.00
D6	2.25	—	1.00 Si	12.00	—	1.00	—	—
D7[b]	2.35	—	—	12.00	4.00	—	1.00	—
				CHROMIUM HOT-WORK TOOL STEELS				
H11	0.35	—	—	5.00	0.40	—	1.50	—
H12*	0.35	—	—	5.00	0.40	1.50	1.50	—
H13*[b]	0.35	—	—	5.00	1.00	—	1.50	—
H14	0.40	—	—	5.00	—	5.00	—	—

Table **20-8** *(cont.)*

DESIGNATION	C	Mn	Si OR Ni	Cr	V	W	Mo	Co
				CHROMIUM HOT-WORK TOOL STEELS				
H15	0.40	—	—	5.00	—	—	5.00	—
H16	0.55	—	—	7.00	—	7.00	—	—
				TUNGSTEN HOT-WORK TOOL STEELS				
H20	0.35	—	—	2.00	—	9.00	—	—
H21*	0.35	—	—	3.50	—	9.50	—	—
H22	0.35	—	—	2.00	—	11.00	—	—
H23	0.30	—	—	12.00	—	12.00	—	—
H24	0.45	—	—	3.00	—	15.00	—	—
H25	0.25	—	—	4.00	—	15.00	—	—
H26	0.50	—	—	4.00	1.00	18.00	—	—
				MOLYBDENUM HOT-WORK TOOL STEELS				
H41	0.65	—	—	4.00	1.00	1.50	8.00	—
H42	0.60	—	—	4.00	2.00	6.00	5.00	—
H43	0.55	—	—	4.00	2.00	—	8.00	—
				TUNGSTEN HIGH-SPEED TOOL STEELS				
T1*[b]	0.70	—	—	4.00	1.00	18.00	—	—
T2[b]	0.85	—	—	4.00	2.00	18.00	—	—
T3[b]	1.05	—	—	4.00	3.00	18.00	—	—
T4	0.75	—	—	4.00	1.00	18.00	—	5.00
T5	0.80	—	—	4.00	2.00	18.00	—	8.00
T7	0.75	—	—	4.00	2.00	14.00	—	—
T8	0.80	—	—	4.00	2.00	14.00	—	5.00
T15[b]	1.50	—	—	4.00	5.00	12.00	—	5.00
				MOLYBDENUM HIGH-SPEED TOOL STEELS				
M1*[b]	0.80	—	—	4.00	1.00	1.50	8.50	—
M2*[b]	0.85	—	—	4.00	2.00	6.25	5.00	—
M3*[bc]	1.00	—	—	4.00	2.40	6.00	5.00	—
M4	1.30	—	—	4.00	4.00	5.50	4.50	—
M6	0.80	—	—	4.00	1.50	4.00	5.00	12.00
M7	1.00	—	—	4.00	2.00	1.75	8.75	—
M10*[b]	0.85	—	—	4.00	2.00	—	8.00	—
M15	1.50	—	—	4.00	5.00	6.50	3.50	5.00
M30	0.80	—	—	4.00	1.25	2.00	8.00	5.00
M33	0.90	—	—	3.75	1.15	1.75	9.50	8.25
M34	0.90	—	—	4.00	2.00	2.00	8.00	8.00
M35	0.80	—	—	4.00	2.00	6.00	5.00	5.00
M36	0.80	—	—	4.00	2.00	6.00	5.00	8.00

DESIGNATION	C	Mn	Si OR Ni	Cr	V	W	Mo	Co
LOW-ALLOY SPECIAL-PURPOSE TOOL STEELS								
L1	1.00	—	—	1.25	—	—	—	—
L2	0.50 to 1.10[a]	—	—	1.00	0.20	—	—	—
L3	1.00	—	—	1.50	0.20	—	—	—
L4	1.00	0.60	—	1.50	0.20	—	—	—
L5	1.00	1.00	—	1.00	—	—	0.25	—
L6	0.70	—	1.50 Ni	0.75	—	—	0.25 opt	—
L7	1.00	0.35	—	1.40	—	—	0.40	—
CARBON-TUNGSTEN TOOL STEELS								
F1	1.00	—	—	—	—	1.25	—	—
F2	1.25	—	—	—	—	3.50	—	—
F3	1.25	—	—	0.75	—	3.50	—	—
LOW-CARBON MOLD STEELS								
P1	0.10 max	—	—	—	—	—	—	—
P2	0.07 max	—	0.50 Ni	1.25	—	—	0.20	—
P3	0.10 max	—	1.25 Ni	0.60	—	—	—	—
P4	0.07 max	—	—	5.00	—	—	—	—
P5	0.10 max	—	—	2.25	—	—	—	—
P6	0.10	—	3.50 Ni	1.50	0.20	—	—	—
P20	0.30	—	—	0.75	—	—	0.25	—
PPT	0.20	1.20Al	4.00 Ni	—	—	—	—	—
OTHER ALLOY TOOL STEELS[d]								
6G	0.55	0.80	0.25 Si	1.00	0.10	—	0.45	—
6F2	0.55	0.75	0.25 Si, 1.00 Ni	1.00	0.10 opt	—	0.30	—
6F3	0.55	0.60	0.85 Si, 1.80 Ni	1.00	0.10 opt	—	0.75	—
6F4	0.20	0.70	0.25 Si, 3.00 Ni	—	—	—	3.35	—
6F5	0.55	1.00	1.00 Si, 2.70 Ni	0.50	0.10	—	0.50	—
6F6	0.50	—	1.50 Si	1.50	—	—	0.20	—
6F7	0.40	0.35	4.25 Ni	1.50	—	—	0.75	—
6H1	0.55	—	—	4.00	0.85	—	0.45	—
6H2	0.55	0.40	1.10 Si	5.00	1.00	—	1.50	—

*Stocked in almost every warehousing district and made by the majority of tool steel producers. [a]Various carbon contents are available in 0.10% ranges. [b]Available as free-cutting grade. [c]Available with vanadium contents of 2.40 or 3.00%. [d]The designations of these steels are similar to those used in the 1948 Metals Handbook, except they were previously written with Roman numberals (VI F2, etc). Neither AISI nor SAE has assigned type numbers to these steels.

Source: By permission, from *Metals Handbook* Volume 2, Copyright American Society for Metals, 1961– .

Table **20-9** Recommended preweld and postweld heat treatments for tool and die steels

STEEL (AISI TYPE)	TYPE OF TOOL STEEL ELECTRODE[a]	ANNEALED BASE METAL		
		ANNEALING TEMPERATURE, F	*PREHEAT AND POSTHEAT TEMPERATURE, F*	*HARDENING TEMPERATURE, F*
W1, W2	Water hardening[c]	1375 to 1425	250 to 450	1375 to 1475
S1	Hot work[d]	1475	300 to 500	1750
S5	Hot work[d]	1450	300 to 500	1625
S7	Hot work[d]	1500 to 1550	300 to 500	1725
O1	Oil hardening[g]	1450	300 to 400	1475
O6	Oil hardening[g]	1425 to 1450	300 to 400	1450 to 1500
A2	Air hardening[h]	1650	300 to 500	1775
A4	Air hardening[h]	1425	300 to 500	1550
D2	Air hardening[h]	1650	700 to 900	1850
H11, H12, H13	Hot work[d]	1600[j]	900 to 1200	1850
M1, M2, M10	High speed[k]	1550[j]	950 to 1100	[m]

[a]Nominal compositions of weld deposits are footnoted by type. The compositions of proprietary electrodes vary. [b]As deposited after postheat. [c]Deposit: 0.95 C, 0.20 Si, 0.30 Mn, 0.20 V. [d]Deposit: 0.33 C, 1.00 Si, 0.40 Mn, 5.00 Cr, 1.35 Mo, 1.25 W. [e]Oil may also be used. [f]Double temper recommended. [g]Deposit: 0.92 C, 0.30 Si, 1.28 Mn, 0.50 Cr, 0.50 W. [h]Deposit: 0.95 C, 0.30 Si, 0.40 Mn, 5.25 Cr, 1.10 Mo, 0.25 V. [j]For H12 and M2 steels, anneal at 1625 F. [k]Proprietary compositions. [m]2240 F for M1, 2260 F for M2, 2215 F for M10.

Source: By permission, from *Metals Handbook* Volume 6, Copyright American Society for Metals, 1971.

Although all tool steels are susceptible to decarburization, the shock-resisting and high-speed tool steels are more susceptible than the other steels. Therefore, it is recommended that these steels be preheated in furnaces using an oxidizing atmosphere and operating at around 1500°F. The oxidizing atmosphere forms a thin, adherent scale on the steel *that appears to partially protect it from decarburization.*

Oil-hardening (O) steels distort very little during preweld and postweld heat treatment. They have (as is the case with water-hardening steels) low red-hardness and are mainly used for cold-working tools and die sections with sharp corners.

Air-hardening (A) steels are nondeforming tool steels used for applications where minimum size change and cracking can be tolerated. These applications include gages, blanking and forming punches and dies, forming rolls, and threading dies. The A7 grade has medium red-hardness, permitting its use in plastic molds. Cold-work (D) steels are used in most applications where precision,

			Hardened base metal	
Quenching medium	Tempering temperature, F	Resulting hardness, Rockwell C	Preheat and postheat temperature, F	Rockwell C hardness[b]
Water	300 to 650	54 to 65	250 to 450	56 to 62
Oil	300 to 500	54 to 57	300 to 500	52 to 56
Oil	500 min	55 to 59	300 to 500	52 to 56
Air[e]	400 to 425[f]	56 to 58	300 to 500	52 to 56
Oil	300 to 450	61 to 63	300 to 400	56 to 62
Oil	300 to 450	61 to 63	300 to 400	56 to 62
Air	350 to 400[f]	60 to 61	300 to 400	56 to 58
Air	350 to 400[f]	60 to 61	300 to 400	60 to 62
Air	900 to 925[f]	58 to 60	700 to 900	58 to 60
Air	1000 to 1150[f]	40 to 50	700 to 1000	46 to 54
Salt[e]	1000 to 1050[f]	65 to 66	950 to 1050	60 to 63

dimensional stability, outstanding wear resistance, and long life are required, such as in lamination dies and gages.

Hot-work (H) steels are air-hardening, and used on high-impact tools, such as forging dies, high-temperature mandrels, and hot shears. Because these steels are subject to cracking from thermal shock, interpass temperatures must be maintained and the steel must be allowed to cool in the furnace.

The high-speed (T and M) steels have high red-hardness and little or no distortion during heat treatment. They are widely used in applications requiring a balance of high red-hardness, abrasion resistance, and toughness, such as in lathe-tool bits, twist drills, milling cutters, taps, reamers, and others. To minimize decarburization, these steels should be preheated in oxidizing-atmosphere furnaces operating at around 1500°F.

The special-purpose (L) steels are oil-hardening and offer only a fair amount of resistance to distortion as a result of heating and cooling. They are frequently substituted for the higher-cost and nondifficult-to-machine tool steels in applications where high wear resistance with good toughness is required, as in bearings, rollers, clutch plates, high-wear springs, feed fingers, and chuck parts.

The carbon-tungsten (F) steels are shallow hardening, have high wear resistance, low red-hardness, and are relatively brittle. Because of these characteristics, they are used mainly for such parts as paper-cutting knives, wire drawing dies, plug gages, forming tools, and brass-cutting tools.

The low-carbon mold (P) steels are alloy steels produced to tool-steel quality that can be hardened by carburizing. They are readily weldable in the annealed state and possess poor red-hardness and low wear resistance (unless carburized and heat treated). Because of these characteristics, the steels are used almost exclusively for low-temperature die castings, dies, and molds for injection or compression molding of plastics.

Groups 6G, 6F, and 6H steels are multialloy deep-hardening steels of medium-carbon content, characterized by high toughness and, in some instances (6F4), good heat resistance. The steels are oil-hardening, offer fairly good resistance to deformation during heat treatment, and are used extensively for hot-upset forging dies and dies for hammer and press forging.

As a group, tool and die steels are difficult to weld; however, they are usually welded for one of the following purposes:

1. Assembly of components into a tool.

2. Fabrication of composite tools by depositing an overlay of tool-steel weld metal on specific areas (hard-surfacing) of a carbon steel or a less highly alloyed steel (a shear blank, for example).

3. Rebuilding of worn surfaces and edges.

4. Alteration of a tool or die to meet a change in design of the product being manufactured using the tool.

5. Repair of cracked or otherwise damaged tools.

Whenever possible, tool and die steels should be repair welded in the hardened and tempered condition. The oxyacetylene process, due to its slowness, introduces too much heat into the base metal, causing distortion, excessive softening of hardened metal or excessive embrittlement, and cracking of annealed metal. Therefore, the oxyacetylene process *should not be used to weld these steels.* However, regardless of the welding process (other than oxyacetylene), *peening of each weld bead immediately after welding will minimize the formation of shrinkage cracks.* Maintenance of interpass temperatures will also help in minimizing the formation of shrinkage cracks.

20-4 CAST IRON

The term *cast iron* describes a wide variety of iron-base materials containing carbon, silicon, manganese, phosphorus, sulfur, nickel, molybdenum, titanium, vanadium, chromium, magnesium, copper, and aluminum. For the purposes of this book, the term *cast iron,* however, will apply only to gray malleable, nodular, or austenic irons. White iron will not be discussed because it is considered "unweldable."

Table **20-10** Minimum transverse breaking loads (ASTM A48)[a]

ASTM CLASS	MINIMUM TENSILE STRENGTH (psi)	MINIMUM TRANSVERSE BREAKING LOAD (lb)		
		0.875-IN. DIAM 12-IN. SUPPORTS	1.2-IN. DIAM 18-IN. SUPPORTS	2.0-IN. DIAM 24-IN. SUPPORTS
20	20,000	900	1800	6,000
25	25,000	1025	2000	6,800
30	30,000	1150	2200	7,600
35	35,000	1275	2400	8,300
40	40,000	1400	2600	9,100
50	50,000	1675	3000	10,300
60	60,000	1925	3400	12,500

[a]For separately cast test specimens. Included in specifications only by agreement between manufacturer and purchaser.

Source: By permission, from *Metals Handbook* Volume 1, Copyright American Society for Metals, 1961–

Two systems of classifying gray irons are in use: the system developed by the American Society for Testing and Materials (ASTM) and that developed by the Society of Automotive Engineers (SAE). In the ASTM system, all gray irons are classified (Table 20-10) by minimum tensile strength and section thickness.

The SAE system (Table 20-11) is more specific than the ASTM system in that it uses only one section thickness. Gray cast irons are used for applications such as window-sash weights, elevator counterweights, industrial furnace doors, low-pressure valves, culvert pipes, machine parts, and most other applications where minimum cost is a factor and low pressures and static loading conditions are the rule.

There are two types of malleable iron: the *black hearth* type, so-called because when fractured, the fracture has a dark appearance; and the *white hearth,* so-called because when fractured, the fracture

Table **20-11** Mechanical properties of SAE class automotive-type gray iron

SAE NO.	BRINELL HARDNESS NUMBER	MINIMUM TRANSVERSE LOAD (lb)	MINIMUM DEFLECTION (in.)	MINIMUM TENSILE STRENGTH (psi)
110	187 max	1800	0.15	20,000
111	170 to 223	2200	0.20	30,000
120	187 to 241	2400	0.24	35,000
121	202 to 255	2600	0.27	40,000
122	217 to 269	2800	0.30	45,000

Source: *SAE Handbook*, 1960, page 135. Properties determined from arbitration test bar (1.2-in. diam) as cast or stress relieved at 1050 F max.

will be light in color, meaning that the iron was decarburized *(lost carbon due to burning)* during the pouring process. Only the black hearth is produced in the USA, and while both types are produced in Europe, the black hearth is preferred there also.

Malleable iron is produced to one of three existing grades, depending on the melting practice employed and the applicable ASTM specification for the casting (Table 20-12). Malleable irons are commonly used to make such automotive parts as differential cases, steering knuckles, steering gear housings, leaf-spring hangers, universal joint yokes, automatic transmission parts, rocker arms, wheel hubs, chain links, electrical-pole line hardware, and other parts requiring the section thickness and properties obtainable in these materials.

Nodular iron, also called *ductile iron,* is made by the same processes and with the same materials as gray iron, except that magne-

Table **20-12** Specifications for pearlitic malleable iron

GRADE	MINIMUM TENSILE STRENGTH (psi)	MINIMUM YIELD STRENGTH (psi)	MINIMUM ELONGATION (% in 2 in.)	TYPICAL BRINELL HARDNESS RANGE
		ASTM A220-55T		
45010	65,000	45,000	10	163 to 207
45007	68,000	45,000	7	163 to 217
48004	70,000	48,000	4	163 to 228
50007	75,000	50,000	7	179 to 228
53004	80,000	53,000	4	197 to 241
60003	80,000	60,000	3	197 to 255
80002	100,000	80,000	2	241 to 269
		SOCIETY OF AUTOMOTIVE ENGINEERS		
43010	60,000	43,000	10	163 to 207
48005	70,000	48,000	5	179 to 228
53004	80,000	53,000	4	197 to 241
60003	80,000	60,000	3	197 to 241
70002	90,000	70,000	2	241 to 285
		FEDERAL SPECIFICATIONS MIL-I-11444		
Class 1 (70002)	90,000	70,000	2	241 to 285
Class 2 (60003)	80,000	60,000	3	197 to 241
Class 3 (53004)	80,000	53,000	4	197 to 241
Class 4 (48005)	70,000	48,000	5	179 to 228
Class 5 (43010)	60,000	43,000	10	163 to 207

Source: By permission, from *Metals Handbook* Volume 1, Copyright American Society for Metals, 1961–

sium and/or cesium is added to the melt. These alloying elements cause the carbon to precipitate as tiny balls (nodules) or spheres of graphite within a matrix (base or net) that is basically pearlite, although usually some ferrite and cementite are also present.

Specifications have been developed for various classes of nodular iron castings. ASTM and military specifications are listed in Table 20-13. Nodular iron castings are used for such applications as generator shafts having diameters of 10 and 20 inches and lengths of 7 to 12 feet, respectively, centrifugal pump castings, and other applications requiring thermal shock resistance up to 1600°F.

Table **20-13** Specifications for nodular iron

CLASS	MINIMUM TENSILE STRENGTH (psi)	MINIMUM YIELD STRENGTH (psi)	MINIMUM ELONGATION (% in 2 in.)	COMPOSITION REQUIREMENT	CONDITION
			ASTM A339-55		
80-60-03	80,000[a]	60,000[b]	3.0	None	As cast
60-45-10	60,000[a]	45,000[b]	10.0	None	Usually annealed
			ASTM A396-58		
120-90-02	120,000[a]	90,000[b]	2.0	None	Heat treated
100-70-03	100,000[a]	70,000[b]	3.0	None	Heat treated
			ASTM A395-56T		
60-45-15	60,000[a]	45,000[b]	15.0	[e]	Ferritized by annealing
			MIL-I-17166A		
60-40-15	60,000[a]	40,000[b]	15.0	[f]	Ferritized by annealing to 190 Bhn max[g]
			MIL-I-11466		
Class 1	120,000[c]	90,000[d]	2.0	None	Usually heat treated[g]
Class 2	100,000[c]	75,000[d]	4.0	None	Usually heat treated[g]
Class 3	85,000[c]	60,000[d]	6.0	None	Usually heat treated[g]
Class 4	80,000[c]	60,000[d]	3.0	None	Ferritized by annealing[g]
Class 5	60,000[c]	45,000[d]	10.0	None	Ferritized by annealing[g]
Class 6	60,000[c]	40,000[d]	18.0	None	Ferritized by annealing[g]

[a]Test specimen to be machined from 1-in. keel block or Y-block in ½, 1 or 3-in. size. [b]Yield strength at 0.2% offset. [c]Test specimens to be machined from ¾, 1 or 3-in. Y-block, in accordance with critical section of casting. [d]Yield strength at 0.1% offset or extension-underload method. [e]3.00% C min, 2.75% Si max, 0.08% P max. [f]Same as ASTM A395-56T plus 4.5% carbon equivalent maximum. Applies to castings with sections 2 in. and over. Carbon equivalent = total carbon plus one third of the silicon. [g]Metallographic test required for each lot of castings.

Source: By permission, from *Metals Handbook* Volume 1, Copyright American Society for Metals, 1961– .

Cast irons are readily weldable by the standard welding processes. The welding procedures are designed to restrict penetration to the minimum depth required for fusion, thereby minimizing or preventing the transformation of the base metal into a brittle zone at the weld, caused by rapid freezing of weld metal and rapid cooling of the weld metal in the heat-affected zone (HAZ). High preheats produce softer, less brittle microstructures than do low preheats. With high preheats, however, welding is more difficult. The casting must be insulated from rapid heat loss, so the preheat temperature can be maintained throughout the welding operation. The heat input from welding can be used to help maintain interpass temperatures in small- and medium-size castings. In large castings, steep thermal gradients should be avoided to prevent cracking. To ensure proper control, the temperature should be measured by contact pyrometers or temperature-indicating crayons at or near the weld zone, and at as many other places as the casting size and heating method may indicate.

During welding of large preheated castings, the welder must be protected from the heat by exposing only the area being welded at the time, and by changing welders if the operation will take more than 2 hours.

In addition to preheating, it is important that the weld groove be prepared correctly. For instance, for butt welding with the shielded metal–arc process or to repair a defective casting, a groove (extending beyond an existing crack) with a 60-degree included angle is prepared by chipping, grinding, or machining. Care must be exercised to allow a root opening wide enough for uninterrupted electrode manipulation and to allow fusion with the root faces and backing plate (if used). On the other hand, *if the weld is to be made by the oxyacetylene process, a 90-degree included groove angle is satisfactory if a double-V groove is used.* Where the weld can be made from one side, the included groove angle may be increased to 120 degrees.

20-5 CAST STEELS

In general, cast steel has the same chemical composition as rolled steel, except that it is poured into a mold in the desired shape instead of being poured into ingots and rolled. Steel castings are usually purchased to meet specified mechanical properties. Table 20-14 lists some standard ASTM, SAE, and government specifications.

Among the most commonly selected grades of steel castings are (a) low-carbon annealed steel corresponding to QQ-S-681 class 2, ASTM 65-35, or SAE 0030, and (b) a higher-strength, often-alloyed or fully heat-treated steel, or both, similar to QQ-S-681 class 4C2, ASTM 105-85, or SAE 0105. Cast steels have the same weldability

Table **20-14** Summary of specification requirements for steel castings

CLASS OR GRADE	TENSILE STRENGTH (min psi)	YIELD STRENGTH (min psi)	ELONGATION IN 2 IN. (min %)	REDUCTION OF AREA (min %)	CHEMICAL RESTRICTIONS, MAX[a] C	CHEMICAL RESTRICTIONS, MAX[a] Mn
FEDERAL SPECIFICATION QQ-S-681						
X	—	—	—	—	0.30	1.00
0	—	—	—	—	0.45	1.00
	60,000	30,000	24	35	0.30	0.60
2	65,000	35,000	20	30	0.35	0.70
3	80,000	40,000	17	25	0.50	—
4A1	75,000	40,000	24	35	—	—
4A2	85,000	53,000	22	35	—	—
4B1	85,000	55,000	22	40	—	—
4B2	90,000	60,000	22	45	—	—
4B3	100,000	65,000	17	30	—	—
4C1	90,000	65,000	20	45	—	—
4C2	105,000	85,000	15	30	—	—
4C3	120,000	100,000	12	30	—	—
4C4	150,000	125,000	10	25	—	—
ASTM SPECIFICATION A27-58[b]						
N-1	—	—	—	—	0.25	0.75
N-2	—	—	—	—	0.35	0.60
N-3	—	—	—	—	—	1.00
U-60-30	60,000	30,000	22	30	0.25	0.75
60-30	60,000	30,000	24	35	0.30	0.60
65-30	65,000	30,000	20	30	—	—
65-35	65,000	35,000	24	35	0.30	0.70
70-36	70,000	36,000	22	30	0.35	0.70
70-40	70,000	40,000	22	30	0.25	1.20
ASTM SPECIFICATION A148-58						
80-40	80,000	40,000	18	30	—	—
80-50	80,000	50,000	22	35	—	—
90-60	90,000	60,000	20	40	—	—
105-85	105,000	85,000	17	35	—	—
120-95	120,000	95,000	14	30	—	—
150-125	150,000	125,000	9	22	—	—
175-145	175,000	145,000	6	12	—	—
SAE SPECIFICATIONS[c,d]						
0022	—	—	—	—	0.12 to 0.22	0.50 to 0.90
0030[g]	65,000	35,000	24	35	0.30	0.70
0050[h]	85,000	45,000	16	24	0.40 to 0.50	0.50 to 0.90[e]

Table **20-14** *(cont.)*

CLASS OR GRADE	TENSILE STRENGTH (min psi)	YIELD STRENGTH (min psi)	ELONGATION IN 2 IN. (min %)	REDUCTION OF AREA (min %)	CHEMICAL RESTRICTIONS, MAX[a]	
					C	Mn
			SAE SPECIFICATIONS[c,d]			
0050[i]	100,000	70,000	10	15	0.40 to 0.50	0.50 to 0.90[f]
080[j]	80,000	40,000	18	30	—	—
090[k]	90,000	60,000	20	40	—	—
0105[l]	105,000	85,000	17	35	—	—
0120[m]	120,000	100,000	14	30	—	—
0150[n]	150,000	125,000	9	22	—	—
0175[o]	175,000	145,000	6	12	—	—
HA[p]	—	—	—	—	0.25 to 0.34	—
HB[p]	—	—	—	—	0.25 to 0.34	—
HC[p]	—	—	—	—	0.25 to 0.34	—

[a]Carbon and manganese are maximum limits unless a range is given. All specifications restrict phosphorus to 0.05% and sulfur to 0.06% max. Silicon and alloying elements are restricted on some grades. [b]For each reduction of 0.01% C below the maximum specified, an increase of 0.04% Mn above the maximum specified will be permitted to a maximum of 1.40% for grade 70-40 and 1.00% for the other grades. [c]For each reduction of 0.01% C below the maximum specified, an increase of 0.04% Mn above the maximum specified will be permitted to a maximum of 1% Mn. [d]Hardness values given in footnotes g through o are nominal and applicable to casting sections not over 3 in. [e]Normalized or normalized and tempered. [f]Quenched and tempered. [g]131 Bhn. [h]170 Bhn. [i]207 Bhn. [j]163 Bhn. [k]187 Bhn. [l]217 Bhn. [m]248 Bhn. [n]311 Bhn. [o]363 Bhn. [p]Purchased on the basis of hardenability. Manganese and other elements added as required.

Source: By permission, from *Metals Handbook* Volume 1, Copyright American Society for Metals, 1961–.

characteristics as rolled or drawn steels of the same composition, but do not exhibit the same effects of directionality on mechanical properties that are typical of drawn or rolled steels.

20-6 STAINLESS STEELS

The descriptive words *stainless steel* are applied to many iron-base alloys, all of which contain at least 12% chromium, with or without additions of other alloying elements. The outstanding property of stainless steels is their resistance to corrosion in numerous, but not all, corrosive environments. In addition, and very important, is the exceptional adaptability of these steels to cold- and hot-forming processes and their ability to develop high tensile and creep strengths up to 350,000 and 25,000 psi, respectively, at 1000°F. The composition of stainless steels has been standardized, and each alloy has been assigned a specific AISI type number (Table 20-15).

Table **20-15** AISI stainless-steel weldability

		WELDABILITY			
		ARC	*GAS*	*RESISTANCE*	OUTSTANDING CHARACTERISTICS
Austenitic	301	1	2	1	A general utility stainless steel, easily worked.
	302	1	3	1	Readily fabricated, for decorative or corrosion resistance.
	304, 304LC	1	3	1	A general utility stainless steel, easily worked.
	308	1	3	1	Used where corrosion resistance better than 18-8 is needed.
	309	1	3	1	High scaling resistance and good strength at high temperatures.
	310	1	3	1	More chromium and nickel for greater resistance to scaling in high heat.
	316, 316LC	1	2	1	Excellent resistance to chemical corrosion.
	317	1	2	1	Higher alloy than 316 for better corrosion resistance.
	321	1	2	1	Titanium stabilized to prevent carbide precipitation.
	347	1	2	1	Columbian stabilized to prevent carbide precipitation.
Martensitic	403	2	2	2	Used for forged turbine blades.
	410	2	3	2	General purpose, low priced, heat treatable.
	414	2	3	2	Nickel added, for knife blades, springs.
	416	4	3	4	Free machining.
	420	2	3	2	Higher carbon for cutlery and surgical instruments.
	431	2	2	2	High mechanical properties.
	440A	5	3	4	For instruments, cutlery, valves.
	440B	5	3	4	Higher carbon than 440A.
	440C	5	3	4	Higher carbon than 440A or B for high hardness.
	501	2	3	2	Less resistance to corrosion than chromium nickel types.
	502	2	3	2	
Ferritic	405	2	3	2	Nonhardening when air cooled from high temperatures.
	406	2	3	2	For electrical resistances.
	430	2	3	2	Easily formed alloy, for automobile trim.
	430F	2	3	2	Free machining variety of 430 grade.
	446	2	3	2	High resistance to corrosion and scaling up to 2150F.

1—Readily weldable. 2—Weldable under favorable conditions. 3—Weldable by gas process in 20 ga. or thinner. 4—No welding information available. 5—Weldable with pre- and post-heat.

Source: Courtesy of Welding Data Book/Welding Design and Fabrication Magazine.

In addition to the stainless steels listed in Table 20-15, there are 22 alloys that are recognized as belonging to the stainless-steel group but lacking a standard terminology such as that provided by AISI. The situation is further complicated by the fact that various organizations have provided different names for the same alloy. Also, separate designations are in use for pipe, tubing, welding electrodes, and welding rods. For example, the Aeronautical Materials Specification (AMS) of the Society of Automotive Engineers covers materials for air frames, engines, and missiles. The various branches of the armed services have developed MIL specifications, elaborating both these and the original government QQ-S designations with departmental suffixes.

The stainless steels most commonly welded by standard welding processes are the chromium-nickel and straight-chromium grades.

Stainless steel types 301, 302, 303, and 304 are frequently used for architectural trim, dairy equipment (such as milking apparatus and homogenizing equipment), restaurant and fountain equipment, kitchen ware, screws, and rivets, as well as machine and/or forged parts.

Type 347 is used in corrosive atmospheres between 800 and 1500°F and for applications in exhaust manifolds, space heaters, flash boilers, pressure vessels, fire walls, boiler casings, and others.

Type 309 resists oxidation in temperatures to 2000°F and is used in applications such as boiler baffles, fire-box sheets, oven linings, still-table supports, kiln linings, and others.

Type 310 resists oxidation in temperatures up to 2100°F and is used for furnace linings, heat-treating furnace conveyors and supports, boiler baffles, furnace stacks and dampers, nitrate crystallizing pans, and others.

Type 314 is used extensively to manufacture parts exposed to elevated temperature-oxidation combinations with carburization; for example, boxes used for medium- and long-run pack carburizing of steel parts.

Types 403 and 410 are air-hardening and used for low-priced cutlery, pump parts, oil refinery equipment, coal washers, and turbine blades.

Type 405 is a nonhardening grade used in applications where the air-hardening types (403 and 410) are objectionable.

Type 414 is a hardenable type used to manufacture such items as strings, steel rules, and machine parts.

Type 416 is stainless-steel type 410 except that it was modified for screw machine work, while type 416 Se is similar to type 416 except that selenium has been added to make hot- or cold-forged parts.

Type 420 is similar to type 410 except that its carbon limits are higher. It is used for good-quality cutlery, surgical instruments, valves, and other wear-resistant parts. Type 420F is the free-cutting modification of type 420.

Type 430 is a general-purpose, nonhardenable chromium type used for decorating trim, nitric acid tanks, and annealing baskets. Types 430F and 430FSe are the free-cutting versions of type 430. Type 430FSe is used to make parts by hot or cold forging.

Type 431 is a special-purpose hardenable steel used where particularly high mechanical properties are required, for example, in aircraft fittings, heater bars, paper machinery, and bolts.

Type 440A is used for products requiring higher hardness than that obtainable from type 420. Type 440B is used for applications requiring higher hardenability than that which can be obtained from type 440A.

Type 440C yields the highest hardness of hardenable stainless steels. It is used for ball bearings and bearing races.

Types 500 and 501 are not true stainless steels although sometimes classified as such. They do have a corrosion resistance of four to ten times that of mild steel, and an oxidation resistance from three to eight times that of mild steel. They have considerable resistance to oxidation at temperatures in the range of 1100 to 1200°F. They are used extensively for hot-oil transfer lines and other oil refinery equipment operating at elevated temperatures, furnace tubes, heat exchangers, valves, and high-temperature steam lines.

All the stainless steels described are readily weldable with standard welding processes, but their preweld and postweld heat treatments differ. The austenitic (300 series) stainless steels, except for the free-machining grades, are more weldable than the ferritic martensitic stainless steels. Because the coefficient of thermal expansion of austenitic stainless steel is approximately 50% greater than that of carbon steel and its thermal conductivity is only one-third that of carbon steel, these steels may present some distortion problems unless welding practices for controlling distortion are implemented, such as placing copper chill bars under weld areas, to draw off the heat more quickly, or under welding fixtures to prevent movement of the base metals.

Austenitic stainless steels, when heated in the range 800–1500°F, undergo a migration of chromium that lowers its resistance to corrosion. This is due to the precipitation at grain boundaries of very fine films of chromium-rich carbides containing as much as 90% chromium. Since the chromium is taken from the layer of metal immediately adjacent to the grain boundary, the metal there may have its corrosion resistance seriously lowered. This phenomenon is called *carbide precipitation*, and the type of corrosion that is likely to occur is known as *intergranular corrosion*.

Because of carbide precipitation or the possibility of workpiece warpage or distortion, preheating is not recommended. *Stress relieving* to ensure dimensional stability may, however, be required. Stress relieving can be performed over a wide range of temperatures depending on the amount of relaxation required. Time at temperatures should be about one hour per inch of section thickness at temperatures below 1200°F. Because of the coefficient of expansion and the low thermal conductivity of these steels, they must be allowed to cool in the furnace.

The estimated percentages of residual stress relieved at various temperatures for the previously noted times are

1550 to 1650°F	85% stress relief
1000 to 1200°F	35% stress relief

The hardenable martensitic (400 series) stainless steels can be welded in the annealed, hardened, or hardened and tempered conditions. Regardless of the prior condition of the steel, welding will produce a hardened martensitic zone adjacent to the weld. The hardness of the heat-affected zone (HAZ) depends primarily on the carbon of the base metal. As hardness increases, toughness decreases, and the zone becomes more susceptible to cracking. Preheating and control of interpass temperatures are the most effective means of avoiding cracking. Postweld heat treatment is required to obtain best results.

The following guide can be used to correlate preheating and postweld heat treating with carbon contents and the welding characteristics of these steels.

- *Carbon below 0.10%*. Steels with carbon content this low are not standard. Neither preheating nor postheating is required.
- *Carbon 0.10 to 0.20%*. Preheat to 500°F, weld at this temperature, and cool at room temperature.
- *Carbon 0.20 to 0.50%*. Preheat to 500°F, weld at this temperature, and anneal.
- *Carbon over 0.50%*. Preheat to 500°F, weld with high heat input, and anneal after welding.

If the weldment is to be hardened and tempered immediately after welding, annealing may be omitted. Otherwise, the weldment should be annealed immediately after welding without being allowed to cool to room temperature. The temperatures for subcritical and full annealing of the martensitic steels are listed in Table 20-16.

The nonhardenable ferritic (400 series) stainless steels produce welded joints with lower impact resistance (notch toughness) because of the grain coarsening that occurs at the welding temperature.

The recommended preheating temperature range for ferritic stainless steels is 300 to 450°F. Because steels less than ¼-inch thick are less likely to crack, preheating of these thin sections is optional.

Table **20-16** Temperatures for subcritical and full annealing of martensitic stainless steels

AISI TYPE	SUBCRITICAL ANNEAL TEMPERATURE (°F)	FULL ANNEAL TEMPERATURE (°F)
403, 410, 416	1400	1600
420	1400	1650
414	1250	—
431	1200	—
440A, 440B, 440C	1400	1650

However, steels over ¼-inch thick should be preheated. Although postweld annealing in the range of 1450 to 1550°F is recommended, the process has three major disadvantages:

1. The long time required for the procedure.

2. The need to prevent the formation of scale during heat treatment, or the removal of scale that does form.

3. The need to use fixture to prevent sagging or distortion of the weldment.

Ferritic stainless steels must not be allowed to cool below 1100°F in the furnace, because between 1050 and 750°F these steels are susceptible to what is known as *885°F embrittlement.* The steels may, however, be cooled from 1100°F to room temperature in air or by quenching with water.

REVIEW QUESTIONS

1. By what other names are carbon steels also known?

2. What can happen when cast iron is rapidly cooled after welding? Explain.

3. In what way is manganese similar to carbon when used as an alloying element in steel?

4. Why are rephosporized steels considered difficult to weld?

5. Describe the procedure for minimizing porosity in a weld.

6. Name some applications for such steels as ASTM A36 and ASTM A441.

7. Describe the welding technique to be used to minimize cracking in sections of mild steel thicker than ⅜ inch.

8. To what temperature should sections of mild steel showing a tendency to crack be preheated?

9. What are the carbon limits of medium-carbon steel? How would you determine whether the steel has a tendency to crack?

10. Name three problems that can arise during the welding of high-carbon steels. What procedure(s) would you follow to minimize the occurrence of these problems?

11. What are the recommended preheating temperatures for welding a 3/8-inch-thick section of A441 steel with and without the use of low-hydrogen electrodes and the following welding procedures: SMAW, GMAW Flux Cored–Arc.

12. What steel would most likely be used for the following applications?

a. Pressure vessels.
b. Buildings.
c. TV transmission towers.
d. Plowshares.
e. Hydrospace vehicle hulls.
f. Cams.

13. List seven major AISI tool and die steel groups.

14. To what temperature and in what type of furnace atmosphere should shock-resisting and high-speed steel be heated prior to welding?

15. What is the tensile strength and Bhn of a steel having a Rockwell hardness of C 25?

16. Describe a step-by-step procedure by which you would test a piece of steel in the shop (without the use of a special equipment) to determine whether it is quarter hard.

17. What should be done if, after testing, it is found that a 1/2-inch-thick piece of SAE 1035 steel developed internal cracks in the weld during welding?

18. Based on carbon content, the AWS classes steels into four groups. Into which of the four groups would a piece of low-carbon steel fall?

19. List four problems that can arise during the welding of high-carbon steels.

20. The AWS classes low-alloy steels into four groups. A pressure vessel that must withstand sudden impacts exceeding 60,000 psi would be most likely made of steel belonging to which group?

21. Name the three steels that are commonly used in the construction of supersonic-aircraft frames.

22. To what temperature should a piece of 2-inch-thick SAE 3140 steel be preheated? How long should it be kept at the preheat temperature to ensure that the part has been heated all the way through?

23. Name the SAE–AISI type tool steel used to make a blade for a sheet-metal shear.

24. In what heat-treated condition should tool and die steels be welded?

25. Why should tool and die steels not be welded with the oxyacetylene process?

26. Which of the cast irons is considered to be unweldable?

27. What is the minimum yield strength of a class 60-45-10 nodular cast iron?

28. What is a likely cause of intergranular corrosion in stainless steel?

29. A broken stainless steel knife is to be repaired. Can the knife be welded without first being annealed? Of what type of stainless steel would the knife most probably be made?

CHAPTER 21

NONFERROUS METALS: CLASSIFICATION, APPLICATION, AND WELDABILITY

The commonly welded nonferrous metals are aluminum and aluminum alloys, copper and copper alloys, nickel and high-nickel alloys, magnesium and magnesium alloys, lead, zinc, titanium and titanium alloys, and reactive, refractory, and precious metals. Welding procedures for all these metals are described in the following sections.

21-1 ALUMINUM AND ALUMINUM ALLOYS

There are two classes of aluminum alloys: cast and wrought. Within the two classes, specific aluminum alloys are identified by a numeric code. This code is shown in Table 21-1. In addition to the numeric designation, wrought alloys are also identified by their temper. The designations used to identify the temper of aluminum alloys are shown in Table 21-2. The alpha-numeric code to designate temper follows the alloy designation with a dash, such as the T6 in 6061-T6. The designation 6061-T6 means that the major alloying elements (Table 21-1) of the alloy are magnesium and silicon and that the alloy has been solution heat treated by artificial aging.

The weldability ratings of weldable, commercially available aluminum alloys are listed in Table 21-3. Of the high-strength, heat-treatable aluminum alloys (7000 series), alloy 7039, though not listed in the table and not quite as strong as alloy 7075, is the all-around high-strength alloy because it is readily weldable by any of

Table **21-1** Aluminum alloy designations[a]

DESIGNATIONS FOR ALLOY GROUPS		
Aluminum	99.00% minimum and greater	1XXX
Aluminum alloys grouped by major alloying elements	Copper	2XXX
	Manganese	3XXX
	Silicon	4XXX
	Magnesium	5XXX
	Magnesium and silicon	6XXX
	Zinc	7XXX
	Other element	8XXX
Unused series		9XXX

[a]The aluminum alloy designation system was adopted by the Aluminum Association on October 1, 1954. Under this system aluminum and its alloys are designated by a four-digit number. The first digit indicates the alloy group. The last two digits identify the alloy or the aluminum purity. The second digit indicates modifications of the original alloy except that for aluminum of greater than 99% purity, a zero indicates no control of impurities and an integer indicates specific control.

Table **21-2** Aluminum temper designations[a]

-F As fabricated.
-O Annealed and recrystallized (wrought only).
-H Strain hardened (wrought only).
- H1 Strain hardened only. The number following this designation indicates the degree of strain hardening.
  - H111 Strain hardened less than the amount required for a controlled H11 temper.
  - H112 Acquires some temper from shaping processes not having special control over the amount of strain hardening or thermal treatment, but for which there are mechanical property limits or required mechanical property testing.
- H2 Strain hardened and then partially annealed. The number following this designation indicates the degree of strain hardening remaining after the product has been partially annealed.
- H3 Strain hardened and then stabilized. The number following this designation indicates the degree of strain hardening remaining after the product has been strain hardened a specific amount and then stabilized.

Table **21-2** (*cont.*)

H311 Strain hardened less than the amount required for a controlled H31 temper.

H321 Strain hardened less than the amount required for a controlled H32 temper.

-T Thermally treated to produce stable tempers other than F, O or H.

-T1, Partially solution heat treated and naturally aged to a substantial stable condition.

-T2, Annealed (cast products only).

-T3, Solution heat treated, then cold worked.

-T4, Solution heat treated and naturally aged to a substantially stable condition.

-T5, Partially solution heat treated, then artificially aged.

-T6, Solution heat treated, then artificially aged.

-T7, Solution heat treated, then stabilized.

-T8, Solution heat treated, cold worked, then artificially aged.

-T9, Solution heat treated, artificially aged, then cold worked.

-T10, Partially solution heat treated, artificially aged, then cold worked.

[a]The temper designation follows and is separated from the alloy designation by a dash, e.g. 6061-T6. It indicates mechanical or thermal treatment. Additional digits to the right indicate variations in basic treatment.

Table **21-3** Weldability ratings of wrought-aluminum alloys

ALLOY	GAS	ARC WITH INERT GAS	RESISTANCE WELDING	PRESSURE WELDING	BRAZING	SOLDERING WITH FLUX *LOW TEMP.*	SOLDERING WITH FLUX *HIGH TEMP.*	ULTIMATE TENSILE STR. (PSI)	YIELD STRENGTH TENSION (PSI)
1060-O	A	A	B	A	A	A	A	10,000	4,000
1060-H14	A	A	A	A	A	A	A	14,000	13,000
EC-O	A	A	B	A	A	A	A	12,000	4,000
EC-H19	A	A	A	A	A	A	A	27,000	24,000
1100-O	A	A	B	A	A	A	A	13,000	5,000
1100-H18	A	A	A	A	A	A	A	24,000	22,000
3003-O	A	A	B	A	A	A	A	16,000	6,000
3003-H18	A	A	A	A	A	A	A	29,000	27,000
Alclad 3003-O	A	A	B	A	A	A	A	16,000	6,000
Alclad 3003-H18	A	A	A	A	A	A	A	29,000	27,000
3004-O	A	A	B	A	B	B	A	26,000	10,000
3004-H38	A	A	A	B	B	B	A	41,000	36,000

Table **21-3** (*cont.*)

ALLOY	GAS	ARC WITH INERT GAS	RESISTANCE WELDING	PRESSURE WELDING	BRAZING	SOLDERING WITH FLUX LOW TEMP.	SOLDERING WITH FLUX HIGH TEMP.	ULTIMATE TENSILE STR. (PSI)	YIELD STRENGTH TENSION (PSI)
2011-T3	D	D	D	D	D	D	D	55,000	43,000
2011-T8	D	D	D	D	D	D	D	59,000	45,000
2014-T4	D	C	B	C	D	D	D	62,000	42,000
2014-T6	D	C	B	D	D	D	D	70,000	60,000
Alclad 2014-T3	D	B	A	C	D	B	C	63,000	40,000
Alclad 2014-T6	D	B	A	C	D	B	C	68,000	60,000
2017-T4	D	C	B	D	D	D	D	62,000	40,000
2018-T61	D	C	B	D	D	D	D	61,000	46,000
2024-T3	D	C	B	C	D	D	D	70,000	50,000
2024-T36	D	C	B	C	D	D	D	72,000	57,000
Alclad 2024-T3	D	B	A	C	D	B	C	65,000	45,000
Alclad 2024-T36	D	B	A	C	D	B	C	67,000	53,000
2219-T31	D	A	A	C	D	D	D	50,000	37,000
2219-T81	D	A	A	C	D	D	D	70,000	53,000
4032-T6	D	B	C	C	D	D	D	55,000	46,000
5005-O	A	A	B	A	B	B	A	18,000	6,000
5005-H38	A	A	A	B	B	B	A	29,000	27,000
5050-O	A	A	B	A	B	C	B	21,000	8,000
5050-H38	A	A	A	B	B	C	B	32,000	29,000
5052-O	A	A	B	A	C	C	C	28,000	13,000
5052-H38	A	A	A	B	C	C	C	42,000	37,000
5056-O	C	A	B	B	D	D	D	42,000	22,000
5056-H38	C	A	A	C	D	D	D	60,000	50,000
5083-O	C	A	B	B	D	D	D	42,000	21,000
5083-H113	C	A	A	C	D	D	D	46,000	33,000
5086-O	C	A	B	B	D	D	D	38,000	17,000
5086-H34	C	A	A	C	D	D	D	47,000	37,000
5154-O	C	A	B	A	D	D	D	35,000	17,000
5154-H34	C	A	A	B	D	D	D	42,000	33,000
5154-H38	C	A	A	B	D	D	D	48,000	39,000
5454-O	C	A	B	B	D	D	D	36,000	17,000
5454-H34	C	A	A	B	D	D	D	44,000	35,000
5456-O	C	A	B	B	D	D	D	45,000	23,000
5456-H321	C	A	A	C	D	D	D	51,000	37,000
6053-T4	A	A	A	B	A	B	B	30,000	20,000
6053-T6	A	A	A	B	A	B	B	37,000	32,000
6061-T4	A	A	A	B	A	B	B	35,000	21,000
6061-T6	A	A	A	B	A	B	B	45,000	40,000
6062-T4	A	A	A	B	A	B	B	35,000	21,000

Table **21-3** (*cont.*)

ALLOY	GAS	ARC WITH INERT GAS	RESISTANCE WELDING	PRESSURE WELDING	BRAZING	SOLDERING WITH FLUX *LOW TEMP.*	*HIGH TEMP.*	ULTIMATE TENSILE STR. (PSI)	YIELD STRENGTH TENSION (PSI)
6062-T6	A	A	A	B	A	B	B	45,000	40,000
6063-T5	A	A	A	B	A	B	B	27,000	21,000
6070-T4	D	A	A	B	B	A	A	49,000	30,000
6070-T6	D	A	A	B	B	A	A	57,000	52,000
6071-T4	D	A	A	B	B	A	A	49,000	30,000
6071-T6	D	A	A	B	B	A	A	57,000	52,000
6151-T6	A	A	A	B	B	B	B	48,000	43,000
6951-O	A	A	B	A	A	B	A	16,000	7,000
6951-T6	A	A	A	B	A	B	A	—	—
X7006-T63	D	A	B	D	C	D	C	61,000	55,000
7075-T6	D	C	B	D	D	D	D	83,000	73,000
Alclad 7075-T6	D	C	B	D	D	B	C	76,000	67,000

Note: Ratings, A, B, C and D are defined:
A. Generally weldable by usual methods.
B. Weldable with special technique or on specific applications. Develop welding procedure and weld performance.
C. Limited weldability because of crack sensitivity or loss in resistance to corrosion and mechanical properties.
D. Welding methods have not been developed.

Source: Welding Data Book/Welding Design and Fabrication Magazine.

the fusion processes. In the T61 temper, this alloy, in thicknesses of $3/16$ inch or more, is suitable for welded pressure vessels for use at from −200 to −300°F under *ASME Boiler and Pressure Vessel Code*, Section VII. Alloy 7039 is also used for armor plate under the military MIL-A-46063 rating.

Preweld cleaning of aluminum is essential for optimum weld quality. Precleaning requirements are especially stringent prior to straight-polarity dc gas tungsten–arc welding, because under such conditions the arc exerts no cleaning action. However, the highest quality welds are not always needed. Where service requirements permit, many aluminum parts are welded with no preweld cleaning at all.

Surface contaminants that should be removed from the base metal include dirt, metal particles, oil and grease, paint, moisture, and heavy oxide coatings. Another source of contamination is oxide film on the filler metal. Base metals such as 1100 and 3003 have a relatively thin oxide coating as fabricated, and the 5000 and 6000 series alloys generally have a thick, dark oxide coating. The thicker the oxide, the greater its adverse effect on the weld-metal flow and solidification, and the greater the risk of porosity. For best results, all cleaning and oxide removal should be done within 24 hours for resistance welding and within two or three days for fusion welding, providing the surface to be welded is wire-brushed immediately before welding. For wire-brushing, stainless steel bristles are preferred to carbon-steel bristles because the iron deposits left by the carbon steel may rust in the presence of moisture. Also, large amounts of iron oxide may result in weld-metal inclusions. Carbon-steel bristles are satisfactory if they are grease-free and both they and the work are kept dry. A summary of the general cleaning procedures is given in Table 21-4.

The joints most commonly used to weld aluminum are shown in Figure 21-1 the applicable welding conditions are given in Table 21-5. To obtain a joint strength equal to about 20% of the base metal, lap joints and T joints should be made by using a small, continuous fillet weld on each side of the joint (Figure 21-2).

Square edges for butt joints in materials up to ⅜ inch thick can be made by shearing. However, the shear blades, as well as the aluminum to be sheared, should be cleaned of excessive oil and/or grease prior to shearing. Sections thicker than ⅜ inch can be prepared by lubricating the section to be cut with stick wax and degreasing the sawed-off ends afterwards.

V-groove and other-shaped butt-joint edges can be produced by machining with cutting tools specifically designed for machining aluminum. Lubricants are generally not required for preparing butt joints by methods other than shearing and band sawing.

Usually only heat-treatable aluminum alloys are preheated; however, when moisture is present in the area of the joint, non-heat-treatable alloys are also preheated to eliminate the moisture.

Preheating of parts to be joined to temperatures up to 400°F is usually sufficient. Because of the relatively high thermal-expansion coefficient of aluminum alloys, the use of properly designed welding jigs and fixtures is extremely important to minimize the amount of distortion resulting from the heat put into the material during welding and its subsequent cooling. It should be remembered that molten aluminum will shrink about three times as much as a similar volume of steel. Sometimes, when it is not possible to use heat sinks or chill bars, other means of cooling (such as dry ice) should be used to isolate the weld zone and minimize distortion.

To obtain good penetration with the GMAW (MIG) welding process, only reverse-polarity direct current should be used. The MIG process can be used to make welds meeting the requirements of the *ASME Boiler and Pressure Vessel Code* in sections ranging from ⅛ to several inches in thickness. Satisfactory welds in sections less than ⅛ inch thick can be made by using pulsed-current power

Table **21-4** Common methods[a] for cleaning aluminum surfaces for welding

TYPE OF SOLUTION	CONCENTRATION	TEMPERATURE	TYPE OF CONTAINER	PROCEDURE	PURPOSE
Technical grade nitric acid	50% water 50% nitric acid, technical grade	Room	Stainless steel-347	Immersion 15 min. Rinse in cold water. Rinse in hot water & dry	For removing thin oxide film for fusion welding.
1. Sodium hydroxide (caustic soda) followed by	1. 5%	1. 160°F	1. Mild steel	1. Immersion 10–60 seconds. Rinse in cold water.	Removes thick oxide film for all welding and brazing processes.
2. Technical grade nitric acid	2. Concentrated (use as received)	2. Room	2. Stainless steel-347	2. Immerse for 30 seconds. Rinse in cold water. Rinse in hot water & dry.	
Sulfuric-chromic	H_2SO—1 gal. CrO_3—45 oz. water—9 gal.	160–180°F	Antimonal lead-lined steel tank	Dip for 2–3 min. Rinse in cold water. Rinse in hot water & dry.	For removal of heat-treatment & annealing films and stains & for stripping oxide coatings.
Phosphoric-chromic	H_3PO_3(75%) 3.5 gal. CrO_4—1.75 lbs. water—10 gal.	200°F	Stainless steel-347	Dip for 5–10 min. Rinse in cold water. Rinse in hot water & dry.	For removing anodic coatings.

[a]There are many proprietary materials and methods available for removing aluminum oxides. Most of these are as efficient as the preparations listed.

Source: American Welding Society, *Welding Handbook*, 6th. ed., section 4, p. 69.77.

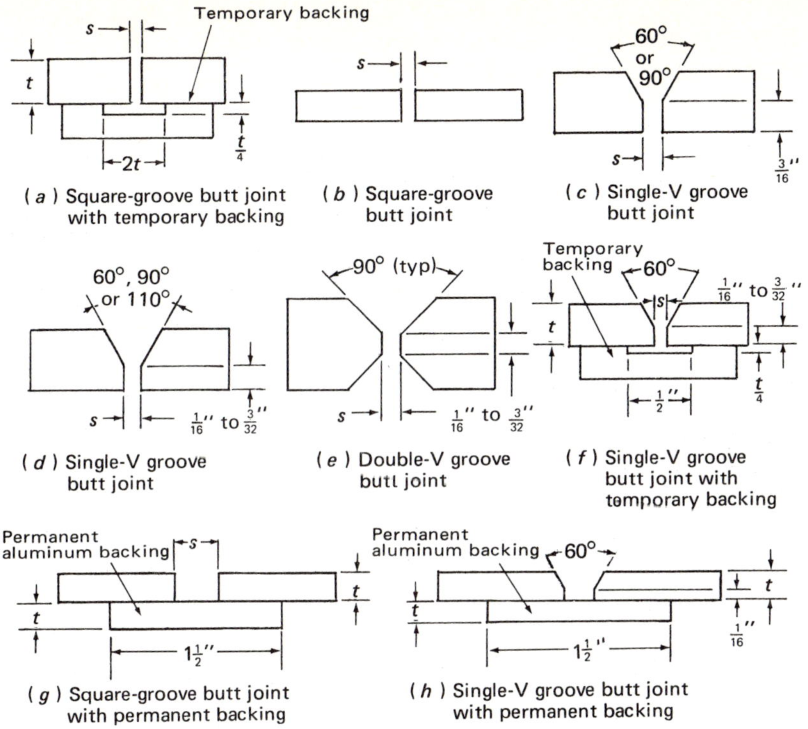
Temporary backing
s
t
2t
t/4
(a) Square-groove butt joint with temporary backing
s
(b) Square-groove butt joint
60° or 90°
s
3/16"
(c) Single-V groove butt joint
60°, 90° or 110°
s
1/16" to 3/32"
(d) Single-V groove butt joint
90° (typ)
s
1/16" to 3/32"
(e) Double-V groove butt joint
Temporary backing
60°
s
1/16" to 3/32"
t
t/4
1/2"
(f) Single-V groove butt joint with temporary backing
Permanent aluminum backing
s
t
1 1/2"
(g) Square-groove butt joint with permanent backing
Permanent aluminum backing
60°
t
1/16"
1 1/2"
(h) Single-V groove butt joint with permanent backing

Figure **21-1** Joints most commonly used in aluminum welding. (By permission, from *Metals Handbook* Volume 6, Copyright American Society for Metals, 1971.)

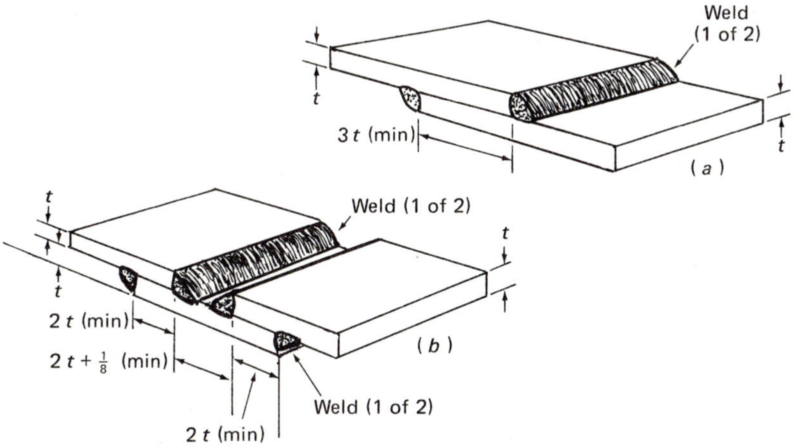
Weld (1 of 2)
t
3 t (min)
t
(a)
t
Weld (1 of 2)
t
t
2 t (min)
2 t + 1/8 (min)
(b)
Weld (1 of 2)
2 t (min)

Figure **21-2** (*a*) Lap joint designed for maximum strength. (*b*) T joint designed for strength. (By permission, from *Metals Handbook* Volume 6, Copyright American Society for Metals, 1971.)

Table **21-5** Recommended conditions for the gas metal—arc welding (dcrp) and gas tungsten—arc welding (ac) of aluminum alloys. Joints *b, c, d, e,* and *f* should be back gouged to solid weld metal before applying a pass on the root side (refer to Figure 21-1)

	SEMIAUTOMATIC GAS METAL–ARC WELDING			MANUAL GAS TUNGSTEN–ARC WELDING		
METAL THICKNESS, *t* (in.)	WELDING POSITION[a]	JOINT DESIGN[b]	ROOT OPENING, s, (in.)	WELDING POSITION[a]	JOINT DESIGN	ROOT OPENING, s (in.)
1/16	F	*a, g*	0–3/32	F,H,V,O	2	0
3/32	F	*a*	0	F,H,V	2	0
	F, H,V,O	*g*	1/8	O	2	0
1/8	F,H,V	*a*	0–3/32	F	2	0
	F,H,V,O	*g*	3/16	H,V,O	2	0
3/16	F,H,V	*b*	0	F	4–60°	0–1/8
	F,H,V,O	*f*	0–1/16	H	4–90°	0–3/32
	F,V	*h*	3/32–3/16	V	4–60°	0–3/32
	H,O	*h*	3/16	O	4–110°	0–3/32
1/4	F	*b*	0	F	4–60°	0–1/8
	F,H,V,O	*f*	0–3/32	H	4–90°	0–3/32
	F,V	*h*	1/8–1/4	V	4–60°	0–3/32
	H,O	*h*	1/4	O	4–110°	0–3/32
3/8	F	*c*–90°	0–3/32	F	4–60°	0–1/8
	F	*f*	0–3/32	F	5	0–3/32
	H,V,O	*f*	0–3/32	V	4–60°	0–3/32
	F,V	*h*	1/4–3/8	H,V,O	5	0–3/32
	H	*h*	3/8	H	4–90°	0–3/32
	O	*h*	3/8	O	4–110°	0–3/32
3/4	F	*c*–60°	0–3/32	—	—	—
	F	*h*	0–1/8	—	—	—
	H,V,O	*h*	0–1/16	—	—	—
	F,H,V,O	*h*	0–1/16	—	—	—

[a]F = flat; H = horizontal; V = vertical; O = overhead. [b]For design 8, $t_1 = t$ for t less than 3/8 in., and t_1 = 3/8 in. for t greater than 3/8 in.

Source: By permission, from *Metals Handbook* Volume 6, Copyright American Society for Metals, 1971.

supplies. Backing bars of either the permanent or temporary type must also be used with this process.

Carbon steel is the most commonly used backing-bar material, though stainless steel is used when lower thermal conductivity is required in the backing. Temporary backing bars (Figure 21-1) should be grooved.

The manual GTAW (TIG) process can be used to weld sections ranging in thickness from 0.040 to 3/8 inch, and the automatic TIG process can be used to weld sections ranging from 0.010 to 1 inch thick. Either ac or dc within the direct- or reverse-polarity modes can be used.

Although backing bars are commonly used, they are not necessary when the manual TIG process is used to make butt welds from one side only. They are also not necessary in automatic straight-polarity dc welding of square-groove butt joints from one side only. Typical conditions for semiautomatic welding of aluminum with the GMAW process are given in Table 21-6.

In addition to the GMAW and GTAW processes, aluminum alloys are sometimes welded by the shielded metal–arc, stud, percussion, and oxyfuel welding processes.

Shielded metal–arc welding is used primarily in small shops for miscellaneous repair work in noncritical applications. A flux-covered aluminum-alloy electrode is used. The flux combines with aluminum oxide to form a slag, which must be removed after each pass. Weld soundness and surface smoothness are poor. The process is limited to butt welds in ⅛-inch and thicker aluminum. AWS A5.3-69 includes two covered electrodes: one wire-core wire corresponding to ER1100 and one to ER4043.

Stud welding of aluminum alloys is generally done by the capacitor-discharge method, rather than the arc method.

Percussion welding of aluminum alloys is used principally for joining wires. This method is sucessfully used to weld numerous dissimilar metals, including aluminum to copper, aluminum to steel, and copper to steel.

Although industrially obsolete, the oxyfuel-gas welding process is still employed in some furniture and decorative product applications where smoothness and appearance of the joint are important, or in locations with no access to electric power. Sections ranging in thickness from $\frac{1}{32}$ to 1 inch can be welded by this process. Although oxybutane, oxypropylene, or oxy–natural gas can be used, they are—because of their low heat value—used mainly for welding sections ranging in thickness from $\frac{1}{32}$ to $\frac{1}{16}$ inch, and for preheating the thicker sections that are welded with an oxyacetylene flame. If the oxyfuel-gas method is used, the flame should be neutral or slightly reducing. A flux must also be used.

Initially the flame should be held perpendicular to the work and moved in a circular fashion to preheat the adjacent metal. After uniformly preheating both edges of the joint, the flame is held over the spot where the weld is to begin. If the flame is held over one spot for a moment or so, the operator will see a small blister form on the metal surface. At this point, the end of the filler rod should be placed in the small blister or puddle.

After welding has begun, the angle of the torch should be decreased by pointing the flame in the direction of travel. This

Table **21-6** Typical conditions for semiautomatic welding of aluminum alloys

		ELECTRODE WIRE			ARGON		ARC			WELDING	
METAL THICKNESS (in.)	WELDING POSITION[a]	DIAMETER (in.)	FEED (ipm)	USED PER 100 FT. LB	FLOW RATE, CFH	USED PER 100 FT. CU FT	CURRENT (dcrp), AMP	VOLTAGE, (v)	TIME PER 100 FT HR[b]	SPEED (ipm)	NUMBER OF PASSES
BUTT JOINTS											
1/8	F	3/64	175	2	30	34	110	20	1.14	24	1
	H, V	3/64	170	2	30	35	100	20	1.17	24	1
	O	3/64	170	2½	40	58	105	20	1.46	24	1
3/16	F	3/64	235	4½	30	57	170	20	1.90	24	1
	H, V	3/64	215	4½	35	75	150	20	2.08	20	1
	O	3/64	225	5	40	90	160	20	2.21	18	1
1/4	F	1/16	170	8	40	105	200	25	2.63	24	1
	H, V	1/16	150	8	45	255	170	25	2.98	24	3
	O	1/16	160	10	50	175	180	29	3.49	24	3
3/8	F	1/16	265	18	50	190	290	25	3.80	24	2
	F	1/16	250	15	50	170	275	29	3.35	24	2
	H, V	1/16	160	18	50	315	190	25	6.29	24	2
	H, V	1/16	150	15	50	280	170	29	5.29	24	2
	O	1/16	170	23	50	380	200	25–29	7.56	24	5
1/2	F	3/32	130	31	50	295	290	25–31	5.87	16	2
	F	3/32	140	30	50	265	320	25–31	5.27	16	2
	F	3/32	130	29	50	275	300	25–31	5.49	16	3
	H, V	1/16	—	31	50	—	215	25–29	—	12	3
	H, V	1/16	160	29	50	505	190	25–29	10.13	12	2
	O	1/16	200	31	80	695	225	25–29	8.66	18	8
3/4	F	3/32	150	62	60	610	350	25–29	10.17	16	4
	F	3/32	145	72	60	735	330	25–29	12.22	16	4
	H, V	1/16	225	62	60	925	250	25–29	15.47	8	4
	H, V	1/16	215	72	60	1125	240	25–29	18.71	8	4
	O	1/16	225	62	80	1235	250	25–29	15.42	18	12
1	F	3/32	170	105	60	910	400	25–31	15.20	12	4
	F	3/32	165	85	60	760	380	25–31	12.68	12	6
	H, V	1/16	225	105	60	1565	250	25–29	26.11	6	4
	H, V	1/16	215	95	60	1480	240	25–29	24.69	6	6
	O	1/16	250	105	80	1880	275	25–29	23.48	18	15

1½	F	3/32	180	200	80	2185	425	25–31	27.33	12	10
	H	1/16	255	105	80	1840	280	25–31	23.02	24	24
	H	3/32	150	105	80	1380	350	25–31	17.22	24	14
	V	1/16	225	200	80	3980	260	25–31	49.73	24	20
	V	3/32	145	200	80	2715	330	25–31	33.93	24	12
2	F	3/32	180	335	80	3665	425	25–31	45.79	12	12
	H	1/16	215	185	80	3845	245	25–31	48.08	24	30
	H	3/32	170	185	80	2140	425	25–31	26.78	24	24
	V	1/16	215	300	80	6240	240	25–31	77.97	20	26
	V	3/32	150	300	80	3935	350	25–31	49.19	20	15
2½	F	3/32	180	350	80	3825	425	25–31	47.83	12	14
	H	1/16	215	270	80	5615	245	25–31	70.18	24	32
	H	3/32	150	270	80	3540	350	25–31	44.27	24	26
3	F	3/32	190	500	80	5180	450	25–31	64.77	20	30

T AND LAP JOINTS

1/8	F	3/64	190	2	30	31	125	20	1.04	30	1
	H, V	3/64	180	2	30	33	115	20	1.10	24	1
	O	3/64	175	2	40	45	110	20	1.13	24	1
3/16	F	3/64	255	4½	30	55	190	20	1.75	24	1
	H, V	3/64	230	4½	35	70	165	20	1.94	20	1
	O	3/64	245	4½	40	75	180	20	1.82	20	1
1/4	F	1/16	195	7	40	80	225	25 to 29	2.01	24	1
	H, V	1/16	170	7	45	105	200	25 to 29	2.30	20	1
	O	1/16	170	7	50	115	200	25 to 29	2.30	20	1
3/8	F	1/16	275	17	50	175	300	25 to 29	3.45	30	3
	H, V	1/16	170	17	50	280	200	25 to 29	5.59	24	3
	O	1/16	195	17	60	290	220	25 to 29	4.87	24	3
1/2	F	3/32	145	30	60	305	340	25 to 31	5.09	16	3
	H, V	1/16	200	30	60	505	225	25 to 29	8.39	12	3
	O	1/16	205	30	60	655	230	25 to 29	8.18	18	5
3/4	F	3/32	160	66	60	610	375	25 to 31	10.15	16	4
	H, V	1/16	235	66	60	945	260	25 to 29	15.71	8	4
	O	1/16	250	66	80	1180	275	25 to 29	14.76	18	10
1	F	3/32	180	120	60	985	425	25 to 31	16.40	8	4
	H,V	1/16	235	120	80	1715	260	25 to 29	28.56	6	6
	O	1/16	265	120	80	2025	290	25 to 29	25.31	18	14

Table **21-6** (*cont.*)

		ELECTRODE WIRE		ARGON	ARC		WELDING	
METAL THICKNESS (in.)	WELDING POSITION[a]	DIAMETER (in.)	USED PER 100 FT. LB	FLOW RATE, CFH	CURRENT (dcrp), AMP	VOLTAGE, (v)	SPEED (ipm)	NUMBER OF PASSES
				CORNER JOINTS				
1/8	F	3/64	2	30	110	20	30	1
	H, V	3/64	2	30	100	20	24	1
	O	3/64	2	40	100	20	24	1
3/16	F	3/64	4½	30	170	20	30	1
	H, V	3/64	4½	35	150	20	24	1
	O	3/64	4½	40	160	20	24	1
1/4	F	1/16	7	40	200	25 to 29	30	1
	H, V	1/16	7	45	170	25 to 29	24	1
	O	1/16	7	50	180	25 to 29	24	1
3/8	F	1/16	17	50	250	25 to 29	30	3
	H, V	1/16	17	50	170	25 to 29	24	3
	O	1/16	17	60	180	25 to 29	24	3
1/2	F	3/32	30	50	290	25 to 31	16	3
	H, V	1/16	30	50	190	25 to 29	12	3
	O	1/16	30	70	200	25 to 29	18	5
3/4	F	3/32	66	60	310	25 to 31	16	4
	H, V	1/16	66	60	220	25 to 29	8	4

EDGE JOINTS

1/8	F	3/64	4	30	110	20	30	1
	H, V	3/64	4	30	100	20	24	1
	O	3/64	4	40	100	20	24	1
3/16	F	3/64	8	30	170	20	30	1
	H, V	3/64	8	35	150	20	24	1
	O	3/64	8	40	160	20	24	1
1/4	F	1/16	15	40	200	25 to 29	30	1
	H, V	1/16	15	45	170	25 to 29	24	1
	O	1/16	15	50	180	25 to 29	24	1
3/8	F	1/16	34	50	250	25 to 29	30	3
	H, V	1/16	34	50	170	25 to 29	24	3
	O	1/16	34	60	180	25 to 29	24	3
1/2	F	3/32	60	50	290	25 to 31	16	3
	H, V	1/16	60	50	190	25 to 29	12	3
	O	1/16	60	70	200	25 to 29	18	5

[a]F = flat; H = horizontal; V = vertical; O = overhead. [b]Based on 100% arc efficiency at electrode-wire feed shown.

Source: By permission, from *Metals Handbook* Volume 6, Copyright American Society for Metals, 1971.

preheats the edges just ahead of the puddle as the torch is moved along the joint. The angle is decreased more for thin than for thick material, because welding progresses faster on thin material.

As welding progresses, the tip of the filler rod is moved in and out of the puddle at a 30- to 45-degree angle. The forehand technique is used whenever possible. The center of the flame should never touch the molten weld metal, but should be kept from $\frac{1}{16}$ to $\frac{1}{4}$ inch away at all times.

Also, as welding progresses across the joint, it helps to use the torch in an oscillating motion to melt both sides of the weld joint simultaneously. The oscillating frequency depends on the gage of the material and somewhat upon the size of the tip and flame adjustment.

The weld defects most commonly present in aluminum welds include: crater cracks, longitudinal cracks, incomplete fusion, incomplete penetration, porosity, and undercutting. Each one of these faults, along with its most probable cause and possible means of correction, is discussed in the following paragraphs.

Crater cracks are small cracks or crow's-foot-type defects that occur during solidification. These cracks may be small but are very serious, since they usually occur at the end of a weld where stress concentration or *end effect* is most pronounced.

Crater cracks can be minimized by breaking or restarting the weld several times so that the pipe formed in the crater may be flued. Runout tabs are often used to prevent crater cracks from occurring. Great care must be taken to determine if crater cracks are present. If found, they should be chipped out and removed before welding. The presence of crater cracks can be determined by x-ray, ultrasonic, guided-bend, and/or metallographic tests.

Longitudinal cracks can sometimes occur when the metal is passing between the liquidus and solidus temperatures. These cracks usually occur from the use of incorrect filler metal, incorrect welding speed, incorrect edge preparation, or improper joint spacing.

Incomplete penetration is usually caused by too high a welding current, not enough edge preparation or joint spacing, or too high a welding-travel speed for the heat input employed.

Incomplete fusion occurs when the refractory aluminum-oxide film on the surface is not completely removed by good cleaning procedures prior to welding.

Porosity is due to entrapped hydrogen. It occurs most frequently in GMAW because of the high solidification characteristic of this process.

Undercutting is a most serious defect because it reduces the cross-sectional area of the weld zone and thereby its load-carrying capacity. It occurs most frequently when improper welding techniques or procedures are used.

Inclusions may be of two types, metallic and nonmetallic. In GTAW, the use of excessive current for a given electrode size will cause deposition of tungsten in the weld. Improperly adjusted high frequency and rectification of the alternating current may also cause this defect. Tungsten inclusions can also occur when the weld is started with a cold electrode. It should be noted that fine, widely scattered particles of tungsten have little effect on the properties of a GTA weld; however, many codes require that welds be essentially free of this defect. Improper use of wire brushes or abrasive wheels, improper removal of oxides, or improper flux-shielded metal–arc welding procedures can also cause the deposit of inclusions in the weld. The procedures usually employed to detect these defects are described in Chapter 22.

21-2 COPPER AND COPPER ALLOYS

As with aluminum, copper alloys are available in the cast or wrought forms. Some of the typical applications for the various copper alloys are given in Table 21-7. The two forms consist of about 250 different alloys and have been arranged into alloy families.

Each alloy family and each alloy within a family is referred to by a three-digit numerical code (Table 21-8). The designation system was developed by the copper and brass industry in the United States and is now used by the U.S. government, the American Society for Testing and Materials (ASTM), the Society of Automotive Engineers (SAE), and nearly all producers of copper and copper alloys in North America. The designation system is administered by the Copper Development Association (CDA).

Since the system is of relatively recent origin, many of the alloys—especially flat products and rod, wire, tube, pipe, and structural shapes that conform to specifications of the ASTM—are still known by their original trade names. When this is the case, the trade name is usually followed by numerals and/or letters. For example, alloy no. 101 is also known by the designation *OF E,* which means that it is an oxygen-free electronic copper alloy. Alloy no. 230 was previously (and may still be) known under the trade name *red brass, 85%*. This means that the alloy contains 85% copper.

Acceptable representative parameters for welding the various kinds of copper and copper alloys are given in Figure 21-3 and Tables 21-9 and 21-10.

In addition to observing the given parameters, certain manipulative techniques must be followed to obtain good welds. The manipulative techniques are discussed on a process-by-process basis beginning with the coated (shielded) metal–arc welding process.

Warning: Inhalation of copper fumes in amounts exceeding 0.1 mg/m^3 (threshold limit) will produce a severe condition of chills,

Table **21-7** Typical applications for coppers and copper alloys

CLASSIFICATION	TYPICAL APPLICATIONS
	WROUGHT ALLOYS
Oxygen-free copper	Bus bars and conductors, wave guides, applications requiring welding or brazing
Tough-pitch coppers	Roofing, automotive radiators, bus bars, wire, switches, kettles, printing rolls
Phosphorus deoxidized copper	Air conditioners, plumbing tubes, chemical process vessels, hydraulic lines
High-coppers	Springs, electrical connectors, valves, pump parts, welding equipment
Red brasses	Architectural applications, plumbing tube, marine hardware, jewelry, condenser and heat exchanger tubes
Yellow brasses	Automotive radiators, pump components, ammunition, plumbing accessories, architectural components, condenser and heat exchanger tubes
Tin-brasses	Aircraft and marine hardware, condenser plates, evaporators and heat exchanges
Tin (phosphor) bronzes	Chemical processing hardware, bushings, bearings, springs, switches, electrical contacts, welding rods
Aluminum bronzes	Nuts and bolts, chemical process vessels, condenser tubes and piping, marine hardware, welding and surfacing applications
Silicon bronzes	Heat exchanger tubes, hydraulic lines, screws and clamps, welding rod, bushings and bearings
Alloy brasses	Forgings, condenser and heat exchanger tubes
Copper-nickels	Desalination equipment, marine piping, condenser, evaporator and heat exchanger piping, antifouling surfaces
Nickel-silvers	Fasteners, optical goods, architectural hardware, springs
	CAST ALLOYS
High-coppers	Electrical and thermal conductors, applications requiring corrosion resistance
Beryllium-coppers	Electrical applications requiring high strength and high hardness, resistance welding electrodes, nonsparking tools, molds, pump parts, bearings, bushings
Red brasses	Valves, plumbing goods, pump parts, fittings
Yellow brasses	Plumbing valves and fittings, ornamental items

Table **21-7** (*cont.*)

CLASSIFICATION	TYPICAL APPLICATIONS
	CAST ALLOYS
Manganese bronzes	Marine castings, gears, gun mounts, bushings, bearings
Silicon bronzes	Bearings, impellers, pump and valve components, marine fittings
Tin bronzes	Bearings and bushings, impellers, valve components, gears, fittings, marine hardware, bridge plates
Nickel-tin bronzes	Structural components, gears, valve seats, pump rods, bearings, bushings, pumps
Aluminum bronzes	Marine equipment, nuts, gears, wear resistant components, acid resistant hardware valves, pump casings
Nickel-aluminum bronzes	Propeller hub and blades, marine hardware
Copper-nickels	Marine hardware, valves, fittings
Nickel-silvers	Valves, statuary, ornamental applications, sanitary fittings

Source: American Welding Society, *Welding Handbook*, 6th ed., section 4, p. 68.5.

Table **21-8** Three-digit numerical code for copper and copper alloys

NUMBER SERIES	DESCRIPTION	COMPOSITION RANGES	REPRESENTATIVE ELECTRICAL CONDUCTIVITIES (% IACS)
		WROUGHT COPPERS[a]	
101 to 107	Oxygen-free	99.95% Cu or better	>100
109 to 142	Tough-pitch and deoxidized	Contain oxygen or deoxidizers	80 to 100
145 to 147	Free-machining	Small additions of S, Te etc.	>90
150 to 194	High-copper alloys	Neighborhood of 1 or 2% additions of Cd, Be, Cr, Co, Fe, Ni, Zn and/or Sn	20 to 85
205 to 240	Red brasses	Up to 20% Zn	35 to 60
250 to 298	Yellow brasses	From 25 to 50% Zn	25 to 35
310 to 385	Leaded brasses	From 10 to 45% Zn and up to 4.5% Pb	25 to 45
405 to 485	Tin brasses	To 5.5% Sn, to 48% Zn	25 to 30
502 to 529	Copper-tin alloys (phosphor bronzes)	From 1 to 11% Sn	10–50

Table **21-8** (*cont.*)

NUMBER SERIES	DESCRIPTION	COMPOSITION RANGES	REPRESENTATIVE ELECTRICAL CONDUCTIVITIES (% IACS)
		WROUGHT COPPERS[a]	
532 to 546	Leaded phosphor bronzes	1 to 4% Pb, about 5% Sn, some with additions of Zn	10–20
606 to 642	Aluminum bronzes	From 2.6 to 13% Al, to 5% Fe, some with additions of Si or Ni	10–20
647 to 661	Silicon bronzes	From 1 to 3.5% Si, some with Mn, Si, or Sn	7–12
665 to 697	Alloy brasses	Zinc containing alloys with additions of Ni, Sn, Mn, Al, and Si	20–25
701 to 720	Copper-nickels	From 2 to 40% Ni, additions of Fe, Be, Mn, or Cr	4–10
732 to 798	Nickel-silvers	From about 43 to 73% Cu, from 7 to 23% Ni, some with Pb or Mn, bal. Zn	5–10
		CAST COPPERS[b]	
801 to 811	Coppers	Minimum of 99.70% Cu & Ag	92–100
813 to 828	High copper alloys	Additions of up to about 2.5% Be, Co, Si, Ni and/or Cr	20–80
833 to 838	Red brasses	83 to 93% Cu, to 12% Zn with lesser amounts of Sn, Pb	15–40
842 to 848	Semi-red brasses	76 to 80% Cu, 8 to 15% Zn with lesser amounts of Sn, Pb	15–20
852 to 858	Yellow brasses	57 to 72% Cu, bal. primarily Zn, 1 to 2% Sn, Pb, Ni or Al	18–28
861 to 868	High-strength yellow brasses	55 to 67% Cu, additions of Fe, Ni, Mn, Al, bal. Zn	7–22
872 to 879	Silicon brasses and silicon bronzes	65 to 90% Cu, about 3 to 5% Si, some with large amounts of Zn	6–15
902 to 945	Tin bronzes	3 to 19% Sn, some with large amounts of Pb, less Zn, Ni	7–15
947, 948	Nickel-tin bronzes	About 5% Sn and 5% Ni, to 2.5% Zn, Alloy 948 has 1% Pb	12

Table **21-8** (*cont.*)

NUMBER SERIES	DESCRIPTION	COMPOSITION RANGES	REPRESENTATIVE ELECTRICAL CONDUCTIVITIES (% IACS)
		CAST COPPERS[b]	
952 to 958	Aluminum bronzes	7 to 11% Al, at least 71% Cu, bal. Ni, Fe, Mn and/or Si	3–13
962 to 966	Copper-nickels	10 to 31% Ni, about 1% additions of Fe, Cb, Si, Mn and/or Si	4–11
973 to 978	Nickel-silvers	55 to 65% Cu, Pb and Sn additions, 12 to 25% Ni, bal. Zn	4–5

[a]For specific compositions and properties see Standards Handbook No. 2, Copper Development Association, N.Y.

[b]For specific compositions and properties see Standards Handbook No. 7, Copper Development Association, N.Y.

Source: American Welding Society, *Welding Handbook*, 6th ed., section 4, p. 68.4.

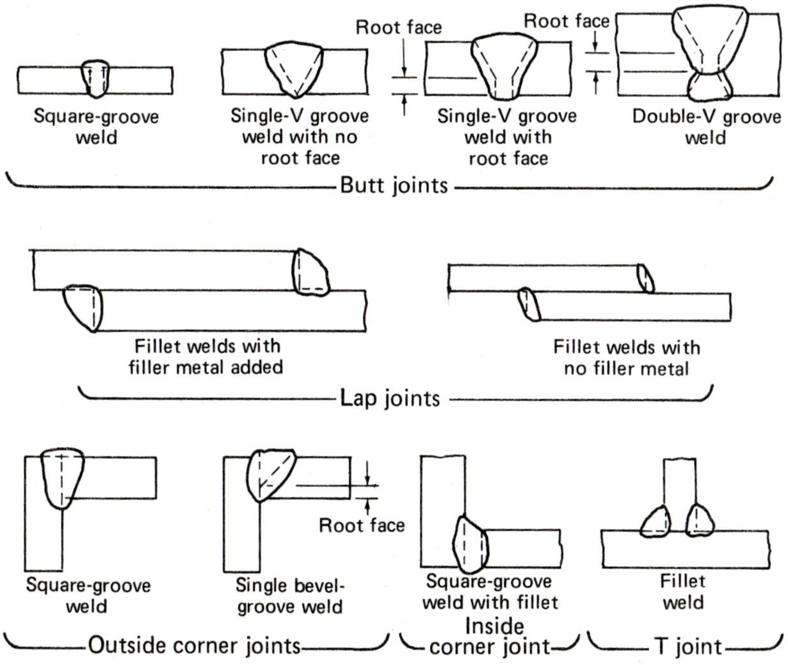

Root face
Root face
Square-groove weld
Single-V groove weld with no root face
Single-V groove weld with root face
Double-V groove weld
Butt joints
Fillet welds with filler metal added
Fillet welds with no filler metal
Lap joints
Root face
Square-groove weld
Single bevel-groove weld
Square-groove weld with fillet
Fillet weld
Outside corner joints
Inside corner joint
T joint

Figure **21-3** Groove configurations for arc welding copper and copper alloys. (By permission, from *Metals Handbook* Volume 6, Copyright American Society for Metals, 1971.)

Table **21-9** Nominal conditions for gas tungsten–arc welding of commercial coppers[a]
(Using EWTH-2 electrodes, RCu welding rod, and straight-polarity direct current)

WORK-METAL THICKNESS (in.)	ROOT OPENING (in.)[b]	ELECTRODE DIAMETER (in.)	DIAMETER OF WELDING ROD (in.)	SHIELDING GAS[c]	GAS-FLOW RATE (cfh)	CURRENT (amp)	TRAVEL SPEED (ipm)	NUMBER OF PASSES	PREHEAT TEMPERATURE, F
BUTT JOINTS—SQUARE GROOVE									
1/16	0	1/16	None used	Argon	15	110 to 140	12	1	None
1/8	0	3/32	None used	Argon	15	175 to 225	11	1	None
1/8	1/8	3/32	3/32, 1/8	Argon	15	175 to 225	11	1	None
3/16	3/16	1/8	1/8	Helium	30	190 to 225	10	1	200
BUTT JOINTS—60° SINGLE-V GROOVE, 1/16-IN. ROOT FACE									
1/4	1/16 max	1/8	1/8	Helium	30	225 to 260	9	1	300
3/8	1/16 max	3/16	3/16	Helium	40	280 to 320	—	2	500
BUTT JOINTS—60° DOUBLE-V GROOVE, 1/8-IN. ROOT FACE[d]									
1/2	1/16 max	3/16, 1/4	1/4	Helium	40	375 to 525	—	3	500
LAP JOINTS—FILLET WELDED[e]									
1/16	0	1/16	1/16	Argon	15	130 to 150	10	1	None
1/8	0	3/32	3/32, 1/8	Argon	15	200 to 250	9	1	None
3/16	0	1/8	1/8	Helium	30	205 to 250	8	1	200
1/4	0	1/8	1/8	Helium	30	250 to 280	7	1	300
3/8	0	3/16	3/16	Helium	40	300 to 340	—	3	500

				OUTSIDE CORNER JOINTS—SQUARE GROOVE					
1/8	1/8 max	3/32	3/32, 1/8	Argon	15	175 to 225	11	1	None
3/16	3/16 max	1/8	1/8	Helium	30	190 to 225	10	1	200
1/4	3/16 max	1/8	1/8	Helium	30	225 to 260	9	1	300
3/8	1/4 max	3/16	3/16	Helium	40	280 to 320	—	2	500
				OUTSIDE CORNER JOINTS—50° SINGLE-BEVEL GROOVE, 1/16-IN. ROOT FACE					
3/16	1/16 max	1/8	1/8	Helium	30	205 to 250	8	1	200
1/4	1/16 max	1/8	1/8	Helium	30	250 to 280	7	1	300
3/8	1/16 max	3/16	3/16	Helium	40	300 to 340	—	3	500
				INSIDE CORNER JOINTS—SQUARE GROOVE, FILLET WELDED					
1/8	1/8 max	3/32	3/32, 1/8	Argon	15	200 to 250	9	1	None
				T JOINTS—FILLET WELDED					
1/8	1/16 max	3/32	3/32, 1/8	Argon	15	200 to 250	9	1	None
3/16	1/16 max	1/8	1/8	Helium	30	205 to 250	8	1	200
1/4	1/16 max	1/8	1/8	Helium	30	250 to 280	7	1	300
3/8	1/16 max	3/16	3/16	Helium	40	3C0 to 340	—	3	500

[a]The data in this table are intended to serve as starting points for the establishment of optimum joint design and conditions, for welding of parts on which previous experience is lacking; they are subject to adjustments as necessary to meet the special requirements of individual applications. [b]Copper, carbon, or graphite backing strips or rings may be used (see text). [c]Mixtures of argon and helium are also used (see text). [d]Depth of back V is 3/8 of stock thickness. [e]Use of filler metal optional for thicknesses of 1/4 in. or less.

Source: By permission, from *Metals Handbook* Volume 6, Copyright American Society for Metals, 1971.

Table **21-10** Nominal conditions for gas metal–arc butt welding of commercial coppers and copper alloys[a]

WELD TYPES FOR BUTT JOINTS	WORK-METAL THICKNESS (in.)	ROOT FACE (in.)	ROOT OPENING (in.)	ELECTRODE	ELECTRODE-WIRE DIAMETER (in.)	SHIELDING GAS	GAS-FLOW RATE (cfh)	CURRENT (DCRP) (amp)	VOLTAGE (v)	TRAVEL SPEED (ipm)	NUMBER OF PASSES	PREHEAT TEMPERATURE, F
COMMERCIAL COPPERS												
Square groove[b]	1/8	1/8	0	ECu	1/16	Argon	30	310	27	30	1	None
Square groove[c]	1/8	1/8	0–1/16	ECu	1/16	Argon[d]	30–35	325–350	28–33	—	1	None
Square groove	1/4	1/4	0	ECu	3/32	Argon	30	460	26	20	2	200
	1/4	1/4	0	ECu	3/32	Argon	30	500	27	20	1	200
75–90° single-V	1/4	1/8	0–1/8	ECu	1/16	Argon[d]	30–35	400–425	32–36	—	2	400–500
groove[c]	1/2	0–1/8	0–1/8	ECu	1/16	Argon[d]	30–35	425–450	35–40	—	4	800–900
90° single-V	3/8	3/16	0	ECu	3/32	Argon	30	500	27	14	[e]	400
groove	3/8	3/16	0	ECu	3/32	Argon	30	550	27	14	[e]	400
	1/2	1/4	0	ECu	3/32	Argon	30	540	27	12	[e]	400
	1/2	1/4	0	ECu	3/32	Argon	30	600	27	10	[e]	400
ALLOY 175 (HIGH-CONDUCTIVITY BERYLLIUM COPPER)[f]												
90° single-V groove	1/4–1/2	1/32	—	Alloy 175	0.045	A-He	30	200–240	—	—	3–4[g]	600
	3/4	1/32	—	Alloy 175	0.045	A-He	30	200–240	—	—	6[g]	900
ALLOYS 170 AND 172 (HIGH-STRENGTH BERYLLIUM COPPERS)[f]												
90° single-V groove	1/4–1/2	1/32–1/16	—	Alloy 170, 172	0.045	A-He	45	175–200	—	—	3–4[h]	300–400
30° double-U groove[j]	3/4–1 1/2	1/16	—	Alloy 170, 172	1/16	A-He	60	325–350	—	—	10–20[k]	300–400

LOW-ZINC BRASSES												
Square groove[c]	1/8	1/8	0	ECuSi	1/16	Argon	30	275–285	25–28	—	1	None
	1/8	1/8	0	ECuSN-C	1/16	Helium	35	275–285	25–28	—	1	None
60° single-V	3/8	0	1/8	ECuSi	1/16	Argon	30	275–285	25–28	—	2	None
groove[c]	1/2	0	1/8	ECuSi	1/16	Argon	30	275–285	25–28	—	4	None
70° single-V	3/8	0	1/8	ECuSn-C	1/16	Helium	35	275–285	25–28	—	2	500[m]
groove[c]	1/2	0	1/8	ECuSn-C	1/16	Helium	35	275–285	25–28	—	4	500[m]
HIGH-ZINC BRASSES, TIN BRASSES, SPECIAL BRASSES, NICKEL SILVERS												
Square groove[c]	1/8	1/8	0	ECuSn-C	1/16	Argon	30	275–285	25–28	—	1	None
70° single-V	3/8	0	1/8	ECuSn-C	1/16	Argon	30	275–285	25–28	—	2	None
groove[c]	1/2	0	1/8	ECuSn-C	1/16	Argon	30	275–285	25–28	—	4	None
PHOSPHOR BRONZES[n]												
90° single-V	3/8	0	1/8	ECuSn-A[p]	1/16	Helium	35	275–285	25–28	—	3–4[q]	200–300
groove[c]	1/2	0	1/8	ECuSn-A[p]	1/16	Helium	35	275–285	25–28	—	5–6[q]	350–400
ALUMINUM BRONZES[r]												
Square groove[s]	1/8	1/8	0	ECuAl-A2	1/16	Argon	30	280–290	27–30	—	1	None
60–70° single-V	3/8	0	1/8	ECuAl-A2	1/16	Argon	30	280–290	27–30	—	2	None
groove[c]	1/2	0	1/8	ECuAl-A2	1/16	Argon	30	280–290	27–30	—	3	Slight
SILICON BRONZES[t]												
Square groove[u]	1/8	1/8	0	ECuSi	1/16	Argon	30	260–270	27–30	8 min	1	None
60° single-V	3/8	0	1/8	ECuSi	1/16	Argon	30	260–270	27–30	8 min	2	None
groove[c]	1/2	0	1/8	ECuSi	1/16	Argon	30	260–270	27–30	8 min	3	None

Table **21-10** *(cont.)*

WELD TYPES FOR BUTT JOINTS	WORK-METAL THICKNESS (in.)	ROOT FACE (in.)	ROOT OPENING (in.)	ELECTRODE	ELECTRODE-WIRE DIAMETER (in.)	SHIELDING GAS	GAS-FLOW RATE (cfh)	CURRENT (DCRP) (amp)	VOLTAGE (v)	TRAVEL SPEED (ipm)	NUMBER OF PASSES	PREHEAT TEMPERATURE, F
COPPER NICKELS												
Square groove[c]	1/8	1/8	0	ECuNi	1/16	Argon	30	280	27–30	—	1	None
60–80° single-V	3/8	0–1/32	1/8–1/4	ECuNi	1/16	Argon	30	280	27–30	—	2	None
groove[c]	1/2	0–1/32	1/8–1/4	ECuNi	1/16	Argon	30	280	27–30	—	4	None
COMMERCIAL COPPERS TO STEEL												
70–80° single-V groove	3/8	1/16	1/8	ERNi-3	1/16	Argon	60	375	29–31	—	4	800–1000
COPPER NICKEL TO STEEL												
70–80° single-V groove	3/8	1/16	1/8	ERNi-3	1/16	Argon	60	375	29–31	—	4	150 max
ALUMINUM BRONZE TO STEEL[v]												
60° single-V groove	3/8	0	5/16–3/8	ECuAl-A2	1/16	Argon	30	270–280	25–27	—	6	300–500
SILICON BRONZE TO STEEL[w]												
60° single-V groove	3/8	0	5/16–3/8	ECuAl-A2	1/16	Argon	30	270–280	28–30	—	6	150 max[x]

[a]The data in this table are intended to serve as starting points for the establishment of optimum joint design and conditions for welding of parts on which previous experience is lacking; they are subject to adjustment necessary to meet the requirements of individual applications. Thicknesses up to about 1½ in. are sometimes welded by use of slightly higher current and lower travel speed than shown for a thickness of ½ in. [b]Copper backing. [c]Grooved copper backing. [d]Or 75% argon, 25% helium. [e]Special welding sequence is used; see text. [f]See Table 21-8, page 597 for compositions. [g]The final pass is made on the root side after back chipping. Grind after each pass. [h]The final pass is made on the root side after back chipping. Wire brush after each pass. [j]Similar to the double-V-groove weld shown in Figure 21-3 but with a groove radius of ⅜ in. [k]Several passes are made on the face side, then several on the back side, until the weld is completed. Back chip the root pass before making the first pass on the back side. Wire brush after each pass. [m]Should not be overheated; as little preheat as possible should be used. [n]Welding conditions based on alloys 510, 521 and 524; current is increased or speed decreased for alloy 505. [p]Or ECuSn-C. [q]Hot peening between passes is recommended for maximum strength. [r]Slight preheat may be needed on heavy sections; interpass temperature should not exceed 600 F. [s]With ⅛-by-1-in. aluminum bronze backing. [t]No preheat is used on any thickness; interpass temperature should not exceed 200 F. [u]With ⅛-by-1-in. silicon bronze backing. [v]Steel should be well penetrated; an overlay is not usually needed. [w]Steel should be well penetrated; an overlay should be applied to avoid excessive dilution of the silicon bronze. [x]Except in welding silicon bronze to high-carbon or low-alloy steel, for which preheat temperature is 400 F.

Source: By permission from *Metals Handbook*, Volume 6. Copyright American Society for Metals, 1971.

fever, and nausea beginning very shortly after exposure and lasting longer than 24 hours.

SHIELDED METAL–ARC PROCESS As pointed out in Table 21-10, electrodes such as ECuSn-C and ECuSn-A can be used for welding the *low-zinc brasses*, providing the metal to be welded is preheated to approximately 450°F and the weld metal is applied in narrow, shallow stringer beads.

The *high-zinc copper alloys* can be satisfactorily welded with ECuSn-C electrodes and no preheating. However, the arc should be held directly on the molten weld puddle rather than toward the base metal, and advanced slowly to minimize zinc volatization.

Warning: Inhalation of zinc oxide fumes in amounts exceeding 5 mg/m^3 (threshold limit) will produce a condition of chills, fever, and nausea occurring 4 to 8 hours after exposure and disappearing almost invariably within 24 hours.

Phosphor bronzes are welded interchangeably with ECuSn-C and ECuSn-A electrodes. However, the material to be welded must be preheated to approximately 350°F and welded rapidly with light passes, taking care that the interpass temperature does not exceed the preheat temperature.

Prior to groove welding, the reader should refer to Figure 21-3 to ascertain an acceptable joint configuration. The weaving technique should be used in groove welding, taking care that the width of the weave does not exceed two electrode diameters. If for some reason the weld assembly is not postheated, it is recommended that the joint be peened immediately after welding. Peening helps to break up coarse grain structure. If, however, the welded joint is to be postheated, it should be heated to 900°F, maintained at that temperature until the piece reaches 900°F throughout, removed from the heat, and cooled rapidly and evenly over its entire surface.

Aluminum bronzes, such as alloy numbers 606, 613, 614, and so on, are welded with ECuAl-A2 and ECuAl-B electrodes. Preheating and interpass temperatures equal to those used for the GMAW process (Table 21-10) are also acceptable for the SMAW process. The two alloys do not require a postweld heat treatment. However, aluminum–bronze alloys, such as 618 and 624, that have more than the average 7% aluminum content may require—depending on section thickness—preheating and interpass temperatures up to 1100°F (see the curve for the GMAW process in Figure 21-4) followed by annealing at 1150°F and fan cooling. These higher-aluminum-content alloys are usually welded with ECuAl-C, ECuAl-D, and ECuAl-E electrodes.

Silicon bronzes are welded with joints dimensionally similar to joints used with mild steel of similar thickness. Silicon bronzes are usually welded with ECuSi electrodes. They are *not* preheated, and

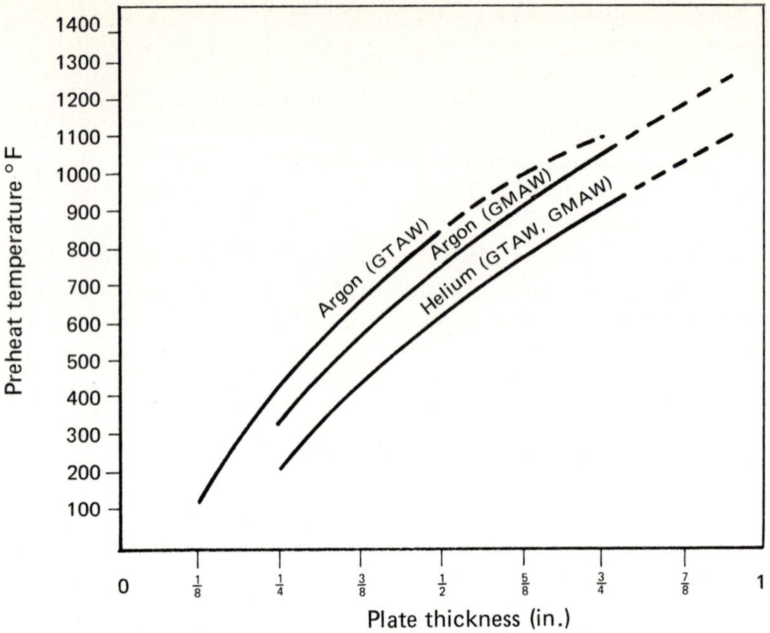
1400
1300
1200
1100
1000
900
800
700
600
500
400
300
200
100
Preheat temperature °F
Argon (GTAW)
Argon (GMAW)
Helium (GTAW, GMAW)
0
1/8
1/4
3/8
1/2
5/8
3/4
7/8
1
Plate thickness (in.)

Figure **21-4** Preheating temperatures for inert-gas welding of copper and copper alloys. (Redrawn by permission from American Welding Society, *Welding Handbook,* 6th ed., 1971, section 4, p. 68.22.)

care should be taken that the interpass temperature does not exceed 200°F. Peening immediately after welding will reduce residual stress and minimize distortion.

Copper-nickel alloys 706, 710, 715, and 725 are as easily welded as mild steels. They are usually welded with the 70-30 copper–nickel ECuNi electrodes, usually with reverse-polarity direct current. These alloys are readily weldable in any welding position; however, best results are obtained in flat-position welding. *Note:* Because the slag produced by this metal is viscous when molten and adherent when cold, special care must be taken to prevent slag entrapment between passes.

Alloy no. 745 was previously (and may still be) known under the trade name *nickel silver, 65-10.* This means that the alloy contains 65% copper and 10% nickel.

If the reader wishes to learn more about these designations, he or she is invited to contact the Copper Development Association, Inc., 405 Lexington Avenue, New York, New York, 10017, to request their standards handbook for wrought, mill, and cast products (see Parts 2 and 7).

The welding processes that can be used to join the different types of coppers and copper alloys are listed in Table 21-11.

Table **21-11** Welding processes that can be used to weld copper and copper alloys (*explanation of terms:* E = excellent; G = good; P = poor; N/R = not recommended)

	APPLICABLE JOINING PROCESSES							
						RESISTANCE		
NUMBER SERIES	SOLDERING	BRAZING	FUEL-GAS	GAS SHIELDED–ARC	COATED METAL–ARC	SPOT	SEAM	BUTT
101–107	E	E	F	G	N/R	N/R	N/R	G
109	E	E	G	G	N/R	N/R	N/R	G
110–116	E	G	N/R	F	N/R	N/R	N/R	G
119–121	E	E	F	E	N/R	N/R	N/R	G
122	E	E	G	E	N/R	N/R	N/R	G
125–130	E	G	N/R	F	N/R	N/R	N/R	G
142, 143	E	E	G	E	N/R	N/R	N/R	G
145	E	G	F	F	N/R	N/R	N/R	F
147	E	E	N/R	N/R	N/R	N/R	N/R	G
150	E	E	—	—	N/R	—	—	G
155	E	E	N/R	N/R	N/R	G	N/R	N/R
162	E	E	G	G	N/R	N/R	N/R	G
165	E	E	N/R	F	N/R	N/R	N/R	E
170–173	G	G	N/R	G	G	G	F	F
175	G	G	N/R	F	F	G	F	F
182–185	G	G	N/R	G	N/R	N/R	N/R	F
187	E	G	—	—	N/R	—	—	F
189	E	E	E	E	G	F	N/R	E
190	E	E	G	G	N/R	F	F	G
191	E	G	F	F	F	P	P	P
192–194	E	E	G	E	N/R	N/R	N/R	G
195	E	E	F	G	F	F	G	G
210, 220	E	E	G	G	N/R	N/R	N/R	G
226, 230, 240	E	E	G	G	N/R	F	N/R	G
260, 268, 270, 280	E	E	G	F	N/R	G	N/R	G
314, 316	E	G	N/R	—	—	—	N/R	G
330	E	G	F	F	N/R	F	N/R	F
332	E	G	—	—	—	—	N/R	F
335	E	G	F	F	N/R	F	N/R	F
340, 342, 353	E	G	N/R	—	—	—	N/R	F
349	E	G	F	F	N/R	F	N/R	F
350, 356, 360	E	G	N/R	—	—	—	N/R	F
365, 366, 367, 368	E	G	F	F	—	—	N/R	F
370	E	G	N/R	—	—	—	N/R	F
377, 378	E	G	N/R	—	—	—	N/R	F

Table **21-11** (*cont.*)

NUMBER SERIES	APPLICABLE JOINING PROCESSES							
						RESISTANCE		
	SOLDERING	BRAZING	FUEL-GAS	GAS SHIELDED–ARC	COATED METAL–ARC	SPOT	SEAM	BUTT
411	E	E	F	G	N/R	F	N/R	G
413	E	E	G	F	N/R	F	N/R	G
425	E	E	G	G	N/R	F	N/R	G
435	E	E	G	F	N/R	G	N/R	G
443, 444, 445	E	E	G	F	N/R	G	N/R	G
464, 467	E	E	G	F	N/R	G	F	G
482, 485	E	G	N/R	—	—	—	N/R	F
505	E	E	F	G	F	N/R	N/R	E
510, 511, 521, 524	E	E	F	G	F	G	F	E
544	E	G	—	—	—	—	N/R	F
608	N/R	F	N/R	G	—	—	—	G
610	F	N/R	N/R	G	G	G	F	G
613	N/R	F	N/R	E	G	G	G	G
614, 618	N/R	F	N/R	G	—	—	—	G
619	F	F	N/R	E	G	G	G	G
623, 624	N/R	F	N/R	G	—	—	—	G
625	N/R	F	N/R	G	G	F	F	F
630	N/R	F	N/R	G	—	—	—	G
632	F	F	N/R	E	G	G	F	G
638	F	F	F	E	—	—	—	E
642	N/R	F	N/R	F	—	—	—	F
651	E	E	G	E	F	E	G	E
655	G	E	G	E	F	E	E	E
667	E	E	G	F	N/R	E	E	E
674	F	G	N/R	F	N/R	G	G	G
675	E	E	G	F	N/R	G	F	G
687	F	G	F	F	N/R	G	F	G
688	F	F	N/R	N/R	N/R	G	G	G
694	E	E	G	N/R	N/R	G	G	G
706	E	E	F	E	G	G	G	E
710	E	E	F	E	G	E	E	E
715	E	E	G	E	—	—	—	E
725	E	E	F	G	G	E	G	E
745–770	E	E	G	F	N/R	G	G	G
782	E	G	N/R	—	—	—	N/R	F

Table **21-11** (*cont.*)

	APPLICABLE JOINING PROCESSES					
NUMBER SERIES	SOLDERING	BRAZING	FUEL-GAS	CARBON-ARC	GAS SHIELDED–ARC	COATED METAL–ARC
801–811	E	E	N/R	F	F	N/R
813, 814, 815	E	G	N/R	F	F	F
817	G	G	N/R	N/R	F	N/R
818, 820	G	G	N/R	F	F	F
821	G	G	N/R	N/R	F	N/R
822	G	G	N/R	F	F	F
824–828	F	F	N/R	F	F	F
833	E	G	N/R	N/R	F	N/R
834	E	E	F	N/R	F	N/R
836–848	E	G	N/R	N/R	N/R	F
852	E	F	F	N/R	N/R	N/R
852	E	E	G	F	G	N/R
852	E	E	G	F	G	N/R
854	E	E	F	P	P	P
855, 857	G	F	N/R	N/R	N/R	N/R
858	G	G	N/R	—	—	N/R
861, 862	P	P	G	N/R	F	G
863	P	—	—	—	P	G
864–868	F	F	P	—	—	P
872	N/R	F	G	P	F	F
874, 975	N/R	G	F	N/R	F	N/R
876	N/R	F	G	P	F	F
878	N/R	F	N/R	—	—	N/R
879	G	G	N/R	—	—	N/R
902, 903, 905	E	G[1]	F	—	—	F
907–913	E	G[1]	F	—	—	F
915	E	G[1]	N/R	—	—	N/R
916, 917	E	G[1]	F	—	—	F
922	E	E[1]	N/R	—	—	N/R
923–932	E	G[1]	N/R	—	—	N/R
934, 935, 937	G	G[1]	N/R	—	—	N/R
938, 939, 943	G	P[1]	N/R	—	—	N/R
944	G	G[1]	N/R	—	—	N/R
945	G	P[1]	N/R	—	—	N/R
947	E	E[1]	F	N/R	G	G

Table **21-11** (*cont.*)

NUMBER SERIES	SOLDERING	BRAZING	FUEL-GAS	CARBON-ARC	GAS SHIELDED–ARC	COATED METAL–ARC
	APPLICABLE JOINING PROCESSES					
948	E	G[1]	N/R	—	—	N/R
952, 953, 954	G	G	G	E	E	G
955	G	F	N/R	G	G	G
956	G	G	N/R	F	G	F
958	G	F	N/R	P	G	G
962	E	E	N/R	N/R	F[2]	G[2]
963	E	E	N/R	N/R	F[2]	F[2]
964	E	E	N/R	N/R	G[2]	G[2]
966	E	E	G	F	F	F
973–978	E	E	N/R	—	—	N/R
993	N/R	G	N/R	P	G	G

[1]Since brazing is performed at temperatures within the hot-short range, strain must be avoided during the brazing and cooling of these alloys.
[2]Use filler metal RCuNi or ECuNi.

GAS-SHIELDED PROCESSES The gas-shielded processes commonly used to weld copper and its alloys are (1) the argon-GTAW process, (2) the argon-GMAW process, and (3) the helium GTAW and GMAW processes. The application of each process will be described separately in the following paragraphs.

The GTAW process is used most frequently for welding copper-alloy sections up to ½ inch thick. No filler metal is used for sections ⅛ inch thick or less. Table 21-9 and Tables 21-12 through 21-17 (pages 614–619) list the nominal conditions for GTAW welding of commercial copper and copper alloys.

Alternating current stabilized by high frequency is used on beryllium coppers and aluminum bronzes. All other copper alloys are welded with dc straight polarity.

Warning: Airborne particles of beryllium are a health hazard when inhaled. The recommendations on respiratory protection and ventilation given in Chapter 13 of this text should be followed strictly when welding beryllium coppers. The booklet *Beryllium and Its Compounds*, published as a hygiene guide by the American Industrial Hygiene Association, should also be consulted if more specific information on this subject is desired.

The GMAW process is almost invariably used for welding copper and copper-alloy sections ½ inch or more in thickness. The nominal welding conditions (parameters) applicable to this process are listed in Tables 21-10, 21-14, and 21-19. Square-groove butt joints can be used in materials up to ¼ inch thick, providing one pass per side is used for this weldment. In welding the single-V-groove joints referred to in Table 21-12 for thicknesses of ⅜ to ½ inch, the metal is deposited on one side in three or more passes, and the root pass is back gouged to sound metal before the last pass is applied to the back of the joint. Sections thicker than ½ inch should be prepared with a double V or a double-V groove and welded with alternate passes applied to opposite sides of the joint, if readily accessible, to minimize distortion.

Welds in high-conductivity beryllium copper (alloy 175, Table 21-10) thicker than 0.090 inch that are to be heat treated, and welds on heat-treated alloy 175 thicker than ¼ inch, are made with the GMAW process. *Note:* Sections of this alloy thicker than ¾ inch should not be welded by this process. When ECuAl-B filler metal is used, joint strength can be increased by aging the alloy for 3 hours at 900°F.

Welding conditions for high-strength beryllium copper alloys number 170 (1.7% Be) and 172 (1.9 percent Be) are also shown in Table 21-10. It should be pointed out that, unlike with high-conductivity copper alloys, the gas tungsten–arc (TIG) process can be used to weld high-strength beryllium coppers thicker than ½ inch. (Table 21-13)

The low-zinc brasses (alloy numbers 210, 220, 230, and 240) are readily weldable with either the GMAW (MIG) or the GTAW (TIG) processes.

To minimize zinc fuming (which is a health hazard), the arc is struck and maintained on the filler metal (RCuSn-C or RCuSi for the TIG process and ECuSi-A or ECuSn-C for the MIG process) and the base metal is heated by conduction from the weld puddle. Sections thicker than ¼ inch should be preheated to 400°F; sections thicker than ⅜ inch are usually welded with the MIG process. Although argon can be used as a shielding gas, adding up to 75% helium to the argon will provide a hotter arc.

ECuSn-A, ECuSn-C, or ECuSi-A electrodes are used for welding the low-zinc brasses with the SMAW (shielded metal–arc) process. Preheating the part to be welded and maintaining interpass temperatures of about 400°F will help obtain adequate penetration. All welding should be done in the flat position with backing strips. High welding speeds and welding currents on the high side of the recommended range should be used with thin stringer beads.

Fuel-gas welding of the low-zinc brasses can be done in all welding positions. The best corrosion resistance is obtained when RCuSi-A is used as the filler metal. Joints may also be made with RCuZn-A and RCuZn-C brazing rods using an AWS type-5 flux. However, joints made with these rods will not be as corrosion resistant as the base metal.

Nominal conditions for welding phosphor–bronze alloys 510, 521, and 524 are shown in Tables 21-18 and 21-19. Alloy number 505, which has a higher thermal conductivity than the other three alloys in this group, should be welded using either a higher welding current or a lower welding speed than recommended in Table 21-14. If the metal is not thicker than ⅛ inch, it is not necessary to utilize filler metal; however, sections ⅛ inch thick or thicker are welded using RCuSn-A and RCuSi-S as filler metals and are preheated to 400°F, taking care that interpass temperatures do not exceed the preheating temperature. *Note:* Because preheating increases the susceptibility of the weld to hot-short cracking and to large columnar grain growth, it is common practice to weld with a stringer bead and to peen between weld beads.

Aluminum bronze alloys (Table 21-15) containing less than 10% aluminum do not require preheating. Their interpass temperatures should not exceed 300°F and the completed weldment should be allowed to cool slowly in air. Iron-bearing alloys with aluminum content from 10 to 13% require preheat temperature and interpass temperature of 500°F, and the completed weldment should be cooled in air with the aid of a fan of appropriate capacity. Iron-bearing alloys containing more than 13% aluminum require preheat and interpass temperatures of 1150°F. The completed weldment should be cooled rapidly using an air blast or a fan of appropriate capacity.

Representative parameters for welding silicon bronzes are given in Table 21-16. Although the GTAW, GMAW, and fuel-gas welding processes may be used for welding these alloys, experience has shown that the GTAW process is the most desirable of the three. Because silicon–bronze alloys tend to form hot-short cracks, they should not be preheated, and the interpass temperatures must be kept below 200°F.

All silicon bronzes can be readily welded with ECuSi and ECuSi-A electrodes or welding rods. Straight-polarity dc is preferred, but alternating current may also be used.

If the fuel-gas process is used, filler metal designated RCuSi or RCuSi-A should be used, together with a neutral or slightly oxidizing flame and one of the following fluxes in either powder or paste form.

- Flux no. 1. 90% fused borax, 10% sodium fluoride
- Flux no. 2. 72% boric acid, 15% anhydrous sodium phosphate, 13% sodium chloride.
- Flux no. 3. Fused borax, sodium fluoride, manganese boride, barium carbonate (equal parts by weight)

If a paste-type flux is desired, methylated spirits should be added to the powder.

Copper-nickel alloys can be welded by nearly all arc-welding processes. Electrodes of the 70-30 copper-nickel composition (CuNi) type are used for all copper–nickel alloys. Representative parameters that can be used to obtain acceptable welds are listed in Tables 21-10 and 21-17.

Straight-polarity dc is used for the GTAW process; reverse-polarity dc is used for the GMAW and SMAW processes. A weave bead is generally preferred to ensure satisfactory bead contour and wash; but it is important that the weave not exceed three times the core-wire diameter. If necessary, a straight drag or string bead with a minimum of weaving may be used in the bottom of a deep groove.

Parameters for obtaining acceptable welds when joining unlike copper alloys and when welding copper alloys to dissimilar metals are given in Tables 21-18 and 21-19.

When joining a copper alloy to a ferrous or nickel alloy, care must be taken to prevent dilution and iron pick-up, respectively. The most serious effect of excessive dilution is unsoundness in the form of shrinkage cracks, which can start in the weld (filler) metal and continue into the base metal. The method most commonly used to prevent dilution and iron pick-up when MIG welding copper, aluminum bronzes, or copper-zinc alloys to low-carbon steel, alloy steel, stainless steel, cast iron, or nickel alloys is to braze weld one side of the joint and weld the other side. Other copper alloys are joined to ferrous metals and nickel alloys by means of the overlay method. The *overlay method* consists of applying a layer (coat) of a metal, selected for its ability to act as a buffer between two incompatible metals, to one or both sides of a joint. The buffer metal minimizes or eliminates the mixing (diffusion) of the incompatible metals. The electrode for welding the remainder of the joint can then be selected on the basis of its compatibility with the overlay metal(s) rather than its compatibility with the base metal.

(*Text continues on page 624*)

Table **21-12** Filler metals and preheat and interpass temperatures for TIG welding of coppers and copper alloys to dissimilar metals[a]

ONE METAL TO BE WELDED	FILLER METALS (AND PREHEAT AND INTERPASS TEMPERATURES) FOR WELDING METAL IN COLUMN 1 TO:				
	COPPERS	*PHOSPHOR BRONZES*	*ALUMINUM BRONZES*	*SILICON BRONZES*	*COPPER NICKELS*
	COPPER ALLOYS				
Low-zinc brasses	ECuSn-C[b] or RCu (1000 F)	—	—	—	—
Phosphor bronzes	ECuSn-C[b] or RCu (1000 F)	—	—	—	—
Aluminum bronzes	RCuAl-A2 (1000 F)	RCuAl-A2 or ECuSn-C[b] (400 F)	—	—	—
Silicon bronzes	ECuSN-C[b] or RCu (1000 F)	RCuSi-A (150 F max)	RCuAl-A2 (150 F max)	—	—
Copper nickels	RCuAl-A2 or RCu or RCuNi (1000 F)	ECuSn-C[b] (150 F max)	RCuAl-A2 (150 F max)	RCuAl-A2 (150 F max)	—
	NICKEL ALLOYS				
Nickel and Ni-Cu alloys	RCuNi or ERNiCu-7 (1000 F)	[c]	[c]	[c]	RCuNi or ERNiCu-7 (150 F max)
Ni-Cr, Ni-Fe and Ni-Cr-Fe alloys	ERNi-3 (1000 F)	[c]	[c]	[c]	ERNi-3 (150 F max)
	STEELS				
Low-carbon steel	RCuAl-A2 or RCu or ERNi-3 (1000 F)	ECuSn-C[b] (400 F)	RCuAl-A2 (300 F)	RCuAl-A2 (150 F max)	RCuAl-A2 or ERNi-3 (150 F max)
Medium-carbon steel	RCuAl-A2 or RCu or ERNi-3 (1000 F)	ECuSn-C[b] (400 F)	RCuAl-A2 (400 F)	RCuAl-A2 (150 F max)	RCuAl-A2 or ERNi-3 (150 F max)
High-carbon steel	RCuAl-A2 or RCu or ERNi-3 (1000 F)	ECuSn-C[b] (500 F)	RCuAl-A2 (500 F)	RCuAl-A2 (400 F)	RCuAl-A2 or ERNi-3 (150 F max)

Table **21-12** *(cont.)*

ONE METAL TO BE WELDED	FILLER METALS (AND PREHEAT AND INTERPASS TEMPERATURES) FOR WELDING METALS IN COLUMN 1 TO:				
	COPPERS	*PHOSPHOR BRONZES*	*ALUMINUM BRONZES*	*SILICON BRONZES*	*COPPER NICKELS*
			STEELS		
Low-alloy steel	RCuAl-A2 or RCu or ERNi-3 (1000 F)	ECuSn-C[b] (500 F)	RCuAl-A2 (500 F)	RCuAl-A2 (400 F)	RCuAl-A2 or ERNi-3 (150 F max)
Stainless steel	RCuAl-A2 or RCu or ERNi-3 (1000 F)	ECuSn-C[b] (400 F)	RCuAl-A2 (150 F max)	RCuAl-A2 (150 F max)	RCuAl-A2 or ERNi-3 (150 F max)
			CAST IRONS		
Gray and malleable irons	RCuAl-A2 or RCu (1000 F)	ECuSn-C[b] (400 F)	RCuAl-A2 (400 F)	RCuAl-A2 or RCuSi-A (300 F)	RCuAl-A2 (150 F max)
Ductile iron	RCuAl-A2 or RCu (1000 F)	ECuSn-C[b] (400 F)	RCuAl-A2 (150 F max)	RCuAl-A2 or RCuSi-A (150 F max)	RCuAl-A2 (150 F max)

[a]Filler-metal selections shown in table are based on weldability, except where mechanical properties are usually more important. Preheating is ordinarily used only when at least one member is thicker than about ⅛ in. or is highly conductive (see text). Preheat and interpass temperatures are subject to adjustment on the basis of the size and shape of the weldment. [b]ECuSn-C is classified by AWS as an electrode wire for gas metal–arc welding, but is used also as filler wire in gas tungsten–arc welding. [c]These combinations of work metals are only infrequently joined by welding; as a starting point in developing welding procedures for joining them, the use of RCuAl-A2 filler metal is recommended, except for welding of combinations that include phosphor bronzes.

Source: By permission, from *Metals Handbook* Volume 6, Copyright American Society for Metals, 1971.

Table **21-13** Nominal conditions for gas tungsten–arc welding of beryllium coppers[a]

WORK-METAL THICKNESS (in.)	BUTT-JOINT GROOVE[b]	ELECTRODE DIAMETER (in.)	CURRENT (amp)[c]	TRAVEL SPEED (ipm)	NUMBER OF PASSES	PREHEAT TEMPERATURE, °F
ALLOY 175 (High-conductivity beryllium copper)						
0 to 0.090	Square	3/32	150	5 to 10	1	None
0.090 to 1/8	90° single-V[d]	3/16	250	5 to 10	1 to 2	None
1/4	90° single-V[d]	3/16	250	5 to 10	4 to 5	800
ALLOYS 170 AND 172 (High-strength beryllium coppers)						
0 to 0.090	Square	3/32	150	5 to10	1	None
0.090 to 1/8	90° single-V[d]	3/32	180	5 to 10	1	None
1/4 to 1/2[e]	90° single-V[d]	3/16	250	5 to 10	3 to 4	300
Over 1/2[e]	90° single-V[d]	3/16	250	5 to 10	5 to 8	400

[a]The data in this table are intended to serve as starting points for the establishment of optimum joint design and conditions for welding of parts for which previous experience is lacking; they are subject to adjustment as necessary to meet the special requirements of individual applications. [b]See Figure 21-3 for illustrations of joints. [c]High-frequency-stabilized alternating current is preferred; straight-polarity direct current, with a thoriated tungsten electrode, is suitable under some conditions (see text). [d]Maximum root face is 1/16 in. [e]Gas tungsten–arc welding is used on these thicknesses only when gas metal–arc welding cannot be used.

For butt joints having zero root opening; welding with a zirconiated tungsten electrode, filler metal of the same composition as the base metal, argon-helium shielding gas at 25 cfh

Source: By permission, from *Metals Handbook* Volume 6, Copyright American Society for Metals, 1971.

Table **21-14** Parameters used for joining phosphor bronzes by the GMAW process (ECuSn-A or ECuSn-C filler wire and argon at 50 cfh)

METAL THICKNESS (in.)	JOINT PREPARATION	WIRE DIAMETER	JOINT GAP (in.)	VOLTS	CURRENT (amps)
1/16	Square butt	0.030-inch	3/64	25–26	130–140
1/8	Square butt	0.035-inch	3/32	26–27	140–160
1/4	V-groove	0.045-inch	1/16	27–28	165–185
1/2	V-groove	1/16-inch	3/32	29–30	315–335
3/4	[1]	5/64-inch	0–3/32	31–32	365–385
1	[1]	3/32-inch	0–3/32	33–34	440–460

[1]Double V or single V with approximately 3/32-in. root face and 1/4-in. radius rather than sharp included angle.

Source: American Welding Society, *Welding Handbook*, 6th ed., section 4, p. 68.33.

Table **21-15** Nominal conditions for gas tungsten–arc welding of aluminum bronzes[a]

WORK-METAL THICKNESS (in.)	ROOT OPENING (in.)	ELECTRODE DIAMETER (in.)[b]	DIAMETER OF WELDING ROD (in.)[c]	FLOW RATE OF ARGON (cfh)	CURRENT (ac, HF-stabilized amp)[d]	NUMBER OF PASSES
			SQUARE-GROOVE BUTT JOINTS			
Up to 1/16	0	1/16	1/16[e]	20 to 30	25 to 80	One
1/16 to 1/8	1/16 max	3/32	1/8	20 to 30	60 to 175	One
1/8	1/8 max	5/32 to 3/16	5/32	30	210	One
			70° SINGLE-V-GROOVE BUTT JOINTS			
3/8	0	5/32 to 3/16	5/32	30	210 to 330	Four
			FILLET-WELDED T JOINTS OR SQUARE-GROOVE INSIDE CORNER JOINTS			
3/8	[f]	5/32 to 3/16	3/32	30	225	Three

[a]The data in this table are intended to serve as starting points for the establishment of optimum joint design and conditions for welding parts on which previous experience is lacking; they are subject to adjustment as necessary to meet the special requirements of individual applications. Preheating is not ordinarily used in welding the thicknesses shown. [b]Zirconiated or unalloyed tungsten electrodes are recommended with high-frequency-stabilized alternating current. [c]Preferred welding rod is RCuAl-A2; otherwise, RCuAl-B or rod of the same composition as the base metal. [d]Straight-polarity direct current can also be used in making single-pass welds; see text. [e]Use of welding rod is optional for thicknesses up to 1/16 in. [f]Zero root opening for T-joints; 3/8 in. max for corner joints.

For joint configurations, see Figure 21-3.

Source: By permission, from *Metals Handbook* Volume 6, Copyright American Society for Metals, 1971.

Table **21-16** Nominal conditions for gas tungsten–arc welding of silicon bronzes[a]

WORK-METAL THICKNESS (in.)	CURRENT (amp)	ELECTRODE DIAMETER (in.)	TRAVEL SPEED (ipm)	DIAMETER OF WELDING ROD (in.)	SHIELDING-GAS FLOW RATE (cfh)	NUMBER OF PASSES
			AUTOMATIC WELDING SQUARE-GROOVE BUTT JOINTS, FLAT POSITION			
0.012 to 0.050	80 to 140	1/8	60 to 80	None used	15 to 35	1
1/16 to 1/8	90 to 210	1/8	45 to 60	None used	15 to 35	1
1/8	250	1/8	18 to 20	1/16[b]	15 to 35	1
			MANUAL WELDING SQUARE-GROOVE BUTT JOINTS, FLAT POSITION			
1/16	100 to 120	1/16	12	1/16	15	1
1/8	130 to 150	1/16	12	3/32	15	1
3/16	150 to 200	3/32	—	1/8	20	1
1/4	250 to 300	1/8	—	1/8, 3/16	20	1
1/4	150 to 200	3/32	—	1/8, 3/16	20	3
			SQUARE-GROOVE BUTT JOINTS, VERTICAL AND OVERHEAD POSITIONS			
1/16	90 to 110	1/16	—	1/16	15	1
1/8	120 to 140	1/16	—	3/32	15	1

Table **21-16** (*cont.*)

WORK-METAL THICKNESS (in.)	CURRENT (amp)	ELECTRODE DIAMETER (in.)	TRAVEL SPEED (ipm)	DIAMETER OF WELDING ROD (in.)	SHIELDING-GAS FLOW RATE (cfh)	NUMBER OF PASSES
MANUAL WELDING 60° SINGLE-V-GROOVE BUTT JOINTS, FLAT POSITION						
3/8	230 to 280	1/8	—	1/8, 3/16	20	3 to 4
1/2	250 to 300	1/8	—	1/8, 3/16	20	4 to 5
3/4[c]	300 to 350	1/8	—	3/16	20	9 to 10
1[c]	300 to 350	1/8	—	3/16, 1/4	20	13
FILLET-WELDED LAP JOINTS, FLAT POSITION						
1/16	110 to 130	1/16	10	1/16	15	1
1/8	140 to 160	1/16, 3/32	10	3/32	15	1
3/16	175 to 225	3/32	—	1/8	20	1
1/4	175 to 225	3/32	—	1/8, 3/16	20	3
3/8	250 to 300	1/8	—	1/8, 3/16	20	3
1/2	275 to 325	1/8	—	1/8, 3/16	20	6
3/4[c]	300 to 350	1/8	—	3/16	20	12
1[c]	325 to 350	1/8	—	1/4	20	16
FILLET-WELDED LAP JOINTS, VERTICAL AND OVERHEAD POSITIONS						
1/16	100 to 120	1/16	—	1/16	15	1
1/8	130 to 150	1/16, 3/32	—	3/32	15	1
SQUARE-GROOVE OUTSIDE CORNER JOINTS, FLAT POSITION						
1/16	100 to 130	1/16	12	1/16	15	1
1/8	130 to 150	1/16	12	3/32	15	1
3/16	150 to 200	3/32	—	1/8	20	1
SQUARE-GROOVE OUTSIDE CORNER JOINTS, VERTICAL AND OVERHEAD POSITIONS						
1/16	90 to 110	1/16	—	1/16	15	1
1/8	120 to 140	1/16	—	3/32	15	1
50° SINGLE-BEVEL-GROOVE OUTSIDE CORNER JOINTS, FLAT POSITION[d]						
1/4	175 to 225	3/32	—	1/8, 3/16	20	3
3/8	230 to 280	1/8	—	1/8, 3/16	20	3
1/2	275 to 325	1/8	—	1/8, 3/16	20	7
3/4[c]	300 to 350	1/8	—	3/16	20	14
1[c]	325 to 350	1/8	—	3/16, 1/4	20	20
FILLET-WELDED SQUARE-GROOVE INSIDE CORNER JOINTS, FLAT POSITION[e]						
1/16	110 to 130	1/16	10	1/16	15	1
1/8	140 to 150	1/16, 3/32	10	3/32	15	1
3/16	175 to 225	3/32	—	1/8	20	1

Table **21-16** (*cont.*)

WORK-METAL THICKNESS (in.)	CURRENT (amp)	ELECTRODE DIAMETER (in.)	TRAVEL SPEED (ipm)	DIAMETER OF WELDING ROD (in.)	SHIELDING-GAS FLOW RATE (cfh)	NUMBER OF PASSES
MANUAL WELDING FILLET-WELDED T JOINTS, FLAT POSITION						
1/16	110 to 130	1/16	10	1/16	15	1
1/8	140 to 160	1/16, 3/32	10	3/32	15	1
3/16	175 to 225	3/32	—	1/8	20	1
1/4	175 to 225	3/32	—	1/8, 3/16	20	3
3/8	230 to 280	1/8	—	1/8, 3/16	20	3
1/2	275 to 325	1/8	—	1/8, 3/16	20	7
3/4[c]	300 to 350	1/8	—	3/16	20	14
1[c]	325 to 350	1/8	—	3/16, 1/4	20	20

[a]The data in this table are intended to serve as starting points for the establishment of optimum joint design and conditions for welding parts on which previous experience is lacking; they are subject to adjustment as necessary to meet the special requirements of individual applications. [b]Wire-feed rate, 115 to 125 in. per minute. [c]Thicknesses greater than about 1/2 in. are gas tungsten-arc welded only when it is not practicable to use gas metal-arc welding. [d]Root face is 1/16 in. for thicknesses of 1/2 in. or less, and 1/8 in. for thicknesses greater than 1/2 in. [e]Maximum root opening = *t* (work-metal thickness).

Using zero root opening, no preheat, EWTh-2 electrodes, RCuSi-A welding rod, argon shielding gas, and straight-polarity direct current; for joint configurations, see Figure 21-3.

Source: By permission, from *Metals Handbook* Volume 6, Copyright American Society for Metals, 1971.

Table **21-17** Nominal conditions for gas tungsten–arc butt welding of copper nickels[a]

WORK-METAL THICKNESS (in.)	BUTT-JOINT GROOVE[b]	CURRENT (dcsp, amp)	ELECTRODE DIAMETER (in.)[c]	TRAVEL SPEED (ipm)	DIAMETER OF RCuNi WELDING ROD (in.)[d]	FLOW RATE OF ARGON (cfh)	NUMBER OF PASSES
AUTOMATIC WELDING OF ALLOY 706 (Copper nickel, 10%)							
1/8	Square	310 to 320	3/16	15 to 18	1/16	25–30	1
MANUAL WELDING OF ALLOY 706 (Copper nickel, 10%)							
0 to 1/8	Square	300 to 310	3/16	5	1/8	25–30	1
1/8 to 3/8	70–80° single-V	300 to 310[e]	3/16	6	1/8, 3/16	25–30	2 to 4
MANUAL WELDING OF ALLOY 715 (Copper nickel, 30%)							
0 to 1/8	Square	270 to 290	3/16	5	1/8	25–30	1
1/8 to 3/8	70–80° single-V	270 to 290[e]	3/16	6	5/32	25–30	4

[a]The data in this table are intended to serve as starting points for the establishment of optimum joint design and conditions for welding parts on which previous experience is lacking; they are subject to adjustment as necessary to meet the special requirements of individual applications. Root opening is zero. Preheating is not needed. [b]See Figure 21-3 for illustrations of joints. [c]Preferred electrode material is EWTh-2. [d]Filler metal (RCuNi) must be used on all welded joints. [e]Current should be increased in equal increments with each pass, up to a maximum of about 375 amp, with larger welding rods.

Source: By permission, from *Metals Handbook* Volume 6, Copyright American Society for Metals, 1971.

Table **21-18** Filler metals and preheat and interpass temperatures for TIG welding of coppers and copper alloys to dissimilar metals[a]

ONE METAL TO BE WELDED	FILLER METALS (AND PREHEAT AND INTERPASS TEMPERATURES) FOR WELDING METAL IN COLUMN 1 TO:				
	COPPERS	*PHOSPHOR BRONZES*	*ALUMINUM BRONZES*	*SILICON BRONZES*	*COPPER NICKELS*
		COPPER ALLOYS			
Low-zinc brasses	ECuSn-C[b] or RCu (1000 F)	—	—	—	—
Phosphor bronzes	ECuSn-C[b] or RCu (1000 F)	—	—	—	—
Aluminum bronzes	RCuAl-A2 (1000 F)	RCuAl-A2 or ECuSn-C[b] (400 F)	—	—	—
Silicon bronzes	ECuSn-C[b] or RCu (1000 F)	RCuSi-A (150 F max)	RCuAl-A2 (150 F max)	—	—
Copper nickels	RCuAl-A2 or RCu or RCuNi (1000 F)	ECuSn-C[b] (150 F max)	RCuAl-A2 (150 F max)	RCuAl-A2 (150 F max)	—
		NICKEL ALLOYS			
Nickel and Ni-Cu alloys	RCuNi or ERNiCu-7 (1000 F)	[c]	[c]	[c]	RCuNi or ERNiCu-7 (150 F max)
Ni-Cr, Ni-Fe and Ni-Cr-Fe alloys	ERNi-3 (1000 F)	[c]	[c]	[c]	ERNi-3 (150 F max)
		STEELS			
Low-carbon steel	RCuAl-A2 or RCu or ERNi-3 (1000 F)	ECuSn-C[b] (400 F)	RCuAl-A2 (300 F)	RCuAl-A2 (150 F max)	RCuAl-A2 or ERNi-3 (150 F max)
Medium-carbon steel	RCuAl-A2 or RCu or ERNi-3 (1000 F)	ECuSn-C[b] (400 F)	RCuAl-A2 (400 F)	RCuAl-A2 (150 F max)	RCuAl-A2 or ERNi-3 (150 F max)
High-carbon steel	RCuAl-A2 or RCu or ERNi-3 (1000 F)	ECuSn-C[b] (500 F)	RCuAl-A2 (500 F)	RCuAl-A2 (400 F)	RCuAl-A2 or ERNi-3 (150 F max)

Table **21-18** *(cont.)*

ONE METAL TO BE WELDED	FILLER METALS (AND PREHEAT AND INTERPASS TEMPERATURES) FOR WELDING METAL IN COLUMN 1 TO:				
	COPPERS	*PHOSPHOR BRONZES*	*ALUMINUM BRONZES*	*SILICON BRONZES*	*COPPER NICKELS*
			STEELS		
Low-alloy steel	RCuAl-A2 or RCu or ERNi-3 (1000 F)	ECuSn-C[b] (500 F)	RCuAl-A2 (500 F)	RCuAl-A2 (400 F)	RCuAl-A2 or ERNi-3 (150 F max)
Stainless steel	RCuAl-A2 or RCu or ERNi-3 (1000 F)	ECuSn-C[b] (400 F)	RCuAl-A2 (150 F max)	RCuAl-A2 (150 F max)	RCuAl-A2 or ERNi-3 (150 F max)
			CAST IRONS		
Gray and malleable irons	RCuAl-A2 or RCu (1000 F)	ECuSn-C[b] (400 F)	RCuAl-A2 (400 F)	RCuAl-A2 or RCuSi-A (300 F)	RCuAl-A2 (150 F max)
Ductile iron	RCuAl-A2 or RCu (1000 F)	ECuSn-C[b] (400 F)	RCuAl-A2 (150 F max)	RCuAl-A2 or RCuSi-A (150 Fmax)	RCuAl-A2 (150 F max)

[a]Filler-metal selections shown in table are based on weldability, except where mechanical properties are usually more important. Preheating is ordinarily used only when at least one member is thicker than about ⅛ in. or is highly conductive (see text). Preheat and interpass temperatures are subject to adjustment on the basis of the size and shape of the weldment. [b]ECuSn-C is classified by AWS as an electrode wire for gas metal-arc welding, but is used also as filler wire in gas tungsten-arc welding. [c]These combinations of work metals are only infrequently joined by welding; as a starting point in developing welding procedures for joining them, the use of RCuAl-A2 filler metal is recommended, except for welding of combinations that include phosphor bronzes.

Source: By permission, from *Metals Handbook* Volume 6, Copyright American Society for Metals, 1971.

Table **21-19** Electrodes and preheat and interpass temperatures used in gas metal–arc welding of coppers and copper alloys to dissimilar metals[a]

ONE METAL TO BE WELDED	ELECTRODES (AND PREHEAT AND INTERPASS TEMPERATURES) FOR WELDING METAL IN COLUMN 1 TO:						
	COPPERS	*LOW-ZINC BRASSES*	*HIGH-ZINC BRASSES, TIN BRASSES, SPECIAL BRASSES*	*PHOSPHOR BRONZES*	*ALUMINUM BRONZES*	*SILICON BRONZES*	*COPPER NICKELS*
COPPER ALLOYS							
Low-zinc brasses	ECuSn-C or ECu (1000 F)	—	—	—	—	—	—
High-zinc brasses, tin brasses, special brasses	ECuSi or ECuSn-C or ECu (1000 F)	ECuSn-C (600 F)	—	—	—	—	—
Phosphor bronzes	ECuSn-C or ECu (1000 F)	ECuSn-C (500 F)	ECuSn-C (600 F)	—	—	—	—
Aluminum bronzes	ECuAl-A2 (1000 F)	ECuAl-A2 (600 F)	ECuAl-A2 (600 F)	ECuAl-A2 or ECuSn-C (400 F)	—	—	—
Silicon bronzes	ECuSn-C or ECu (1000 F)	ECuAl-A2 or ECuSi (150 F max)	ECuAl-A2 or ECuSi (150 F max)	ECuSi (150 F max)	ECuAl-A2 (150 F max)	—	—
Copper nickels	ECuAl-A2 or ECuNi or ECu (1000 F)	ECuAl-A2 (150 F max)	ECuAl-A2 (150 F max)	ECuSn-C (150 F max)	ECuAl-A2 (150 F max)	ECuAl-A2 (150 F max)	—
NICKEL ALLOYS							
Nickel and Ni-Cu alloys	ECuNi or ERNiCu-7 (1000 F)	[b]	[b]	[b]	[b]	[b]	ECuNi or ERNiCu-7 (150 F max)
Ni-Cr, Ni-Fe and Ni-Cr-Fe alloys	ERNi-3 (1000 F)	[b]	[b]	[b]	[b]	[b]	ERNi-3 (150 F max)

			STEELS				
Low-carbon steel	ECuAl-A2 or ECu or ERNi-3 (1000 F)	ECuSn-C (600 F)	ECuAl-A2 (500 F)	ECuSn-C (400 F)	ECuAl-A2 (300 F)	ECuAl-A2 (150 F max)	ECuAl-A2 or ERNi-3 (150 F max)
Medium-carbon steel	ECuAl-A2 or ECu or ERNi-3 (1000 F)	ECuAl-A2 (600 F)	ECuAl-A2 (500 F)	ECuSn-C (400 F)	ECuAl-A2 (400 F)	ECuAl-A2 (150 F max)	ECuAl-A2 or ERNi-3 (150 F max)
High-carbon steel	ECuAl-A2 or ECu or ERNi-3 (1000 F)	ECuAl-A2 (600 F)	ECuAl-A2 (500 F)	ECuSn-C (500 F)	ECuAl-A2 (500 F)	ECuAl-A2 (400 F)	ECuAl-A2 or ERNi-3 (150 F max)
Low-alloy steel	ECuAl-A2 or ECu or ERNi-3 (1000 F)	ECuAl-A2 (600 F)	ECuAl-A2 (600 F)	ECuSn-C (500 F)	ECuAl-A2 (500 F)	ECuAl-A2 (400 F)	ECuAl-A2 or ERNi-3 (150 F max)
Stainless steel	ECuAl-A2 or ECu or ERNi-3 (1000 F)	ECuAl-A2 or ECuSn-C (600 F)	ECuAl-A2 (600 F)	ECuSn-C (400 F)	ECuAl-A2 (150 F max)	ECuAl-A2 (150 F max)	ECuAl-A2 or ERNi-3 (150 F max)
			CAST IRONS				
Gray and malleable irons	ECuAl-A2 or ECu (1000 F)	ECuAl-A2 or ECuSn-C (600 F)	ECuAl-A2 (600 F)	ECuSn-C (400 F)	ECuAl-A2 (400 F)	ECuAl-A2 or ECuSi (300 F)	ECuAl-A2 or ECuNi (150 F max)
Ductile iron	ECuAl-A2 or ECu (1000 F)	ECuAl-A2 (600 F)	ECuAl-A2 (600 F)	ECuSn-C (400 F)	ECuAl-A2 (150 F max)	ECuAl-A2 or ECuSi (150 F max)	ECuAl-A2 or ECuNi (150 F max)

[a]Electrode selections in table are based on weldability, except where mechanical properties are usually more important. Preheating is usually used only when at least one member is thicker than 1/8 in. or is highly conductive; see text. Preheat and interpass temperatures are subject to adjustment based on size and shape of weldment. [b]These combinations are seldom welded; as a starting point in developing welding procedures, use of ECuAl-A2 electrodes is recommended, except for combinations including phosphor bronzes.

Source: By permission, from *Metals Handbook* Volume 6, Copyright American Society for Metals, 1971.

In addition to the data presented in Tables 21-18 and 21-19, it should be pointed out that:

1. When silicon bronzes are joined to ferrous or nickel alloys, the silicon bronze side of the joint is usually overlaid with ECuAl-A2, the filler metal which will be used to weld the joint.

2. Overlaying is usually unnecessary in welding copper nickels or aluminum bronzes to steel. However, when welding these alloys to ferrous metals, care must be taken to ensure that the ferrous metal is well penetrated, otherwise the joint will only be as strong as a brazed joint and may come apart under severe stress at the ferrous interface.

3. Commercial coppers can also be joined to ferrous metals without the use of a buffer metal if filler metal ERNi-3 is used.

4. The overlay (weld deposit) should be at least ⅛ inch thick when welding nickel alloys to coppers and copper nickels. Either ERNi-3 (nickel), ERNiCu-7 (nickel-copper), or ECuNi (copper-nickel) can be used as filler metals.

21-3 MAGNESIUM AND MAGNESIUM ALLOYS

Magnesium and its alloys have the least density (weight per unit volume) of the commercial metals. Some comparative values of the properties of pure magnesium and other metals are given in Table 21-20.

Magnesium alloys are designated by the system shown in Table 21-21. Using the table and alloy AZ91C-T6, let's examine the system.

The first part of the system indicates that aluminum and zinc are the two principal alloying elements in our alloy. The second part of

Table **21-20** Comparative properties of magnesium and other metals

				APPROXIMATE RATIOS, MAGNESIUM = 1		
	APPROX. MELT. PT. °F	WEIGHT (lb./cu. in.)	WEIGHT (lb./cu. ft.)	*WEIGHT RATIO*	*THERMAL CONDUCTIVITY RATIO*	*EXPANSION RATIO*
Magnesium	1204	0.063	109	1.0	1.0	1.0
Aluminum	1215	0.098	170	1.55	1.4	0.9
Copper	1980	0.323	560	5.1	2.5	0.62
Steel	2700	0.284	490	4.5	0.5	0.45
18-8 Stainless	2600	0.286	495	4.56	0.17	0.63
Nickel	2646	0.322	560	5.1	0.38	0.52

Source: *Metals and How to Weld Them*, p. 277. Courtesy James F. Lincoln Arc Welding Foundation, Cleveland Ohio.

Table **21-21** Standard four-part system for alloy designation (ASTM)[a]

FIRST PART	SECOND PART	THIRD PART	FOURTH PART[b]
Indicates the two principal alloying elements	Indicates the amounts of the two principal alloying elements	Distinguishes between different alloys with the same percentages of the two principal alloying elements	Indicates condition and properties
Consists of two code letters representing the two main alloying elements arranged in order of decreasing percentage (or alphabetically if percentages are equal)	Consists of two numbers corresponding to rounded-off percentages of the two main alloying elements and arranged in same order as alloy designations in first part	Consists of a letter of the alphabet assigned in order as compositions become standard	Consists of a letter followed by a number (separated from the third part of the designation by a hyphen)
A—Aluminum B—Bismuth C—Copper D—Cadmium E—Rare Earth F—Iron H—Thorium K—Zirconium L—Beryllium M—Manganese N—Nickel P—Lead Q—Silver R—Chromium S—Silicon T—Tin Z—Zinc	Whole numbers	Letters of alphabet except I and O	F—As fabricated O—Annealed H10 and H11—Slightly strain hardened H23, H24 and H26—Strain hardened and partially annealed T4—Solution heat treated T5—Artificially aged only T6—Solution heat treated and artificially aged

Note: As an example of a typical four-part designation, AZ91C-T6 is explained in the text.

[a]This system is standard for both magnesium and aluminum alloys; thus, designation should be preceded by the name of the base metal unless base metal is obvious. [b]For a complete explanation of the system, see the article on temper designations in the Aluminum section. In this table, only the designations most commonly used in magnesium standards are given.

Source: By permission, from *Metals Handbook* Volume 1, Copyright American Society for Metals, 1961– .

the designation, 91, means that aluminum and zinc are present in rounded-off percentages of 9 and 1%, respectively. The third part, C (the *third* letter of the alphabet), indicates that this is the third alloy standardized with 9% aluminum and 1% zinc as the principal alloying additions. The fourth part, T6, notes, as in the aluminum system (Table 21-2), that the alloy is solution treated and artificially aged.

Although the possibility of a self-ignited fire when welding magnesium alloys above foil thickness is extremely remote, welders inexperienced in welding magnesium alloys believe the opposite. Magnesium alloys will not ignite until the temperature of fusion (1200°F) is reached; and if they do ignite, they will keep on burning only if the temperature of fusion is maintained in the metal. Graphite-base powder CG-11 or proprietary salt-base powder are both recommended by Underwriter's Laboratory to put out magnesium fires; therefore, they should be conveniently located. The following further precautions should also be taken:

1. Special care should be taken to prevent fires during the preparation of grooves, as magnesium fires usually occur in accumulations of grinding dust or machine chips, and in band saws used alternately on magnesium alloys and ferrous metals. If large amounts of fine particles are produced, they should be collected in a specially designed water-wash–type dust collector.

2. Special precautions pertaining to the handling of wet magnesium fires must be taken.

Magnesium alloys are usually supplied by the mill with either a protective oil coating, an acid-pickled surface, or a chromate-coated surface. The protective coating must be removed by degreasing and/or alkaline cleaning, followed by mechanical cleaning, before welding.

21-4 NICKEL AND HIGH-NICKEL ALLOYS

For many years, nickel alloys were identified by trade names or by military or technical-society material specifications. More recently, however, manufacturers of the wrought nickel alloys have used the three-digit ASTM numbering system shown in Table 21-22. In practice, manufacturers usually apply their particular trade name ahead of the three-digit number (see Table 21-23).

In general, nickel and its alloys are specified for two principal types of application:

1. Where nickel performs the function of corrosion-resistant structural materials.

2. Where physical properties of nickel and its alloys play the dominant role.

Table **21-22** ASTM three-digit numbering system for nickel and nickel alloys

ASTM SERIES	ALLOY GROUP
200	Nickel, solid-solution
300	Nickel, precipitation-hardenable
400	Nickel-copper, solid-solution
500	Nickel-copper, precipitation-hardenable
600	Nickel-chromium, solid-solution
700	Nickel-chromium, precipitation-hardenable
800	Nickel-iron-chromium, solid-solution
900	Nickel-iron-chromium, precipitation-hardenable

Table **21-23** Nominal compositions of weldable wrought nickel and nickel alloys

ALLOY	COMPOSITION
	NICKEL AND SOLID-SOLUTION ALLOYS
Nickel 200	99.5 Ni, 0.06 C, 0.25 Mn, 0.15 Fe
Nickel 201	99.5 Ni, 0.01 C, 0.20 Mn, 0.15 Fe
Nickel 205	99.5 Ni, 0.06 C, 0.20 Mn, 0.10 Fe, 0.04 Mg
Nickel 211	95.0 Ni, 0.10 C, 4.75 Mn, 0.05 Fe
Nickel 220	99.5 Ni, 0.06 C, 0.12 Mn, 0.05 Fe, 0.04 Mg
Nickel 230	99.5 Ni, 0.09 C, 0.10 Mn, 0.05 Fe, 0.06 Mg
Nickel 233	99.5 Ni, 0.09 C, 0.18 Mn, 0.05 Fe, 0.07 Mg
Nickel 270	99.98 Ni, 0.01 C
Monel 400	66.0 Ni, 31.5 Cu, 0.90 Mn, 1.35 Fe
Monel 401	44.5 Ni, 53.0 Cu, 1.70 Mn, 0.20 Fe, 0.50 Co
Monel 404	55.0 Ni, 44.0 Cu
Monel R-405	66.0 Ni, 31.5 Cu, 1.35 Fe
	PRECIPITATION-HARDENABLE ALLOYS
Monel K-500	65.0 Ni, 29.50 Cu, 0.60 Mn, 1.00 Fe, 0.60 Ti, 2.73 Al
Monel 502	66.5 Ni, 28.0 Cu, 0.75 Mn, 1.0 Fe, 0.25 Ti, 3.0 Al
	SPECIAL-PURPOSE ALLOYS
4-79 Moly-Permalloy	79 Ni, 4 Mo, rem Fe
Mumetal (AMS 7701)	80 Ni, 15 Fe, rem Cu, Mo or Cr
Nichrome	57 Ni, 16 Cr, 27 Fe
Invar 36	35.5 Ni, 0.18 C, 0.42 Mn, rem Fe

Source: By permission, from *Metals Handbook* Volume 6, Copyright American Society for Metals, 1971.

SOLID-SOLUTION STRENGTHENED ALLOYS 200-series-nickel solid-solution alloys are used for applications involving exposures to temperatures above 600°F (315°C), such as food-processing equipment, acid piping and handling equipment, shipping containers for chemicals, and operating metal parts for electrical equipment.

400-series nickel-copper solid-solution alloys are used for roofs, gutters and flashing, propellers and propeller shafts, pump shafts, impellers, condenser tubes in sea-water service, coils in zinc-chloride evaporators, pulp washers, thickeners, screens in sulfide pulp mills, smoke-scrubbing systems, crude petroleum stills, processing vessels, piping, boiler-feed water heaters, other heat exchangers, marine fixtures and fasteners, and pharmaceutical, electrical, laundry, and textile equipment.

Nickel-molybdenum alloys, such as Hastelloy B and N, are extensively used for equipment designed to perform in corrosive environments at temperatures between 1400 and 2000°F. Alloy W is used as a filler metal for joining dissimilar metals and has good corrosion and oxidation resistances.

Nickel-chromium-molybdenum alloys, such as Hastelloy C, C-276, F, G, and X, have good corrosion resistance and high-temperature properties. They find applications in the linings of sulfite pulp digesters, in equipment for handling ammonia and magnesia-based pulp liquors, industrial furnace parts, jet engine tail pipes, afterburners, and turbine blades and vanes, as well as space heaters.

Of the nickel-silicon alloys, the three most widely used are alloys 600, 601, and 625. Alloy 600 contains approximately 77% nickel, 16% chromium, and 8% iron. The alloy has good corrosion resistance at elevated temperatures, along with good high-temperature strength. Because of its resistance to chloride-ion stress-corrosion cracking and corrosion by high-purity water, alloy 600 is often used in nuclear reactors. Alloy 601 is a versatile high-temperature alloy that has shown outstanding resistance to oxidation and scaling at temperatures as high as 2200°F. Nickel-chromium alloy 625 has additions of 9% molybdenum and 4% columbium, which considerably increase the room- and high-temperature strength and corrosion resistance. Alloy 625 has exceptionally good weldability, along with good formability.

The 50%-nickel, 50%-chromium alloy is another material in this group, but it is actually a two-phase alloy rather than a solid-solution alloy. The cast version of this alloy is of increasing commercial importance because of its superior high-temperature resistance to high-sulfur and high-vanadium fuels. The material specification for casting is ASTM-A-560-66 grade 50Cr-50Ni.

Of the nickel-iron-chromium alloys, alloys 800, 825, and 20Cb are the most widely used. Alloy 800 is used extensively for high-temperature applications because of its strength and excellent oxidation and carburization resistance. Alloy 825 and alloy 20Cb are used extensively in corrosive environments because of their resistance to chloride-ion stress-corrosion cracking and to reducing acids.

The principal alloys in the nickel-molybdenum group are Hastelloy B, N, and W. They are nickel-base alloys and contain from 16 to 28% molybdenum and lesser amounts of chromium and iron. The alloys are primarily used for their corrosion resistance. Alloy B has excellent resistance to hydrochloric and other acids. Alloy N was developed for resistance to molten fluoride salts and has good resistance to the many acids. Alloy W is a filler metal for joining dissimilar metals and has good corrosion and oxidation resistance.

Included in the nickel-chromium-molybdenum group are Hastelloy alloys C, C-276, F, G, and X. Alloy C has good corrosion resistance and high-temperature properties. Alloy C-276 is similar to alloy C, but has lower carbon and silicon contents to reduce the formation of grain-boundary precipitates and to enable the alloy to be used in the as-welded condition. Alloy F resists corrosion by both oxidizing and reducing agents. Alloy G has excellent resistance to hot sulfuric, phosphoric, and nitric acids. Alloy X has high strength and oxidation resistance at temperatures up to 2200°F.

PRECIPITATION-HARDENABLE ALLOYS Of the two alloys in the 200 series, Permanickel and Duranickel, only Duranickel (alloy 301) can be welded. Duranickel is used for parts that must operate at very high temperatures, such as extrusion presses and glass molds. It is also used for springs and diaphragms.

There are three alloys in the 500 series (nickel-copper alloys)—monel K-500, monel 502, and monel 505. Monel 502 can only be repair welded, while monel 505 has only limited weldability. The two alloys are used where high strength and hardness, in addition to corrosion resistance, are required. Such parts include valves, pump parts, propellor shafts, marine fixtures, fasteners, crude petroleum stills, heat exchangers, screw-machine parts, oil-well equipment, and springs.

The six weldable alloys in the 700 series (nickel-chromium alloys) have specific applications. Inconel 706 is used for turbine parts; Inconel 718 and Inconel X-750 are used for aircraft turbine parts, pumps, and rocket motors; Inconel 722 is used for industrial turbine frames; Inconel 751 is used for diesel exhaust valves.

The two alloys in the 900 series (nickel-iron-chromium) are used to make aircraft turbine parts.

The welding processes commonly used for nickel and high-nickel alloys are listed in Table 21-24. Because nickel and nickel alloys are susceptible to embrittlement by lead, sulfur, phosphorus, and some low-melting-point metals that may be present in grease, oil, paint, marking crayons and inks, forming lubricants, cutting fluids, shop

Table **21-24** Joining processes applicable to nickel alloys

ALLOY	SHIELDED METAL–ARC	GAS TUNGSTEN ARC	GAS METAL ARC	SUBMERGED ARC	ELECTRON BEAM	OXYACETYLENE	BRAZING
Nickel 200	X	X	X	X	X	X	X
Nickel 201	X	X	X	X	X	—	X
MONEL alloy 400	X	X	X	X	X	X	X
MONEL alloy 401	—	X	—	—	X	—	X
MONEL alloy 404	—	X	—	—	—	X	X
MONEL alloy R-405	X	X	X	—	X	X	X
MONEL alloy K-500	X	X	X	—	X	X	X
MONEL alloy 502	X	X	X	—	X	X	X
INCONEL alloy 600	X	X	X	X	X	X	X
INCONEL alloy 601	X	X	X	X	X	X	X
INCONEL alloy 625	X	X	X	X	X	—	X
INCONEL alloy 706	—	X	—	—	X	—	X
INCONEL alloy 718	—	X	X	—	X	—	X
INCONEL alloy X-750	—	X	—	—	X	—	X
INCOLOY alloy 800	X	X	X	X	X	X	X
INCOLOY alloy 825	X	X	X	—	X	—	X
INCOLOY alloy 901	—	X	—	—	X	—	X
Alloy 713 C	—	—	—	—	—	—	X
UDIMET 500	X	X	—	—	X	—	X
UDIMET 700	—	X	—	—	X	—	X
RENE 41	—	X	—	—	X	—	X
ASTROLOY	—	X	—	—	X	—	X
WASPALOY	—	X	—	—	X	—	X
CARPENTER 20Cb3	X	X	X	—	X	—	X
HASTELLOY alloy B	X	X	X	—	X	—	X
HASTELLOY alloy C	X	X	X	—	X	—	X
HASTELLOY alloy C-276	X	X	X	—	X	—	X
HASTELLOY alloy D	—	—	—	—	—	X	X
HASTELLOY alloy F	X	X	—	—	X	—	X
HASTELLOY alloy G	X	X	X	—	—	—	X
HASTELLOY alloy N	X	X	—	—	X	—	X
HASTELLOY alloy R-235	X	X	X	—	X	—	X
HASTELLOY alloy X	X	X	X	—	X	—	X
Alloy IN100	—	—	—	—	—	—	X

X = Weldable by this method.
— = Not weldable by this method or no information available; consult manufacturer.

Source: American Welding Society, *Welding Handbook*, 6th ed., section 4, p. 67.8.

dirt, and processing chemicals, it is essential that both sides of a joint to be welded be completely clean.

Shop dirt can be removed by vapor degreasing or by swabbing with acetone or other nontoxic solvents. Paint and other materials that cannot be dissolved with cleaning fluids may require the use of methylene chloride, alkaline cleaners, or special proprietary compounds. Oxides can be removed by wire brushing, and imbedded particles can be removed by grinding or abrasive blasting, followed by a swabbing with 10% (by volume) hydrochloric acid solution, followed by a thorough wash with tapwater.

Joint designs for welding nickel alloys should be tailored to the particular joining process. As the molten nickel alloy does not spread and wet the base metal as readily as carbon steel and stainless steel, the joints must be sufficiently open to permit manipulation of the filler metal and placement of the weld bead. Typical weld joint configurations are shown in Figure 21-5. The appropriate dimensions and amounts of filler metal and electrode required appear in Table 21-25. To control the magnitude of the residual and processing stresses, precipitation-hardening alloys are welded in the stress-relieved condition, or are sequence welded, or are reannealed during fabrication. *Note:* To avoid prolonged exposure of the weld to temperatures within the precipitation-hardening range, the part should be heated in a furnace that has been preheated to about 300°F (149°C) above the solution-treatment temperature (annealing temperature).

Solid-solution-strengthened alloys are welded in the solution-treated condition. Postweld heat treatment includes resolution and an aging treatment.

General conditions for GMAW (MIG) welding of nickel-base alloys are listed in Table 21-26. For shielded metal–arc (SMAW) welding of nickel and nickel alloys, reverse-polarity dc is used. The flat position should be used whenever possible. For vertical and overhead welding the arc should be slightly shorter than for flat-position welding, and the current should be approximately 10% lower (Table 21-27).

Single-pass welds can be made with the stringer technique. The weaving technique should be used on all sections requiring more than one pass. When the weave is used, it should be no wider than about three times the electrode diameter. The following precautions should be maintained:

1. If spatter occurs, it is an indication of one or a combination of these conditions: (a) the arc is too long; (b) the amperage is too high; or (c) straight polarity is being used.

2. If arc blow occurs, change the location of the ground clamp on the workpiece or the direction of the electrical path to the arc (you may be using straight polarity).

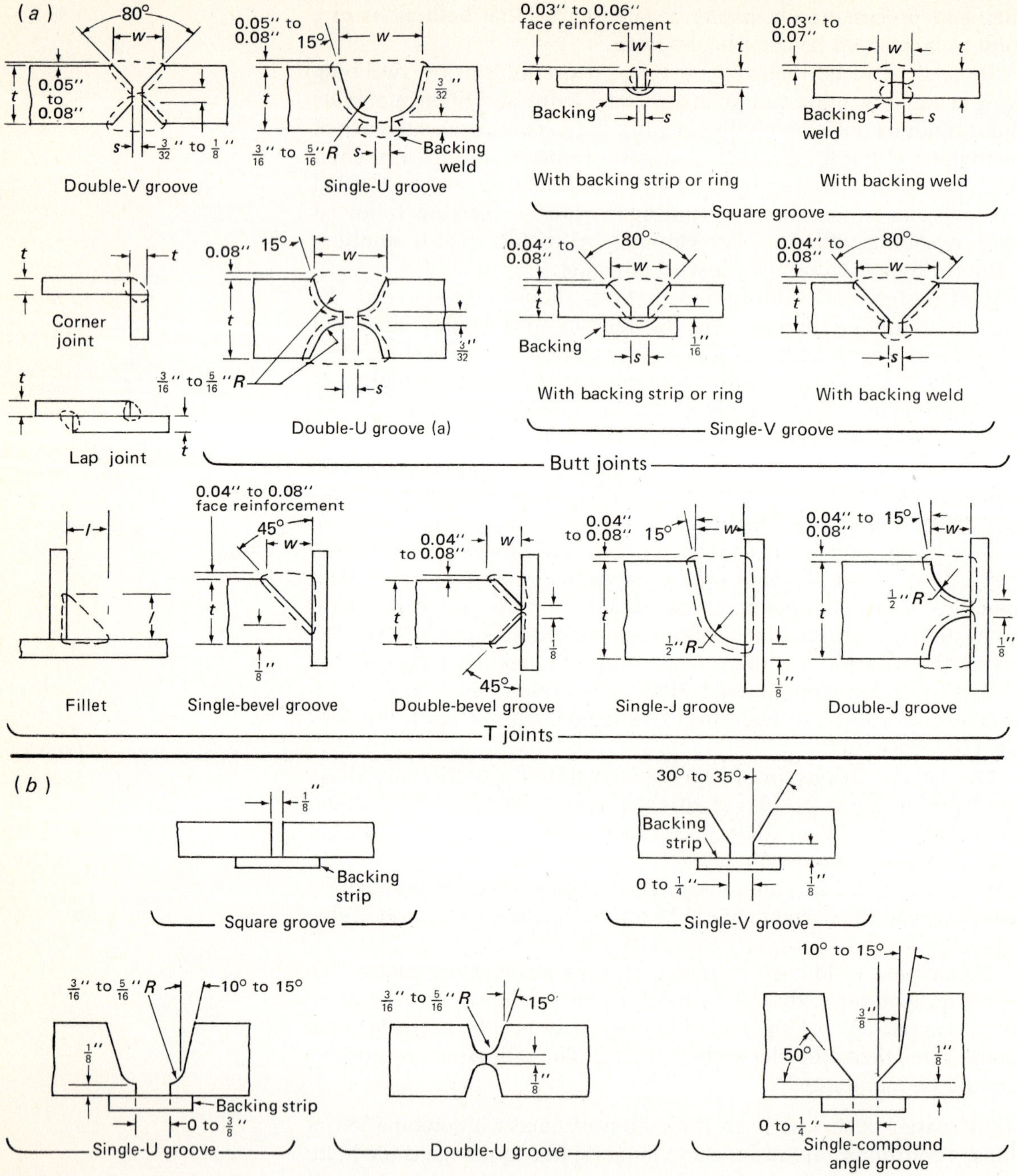
(a)
Double-V groove
Single-U groove
Backing weld
With backing strip or ring
With backing weld
Square groove
Corner joint
Lap joint
Double-U groove (a)
Single-V groove
Butt joints
Fillet
Single-bevel groove
Double-bevel groove
Single-J groove
Double-J groove
T joints
(b)
Backing strip
Square groove
Single-V groove
Single-U groove
Double-U groove
Single-compound angle groove

Figure **21-5** (*a*) Typical joint designs for gas tungsten–arc, gas metal–arc, and shielded metal–arc welding of nickel-base alloys. (*b*) Joint designs for submerged metal–arc welding of nickel-base alloys. (By permission, from *Metals Handbook* Volume 6, Copyright American Society for Metals, 1971.)

Table **21-25** Dimensions of grooves, amount of filler metal, and consumable electrode required for nickel-base heat-resisting alloys

BASE-METAL THICKNESS (in.)	WIDTH OF GROOVE OR BEAD (*w*), (in.)	MAXIMUM ROOT OPENING (*s*), (in.)	APPROXIMATE AMOUNT OF METAL DEPOSITED PER LINEAR FOOT		APPROX WEIGHT OF ELECTRODE REQUIRED PER LINEAR FT. LB[a]
			CU IN.	*LB*	
SQUARE-GROOVE BUTT JOINT WITH BACKING STRIP OR RING					
0.037	1/8	0	0.07	0.02	0.025
0.050	5/32	0	0.13	0.04	0.05
0.062	3/16	0	0.13	0.04	0.06
0.093	3/16 to 1/4	0 to 1/32	0.18	0.06	0.08
0.125	1/4	1/32 to 1/16	0.22	0.07	0.09
SQUARE-GROOVE BUTT JOINT WITH BACKING WELD					
1/8	1/4	0 to 1/32	0.35	0.11	0.15
5/16	3/8	1/32 to 1/16	0.74	0.24	0.32
1/4	7/16	1/16 to 3/32	0.97	0.31	0.42
SINGLE-V-GROOVE BUTT JOINT WITH BACKING STRIP OR RING					
3/16	0.35	1/8	0.72	0.227	0.31
1/4	0.51	3/16	1.39	0.443	0.61
5/16	0.61	3/16	1.84	0.582	0.80
3/8	0.71	3/16	2.36	0.745	1.02
1/2	0.91	3/16	3.68	1.16	1.59
5/8	1.16	3/16	5.10	1.61	2.21
SINGLE-V-GROOVE BUTT JOINT WITH BACKING WELD					
1/4	0.41	3/32	1.33	0.42	0.58
5/16	0.51	3/32	1.71	0.54	0.74
3/8	0.65	1/8	2.30	0.73	1.00
1/2	0.85	1/8	3.85	1.21	1.67
5/8	1.06	1/8	4.63	1.46	2.00
DOUBLE-V-GROOVE BUTT JOINT					
1/2	0.40	1/8	2.65	0.89	1.16
5/8	0.49	1/8	3.45	1.08	1.48
3/4	0.62	1/8	4.60	1.46	2.00
1	0.81	1/8	7.70	2.42	3.34
1 1/4	1.03	1/8	9.26	2.92	4.00
SINGLE-U-GROOVE BUTT JOINT[b]					
1/2	0.679	1/8	3.27	1.03	1.41
5/8	0.745	1/8	4.37	1.38	1.90

Table **21-25** *(cont.)*

BASE-METAL THICKNESS (in.)	WIDTH OF GROOVE OR BEAD (*w*), (in.)	MAXIMUM ROOT OPENING (s), (in.)	APPROXIMATE AMOUNT OF METAL DEPOSITED PER LINEAR FOOT		APPROX WEIGHT OF ELECTRODE REQUIRED PER LINEAR FT.
			CU IN.	*LB*	LB[a]
SINGLE-U-GROOVE BUTT JOINT[b]					
3/4	0.813	1/8	5.33	1.68	2.30
1	0.957	1/8	8.35	2.63	3.60
1 1/4	1.073	1/8	11.48	3.62	4.96
1 1/2	1.215	1/8	15.16	4.79	6.55
1 3/4	1.349	1/8	18.90	5.98	8.19
2	1.485	1/8	23.45	7.40	10.12
DOUBLE-U-GROOVE BUTT JOINT[b]					
1	0.679	1/8	6.54	2.06	2.82
1 1/4	0.745	1/8	8.74	2.76	3.80
1 1/2	0.813	1/8	10.66	3.36	4.60
2	0.957	1/8	16.66	5.26	7.20
2 1/2	1.073	1/8	22.96	7.24	9.92
CORNER AND LAP JOINT					
1/16	—	—	0.05	0.02	0.04
1/8	—	—	0.15	0.05	0.07
3/16	—	—	0.33	0.10	0.14
1/4	—	—	0.59	0.19	0.26
3/8	—	—	1.32	0.42	0.57
1/2	—	—	2.35	0.74	1.02
T-JOINT WITH FILLET WELD					
	—	1/8	0.09	0.03	0.04
	—	3/16	0.22	0.07	0.10
	—	1/4	0.38	0.12	0.16
	—	5/16	0.59	0.19	0.26
	—	3/8	0.84	0.27	0.37
	—	1/2	1.50	0.47	0.64
	—	5/8	2.34	0.74	1.01
	—	3/4	3.38	1.07	1.46
	—	1	6.00	1.90	2.60
SINGLE-BEVEL-GROOVE T-JOINT					
1/4	0.125	—	0.22	0.07	0.09
5/16	0.188	—	0.40	0.13	0.17
3/8	0.250	—	0.61	0.19	0.26
1/2	0.375	—	1.21	0.38	0.52

Table **21-25** (*cont.*)

BASE-METAL THICKNESS (in.)	WIDTH OF GROOVE OR BEAD (*w*), (in.)	MAXIMUM ROOT OPENING (*s*), (in.)	APPROXIMATE AMOUNT OF METAL DEPOSITED PER LINEAR FOOT		APPROX WEIGHT OF ELECTRODE REQUIRED PER LINEAR FT. LB[a]
			CU IN.	*LB*	
SINGLE-BEVEL-GROOVE T-JOINT					
5/8	0.500	—	1.98	0.63	0.86
3/4	0.625	—	2.95	0.93	1.28
1	0.875	—	5.57	1.77	2.42
DOUBLE-BEVEL-GROOVE T-JOINT					
1/2	0.188	—	0.78	0.25	0.34
5/8	0.250	—	1.24	0.39	0.54
3/4	0.313	—	1.78	0.56	0.77
1	0.438	—	3.13	0.99	1.36
1 1/4	0.563	—	4.87	1.54	2.15
1 1/2	0.688	—	7.00	2.21	3.03
1 3/4	0.813	—	9.47	3.00	4.09
2	0.938	—	12.33	3.90	5.35
SINGLE-J-GROOVE T-JOINT					
1	0.625	—	5.64	1.78	2.4
1 1/4	0.719	—	7.91	2.50	3.4
1 1/2	0.781	—	10.20	3.23	4.4
1 3/4	0.875	—	12.95	4.09	5.6
2	0.969	—	15.60	4.93	6.8
2 1/4	1.031	—	18.35	5.80	8.0
2 1/2	1.094	—	21.95	6.94	9.5
DOUBLE-J-GROOVE T-JOINT					
1	0.500	—	4.67	1.48	2.0
1 1/4	0.563	—	6.90	1.90	2.6
1 1/2	0.594	—	8.10	2.56	3.5
1 3/4	0.625	—	9.83	3.11	4.3
2	0.656	—	12.06	3.81	5.2
2 1/4	0.688	—	14.29	4.51	6.2
2 1/2	0.750	—	16.68	5.27	7.2

[a]To obtain linear feet of weld per pound of consumable electrode, take the reciprocal of pounds per linear foot. If the underside of the first bead is chipped out and welded, add 0.21 lb of metal deposited (equivalent to 0.29 lb of consumable electrode). [b]For gas metal–arc welding (except with the short-circuiting arc), foot radius should be one-half the value shown and bevel angle should be twice as great.

Source: By permission, from *Metals Handbook* Volume 6, Copyright American Society for Metals, 1971.

3. To reduce the probability of crater oxidation due to arc interruption, and to prepare for restriking the arc by developing a rolled edge, the arc should be shortened and the rate of travel increased before breaking the arc.

4. A good technique for restriking the arc is to restrike it at about 1 inch behind the crater on top of the previous pass. The excess metal can be removed by grinding after the weld is completed.

Solid-solution alloys up to 3 inches thick can be welded by the submerged-arc process. Some joint designs used in this process are shown in Figure 21-5.

Direct current with either straight or reverse polarity can be used, but reverse polarity is preferred because it produces a flatter bead with deeper penetration and at lower arc voltage than straight polarity (30 amps versus 35 amps, respectively).

The oxyacetylene process uses a slightly reducing flame with a tip of the same size or one size larger than that for steel. Because generator-produced acetylene may contain sulfur, only clean, bottled acetylene should be used. Commercially pure nickel can be oxyacetylene welded without flux, but other alloys cannot.

Welding of a nickel alloy to steel or to a different nickel alloy can be accomplished after giving due consideration to such factors as the difference in the thermal coefficient of expansion between the filler

Table **21-26** Conditions for GMAW welding of nickel and nickel-based alloys

BASE METAL	FILLER METAL	PEAK VOLTAGE	AVE. VOLTAGE	AVE. CURRENT (amp)	WIRE FEED (ipm)
MIG PROCESS, SPRAY TRANSFER (Argon shielding gas flow, 60 cfh. Weld position, flat. Filler metal diameter, 0.062 in.)					
Nickel 200	ERNi-3	—	29–31	375	205
MONEL alloy 400	ERNiCu-7	—	28–30	290	200
INCONEL alloy 600	ERNiCr-3	—	28–30	265	200
SHORT-CIRCUITING ARC PROCESS (Argon-helium shielding gas flow, 50 cfh. Weld position, vertical. Filler metal diameter, 0.035 in.)					
Nickel 200	ERNi-3	—	20–21	160	360
MONEL alloy 400	ERNiCu-7	—	16–18	130–135	275–290
INCONEL alloy 600	ERNiCr-3	—	16–18	120–130	270–290
PULSING-ARC PROCESS (Argon or argon-helium shielding gas flow, 25–35 cfh. Weld position, vertical. Filler metal diameter, 0.045 in.)					
Nickel 200	ERNi-3	46	21–22	150	160
MONEL alloy 400	ERNiCu-7	40	21–22	110	140
INCONEL alloy 600	ERNiCr-3	44	20–22	90–120	140

Source: American Welding Society, *Welding Handbook*, 6th ed., section 4, p. 67.19.

Table **21-27** Recommended electrode diameter and welding current for shielded metal–arc welding of various thicknesses of high-nickel alloys and nickel-copper alloys in the flat position

BASE-METAL THICKNESS (in.)	ELECTRODE DIAMETER[a] (in.)	CURRENT[b] (amp)
	HIGH-NICKEL ALLOYS	
0.037	3/32	c
0.043	3/32	c
0.050	3/32	c
0.062	3/32	75
0.078	3/32	80
0.093	3/32	85
0.109	1/8	105
0.125	1/8	105
0.125	3/32 5/32	80–150
0.140	5/32	130
0.156	5/32	135
0.187[d]	5/32	150
	NICKEL-COPPER ALLOYS	
0.037	3/32	c
0.043	3/32	c
0.050	3/32	c
0.062	3/32	50
0.078	3/32	55
0.093	3/32	60
0.109	3/32	60
0.109	1/8	65
0.125	3/32–5/32	60–140
0.140	3/32–5/32	60–140
0.156	3/32–5/32	60–140
0.250	3/32–5/32	60–140
0.375	3/32–3/16	60–180
0.500[d]	3/32–3/16	60–180

[a]Where a range is shown, the smaller-diameter electrodes are used for the first passes at the bottom of the groove, and the joints are completed with the larger-diameter electrodes. [b]Current should be in the range recommended by the electrode manufacturer. [c]Use minimum amperage at which arc control can be maintained. [d]And thicker.

Source: By permission, from *Metals Handbook* Volume 6, Copyright American Society for Metals, 1971.

metal and the base metals, the effects of dilution at both base-metal/weld interfaces, and the possibility of changes in the solid state (change in microstructure) after long service at high temperatures.

The spray or pulsed-arc mode of metal transfer should be used when welding with the GMAW (MIG) process. When using the shielded metal–arc process, the current should be maintained near the middle range recommended for the electrode being used. To keep dilution to a minimum with the SMAW process, the electrode should be manipulated so as to dissipate the arc force on the already-deposited metal.

21-5 LEAD AND LEAD ALLOYS

The available commercial grades of lead are classified under ASTM specifications B 29-55 and B 325-58T.

Two methods are used to designate lead alloys: the ASTM method, which is based on major alloying elements present in the alloy, and the traditional method. The traditional method uses colloquial terms to designate the alloy. Table 21-28 lists the alloys by their ASTM designations and their colloquial names, and keys the alloys to their principal uses.

Joints in lead are made by *lead burning,* which is the colloquial term for the fusion-welding process when it is used to join lead. Three fuel gases are commonly used to burn lead: acetylene, natural gas, and hydrogen, each combined with oxygen. Although oxyhydrogen and oxyacetylene mixtures can be used for all-position welding, the oxyhydrogen mixture is preferred by most lead welders because the flame can be more precisely controlled. The oxy–natural gas mixture is recommended for flat-position welding only.

The joints to be welded should be prepared as shown in Figure 21-6. The weld should be made using a neutral flame and gas pressures varying from 1½ to 5 psi, depending on the thickness of the section to be welded. Figures 21-7 and 21-8 illustrate some of the manipulative techniques that have proven successful.

Because pure lead does not readily wet or alloy with other metals, it is necessary to mechanically or chemically clean the surface of the metal (usually steel) that is to be leaded and then apply a flux of zinc-ammonium chloride and stannous chloride to the surface. The flux galvanizes the steel with a layer of tin. The lead is then applied simultaneously to the galvanized surface and the lead filler metal. The following procedures should be followed:

1. Experience has proven that three is the usual minimum number of passes to build the surface to a thickness of ¼ inch. A proportionately greater number of passes is required to build up a thicker layer.

2. After each pass, the lead should be carefully examined for pin holes and other faults. If any pin holes and/or faults exist, they should be repaired by scraping down to the steel, refluxing the scraped area, and filling it with lead.

3. When the required thickness is obtained, the lead may be smoothed by machining or scraping.

Warning: Because of the toxic nature of lead, lead welding should be performed in well-ventilated locations to prevent inhalation of lead oxide fumes. Welders working in confined areas, such as the bottom of a deep, open tank, should be provided with a positive air supply directed into the tank by a fan or blower to a position below the breathing level. In enclosed areas, such as pressure vessels or closed tanks, each worker should be required to wear an approved air-supplied respirator or mask.

Table **21-28** Lead and its alloys—cast, wrought

	TYPE				
	CHEMICAL LEAD	*COMMON LEAD*	*TELLURIUM LEAD*	*LEAD-0.08 Ca*[a]	*LEAD 1 Sn-0.08 CA*[ab]
Density, lb/cu in.	0.41	0.41	0.41	0.41	0.41
Melting point, F	618	621	617	619–621	610–640
Ther cond (212 F), Btu/hr/sq ft/°F/ft	19.6	19.6	19.3	—	—
Coef of ther exp., per °F	16.3×10^{-6}	16.3×10^{-6}	16×10^{-6}	—	—
Specific heat (32 F), Btu/lb/°F	0.031	0.031	0.031	0.032	0.032
Mod of elast in ten, psi	2×10^6	2×10^6	2×10^6	—	—
Tensile strength, 1000 psi					
Rolled	2.4	2.1	2.8	5.5	9.0
Extruded	2.5	2.0	—	4.0	—
Chill cast	2.6	2.0	3.0	4.5	—
Yld str, (0.5% Off) 1000 psi					
Rolled	1.6	—	—	5.0	8.0
Extruded	—	—	1500	—	—
Elong (in 2 in.), %					
Rolled	51	43	47	35	20
Extruded	50	—	—	60	20
Chill cast	45	47	45	40	35
Hardness (Brinell)					
Rolled	4.7	—	5.5	12	17
Extruded	—	—	6	—	—
Chill cast	5.2	4.2	5.7	9	17

Table **21-28** *(cont.)*

	TYPE				
	CHEMICAL LEAD	*COMMON LEAD*	*TELLURIUM LEAD*	*LEAD-0.08 Ca*[a]	*LEAD 1 Sn-0.08 CA*[ab]
Impact str (Charpy, chill cast), ft-lb	—	10	—	—	—
Endur limit (10^7 cycles, extr & aged), psi	725	470	1000	—	—
Shear str (chill cast), psi	—	1820	—	—	—
Creep str (0.1% per yr, rolled, 85 F), psi	300	250	300	600[c]	> 2000[d]
Casting temp range, F	790–850	790–850	790–850	750–850	750–850
Joining	Soft sold with 50–50 or 40–60 solder using rosin or stearic acid flux. Oxyhyd weld (lead burning); slightly reducing flame; no flux			Soft sold with Pb-Sn or Pb-Sb stick, $ZnCl_2$ flux. Oxydyd weld; filler metal, no flux	
Corrosion resistance	Resists sulfuric, sulfurous, phosphoric and chromic acids. Attacked by acetic, formic and nitric acids. Resistant to atmosphere and fresh and salt water				
Available forms	Castings, rolled and extruded shapes, sheet			Strip	
Uses	Nuclear reflectors and shields; anodes for cathodic protection			Similar to soft lead and Pb-Sb alloys and for uses requiring machinability, higher strength and creep resistance	
	Chemical apparatus	Batteries, cable sheath, ammo, calking, coatings	Chemical apparatus		

[a]Alloy represents median calcium content for family of alloys containing 0.06 to 0.10% Ca. [b]Alloy represents median tin content for family of alloys containing 0.5 to 1.5% Sn. [c]For 0.5% ext/yr. [d]Negligible creep after 10,000 hrs at 2000 psi.

Table **21-28** *(cont.)*

	TYPE				
	1% Sb-LEAD	4% Sb-LEAD	6% Sb-LEAD	8% Sb-LEAD	9% Sb-LEAD
Density, lb/cu in.	0.406	0.398	0.393	0.388	0.385
Melting temp range, F	608–595	570–486	545–486	520–486	509–486
Ther cond (212 F), Btu/hr/sq ft/°F/ft	19	18	17	16	16
Coef of ther exp (68-212 F), per °F	16×10^{-6}	15.5×10^{-6}	15.4×10^{-6}	14.5×10^{-6}	14.4×10^{-6}
Spec ht, Btu/lb/°F	0.031	0.032	0.032	0.032	0.032
Ten str, 1000 psi					
Rolled	3.0	4.0	4.2	4.6	4.7
Extruded	2.9	3.1	3.3	3.3	—
Chill cast	3.4	5.6	6.8	7.4	7.4
Elong (in 2 in.), %					
Rolled	50	50	50	30	20
Extruded	58	58	65	75	—
Chill cast	16	22	24	19	—
Hardness (Brinell)					
Rolled	6	8	9	9	9
Extruded	5.1	8.9	10.7	12.4	—
Chill cast	7	10	11.8	13.3	15.4
Endur limit, psi[b]					
Rolled	1200	1500	1500	1750	1800
Extruded	—	—	1200	—	—
Chill cast	—	—	2500	—	2700
Creep str (86 F)[c]				—	
Rolled	250	250	400	400	400
Extruded	350	210	—	—	—
Casting Temp Range, F	750–925	750–925	750–850	750–925	750–925
Joining	Soft solder with 50–50 or 40–60 solder using rosin or stearic acid flux. Oxyhydrogen welding (lead burning): no flux, slightly reducing flame				
Corrosion resistance	Similar to soft lead (see above)				
Uses	Nuclear reflectors and shields; anodes for cathodic protection				
	Cable sheathing	Rolled sheet for roofing and flashing; extruded pipe for corrosion resistance applications requiring greater strength than soft lead; battery grids			

The mechanical properties of antimony-lead alloys are influenced by thermal history, amount of deformation, and period of aging. But since commercially rolled products usually have attained a stable condition if they are allowed to cool normally to room temperature, the values given are reasonable for design purposes. [b]2×10^7 cycles. [c]1% extension in 10,000 hr.

Source: *Materials Selector, Materials Engineering,* a Penton/DPC Reinhold Publication.

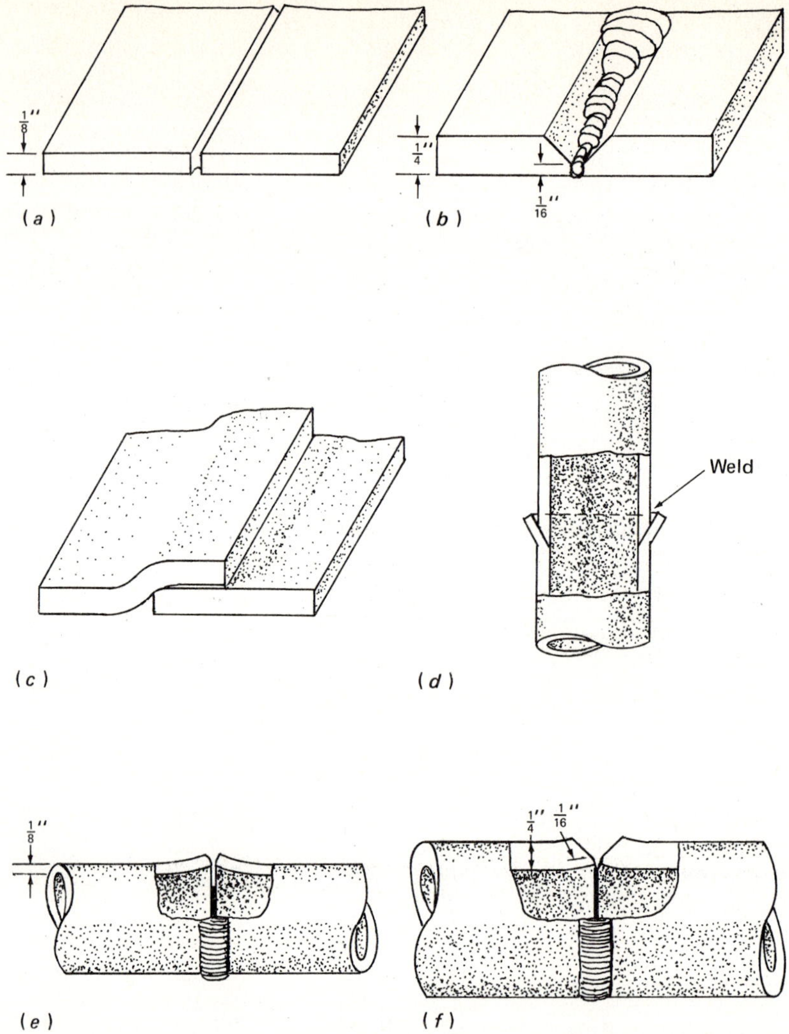
1"/8
(a)
1"/4
1"/16
(b)
(c)
Weld
(d)
1"/8
(e)
1"/4
1"/16
(f)

Figure **21-6** Typical joints for use in lead welding: (*a*) butt joint for less than ⅛-inch sheet; (*b*) butt joint for greater than ⅛-inch sheet; (*c*) lap joint; (*d*) cup joint; (*e*) butt joint for less than ⅛-inch-thick pipe; (*f*) butt joint for greater than ⅛-inch-thick pipe. (Redrawn by permission from American Welding Society, *Welding Handbook*, 6th ed., 1972, section 4, pp. 71.4, 71.5.)

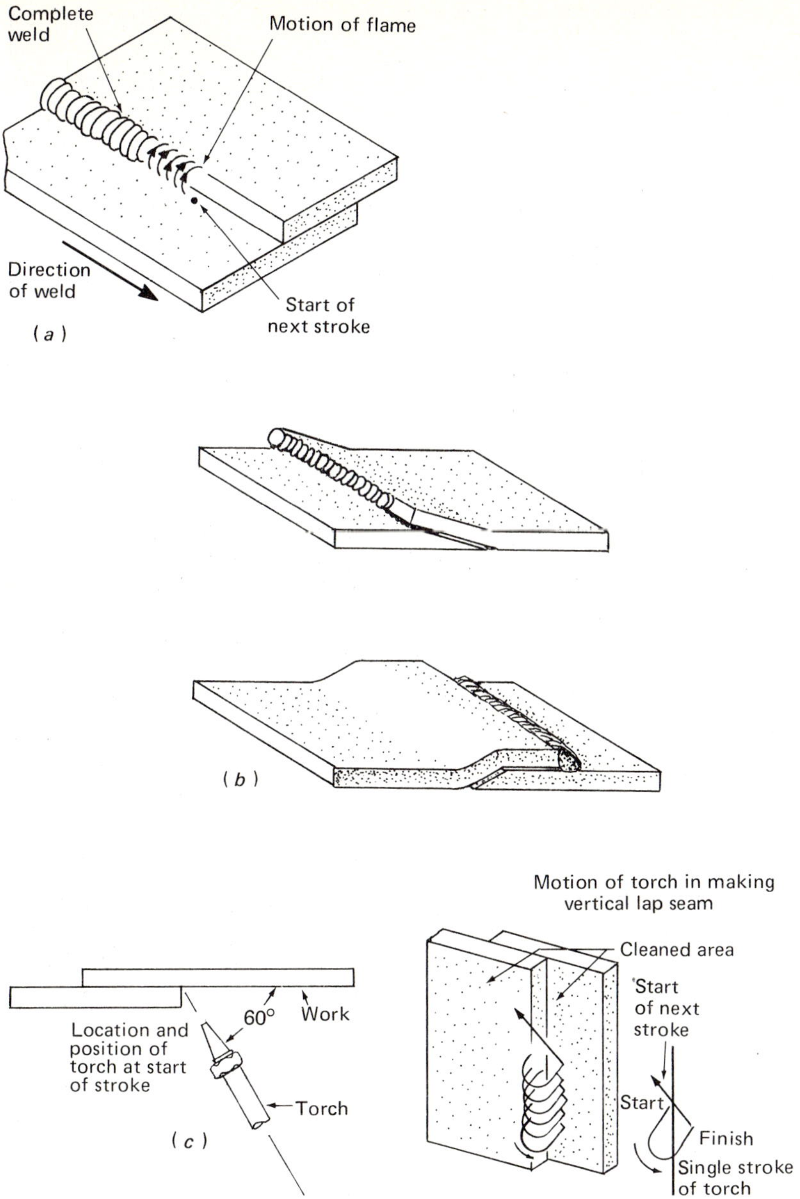
Complete
weld
Motion of flame
Direction
of weld
Start of
next stroke
(a)
(b)
Motion of torch in making
vertical lap seam
Cleaned area
Start
of next
stroke
60°
Work
Location and
position of
torch at start
of stroke
Torch
(c)
Start
Finish
Single stroke
of torch

Figure **21-7** Techniques for lead welding: (*a*) lap joint in flat position; (*b*) overhand and underhand techniques for horizontal position; and (*c*) lap joint in vertical position. (Redrawn by permission from American Welding Society, *Welding Handbook,* 6th ed., 1972, section 4, p. 71.7.)

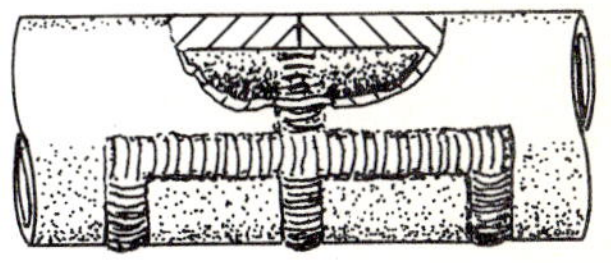

Figure **21-8** Technique for welding lead pipe in the horizontal position. (Redrawn by permission from American Welding Society, *Welding Handbook,* 6th ed., 1972, section 4, p. 71.7.)

21-6 ZINC AND ZINC ALLOYS

Commercial grades of zinc conform to ASTM specifications B6, B69, B86, and B240. Table 21-29 shows the compositions of some typical zinc alloys, together with their weldabilities and common uses.

Satisfactory welds can be made in any of the listed zinc alloys in thicknesses up to 0.125 inch with any of the oxyfuel-gas processes. Sheets less than 0.040 inch thick are flanged and welded without the addition of filler metal. Square-groove butt joints are welded in materials up to 0.125 inch thick. Butt joints over 0.063 inch thick should be welded with a gap equal to the metal thickness between edges. Thicker zinc sheets should be beveled to form a 70- to 90-degree included angle, with little or no land. Lap joints may also be used, with a fillet weld generally made on both sides of the joint. The joint area must be thoroughly cleaned and abraded. Tack welding before finish welding is usually desirable. A small torch tip should be used; for example, a tip suitable for welding 0.031-inch steel is used for 0.125-inch zinc. To minimize surface oxides, the flame should be neutral or slightly reducing.

Filler metals may be either zinc or the same composition as the base metal. Filler-rod diameters should be approximately ⅔ the thickness of the metal being welded, to a maximum of 0.16 inch. The tendency for overheating the base metal may be minimized by using larger-diameter rods or by directing the flame onto the filler rod.

Since holes may be formed if the torch is held vertically, the flame should be maintained at an angle of 15 to 45 degrees to the work, depending on the sheet thickness; the thinner the material, the smaller the angle. In groove welding, the flame is directed toward the rod, which is raised out of the flame after each drop is deposited. The torch should be oscillated to prevent burning through the sheet. Backhand or forehand welding techniques are satisfactory. The joint must have adequate support to prevent excessive drop-through. Backup materials may be asbestos, graphite, fire clay, or plaster of paris.

To improve strength and ductility, the weld joint may be peened. This operation, however, must be done between 200 and 300°F, since peening at room temperature or above 300°F might crack the weld.

The gas tungsten–arc welding process will produce uniform penetration and satisfactory welds in zinc alloys. Argon shielding gas is recommended because it allows the use of lower voltages and longer arc lengths. High-frequency current superimposed over a 60-cycle-ac power source is necessary to overcome interference with arc stability caused by the presence of titanium, aluminum, and zinc oxides. *Note:* Manual arc welding requires a highly skilled welder. Automatic welding, where practical, is desirable to maintain close control of the arc length and travel speed. Inert gas backing in a plenum chamber on the root side of the joint is recommended to overcome the tendency toward lack of fusion at the butting edges of the root, a condition caused by the formation of oxides when plain, refractory, or copper backup materials are used.

GTAW (TIG) welding of zinc-copper-titanium alloys requires the use of a filler metal that will produce grain refinement in the deposit without appreciably reducing the ductility by a precipitation-hardening reaction during cooling. Titanium content in the filler metal produces grain refinement in the cast structure, but may also cause precipitation reaction, which decreases the ductility of the weld deposit. Titanium contents of the deposit should not exceed 0.12% to be effective. Filler metal should contain at least 0.10% titanium and have the same copper content as the base metal. This composition will produce the strongest and most ductile weld deposit. Typical GTAW welding schedules for zinc-copper-titanium alloys are shown in Table 21-30.

Cast zinc alloys as thin as 0.033 inch can be successfully repair welded with the fuel-gas processes. Joint preparation of a break in a casting consists of removing all grease, oil, dirt, oxides, and any nickel or chrome plating from the weld area. After cleaning, the break is grooved through the full thickness of the casting to an included angle of 45 degrees.

If the casting is large and complicated, the whole structure must be preheated to approximately 250°F, because localized heating may cause serious distortion. The joint area should be further heated with the torch until fine droplets appear on the surface at the joint area. Extra heating must be confined to the repair area. Should these droplets appear beyond the joint area, the part has been overheated and may be destroyed. The gas torch should be adjusted to give a soft, lazy, soot-free reducing flame. The part must be properly supported during welding with either damp shredded asbestos, carbon paste, fire clay, or plaster.

A filler metal of the composition of AC43A alloy (Al 3.5–4.3; Mg 0.02–0.03; Cu 2.5–3.5; Zn bal.) is often used because it has a low melting point. The preheat is maintained during welding by occassionally sweeping the flame over the casting. The filler rod, or a small puddling spade, may be used to break up and scrape away surface oxides as filler metal is slowly added. The weld may be shaped with the puddling spade while still in a mushy condition.

Good welds can be obtained in galvanized (zinc-coated) steel with the oxyfuel-gas, SMAW, GMAW (CO_2), and submerged-arc processes. Procedures for shielded metal–arc welding of galvanized steel with both rutile and basic coated mild electrodes are similar to the procedures used for welding uncoated steel.

Table **21-29** Zinc and zinc alloys

	TYPE					
	COMMERCIAL ROLLED ZINC (DEEP DRAWING)	COMMERCIAL ROLLED ZINC	COMMERCIAL ROLLED ZINC (HIGHER Pb, Cd)	COPPER HARDENED ROLLED ZINC ALLOY	ROLLED ZINC ALLOY (Cu, Mg)	ROLLED ZINC ALLOY (Cu, Ti)
Composition, %	Pb 0.10 max, Zn bal	Pb 0.05–0.10, Cd 0.05–0.08, Zn bal	Pb 0.25–0.50, Cd 0.25–0.45, Zn bal	Cu 0.85–1.25, Zn bal	Cu 0.85–1.25, Mg 0.006–0.016, Zn bal	Cu 0.50–1.5, Ti 0.12–0.50, Zn bal
PHYSICAL PROPERTIES						
Density, lb/cu in	0.258	0.258	0.258	0.259	0.259	0.258–0.259
Melting point, F	786	786	786	792	792	792
Ther cond (64 F), Btu/hr/sq ft/°F/ft	62.2	62.2	—	—	60.5	60.5
Coef of ther exp, per °F						
With grain	18.1×10^{-6}	18.1×10^{-6}	18.8×10^{-6}	—	19.3×10^{-6}	$13.0–13.8 \times 10^{-6}$
Across grain	12.8×10^{-6}	12.8×10^{-6}	13.0×10^{-6}	—	11.7×10^{-6}	$9.8–10.8 \times 10^{-6}$
Spec ht (68–212 F), Btu/lb/°F	0.094	0.094	0.094	0.0957	0.0957	0.096
Elec res (68 F), microhm-cm						
Hot rolled	6.06	—	—	6.22	6.31	6.24
Cold rolled	6.10	—	—	—	—	—
MECHANICAL PROPERTIES[a]						
Ten str, 1000 psi						
Hot rolled	19.5, 23	21, 25	23, 29	24, 30	29, 40	24–32, 32–42
Cold rolled	21, 27	22, 29	25, 31	31, 40	36, 46	29, 37
Elong (in 2 in.), %						
Hot rolled	65, 50	52, 30	50, 32	50, 35	20, 10	14–38, 21–22
Cold rolled	50, 40	40, 30	45, 28	44, 30	25, 10	44, 60
Hardness (Brinell)						
Hot rolled	38	43	47	52	61	—
Endur limit (hot rolled), 1000 psi	2.5	3.8	4.1	6.1	6.8	—
Creep rate (12,000 psi, 77 F), days/%						
Hot rolled	—	—	—	0.15	—	115.0

FABRICATING PROPERTIES						
Hot working temp range, F	248–527	248–527	248–437	447–572	447–572	382–572
Melting range, F	887–977	887–977	887–977	887–977	887–977	887–977
Ingot casting range, F	815–905	815–905	815–905	815–905	815–905	815–905
Annealing temp, F	—	—	221	347	347	475–482
Machinability	Good	Good	Good	Good	Good	Good
Joining						
Torch welding	Poor to fair	Poor to fair	Poor to fair	Poor to fair	Poor to fair	Good
Single impulse resistance welding	Poor	Poor	Poor	Poor	Poor	Good
Multiple impulse resistance welding	Fair to good	Fair to good	Fair to good	Fair to good	Fair to good	Good
Soldering	Good	Good	Good	Good	Good	Good
Common processes	Drawing, bending, roll forming, stamping, swaging, coining, extruding, spinning					
Corrosion resistance	Excellent resistance to both metropolitan and rural atmospheric corrosion (penetration in in. per yr is 0.000064 in Palmerton, Pa., and 0.00028 in New York City); also hot soapy water, printing inks, trichloroethylene, carbon tetrachloride, dry illuminating gas, and moisture- and acid-free hydrocarbons. Fair resistance to pure ethyl and methyl alcohols, glycerine, water, petroleum products. Poor resistance to steam, spray insectides, animal oils, strong acids and bases, and mixtures of glycerine or alcohol and water					
Available forms	Rolled plate, strip and sheet; extruded rod and shapes; drawn rod and wire					
Uses	Dry batteries, eyelets and grommets, address plates, flashing, weatherstrip, laundry tags, novelties, lithoplates, condenser cans, embossing tape, leaders and gutters, corrugated roofing					
	Parts requiring no rigidity	Parts requiring some rigidity		Parts requiring maximum rigidity	Parts requiring maximum rigidity and some creep resistance	Parts requiring maximum rigidity and creep strength; low thermal expansion; resistance to grain growth

[a] Two values represent properties parallel to grain and perpendicular to grain, in that order.

Table **21-29** (*cont.*)

	TYPE				
	EXTRUDED, FORGED ZINC ALLOY (Cu, Ti)	*EXTRUDED, FORGED ZINC ALLOY (Cu, Mn, Ti)*	*FORGED ZINC ALLOY (Al, Cu)*	*FORGED ZINC ALLOY (Al, Cu)*	*SUPER-PLASTIC ZINC*[b]
Composition, %	Cu 1.0, Ti < 1, Zn bal	Cu 1.0, Mn 1.0, Ti < 1, Zn bal	Al 12.0, Cu < 1, Zn bal	Al 14.5, Cu < 1, Zn bal	Al, 22, Zn bal
PHYSICAL PROPERTIES					
Density, lb/in.^{3}	0.258	0.259	0.217	0.210	0.188
Melting point, F	786–792	786–792	718–810	718–815	—
Coef of ther exp, per °F	$8–12 \times 10^{-6}$	$6–12 \times 10^{-6}$	$6–12 \times 10^{-6}$	$6–13 \times 10^{-6}$	$11.2–15.6 \times 10^{-6}$
Elec cond, % IACS	28	17	25	25	28–34
MECHANICAL PROPERTIES					
Mod of elas, 10^6 psi	13	13	14	14	$6.2–13.5 \times 10^6$
Ten str, 1000 psi	35	48	50–72	50–75	28–72
Proof stress (0.2%), 1000 psi	23	35	45–60	45–65	23–61[c]
Elong (in 2 in.), %	30	20	13–18	14–20	7–98
Hardness, Brinell	65	90	100–120	100–125	46–87[d]
Shear str, 1000 psi	25	33	36	36	—
Creep str,[a] 1000 psi	10	15	3	3	0.19–10[e]
FABRICATION PROPERTIES					
Cold & hot workability	Excellent	Excellent	Excellent	Excellent	Excellent[f]
Forging temp, F	580	580	580	580	—
Machinability	Good	Good	Excellent	Excellent	—
Soldering	Excellent	Excellent	Fair	Fair	—
Welding					
Torch	Fair	Fair	Fair	Fair	—
Resistance	Good	Good	Good	Good	—
TIG	Good	Good	Good	Good	—
Corrosion resistance	Good resistance to hot soapy water, printer's ink, trichloroethylene, carbon tetrachloride, dry illuminating gas, moisture and acid-free hydrocarbons, methyl alcohol, glycerine, petroleum products, rural and industrial atmospheres. The titanium-bearing alloys are not susceptible to intergranular corrosion.				
Available forms	Extruded or continuously cast stock for subsequent forging; extrusions, forgings.				Sheet, strip
Uses	Industrial and commercial hardware, automotive fittings, marine hardware, machinery and electrical components, appliance parts, sporting goods.		Industrial and commercial hardware, electrical and machinery components, sporting goods, bearings, decorative applications.		Parts requiring severely formed/ drawn shapes.

[a]Extrapolated values of allowable creep stress to cause a steady state creep rate of 1% in 10 yrs. [b]Properties are ranges for three grades, having 782m-22Al base composition, in as-rolled and annealed plus air-cooled conditions. [c]0.2% offset yield strength. [d]Rockwell (15-T). [e]Stress for creep rate of 0.01%/1000 hr; 1% in 11.4 yr. [f]Alloys are superplastic at 500 F.

	TYPE					
	ALLOY AG40A (XXII)[a]	*ALLOY AC41A (XXV)*[a]	*SLUSH CASTING ALLOY*[b]	*SLUSH CASTING ALLOY (UNBREAKABLE METAL)*[b]	*ILZRO 12 FOUNDRY ALLOY*	*SAE 903 ZAMAK 7*
Composition, %	Al 3.5–4.3, Mg 0.03–0.08, Zn[c] bal	Al 3.5–4.3, Cu 0.75–1.25, Mg 0.03–0.08, Zn[c] bal	Al 4.5–5.0, Cu 0.2–0.3, Zn[c] bal	Al 5.25–5.75, Zn[c] bal	Al 11–13, Cu 0.5–1.25, Mg 0.01–0.03, Zn bal	Al 3.9–4.3 Cu 0.25–0.75 Mg 0.020–0.05
PHYSICAL PROPERTIES						
Density, lb/cu in.	0.24	0.24	—	—	0.216	0.238
Melting point, F	728	727	734	743	716–810	717
Ther cond (158–284 F), Btu/hr/sq ft/°F/ft	65.3	62.9	—	—	—	—
Coef of ther exp (68–212 F), per °F	15.2×10^{-6}	15.2×10^{-6}	—	—	15.5×10^{-6}	15.2×10^{-6}
Spec ht (68–212 F), Btu/lb/°F	0.10	0.10	—	—	—	0.10
Elec res (68 F), microhm-cm	6.37	6.54	—	—	—	—
MECHANICAL PROPERTIES[d]						
Ten str, 1000 psi						
Die cast	41	47.6	—	—	41–45[g]	41
Chill cast	—	—	28.0	25.0	50–53	—
Elong (in 2 in.), %						
Die cast	10	7	—	—	2–3.5[g]	10
Chill cast	—	—	—	1	5–7	—
Hardness (Brinell, die cast)	82	91	—	—	95–113[h]	82
Impact str (Charpy), ft-lb						
Die cast	43	48	—	—	—	43
Chill cast	—	—	3	1	—	—
Endur limit (10^8 cycles, die cast), 1000 psi	6.9	8.2	—	—	—	—
Compr yld str (die cast), 1000 psi	60	87	—	—	—	60
Shear str (die cast), 1000 psi	31	38	—	—	32[h]	31
FABRICATING PROPERTIES						
Melting temp range, F	728–932	727–932	—	—	890–970[i]	—
Die casting temp range, F	740–800	740–800	—	—	—	740–800
Solidification shrinkage, %	1.17	1.17	—	—	—	—
Machinability	Good	Good	Good	Good	Excellent	—

Table **21-29** (*cont.*)

	TYPE					
	ALLOY AG40A (XXII)[a]	*ALLOY AC41A (XXV)*[a]	*SLUSH CASTING ALLOY*[b]	*SLUSH CASTING ALLOY (UNBREAKABLE METAL)*[b]	*ILZRO 12 FOUNDRY ALLOY*	*SAE 903 ZAMAK 7*
Joining						
Torch welding	Poor to fair	Poor to fair	—	—	—	—
Single impulse resistance welding	Poor	Poor	—	—	—	—
Multiple impulse resistance welding	Fair to good	Fair to good	—	—	—	—
Soldering	Poor[e]	Poor[e]	—	—	—	—
Common fabrication processes	Welding, soldering, machining, riveting, spinning, cold swaging		Welding, soldering, machining		— —	— —
Corrosion resistance	Excellent resistance to both metropolitan and rural atmospheric corrosion[f]; also hot soapy water, printing inks, trichloroethylene, carbon tetrachloride, dry illuminating gas, and moisture- and acid-free hydrocarbons. Fair resistance to pure ethyl and methyl alcohols, glycerine, water and petroleum products. Poor resistance to steam, spray insecticides, animal oils, strong acids and bases, and mixtures of glycerine or alcohol and water					
Uses	Automotive parts, household utensils, office equipment, building hardware, padlocks, toys, novelties, drop hammer dies (XXV)		Slush and permanent mold castings, principally for lighting fixtures		Gravity casting alloy for prototype die castings	Die castings, general purpose

[a]ASTM B86-64. [b]Because of their limited use, few data are available on these alloys. [c]Special high grade zinc is required. [d]Based on a ¼-in. section for die cast alloys; a ½-in. section for chill cast alloys. [e]Cadmium-zinc and lead-tin solders diffuse into the casting, promoting subsurface attack. Castings must be nickel plated to be joined by lead-tin solders. [f]Penetration in in. per yr in Palmerton, Pa., and New York City, respectively, is 0.78×10^{-4} and 2.8×10^{-4} for XXIII, and 0.63×10^{-4} and 2.8×10^{-4} for XXV. [g]Sand cast. [h]Chill and sand cast. [i]Pouring temperature.

Source: *Materials Selector, Materials Engineering,* a Penton/DPC Reinhold Publication.

Table **21-30** Gas tungsten–arc welding schedules for zinc–copper–titanium wrought zinc alloys

THICKNESS (in.)	AC CURRENT (amps)	ARC VOLTAGE	ARC LENGTH (in.)	TUNGSTEN ELECTRODE *DIAMETER* *(in.)*	TUNGSTEN ELECTRODE *STICK-OUT* *(in.)*	ARGON FLOW (cfh)	TRAVEL SPEED (imp)	FILLER METAL
AUTOMATIC WELDING—BUTT JOINT*								
0.024	54	10.5	0.062	0.040	3/16	8	48	None
0.062	98	17–18	0.075	0.062		25	52	
0.075	125–130	11.5	0.093	0.062	5/16	18	30	Base Metal
0.125	172–180	13–15	0.125	0.125	3/8	25	20	Pure Zinc
AUTOMATIC WELDING—FILLET OVERLAP								
0.062	94	17–18	0.093			25	31	None
0.075	105	20	0.093†	0.062	5/16	18	30	None
0.125	140	17–18	0.156†			25	19	None
MANUAL WELDING—BUTT JOINT								
0.040	30		0.062	0.062	5/16	25	6	Base Metal
0.075	150		0.125	0.062	5/16	25	18	Base Metal

*Square butt joint preparation except 0.125 inch which is 60° V groove.
†Measured from bottom plate.

Source: American Welding Society, *Welding Handbook,* 6th ed., section 4, p. 72.8.

On butt joints, gaps should be wider when welding zinc-coated steel, to compensate for lower weld penetration. For example, in square-edge butt joints in 1/8-inch galvanized steel welded from one side, a gap of 3/32 inch is necessary for complete penetration. This compares with a 1/16-inch gap in uncoated steel.

The welding electrode should be repeatedly moved ahead of the weld pool (1/8 to 5/16 inch) and then back as the weld advances. This action will tend to volatize the zinc ahead of the bead and lessen the chance for porosity or gas pockets. The "whipping" motion of the electrode results in reduced travel speed (10 to 20%) as compared with welding uncoated steel.

Butt welds made in the horizontal or overhead position on galvanized steel can be made at speeds similar to those for welds on uncoated steel. When multipass welds are made to fill the joint, the usual procedure for 1/4-inch and thicker plate is only for the first pass to be deposited with reduced travel speed (due to the whipping motion). Subsequent passes can generally be made at the same speed as on uncoated steel.

Electrode materials other than mild steel are often used to weld galvanized steel. The most common of these electrodes are phosphor bronze and aluminum bronze. These are low-melting (1875 to 1905°F) alloys that minimize zinc-coating burn-off adjacent to the weld area and provide excellent corrosion resistance.

The procedure for making T joints (fillet welds) is essentially the same as that used to make butt joints. Best results are obtained with the EXX18-type electrode. If the EXX18-type electrode is not available and the plain organic or rutile coated electrode is to be used, it is recommended that the weld be made with either a sideways weaving motion or a back-and-forth whipping motion. This technique helps minimize undercutting the edge of the weld on the vertical plane.

Although the CO_2 short-circuit transfer process has been found to produce the most satisfactory welds, it also produces spatter. To overcome the spatter problem, it is recommended that the area around the weld and the nozzle of the welding gun be coated with antispatter compound. Although joints in sections up to ¼ inch thick give clear radiographs, welds made in sections thicker than ¼ inch or with filler metal having a silicon content of 0.2% or more have a tendency toward intergranular cracking. To minimize the chances of this, it is recommended that:

1. The edge be prepared such that it has either a single- or double-bevel gas cut or a machined cut, either before or after galvanizing.

2. The zinc from both joint surfaces be removed by burning with an oxyfuel-gas torch or by shot-blasting with portable equipment.

3. A suitable gap be maintained between the standing plate and the horizontal plate of a T joint. This can be done by using a 1/16-inch parallel gap or a wedge-shaped gap produced by a 15-degree angle on the edge of the standing plate. A bevel angle of 15 degrees could be formed on the edge of the plate by oxygen cutting or machining, and it would be effective whether it was done before or after galvanizing.

4. Procedural tests of the electrodes be made. Both weld-metal silicon content and penetration characteristics of the electrode appear to be important; thus, it is not possible to select electrodes for freedom from zinc penetration of fillet welds on the basis of silicon content alone. *Note:* Until experience is built up with particular electrodes, or until electrodes are developed specifically for welding galvanized steel, it is essential that simple procedural tests be made to find out which electrodes to use for making fillet welds in galvanized steel.

The submerged-arc process can be used to produce square-butt and T joints with a minimum of porosity in sections up to ½ inch thick, providing that in the case of butt joints, the joint is positioned

in such a manner as to ensure free access of air to the underside of the joint and a gap of 1/16 to 1/8 inch is maintained for making either type of joint.

To prevent corrosion of the base metal in the area of the weld, it is recommended that the following be done as soon as possible after welding:

1. Use a brush (preferably powered) to remove all slag.

2. Use a brush (preferably powered) to remove all loose particles from the HAZ.

3. Deposit a layer of zinc on the joint and the HAZ by rubbing the areas with any one of a number of zinc-alloy sticks specifically made for that purpose.

Warning: Zinc fumes are a potential health hazard, so specific precautions as to ventilation are required. Consult Chapter 13 and the USA Standard 249.1 (latest edition) *Safety in Welding and Cutting.*

Metal fume fever, more commonly called *zinc chills,* spelter shakes, or brass founders ague, may follow exposure to zinc fumes released during welding operations. The chills are caused by colloidal zinc oxide (0.3 to 0.4 micron in diameter) penetrating the lungs. Larger particles of zinc oxide adhere to the trachea and have no symptoms.

This fever is an acute self-limiting condition without known complications, after-effects, or chronic form. The illness begins a few hours after exposure and may cause a sweet taste in the mouth, dryness of the throat, coughing, fatigue, yawning, weakness, head and body aches, nausea, vomiting, chills, and fever (rarely exceeding 102°F). A second attack seldom occurs during repeated exposures unless there has been an interval of several days between exposures.

Threshold-limit values (TLV) of zinc oxide, as published by the American Conference of Governmental Industrial Hygienists, are on the order of 5.0 mg/m^3. These values are published yearly by this group and may change from year to year.

21-7 TITANIUM AND TITANIUM ALLOYS

Titanium and its alloys are produced commercially in conformance with AMS, ASTM, and military specifications. The three classes of titanium alloys and the three grades of unalloyed titaniums are listed by alloy number, standard designation, available forms, and application in Table 21-31.

The GTAW (TIG) process is most frequently used for welding titanium and its alloys. Sections up to 0.10 inch thick can be square-butt welded without the use of a filler metal. Unless the weld is done

Table **21-31** Titanium and titanium alloys—wrought

	TYPE				
	Ti-8Al-1Mo-1V	*Ti-6Al-4V*	*Ti-6Al-4V ELI*	*Ti-8Mn*	*Ti-6Al-2Sn-4Zr-6Mo*
PHYSICAL PROPERTIES					
Density, lb/cu in.	0.158	0.160	0.160	0.171	0.168
Melting temp,[a] F	—	3000	3000	2730–2970	—
Ther cond (RT), Btu/hr/sq ft/°F/ft	—	4.2	—	6.3	—
Coef of ther exp (RT-1000 F), per °F × 10^{-6}	5.6	5.3	5.3	6.0	5.3
Spec heat (RT), Btu/lb/°F	—	0.135	0.135	0.118	—
Elec res (RT), microhm-cm	199	171	171	92	—
Magnetic perm (20 oersteds)[b]	1.00005	1.00005	1.00005	1.00005	—
MECHANICAL PROPERTIES[cd]					
Mod of elast in tension, 10^6 psi					
Room temp	18.0–18.5	16.5	16.5	16.4	16.5
600 F	—	13.5	13.5	14.4	—
Shear mod, 10^6 psi	—	6.1	6.1	7.0	—
Tensile strength, 1000 psi					
Room temp	135–160	138, 170[e]	135	137	184
800 F[f]	96[n]	90, 130[e]	220 (−320 F)	80	138
Yield strength, 1000 psi					
Room temp	125–150	128, 155[e]	127	125	170
800 F[f]	72[n]	78, 100[e]	205 (−320 F)	59	110
Elongation (in 2 in.),%					
Room temp	10–18	12, 8[e]	15	15	10
800 F[f]	17[n]	18, 8[e]	13 (−320 F)	15	19
Compressive yield strength, 1000 psi					
Room temp	—	—	—	120-140	
800 F	—	—	—	55-65	
Hardness, Rockwell	$36R_C$	$36R_C$[k]	$36R_C$[k]	$28–36R_C$	42
Impact strength (Charpy-V), ft-lb	15–25[k]	10–20[k]	10[k]	—	
Fatigue strength (10^7 cyc),[h] 1000 psi	—	75,[k] 92[ke]	—	90[o]	
Shear strength, 1000 psi	98	100–110[e]	—	100–105	

Corrosion resistance	Titanium and its alloys have outstanding corrosion resistance to most media including nitric acid in all conc to boiling pt; inorganic salts (except aluminum chloride), particularly to pitting attack by chloride solutions (sea water); and to alkalis in all conc to boiling pt (except boiling conc potassium hydroxide). They also have generally good resistance to organic salts (borderline passivity to formic and trichloracetic acids) and dilute, low temperature sulfuric, hydrochloric and phosphoric acids (solution inhibitors permit use at higher temperatures and conc). They are rapidly attacked by hydrofluoric acid and may have pyrophoric reactions with fuming nitric acid containing less than 2% water or more than 6% nitrogen dioxide, liquid oxygen on impact, anhydrous liquid or gaseous chlorine, liquid bromine, hot gaseous fluorine and oxygen enriched atmospheres. Stress corrosion may occur in some alloys if chloride salts are present on stressed parts subsequently subjected to high temperatures. Oxidation resistance is high to 1000 F				
FABRICATING PROPERTIES					
Annealing temp, F	1450–1850[v]	1300–1550	1300–1550	1250–1300	1350–1650
Stress relieving temp, F	1100–1200	900–1200	900–1200	900–1100	1100–1250
Solution temp,[r] F	NHT	1550–1750	[w]	[w]	1525–1675
Aging temp,[r] F	NHT	900–1000	[w]	[w]	1100–1250
Forging temp,[s] F	1850–1950	1400–1900	1400–1900	[w]	1550–1750
Formability[t]	3T–5T	3T–5T	3T–5T	2.5T–4T	—
Weldability[u]	W	W	W	[w]	LW
Available forms[x]	B,b,P,S,s, W,E	B,b,P,S,s, W,E	B,b,P,S,s,T, W,E	P,S,s	B,b,S,P
Uses and remarks	High strength and creep resistance up to about 900 F; aircraft and jet engine components	Most versatile alloy; aircraft and jet engine parts; pressure vessels; rocket motor cases; chemical processing equipment; marine parts	Excellent cryogenic properties down to −320 F; pressure vessels for temperatures down to −320 F	Aircraft sheet parts such as skins, stiffeners, ribs and webs	Jet engine parts

[a]Single values are approximated. [b]Approximate value. [c]Sheet material unless otherwise noted. [d]Annealed condition unless otherwise noted. [e]Aged. [f]Unless otherwise noted. [g]Properties shown are ranges for five grades of unalloyed titanium. [h]Rotating beam test unless otherwise noted. [i]For 99% Ti (90,000 psi tensile strength) at $K_t = 1$ [j]For high strength grades. [k]Bar. [l]Annealed or aged. [m]Direct axial, $K_t = 1$ (alternating stress/mean stress ratio = 0.9). [n]For 135,000 psi annealed sheet. [o]Direct axial, $K_t = 1$ (alternating stress/mean stress ratio = infinity). [p]Direct axial, $K_t = 1$ (alternating stress/mean stress ratio = 0.6). [q]Threaded bolts. [r]NHT = hot heat treatable. [s]Range set by lowest finish forging and highest starting forging temperatures. [t]Minimum bend-to-thickness (T) ratio for sheet; heat usually required. [u]FW = fully weldable, W = weldable, LW = limited weldability. [v]Various treatments possible. [w]Not recommended. [x]B = billet, b = bar, P = plate, S = sheet, s = strip, T = tubing, W = wire, E = extrusions.

Table **21-31** *(cont.)*

	TYPE					
	Unalloyed 98.9-99.5Ti[g]	*Ti-0.15-0.2Pd*	*Ti-5Al-2.5Sn*	*Ti-5Al-2.5Sn ELI*	*Ti-6Al-2Cb-1Ta-1Mo*	*Ti-5Al-6Sn-6Sn-2Zr-1Mo*
PHYSICAL PROPERTIES						
Density, lb/cu in.	0.163–0.164	0.163	0.162	0.161	0.162	0.163
Melting temp,[a] F	3000–3040	3000–3100	2820–3000	2820–3000	3000	—
Ther cond (RT), Btu/hr/sq ft/°F/ft	9.0–11.5	9.5	4.5	4.5	—	—
Coef of ther exp (RT-1000 F), per °F × 10^{-6}	5.1–5.5	5.4	5.3	5.4	—	—
Spec heat (RT), Btu/lb/°F	0.124–0.129	0.125	0.125	0.125	—	—
Elec res (RT), microhm-cm	48–60	57	157	180	—	—
Magnetic perm (20 oersteds)[b]	1.00005	1.00005	1.00005	1.00005	1.00005	—
MECHANICAL PROPERTIES[cd]						
Mod of elast in tension, 10^6 psi						
Room temp	14.9–15.5	14.9	16.0	16.0	—	—
600 F	12.1–12.6	12.3	13.4	13.4	—	—
Shear mod, 10^6 psi	6.5	6.5	7.0	—	—	—
Tensile strength, 1000 psi						
Room temp	38–100	62	125	110	110–115	145[k]
800 F[f]	20–47 (600 F)	28 (600 F)	82 (600 F)	229 (−423 F)	75	103 (900 F)[k]
Yield strength, 1000 psi						
Room temp	27–85	46	117	95	95–100	126[k]
800 F[f]	10–30 (600 F)	13 (600 F)	65 (600 F)	206 (−423 F)	60	78 (900 F)[k]
Elongation (in 2 in.),%						
Room temp	17–30	27	18	20	10	12[k]
800 F[f]	25–50 (600 F)	30 (600 F)	19 (600 F)	15 (−423 F)	15	13 (900 F)[k]
Compressive yield strength, 1000 psi						
Room temp	—	—	110–120	—	—	—
800 F	—	—	60	—	—	—
Hardness, Rockwell	70–100 R_B	70 R_B	36 R_C	36 R_C	30 R_C	—
Impact strength (charpy-V), ft-lb	11–40	—	19[k]	19[k]	—	—
Fatigue strength (10^7 cyc),[h] 1000 psi	63[i]	—	93[m]	—	—	—
Shear strength, 1000 psi	50–60[j]	—	100–110	—	—	—

Corrosion resistance	Titanium and its alloys have outstanding corrosion resistance to most media including nitric acid in all conc to boiling pt; inorganic salts (except aluminum chloride), particularly to pitting attack by chloride solutions (sea water); and to alkalis in all conc to boiling pt (except boiling conc potassium hydroxide). They also have generally good resistance to organic salts (borderline passivity to formic and trichloracetic acids) and dilute, low temperature sulfuric, hydrochloric and phosphoric acids (solution inhibitors permit use at higher temperatures and conc). They are rapidly attacked by hydrofluoric acid and may have pyrophoric reactions with fuming nitric acid containing less than 2% water or more than 6% nitrogen dioxide, liquid oxygen on impact, anhydrous liquid or gaseous chlorine, liquid bromine, hot gaseous fluorine and oxygen enriched atmospheres. Stress corrosion may occur in some alloys if chlorine salts are present on stressed parts subsequently subjected to high temperatures. Oxidation resistance is high to 1000 F

FABRICATING PROPERTIES

Annealing temp, F	1250–1300	1250–1300	1325–1550	1325–1550	—	—
Stress relieving temp, F	1000–1100	1000–1100	1110–1200	1100–1200	1000–1200	1100–1200
Solution temp,[r] F	NHT	NHT	NHT	NHT	NHT	NHT
Aging temp,[r] F	NHT	NHT	NHT	NHT	NHT	NHT
Forging temp,[s] F	1200–1750	1550–1700	1400–1950	1400–1900	1850–1950	1900–1950
Formability[t]	1T–3T	1T–3T	3T–5T	3T–5T	4T–6T	—
Weldability[u]	FW	FW	W	W	W	W
Available forms[x]	B,b,P,S,s, T,W,E	B,b,P,S,s, T,W,E	B,b,P,S,s, W,E	B,b,P,S,s, W,E	B,b,P,S,s, W	B,b
Uses and remarks	Aircraft skins, webs, stiffeners, engine rings and fasteners; heat exchangers; marine parts; chemical process equipment	Pd addition improves resistance to mildly reducing media, e.g., dilute HCl and H_2SO_4 acids. Chemical processing equipment	Combines high strength and weldability with excellent oxidation resistance and cryogenic properties; aircraft parts; pressure vessels	Excellent cryogenic properties down to −423 F; pressure vessels for temperatures below −320 F	Combines high toughness with strength, weldability and corrosion resistance	High strength and creep resistance to about 900 F; aircraft and jet engine components

[a]Single values are approximated. [b]Approximate value. [c]Sheet material unless otherwise noted. [d]Annealed condition unless otherwise noted. [e]Aged. [f]Unless otherwise noted. [g]Properties shown are ranges for five grades of unalloyed titanium. [h]Rotating beam test unless otherwise noted. [i]For 99% Ti (90,000 psi tensile strength) at K_t = 1 [j]For high strength grades. [k]Bar. [l]Annealed or aged. [m]Direct axial, K_t = 1 (alternating stress/mean stress ratio = 0.9). [n]For 135,000 psi annealed sheet. [o]Direct axial, K_t = 1 (alternating stress/mean stress ratio = infinity). [p]Direct axial, K_t = 1 (alternating stress/mean stress ratio = 0.6). [q]Threaded bolts. [r]NHT = hot heat treatable. [s]Range set by lowest finish forging and highest starting forging temperatures [t]Minimum bend-to-thickness (T) ratio for sheet; heat usually required. [u]FW = fully weldable, W = weldable, LW = limited weldability. [v]Various treatments possible. [w]Not recommended. [x]B = billet, b = bar, P = plate, S = sheet, s = strip, T = tubing, W = wire, E = extrusions.

Table **21-31** (*cont.*)

	TYPE					
	Ti-6Al-6V-2Sn	*Ti-4Al-3Mo-1V*	*Ti-6Al-2Sn-4Zr-2Mo*	*Ti-8Mo-8V-2Fe-3Al*	*Ti-3Al-13V-11Cr*	*Ti-11.5Mo-6Zr-4.5Sn*
PHYSICAL PROPERTIES						
Density, lb/cu in.	0.164	0.163	0.164	0.175	0.76	0.183
Melting temp,[a] F	3100	3100	3000	—	—	—
Ther cond (RT) Btu/hr/sq ft/°F/ft	4.2	3.9	4.1 (212 F)	—	4.0	—
Coef of ther exp (RT-1000 F) per °F × 10^{-6}	5.3	5.5	4.5	—	5.9	—
Spec heat (RT), Btu/lb/°F	0.155	0.132	0.110 (212 F)	—	0.120	—
Elec res (RT), microhm-cm	157	165	190	—	142–153	156
Magnetic perm (20 oersteds)[b]	1.00005	1.00005	—	—	1.00005	—
MECHANICAL PROPERTIES[cd]						
Modulus of elast in tension, 10^6 psi						
Room temp	15.0,[k] 16.5[ke]	16.5	16.5	15.5	14.2, 14.8[e]	11, 15[e]
600 F	13.4,[k] 14.5[ke]	14.0	—	—	13.2, 13.8[e]	—
Shear modulus, 10^6 psi	—	7.0	—	—	6.2	—
Tensile strength, 1000 psi						
Room temp	165,[k] 190[ke]	140, 195[e]	130	190	135, 185[e]	141, 205[e]
800 F[f]	90[k]	145[e]	85	—	115, 160[e]	—
Yield strength, 1000 psi						
Room temp	150,[k] 180[ke]	120, 167[e]	120	180	130, 175[e]	128, 191[e]
800 F[f]	80[k]	115[e]	70	—	100, 120[e]	125[e]
Elongation (in 2 in.), %						
Room temp	15,[k] 10[ke]	15, 6[e]	10	8	16, 8[e]	17, 7[e]
800 F[f]	15[k]	8[e]	20–21	—	18, 12[e]	6[e]
Compressive yield strength, 1000 psi						
Room temp	—	—	135–145	—	—	—
800 F	—	—	84–95	—	—	—
Hardness, Rockwell	35,[k] 48[ke]	32–28 R_C	36 R_C	40	32, 26 R_C	—
Impact strength (Charpy-V), ft-lb	15,[k] 5–10[ke]	—	—	—	5–15[k]	—
Fatigue strength (10^7 cyc),[h] 1000 psi	—	124[p]	70	—	—	—
Shear strength, 1000 psi	105–115[ke]	—	90–96	—	—	—

Corrosion resistance	Titanium and its alloys have outstanding corrosion resistance to most media including nitric acid in all conc to boiling pt; inorganic salts (except aluminum chloride), particularly to pitting attack by chloride solutions (sea water); and to alkalis in all conc to boiling pt (except boiling conc potassium hydroxide). They also have generally good resistance to organic salts (borderline passivity to formic and trichloracetic acids) and dilute, low temperature sulfuric, hydrochloric and phosphoric acids (solution inhibitors permit use at higher temperatures and conc). They are rapidly attacked by hydrofluoric acid and may have pyrophoric reactions with fuming nitric acid containing less than 2% water or more than 6% nitrogen dioxide, liquid oxygen on impact, anhydrous liquid or gaseous chlorine, liquid bromine, hot gaseous fluorine and oxygen enriched atmospheres. Stress corrosion may occur in some alloys if chloride salts are present on stressed parts subsequently subjected to high temperatures. Oxidation resistance is high to 1000 F.					
FABRICATING PROPERTIES						
Annealing temp, F	1300–1400	1225	1300–1550	1450	1400–1500	—
Stress relieving temp, F	1100	1000–1100	900–1200	1000	1300–1450	—
Solution temp,[r] F	1600–1675	1625–1650	—	1450	1400–1500	1350–1450
Aging temp,[r] F	900–1100	925	1000–1100	900–1000	900	900
Forging temp,[a] F	1550–1725	1650–1800	1775–1950	1450–1950	1400–2150	1400–1800
Formability[t]	—	2.5T–4T	4.5–5T	2T–4T	2T–4T	2T–3.5T
Weldability[u]	LW	LW	LW	W	W	W
Available forms[x]	B,b,P,S,w,E	P,S,s	B,b,P,S,s,E	B,b,P,S,s	B,b,P,S,s,w	B,P,S,s,t,w
Uses and remarks	Aircraft, parts; ordnance components; rocket motor cases	Good high temperature strength, stability; aircraft sheet parts—skins, stiffeners, ribs, webs	Good high temperature stability; aircraft and jet engine parts	Airframe parts	Very high strength at room and moderate temperatures; limited stability above 600 F; aircraft parts	Very high room temp str, good stability to 700 F; aircraft parts, fasteners

[a]Single values are approximate. [b]Approximate value. [c]Sheet material unless otherwise noted. [d]Annealed condition unless otherwise noted. [e]Aged. [f]Unless otherwise noted. [g]Properties shown are ranges for five grades of unalloyed titanium. [h]Rotating beam test unless otherwise noted. [i]For 99% Ti (90,000 psi tensile strength) at $K_t = 1$. [j]For high strength grades. [k]Bar. [l]Annealed or aged. [m]Direct axial, $K_t = 1$ (alternating stress/mean stress ratio = 0.9). [n]For 135,000 psi annealed sheet. [o]Direct axial, $K_t = 1$ (alternating stress/mean stress ratio = infinity). [p]Direct axial, $K_t = 1$ (alternating stress/mean stress ratio = 0.6). [q]Threaded bolts. [r]NHT = not heat treatable. [s]Range set by lowest finish forging and highest starting forging temperatures. [t]Minimum bend-to-thickness (T) ratio for sheet; heat usually required. [u]FW = fully weldable, W = weldable, LW = limited weldability. [v]Various treatments possible. [w]Not recommended. [x]B = billet, b= bar, P = plate, S = sheet, s = strip, t = tubing, w = wire, E = extrusions.

Source: *Materials Selector, Materials Engineering,* a Penton/DPC Reinhold Publication, pp. 147–148.

in a chamber filled with an inert gas, such as argon or helium, it is absolutely essential to properly shield the joint from the atmosphere by means of a secondary shielding (Figures 21-9 to 21-11). Primary shielding is provided by cups designed to release a constant gas stream. The largest possible cup size consistent with accessibility and visibility of the joint during welding should be used. Another important aid to welding titanium sheets is the *heat sink* or *chill fixture*. This fixture makes the weld cool faster and reduces the amount of shielding necessary. Heat sinks are usually made of copper or stainless steel. Aluminum may also be used, but care must be exercised because of its low melting point. *Note:* Prior to welding, all jigs, fixtures, clamping equipment, the joint (Table 21-32), and a surface area of at least 1 inch on either side of the joint must be cleaned of oil, grease, and shop dirt. A flowchart of the cleaning processes is shown in Figure 21-12.

Best welding results are obtained with a straight-polarity-dc (SPDC) power supply. Because a current fluctuation greater than 5% may mean the difference between a good and a defective weld, it is recommended that the power supply be rectifier controlled and have a foot-operated current control to facilitate the even decrease of current and the filling-up of the shrinkage crater.

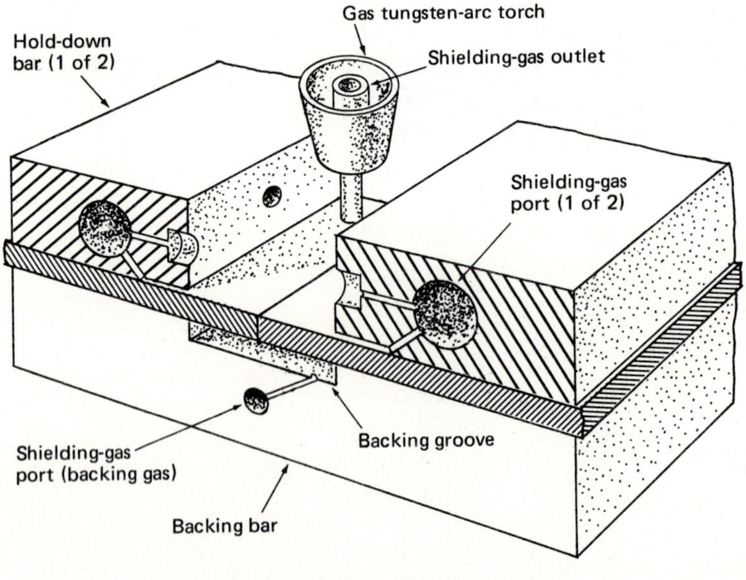
Gas tungsten-arc torch
Hold-down bar (1 of 2)
Shielding-gas outlet
Shielding-gas port (1 of 2)
Shielding-gas port (backing gas)
Backing groove
Backing bar

Figure **21-9** Typical set-up for inert-gas shielding for gas tungsten–arc welding of titanium and its alloys. (By permission, from *Metals Handbook* Volume 6, Copyright American Society for Metals, 1971.)

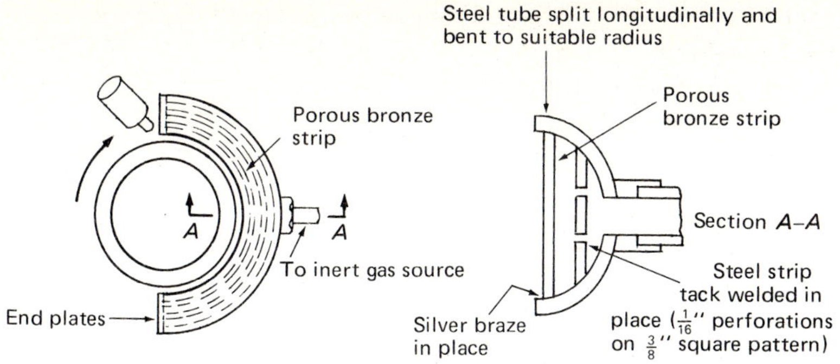
Steel tube split longitudinally and bent to suitable radius
Porous bronze strip
Porous bronze strip
Section A–A
A
A
To inert gas source
Steel strip tack welded in place ($\frac{1}{16}$" perforations on $\frac{3}{8}$" square pattern)
End plates
Silver braze in place

Figure **21-10** Schematic of the trailing gas shield applied to the automatic groove welding of pipe in the horizontal-fixed position. (Redrawn by permission from American Welding Society, *Welding Handbook*, 6th ed., 1972, section 4, p. 73.32.)

In structures where residual stresses (caused by cooling shrinkage) may be a problem, the structure should be heated to the temperature that will produce the required mechanical properties (strength) in the material. Table 21-33 relates titanium and its alloys to recommended heat treatments. *Note:* Before titanium is heated, it should be free of fingerprints, salt, grease, sulfur compounds, and degrease residues. There have been instances where the presence of these agents has led to stress corrosion cracking. After the stress-relief anneal, any furnace discoloration may be removed by a brightening dip in a 2% HF, 20% HNO_3 (160°F) bath. (See page 664.)

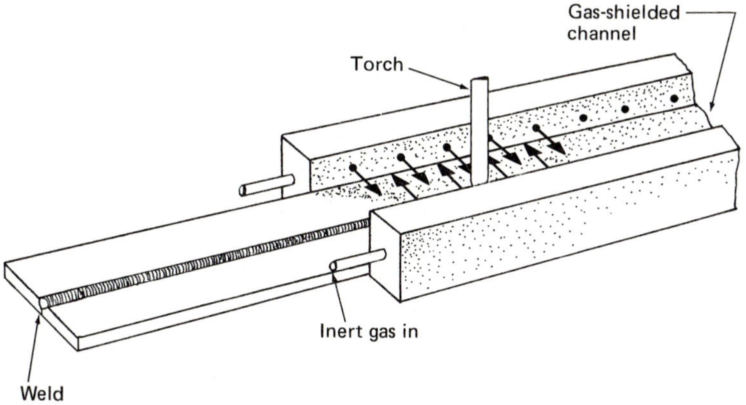
Gas-shielded channel
Torch
Inert gas in
Weld

Figure **21-11** Schematic of the side shield. (Redrawn by permission from American Welding Society, *Welding Handbook*, 6th ed., 1972, section 4, p. 73.33.)

Table **21-32** Dimensions of typical joints for welding titanium and titanium alloys

BASE-METAL THICKNESS, t (in.)	ROOT OPENING (in.)	GROOVE ANGLE (deg)	WELD-BEAD WIDTH (in.)
SQUARE-GROOVE BUTT JOINT			
0.010–0.090	0	—	—
0.031–0.125	0–0.10t	—	—
SINGLE-V-GROOVE BUTT JOINT			
0.062–0.125	0–0.10t	30–60	0.10–0.25T
0.090–0.125	[a]	90	—
0.125–0.250	0–0.10t	30–60	0.10–0.25t
DOUBLE-V-GROOVE BUTT JOINT			
0.250–0.500	0–0.20t	30–120	0.10–0.25t
SINGLE-U-GROOVE BUTT JOINT			
0.250–0.750	0–0.10t	15–30	0.10–0.25t
DOUBLE-U-GROOVE BUTT JOINT			
0.750–1.500	0–0.10t	15–30	0.10–0.25t
FILLET WELD			
0.031–0.125	0–0.10t	0–45	0–0.25t
0.125–0.500	0–0.10t	30–45	0.10–0.25t

[a]Root face, 0.030 in.

Source: By permission, from *Metals Handbook* Volume 6, Copyright American Society for Metals, 1971.

21-8 REACTIVE METALS

Reactive metals are those that react violently (explode) at welding temperatures when combined with the oxygen and nitrogen in the atmosphere. The commercially available, weldable reactive metals are listed in Table 21-34 (page 666).

As indicated in the table, the metals have been welded successfully with the GTAW (TIG) process. The weld must, however, be made in a hooded enclosure (Figure 21-13; page 670) or a glove box with an argon or helium atmosphere.

Manual welding is accomplished with alternating current. Automatic welding is accomplished with straight-polarity dc. Because

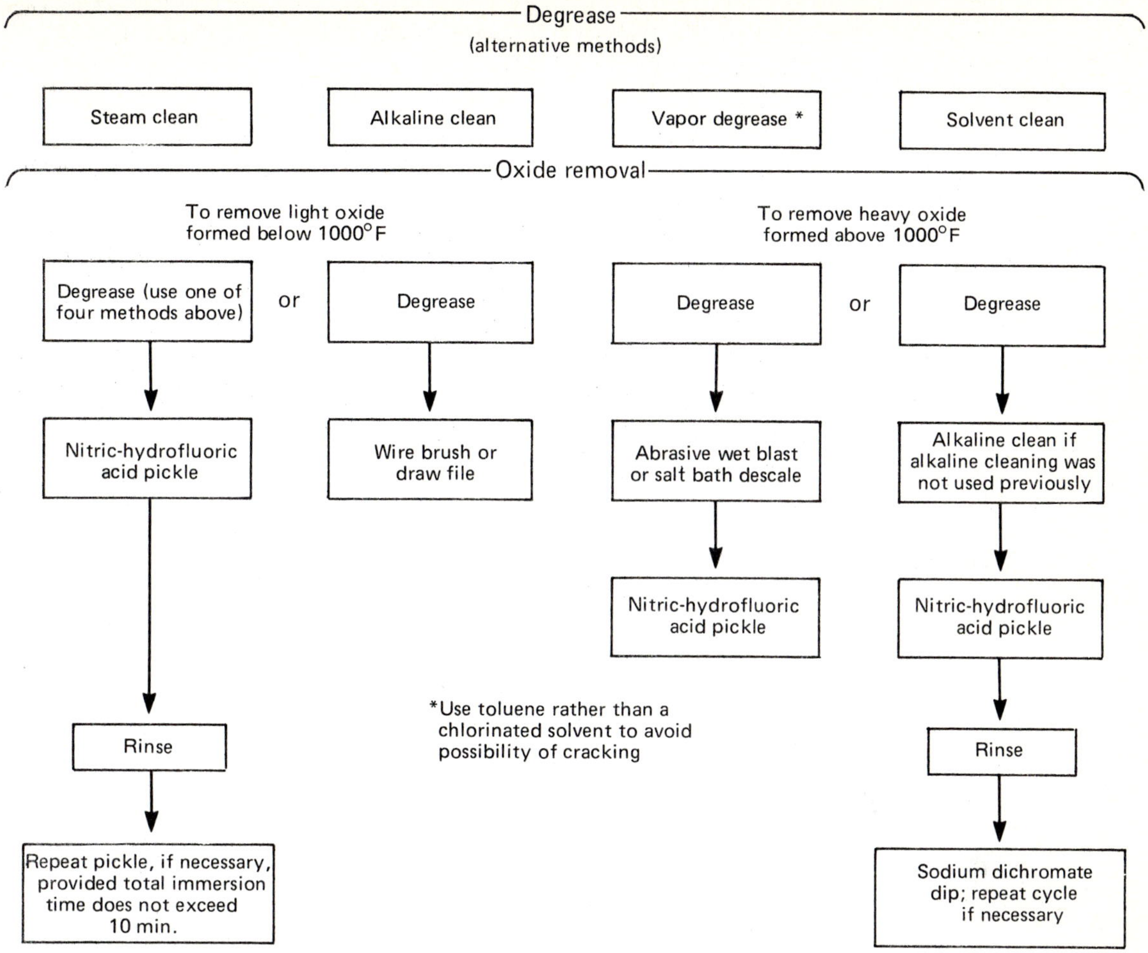
Degrease
(alternative methods)
Steam clean
Alkaline clean
Vapor degrease *
Solvent clean
Oxide removal
To remove light oxide formed below 1000°F
To remove heavy oxide formed above 1000°F
Degrease (use one of four methods above)
or
Degrease
Degrease
or
Degrease
Nitric-hydrofluoric acid pickle
Wire brush or draw file
Abrasive wet blast or salt bath descale
Alkaline clean if alkaline cleaning was not used previously
Nitric-hydrofluoric acid pickle
Nitric-hydrofluoric acid pickle
*Use toluene rather than a chlorinated solvent to avoid possibility of cracking
Rinse
Rinse
Repeat pickle, if necessary, provided total immersion time does not exceed 10 min.
Sodium dichromate dip; repeat cycle if necessary

Figure **21-12** Flow chart of procedures for cleaning titanium alloys. (By permission, from *Metals Handbook* Volume 6, Copyright American Society for Metals, 1971.)

beryllium and zirconium have low ductility, the heat input must be carefully controlled, otherwise thermal-shock cracking will become unavoidable. When available, the electron-beam and solid-state welding processes should be given preference over the GTAW process.

Brazing and braze welding produce better results when joining beryllium to beryllium or zirconium to zirconium or other metals. The filler metal preferred for joining beryllium is Ag 0.5% Li.

Warning: Beryllium fumes are toxic. Before attempting to weld beryllium, the welding area must be properly ventilated.

Table **21-33** Summary of titanium alloys in commercial production

	Nominal Composition Percent	AMS No.	Forms Available[1]	Recommended Heat Treatments[2]: Stress-Relief Annealing	Annealing	Solution Treatment (Aging Treatment)
Commercially Pure Grades	99.5		All	1000 to 1100F, ½Hr., AC	1250 to 1300 F, 2Hr., AC	Not heat treatable
	99.2	4902, 4941, 4942	All	1000 to 1100F, ½Hr., AC	1250 to 1300 F, 2Hr., AC	Not heat treatable
	99.0	4900, 4901, 4921	All	1000 to 1100F, ½Hr., AC	1250 to 1300 F, 2Hr., AC	Not heat treatable
	98.9		B, b, E, W	1000 to 1100F, ½Hr., AC	1250 to 1300 F, 2Hr., AC	Not heat treatable
	0.15 to 0.20 Pd (Balance Ti)		All	1000 to 1100F, ½Hr., AC	1250 to 1300 F, 2Hr., AC	Not heat treatable
Alpha Alloy Grades	5Al–2.5Sn	4910, 4926, 4953, 4966	B, b, E, P, S, s	1000 to 1200F, ¼ to 2Hr., AC	1325 to 1550F, 10Min. to 4Hr., AC	Not heat treatable
	5Al–2.5Sn (low O)	4909, 4924	B, b, E, P, S, s	1000 to 1200F, ¼ to 2Hr., AC	1325 to 1550F, 10Min. to 4Hr., AC	Not heat treatable
	6Al–2Sn-4Zr-2Mo	4975, 4976	B, b, E, P, S, W	1100 to 1200F(1Hr.) AC	(1) 1650F(½Hr.) AC + 1450F (¼Hr.) AC (Sheet) (2) Same as (1) + 1100F(2Hr.) AC (Sheet) (3) 1650 or 1750 (1Hr.) AC + 1100F (8Hrs.) AC	Not heat treatable
	8Al–1Mo–1V	4915, 4916, 4955, 4972, 4973	B, b, E, P, S, W	1100 to 1200F (1Hr.) AC	(1) 1450F (8Hr.) Fc (sheet-heavy sections) (2) Same as (1) + 1450F(¼Hr.) AC (sheet) (3) 1650 or 1850F (1Hr.) WQ or AC + 1100F (8Hr.) AC (heavy sections)	Not heat treatable
	6Al–2Cb–1Ta–1Mo		B, b, E, P, S, W			

ALPHA-BETA ALLOY GRADES	8Mn	4908	P, S	900–1100F, ½ to 2 Hr., AC	1250–1300F (1Hr.) Fc to 1000F	Not recommended
	3Al–2.5V			900–1200F, 1 to 4Hr.	1300F (1Hr.) AC	Not recommended
	4Al–3Mo–1V	4912 4913	P, S	1000–1100F, 1 Hr., AC	1225F (4Hr.) SC to 1050F, AC	1625–1650F, ¼Hr. WQ (925F, 8–12Hr, AC)
	6Al–4V	4911 4928 4935 4954 4965 4967	B, b, P, S, s, E, W	900–1200F, 1 to 4Hr., AC	1300–1550F (1–8Hr.) SC to 1050F, AC	1550–1750F, 5Min-1Hr, WQ (900–1100F, 4–8Hr., AC)
	6Al–4V (low O)	4907 4930	B, b, P, S, E, s, T, W	Same	Same	Not recommended
	6Al–6V–2Sn–1(Fe,Cu)	4918 4971	B, b, P, S, E, W	1100F, 2Hrs., AC	1300–1400F, 1–2Hrs., AC	1600–1650F, 1Hr, WQ (900–1100F, 4–8Hr, AC)
	6Al–2Sn–4Zr–6Mo		B, b, P, S			1600 (1Hr) WQ (1100F, 8Hrs, AC)
BETA ALLOY GRADES	3Al–13V–11Cr	4917	B, b, P, S, s, W			1400–1500, ¼–1Hr, WQ or AC (900F, 2–24Hrs.) AC

[1]B = billet; b = bar; E = extrusion; P = plate; S = sheet; s = strip; T = tubing; W = wire.
[2]AC = air cool; SC = slow cool; FC = furnace cool; WQ = water quench.

Source: American Welding Society, *Welding Handbook*, 6th ed., section 4, p. 73.4.

Table **21-34** Reactive metals hafnium, thorium, uranium, vanadium, beryllium—wrought

	METAL					
	HAFNIUM	*THORIUM*	*DEPLETED URANIUM*	*VANADIUM*	*BERYLLIUM*	*Be-38Al*
PHYSICAL PROPERTIES						
Density, lb/cu in.	0.47	0.42	0.683	0.23	0.067	0.075
Melting point, F	3400	3180	2071	3110	2345	1193
Ther cond (212 F), Btu/hr/sq ft/°F/ft	—	21.4[c]	17.2	—	87	123
Coef of ther exp (70 F), per °F	3.4×10^{-6}	6.2×10^{-6}	7.7×10^{-6}	4.8×10^{6}	6.4×10^{-6}	9×10^{-6}
Spec ht, Btu/lb/°F	0.035	0.03	0.03	0.12	0.45	—
Elec res (68 F), microhm-cm	30[a]	18	25-30	25	4.3	3.0
MECHANICAL PROPERTIES						
Mod of elast in ten, psi	20×10^{6}	10×10^{6}	$20\text{-}30 \times 10^{6}$	20×10^{6}	42×10^{6}	27×10^{6}
Ten str, 1000 psi						
Annealed	77	34	70	72	60-90[e]	55
Cold worked	112[b]	49[b]	90	113	—	—
Yld str, 1000 psi						
Annealed	32	26	35	64	5-20[e]	40
Cold worked	96[b]	45[d]	50	109	—	—
Elong (in 2 in.), %						
Annealed	24	51	5	28	2-5[e]	7-12[e]
Cold worked	10[b]	—	2-10	3	—	—
Hardness (Rockwell)						
Annealed	A58	—	B90	B81	—	7.0
Cold worked	A65[b]	—	—	B93	—	—

FABRICATING PROPERTIES

Annealing temperature, F	1380 in vacuum or inert atm	1380 in vacuum	1100 in vacuum or inert atm	1650 in vacuum or inert atm	1400-2100 in vacuum[e]	950-1100 in vacuum
Workability	Can be hot worked at 1550 F; cold worked 30% between anneals	Can be readily hot or cold worked; fabricated by forging, rolling, swaging, extruding or drawing	Can be forged, rolled, swaged and drawn; heating must be in protective atm	Good cold working properties	Hot worked at 750-1950 F	Can be cold worked slightly; hot worked at 850-1050 F
Machinability	Similar to stainless steel	Can be machined like mild steel with or without cutting fluids	Moderately difficult to machine	Tools similar to those for cold rolled steel	Machines similar to cast iron	Similar to magnesium
Joining	—	Difficult to weld; brazing yields brittle	Can be welded in protective atm or vacuum	Can be welded with heliarc torch under argon	Brazed with zinc, aluminum or silver alloys	Good weldability by TIG or MIG; braze as aluminum
Corrosion resistance	Resistant to oxidizing acids but attacked by hydrofluoric acid	Very poor resistance to atmosphere, water and most reagents	Poor resistance to atmosphere, water and most reagents	Resists sea water; not affected by moderate strength hydrochloric and sulfuric acids; dissolved by any strength nitric acid	Resists atm at ambient temp; attacked by oxygen and nitrogen at elevated temp; attack in fresh water varies with air content	Somewhat similar to beryllium
Available forms	Has been produced in sheet and rod	Has been produced in rod, sheet, thin-walled tube, fine wire, foil	Castings, rolled plate, rod, tube, foil, extrusions	Plate, strip, bar, sheet, wire	Sheet, plate, rod, bar, tube, powder, wire, foil, extrusions	Sheet, rod, plate, bar, foil, extrusions, powder
Uses	Nuclear reactors	Secondary (breeder) reactor fuel	Counterweights, missile ballast, shielding	AEC applications	Nuclear reactors, missiles, aircraft	Nuclear reactors, aircraft, lightweight structures

[a]32 F. [b]20% cold work. [c]70 F. [d]50% cold work. [e]Depends on form.

Table **21-34** (*cont.*)

ZIRCONIUM AND ITS ALLOYS—WROUGHT				
	TYPE			
	COMMERCIAL GRADE[a]	*REACTOR GRADE*	*ZIRCALOY-2*[b]	*ATR*
Composition, %	Hf 2.0, Zr bal	Hf 0.001 max, Zr bal	Sn 1.5, Fe 0.12, Cr 0.10, Ni 0.005, Zr bal	Cu 0.5, Mo 0.5, Zr bal
PHYSICAL PROPERTIES				
Density, lb/cu in.	0.237	0.235	0.237	0.24
Melting point, F	3350	3350	3300	3300
Ther cond (212 F), Btu/hr/sq ft/°F/ft	—	9.6	8.1	—
Coef of ther exp (212 F), per °F	3.1×10^{-6}	3.1×10^{-6}	3.6×10^{-6}	—
Elec res, microhm-cm	40	40	74	—
MECHANICAL PROPERTIES				
Modulus of elasticity, psi	14×10^6	14×10^6	13.8×10^6	14×10^6
Tensile strength (annealed), 1000 psi				
Room temperature	64	35	60	—
600 F	30	19	29	45
Yield strength (0.2% offset; annealed), 1000 psi				
Room temperature	53	15	45	—
600 F	23	10	19	42
Elongation (in 2 in.; annealed), %				
Room temperature	24	32	37	—
600 F	35	52	35	24
Reduction of area (annealed), %				
Room temperature	42	40	45	—
600 F	65	—	60	—
Hardness (Rockwell)	B89	B65	B89	B84

FABRICATING PROPERTIES

Annealing temperature, F[c]	1200-1450	1200-1400	1200-1400	1000-1500
Workability	Zirconium is very ductile and workable and can be fabricated with standard shop equipment (with a few modifications and special techniques, e.g., those used for titanium). Hot working range is 1425-1850 F. Min sheet bend radius is 3-5T			
Joining	May be welded under inert atmosphere; can be brazed and soldered			
Corrosion resistance	Excellent resistance to hydrochloric and nitric acids in all concentrations and temperatures (up to boiling); resists sulphuric acid (up to 55%) to its boiling point; resists alkalis at all concentrations and temperatures. Attacked by hydrofluoric acid and aqua regia		Excellent resistance to steam and pressurized water up to 550 F	Excellent resistance to wet and dry carbon dioxide up to 1000 F
Available forms	Ingot, billet, rod, bar, sheet, strip, tube, and wire. Pipe, fittings, castings, and shapes available on request			
Uses	Chemical plant equipment	Fuel cladding and structural parts in nuclear reactors; flash bulb filler	Fuel cladding and structural parts in water or steam cooled nuclear reactors	Structural parts in gas-cooled nuclear reactors

[a]Another grade (containing 500 ppm Fe + Cr) is available for severe service in hot hydrochloric acid. [b]A similar grade, Zircaloy-4 (Sn 1.5, Fe 0.20, Cr 0.10), picks up less hydrogen in reactor service. [c]Light gage material or finish machined parts should be annealed in a vacuum to prevent surface oxidation.

Source: *Materials Selector, Materials Engineering,* a Penton/DPC Reinhold Publication.

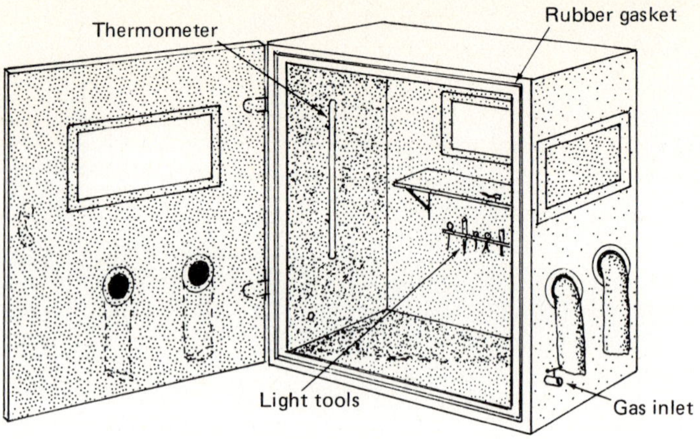

Thermometer
Rubber gasket
Light tools
Gas inlet

(*a*) Simple box chamber

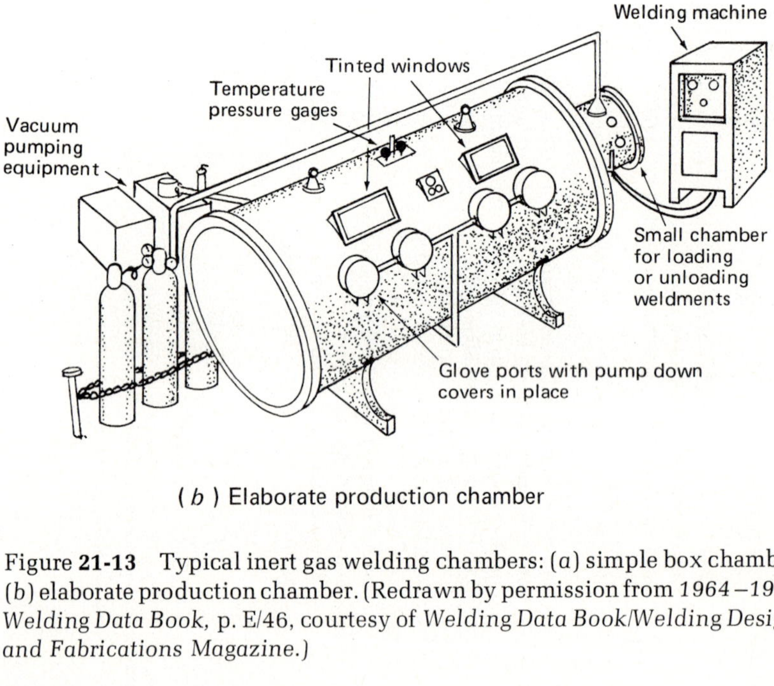

Welding machine
Tinted windows
Temperature
pressure gages
Vacuum
pumping
equipment
Small chamber
for loading
or unloading
weldments
Glove ports with pump down
covers in place

(*b*) Elaborate production chamber

Figure **21-13** Typical inert gas welding chambers: (*a*) simple box chamber; (*b*) elaborate production chamber. (Redrawn by permission from *1964–1965 Welding Data Book,* p. E/46, courtesy of *Welding Data Book/Welding Design and Fabrications Magazine.*)

21-9 REFRACTORY METALS

Refractory metals are those that have melting points above the range of iron, cobalt, and nickel. The weldable, commercially available refractory metals, together with their designations, applications, and weldability, are listed in Table 21-35.

Because the refractory metals are susceptible to weld porosity and sometimes to weld cracking, it is essential that the joint to be welded and the area immediately adjacent to it be wiped with acetone, followed by chemical cleaning. To avoid having the cleaning solution contaminated by metal particles, the solution should be kept in glass or porcelain containers. A satisfactory cleaning procedure, as recommended by the ASM, follows:

1. Immerse for 5 to 10 minutes in a warm (150 to 180°F) solution of 10% (by weight) sodium hydroxide and 5% potassium permanganate in distilled water.
2. Rinse in tap water, scrubbing off loose smut.
3. Immerse for 5 to 10 minutes in a room-temperature solution of 15% (by volume) concentrated sulfuric acid and 15% concentrated hydrochloric acid in distilled water, plus 6 to 10% (by weight) chromic acid.
4. Rinse in tap water and dry in air.

The ASM states that tungsten may be cleaned by etching with a solution of 90% (by volume) concentrated nitric acid and 10% hydrofluoric acid.

Warning: Reaction is vigorous and the noxious gases emitted require forced-draft ventilation (see Chapter 13).

The cleaning procedure for tantalum alloys recommended by the ASM consists of immersing in a hot chromic dip or a chemical polish in a solution of five volumes of 95% sulfuric acid, two volumes of 70% nitric acid, and two volumes of 48% hydrofluoric acid. The ASM recommends the following three solutions as effective for cleaning columbium alloys:

1. 40% (by volume) HNO_3, 5% HF, remainder distilled water
2. 60% (by volume) HNO_3, 40% HF
3. 15% (by volume) H_2SO_4, 22% HF, remainder distilled water

After cleaning, columbium should be handled with plastic gloves. If the cleaned columbium is not welded within a few hours, it should be wrapped in plastic film or similar material. Cleaned columbium that is not to be welded within 24 hours should be stored in a vacuum chamber or in a welding chamber that has an inert gas atmosphere.

Table **21-35** Refractory metals tantalum, tungsten, rhenium, molybdenum—wrought

	TYPE				
	TANTALUM-10W	*AVC W-25Re*	*AVC (70 Mo, 30 W)*	*50Mo-50Re*	*222 (10.5W, 2.4 HF, 0.01C, Ta bal)*
	PHYSICAL PROPERTIES				
Density, lb/cu in.	0.608	0.714	0.43	0.50	0.604
Melting point, F	5516	5650	5150	4620	—
Elec res (70F), microhm-cm	20.0	20.0	5.3	19.4	—
	MECHANICAL PROPERTIES[a]				
Mod of elast in ten, psi					
Room temp	21×10^6	60×10^6	50×10^6	53	—
1600 F	10×10^6	—	—	—	—
2400 F	9×10^6	—	—	—	—
Ten str, 1000 psi					
Room temp	160	242	105	240	115
1600 F	95	148	70	108	—
2400 F	23	100	—	41	58
Yld str, 1000 psi					
Room temp	158	225	95	210	110
1600 F	90	140	85	—	—
2400 F	20	90	—	—	38
Red. of area, %					
Room temp	85+	36	35	40	—
1600 F	—	—	70	—	—
2400 F	—	—	—	—	—
Hardness (VHN)					
Cold worked	—	600	—	600	390
Stress relieved	250	570	225	—	350
Recrystallized	216	450	200	350	280

FABRICATING PROPERTIES

Working temp, F	Cold to 800	1600–2400	1800–2300	Cold	Room temp
Annealing temp, F					
Stress relief	≈2100	≈2400	≈2200	1100–1200	2000
Recrystallization	≈2700	≈3000	≈2600	1500–1650	3000
Machinability	Similar to unalloyed tantalum	EDM; diamond wheel grinding	Similar to molybdenum alloys	Machinable with carbide tools	Excellent
Joining	Similar to unalloyed tantalum	TIG welding	Similar to tungsten	Ductile TIG welds, resistance and fusion welding require oxid prot	Excellent
Corrosion resistance	Oxidation rate approximately two thirds that of unalloyed tantalum	Similar to tungsten	Similar to tungsten and molybdenum; high resistance to molten zinc	Resists H_2SO_4, brine atmos, molten metals and oxides. Rapid oxid over 1100 F	—
Available forms	Sheet, rod	Sheet, strip, bar, wire, tubing	Billet, bar & rod	Sheet, rod, wire, foil, tube, alloyed powder	Billet, rod
Uses	Rocket nozzels, heat shields, high temperature structures (missiles and space craft)	Nuclear fuel sheeting, thermoelectron devices	Rocket nozzle, components subject to molten zinc	Electronic tubes thermocouples, missile, rocket, nuclear and furnace parts	Aerospace applications

[a]Tensile properties are for stress-relieved condition.

Source: *Materials Selector, Materials Engineering*, a Penton/DPC Reinhold Publication, page 143.

The AWS recommends that molybdenum alloys having transition temperatures above the ambient welding temperature be preheated to a temperature above the transition temperature. Tungsten should always be preheated to 900°F.

Tungsten, molybdenum, and tantalum can be welded by the GTAW and EB processes in much the same way as the reactive metals. If proper backing and trailing shields are used, columbium sections up to 0.05 inch thick can be welded without a chamber.

Many brazing alloys are available for brazing tungsten and molybdenum. Columbium and titanium alloys are difficult to braze because the brazing alloy diffuses into the base metal, causing embrittlement. The filler metal to be used depends on the temperature at which the part is to be used. For service at 3500°F, brazing alloys must contain less than 6% Ti and 6% V. For service temperatures below 2800°F, titanium-based filler metals must be used.

21-10 PRECIOUS METALS

The precious metals are known by their generic names. The commonly used precious-metal alloys, together with their applications and weldability, are listed in Table 21-36.

With few exceptions, no special welding techniques are required to weld the precious metals. The two exceptions are (1) pure platinum can be joined by hammer welding in air at 1832 to 2192°F, and (2) a neutral or slightly oxidizing oxyacetylene flame should be used to join palladium.

Table **21-36** Precious metals—wrought

	METAL			
	GOLD[a]	*SILVER*	*PLATINUM*[b]	*PALLADIUM*
PHYSICAL PROPERTIES				
Density, lb/cu in	0.698	0.379	0.775	0.434
Melting point, F	1945	1761	3217	2826
Ther cond (212 F), Btu/hr/sq ft/°F/ft	172	242	42	41
Coef of ther exp, per °F	7.9×10^{-6}	10.9×10^{-6}	4.9×10^{-6}	6.5×10^{-6}
Specific heat, Btu/lb/°F	0.031	0.056	0.031	0.058
Elec res (32 F), microhm-cm	2.19	1.47	9.83	10.0
Temp coef of res (0–100 C), per °C[c]	0.003982	0.0041	0.003927	0.003802
Emf vs platinum (at 100 C), mv	0.78	0.74	0	−0.57
Mag suscept (18 C), mass units	—	—	1.10×10^{-6}	5.4×10^{-6}
Thermionic emission electron volts	4.9	4.5	5.32	4.99
Ther neutron capture cross section, barns/atom	94	60	8.1	8.0
MECHANICAL PROPERTIES				
Mod of elast in tension, psi	12×10^{6}	11×10^{6}	25×10^{6}	18×10^{6}
Tensile str, 1000 psi				
Annealed	19	22	18–21	20–28
Cold rolled, 50%	32	54	28–30	47
Yield str, 1000 psi				
Annealed	Nil	8	2–5.5	5
Cold worked	30	44	27	30
Elong (in 2 in.), %				
Annealed	45	48	30–40	24–40
Cold worked	4	2.5	2.5–3.5	1.5
Hardness (Vickers)[d]				
Annealed	—	—	40	40
Cold worked	—	—	100	100

Table **21-36** (*cont.*)

	METAL			
	GOLD[a]	SILVER	PLATINUM[b]	PALLADIUM
FABRICATING PROPERTIES				
Annealing temp, F	—	400–600	1475–2000[e]	1475
Hot working temp range, F	Any to melting point	—	Depends on cold work, purity	Depends on cold work, purity
Max red between anneals, %	Apparently unlimited	—	99	99
Casting temp range, F	2000–2370	2000	3300	3000
Joining	Braze with silver solder, no flux, any flame. Can be resistance welded by any method. Oxyacetylene weld with no flux, any flame	Braze with silver solder. Can be resistance welded	Braze with fine gold or white platinum solder. Hammer weld at 1800 F. Can be resistance, oxyacetylene, or arc welded	Braze with oxyacetylene torch using platinum solders. Can be resistance welded
Corrosion resistance	Does not oxidize when heated in air. Resists alkalis, salts, most acids. Not attacked by oxygen or sulfur. Rapidly attacked by chlorine and bromine	Does not oxidize when heated in air. Resists most dilute mineral acids and alkalis. Attacked rapidly by sulfur-bearing gases	Excellent oxidation resistance. Resists reducing or oxidizing acids alone, but is dissolved by aqua regia	Oxidizes when heated in air. Resists hydrofluoric, acetic and phosphoric acids. Attacked by nitric and sulfuric acids; and bromine and iodine
Available forms	Foil, rod, wire, sheet, tubing, powder	Sheet, strip, rod, wire, tubing, powder	Foil, sheet, wire, tubing, powder	Sheet, foil, wire, tubing, powder
Uses	Lining of chemical equipment, high melting solder alloys for electrical and chemical purposes, jewelry, dentistry	Electrical contacts, corrosion resisting equipment, bearings, photography supplies; alloying for coinage, brazing alloys, jewelry	Chemical equipment, electrical contacts, catalysts, laboratory equipment, jewelry, glass, extrusion dies, thermocouples	Electrical contacts, catalysts, production of pure hydrogen, jewelry, dental alloys

[a]Gold is generally produced in three grades: proof gold, 99.99% Au; refined gold, 99.95–99.98% Au; and 99.5% Au, which is accepted by the U.S. Mint without penalty.
[b]Grades of platinum metals available: Commercial Grade, 99.8% min purity; C.P. Grade, 99.98% min purity; Standard Thermocouple Grade, 99.99% min purity; Premium Thermocouple Platinum, 99.999% min purity; and Spectrographically Standardized Grade.
[c]Temperature coefficient is for metal of highest available quality.
[d]Refer to pp 354–357 for Hardness of Electrodeposited Metals.
[e]Can be cold worked.

Source: *Materials Selector, Materials Engineering,* a Penton/DPC Reinhold Publication.

REVIEW QUESTIONS

1. Name the composition and heat treatment of the aluminum alloy whose designation is:

a. 2024-T86.
b. 3003-H14.
c. 6061-T6.
d. 7075-O.

2. Refer to Table 4-8 and select the most suitable filler metal to obtain:

a. Best corrosion resistance.
b. Best color match after anodizing.
c. Ease of welding.

for arc welding the following aluminum alloy combinations: 5086 to 5154; 3003 to alclad 3004; alclad 3004 to 5050.

3. Which aluminum alloy is used for making armor-plated wheels, as used in tracked military vehicles?

4. What is the composition of the aluminum cleaning solution that can be stored in a mild-steel container?

5. Sketch the standard AWS butt-joint design recommended for arc welding ¼-inch-thick aluminum sections.

6. How would you prepare the joint designed in question 5 above, and what tools should be used?

7. Why are backing bars used? Of what material are they usually made?

8. Name the combustion gas preferred for fuel-gas welding of aluminum sections ¼ inch thick or thicker.

9. What type of flame should be used for fuel-gas welding of aluminum?

10. What is (or was) the colloquial name of red brass, 85%?

11. Which of the copper alloys listed in column I *should not* be welded by the processes identified in column II?

Column I	*Column II*
858	OAW
191	SMAW
782	TB
876	GMAW
993	GTAW
150	
524	

12. Name two typical applications for the following copper alloys:

a. Wrought alloys
    (1) Yellow brass

(2) Silicon bronze
(3) Copper-nickels

b. Cast alloys
(1) Red brass
(2) Aluminum bronze
(3) Beryllium copper

13. What surface impurities can be removed from magnesium with an alkaline cleaner?

14. Name the temper resulting from postweld heat treatment of the following magnesium alloys:

a. EK 41A-T4.
b. ZH 62A-F.
c. AZ 63A-T4.

15. To what nickel-alloy group do the 400-series nickel alloys belong?

16. Of what nickel alloy would linings used for pulp digesters used in paper mills be made?

17. Refer to Table 21-25 and identify the welding processes that *can* be used to weld:

a. UDIMET 500.
b. Alloy IN 100.
c. Monel Alloy 401.

18. Sketch the standard AWS joint design for a butt joint to be used for SMAW a ½-inch-thick section of nickel-alloy 600. What is the composition of the nickel alloy?

19. What are the conditions for MIG butt-welding a ½-inch-thick section of Inconel 600?

20. Name three welding conditions that could cause spatter when welding nickel alloys.

21. List the filler metal that should be used for SMAW and TIG welding the following nickel alloy combinations:

a. Monel 400 to stainless steel.
b. Hastelloy alloy N to Inconel 600.
c. Nickel 200 to carbon steel.

22. What gas mixture is preferred for welding lead? Why?

23. How should a confined, deep area be ventilated for welding lead?

24. What kind of oxyfuel-gas flame should be utilized to weld zinc alloys?

25. What are the welding conditions (machine settings or parameters) for automatic TIG welding a butt joint in a 1/16-inch-thick zinc-alloy plate?

26. Why should the weld in welded zinc joints be peened? What must the temperature range of the joint be while it is being peened? Why?

27. What filler metal is preferred for arc welding galvanized steel? Why?

28. Describe three welding techniques that should be followed to minimize intergranular cracking of welded zinc joints.

29. When must a secondary shielding be provided for welding titanium alloys?

30. Describe a simple method for: (a) degreasing titanium and (b) removing oxide formed on the titanium at temperatures below 1000°F.

31. Name the materials used to make heat sinks for welding titanium.

32. What are "reactive metals"?

33. What type of ventilation is required when cleaning tungsten?

34. Are special techniques required for welding any precious metals? If so, describe them.

PART 5

QUALITY CONTROL IN WELDING

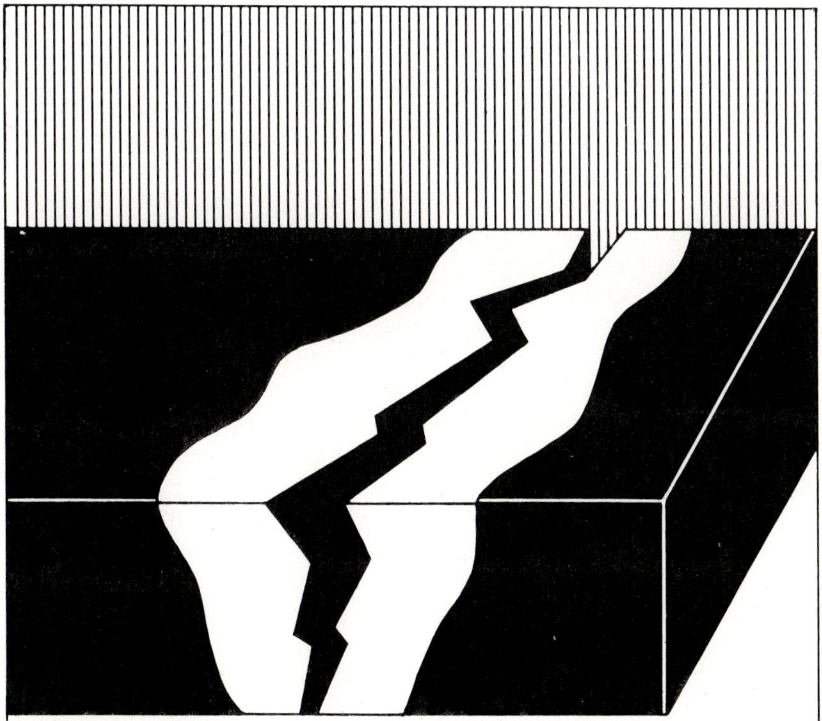

CHAPTER 22

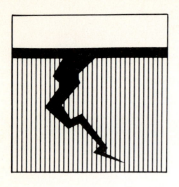

WELD QUALITY AND ITS EVALUATION BY NONDESTRUCTIVE AND DESTRUCTIVE TESTING METHODS

Weld quality is a relative term which depends on the final use to which the weldment is put. Quality welds can be classified as "good welds" and "too-good welds." Good welds are those that meet appearance requirements and will perform as intended until removed from service at the users' choice. Welds that are too good are those that have been made under high quality-control situations and will add nothing but production cost to the weld. Insisting on any method of inspection that serves no useful purpose is unnecessary and wasteful.

The first step in truly *controlling* quality is for the design or manufacturing engineering department to determine the degree of quality required for any particular weld based on performance and (sometimes) appearance of the weld (see Figure 22-1).

A welding procedure sheet [Figure 22-1(*c*), page 685] is prepared after the degree of quality of the welds has been determined. If, after testing a number of welds, it is found that the welds are continuously performing in excess of the standards, the welding procedure or even the joint design itself should be changed.

Regardless of the joint configuration or process used, welding procedures should be qualified prior to use. If it is not possible to try out the procedure on actual preproduction parts, mock-ups of the joint in question, using the same size, type, and shape of the workpiece and filler metal, should be welded. If the welder who is to do the production job also qualifies the procedure, the quality of the welds produced can also be used as a final check on the welder's ability (welder certification).

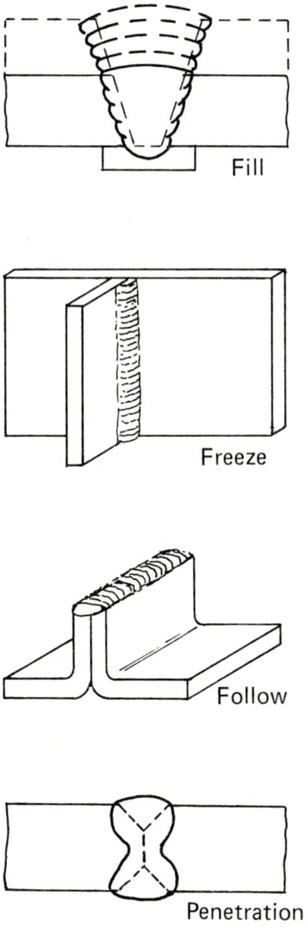
Fill
Freeze
Follow
Penetration

Fast-fill: A high deposition rate.

Fast-freeze: An out-of-position joint, where control of molten crater is important.

Fast-follow: Minimal weld metal required, high arc speed, and usually very small welds.

Penetration: Joint requires deep penetration into the base metal.

Figure **22-1** (*a*) Welding process and procedure selection: any joint can be categorized in terms of its need for these two factors. (Redrawn by permission from *The Procedure Handbook of Arc Welding*, 11th ed., Lincoln Electric Company, Cleveland, Ohio, p. 5.5-3.)

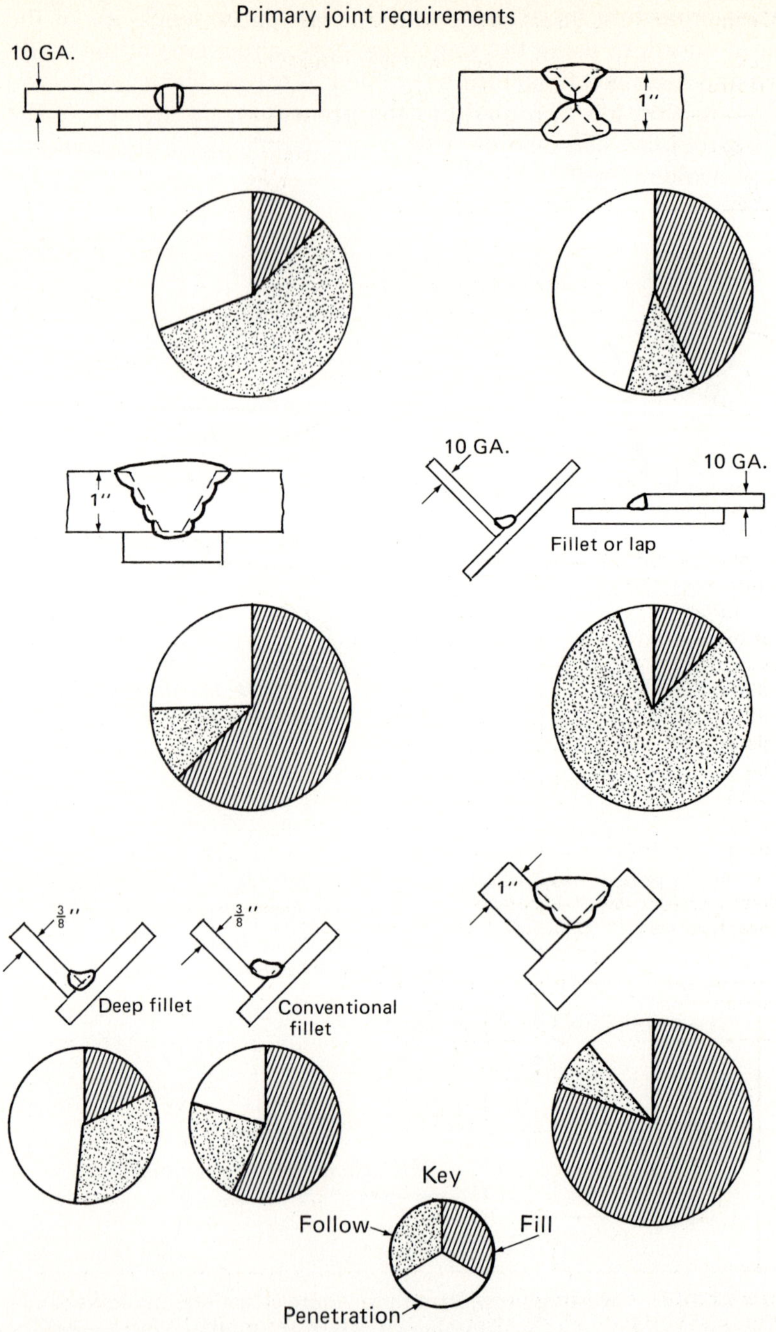
Primary joint requirements
10 GA.
1''
1''
10 GA.
10 GA.
Fillet or lap
3/8''
3/8''
Deep fillet
Conventional fillet
1''
Key
Follow
Fill
Penetration

Figure **22-1** (*b*) Some typical joints and their requirements (Redrawn by permission from *The Procedure Handbook of Arc Welding*, 11th ed., Lincoln Electric Company, Cleveland, Ohio, p. 5.5-3.)

Define joint requirements and performance characteristics

Stick	MIG	Self-shielded	Sub-arc
Electrode E______	Cored wire ___ Solid wire ___	Flux-cored ______	______

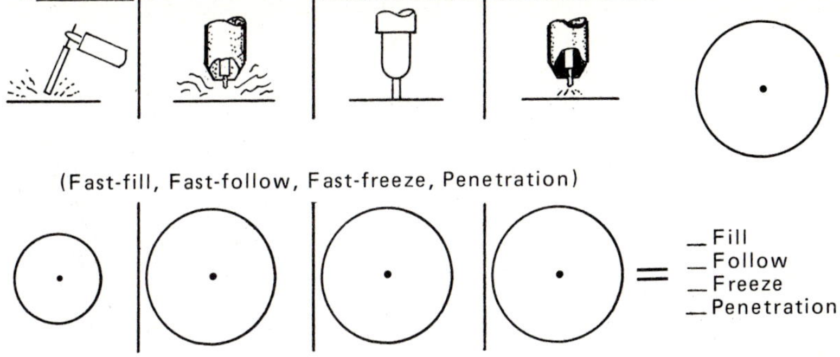

(Fast-fill, Fast-follow, Fast-freeze, Penetration)
=
_ Fill
_ Follow
_ Freeze
_ Penetration

Material specification ....
Welding process ....
Manual or machine ....
Position of welding ....
Filler metal specification ....
Filler metal classification ....
Flux ....
Weld metal grade ....
Shielding gas .... Flow ....
Single or multiple pass ....
Single or multiple arc ....
Welding current ....
Polarity ....
Welding progression ....
Root treatment ....
Preheat and interpass temperature ....
Postheat treatment ....

Welding procedure

Pass no.	Electrode size	Welding current		Travel speed	Joint detail
		Amperes	Volts		

This procedure may vary due to fabrication sequence, fit-up, pass size, etc. within the limitation of variables given in applicable welding codes.

Figure **22-1** (*c*) A welding procedure sheet. (Adapted by permission from *The Procedure Handbook of Arc Welding*, 11th ed., Lincoln Electric Company, Cleveland, Ohio, p. 5.5-3.)

22-1 SOME OBSERVATIONS ON WELD DEFECTS

Much of what an inspector is required to pass judgment on does not resolve itself into a "go" or "no-go" situation. Most x-ray inspection is based on standards that give the inspector a means of comparison as a basis for acceptance or rejection, and usually are not unrealistic in their demands. The quality demanded by the standards is obtainable. On the other hand, the standards are not loose enough to permit defects that would likely lead to in-service weld failure.

Problems arise, however, when a demand is made for much clearer x rays than the code requires. In some cases, inspectors have insisted that they want perfect x rays regardless of what the code says. There is doubt that this attitude would prevail if these same people were aware of the facts illustrated in Figures 22-2 and 22-3. Also, it is an established fact, and one worthy of inspection recognition, that scattered porosity well in excess of that permitted by some codes does not detract from the strength of the joint.

The individual welding-machine operator has a very great responsibility when x-ray quality is required. In addition to being certain that the prescribed procedures for producing x-ray results are met, the welder must be meticulous in making every bead as sound as possible. The ordinary effort is not good enough. Welding for x-ray quality literally requires a change of attitude on the part of the welder.

One defect an inspector is required to look for is *undercutting*. Generally this is not troublesome. It is almost always prevalent to some degree on the vertical leg of large horizontal fillets. But the mere fact of its existence does not mean that the weld is totally unacceptable.

Undercut serves no useful purpose; therefore, no reasonable effort should be spared to eliminate it or at least minimize it. The greatest

Figure **22-2** Weld samples made with progressive degrees of lack of fusion. (Redrawn by permission from *The Procedure Handbook of Arc Welding*, 11th ed., Lincoln Electric Company, Cleveland, Ohio, p. 11.1-4.)

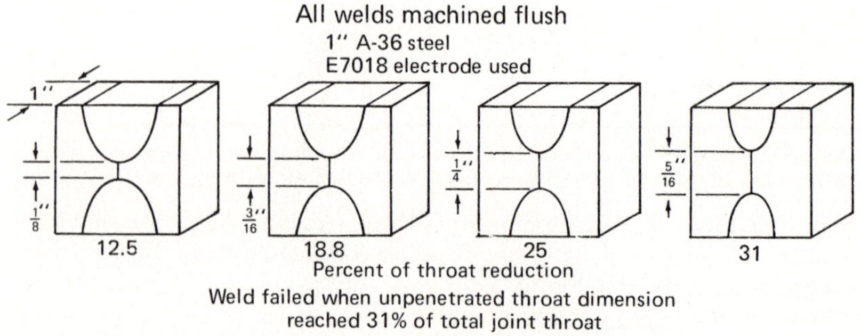
All welds machined flush
1" A-36 steel
E7018 electrode used
1"
1/8"
3/16"
1/4"
5/16"
12.5
18.8
25
31
Percent of throat reduction
Weld failed when unpenetrated throat dimension
reached 31% of total joint throat

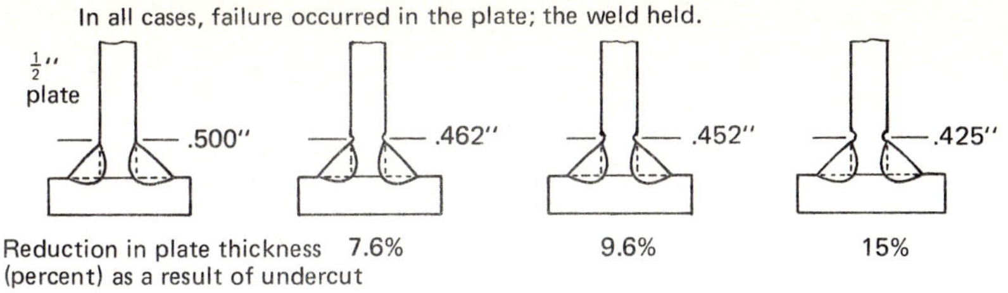
In all cases, failure occurred in the plate; the weld held.
½″ plate
.500″
.462″
.452″
.425″
Reduction in plate thickness (percent) as a result of undercut
7.6%
9.6%
15%

Figure **22-3** Undercut is not always objectionable. (Redrawn by permission from *The Procedure Handbook of Arc Welding,* 11th ed., Lincoln Electric Company, Cleveland, Ohio, p. 11.1-4.)

control generally lies within the area of procedure details or in the skill of the individual welder. But despite its undesirable nature, the presence of undercut, except under fatigue loading, is not nearly as damaging as one might suspect, as illustrated by Figure 22-4. Here undercut has been deliberately increased to a ridiculous degree on both sides of a fillet weld, yet the strength of the joint under static loading has not been impaired. Obviously, it does not make sense to demand repair of a little undercut on the average weldment.

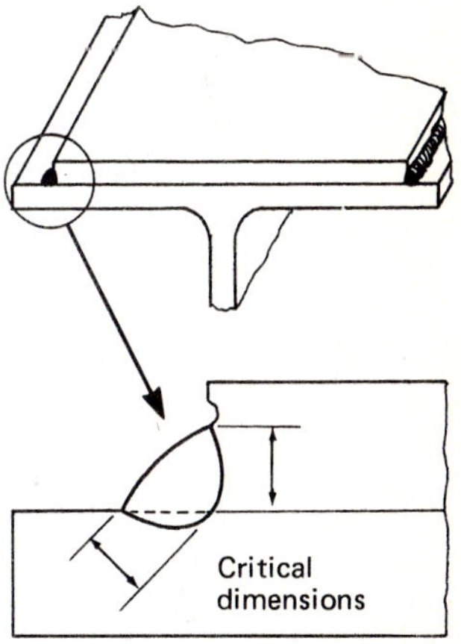
Critical dimensions

Figure **22-4** Undercut on a welded cover plate has no effect on strength of member. (Redrawn by permission from *The Procedure Handbook of Arc Welding,* 11th ed., Lincoln Electric Company, Cleveland, Ohio, p. 11.1-5.)

Another questionable defect controlled by standards is the *weld crater.* Any arc-welding process or procedure produces a weld crater. The ability to fill this crater at the termination of a weld varies considerably with the welding current, weld size, and other factors. If these limitations are understood and service requirements of the joint are considered, it should not be difficult to decide whether a crater must be filled.

Most codes make a general statement regarding crater filling. The *AWS Structural Code* states, "All craters shall be filled to the full cross section of the weld."

This statement appears harmless enough and, in fact, presents no problem on long, continuous welds. In this case, the start of each weld fills the crater left by the previous weld—a technique that any welder can easily master. The only crater requiring special attention would be the final one at the end of the joint.

What happens on intermittent welds is entirely different. An example is the production of 5⁄16-inch fillet welds, each 3 inches long and spaced on 12-inch centers. A typical production welding procedure would specify the use of ¼-inch electrode. But any procedure that will efficiently produce a sound weld of the required size will most certainly develop a pronounced crater that is very difficult to eliminate.

Where intermittent welds are acceptable, it would be more logical to require the weld length to be the total length, exclusive of the crater. Thus if 6-inch welds are required, they should measure 6 inches plus the crater.

The presence of defects that affect service performance is, in most instances, more important than those that affect appearance. The welding defects shown in Figures 22-5 and 22-6 are described in the following paragraphs.

POROSITY The term *porosity* is used to describe the globular voids, free of any solid material, which are frequently found in welds [Figure 22-5(g)]. In reality, the voids are a form of inclusion resulting from the chemical reactions that take place during welding. They differ from slag inclusions in that they contain gases rather than solids.

The gases forming the voids are derived from gases released by the cooling weld metal because of reduced solubility as the temperature drops and by chemical reactions within the weld.

Porosity may be prevented by avoiding excessive currents or excessive arc lengths. High consumption of the deoxidizing elements of the electrode coating may take place during deposition if excessive currents and arc lengths are used, leaving insufficient

Figure **22-5** Different types of weld defects. (Redrawn by permission from American Welding Society, *Welding Inspection*, Publication No. B1.1-45, pp. 44–49.)

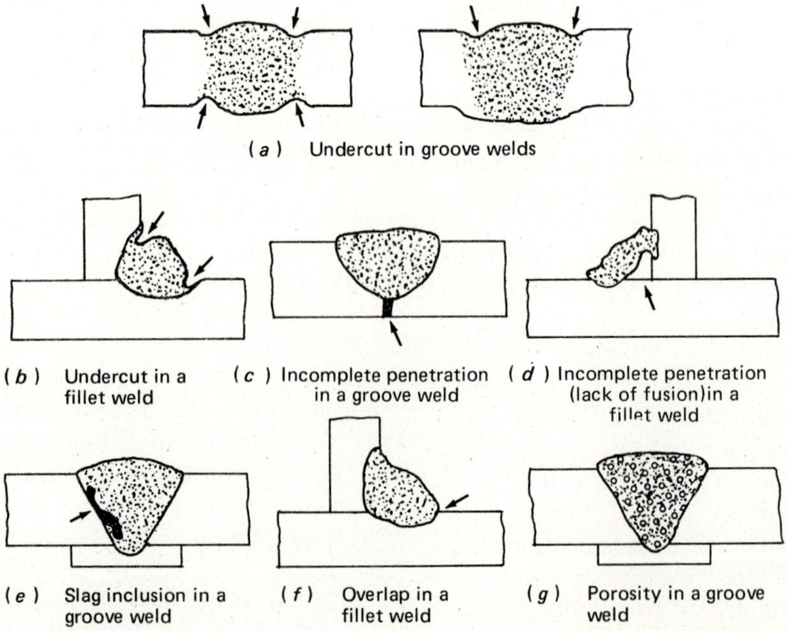

(a) Undercut in groove welds
(b) Undercut in a fillet weld
(c) Incomplete penetration in a groove weld
(d) Incomplete penetration (lack of fusion) in a fillet weld
(e) Slag inclusion in a groove weld
(f) Overlap in a fillet weld
(g) Porosity in a groove weld

quantities available to combine with the gases in the molten metal during cooling.

Porosity may generally be classified in one of three ways. *Uniformly scattered porosity* is such that the cavities are scattered more or less uniformly throughout the volume of the weld metal. The individual cavities may vary from almost microscopic to ⅛ inch or more in size. However, for a given welding condition they tend to be predominantly of one size.

It often happens that cavities occur in groups or clusters separated by considerable lengths of weld metal free of porosity. This is known as *group porosity.* Such groups are often associated with changes in welding conditions as, for example, changes in arc conditions when welding is stopped and started, such as in replacing an electrode.

Linear porosity occurs in the root pass and is often regarded as a special case of incomplete penetration. It is usually defined as a condition in which three or more cavities having an average diameter of not less than 1/16 inch are distributed in a line parallel to the longitudinal axis of the weld, the average distance of closest approach of any three or more cavities being not more than ¾ inch nor less than 1/16 inch.

NONMETALLIC INCLUSIONS This term is used to describe the oxides and other nonmetallic solids that are sometimes found as elongated and globular inclusions in welds [Figure 22-5(*e*)]. During the deposition and subsequent solidification of weld metal, many chemical reactions occur between materials (flux) or with the slag produced. Some of the products of these reactions are nonmetallic compounds soluble only to a slight degree in the molten metal. Due to their lower specific gravity, they tend to seek the upper surface of the molten metal unless restrained from doing so.

When welding by the shielded metal–arc process, slag may be formed and forced below the surface of the molten metal by the stirring action of the arc. Slag may also flow ahead of the arc, causing the metal to be deposited over it. Once slag is present in the molten metal, from any cause, it tends to rise to the surface by virtue of its lower density. A number of factors, such as high viscosity of the weld metal, rapid solidification, or too low a temperature, may prevent its release.

When weld metal is deposited by the gas metal–arc process over a sharp V-shaped recess, such as that produced by a diamond-pointed tool, slag will frequently be found trapped in the weld. Under such conditions, the arc may fail to heat the bottom of the recess to a sufficiently high temperature to permit the slag to float to the surface. A similar condition exists if a sharp recess is present as a result of undercutting or excessive convexity of the previous bead. Slag inclusions of this type are usually elongated and, if individual inclusions are of considerable size, or if inclusions are closely spaced, the strength of the joint may be materially reduced.

Slag forced into the molten metal by the arc or formed there by chemical reactions usually appears as finely divided or globular inclusions. Inclusions of this type are likely to be a particular problem in overhead welding.

The majority of slag inclusions may be prevented by proper preparation of the groove before each bead is deposited, taking care to correct contours that will be difficult to penetrate fully with the arc. Obviously, the release of slag from the molten weld metal will be aided by all factors that tend to make the metal less viscous or retard its solidification, such as preheating and high heat input per inch per unit time.

Slag in the root area results from the electrode being so large that the arc strikes the side of the groove instead of the root. The slag may roll down into the root opening, or it may be trapped in the metal of the root layer.

Another source of root-area slag lies in imperfect grinding or chipping preparatory to making the root pass on the opposite side of the plate. The initial layer of the weld on the first side of the plate does not necessarily have to fuse through the complete depth of the root opening. Even if it does fuse through, it will carry most of the slag with it. In addition, the unprotected metal that comes through will be badly oxidized. The slag, oxide, and irregular protrusions of weld metal should be removed by chipping or grinding to clean the metal before welding is started on the reverse side. If the slag is not completely removed in this way, it will remain in the penetration zone.

In removing slag preparatory to depositing subsequent passes, some may remain in the corner between the weld bead and the groove face. While in most cases this slag will be melted and rise to the surface, it may not, but rather remain as an elongated inclusion in the fusion zone. This is called *fusion-zone slag.*

CRACKING Cracking of welded joints results from the presence of multidirectional, localized stresses that at some point exceed the ultimate strength of the metal. When cracks occur during, or as a result of, welding, usually little deformation of the workpiece is apparent. Although cracks that appear in welded joints during welding are rarely caused by multidirectional stresses alone, experience has indicated that in the case of heavy structures (such as machine bases) they cause failure with little or no deformation when additional load is applied. To avoid this undesirable condition, stress-relief heat treatment is usually specified for thick sections.

An unfused area at the root of a weld may result in cracks without appreciable deformation if this area is subjected to tensile stress. In welding two plates together, the root of the weld is subjected to tensile stress as successive layers are deposited and, as already stated, a partially fused root will frequently permit a crack to start that may progress through practically the entire thickness of the weld.

After a welded joint has cooled, cracking is more likely to occur if the metal is either hard or brittle. A ductile material will withstand stress concentrations that might cause a hard or brittle material to fail.

WELD-METAL CRACKING The ability of the deposit to remain intact under the stress system imposed during the welding operation is a function of the composition and structure of the weld metal. Weld-metal cracking is more likely to occur in the first layer of the weld than elsewhere and, unless repaired, it will usually continue through the other layers as they are deposited. This tendency to continue into succeeding layers is either materially reduced or eliminated with austenitic weld metal. When the problem of cracking of the first layer of weld metal is encountered, improvement may be obtained by one or more of the following modifications:

1. Change in the electrode manipulation or electrical conditions, which will change the contour or composition of the deposit.

2. Decrease in the travel speed to increase the thickness of deposit, thereby providing more weld metal to resist the stresses being imposed.

3. Use of preheat to modify the intensity of the stress system being imposed.

Three different types of cracks, illustrated in Figure 22-6, can occur in weld metal, as follows.

Transverse weld cracks are perpendicular to the axis of the weld and, in some cases, have been observed to extend beyond the weld into the plate metal. This type of crack is more common in joints having a high degree of restraint.

Longitudinal weld cracks are predominantly within the weld metal and are usually confined to the center of the weld. Such cracks may occur as the extension of crater cracks formed at the end of the weld. They may also occur as the extension through successive layers of a crack that existed in the first layer deposited. If a crack is formed in the first layer and is not removed or completely remelted when the subsequent layer is deposited, it tends to progress into the layer above, then into the next adjacent layer, and finally to the surface. The final extension of the crack to the weld surface may occur during the cooling of the weld after welding operations have been completed. The only way to remedy this condition for any

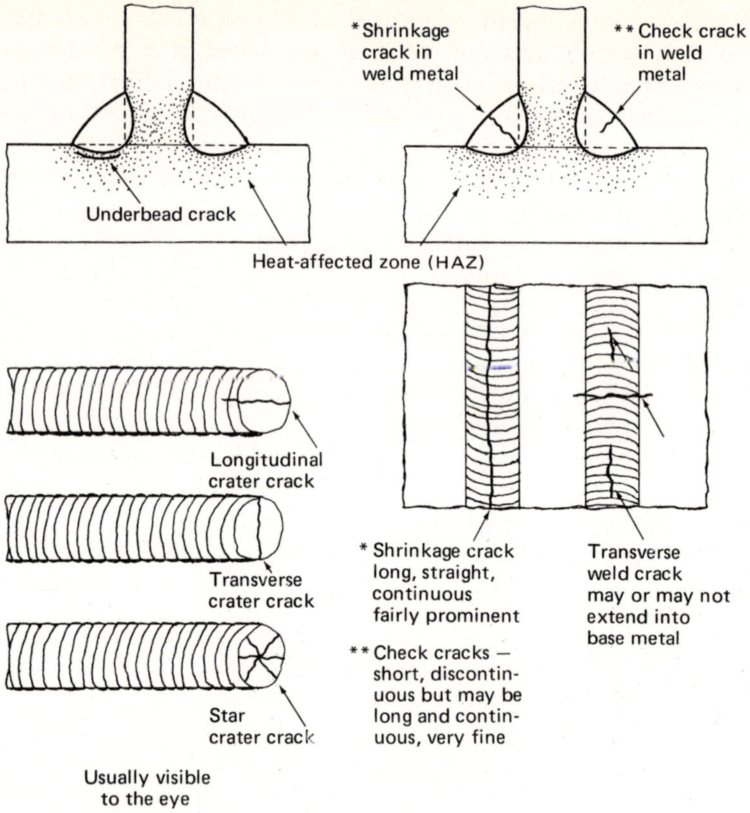

*Shrinkage crack in weld metal
**Check crack in weld metal
Underbead crack
Heat-affected zone (HAZ)
Longitudinal crater crack
Transverse crater crack
Star crater crack
Usually visible to the eye
* Shrinkage crack long, straight, continuous fairly prominent
** Check cracks — short, discontinuous but may be long and continuous, very fine
Transverse weld crack may or may not extend into base metal

Figure **22-6** Three basic kinds of weld cracks. (Redrawn by permission from American Welding Society, *Welding Inspection*, Publication No. B1.1-45, p. 52.)

given filler-metal/base-metal combination is to change all the welding machine settings.

Whenever the welding operation is interrupted, there is a tendency for the formation of a crack in the crater. *Crater cracks* are usually star-shaped and proceed only to the edge of the crater. However, they may be starting points for longitudinal weld cracks, particularly when they occur in the crater that is formed at the end of the weld. When the crater is formed at other places—as, for instance, when an electrode is changed—the crack is usually welded when the operation is resumed. This, however, does not always occur, and sometimes fine, star-shaped cracks may be observed in several places.

The previous paragraphs dealt with defects in the deposited weld metal. The following paragraphs will describe the defects that can occur in the base metal of workpieces.

BASE-METAL CRACKING This type of cracking, shown at the top of Figure 22-6, usually longitudinal in nature, occurs within the heat-affected zone of the metal being welded and is almost always associated with hardenable materials. Hardness and brittleness in the heat-affected zone in welded joints are metallurgical defects produced by the thermal cycle of welding, and are among the main causes of cracking.

In the case of low-carbon, medium-carbon, and low-alloy steels, the hardness and the ability to deform without rupture depend on the alloy group to which the steel belongs, and upon the rate of cooling from the elevated temperatures produced by the welding operation. The rate of cooling depends upon a number of physical factors, such as the temperature, the thickness, and the thermal conductivity of the base metal; the heat input per unit time at a given section of the weld; and the atmospheric temperature. With a given cooling rate, the low-carbon steels harden considerably less than the medium-carbon steels. Low-alloy construction steels, such as SAE 950, exhibit a wider variation in their hardening characteristics, and some of them may be similar to low-carbon steel, while others react like medium-carbon steel. The austenitic and ferritic stainless steels and the martensitic-alloy steels behave similarly to the medium-carbon and low-alloy groups, except that they harden to a greater degree with a given cooling rate. Neither the austenitic steels, of which the common 18% chromium, 8% nickel stainless steel is an example, nor the ferritic stainless steels, of which the low-carbon straight-chromium steels (or irons) containing 18% or more chromium are an example, harden upon quenching from elevated temperatures. However, the ferritic stainless steels are, in general, rendered brittle (but not hard) by welding operations.

The metallurgical characteristics of the metals involved are very important. Since ductility usually decreases with increasing hardness, base-metal cracking has been associated with lack of ductility in the heat-affected zone. This is not the whole answer, however, for it has been established that different heats (batches of steel) of the same steel of equal hardenability vary appreciably in cracking tendency. The characteristics of the electrode coating also have a considerable effect on the tendency toward heat-affected-zone cracking.

Hardenable steels are usually more difficult to weld because of the following two reasons:

1. Variations in the metallographic structure of the heat-affected zone can occur with variations in the cooling rate, resulting in differences in mechanical characteristics.

2. Such steels are usually used because of their higher tensile properties, and so can be likened to increasing the thickness of a mild steel.

The two types of base-metal cracking that occur in hardenable steels are (1) transverse and (2) longitudinal. These types of cracking can be minimized by:

1. The use of a suitable preheat;
2. an increase in heat input that will retard cooling rate; and
3. selection of the best available filler metal.

Transverse base-metal cracks are transverse to the direction of welding. They are usually associated with fillet welds on steels of high hardenability, where the distance between the edge of the weld and the exposed edge of one plate is relatively small. Such cracks usually cannot be detected until the weldment has cooled to room temperature.

Longitudinal base-metal cracks are parallel to the weld and are in the base metal. They may be extensions of fusion-zone cracks. For fillet welds, longitudinal base-metal cracks may be divided into two types.

1. *Toe cracks* proceed from the toe of the fillet weld through the base metal, often starting from the undercuts.
2. *Root cracks* proceed from the root of the fillet weld and progress through the base metal, and are usually evident on the opposite side of the plate.

In the case of groove welds, cracks are more likely to occur in the heat-affected zone adjacent to the weld. Cracks may also occur at the edge of the weld in the fusion zone between the weld and base metals. Usually this type of crack is associated with steels of high hardenability when weld metal and plate metal are entirely different in composition, thereby promoting the formation of alloys of unpredictable properties in this zone.

INCOMPLETE FUSION This term is sometimes applied to the conditions later referred to as *incomplete penetration*. It is used here in a more restricted sense to describe the failure to fuse together adjacent layers of weld metal or adjacent weld and base metal [Figure 22-5(*d*)]. This failure to obtain fusion may occur at any point in the welding groove.

Lack of fusion may be caused by (1) failure to raise the temperature of the base metal at the weld and immediately adjacent thereto and the temperature of the previously deposited weld metal to the melting point, or (2) by failure to dissolve (because of improper fluxing) the oxides or other foreign material present on the surfaces to which the deposited metal must fuse.

It should not be inferred from this brief discussion that it is necessary to melt an appreciable portion of the sidewalls of the groove in order to be certain of securing proper fusion. It is only necessary to bring the surface of the base metal to the fusion temperature to obtain metallurgical continuity of the base and weld metal.

Lack of fusion is best avoided by ascertaining that the surfaces to be welded are free of injurious foreign material and by the use of welding machine operators who have adequately demonstrated their ability to make sound welds.

INCOMPLETE PENETRATION This term is used to describe the failure of the deposited metal and base metal to fuse integrally at the root of the weld. It may be caused by the failure of a groove-weld root face to reach fusion temperature for its entire depth [Figure 22-5(*c*)], or the failure of the weld metal to reach the root of a fillet weld, leaving a void caused by bridging of the weld metal from one member to the other [Figure 22-5(*d*)].

Although incomplete penetration may, in a few cases, be due to failure to dissolve or flux surface oxides and impurities, the heat-transfer conditions existing at the joint are a more frequent source of this defect. If the areas of base metal that first reach fusion temperatures are above the root, molten metal may bridge between these areas and screen off the arc before the base metal at the root melts. In metal-arc welding, the arc will establish itself between the electrode and the closest part of the base metal. All other areas of the base metal will receive heat principally by conduction. If the portion of the base metal closest to the electrode is a considerable distance from the root, the conduction of heat may be insufficient to attain fusion temperature at the root.

Incomplete penetration is undesirable particularly if the root of the weld is subject to either direct tension or bending stresses. The unfused area permits stress concentrations that may result in failure without appreciable deformation. Even though the service stresses in the structure may not involve tension or bending at this point, the shrinkage stresses and consequent distortion of the parts during welding will frequently cause a crack to initiate at the unfused area. Such cracks may progress as successive beads are deposited until they extend through almost the entire thickness of the weld.

The most frequent cause of this type of defect is a groove design unsuitable for the welding process used or for the conditions of actual construction. When a groove is welded from one side only, complete penetration is not likely to be obtained consistently with the metal-arc process if the root opening is adequate. This is also true if the root opening is too small, or if the included angle of a V-type groove is too small. Any of these factors will make it difficult to reproduce test results under conditions of actual construction. If the design is known to be adequate, incomplete penetration may result from the use of too large an electrode, an abnormally high rate of travel, or insufficient welding current.

UNDERCUTTING This term is used to describe either (1) the melting away of the sidewall of a welding groove at the edge of a layer or bead, thus forming a sharp recess in the sidewall in the area to which the next layer or bead must fuse; or (2) the reduction in base-metal thickness at the line where the last bead is fused to the surface [Figures 22-5(*a*) and (*b*)].

Undercutting of both types is usually due to the technique employed by the operator. Certain electrodes, too high a current, or too long an arc may increase the tendency to undercut. However, different types of electrodes show widely varying characteristics in this respect. With some electrodes, even the most skilled operator may be unable to avoid undercutting under certain conditions. The conditions referred to primarily involve position and accessibility of the joint, although magnetic arc blow may also be a factor.

Undercutting of the sidewalls of a welding groove will in no way affect the completed weld if care is taken to properly correct the conditions before depositing the next bead. This is best accomplished by using a well-rounded chipping tool or grinding wheel to eliminate the sharp recess that might serve to trap slag. However, if an operator is sufficiently experienced and knows to what extent subsequent beads will penetrate, chipping is not always necessary.

Undercutting at the surface should not be permitted in an aggravated form, as it may materially reduce the strength of the joint, particularly its resistance to fatigue stresses. Fortunately, this type of undercutting is always readily detectable by a visual examination of the surface of the completed weld and can usually be corrected by additional metal deposition.

The Society for Nondestructive Testing (SNT) defines *nondestructive testing* as a group of tests used to detect defects or flaws in metals, but leaving the specimen fit to perform the task for which it was made after testing. The principal nondestructive testing (inspection) methods for welds are listed in Table 22-1, and are described in the following sections.

22-2 VISUAL INSPECTION

This method is the most extensively used of any method of inspection because it is easy to apply, quick, relatively inexpensive, and gives very important information with regard to the general conformity of the weldment to specification requirements. Visual inspection is performed prior to welding, during the welding operation, and after welding has been completed.

Prior to welding, the inspector checks the material to be welded for such defects as scabs, seams, scale, plate laminations, and plate dimensions. After assembling the parts to be welded, the inspector can note any incorrect root openings, improper edge preparation,

Table **22-1** Principal nondestructive testing methods

INSPECTION METHOD	EQUIPMENT REQUIRED	ENABLES DETECTION OF	ADVANTAGES	LIMITATIONS	REMARKS
Visual	Magnifying glass Weld-size gauge Pocket rule Straight edge Workmanship standards	Surface flaws—cracks, porosity, unfilled craters, slag inclusions. Warpage, underwelding, overwelding, poorly formed beads, misalignments, improper fitup.	Low cost Can be applied while work is in process, permitting correction of faults. Gives indication of incorrect procedures.	Applicable to surface defects only. Provides no permanent record.	Should always be the primary method of inspection, no matter what other techniques are required. Is the only "productive" type of inspection. Is the necessary function of everyone who in any way contributes to the making of the weld.
Radiographic	Commercial x-ray or gamma units, made especially for inspecting welds, castings, and forgings. Film and processing facilities. Fluoroscopic viewing equipment.	Interior macroscopic flaws—cracks, porosity, blow holes, nonmetallic inclusions, incomplete root penetration, undercutting, icicles, and burnthrough.	When the indications are recorded on film, gives a permanent record. When viewed on a fluoroscopic screen, a low-cost method of internal inspection.	Requires skill in choosing angles of exposure, operating equipment, and interpreting indications. Requires safety precautions. Not generally suitable for fillet-weld inspection.	X-ray inspection is required by many codes and specifications. Useful in qualification of weldors and welding processes. Because of cost, its use should be limited to those areas where other methods will not provide the assurance required.
Magnetic-particle	Special commercial equipment.	Excellent for detecting surface discontinuities—especially surface cracks.	Simpler to use than radiographic inspection.	Applicable to ferromagnetic materials only.	Elongated defects parallel to the magnetic field may not give pattern; for this

Table **22-1** (*cont.*)

INSPECTION METHOD	EQUIPMENT REQUIRED	ENABLES DETECTION OF	ADVANTAGES	LIMITATIONS	REMARKS
Magnetic-particle	Magnetic powders—dry or wet form; may be fluorescent for viewing under ultraviolet light.		Permits controlled sensitivity. Relatively low-cost method.	Requires skill in interpretation of indications and recognition of irrelevant patterns. Difficult to use on rough surfaces.	reason the field should be applied from two directions at or near right angles to each other.
Liquid-penetrant	Commercial kits, containing fluorescent or dye penetrants and developers. Application equipment for the developer. A source of ultraviolet light—if fluorescent method is used.	Surface cracks not readily visible to the unaided eye. Excellent for locating leaks in weldments.	Applicable to magnetic, nonmagnetic materials. Easy to use. Low cost.	Only surface defects are detectable. Cannot be used effectively on hot assemblies.	In thin-walled vessels, will reveal leaks not ordinarily located by usual air tests. Irrelevant surface conditions (smoke, slag) may give misleading indications.
Ultrasonic	Special commercial equipment, either of the pulse-echo or transmission type. Standard reference patterns for interpretation of RF or video patterns.	Surface and subsurface flaws, including those too small to be detected by other methods. Especially for detecting subsurface lamination-like defects.	Very sensitive. Permits probing of joints inaccessible to radiography.	Requires high degree of skill in interpreting pulse-echo patterns. Permanent record is not readily obtained.	Pulse-echo equipment is highly developed for weld inspection purposes. The transmission-type equipment simplifies pattern interpretation where it is applicable.

Source: *Procedure Handbook of Arc Welding* (Cleveland, Ohio: Lincoln Electric Company), pp. 11.2–11.4.

and other features of joint preparation that might affect the quality of the welded joint. For example, the applicable drawing may indicate a double-V groove, whereas a single-V groove has actually been made. Also, the groove angle may be 30 degrees, when 45 degrees is required. Oxygen cutting irregularities and the presence of excessive scale are other factors that the inspector should check, since they, too, may affect the quality of the finished joint.

During the welding operation, the inspector makes sure that all requirements of the written procedure sheet are complied with. Where multiple pass welds are made, the inspector may make use of a *workmanship standard*. Figure 22-7 shows how such standards may be prepared. These are sections of joints similar to those in manufacturing, in which the proportions of successive weld layers are shown. Each layer of the production weld may be compared with corresponding layers of the workmanship standard as a guide.

Figure **22-7** Workmanship standards: (*a*) groove welds; (*b*) fillet welds.

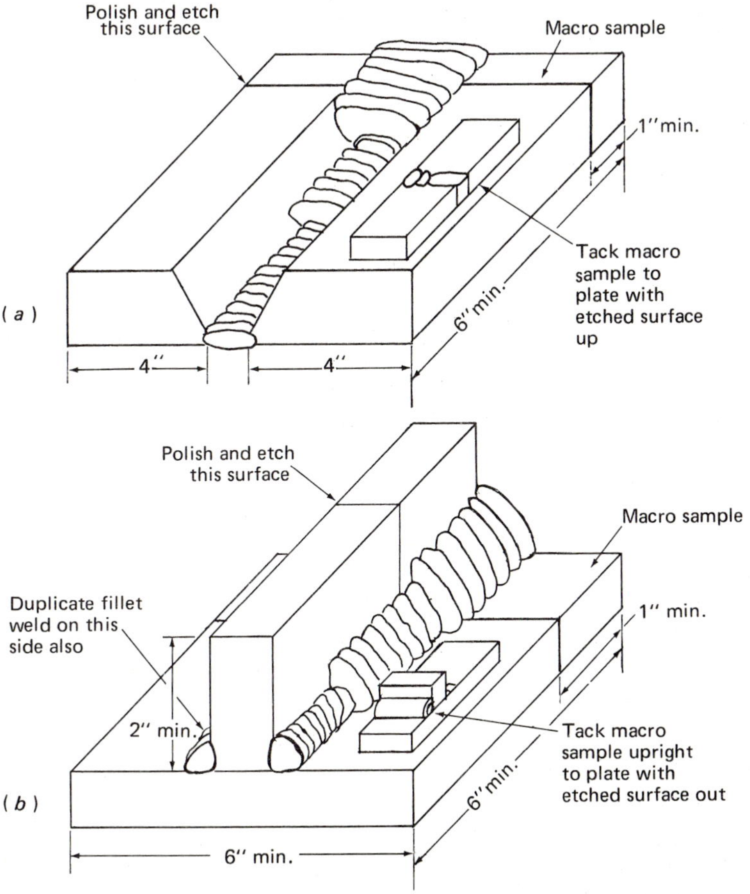

Polish and etch this surface
Macro sample
1" min.
Tack macro sample to plate with etched surface up
(a)
4"
4"
6" min.
Polish and etch this surface
Macro sample
1" min.
Duplicate fillet weld on this side also
2" min.
Tack macro sample upright to plate with etched surface out
(b)
6" min.
6" min.

The first layer, or *root pass,* is the most important from the point of view of final soundness. Because of the geometry of the joint, the relatively large volume of base metal compared to that of the root-pass weld metal, the fact that the plate may be cold, and the possibility that the arc may not strike into the root, the root-pass metal quickly freezes and tends to trap slag or gas that is difficult to remove in making subsequent passes. In addition, this pass is particularly susceptible to cracking. Such cracks may persist and extend to subsequent layers. Inspection of this pass should be carried out meticulously. A workmanship standard is very useful in this connection. Radiographic inspection may give evidence as to conditions in the root pass, thus serving as a check on visual inspection.

Inspection of the root pass offers an additional opportunity to inspect for plate laminations, since they tend to open up because of the heat incident to the welding operation. In the case of groove welds, slag from the root pass on one side of the plate drips through to form slag deposits on the other side. Such deposits are usually chipped or ground out. Where removal is imperfect, slag will be found in the root of the finished weld. Visual inspection is carried out to ensure adequate chipping or grinding.

After welding has been completed, the inspector usually checks the welded assembly for such items as:

1. Dimensional accuracy of the weldment (including warpage).

2. Conformity to drawing requirements. (This involves determination to whether all required welding has been done and whether finished welds conform with regard to size and contour.)

3. Acceptability of welds with regard to appearance, including such items as regularity, surface roughness, and weld spatter.

4. The presence of unfilled craters, pock marks, undercuts, overlaps, and cracks.

The size and contour of fillet welds are usually checked with a *weld gage,* such as the one shown in Figure 22-8(*a*). This gage is only used for fillet welds. The size of the weld is defined in terms of the length of the leg. The gage will determine whether or not the size is within allowable limits and whether there is excessive concavity or convexity. Such a gage is made for use on joints between surfaces that are perpendicular or nearly so. Special gages may be made for surfaces that are at acute or obtuse angles.

For groove welds, the following specifications are considered. The width of a finished groove weld should be in accordance with the required groove angle, root face, and root opening. The height of reinforcement should be consistent with specified requirements; when not specified, the inspector may have to rely on judgment, guided by what he or she considers good welding practice.

To determine the conformity of welds to appearance implies the use of visual standards or of sample weldments submitted by the purchaser or contractor. Requirements as to surface appearance differ widely. Sometimes a smooth weld strictly uniform in size and contour is required because the weld forms part of the exposed surface of the weldment and good appearance is desired for esthetic reasons.

Figure **22-8** (*a*) Weld gage for fillet welds only. (*b*) Acceptable and defective fillet-weld profiles. (Redrawn by permission from American Welding Society, *Welding Inspection*, Publication No. B1.1-45, p. 43.)

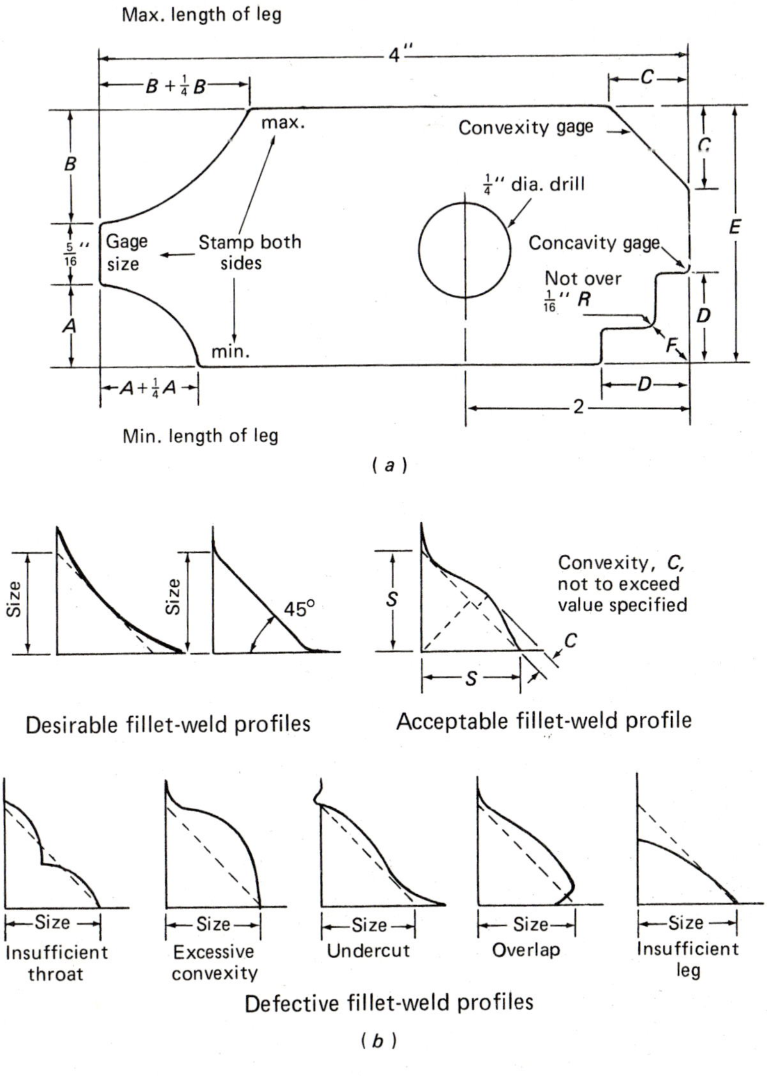

Max. length of leg
4"
B + ¼B
C
max.
Convexity gage
B
C
¼" dia. drill
E
5/16"
Gage size
Stamp both sides
Concavity gage
Not over 1/16" R
A
D
min.
F
A+¼A
D
2
Min. length of leg
(a)
Size
Size
45°
S
Convexity, C, not to exceed value specified
C
S
Desirable fillet-weld profiles
Acceptable fillet-weld profile
Size
Size
Size
Size
Size
Insufficient throat
Excessive convexity
Undercut
Overlap
Insufficient leg
Defective fillet-weld profiles
(b)

22-3 RADIOGRAPHIC INSPECTION

This test method makes use of the ability of short-wave radiations, such as x rays or gamma rays, to penetrate objects opaque to ordinary light. These rays have this property because of their shorter wavelength, as compared to light. In general, the shorter the wavelength the greater the penetrating power. Not all of the radiation penetrates the weld—some is absorbed. The amount of the absorption is a function of the density and thickness of the weld. Should there be a cavity, such as a blowhole, in the weld interior, the beam of radiation will have less metal to pass through than in a sound weld. Consequently, there will be a variation, if measured or recorded on a film sensitive to the radiation, which gives rise to an image indicating the presence of the defect. The image is an x-ray shadow of the interior defect. Such a shadow picture is called a *radiograph*. A successful radiograph will be one which has so faithfully recorded the x-ray image that the presence or absence of a defect in a weld is established; and, if present, its size, shape, and location are clearly defined.

The x ray is the most successful and reliable method for the nondestructive testing of welds. However, like most tools, it has certain limitations; and its correct application and interpretation require a technical knowledge of the method, a reasonable conception of the type of defects disclosed, and a knowledge of the relationship between the defects and the applicable specification.

To reduce the chances of misinterpretation of radiographs that lack sharpness and contrast, a gage known as a *penetrameter* is used on the side of the weld away from the film. Figure 22-9(*a*) shows the dimensions of a standard ASTM-type penetrameter. The penetrameter specified by the *ASME Boiler and Pressure Vessel Code* consists of a thin strip of metal with the same adsorption characteristics as the weld metal. When a weld is to be x rayed, a penetrameter with a thickness equal to or less than 2% of the weld thickness is selected and placed alongside the weld to be x rayed [Figure 22-9(*b*)]. A lead numeral at one end shows the thickness of weld for which the penetrameter is to be used [Figure 22-9(*c*)]. For reference purposes, three holes are drilled in the face of the gage, the diameters of which are multiples of the penetrameter thickness. The appearance of the penetrameter image on the radiograph film tells the observer whether the minimum of 2% sensitivity and adequate sharpness and contrast have been attained. In the hands of a skilled inspector, the penetrameter also gives other items of information. Sharp image delineation gives assurance that the radiographic procedure is correct. The presence of a penetrameter image is also evidence that can be presented at any later time to prove that the weld was executed properly.

(a)

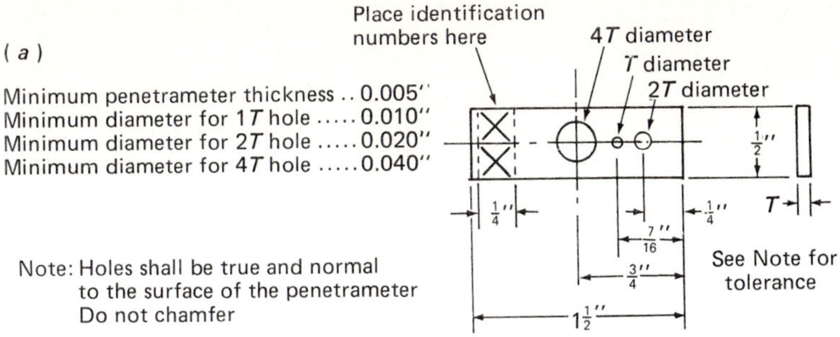
Place identification numbers here
4T diameter
T diameter
2T diameter
Minimum penetrameter thickness .. 0.005''
Minimum diameter for 1T hole ..... 0.010''
Minimum diameter for 2T hole ..... 0.020''
Minimum diameter for 4T hole ..... 0.040''
1/2''
1/4''
1/4''
T
7/16''
3/4''
1 1/2''
See Note for tolerance
Note: Holes shall be true and normal to the surface of the penetrameter Do not chamfer

(b)

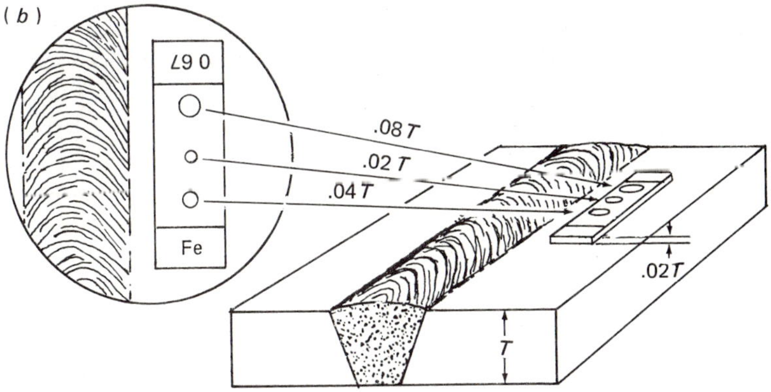
0 67
Fe
.08 T
.02 T
.04 T
.02T
T

(c)

JE
X 566

Figure **22-9** (*a*) Design of penetrameter for thickness from 0.005 inch up to and including 0.050 inch. (*b*) Placement of penetrameter alongside of a weld. (*c*) Some typical penetrameters. (Redrawn by permission from *The Procedure Handbook of Arc Welding*, 11th ed., Lincoln Electric Company, Cleveland, Ohio, p. 11.2-7.)

COMMERCIAL
XE-7680

Figure **22-10** X-raying a section of a pipeline. (Courtesy of the Lincoln Electric Company, Cleveland, Ohio.)

It should be pointed out that a radiograph compresses all the indications of defects that occur throughout the weld into one plane. Thus, the radiograph tends to give an exaggerated impression of scattered types of defects, such as porosity or inclusions, and unless allowance is made for this fact, a weld that is entirely adequate for its function could be ruled defective. The angle of exposure also has an influence on the radiograph.

A typical arrangement for radiography of a pipeline is shown in Figure 22-10. As seen in the photograph, radiographic inspection should be done using only qualified inspection procedures and personnel. Methods of radiographic inspection are covered by ASTM specification E 94, *Radiographic Testing.*

Warning: Radiation from x-ray machines or radioisotope sources can be damaging to body tissue when the exposure is excessive. Safety precautions must be taken. The *American Standard Safety Code for the Industrial Use of X-rays* should be consulted for this purpose.

22-4 MAGNETIC-PARTICLE INSPECTION

Magnetic-particle inspection is a method of locating and defining discontinuities in magnetic materials. It is excellent for detecting surface defects in welds, revealing discontinuities that are too fine to

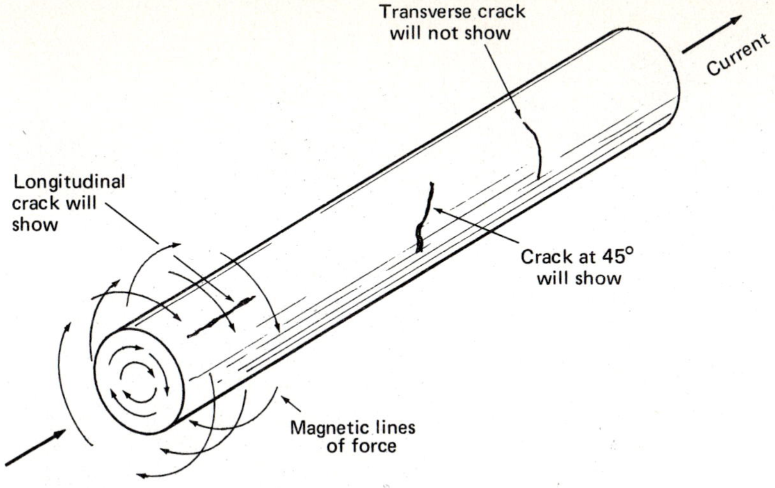
Transverse crack
will not show
Current
Longitudinal
crack will
show
Crack at 45°
will show
Magnetic lines
of force

Figure **22-11** Flux is produced at right angles to the flow of current

be seen with the naked eye. With special equipment, it can also be used to detect defects that are close to the surface.

There are two magnetic inspection methods: *circular magnetization* (Figure 22-11) and *longitudinal magnetization* (Figure 22-12). When using the circular method, probes (Figure 22-13) are usually placed on each side of the area to be inspected and high amperage passed through the workpiece. A magnetic flux is produced at right angles to the flow of current, which may be represented (as in Figure 22-11) by circular lines of force within the workpiece. When these

Figure **22-12** Part magnetized by placement inside a solenoid

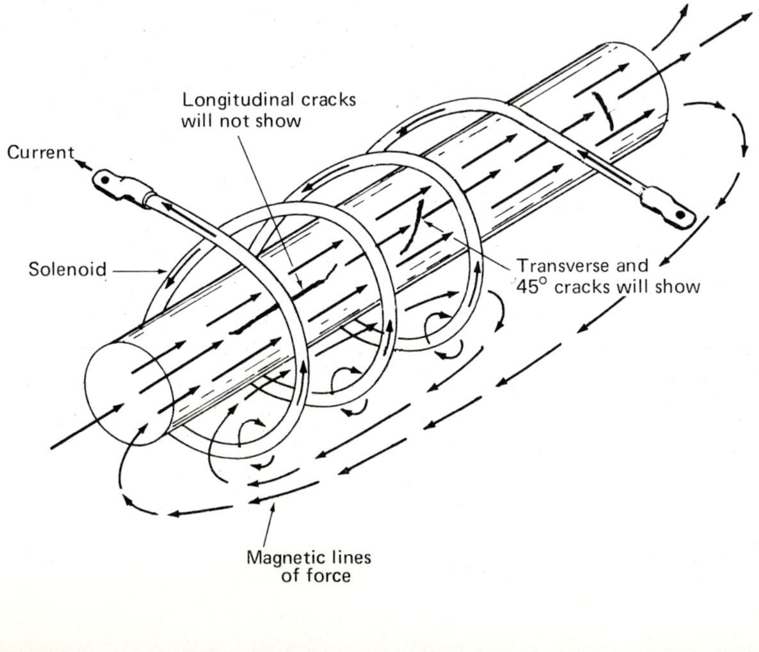
Longitudinal cracks
will not show
Current
Solenoid
Transverse and
45° cracks will show
Magnetic lines
of force

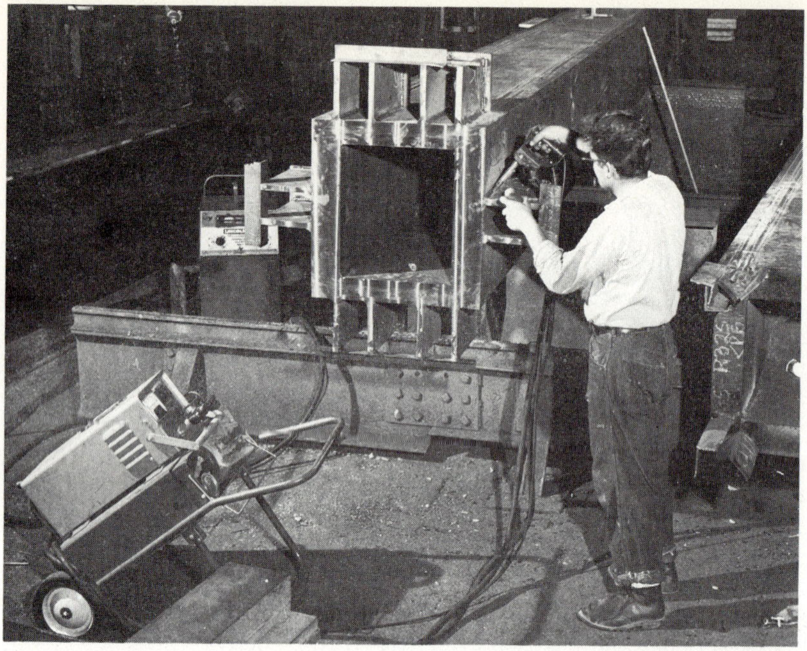

Figure **22-13** Circular magnetization method for inspection of welds in steel building columns. (Courtesy of the Lincoln Electric Company, Cleveland, Ohio.)

lines of force encounter a discontinuity, such as a longitudinal crack, they are diverted and leak through the surface, creating magnetic poles or points of attraction. A magnetic powder dusted onto the surface will cling to the leakage area more tenaciously than elsewhere, forming an indication of the discontinuity .

A workpiece can also be magnetized by putting it inside a solenoid, as in Figure 22-12. In this case, the magnetic lines of force are longitudinal and parallel with the workpiece. Transverse cracks show up under this arrangement.

The magnetic-particle inspection method is much simpler to use than radiographic inspection, but it has its limitations. It is applicable to ferromagnetic materials only. It cannot be used with austenitic steels. A joint between a base metal and a weld metal of different magnetic characteristics will create magnetic discontinuities, which may produce indications interpretable as unsoundness even though the joint is entirely sound. On the other hand, a true defect can be obscured by the powder clinging over the harmless magnetic discontinuity. The sensitivity of the method lessens with decrease in defect size. Sensitivity is less with round forms, such as gas pockets, and best with elongated forms, such as cracks.

To have external leakage, the magnetic field must be distorted sufficiently. Fine, elongated discontinuities, such as hairline cracks, seams, or inclusions, that are "strung" parallel with the magnetic field, will not distort it sufficiently to develop leakage. Thus, no indication will result. By changing the direction of the field, however, the indications can be developed. To be certain that discontinuities are detected, it is advisable to apply the field from two directions, preferably at right angles to each other.

Pieces to be inspected must be clean and dry. Wire brushing and sand blasting are satisfactory methods for cleaning welds. Surface roughness decreases the sensitivity, tends to distort the magnetic field, and interferes mechanically with formation of the powder pattern.

The shape, sharpness of outline, and width and height to which the particles have built up are features used for identifying discontinuities. When unusual patterns are produced, other test methods may be required to establish identity. Once a pattern has been interpreted by correlating it with the identification established by other methods, the interpretation can be applied to similar indications on other parts. Representative indications are shown in Figure 22-14.

Since a powder pattern results from various types of discontinuities in the magnetic field, it is easy to mistake an irrelevant indication for a defect. As noted previously, a change in the magnetic characteristics of materials can create an irrelevant indication in the area of interface. A change in section or a hole drilled in the part will

Figure **22-14** Representative indications of weld defects detected by the magnetic-particle inspection method. Redrawn by permission from *The Procedure Handbook of Arc Welding*, 11th ed., Lincoln Electric Company, Cleveland, Ohio, p. 11.2-12.)

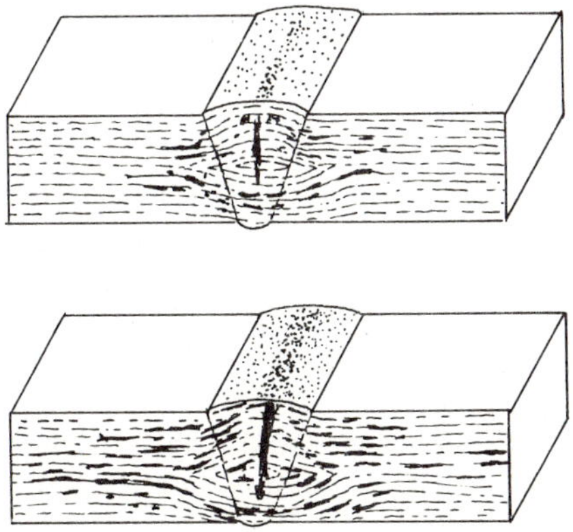

Surface cracks give powder patterns that are sharply defined, tightly held, and usually built up heavily. The deeper the crack, the heavier the build-up of powder.

Subsurface cracks produce a less sharply defined, fuzzy pattern. The powder is also less tightly adherent.

Crater cracks may give a single line, running in almost any direction, or a pattern of lines. The pattern is small and occurs near the end of the weld.

Subsurface porosity gives weak, poorly defined patterns, as do *slag inclusions*.

tend to produce indications that have no significance with respect to weld soundness. These patterns are usually readily recognized for what they represent by their shapes and locations in the part. Abrupt changes in magnetic properties may occur at the edge of the heat-affected zone. The pattern will be fuzzy and diffused, running along the base metal close to the edge of the weld. It resembles the pattern caused by undercutting, the difference being that the particles are much less adherent. If a part is heat-treated or stress-relieved before inspection, the magnetic characteristics of the heat-affected zone are restored and no indications will appear.

In the shop, interpretation of magnetic-particle inspection patterns can be guided by rules or standards based on experience or laboratory tests. Assuming that the inspector has ruled out the possibility of a false indication and has properly identified the indicated defect, he or she must then decide whether to pass or reject the part or require its repair. All defects do not affect the integrity of the part in service. Thus slag inclusions and porosity may have no bearing on the serviceability of the weld. Surface cracks revealed by magnetic-particle inspection, however, should be considered potential stress risers or focal points for fatigue and corrosion.

The equipment for magnetic particle inspection is relatively simple (Figure 22-13). Commercial units, portable and stationary, provide for nearly every situation where the method is applicable. The manufacturer's instructions and recommendations for use of the equipment should be read carefully. The units may provide magnetization by direct, alternating, or rectified currents or combinations thereof. Portable equipment, making use of electromagnets and permanent magnets, is also available.

Direct or rectified current is required for deep penetration. Alternating current magnetizes only the surface and thus is limited to surface inspection. Full-wave, three-phase current produces results comparable to those produced by battery direct current. Half-wave, single-phase rerectified current gives maximum sensitivity. High-amperage, low-voltage current is normally used in all magnetic-particle testing to limit arcing or burning on the test piece.

Magnetic powders may be applied by either the dry or wet methods. Dry powder is uniformly dusted over the work with a spray gun, dusting bag, or atomizer. The finely divided magnetic particles are coated to give them greater mobility, and are available in gray, black, and red colors. It is desirable that the particles impinge on the surface at low velocity and with just enough residual force after impact to move them to possible sites of leakage. Excess powder is removed with a light stream of air.

In the wet method, very fine red or black particles are suspended in water or light petroleum distillate. Powders for liquid suspension come from the manufacturer in either paste or dry form, prepared for use in water or oil baths. After the suspension has been made—in accordance with the manufacturer's instructions—it is flowed or sprayed onto the surface to be inspected, or the piece may be dipped into the liquid. The wet method is more sensitive than the dry method, since extremely fine particles may be used; this enables detection of exceedingly fine defects. Red particles improve visibility on dark surfaces. When the particle coating is a dye that fluoresces under ultraviolet light, sensitivity is further increased. Fluorescent powders are excellent for locating discontinuities in corners, keyways, splines, deep holes, and similar locations.

The techniques for creating a magnetic field in workpieces of various sizes and shapes, the sequence of operations in magnetizing and applying magnetic particles, adjustment of current to bring out desired results, the practical modes for orienting magnetic fields to produce or better delineate indications—all are important to the successful use of this inspection method. The equipment manufacturers' literature and various ASTM and other specifications (such as ASTM E-109-57T, *Method for Dry-Powder Magnetic Particle Inspection* and ASTM E-138-58T, *Method for Wet Magnetic Particle Inspection*) should be consulted for operational details.

Magnetic-particle inspection is applied to many types of welding in production practice. The dry-powder method is especially popular for heavy weldments. Many steel weldments in aircraft manufacture are inspected by the wet method, using direct current. Airframe parts are subjected to fatigue conditions, which means that surface cracks cannot be tolerated. Since these weldments are relatively thin, magnetic-particle inspection will usually detect subsurface defects as well as indicate the finest surface cracks.

22-5 LIQUID-PENETRANT INSPECTION

Liquid-penetrant inspection is a nondestructive method for locating surface cracks and pinholes invisible to the naked eye. It is a favored technique for locating leaks in welds, and can be applied where magnetic particle inspection is useless, such as with austenitic stainless steels or nonferrous metals. Two types of penetrant inspection, which define the penetrating substance, are used—fluorescent and dye.

With *fluorescent-penetrant inspection*, highly fluorescent liquid with good penetrating qualities is applied to the surface of the part to be examined. Capillary action draws the liquid into the surface openings. The excess liquid is then removed from the part, a so-called developer is used to draw the penetrant to the surface, and the resulting indication is viewed by ultraviolet ("black") light. The high contrast between the fluorescent material and the background makes possible the detection of minute traces of penetrant.

Since penetration of minute openings is involved, the part to be inspected must be thoroughly clean and dry. Any foreign matter could close the openings, leading to false conclusions. The penetrant is applied by dipping, spraying, or brushing. Time must be allowed for absorption of the material into the discontinuities—up to an hour or more in very exacting work.

When penetration is complete—or assumed to be adequate in accordance with testing specifications—the excess material is removed from the surface. If the penetrant is designed for water wash, a low-pressure water spray is used. Some commercial penetrant systems, however, require a solvent wash or what is called a *post emulsifier*. With the latter, an emulsifier is applied to the part and allowed to remain on it for one to four minutes before the water spray. The instructions for use of emulsifiers and for the washing operation must be followed closely, since only excess penetrant should be removed.

After the wash, the parts are dried if a dry developer is to be used. Hot air may be used to accelerate drying. A dry developer is applied with a powder gun or spray bulb, or by dipping. The developer draws the penetrant from the defects, making it accessible for viewing by ultraviolet light. If only large discontinuities are sought, a developer may not be required to make the indications visible. If a wet developer is used, drying of the part after removal of excess penetrant is not required. Wet developer is applied in the form of a colloidal water suspension by dipping or spraying, after which the part is dried by hot air.

Under ultraviolet light, the indications fluoresce brilliantly. The extent and depth of the discontinuity can be gauged by the width and length of the indication and the amount of penetrant bleeding to the surface. The darker the room in which it is viewed, the more brilliant the fluorescence and more easily very small indications are observed.

Dye-penetrant inspection is similar to fluorescent penetrant inspection, except that dyes visible under ordinary light are used. By eliminating the need for ultraviolet light, greater portability in equipment is achieved.

Liquid-penetrant inspection is widely used for leak detection. A common procedure is to apply fluorescent material to one side of a joint by brushing or spraying, allow adequate time for capillary action, and then view the other side of the joint with ultraviolet light. Dry developer may be used on the side being inspected to intensify the indications. In thin-walled vessels, this technique will show up leaks that are not ordinarily located by the usual air test with pressures of 5 to 20 psi. The sensitivity of the leak test decreases, however, when the wall thickness is over ¼ inch.

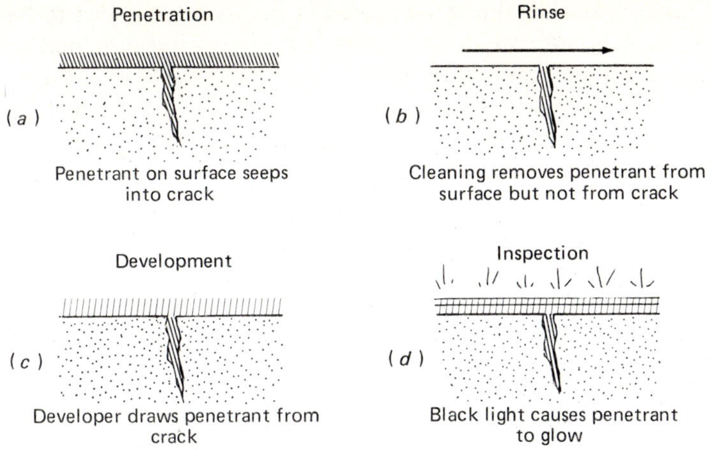
Penetration
(a)
Penetrant on surface seeps into crack
Rinse
(b)
Cleaning removes penetrant from surface but not from crack
Development
(c)
Developer draws penetrant from crack
Inspection
(d)
Black light causes penetrant to glow

Figure **22-15** Typical defect shown by the fluorescent-penetrant method

Penetrant inspection is also widely used in the inspection of large and small weldments for cracks and porosity when the materials are nonmagnetic. It must be remembered that only surface defects are revealed by this method. Figure 22-15 shows a typical defect indication using fluorescent penetrant.

22-6 ULTRASONIC INSPECTION

Ultrasonic inspection is a supersensitive method of detecting, locating, and measuring both surface and subsurface defects in metals. Flaws that cannot be discovered by other methods, and even cracks small enough to be termed microseparations, may be detected. In the practical inspection of welds, the sensitivity of the process is often curbed by designing or setting the equipment to give a response equivalent to a sensitivity of 2% of the metal thickness, thus giving results comparable with those obtained in radiographic inspection.

Ultrasonic inspection is based on the fact that a discontinuity or density change will act as a reflector for high-frequency vibrations propagated through metal. The searching unit of the pulse-echo–type ultrasonic equipment contains a crystal of quartz (or other piezoelectric material), which changes its dimensions with an applied electromotive force. By using alternating current, the dimensional changes are alternately in one direction, then the other, the rapidity of change varying with the frequency of the applied electromotive force. This sets the crystal to vibrating rapidly, and it imparts mechanical vibrations of the same frequency into materials it contacts.

When an ultrasonic probe is held against metal, the vibrational waves are propagated through the material until a discontinuity or change of density is reached. At these points, some of the vibrational energy is reflected back. If, in the meantime, the current that caused the vibration has been shut off, the quartz crystal (probe) can now act as a receiver to pick up the reflected energy. The reflected vibration causes pressure on the quartz crystal, which results in the generation of an electric current. Fed to a cathode-ray tube (CRT), this current produces vertical deflections in the horizontal base line. The pattern on the face of the tube is thus a representation of the reflected signal—and of the defect. The cycle of transmitting and receiving is repeated at a rate of 60 to 1000 times per second.

Two types of CRT presentation are available—radio frequency (RF) and video. Most commercial units (Figure 22-16) present the video pattern, although many inspectors believe RF provides more useful information on flaw identification. Figure 22-17 shows the video patterns produced on the cathode-ray tube by different types

Figure **22-16** A portable, battery-powered, ultrasonic testing machine. (Courtesy of the Lincoln Electric Company, Cleveland, Ohio.)

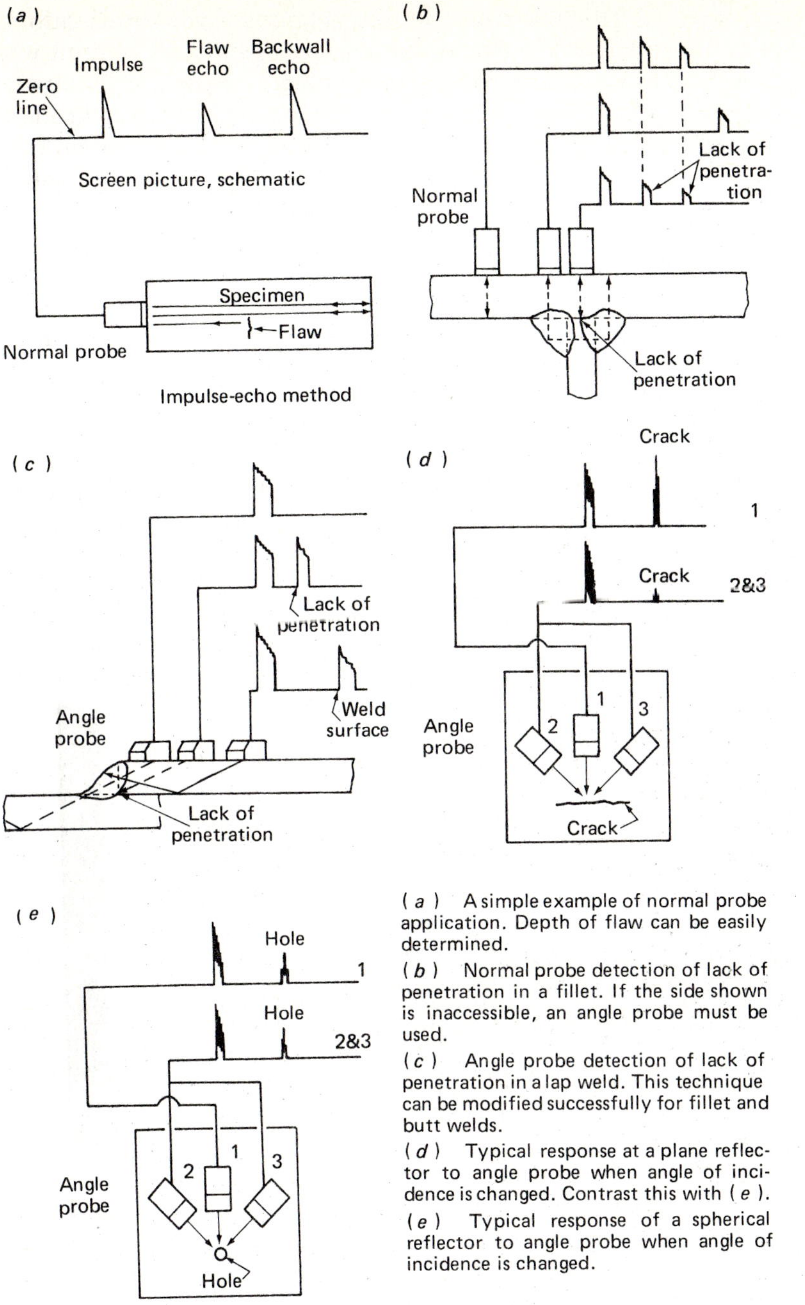

(a)
Impulse
Flaw echo
Backwall echo
Zero line
Screen picture, schematic
Specimen
Flaw
Normal probe
Impulse-echo method
(b)
Lack of penetra-tion
Normal probe
Lack of penetration
(c)
Lack of penetration
Weld surface
Angle probe
Lack of penetration
(d)
Crack
1
Crack
2&3
Angle probe
2
1
3
Crack
(e)
Hole
1
Hole
2&3
Angle probe
2
1
3
Hole

(*a*) A simple example of normal probe application. Depth of flaw can be easily determined.

(*b*) Normal probe detection of lack of penetration in a fillet. If the side shown is inaccessible, an angle probe must be used.

(*c*) Angle probe detection of lack of penetration in a lap weld. This technique can be modified successfully for fillet and butt welds.

(*d*) Typical response at a plane reflector to angle probe when angle of incidence is changed. Contrast this with (*e*).

(*e*) Typical response of a spherical reflector to angle probe when angle of incidence is changed.

Figure **22-17** Typical video patterns produced on the cathode-ray tube by the ultrasonic testing method. (Redrawn by permission from *The Procedure Handbook of Arc Welding,* 11th ed., Lincoln Electric Company, Cleveland, Ohio, p. 11.2-15.)

of defects. It can readily be seen that expertise in interpretation of the deviations from the horizontal oscilloscope line is required, and a similar degree of expertise is needed for operation of the equipment. Interpretations are based on standard patterns made from reference plates, which are prepared in accordance with procedures approved by the ASTM, ASME, and others. Battery-operated pulse-echo ultrasonic equipment is available for field use.

Another type of ultrasonic inspection equipment gives what is similar to a "television picture" of the defect. Although not as popular as pulse-echo equipment, it is believed to have the potential for simplifying pattern interpretation where applicable, since the pattern is essentially a plane image. With this type of equipment, the ultrasonic energy from a generator is transmitted through the specimen, much as x rays are transmitted in radiographic inspection. The transmitted energy impinges on a piezoelectric target, producing varying potentials on its face that are characteristics of the flaws in the specimen. The target is scanned with an electron beam which is modulated by the varying surface potentials. This enables the development of a picture—a representation of the flaw—on a monitoring screen.

22-7 OTHER NONDESTRUCTIVE WELD-TESTING METHODS

Under this heading come the following tests: (1) chemical, (2) metallographic, (3) hardness, and (4) mechanical. Most of these tests are of little importance to the welder and are beyond the scope of this book. If the reader wishes to obtain more detailed information on these tests than is given in the following paragraphs, she or he should refer to the following sources: (1) the author, (2) the Society for Nondestructive Testing (USA), (3) the American Society for Testing and Materials, (4) the Nondestructive Testing and Information Center in San Antonio, Texas.

Chemical tests are most frequently used to determine the corrosion resistance of metals. Since weldments must operate in all types of corrosive liquids, solids, and gases, a large number of corrosion tests have been devised to see if the material will meet service conditions. Corrosion tests are required by very few codes and specifications, because most metals are selected for specific use by previous experience. The chemical industry has developed numerous corrosion tests that are usually required on both base metal and welded joints of the weldment. Because welds are made in so many different metals and their alloys, anything other than a reference to the American Society for Testing and Materials (ASTM) *Standards*, Part 1, Metals, which describes the salt spray test, is beyond the scope of this book.

Metallographic tests are not often required in welding specifications. They are carried out for such purposes as the determination of the distribution of nonmetallic inclusions, the number of weld passes, grain structure in the weld and fusion zone, and the extent and structure of the heat-affected zone.

Metallographic tests may merely involve visual examination, in which case the specimens, called *macrospecimens,* are etched to bring out the structure and examined by the unaided eye or at low magnifications; or they may involve microscopic examination, in which case the specimens, called *microspecimens,* are prepared and etched for examination under the microscope at high magnifications.

Samples for chemical examination may be obtained by sectioning the weld by means of a core drill, or by trepanning. In both of these cases the drilled hole or the trepanned cut may be rewelded and the weldment thus rendered fit for service. Such tests may be regarded as nondestructive.

There are two general types of examination: *macro,* in which the sample is etched to reveal the structure that can be studied by the unaided vision or with the assistance of a low-power (5× or 10×) magnifying glass; and *micro,* in which the cut surface is etched to reveal fine structural details that can be seen only with the aid of a metallurgical microscope. For macroscopic examination, the surface to be examined may be prepared by one of the following methods:

1. Without requiring any finishing or other preparation, place the part in a boiling 50% solution of hydrochloric (muriatic) acid until there is a clear definition of the structure of the weld. This requires approximately one-half hour.

2. Grind and smooth the specimens with silicon-carbide paper, then etch by treating with a solution of one part ammonium persulphate (solid) and nine parts water by weight. The solution should be used at room temperature and be applied by vigorously rubbing the surface to be etched with a piece of cotton kept saturated with the solution. The etching process should be continued until there is a clear definition of the structure of the weld.

3. Grind and smooth the specimens with silicon-carbide paper, then etch by treating with a solution of one part powdered iodine (solid form), two parts powdered potassium iodide, and three parts water (all by weight). The solutions should be used at room temperature and should be brushed onto the surface to be etched until there is a clear definition of the weld structure.

After being etched, the specimens are washed in clear water and the excess water removed. They are then immersed in ethyl alcohol and allowed to dry. The etched surfaces may then be preserved by coating with a thin, clear lacquer.

The specifications under which the weldment is constructed will define the type, sizes, and number of defects allowable for the particular type of welded joint. Where a defect is located, additional specimens are usually cut at certain intervals on each side of the defective segment, wherever the joint length will permit. If additional defects are found, more specimens usually are cut until the limit of the defective welding has been definitely established.

In the examination for exceedingly small defects or for grain structure at high magnification, specimens may be cut from the actual weldment, as above, or from welded test samples. The samples are prepared with a highly polished, mirror-like surface and etched for examination under high-power microscopes to show the structure of the base metal, heat-affected zone, fusion zone, and weld metal. These specimens are called *microspecimens,* and are most often specified for alloy steels or nonferrous metals. A trained metallographer can read much from microscopic examinations. The procedure is complicated, and a considerable amount of skill is necessary to properly polish the samples and to use the proper etchants and technique to show what is desired. (For a detailed discussion of microscopic examination procedures, refer to ASTM *Standards,* Part 1, Metals.)

Hardness tests of a welded joint are affected by the chemical analysis (alloying elements and percentage of each) of the filler metal, while tests of the heat-affected zone are affected by the chemical analysis of the base metal, cold working of the base metal, heat treatment, and many other factors. Four hardness tests are commonly used by inspectors to determine the hardness of a welded joint: (1) Brinell, (2) Rockwell, (3) Vickers or diamond pyramid, and (4) Shore scleroscope. Depending on the size and location of the weld to be tested, the test may be performed with either portable or bench-mounted hardness testers.

Each of the hardness tests supplements the others. The impression obtained with the Brinell tester is large and can only be used for obtaining hardness values over a relatively large area, such as the face of a weld or the base metal. The Rockwell and Vickers tests can be used to survey the hardness of such small zones as the cross-section of a weld, the heat-affected zone, or an individual bead.

As discussed in Chapter 18, the Rockwell hardness measures the depth of residual penetration made by a small, hardened steel ball or a diamond cone. The test is performed by applying a minor load of 10 kg, which seats the penetrator in the surface of the specimen and holds it in position. The dial is turned to the point marked *Set,* and the major load, which varies with the hardness expected, is then applied. After the pointer comes to rest, the major load is released,

leaving the minor load still on. The Rockwell hardness number is read directly on the dial. The B- and C-dial scales are the ones most commonly used.

Before performing the test, one should know the hardness expected [this can be obtained by converting the tensile strength of the metal to be tested to a Rockwell hardness number (see Table 19-1)]. Then make certain that the surfaces of the material to be tested have been properly ground, to enable the specimen to seat solidly with no rocking. The thickness of the specimen should be such that the penetrator will not go through it. The specimen thickness will vary with the type of metal tested.

Softer material requires greater specimen thickness, or lighter load, or both. Results from tests on a curved surface may be in error and should not be reported without stating the radius of curvature. In testing round specimens (cylinders) along their lengths, the effect of curvature may be eliminated by filing a small flat along its length. Impression should not be made within about two ball-penetrator diameters of the edge of the specimen, or closer than that to each other. When changing from one scale to another, a standard test block of a specimen hardness in the range of the new scale should be used to check whether all the required changes have been made and whether the penetrator has properly seated. Care should be taken not to damage the penetrator or the anvil by forcing them together when a specimen is not in the machine. The minor load should be carefully applied so as not to overshoot the mark. The loading lever should be brought back gently. The ball penetrator tends to become flattened by use, especially in testing hardened steels, and should be occasionally checked and replaced when necessary. In the same manner, the cone penetrator should be examined with a 5× or 10× magnifying glass and replaced when found to be blunted or chipped.

The Vickers hardness test consists of impressing a diamond penetrator into the surface of a specimen under a predetermined load. The ratio of the impressed load to the area of the resulting indentation gives the Vickers hardness number. The application and removal of the load, after a predetermined interval, are controlled automatically. Since the Vickers indenter is a diamond, it can be used in testing the hardest steels and remains practically undeformed. The load is light, varying from 1 to 120 kg according to requirements. A normal loading of 30 kg is used for homogeneous materials; 10 kg is used for soft, thin, or surface-hardened materials. One movement of a starting handle releases the mechanism, and the depression of a foot pedal restores it to its original postion. An audible click informs the operator of the duration of the test. It is recommended that the time of load application be standardized at 10 seconds.

Upon lowering the stage, the measuring microscope is swung into a position over the impression. The impression appears as a dark square on a light background. The measurements are taken across the diagonals of the square between the knife edges. There are three knife edges: one is fixed, another movable by means of a micrometer screw connected to the counting mechanism, and the third is provided for rapid readings to specified limits. The left-hand corner of the impression is set to correspond to the fixed knife edge, the movable knife edge adjusted to coincide with the right-hand corner of the impression, and the reading obtained directly from the indicator. The actual hardness figure is then found from a chart or calculated by a simple formula.

Hardness numbers obtained with a pyramid indenter are practically constant, irrespective of the load applied. The Vickers and Brinell hardness values on steel are practically identical up to a hardness of about 300. At higher hardness values, the Brinell number falls progressively lower than the Vickers number and is not reliable above about 600 Brinell hardness, even with specially hardened balls. This irregularity is caused by the flattening of the steel ball under the heavy loads required for testing hard materials, whereas the diamond shows no distortion. As Vickers impressions are much smaller than those obtained when using Brinell or Rockwell testers, the surfaces to be tested should be prepared much more carefully. Although the Vickers instrument is definitely a machine of precision, it requires care in operation rather than any particular skill.

Mechanical tests other than those generally performed on welds as part of a welder-certification test consist principally of impact and proof tests.

Impact tests are performed on those welds that must often withstand impact or sudden applied loads. The velocity of the impact, the presence of sharp notches, and the temperature of the metal are all factors that contribute to the metal's ability to withstand shock. As metals respond differently to these factors, it is usual to test metals under two or more related conditions.

The impact test specimen is removed from a sample weld joint. After removal, and depending on the type of test to be performed—simple-beam (Charpy); cantilever-beam (Izod); or tension-impact (Charpy tension)—the specimen is prepared so that its dimensions conform to those shown in Figure 18-11. Depending on type, the prepared specimens are tested as follows.

The notched face of a Charpy specimen is placed on two anvils, with the notch centered between the anvils and pointing away from the pendulum. The specimen is struck by the top of the pendulum at a point opposite the notch. The Izod specimen is inserted vertically into a slot, provided especially for that purpose, located between the anvils. The specimen is inserted into the slot until the notch is level with the jaws of the vise. This is tested with a gage especially provided for that purpose. The free end is struck by the top of the pendulum. If the specimen is to be used for the tension-impact test, it is attached to the pendulum and carries an especially provided tip that engages the anvils. The direction of the force is parallel to the axis of the specimen, while in the two previous tests it is perpendicular.

Proof testing involves applying a load or pressure equal to or greater than that expected in service, but not great enough to damage the product. It is a method of assuring that the product (pipes and tanks) is serviceably sound at the time of the test. Destructive testing is applied only to prototypes or samples taken from production in some predetermined statistical sequence.

Proof testing is common with welded vessels. The test may be run for tightness of the enclosure, in which case the working pressure will likely be used. If the test is run to confirm strength or safety, the test pressure will usually exceed the working pressure. Judgment enters into specifying the pressure intended to reveal defects that later may cause failure in service. Usually, some multiple of the design value is selected. The passing of a proof test in which the weldment is subjected to pressures several times those of normal service does not mean that working pressures of that magnitude can be used. Experience indicates that failures can occur later at much lower pressures as a result of fatigue, corrosion, or stress risers.

To avoid damage to the weldment, a proof test must always be run at something less than the yield strength of the weakest component. It is usually desirable to limit the pressure or force to 75% of the calculated yield point. Sometimes the combination of residual stresses with the proof-test stress may cause some local yielding without damage. The test, in fact, may act, to a limited extent, as a stress-relief operation.

Air, water, oil, and gases may be used to supply the pressure when proof testing a vessel. Leaks are shown by pressure drops or detected visually. Light-viscous oil will penetrate leaks that will not pass water; air will leak where oil will not; and hydrogen will leak where air will not. (Hydrogen is highly explosive and should never be used where there is a possibility of a flame or spark.) On a clean weld, leaking oil or water is easily observed. Air or gas pressure will show a drop if a leak exists. Soap solution, spread over a weld, will reveal the presence of a leak by bubble formation.

22-8 DESTRUCTIVE TESTING

Destructive testing gives an absolute measure of the strength of the sample tested. Assuming that materials and the method of fabrication are uniform, it is reasonable to infer that the sample is representative of all the units. Periodic testing of specimens tends to lend validity to this inference, providing the results are similar. Usually the destructive force is applied in a way that simulates service conditions.

Any testing procedure based on a sampling may permit defective work to pass inspection. Weight must be given to the penalty resulting therefrom. When safety is an important consideration, destructive techniques with samples do not give the assurance of quality needed. On the other hand, when safety is not a factor and the unit cost of the product is low, destructive testing with samples may be more economical and informative than nondestructive testing.

REVIEW QUESTIONS

1. Why is *weld quality* a relative term?

2. Differentiate between *welding process* and *welding procedure.*

3. Under what conditions can welding processes, procedures, and or joint configurations other than those assigned prequalified status by the AWS be used?

4. Is it possible for a weld found to be of "poor" quality by standards of inspection and judgment to be "acceptable"? Explain.

5. Must undercut welds always be rejected? Explain.

6. What does the *AWS Structural Code* state in regard to weld craters?

7. Does a single qualification test enable a welder to weld any and all work requiring the services of a certified welder? Explain.

8. Either identify by name or sketch:

a. Three base metal defects that are detectable by inspection performed prior to welding.

b. Two sources of root-area slag.

c. Two causes of weld-joint cracking in ferrous metals.

d. Two techniques that can be used to minimize weld-metal cracking.

e. The document against which a welding inspector would check an assembly set-up ready to be welded before the actual welding operation is begun.

f. The location and usual type of cracking that occurs in the base metal as the weld cools.

g. Two techniques that can be used to minimize weld cracking.

h. The type(s) of stainless steel that *cannot* be inspected by means of the magnetic-particle method.

i. The nondestructive testing method that is generally unsuitable for fillet-weld inspection.

j. The test method commonly used to determine the grain structure of weld metal and the HAZ.

k. The hardness test in which the hardness numbers are based on the difference in the depths of penetration between the major and the minor loads.

l. The organization(s) that promulgate(s) procedures for the preparation of standard reference plates used to assist in the interpretation of weld-defect patterns obtained by use of the x-ray and ultrasonic test methods.

m. The directionality of cracks that can be detected with (1) circular magnetization, and (2) longitudinal magnetization.

9. Describe the principles of (a) ultrasonic inspection, and (b) fluorescent-particle inspection.

10. Differentiate between base-metal cracking and weld-metal cracking. Why and when is each type of crack most likely to occur?

11. For what purpose(s) are the following tests performed?

a. Proof-testing.

b. Salt-spray.

12. For what purpose(s) are the following devices used?

a. Workmanship standard.

b. Weld gage.

CHAPTER 23
PROCEDURE QUALIFICATION AND WELDER CERTIFICATION

There is a popular misconception that if a welder or welding-machine operator passes a certain test, he or she receives a card that entitles him or her to do any type of welding anywhere in the country—this is not true. It is true that welders qualified under structural, pressure, or pipe-line welding under certain codes of the AWS, ASME, and the API are often called *certified*, but this applies only to the particular work for which they are qualified, and usually in one plant (company) only. There is no such a thing as a certified welder or welding-machine operator who is able to weld on any and all types of code work because of passing a single qualification test.

Another popular misconception is that once certified a welder remains permanently certified. This is not true, either. For instance, the New York State Department of Transportation states that the card (Figure 23-1) specifies the limits of qualification tested. A welder qualifying for unlimited thickness in the overhead position is, however, automatically qualified for everything else. The card is valid for a period of three years (provided that a minimum of six working days are recorded during each six-month period) and has the expiration date indicated on the card. Certificate renewal is the responsibility of the welder, who must submit application with employer statements verifying maintenance of qualifying skills and employment.

Generally speaking, there are three different qualification tests:

1. A test to qualify the welding procedure to be used for a specific welding project. This is called *procedure qualification.*

TO: MAIN OFFICE REGIONAL OFFICE	**ENGINEERING INSTRUCTION** NEW YORK STATE DEPARTMENT OF TRANSPORTATION
	SUBJECT: WELDER CERTIFICATION Subject Code: 7.35
Distribution: [G] Main Office [G] Regions [ ] Special	Code: EI 75-1
APPROVED: R. N. Kamp DEPUTY CHIEF ENGINEER (STRUCTURES)	Date: 1/2/75 Supersedes: EI 72-8

Welder certification shall continue to be administered in accordance with the provisions of the Contract Documents. Welder qualification tests conducted by Regional Offices shall be performed in accordance with the instructions included in the N. Y. S. Steel Construction Manual, Section 402.

A welder holding a 1974 NYS-DOT Welder Certification Card will be eligible for recertification without retesting provided he is recommended for recertification by a State representative knowledgeable in welding who can state from personal experience that the welder has performed acceptable work using the welding process for which he is qualified within the six month period preceding recommendation of recertification.

Beginning January 1, 1975, NYS-DOT Welder Qualification Certificates will be issued in place of the current Welder Certification Cards. These certificates will be valid for a period of three years and will have the expiration date indicated on the document. Each welder will be issued a card entitled "Work Record" that must accompany the Welder Qualification Certificate. The Work Record and the Certificate must be presented to the Engineer before beginning work. A State of New York Engineer-In-Charge or a licensed Professional Engineer representing the owner at the construction site must sign the Work Record at least once every six months for the Certificate to remain valid. Sufficient signatures must be recorded to represent a minimum of six working days using each qualified process during each six month period. The Engineer signing the Work Record shall have personal knowledge of the welder's satisfactory performance using the welding process described.

Welders who have had a lapse of experience in excess of six months will be recertified on the basis of Standard Welder Qualification Tests, except that welders previously qualified on the basis of qualification tests made with 1" thick test plates shall have the option of using either 3/8" or 1" thick test plates to be requalified to weld steel of unlimited thickness.

The Welder Qualification certificate may be revoked at any time that there is a specific reason to question the welder's ability. When this occurs, a complete retest is required for recertification.

Figure **23-1** New York State Department of Transportation certified welder certificate and card. (Courtesy of the New York State Department of Transportation.)

This certificate must be presented to the Engineer before beginning work. A State of New York Engineer-In-Charge or licensed Professional Engineer must sign this certificate at least once every six months for the certificate to remain valid. Sufficient signatures must be recorded to represent a minimum of 6 working days for each 6 month period. The Engineer signing this card shall have personal knowledge of the welders satisfactory performance using the welding process described.

Engineers Signature	N.Y.S. Region or P.E. No. & State	Date	No. Days Service	Location of work

Figure **23-1** (cont.)

STATE OF NEW YORK
DEPARTMENT OF TRANSPORTATION

No. ______________________

Expiration Date: ______________

____

Social Security No. ______________________
has qualified for the following welding processes to the extent shown below:

☐ Manual Shielded Metal Arc Welding - Low Hydrogen
☐ Semiauto. Flux Cored Arc Welding (with CO_2 shield)
☐ Semiauto. Submerged Arc Welding (Wire Dia.______)

______________	______________________________
Date of Issue	Asst. Deputy Chief Engineer (Structures)

No. ______________

____
Signature of Welder

Qualification is granted on the basis of X-Ray examination of test welds interpreted to the requirements of the New York State Steel Construction Manual, Section VI.

NOT VALID WHERE PUNCHED

WELDING PROCESS	FILLET WELDS				GROOVE WELDS: 3/4" max. thickness				GROOVE WELDS: unlimited thickness			
	Position				Position				Position			
Manual	F	H	V	O	F	H	V	O	F	H	V	O
FCAW-CO^2	F	H	V	O	F	H	V	O	F	H	V	O
Sub-Arc	F	H	[redacted]		F	[redacted]			F	[redacted]		

Figure **23-1** (cont.)

2. A test of the welding proficiency of the operator with hand-held welding equipment (the stick-electrode holder or the semiautomatic welding gun). This is referred to as *welder qualification.*

3. A test of the proficiency of the operator with fully automatic welding equipment—referred to as *welding-operator qualification.*

Although many code-writing institutions have developed their own qualification tests, the tests developed by the AWS have been included by most U.S. codes and specifications concerned with welded structures. Because of this, only the procedure qualification, welder qualification, and welding-operator qualification tests developed by the AWS are discussed in this chapter.

23-1 HOW TO PASS QUALIFICATION TESTS

While the AWS tests are designed to separate capable welders from amateurs, many professional welders have failed for reasons not related to their welding ability. The *AWS Structural Code* minimizes such possibilities by making the test a more positive demonstration of weld quality and welding skill. However, it is still possible to fail the test because of poor plate quality, improperly prepared samples, or incorrect interpretation of the results. This is exasperating to the welder and expensive to his or her employer. It is to everyone's advantage to see that the tests are run properly, so that the results are an accurate measure of welding ability.

Before taking a test, one should make certain that the appropriate test details for the type of steel, plate thickness, joint, and welding position to be used have been selected (see Figure 23-2, pages 727–730). Then the strength of the plate should be checked. Face- and root-bend tests are designed so that both plate and weld stretch during the test. If the plate has a substantially higher strength than the weld metal (and this is possible with steel purchased as mild steel), the plate will not stretch sufficiently, thereby forcing the weld metal to stretch beyond its yield and crack.

As previously indicated, one vertical and one overhead test plate can qualify a manual-electrode or semiautomatic welder for unlimited welding under the AWS Structural Code. Since it permits using the welder throughout the shop without restrictions, most testing is done in these two positions.

For unlimited manual-electrode qualification, a low-hydrogen electrode (usually E7018) is required. EXX10 or EXX11 electrodes are also frequently used but do not qualify the welder to work with low-hydrogen electrodes. Low-hydrogen electrodes require different techniques than EXX10 or EXX11, and welders should be trained with low-hydrogen techniques before testing, even if they are expert EXX10 welders.

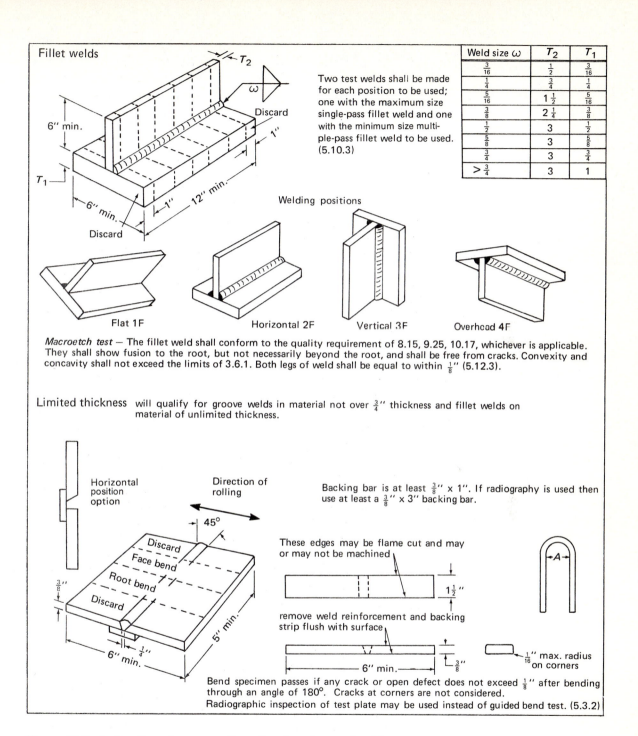

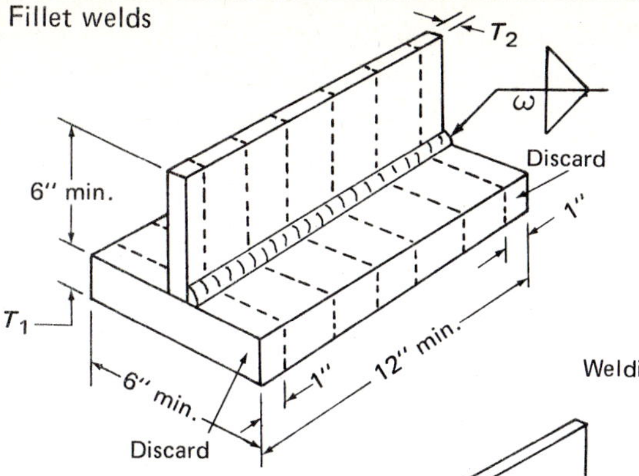
Fillet welds
T_2
ω
Discard
1"
6" min.
T_1
12" min.
6" min.
Discard

Two test welds shall be made for each position to be used; one with the maximum size single-pass fillet weld and one with the minimum size multi-ple-pass fillet weld to be used. (5.10.3)

Weld size ω	T_2	T_1
$\frac{3}{16}$	$\frac{1}{2}$	$\frac{3}{16}$
$\frac{1}{4}$	$\frac{3}{4}$	$\frac{1}{4}$
$\frac{5}{16}$	$1\frac{1}{2}$	$\frac{5}{16}$
$\frac{3}{8}$	$2\frac{1}{4}$	$\frac{3}{8}$
$\frac{1}{2}$	3	$\frac{1}{2}$
$\frac{5}{8}$	3	$\frac{5}{8}$
$\frac{3}{4}$	3	$\frac{3}{4}$
$> \frac{3}{4}$	3	1

Welding positions

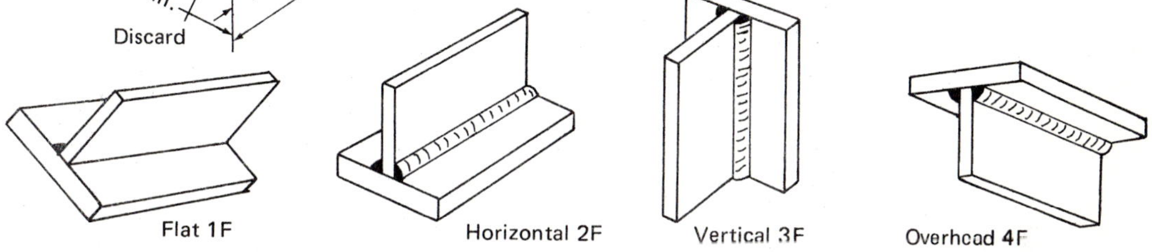
Flat 1F
Horizontal 2F
Vertical 3F
Overhead 4F

Macroetch test — The fillet weld shall conform to the quality requirement of 8.15, 9.25, 10.17, whichever is applicable. They shall show fusion to the root, but not necessarily beyond the root, and shall be free from cracks. Convexity and concavity shall not exceed the limits of 3.6.1. Both legs of weld shall be equal to within $\frac{1}{8}$" (5.12.3).

Limited thickness will qualify for groove welds in material not over $\frac{3}{4}$" thickness and fillet welds on material of unlimited thickness.

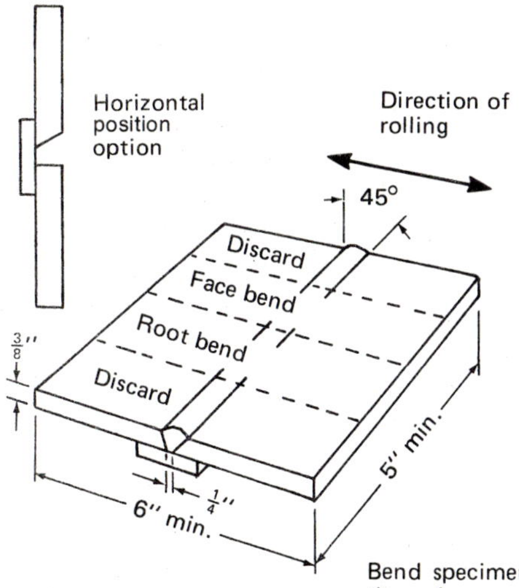
Horizontal position option
Direction of rolling
45°
Discard
Face bend
Root bend
Discard
3/8"
1/4"
6" min.
5" min.

Backing bar is at least $\frac{3}{8}$" x 1". If radiography is used then use at least a $\frac{3}{8}$" x 3" backing bar.

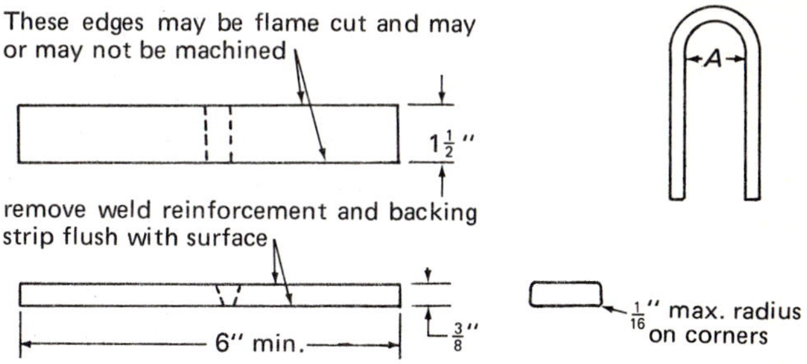
These edges may be flame cut and may or may not be machined
1 1/2"
remove weld reinforcement and backing strip flush with surface
3/8"
6" min.
A
1/16" max. radius on corners

Bend specimen passes if any crack or open defect does not exceed $\frac{1}{8}$" after bending through an angle of 180°. Cracks at corners are not considered.

Radiographic inspection of test plate may be used instead of guided bend test. (5.3.2)

Figure **23-2** Instructions for preparation of various types of weld test specimens used for welder certification. Numbers in parentheses refer to *AWS Structural Welding Code*, 1972 (AWS D1.1-76). (Redrawn by permission from *The Procedure Handbook of Arc Welding*, 11th ed., Lincoln Electric Company, Cleveland, Ohio, pp. 11.3-2, 11.3-3, 11.3-5, 11.3-6.)

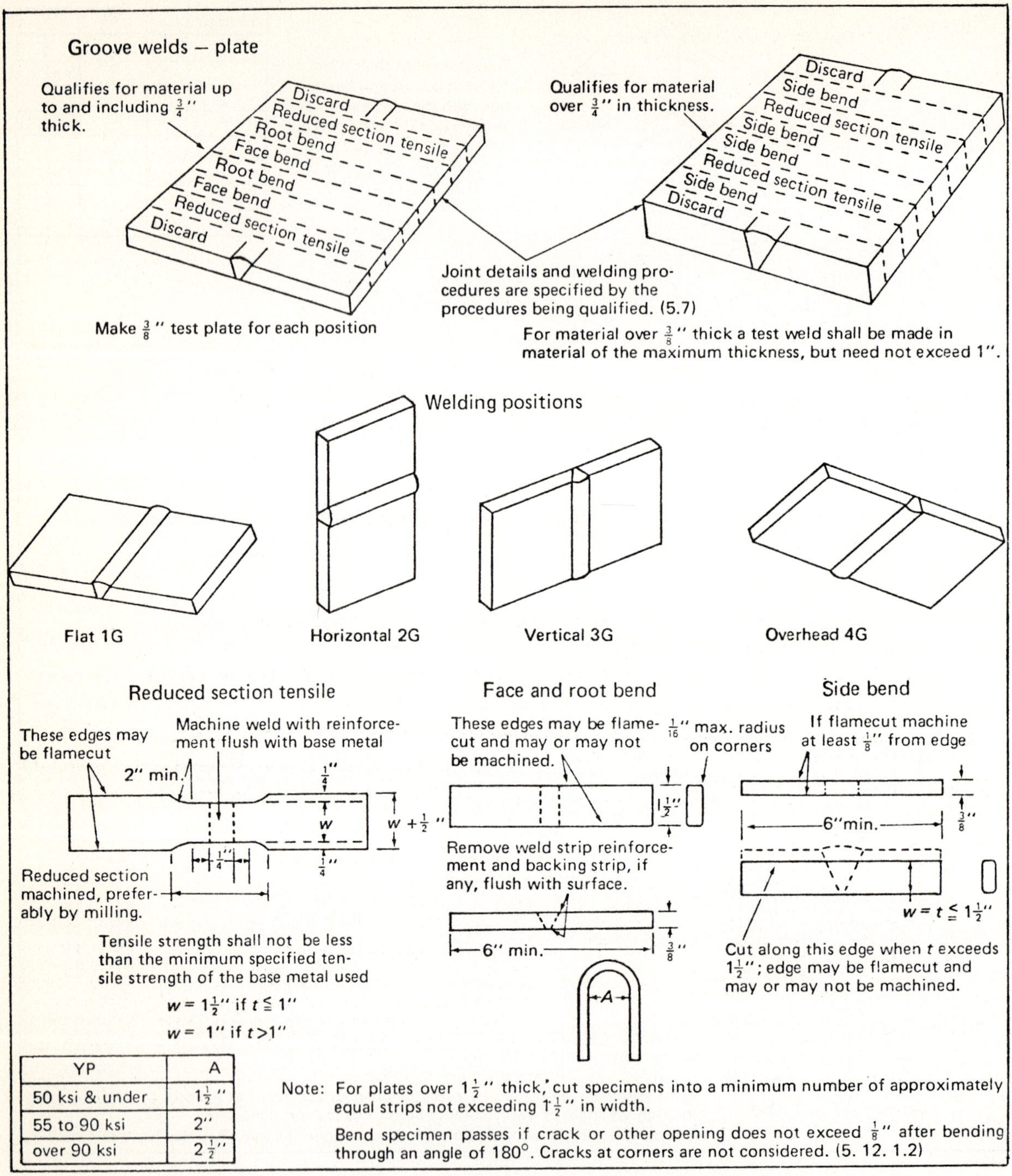
Groove welds – plate
Qualifies for material up to and including $\frac{3}{4}$'' thick.
Discard
Reduced section tensile
Root bend
Face bend
Root bend
Face bend
Reduced section tensile
Discard
Make $\frac{3}{8}$'' test plate for each position
Qualifies for material over $\frac{3}{4}$'' in thickness.
Discard
Side bend
Reduced section tensile
Side bend
Side bend
Reduced section tensile
Side bend
Discard
Joint details and welding procedures are specified by the procedures being qualified. (5.7)
For material over $\frac{3}{8}$'' thick a test weld shall be made in material of the maximum thickness, but need not exceed 1''.
Welding positions
Flat 1G
Horizontal 2G
Vertical 3G
Overhead 4G
Reduced section tensile
These edges may be flamecut
Machine weld with reinforcement flush with base metal
2'' min.
$\frac{1}{4}$''
w
$w + \frac{1}{2}$''
$\frac{1}{4}$''
$\frac{1}{4}$''
Reduced section machined, preferably by milling.
Tensile strength shall not be less than the minimum specified tensile strength of the base metal used
$w = 1\frac{1}{2}$'' if $t \leqq 1$''
$w = 1$'' if $t > 1$''
Face and root bend
These edges may be flamecut and may or may not be machined.
$\frac{1}{16}$'' max. radius on corners
$1\frac{1}{2}$''
Remove weld strip reinforcement and backing strip, if any, flush with surface.
6'' min.
$\frac{3}{8}$''
A
Side bend
If flamecut machine at least $\frac{1}{8}$'' from edge
6''min.
$\frac{3}{8}$''
$w = t \leqq 1\frac{1}{2}$''
Cut along this edge when t exceeds $1\frac{1}{2}$''; edge may be flamecut and may or may not be machined.
Note: For plates over $1\frac{1}{2}$'' thick, cut specimens into a minimum number of approximately equal strips not exceeding $1\frac{1}{2}$'' in width.
Bend specimen passes if crack or other opening does not exceed $\frac{1}{8}$'' after bending through an angle of 180°. Cracks at corners are not considered. (5. 12. 1.2)

YP	A
50 ksi & under	$1\frac{1}{2}$''
55 to 90 ksi	2''
over 90 ksi	$2\frac{1}{2}$''

Figure **23-2** (*cont.*)

Unlimited thickness will qualify for groove or fillet welds in unlimited thickness.

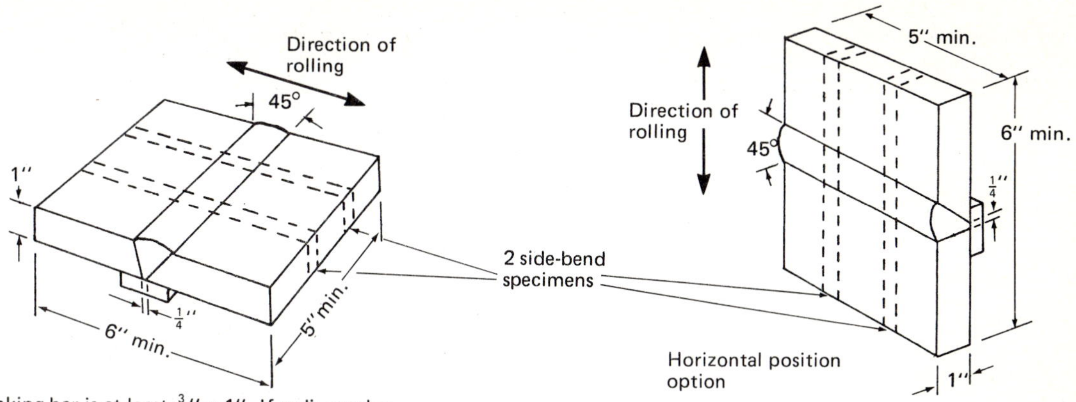
Direction of rolling
45°
1"
1/4"
6" min.
5" min.
2 side-bend specimens
5" min.
Direction of rolling
45°
6" min.
1/4"
1"
Horizontal position option

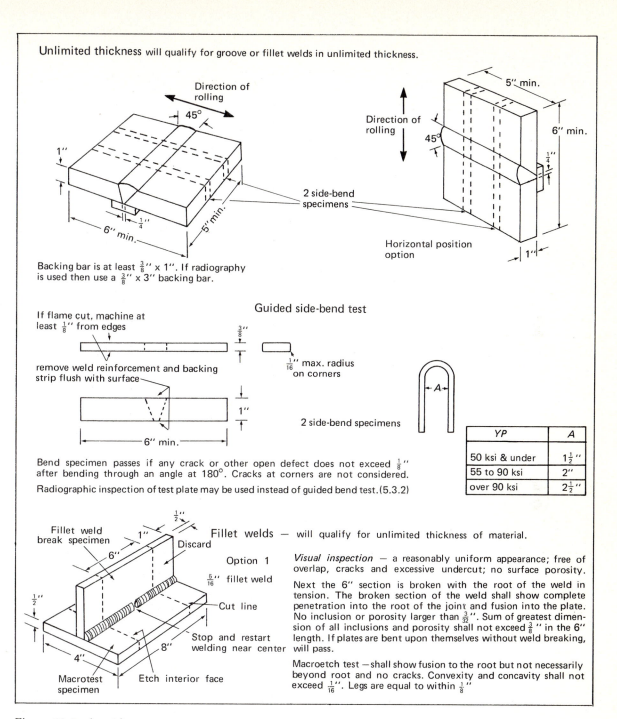

Backing bar is at least $\frac{3}{8}$" x 1". If radiography is used then use a $\frac{3}{8}$" x 3" backing bar.

Guided side-bend test

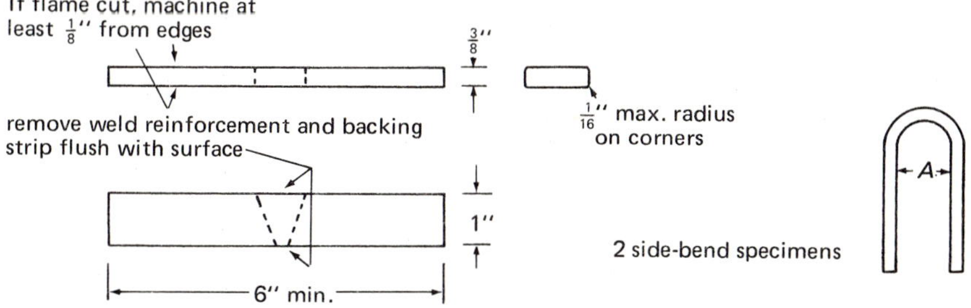
If flame cut, machine at least 1/8" from edges
3/8"
1/16" max. radius on corners
remove weld reinforcement and backing strip flush with surface
1"
6" min.
2 side-bend specimens
A

YP	*A*
50 ksi & under	$1\frac{1}{2}$"
55 to 90 ksi	2"
over 90 ksi	$2\frac{1}{2}$"

Bend specimen passes if any crack or other open defect does not exceed $\frac{1}{8}$" after bending through an angle at 180°. Cracks at corners are not considered.

Radiographic inspection of test plate may be used instead of guided bend test. (5.3.2)

Fillet welds — will qualify for unlimited thickness of material.

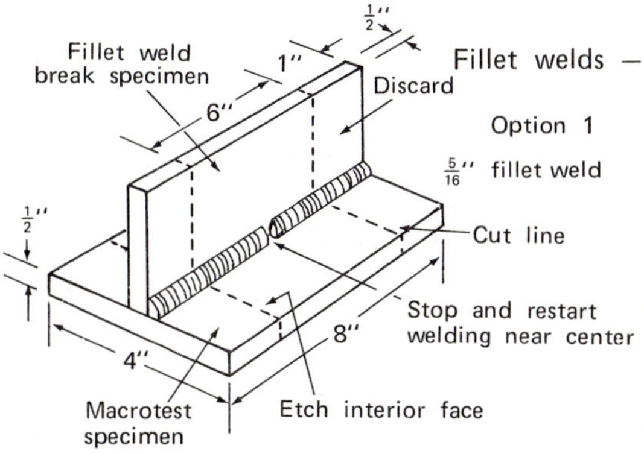
1/2"
Fillet weld break specimen
1"
Discard
6"
Option 1
5/16" fillet weld
1/2"
Cut line
Stop and restart welding near center
8"
4"
Macrotest specimen
Etch interior face

Visual inspection — a reasonably uniform appearance; free of overlap, cracks and excessive undercut; no surface porosity.

Next the 6" section is broken with the root of the weld in tension. The broken section of the weld shall show complete penetration into the root of the joint and fusion into the plate. No inclusion or porosity larger than $\frac{3}{32}$". Sum of greatest dimension of all inclusions and porosity shall not exceed $\frac{3}{8}$" in the 6" length. If plates are bent upon themselves without weld breaking, will pass.

Macroetch test — shall show fusion to the root but not necessarily beyond root and no cracks. Convexity and concavity shall not exceed $\frac{1}{16}$". Legs are equal to within $\frac{1}{8}$"

Figure **23-2** *(cont.)*

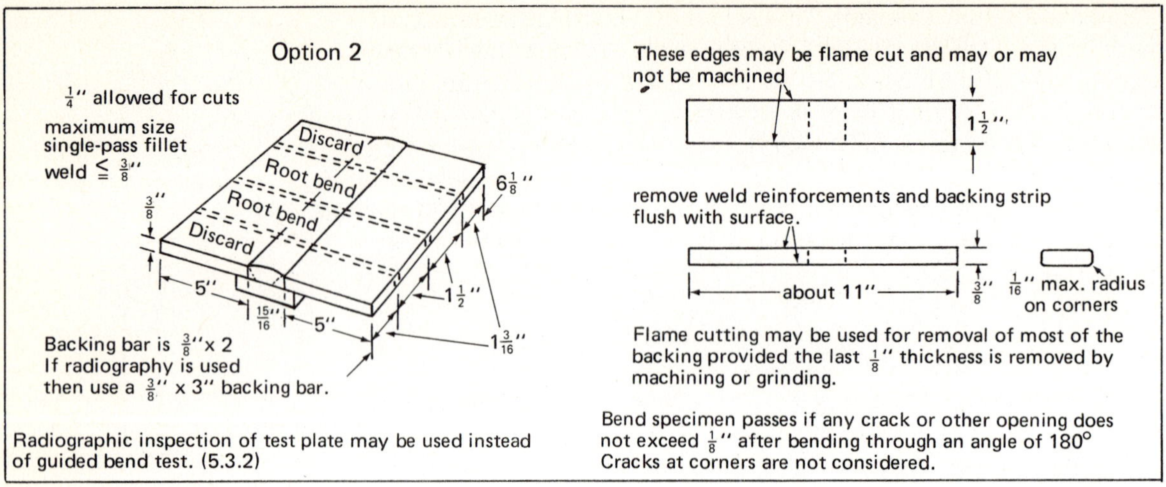
Option 2
1/4" allowed for cuts
maximum size single-pass fillet weld ≦ 3/8"
3/8"
Discard
Root bend
Root bend
Discard
6 1/8"
5"
15/16"
5"
1 1/2"
1 3/16"
Backing bar is 3/8"x 2
If radiography is used then use a 3/8" x 3" backing bar.
Radiographic inspection of test plate may be used instead of guided bend test. (5.3.2)
These edges may be flame cut and may or may not be machined
1 1/2"
remove weld reinforcements and backing strip flush with surface.
about 11"
3/8"
1/16" max. radius on corners
Flame cutting may be used for removal of most of the backing provided the last 1/8" thickness is removed by machining or grinding.
Bend specimen passes if any crack or other opening does not exceed 1/8" after bending through an angle of 180°
Cracks at corners are not considered.

Figure **23-2** (*cont.*)

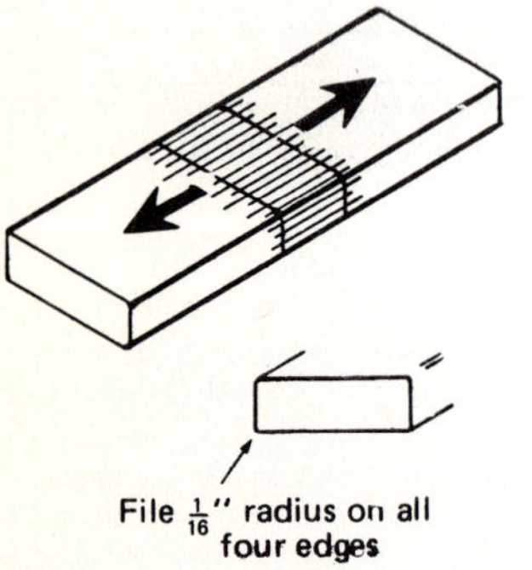
File 1/16" radius on all four edges

Figure **23-3** How to correctly prepare surfaces and edges of a weld test specimen. (Redrawn by permission from *The Procedure Handbook of Arc Welding,* 11th ed., Lincoln Electric Company, Cleveland, Ohio, p. 11.3-10.)

Good penetration and sound weld metal in the root passes are critical to passing any of the tests. To get good penetration, it is generally best to use the highest current one can handle within the recommended range for the electrode. Electrode data tables furnished by the electrode manufacturer should be consulted for electrode current ranges.

Many operators worry too much about appearance on the first beads. Appearance means nothing to the testing machine. But poor penetration and unsound weld metal cause most failures.

When manual-electrode welding, vertical welds should be made with the vertical-up technique. For gas metal–arc or flux-cored arc welding, vertical welds can be made either in the vertical-up or vertical-down direction. However, the welder must use the same direction on the job as was used on the qualification test.

No preheat or postheat treatment is necessary or permissible to pass the test. It is, however, helpful to keep the plates hot during welding and allow them to cool slowly after completion. Interpass temperatures of 300 to 400°F are ideal and can usually be maintained by operating at a normal pace. Never quench the plates in cold water or in any other way accelerate the cooling rate.

Poor specimen preparation can cause sound weld metal to fail. Specimen size, location, and preparation are specified for each test. Even a slight nick across the sample may open up under the severe bending stress of the test, causing failure. Therefore, always grind or machine lengthwise on the specimen, as indicated in Figure 23-3. Always grind or machine both faces of the specimen until the entire bend area is even, leaving no indentations or irregular spots.

Remove all reinforcement. This is part of the test requirement and, more important to the welder, failure to do so can cause the failure of a good weld (Figure 23-4). Be sure that the edges are rounded to a smooth 1/16-inch radius. This can be done quickly with a file, and is a good insurance against failure caused by cracks starting at the sharp corner.

When grinding specimens, do not water quench them when they are hot. Quenching may create tiny surface cracks that become larger during the bend test.

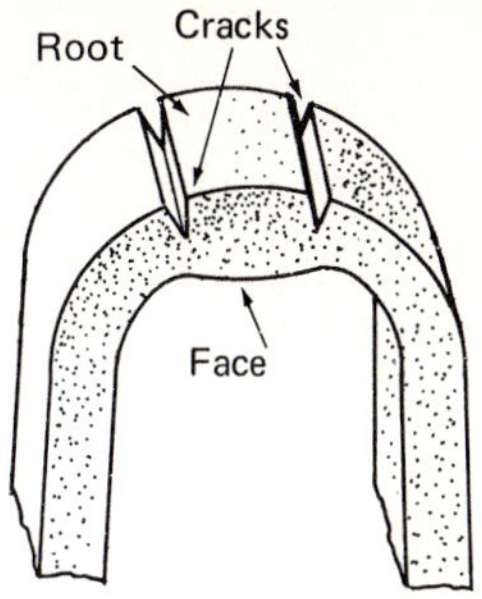
Root
Cracks
Face

Figure **23-4** Failure of a good weld due to nonremoval of weld reinforcement. (Redrawn by permission from *The Procedure Handbook of Arc Welding*, 11th ed., Lincoln Electric Company, Cleveland, Ohio, p. 11.3-10.)

23-2 WELDER QUALIFICATION

Manual-electrode and semiautomatic welders must be qualified for work covered by the *AWS Structural Welding Code*.

A welder is qualified for any of the approved steels if he or she qualifies on any one of them. For manual shielded metal–arc welding the proper electrode classification must be used.

With manual shielded metal–arc welding, qualification with any of the following electrodes will qualify for any electrode having the same or lower group designation, where XX represents the strength level of the electrode (60, 70, 80, 90, 100, 110, and 120 × 10^3 psi):

Group	*AWS Electrode Classification*
F4	EXX15 EXX16 EXX18
F3	EXX10 EXX11
F2	EXX12 EXX13 EXX14
F1	EXX20 EXX24 EXX27 EXX28

A welder qualified with an approved combination of electrode and shielding medium is also qualified for any other approved combination for the same semiautomatic process as used in the qualification test. The types of joints and welding positions qualified with each test are indicated in Table 23-1.

The welder taking and passing a procedure qualification test is also qualified for that process and test position for plates of thickness equal to or less than that of the test plate welded. In the case of 1-inch or other thickness, qualification will be for all thicknesses.

A welder could be qualified for making any type of weld in any position by making two satisfactory groove welds with EXX15, EXX16, or EXX18 electrodes in material prepared (by the welder) with the prescribed bevel angle and root opening in a 1-inch-thick section of any one of the twelve qualifiable steels. One weld would be made in the vertical position and the other in the overhead. Radiographic inspection would eliminate any machining and bending. If no overhead welding will be required on the job, just one test plate in the vertical position could be made.

Table **23-1** Position and type of weld qualified

TEST POSITION	UNLIMITED* & LIMITED** THICKNESS TEST		FILLET WELD TESTS***
Overhead OH	F & OH groove	F, H & OH fillet	F, H & OH fillet
Vertical V	F, H & V groove	F, H & V fillet	F, H & V fillet
Horizontal H	F & H groove	F & H fillet	F & H fillet
Flat F	F groove	F & H fillet	F fillet

* Qualifies for groove and fillet welds on material of unlimited thickness.
** Qualifies for groove welds in material not over ¾-inch thick and fillet welds on material of unlimited thickness.
*** Qualifies for fillet welds on material of unlimited thickness.

Source: *The Procedure Handbook of Arc Welding* (Cleveland, Ohio: Lincoln Electric Company), pp. 11.3–11.4.

Results of welder-qualification tests remain valid indefinitely unless:

1. the welder does not work with the welding process for which qualification has been shown for a period exceeding six months. In this case a requalification test is required on ⅜-inch-thick plate.

2. there is specific reason to question the welder's ability.

If a welder fails the test, he or she may be retested as follows:

1. An immediate retest shall consist of two test welds of each type failed, and all test specimens must pass.

2. A retest can be made if the welder has had further training or practice. In this case a complete retest shall be made.

23-3 RADIOGRAPHIC EXAMINATION OF WELDS

Radiographic examination of test plates for welder qualification may be used in place of the guided-bend test. Radiographic examination, in addition to being more fair (in that it gives a more positive appraisal of the weld), is less expensive, and permits a welder to be tested in the morning and put to work in the afternoon, avoiding the time delays that frustrate employers and employees alike.

The *AWS Structural Code* standards for radiographic weld inspection are applicable to the examination of qualification welds by radiography. They specify that porosity or fusion-type defects less than $\frac{3}{32}$ inch in their greatest dimension are acceptable if the sum of the greatest dimension, when scattered, does not exceed $\frac{3}{8}$ inch in any linear inch of weld. Larger porosity of fusion defects are acceptable when:

1. The greatest dimension of the defect does not exceed $\frac{2}{3}t$ (t = joint or weld throat in inches) or $\frac{3}{4}$ inch.

2. The sum of the greatest dimensions of defects in line does not exceed t in a length of $6t$, or the space between each pair of adjacent defects exceeds three times the greatest dimension of the larger defects. When the length of the weld being examined is less than $6t$, the permissible sum of the greatest dimension of all such defects shall be proportionately less than t.

3. The defect does not lie closer than three times its greatest dimension to the end of a groove-welded joint carrying primary tensile stress.

Piping porosity is acceptable if the sum of the diameters does not exceed $\frac{3}{8}$ inch in any linear inch of weld nor $\frac{3}{4}$ inch in any 12-inch length of weld. Figure 23-5 illustrates graphically these limitations in a groove-welded joint carrying primary tensile stress.

23-4 AWS PREQUALIFIED JOINTS

When the work involves the joints shown in Figure 2-11 and the procedures used are as shown in Figure 23-6, the *AWS Structural Code* does not require procedure qualification tests, but does require the services of a certified welder.

23-5 AWS QUALIFICATION TESTS FOR PIPING AND TUBING

The American Welding Society Standard D10.9 covers the qualification of procedures and personnel for the welding of pipe, tubing, and associated components. The standard provides three levels of qualification, termed *Acceptance Requirements 1, 2,* and *3* (abbreviated AR-1, and AR-2, and AR-3, respectively). AR-1 is the highest level of qualification, followed by AR-2 and AR-3. The differences lie in the type of tests and the acceptance requirements for the sample welds. The welding-procedure qualification tests are shown in Table 23-2.

Each of the three levels of welding-procedure qualification has a corresponding level of welder qualification. Qualification for a given level automatically qualifies for all lower levels. For example, qualification for AR-1 also qualifies for AR-2 and AR-3.

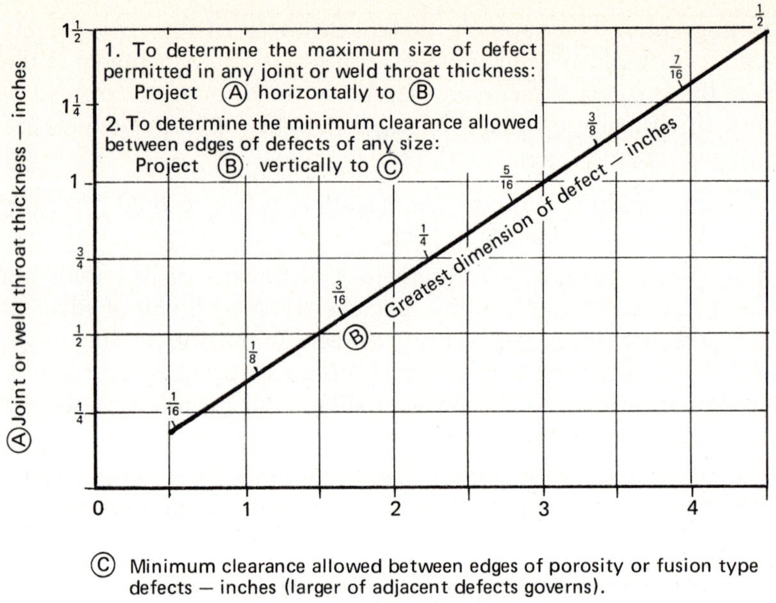

1. To determine the maximum size of defect permitted in any joint or weld throat thickness:
Project (A) horizontally to (B)
2. To determine the minimum clearance allowed between edges of defects of any size:
Project (B) vertically to (C)
(A) Joint or weld throat thickness — inches
Greatest dimension of defect — inches
(C) Minimum clearance allowed between edges of porosity or fusion type defects — inches (larger of adjacent defects governs).

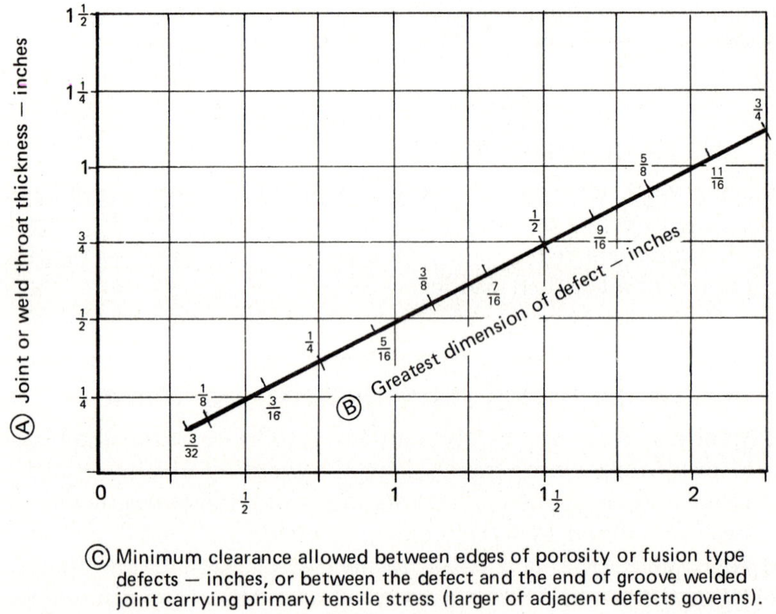

(A) Joint or weld throat thickness — inches
Greatest dimension of defect — inches
(C) Minimum clearance allowed between edges of porosity or fusion type defects — inches, or between the defect and the end of groove welded joint carrying primary tensile stress (larger of adjacent defects governs).

Figure **23-5** Graphs used in determining allowable dimensions of piping porosity in welded joints. (Redrawn by permission from *The Procedure Handbook of Arc Welding*, 11th ed., Lincoln Electric Company, Cleveland, Ohio, p. 11.3-7.)

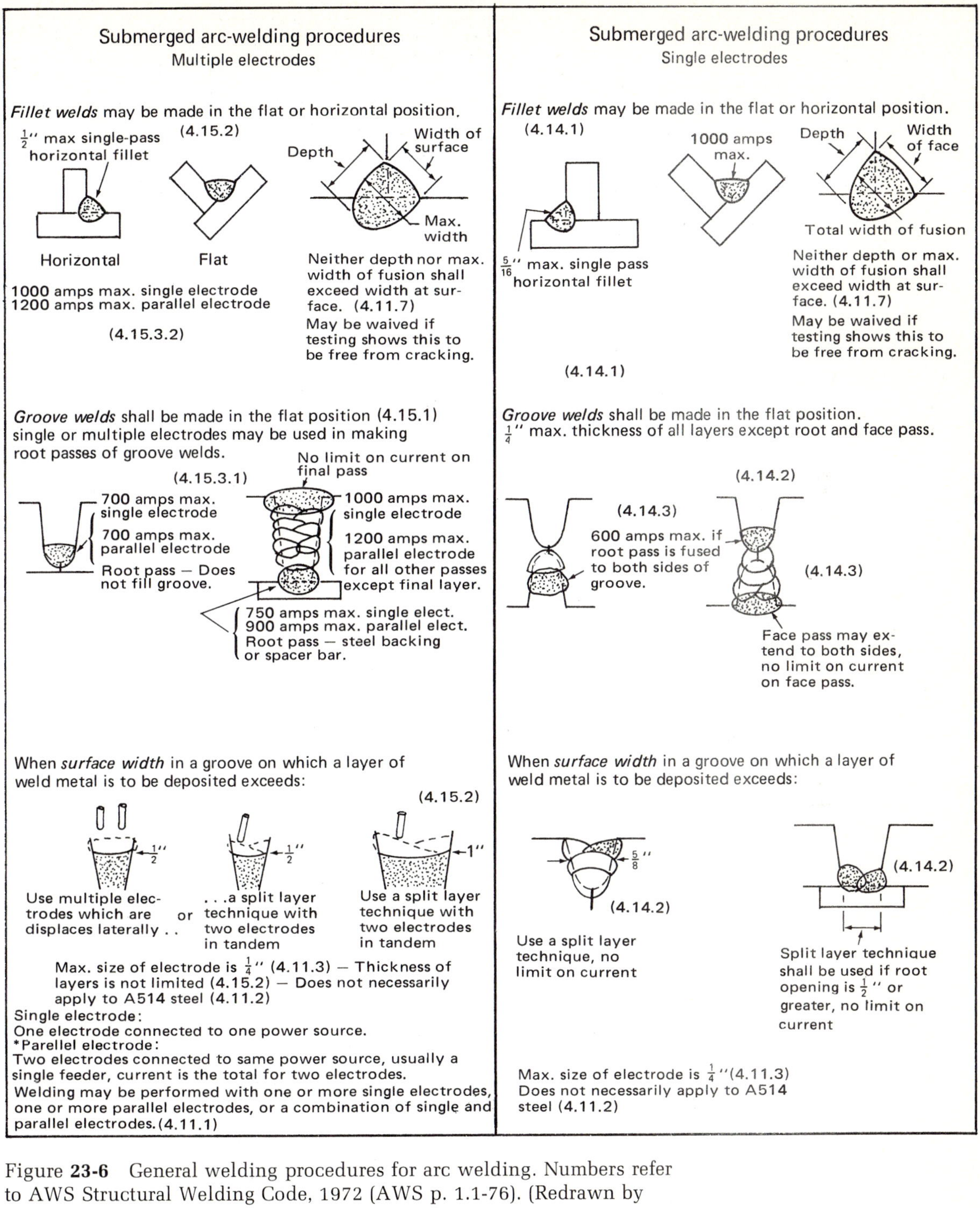

Submerged arc-welding procedures
Multiple electrodes
Fillet welds may be made in the flat or horizontal position.
(4.15.2)
1/2'' max single-pass horizontal fillet
Depth
Width of surface
Max. width
Horizontal
Flat
Neither depth nor max. width of fusion shall exceed width at surface. (4.11.7)
May be waived if testing shows this to be free from cracking.
1000 amps max. single electrode
1200 amps max. parallel electrode
(4.15.3.2)
Groove welds shall be made in the flat position (4.15.1) single or multiple electrodes may be used in making root passes of groove welds.
(4.15.3.1)
No limit on current on final pass
700 amps max. single electrode
700 amps max. parallel electrode
Root pass — Does not fill groove.
1000 amps max. single electrode
1200 amps max. parallel electrode for all other passes except final layer.
750 amps max. single elect.
900 amps max. parallel elect.
Root pass — steel backing or spacer bar.
When surface width in a groove on which a layer of weld metal is to be deposited exceeds:
(4.15.2)
1/2''
1/2''
1''
Use multiple electrodes which are displaces laterally . .
or
. . .a split layer technique with two electrodes in tandem
Use a split layer technique with two electrodes in tandem
Max. size of electrode is 1/4'' (4.11.3) — Thickness of layers is not limited (4.15.2) — Does not necessarily apply to A514 steel (4.11.2)
Single electrode:
One electrode connected to one power source.
*Parellel electrode:
Two electrodes connected to same power source, usually a single feeder, current is the total for two electrodes.
Welding may be performed with one or more single electrodes, one or more parallel electrodes, or a combination of single and parallel electrodes. (4.11.1)
Submerged arc-welding procedures
Single electrodes
Fillet welds may be made in the flat or horizontal position.
(4.14.1)
1000 amps max.
Depth
Width of face
Total width of fusion
5/16'' max. single pass horizontal fillet
Neither depth or max. width of fusion shall exceed width at surface. (4.11.7)
May be waived if testing shows this to be free from cracking.
(4.14.1)
Groove welds shall be made in the flat position.
1/4'' max. thickness of all layers except root and face pass.
(4.14.2)
(4.14.3)
600 amps max. if root pass is fused to both sides of groove.
(4.14.3)
Face pass may extend to both sides, no limit on current on face pass.
When surface width in a groove on which a layer of weld metal is to be deposited exceeds:
5/8''
(4.14.2)
(4.14.2)
Use a split layer technique, no limit on current
Split layer technique shall be used if root opening is 1/2'' or greater, no limit on current
Max. size of electrode is 1/4''(4.11.3)
Does not necessarily apply to A514 steel (4.11.2)

Figure **23-6** General welding procedures for arc welding. Numbers refer to AWS Structural Welding Code, 1972 (AWS p. 1.1-76). (Redrawn by permission from *The Procedure Handbook of Arc Welding*, 11th ed., Lincoln Electric Company, Cleveland, Ohio, p. 11.3-14.)

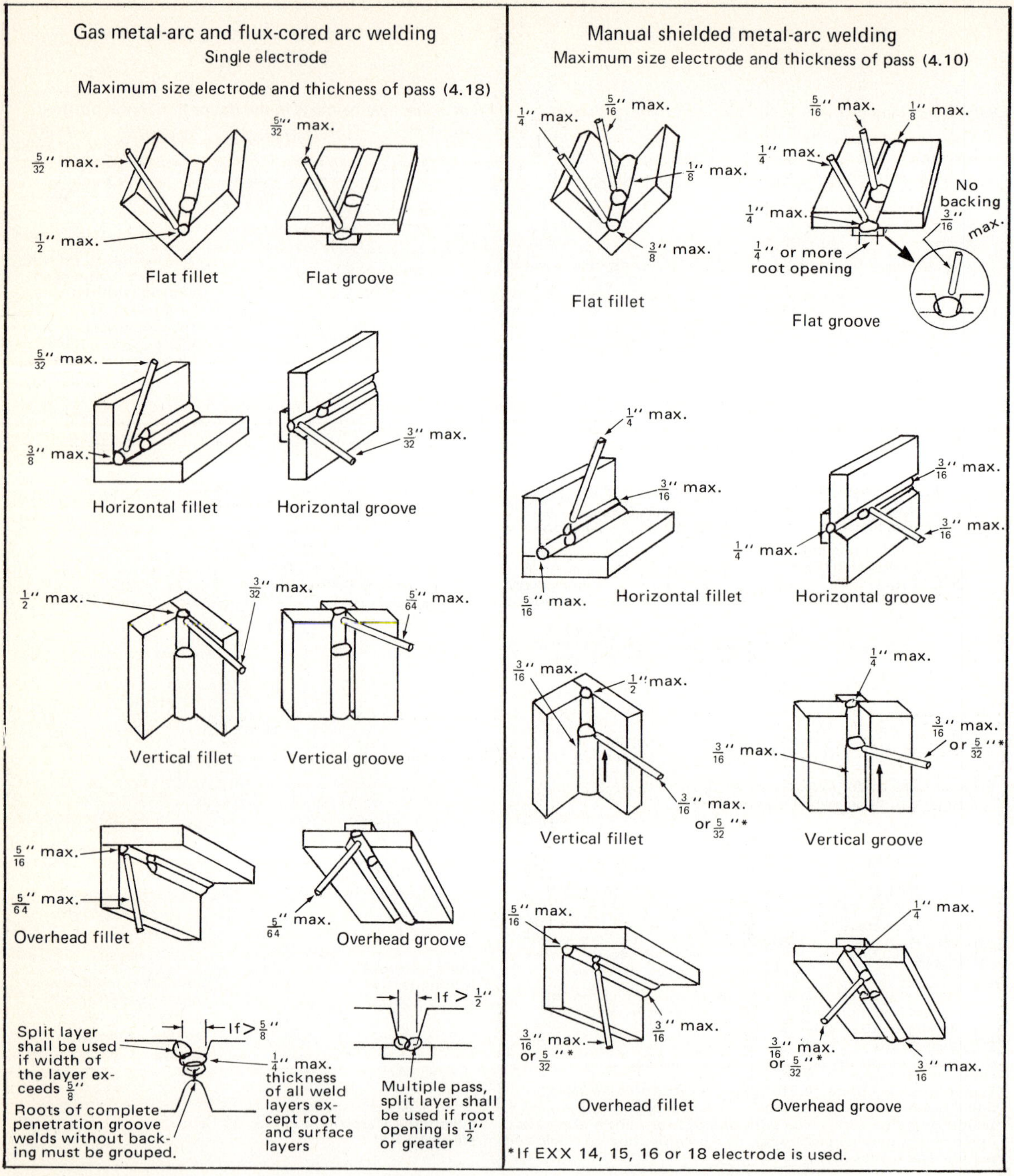

Gas metal-arc and flux-cored arc welding
Single electrode
Maximum size electrode and thickness of pass (4.18)
5/32" max.
5/32" max.
1/2" max.
Flat fillet
Flat groove
5/32" max.
3/8" max.
3/32" max.
Horizontal fillet
Horizontal groove
1/2" max.
3/32" max.
5/64" max.
Vertical fillet
Vertical groove
5/16" max.
5/64" max.
Overhead fillet
5/64" max.
Overhead groove
Split layer shall be used if width of the layer exceeds 5/8"
If > 5/8"
Roots of complete penetration groove welds without backing must be grouped.
1/4" max. thickness of all weld layers except root and surface layers
If > 1/2"
Multiple pass, split layer shall be used if root opening is 1/2" or greater
Manual shielded metal-arc welding
Maximum size electrode and thickness of pass (4.10)
1/4" max.
5/16" max.
1/8" max.
3/8" max.
Flat fillet
5/16" max.
1/8" max.
1/4" max.
1/4" max.
No backing
3/16" max.
1/4" or more root opening
Flat groove
1/4" max.
3/16" max.
5/16" max.
Horizontal fillet
3/16" max.
3/16" max.
1/4" max.
Horizontal groove
3/16" max.
1/2" max.
3/16" max. or 5/32"*
Vertical fillet
1/4" max.
3/16" max.
3/16" max. or 5/32"*
Vertical groove
5/16" max.
3/16" max. or 5/32"*
3/16" max.
Overhead fillet
1/4" max.
3/16" max. or 5/32"*
3/16" max.
Overhead groove
*If EXX 14, 15, 16 or 18 electrode is used.

Figure **23-6** (*cont.*)

Table **23-2** Acceptance requirements tests AR-1, AR-2, and AR-3

AR-1 LEVEL OF WELDING PROCEDURE QUALIFICATION

SAMPLE WELDS WALL THICKNESS ¾″ AND UNDER					TESTS REQUIRED						
		PIPE OR TUBE WALL THICKNESS (t) QUALIFIED						TENSILE TEST	BEND TESTS NUMBER OF SPECIMENS		
PIPE SIZE OF SAMPLE WELD	PIPE OR TUBE SIZE QUALIFIED	MIN.	MAX.	NUMBER OF SAMPLE WELDS PER POSITION	VISUAL INSPECTION	PENETRANT INSPECTION OPTIONAL	RADIOGRAPHY	NUMBER OF SPECIMENS	FACE	ROOT	SIDE
½″ Sch. 40	Through 1½″	.063″	.400″	2	Yes	Yes	Yes	2	0	0	0
2″ Sch. 80	1″ Through 4″	.063″	.674″	2	Yes	Yes	Yes	2	2	2	0
5″ Sch. 80	Over 4″	.187″	.750″	1	Yes	Yes	Yes	2	2	2	0
WALL THICKNESS OVER ¾″											
8″ Sch. 120	5″ and Over	.187″	Any	1	Yes	Yes	Yes	2	0	0	4

AR-2 LEVEL OF WELDING PROCEDURE QUALIFICATION

		PIPE OR TUBE WALL THICKNESS (t) QUALIFIED					TENSILE TEST	BEND TESTS NUMBER OF SPECIMENS		
PIPE SIZE OF SAMPLE WELD	PIPE OR TUBE SIZE QUALIFIED	MIN.	MAX.	NUMBER OF SAMPLE WELDS PER POSITION	VISUAL INSPECTION	RADIOGRAPHY OPTIONAL	NUMBER OF SPECIMENS	FACE	ROOT	SIDE
½″ Sch. 40	Through 1½″	.063″	.400″	2	Yes	Yes	2	0	0	0
2″ Sch. 80	1″ Through 4″	.063″	.674″	2	Yes	Yes	2	2	2	0
5″ Sch. 80	Over 4″	.187″	.750″	1	Yes	Yes	2	2	2	0
WALL THICKNESS OVER ¾″										
8″ Sch. 120	5″ and Over	.187″	Any	1	Yes	Yes	2	0	0	4

AR-3 LEVEL OF WELDING PROCEDURE QUALIFICATION
LIMITED TO PIPE UP TO AND INCLUDING 10″ DIAMETER AND ⅜″ WALL THICKNESS

SAMPLE WELDS WALL THICKNESS ⅜″ AND UNDER					TESTS REQUIRED		
		PIPE OR TUBE WALL THICKNESS (t) QUALIFIED				PRESSURE BEND	
PIPE SIZE OF SAMPLE WELD	PIPE OR TUBE SIZE QUALIFIED	MIN.	MAX.	NUMBER OF SAMPLE WELDS PER POSITION	VISUAL INSPECTION	No. of Tests	MACROSTRUCTURE NO. OF SPECIMENS
½″ Sch. 40	1½″ and Under	1/16″	2 t	1	Yes	1	0
2″ Sch. 40	1″ Through 4″	½ t	2 t	1	Yes	1	0
5″ Sch. 40	4½″ Through 10″	½ t	⅜″	1	Yes	0	4

Source: *The Procedure Handbook of Arc Welding* (Cleveland, Ohio: Lincoln Electric Company), pp. 11.3-19, 11.3-21, 11.3-22.

The AR-1 level of quality is intended to provide the confidence required for pipelines that may be found in nuclear-energy, space, high-pressure, high-temperature, chemical, and gas systems.

The AR-2 level of quality is intended to provide the confidence necessary for some lines that may be found in nuclear-energy, steam, water, petroleum, gas, or chemical systems.

The AR-3 level of quality is intended to provide the confidence needed for lines such as low-pressure heating, air-conditioning, and sanitary water.

The standard provides for qualification with groove and fillet welds. Procedure and welder qualifications with groove welds automatically qualifies for fillet welds. However, if fillet welds alone are required for the specific job, the welding procedure and welder can be qualified for fillet welds only.

Some of the test positions required for the AR-1 level procedure qualification are shown in Figure 23-7. For more detailed information consult AWS Standard D10.9.

23-6 *ASME BOILER AND PRESSURE VESSEL CODE* PROCEDURE QUALIFICATION TESTS

Section IX of the *Boiler and Pressure Vessel Code of the American Society of Mechanical Engineers* provides for qualification tests of procedures and personnel for welding under that code. The following is a condensed and incomplete description of these qualifications. Qualification tests for steels of 85,000 psi or more and for quenched and tempered steels are not covered here. Before making tests, a copy of the complete code should be obtained from the American Society of Mechanical Engineers, Order Department, United Engineering Center, 345 East 47th Street, New York, NY 10017.

Procedure qualification for groove welds in pipe requires that one test assembly be welded for each position shown in Figure 23-8. In position 1G, the pipe axis is horizontal, and the pipe is rotated during welding. Weld metal is deposited from above (flat welding). In position 2G, the pipe axis is horizontal, and the pipe is not rotated (a combination of flat, vertical, and overhead welding).

Qualification in the horizontal, vertical, or overhead position also qualifies for the flat position. Qualification in the horizontal fixed position, 5G, qualifies for flat, vertical, and overhead positions. Qualification in the horizontal, vertical, and overhead positions qualifies for all positions. Procedure qualification on pipe also qualifies for plate, but not vice versa.

Procedure qualification for groove welds in plate requires one test assembly to be welded for each position. Qualifying positions are also shown in Figure 23-8. In position 1G, the plate is in the

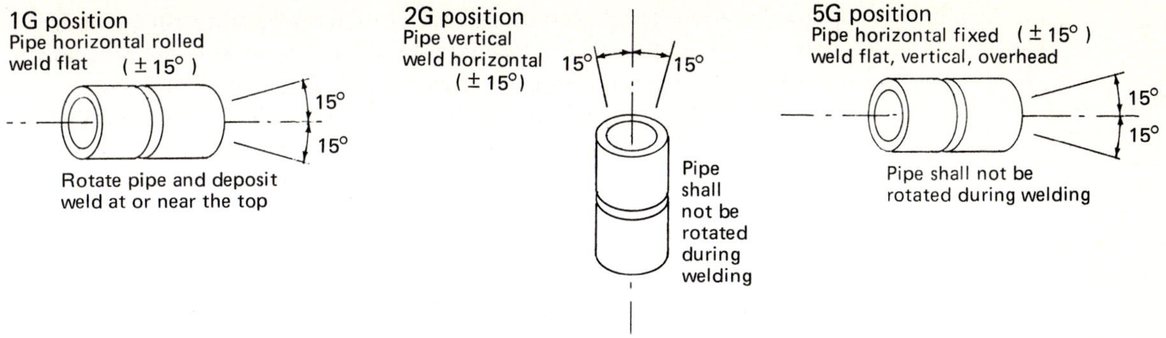
1G position
Pipe horizontal rolled
weld flat (± 15°)
15°
15°
Rotate pipe and deposit
weld at or near the top
2G position
Pipe vertical
weld horizontal
(± 15°)
15°
15°
Pipe
shall
not be
rotated
during
welding
5G position
Pipe horizontal fixed (± 15°)
weld flat, vertical, overhead
15°
15°
Pipe shall not be
rotated during welding

The limits of qualification in a given position are shown below:

1. Qualification in the 1G position qualifies only for this position.
2. Qualification in the 2G position qualifies for welding in the 1G and 2G positions only.
3. Qualification in the 5G position qualifies for welding in the 1G and 5G positions only.
4. Qualification in both the 2G and 5G positions qualifies for welding in all positions.
5. Qualification in the 6G position qualifies for welding in all positions, because it includes elements of those positions.
6. Qualification in a position other than the four standard positions described above is valid only for that position (plus or minus 15 degrees).

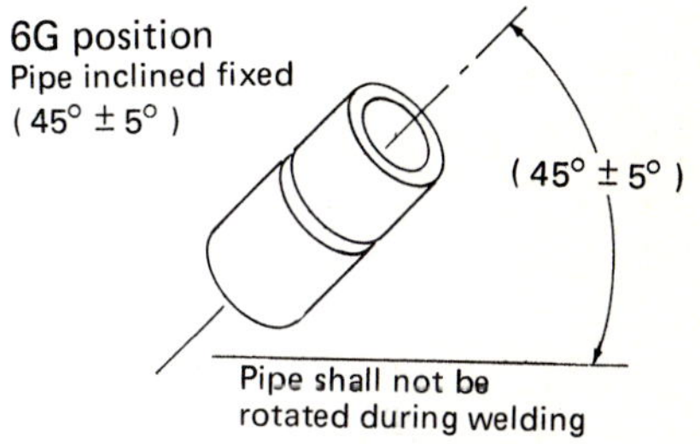
6G position
Pipe inclined fixed
(45° ± 5°)
(45° ± 5°)
Pipe shall not be
rotated during welding

Figure **23-7** Some of the test positions required for procedure qualification at the AR-1 level. (Redrawn by permission from *The Procedure Handbook of Arc Welding*, 11th ed., Lincoln Electric Company, Cleveland, Ohio, p. 11.3-16.)

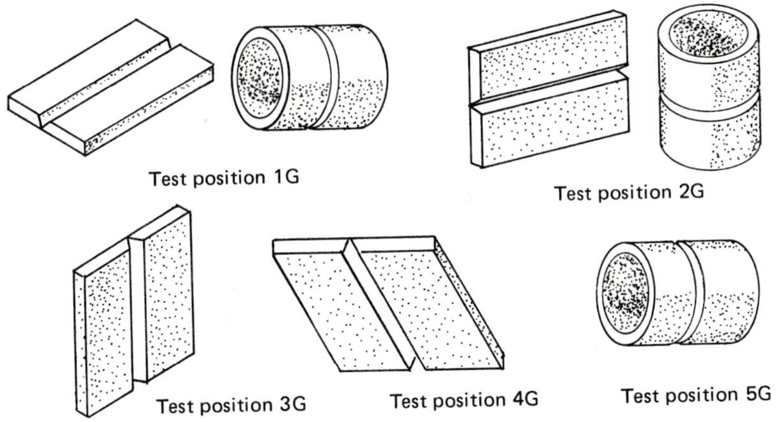
Test position 1G
Test position 2G
Test position 3G
Test position 4G
Test position 5G

Figure **23-8** Test positions used in qualifying procedure for groove welds in pipe and plate. (Redrawn by permission from *The Procedure Handbook of Arc Welding*, 11th ed., Lincoln Electric Company, Cleveland, Ohio, p. 11.3-25.)

horizontal plane and weld metal is deposited from above (flat welding). In position 2G, the plate is in the vertical plane with the axis of the weld horizontal (horizontal welding). In position 3G, the plate is in the vertical plane and the axis of the weld is vertical (vertical welding). In position 4G, the plate is in the horizontal plane and weld metal is deposited from underneath (overhead welding). Groove-weld tests qualify the welding procedure for use with both groove and fillet welds.

Procedure qualification for fillet welds requires one test assembly in each of the positions 1, 2, 3, and 4 F. (F stands for fillet) (Figure 23-9).

Qualification in the horizontal, vertical, or overhead position qualifies also for the flat position. Qualification in the horizontal, vertical, and overhead positions qualifies for all positions.

The base material, filler material, and the welding procedure for the test joint must comply with the job specification. The base material may be either plate or pipe (see Table 23-3). The recommended pipe size is 5 inches in diameter and 3/8 inch in wall thickness, though larger pipe may be used. A smaller pipe size (job size) may be used, but in such cases the procedure must be qualified for thicknesses between 1/2 and twice the wall thickness of the test pipe but not over 3/4 inch.

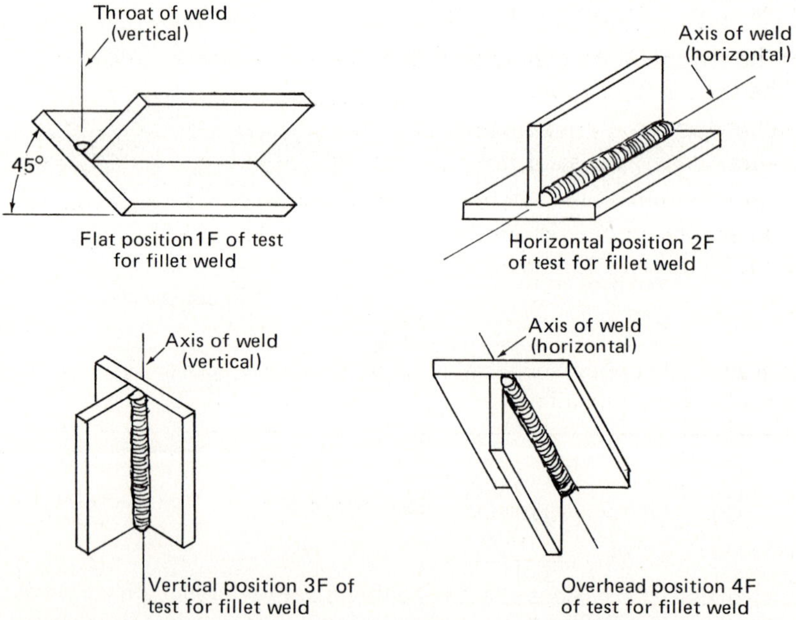
Throat of weld
(vertical)
45°
Flat position 1F of test
for fillet weld
Axis of weld
(horizontal)
Horizontal position 2F
of test for fillet weld
Axis of weld
(vertical)
Vertical position 3F of
test for fillet weld
Axis of weld
(horizontal)
Overhead position 4F
of test for fillet weld

Figure **23-9** Test positions for fillet welds. (Redrawn by permission from *The Procedure Handbook of Arc Welding*, 11th ed., Lincoln Electric Company, Cleveland, Ohio, p. 11.3-25.)

Table **23-3** Procedure qualification: Type, number of test specimens, and range of thickness qualified

THICKNESS t OF TEST PLATE OR PIPE WALL (in.)	RANGE OF THICKNESS (t) QUALIFIED (in.)		TYPE AND NUMBER OF TESTS REQUIRED			
				TRANSVERSE BEND TESTS[5]		
	MIN.[6]	*MAX.*[1,3,6]	*TENSION*	SIDE BEND	FACE BEND	ROOT BEND
1/16 to 3/8 inclusive	1/16	$2\,t^2$	2		2	2
Over 3/8 to 3/4	3/16	2 t	2		2	2
3/4 and over	3/16	2 t	2	4[4]		

Note: for notes 1–6, see text.

Source: *The Procedure Handbook of Arc Welding* (Cleveland, Ohio: Lincoln Electric Company), p. 11.3–26.

The type and number of test specimens for procedure qualification are shown in Tables 23-3 and 23-4. Also shown is the range of thickness that is qualified for use in construction by a given thickness of test plate or pipe used in making the qualification. Test specimens are to be removed as shown in Figure 23-10 for plate and Figure 23-11 for pipe.

The notes to Tables 23-3 and 23-4 are essential to the use of the tables. The applicable notes are:

1. The maximum thickness qualified in gas welding is the thickness of the test plate or pipe.

2. The maximum thickness qualified for pipe smaller than 5 inches is twice the thickness of the pipe but not more than 3/4 inch.

3. For submerged-arc welding and gas metal–arc welding, the thickness limitation for production welding, based on plate thickness t, shall be as follows:

Table **23-4** Procedure qualification: Type, number of test specimens, and range of thickness qualified

THICKNESS t OF TEST PLATE (in.)	RANGE OF THICKNESS (t) QUALIFIED		TYPE AND NUMBER OF TESTS REQUIRED		
				LONGITUDINAL BEND TESTS[5]	
	MIN.[6]	*MAX.*[1,3,6]	*TENSION*	FACE BEND	ROOT BEND
1/16 to 3/8	1/16	2 t	2	2	2
Over 3/8	3/16	2 t	2	2	2

Note: For notes 1–6, see text.

Source: *The Procedure Handbook of Arc Welding* (Cleveland, Ohio: Lincoln Electric Company), p. 11.3–26.

Discard	this piece
Reduced section	tensile specimen
Root bend	specimen
Face bend	specimen
Root bend	specimen
Face bend	specimen
Reduced section	tensile specimen
Discard	this piece

Discard	this piece
Side bend	specimen
Reduced section	tensile specimen
Side bend	specimen
Side bend	specimen
Reduced section	tensile specimen
Side bend	specimen
Discard	this piece

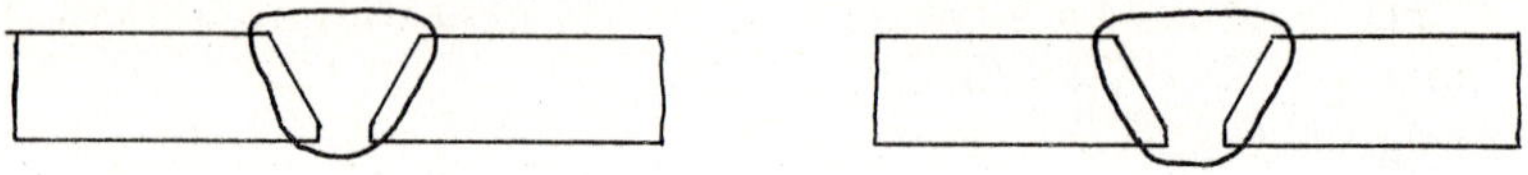

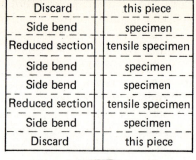

For plate $\frac{1}{16}$ to $\frac{3}{4}$ in. thick.

For plate over $\frac{3}{4}$ in. thick. May be used also for thicknesses from $\frac{3}{8}$ to $\frac{3}{4}$ in.

Figure **23-10** Locations for test specimens that are to be removed from plate. (Redrawn by permission from *The Procedure Handbook of Arc Welding,* 11th ed., Lincoln Electric Company, Cleveland, Ohio, p. 11.3-26.)

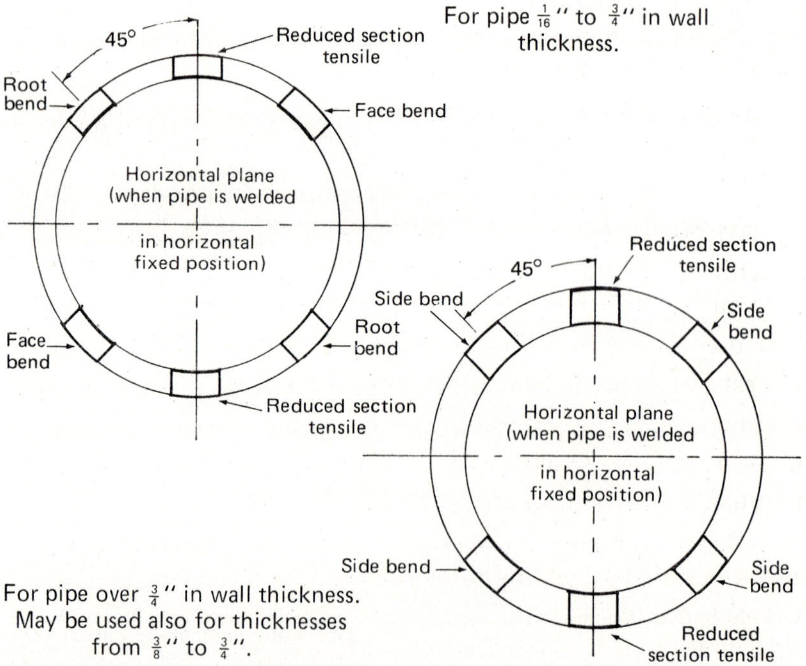
45°
Reduced section tensile
For pipe 1/16" to 3/4" in wall thickness.
Root bend
Face bend
Horizontal plane (when pipe is welded in horizontal fixed position)
Face bend
Root bend
Reduced section tensile
45°
Reduced section tensile
Side bend
Side bend
Horizontal plane (when pipe is welded in horizontal fixed position)
Side bend
Side bend
For pipe over 3/4" in wall thickness. May be used also for thicknesses from 3/8" to 3/4".
Reduced section tensile

Figure **23-11** Locations for test specimens to be removed from pipe. (Redrawn by permission from *The Procedure Handbook of Arc Welding,* 11th ed., Lincoln Electric Company, Cleveland, Ohio, p. 11.3-27.)

a. For single-pass welding with no backing strip or against a metal or ceramic (flux) backing, the maximum thickness welded in production shall not exceed the thickness of the test plate or pipe.

b. If the test plate is welded by a procedure involving one pass from each side, the maximum thickness that may be welded in production shall be $2t$, where $2t$ shall not exceed 2 inches. If sections thicker than 2 inches are to be welded in production, a separate test plate shall be prepared with the thickness not less than the thickness to be used in production.

c. For multiple-pass welding, the thickness limitations refer to the applicable *AWS Structural Code*.

4. Either face and root bends or side bends may be used for thicknesses from ⅜ to ¾ inch.

5. Longitudinal bend tests may be used in lieu of transverse bend tests only for testing material combinations differing markedly in physical bending properties between (a) the two base materials or (b) the weld metal and base materials.

6. For quenched and tempered steels (tensile strength 95,000 psi or higher) of thicknesses less than ⅝ inch, the thickness of the test plate or pipe is the minimum thickness qualified. For test plates or pipe receiving a postweld heat treatment in which the lower critical temperature is exceeded, the maximum thickness qualified is the thickness of the test plate or pipe.

For more details, the student should refer directly to the applicable welding code.

REVIEW QUESTIONS

1. With which AWS electrode must a welder take a certification test to be qualified for welding with any manual electrode?

2. Is it necessary to preheat or postheat test plates? Explain.

3. Test plates should be maintained at what interpass temperature?

4. How must guided-bend test specimens be prepared?

5. Make a free-hand sketch of a root-bend specimen.

6. When and why must a welder be recertified?

7. To what level must a welder be certified and the welding procedure qualified for welding piping and tubing acceptably?

8. What tests are usually required to certify a welder for welding pipes to be used at below-freezing temperatures?

9. Make a free-hand sketch of the end view of a pipe and identify the locations from which specimens must be removed for tensile, root bend, and face bend.

REFERENCE TABLES

Temperature conversion table

459.4 TO 0			0 TO 100						100 TO 1000					
C	F \ C	F	C	F \ C	F	C	F \ C	F	C	F \ C	F	C	F \ C	F
−273	−459.4		−17.8	0	32	10.0	50	122.0	38	100	212	260	500	932
−268	−450		−17.2	1	33.8	10.6	51	123.8	43	110	230	266	510	950
−262	−440		−16.7	2	35.6	11.1	52	125.6	49	120	248	271	520	968
−257	−430		−16.1	3	37.4	11.7	53	127.4	54	130	266	277	530	986
−251	−420		−15.6	4	39.2	12.2	54	129.2	60	140	284	282	540	1004
−246	−410		−15.0	5	41.0	12.8	55	131.0	66	150	302	288	550	1022
−240	−400		−14.4	6	42.8	13.3	56	132.8	71	160	320	293	560	1040
−234	−390		−13.9	7	44.6	13.9	57	134.6	77	170	338	299	570	1058
−229	−380		−13.3	8	46.4	14.4	58	136.4	82	180	356	304	580	1076
−223	−370		−12.8	9	48.2	15.0	59	138.2	88	190	374	310	590	1094
−218	−360		−12.2	10	50.0	15.6	60	140.0	93	200	392	316	600	1112
−212	−350		−11.7	11	51.8	16.1	61	141.8	99	210	410	321	610	1130
−207	−340		−11.1	12	53.6	16.7	62	143.6	100	212	413.6	327	620	1148
−201	−330		−10.6	13	55.4	17.2	63	145.4	104	220	428	332	630	1166
−196	−320		−10.0	14	57.2	17.8	64	147.2	110	230	446	338	640	1184
−190	−310		−9.4	15	59.0	18.3	65	149.0	116	240	464	343	650	1202
−184	−300		−8.9	16	60.8	18.9	66	150.8	121	250	482	349	660	1220
−179	−290		−8.3	17	62.6	19.4	67	152.6	127	260	500	354	670	1238
−173	−280		−7.8	18	64.4	20.0	68	154.4	132	270	518	360	680	1256

−169	−273	−459.4	−7.2	19	66.2	20.6	69	156.2	138	280	536	366	690	1274
−168	−270	−454	−6.7	20	68.0	21.1	70	158.0	143	290	554	371	700	1292
−162	−260	−436	−6.1	21	69.8	21.7	71	159.8	149	300	572	377	710	1310
−157	−250	−418	−5.6	22	71.6	22.2	72	161.6	154	310	590	382	720	1328
−151	−240	−400	−5.0	23	73.4	22.8	73	163.4	160	320	608	388	730	1346
−146	−230	−382	−4.4	24	75.2	23.3	74	165.2	166	330	626	393	740	1364
−140	−220	−364	−3.9	25	77.0	23.9	75	167.0	171	340	644	399	750	1382
−134	−210	−346	−3.3	26	78.8	24.4	76	168.8	177	350	662	404	760	1400
−129	−200	−328	−2.8	27	80.6	25.0	77	170.6	182	360	680	410	770	1418
−123	−190	−310	−2.2	28	82.4	25.6	78	172.4	188	370	698	416	780	1436
−118	−180	−292	−1.7	29	84.2	26.1	79	174.2	193	380	716	421	790	1454
−112	−170	−274	−1.1	30	86.0	26.7	80	176.0	199	390	734	427	800	1472
−107	−160	−256	−.6	31	87.8	27.2	81	177.8	204	400	752	432	810	1490
−101	−150	−238	0	32	89.6	27.8	82	179.6	210	410	770	438	820	1508
−96	−140	−220	.6	33	91.4	28.3	83	181.4	216	420	788	443	830	1526
−90	−130	−202	1.1	34	93.2	28.9	84	183.2	221	430	806	449	840	1544
−84	−120	−184	1.7	35	95.0	29.4	85	185.0	227	440	824	454	850	1562
−79	−110	−166	2.2	36	96.8	30.0	86	186.8	232	450	842	460	860	1580
−73	−100	−148	2.8	37	98.6	30.6	87	188.6	238	460	860	466	870	1598
−68	−90	−130	3.3	38	100.4	31.1	88	190.4	243	470	878	471	880	1616
−62	−80	−112	3.9	39	102.2	31.7	89	192.2	249	480	896	477	890	1634
−57	−70	−94	4.4	40	104.0	32.2	90	194.0	254	490	914	482	900	1652
−51	−60	−76	5.0	41	105.8	32.8	91	195.8				488	910	1670
−46	−50	−58	5.6	42	107.6	33.3	92	197.6				493	920	1688
−40	−40	−40	6.1	43	109.4	33.9	93	199.4				499	930	1706
−34	−30	−22	6.7	44	111.2	34.4	94	201.2				504	940	1724
−29	−20	−4	7.2	45	113.0	35.0	95	203.0				510	950	1742
−23	−10	14	7.8	46	114.8	35.6	96	204.8				516	960	1760
−17.8	0	32	8.3	47	116.6	36.1	97	206.6				521	970	1778
			8.9	48	118.4	36.7	98	208.4				527	980	1796
			9.4	49	120.2	37.2	99	210.2				532	990	1814
						37.8	100	212.0				538	1000	1832

Look up reading in middle column. If in degrees centigrade, read Fahrenheit equivalent in right-hand column; if in degrees Fahrenheit, read centigrade equivalent in left-hand column.

Temperature conversion table *(cont.)*

1000 TO 2000						2000 TO 3000					
C	C/F	F	C	C/F	F	C	C/F	F	C	C/F	F
538	1000	1832	816	1500	2732	1093	2000	3632	1371	2500	4532
543	1010	1850	821	1510	2750	1099	2010	3650	1377	2510	4550
549	1020	1868	827	1520	2768	1104	2020	3668	1382	2520	4568
554	1030	1886	832	1530	2786	1110	2030	3686	1388	2530	4586
560	1040	1904	838	1540	2804	1116	2040	3704	1393	2540	4604
566	1050	1922	843	1550	2822	1121	2050	3722	1399	2550	4622
571	1060	1940	849	1560	2840	1127	2060	3740	1404	2560	4640
577	1070	1958	854	1570	2858	1132	2070	3758	1410	2570	4658
582	1080	1976	860	1580	2876	1138	2080	3776	1416	2580	4676
588	1090	1994	866	1590	2894	1143	2090	3794	1421	2590	4694
593	1100	2012	871	1600	2912	1149	2100	3812	1427	2600	4712
599	1110	2030	877	1610	2930	1154	2110	3830	1432	2610	4730
604	1120	2048	882	1620	2948	1160	2120	3848	1438	2620	4748
610	1130	2066	888	1630	2966	1166	2130	3866	1443	2630	4766
616	1140	2084	893	1640	2984	1171	2140	3884	1449	2640	4784
621	1150	2102	899	1650	3002	1177	2150	3902	1454	2650	4802
627	1160	2120	904	1660	3020	1182	2160	3920	1460	2660	4820
632	1170	2138	910	1670	3038	1188	2170	3938	1466	2670	4838
638	1180	2156	916	1680	3056	1193	2180	3956	1471	2680	4856
643	1190	2174	921	1690	3074	1199	2190	3974	1477	2690	4874
649	1200	2192	927	1700	3092	1204	2200	3992	1482	2700	4892
654	1210	2210	932	1710	3110	1210	2210	4010	1488	2710	4910
660	1220	2228	938	1720	3128	1216	2220	4028	1493	2720	4928

666	1230	2246	943	1730	3146	1221	2230	4046	1499	2730	4946
671	1240	2264	949	1740	3164	1227	2240	4064	1504	2740	4964
677	1250	2282	954	1750	3182	1232	2250	4082	1510	2750	4982
682	1260	2300	960	1760	3200	1238	2260	4100	1516	2760	5000
688	1270	2318	966	1770	3218	1243	2270	4118	1521	2770	5018
693	1280	2336	971	1780	3236	1249	2280	4136	1527	2780	5036
699	1290	2354	977	1790	3254	1254	2290	4154	1532	2790	5054
704	1300	2372	982	1800	3272	1260	2300	4172	1538	2800	5072
710	1310	2390	988	1810	3290	1266	2310	4190	1543	2810	5090
716	1320	2408	993	1820	3308	1271	2320	4208	1549	2820	5108
721	1330	2426	999	1830	3326	1277	2330	4226	1554	2830	5126
727	1340	2444	1004	1840	3344	1282	2340	4244	1560	2840	5144
732	1350	2462	1010	1850	3362	1288	2350	4262	1566	2850	5162
738	1360	2480	1016	1860	3380	1293	2360	4280	1571	2860	5180
743	1370	2498	1021	1870	3398	1299	2370	4298	1577	2870	5198
749	1380	2516	1027	1880	3416	1304	2380	4316	1582	2880	5216
754	1390	2534	1032	1890	3434	1310	2390	4334	1588	2890	5234
760	1400	2552	1038	1900	3452	1316	2400	4352	1593	2900	5252
766	1410	2570	1043	1910	3470	1321	2410	4370	1599	2910	5270
771	1420	2588	1049	1920	3488	1327	2420	4388	1604	2920	5288
777	1430	2606	1054	1930	3506	1332	2430	4406	1610	2930	5306
782	1440	2624	1060	1940	3524	1338	2440	4424	1616	2940	5324
788	1450	2642	1066	1950	3542	1343	2450	4442	1621	2950	5342
793	1460	2660	1071	1960	3560	1349	2460	4460	1627	2960	5360
799	1470	2678	1077	1970	3578	1354	2470	4478	1632	2970	5378
804	1480	2696	1082	1980	3596	1360	2480	4496	1638	2980	5396
810	1490	2714	1088	1990	3614	1366	2490	4514	1643	2990	5414
			1093	2000	3632				1649	3000	5432

Conversion factor table

TO CONVERT	INTO	MULTIPLY BY
amperes/sq in.	amp/sq cm	0.155
amperes/sq in.	amp/sq m	1,550.0
ampere-hr	coulombs	3,600.0
ampere-hr	faradays	0.03731
ampere-hr	gilberts	1.257
ampere-turns/in.	amp-turns/cm	0.3937
Btu	ergs	1.055×10^{10}
Btu	ft-lb	778.3
Btu	gram-cal	252.0
Btu	joules	1,054.8
Btu	kg-cal	0.252
Btu/hr	gram-cal/s	0.07
coulombs	faradays	1.036×10^{-5}
coulombs/sq in.	coulombs/sq meter	1,550.0
cu ft	cu cm	28,320.0
cu ft	liters	28.32
cu ft/min	cu cm/s	472.0
cu ft/min	liters/s	0.472
cu in.	cu cm	16.39
cu in.	liters	0.01639
dyne/cm	erg/sq mm	0.01
erg/s	dyne-cm/s	1.0
ft	cm	30.48
ft	meters	0.3048
ft	mils	1.2×10^{4}
ft of water	atmospheres	0.0295
ft of water	in. of mercury	0.8826
ft of water	kg/sq cm	0.03048
ft/min	cm/s	0.508
ft-candle	Lumen/sq. meter	10.764
ft-lb	joules	1.356
ft-lb	kg-cal	3.24×10^{-4}
ft-lb	kg-meters	0.1383
ft-lb	kilowatt-hr	3.766×10^{-7}
ft-lb/min	kg-cal/min	3.24×10^{-4}
ft-lb/min	kilowatts	2.260×10^{-5}
ft-lb/s	kilowatts	1.356×10^{-3}
gallons	cu cm	3,785.0
gallons	cu meters	3.785×10^{-3}
gallons	liters	3.785
gallons/min	liters/s	0.06308
gausses	lines/sq in.	6.452
gausses	webers/sq cm	6.452×10^{-8}

Conversion factor table (*cont.*)

TO CONVERT	INTO	MULTIPLY BY
gilberts	amp-turns	0.7958
gilberts/cm	amp-turns/cm	0.7958
grams	dynes	980.7
grams	joules/cm	9.807×10^{-5}
grams	joules/meter (newtons)	9.807×10^{-3}
hp (550 ft lb/s)	hp (metric; 542.5 ft lb/s)	1.014
hp	kg-cal/min	10.68
hp	watts	745.7
hp (boiler)	kilowatts	9.803
hp-hr	gram-cal	641,190.0
hp-hr	joules	2.684×10^{6}
in.	cm	2.54
in.	meters	2.54×10^{-2}
in.	mm	25.4
in. of mercury	kg/sq cm	0.03453
in. of water (at 4°C)	kg/sq cm	2.54×10^{-3}
international ampere	ampere (absolute)	0.9998
international volt	joules (absolute)	1.593×10^{-19}
international volt	joules	9.654×10^{4}
international volt	volts (absolute)	1.0003
joules	kg-cal	2.389×10^{-4}
joules	kg-meters	0.102
joules	watt-hr	2.778×10^{-4}
joules/cm	dynes	10^{7}
joules/cm	grams	1.02×10^{4}
joules/cm	joules/meter (newtons)	100.0
joules/cm	lb	22.48
kg	dynes	980,665.0
kg-cal	hp-hr	1.56×10^{-3}
kg-meters	joules	9.804
kilolines	maxwells	1,000.0
kilowatts	kg-cal/min	14.34
lines/sq in.	gausses	0.155
lines/sq in.	webers/sq cm	1.55×10^{-9}
lumens/sq ft.	lumens/sq meter	10.76
maxwells	kilolines	0.001
maxwells	webers	10^{-8}
megalines	maxwells	10^{6}
megohms	microhms	10^{12}
megohms	ohms	100.0
meter-kg	cm-dynes	9.807×10^{7}
meter-kg	cm-grams	10^{5}
microfarads	farads	10^{-6}

Conversion factor table (*cont.*)

TO CONVERT	INTO	MULTIPLY BY
microns	meters	1×10^{-6}
miles (statute)	cm	1.609×10^{5}
miles (statute)	ft	5,280.0
miles (statute)	meters	1,609.0
miles (statute)	yards	1,760.0
miles/hr	cm/s	44.7
miles/hr	km/hr	1.609
miles/hr	meters/min	26.82
miles/hr/s	cm/s/s	44.7
millimicrons	meters	1×10^{-9}
min (angles)	degrees	0.01667
min (angles)	radians	2.909×10^{-4}
min (angles)	s	60.0
nepers	decibels	8.686
newtons	dynes	1×105
ohms (international)	ohms (absolute)	1.0005
ounces	grams	28.349527
ounces	tons (metric)	2.835×10^{-5}
ounces/sq in	dynes/sq cm	4309
parsecs	miles	19×10^{12}
parsecs	km	3.084×10^{13}
poundals	dynes	13,826.0
poundals	grams	14.1
poundals	joules/cm	1.383×10^{-3}
poundals	joules/meter (newtons)	0.1383
lb	dynes	44.4823×10^{4}
lb	grams	453.5924
lb	joules/cm	0.04448
lb	joules/meter (newtons)	4.448
lb-ft	cm-dynes	1.356×10^{7}
lb-ft	cm-grams	13,825.0
lb-ft	meter-kg	0.1383
lb/sq in.	atmospheres	0.06804
lb/sq in.	kg/sq meter	703.1
quarts (liq.)	cu cm	946.4
quarts (liq.)	cu meters	9.464×10^{-4}
quarts (liq.)	liters	0.9463
radians	degrees	57.3
radians	quadrants	0.6366
revolutions	radians	6.283
rev/min	degrees/s	6.0

Conversion factor table *(cont.)*

TO CONVERT	INTO	MULTIPLY BY
rev/min	radians/s	0.1047
rods	meters	5.029
slugs	kg	14.59
sq cm	sq mm	100.0
sq ft	sq cm	929.0
sq ft	sq meters	0.0929
sq in.	sq cm	6.452
sq in.	sq mm	645.2
sq km	sq cm	10^{10}
sq miles	sq km	2.590
sq miles	sq meters	2.590×10^{6}
sq mils	sq cm	6.452×10^{-6}
sq yards	sq cm	8,361.0
sq yards	sq meters	0.8361
sq yards	sq mm	8.361×10^{5}
temperature (°C) + 273	absolute temperature (°C)	1.0
temperature (°C) + 17.78	temperature (°F)	1.8
temperature (°F) + 460	absolute temperature (°F)	1.0
temperature (°F) − 32	temperature (°C)	5/9
tons (long)	kg	1,016.0
tons (short)	kg	907.1848
tons (short)/sq ft	kg/sq meter	9,765.0
volts (absolute)	statvolts	0.003336
volts/in.	volts/cm	0.3937
watts	ergs/s	107.0
watts	hp (metric)	1.36×10^{-3}
watts	kg-cal/min	0.01433
watts	kilowatts	0.001
watts (absolute)	Btu (mean)/min	0.056884
watt-hr	ergs	3.6×10^{10}
watt-hr	kg-cal	0.8605
watt-hr	kg-meters	367.2
watts (international)	watts (absolute)	1.0002
yards	cm	91.44
yards	km	9.144×10^{-4}

GLOSSARY

ABRASION The process of rubbing, grinding, or wearing away by friction.

Ac_{cm}, Ac_1, Ac_3, Ac_4 See *transformation temperature.*

ADHESION Force of attraction between the molecules (or atoms) of two different phases, such as liquid-brazing fillet metal and solid copper, or plated metal and base metal. Contrast with *cohesion.*

AGE HARDENING Hardening by aging, usually after rapid cooling or cold working. See also *aging.*

AGING In a metal or alloy, a change in properties that generally occurs slowly at room temperature and more rapidly at higher temperatures. See also *age hardening.*

AIR-HARDENING STEEL A steel containing sufficient carbon and other alloying elements to fully harden during cooling in air or other gaseous mediums from a temperature above its transformation range. The term should be restricted to steels that are capable of being hardened by cooling in air in fairly large sections, about 2 inches or more in diameter. Same as self-hardening steel.

ALCLAD Composite sheet produced by bonding either corrosion-resistant aluminum alloy or aluminum of high purity to base metal of structurally stronger aluminum alloy.

ALKALINE CLEANER A material blended from alkali hydroxides and such alkaline salts as borates, carbonates, phosphates, and silicates. The cleaning action may be enhanced by the addition of surface-active agents and special solvents.

ALLOTROPY The reversible phenomenon by which certain metals may exist in more than one crystal structure. If not reversible, the phenomenon is termed *polymorphism.*

ALLOY A substance having metallic properties and composed of two or more chemical elements, at least one of which is an elemental metal.

ALLOYING ELEMENT An element added to a metal to effect changes in properties, and which remains within the metal.

ALLOY STEEL Steel containing significant quantities of alloying elements (other than carbon and the commonly accepted amounts of manganese, silicon, sulfur, and phosphorus) added to effect changes in the mechanical or physical properties.

ALL-POSITION ELECTRODE In arc welding, a filler-metal electrode for depositing weld metal in the flat, horizontal, overhead, and vertical positions.

ALL-WELD-METAL TEST SPECIMEN A test specimen wherein the portion being tested is composed wholly of weld metal.

ARC BLOW The swerving of an electric arc from its normal path because of magnetic forces.

ARC BRAZING Brazing with an electric arc, usually with two non-consumable electrodes.

ARC CUTTING Metal cutting with an arc between an electrode and the metal itself. The terms *carbon-arc cutting* and *metal-arc cutting* refer, respectively, to the use of a carbon or a metal electrode.

ARC FURNACE A furnace in which material is heated either directly by an electric arc between an electrode and the work, or indirectly by an arc between two electrodes adjacent to the material.

ARC TIME The amount of time the arc is maintained in making an arc weld.

ARC VOLTAGE The voltage across a welding arc.

ARC WELDING Welding with an electric arc.

AUSTEMPERING Quenching a ferrous alloy from a temperature above the transformation range, in a medium having a rate of heat abstraction high enough to prevent the formation of high-temperature transformation products; then holding the alloy until transformation is complete at a temperature below that of pearlite formation and above that of martensite formation.

AUSTENITE A solid solution of one or more elements in face-centered cubic iron. Unless otherwise designated (as in *nickel* austenite), the solute is generally assumed to be carbon.

AUSTENITIC STEEL An alloy steel whose structure is normally austenitic at room temperature.

AXIS OF WELD A line through the length of a weld perpendicular to the cross-section at its center of gravity.

BACKHAND WELDING Welding in which the back of the principal hand (torch or electrode hand) of the welder faces the direction of travel. It has special significance in gas welding in that it provides postheating. Compare with *forehand welding*.

BACKSTEP SEQUENCE A longitudinal welding sequence in which the direction of general progress is opposite to that of welding the individual increments.

BACK WELD A weld deposited at the back of a single-groove weld.

BARE ELECTRODE A filler-metal arc welding electrode in the form of a wire or rod having no coating other than that incidental to the drawing of the wire.

BEAD WELD A weld composed of one or more string or weave beads deposited on an unbroken surface.

BEAM Directed radiant energy, such as x ray or sonic beam, normally limited in cross-section by the mechanism of generation or by collimating apertures.

BLAST FURNACE A shaft furnace in which solid fuel is burned with an air blast to smelt ore in a continuous operation. Where the temperature must be high, as in the production of pig iron, the air is preheated. Where the temperature can be lower, as in smelting copper, lead, and tin ores, a smaller furnace is economical, and preheating of the blast is not required.

BLASTING Cleaning or finishing metals by impingement with abrasive particles moving at high speed, and usually carried by gas or liquid or thrown from a centrifugal wheel.

BRASS An alloy consisting mainly of copper (over 50%) and zinc, to which smaller amounts of other elements may be added.

BRAZE WELDING Welding in which a groove, fillet, plug, or slot weld is made using a nonferrous filler metal with a melting point lower than that of the base metal but higher than 800°F. The filler metal is not distributed by capillarity.

BRAZING Joining metals by flowing a thin layer, capillary thickness, of nonferrous filler metal into the space between them. Bonding results from the intimate contact produced by the dissolution of a small amount of base metal in the molten *filler metal,* without fusion of the base metal. Sometimes the filler metal is put in place as a thin, solid sheet or as a clad layer, and the composite is heated as in furnace brazing. The term *brazing* is used where the temperature exceeds some arbitrary value, such as 800°F; the term *soldering* is used for temperatures lower than the arbitrary value.

BRAZING FILLER METAL A nonferrous filler metal used in *brazing* and *braze welding.*

BUTT JOINT A joint between two abutting members lying approximately in the same plane. A welded butt joint may contain a variety of grooves. See also *groove weld.*

CARBON-ARC CUTTING Metal cutting by melting with the heat of an arc between a carbon electrode and the base metal.

CARBON-ARC WELDING Welding in which an arc is maintained between a nonconsumable carbon electrode and the work.

CARBON ELECTRODE A carbon or graphite rod used in carbon-arc welding.

CARBON STEEL Steel containing up to about 2% carbon and only residual quantities of other elements, except those added for deoxidation, with silicon usually limited to 0.60% and manganese to about 1.65%. Also termed *plain carbon steel, ordinary steel,* and *straight carbon steel.*

CARBURIZING FLAME A gas flame which will introduce carbon into some heated metals, as during a gas-welding operation. A carburizing flame is a *reducing flame,* but a reducing flame is not necessarily a carburizing flame.

CAST IRON An iron containing carbon in excess of the solubility in the austenite that exists in the alloy at the eutectic temperature. For the various forms of gray cast iron, white cast iron, malleable cast iron, and modular cast iron, the word *cast* is often left out, resulting in *gray iron, white iron, malleable iron* and *nodular iron,* respectively.

CHARPY TEST A pendulum-type, single-blow impact test in which the specimen, usually notched, is supported at both ends as a simple beam and is broken by a falling pendulum. Used as a measure of impact strength or notch toughness.

CONTINUOUS WELD A weld extending continuously from one end of a joint to the other; where the joint is essentially circular, completely around the joint. Contrast with *intermittent weld.*

COPPER BRAZING Brazing with copper as the filler metal.

COVERED ELECTRODE A filler-metal electrode, used in arc welding, consisting of a metal core wire with a relatively thick covering, which provides protection for the molten metal from the atmosphere, improves the properties of the weld metal, and stabilizes the arc. The covering usually consists of mineral or metal powders mixed with cellulose or other binder.

CRATER A depression at the termination of a bead or in the weld pool beneath the electrode.

DECOMPOSITION EFFICIENCY In welding, the ratio of the weight of deposited weld metal to the net weight of electrodes consumed, exclusive of stubs.

DEPOSITION SEQUENCE The order in which the increments of weld metal are deposited.

DEPTH OF FUSION The depth to which the base metal melts during welding.

DOUBLE-BEVEL GROOVE WELD A groove weld in which the joint edge of one member is beveled from both sides.

DOUBLE-J GROOVE WELD A groove weld in which the joint edge of one member is in the form of two J's, one from either side of the member.

DOUBLE-U GROOVE WELD A groove weld in which each joint edge is in the form of two J's or two half-U's, one from either side.

DOUBLE-V GROOVE WELD A groove weld in which each joint edge is beveled from both sides.

DOUBLE-WELDED JOINT A butt, edge, T, corner, or lap joint in which welding has been done from both sides.

DOWNHAND WELDING Same as flat-position welding.

DOWN SLOPE TIME In resistance welding, time associated with current decrease using slope control.

DUTY CYCLE For electric welding equipment, the amount of time that current flows during a specified period. In arc welding, the specified period is 10 minutes.

ELASTICITY That property of a material by virtue of which it tends to recover its original size and shape after deformation.

ELASTIC LIMIT The maximum stress to which a material may be subjected without any permanent strain remaining upon complete release of stress.

ELECTRODE (1) In arc welding, a current-carrying rod which supports the arc between the rod and the work, or between two rods, as in twin carbon-arc welding. It may or may not furnish filler metal. See also *bare electrode, carbon electrode,* and *covered electrode.* (2) In resistance welding, a part of a resistance-welding machine through which current and, in most cases, pressure are applied directly to the work. The electrode may be in the form of a rotating wheel, rotating roll, bar, cylinder, plate, clamp, chuck, or modification thereof. (3) An electrical conductor for leading current into or out of a medium.

ELECTRODE LEAD The electrical conductor between the source of arc welding current and the electrode holder.

ELECTRODE SKID In spot, seam, or projection welding, the sliding of an electrode along the surface of the work.

ELECTROMOTIVE SERIES A list of elements arranged according to their standard electrode potentials. In corrosion studies, the analogous but more practical galvanic series of metals is generally used. The relative position of a given metal is not necessarily the same in the two series.

EQUILIBRIUM DIAGRAM A graphical representation of the temperature, pressure, and composition limits of phase fields in an alloy system as they exist under conditions of complete equilibrium. In metal systems, pressure is usually considered constant.

EUTECTIC (1) An isothermal, reversible reaction in which a liquid solution is converted into two or more intimately mixed solids on cooling, the number of solids formed being the same as the number of components in the system. (2) An alloy having the composition indicated by the eutectic point on an equilibrium diagram. (3) An alloy structure of intermixed solid constituents formed by a eutectic reaction.

EUTECTOID (1) An isothermal, reversible reaction in which a solid solution is converted into two or more intimately mixed solids on cooling, the number of solids formed being the same as the number of components in the system. (2) An alloy having the composition indicated by the eutectoid point on an equilibrium diagram. (3) An alloy structure of intermixed solid constituents formed by a eutectoid reaction.

FILE HARDNESS Hardness as determined by the use of a file of standardized hardness, on the assumption that a material that cannot be cut with the file is as hard as, or harder than, the file. Files covering a range of hardness may be employed.

FILLER METAL Metal added in making a brazed, soldered, or welded joint.

FILLET (1) A radius (curvature) imparted to inside meeting surfaces. (2) A concave cornerpiece used on foundry patterns.

FILLET WELD A weld, approximately triangular in cross-section, joining two surfaces essentially at right angles to each other in a lap, T, or corner joint.

FLAME ANNEALING Annealing in which the heat is applied directly by a flame.

FLAME CLEANING Cleaning metal surfaces of scale, rust, dirt, and moisture by use of a gas flame.

FLAME HARDENING Quench hardening in which the heat is applied directly by a flame.

FLAME STRAIGHTENING Correcting distortion in metal structures by localized heating with a gas flame.

FLASHBACK The recession of a flame into or in back of the interior of a torch.

FLASHING In flash welding, the heating portion of the cycle, consisting of a series of rapidly recurring, localized short circuits followed by molten metal expulsions, during which time the surfaces to be welded are moved toward one another at the proper speed.

FLASH WELDING A resistance butt-welding process in which the weld is produced over the entire abutting surface by pressure and heat, the heat being produced by electric arcs between the members being welded.

FLOW BRAZING Brazing by pouring molten filler metal over a joint.

FLUX-OXYGEN CUTTING Oxygen cutting with the aid of a flux.

FOREHAND WELDING Welding in which the palm of the principal hand (torch or electrode hand) of the welder faces the direction of travel. It has special significance in gas welding in that it provides preheating. Contrast with *backhand welding*.

FORGE DELAY TIME In spot, seam, or projection welding, the time between the start of the welding current or weld interval and the application of forging pressure.

FORGE WELDING Welding hot metal by pressure or blows only.

GAS METAL–ARC WELDING Welding process in which the electrode is a continuous filler metal protected by externally supplied gases.

GAS SHIELDED–ARC WELDING Arc welding in which the arc and molten metal are shielded from the atmosphere by a stream of gas, such as argon, helium, argon-hydrogen mixtures, or carbon dioxide.

GAS TUNGSTEN–ARC WELDING Arc welding in which the arc and molten metal are shielded by externally supplied gases and the necessary filler metal is supplied by welding rods. Electrodes are made of nonconsumable tungsten.

GAS WELDING Welding with the heat from a gas flame.

GROOVE WELD A weld made in the groove between two members. The standard types are: square, single bevel, single-flare bevel, single-flare V, single J, single U, single V, double bevel, double-flare bevel, double-flare V, double J, double U, and double V.

GROSS POROSITY In weld metal or in a casting, pores, gas holes, or globular voids that are larger and in greater number than obtained in good practice.

HALF-LIFE The characteristic time required for half of the nuclei of a radioactive species to disintegrate spontaneously.

HARD CHROMIUM Chromium deposited for engineering purposes, such as increasing the wear resistance of sliding metal surfaces, rather than as a decorative coating. It is usually applied directly to base metal and is customarily thicker than a decorative deposit.

HARDENABILITY In a ferrous alloy, the property that determines the depth and distribution of hardness induced by quenching.

HARD FACING Depositing filler metal on a surface by welding, spraying, or braze welding, for the purpose of resisting abrasion, erosion, wear, galling, and impact.

HARDNESS (1) Resistance of metal to plastic deformation (usually by indentation). However, the term may also refer to stiffness or temper, or to resistance to scratching, abrasion, or cutting. Indentation hardness may be measured by various hardness tests, such as Brinell, Rockwell, and Vickers. (2) For grinding wheels, same as grade.

HARD SOLDERING (obsolete) Formerly referred to a process using materials now called brazing alloys.

HEAT-AFFECTED ZONE The portion of the base metal that was not melted during brazing, cutting, or welding, but whose microstructure and physical properties were altered by the heat.

HEAT TIME In multiple-impulse or seam welding, the time that the current flows during any one impulse.

HEAT TREATMENT Heating and cooling a solid metal or alloy in such a way as to obtain desired conditions or properties. Heating for the sole purpose of hot-working is excluded from the meaning of this definition.

HOMOGENIZING Holding at high temperature to eliminate or decrease chemical segregation by diffusion.

HORIZONTAL-POSITION WELDING (1) Making a fillet weld on the upper side of the intersection of a vertical surface and a horizontal surface. (2) Making a horizontal groove weld on a vertical surface.

HORIZONTAL-ROLLED-POSITION WELDING Topside welding of a butt joint connnecting two horizontal pieces of rotating pipe.

HYDROGEN BRAZING Brazing in a hydrogen atmosphere, usually in a furnace.

IMPACT ENERGY (IMPACT VALUE) The amount of energy required to fracture a material, usually measured by means of an *Izod* or a *Charpy test.* The type of specimen and testing conditions affect the values and therefore should be specified.

IMPACT TEST A test to determine the behavior of materials when subjected to high rates of loading, usually in bending, tension, or torsion. The quantity measured is the energy absorbed in breaking the specimen by a single blow, as in the Charpy or Izod tests.

INDENTATION In a spot, seam, or projection weld, the depression on the exterior surface of the base metal.

INDUCTION BRAZING Brazing with induction heat.

INDUCTION WELDING Welding with induction heat.

INERT-GAS SHIELDED–ARC CUTTING Metal cutting with the heat of an arc in an inert gas such as argon or helium.

INERT-GAS SHIELDED–ARC WELDING Arc welding in an inert gas such as argon or helium.

INTERMITTENT WELD A weld in which the continuity is broken by recurring unwelded spaces.

INTERPASS TEMPERATURE In a multipass weld, the lowest temperature of a pass before the succeeding one is commenced.

IZOD TEST A pendulum type of single-blow impact test in which the specimen, usually notched, is fixed at one end and broken by a falling pendulum. The energy absorbed, as measured by the subsequent rise of the pendulum, is a measure of impact strength or notch toughness.

JOINT The location where two or more members are to be, or have been, fastened together mechanically or by brazing or welding.

JOINT EFFICIENCY The strength of a welded joint expressed as a percentage of the strength of the unwelded base metal.

JOINT PENETRATION The distance that weld metal and fusion extend into a joint.

LAP JOINT A joint made with two overlapping members.

LAYER A stratum of one or more weld beads lying in a plane parallel to the surface from which welding was done.

LEG OF A FILLET WELD *Actual:* The distance from the root of the joint to the toe of a fillet weld. *Nominal:* The length of a side of the largest right triangle that can be inscribed in the cross-section of the weld.

MACHINE WELDING Welding with equipment that performs under the observation and control of an operator. It may or may not perform the loading and unloading of the work.

MACRO-ETCH Etching of a metal surface for accentuation of gross structural details and defects for observation by the unaided eye or at magnifications not exceeding ten diameters.

MACROGRAPH A graphic reproduction of the surface of a prepared specimen at a magnification not exceeding ten diameters. When photographed, the reproduction is known as a photomacrograph.

MACROSCOPIC Visible at magnifications from one to ten diameters.

MAGNETIC-ANALYSIS INSPECTION A nondestructive method of inspection for determining the existence and extent of possible defects in ferromagnetic materials. Finely divided magnetic particles, applied to the magnetized part, are attracted to and outline the pattern of any magnetic-leakage fields created by discontinuities.

MALLEABLE CAST IRON A cast iron made by a prolonged anneal of white cast iron in which decarburization or graphitization, or both, take place to eliminate some or all of the cementite. The graphite is in the form of temper carbon. If decarburization is the predominant reaction, the product will have a light fracture, hence "whiteheart malleable"; otherwise, the fracture will be dark, hence "blackheart malleable." "Pearlitic malleable" is a blackheart variety having a pearlitic matrix, along with perhaps some free ferrite.

MANUAL WELDING Welding wherein the entire welding operation is performed and controlled by hand.

MARTEMPERING Quenching an austenitized ferrous alloy in a medium at a temperature in the upper part of the martensite range, or slightly above that range, and holding it in the medium until the temperature throughout the alloy is substantially uniform. The alloy is then allowed to cool in air through the martensite range.

MARTENSITE (1) In an alloy, a metastable transitional structure intermediate between two allotropic modifications whose abilities to dissolve a given solute differ considerably, the high-temperature phase having the greater solubility. The amount of the high-temperature phase transformed to martensite depends to a large extent upon the temperature attained in cooling, there being a rather distinct beginning temperature. (2) A metastable phase of steel, formed by a transformation of austenite below the M_s (or Ar'') temperature. It is an interstitial, supersaturated solid solution of carbon in iron having a body-centered tetragonal lattice. Its microstructure is characterized by an acicular, or needle-like, pattern.

MASH SEAM WELD A seam weld made in a lap joint, in which the thickness at the lap is reduced plastically to approximately the thickness of one of the lapped parts.

MELTING RATE In electric-arc welding, the weight or length of electrode melted in a unit of time. Sometimes called *melt-off rate* or *burn-off rate.*

METAL-ARC CUTTING Metal cutting with the heat of an arc between a metal electrode and the base metal.

METAL-ARC WELDING Arc welding with metal electrodes. Commonly refers to shielded metal–arc welding using covered electrodes.

METALLIZING (SPRAY METALLIZING) Forming a metallic coating by atomized spraying with molten metal or by vacuum deposition.

METALLOGRAPHY The science dealing with the constitution and structure of metals and alloys as revealed by the unaided eye or by such tools as low-powered magnification, optical microscope, electron microscope, and diffraction or x-ray techniques.

MILD STEEL Carbon steel with a maximum of about 0.25% carbon.

NECKING (1) Reducing the cross-sectional area of metal in a localized area by stretching. (2) Reducing the diameter of a portion of the length of a cylindrical shell or tube.

NEUTRAL FLAME A gas flame in which there is no excess of either fuel or oxygen.

NODULAR CAST IRON A cast iron that has been treated while molten with a master alloy containing an element such as magnesium or cerium to give primary graphite in the spherulitic form.

NONDESTRUCTIVE INSPECTION Inspection, by methods that do not destroy the part, to determine its suitability for use.

OFF TIME In resistance welding, the time that the electrodes are off the work. This term is generally applied where the welding cycle is repetitive.

OVERHEAD-POSITION WELDING Welding that is performed from the underside.

OVERLAP (1) Protrusion of weld metal beyond the bond at the toe of the weld. (2) In spot, seam, or projection welding, the amount one sheet overlays the other.

OXIDIZING FLAME A gas flame produced with excess oxygen.

OXYACETYLENE CUTTING Oxygen cutting in which the initiation temperature is attained with an oxyacetylene flame.

OXYACETYLENE WELDING Welding with an oxyacetylene flame.

OXYGEN CUTTING Metal cutting by directing a stream of oxygen upon a hot metal. The chemical reaction of oxygen and the base metal furnishes heat for the localized melting, hence, cutting.

OXYGEN GOUGING Oxygen cutting in which a chamfer or groove is formed.

OXYGEN LANCE A length of pipe used to convey oxygen to the point of cutting in oxygen-lance cutting.

OXYHYDROGEN CUTTING Oxygen cutting in which the initiation temperature is attained with an oxyhydrogen flame.

OXYHYDROGEN WELDING Welding with an oxyhydrogen flame.

PLUG WELD A circular weld made by either arc or gas welding through one member of a lap or T joint. If a hole is used, it may be only partially filled. Neither a fillet-welded hole nor a spot weld is to be construed as a plug weld.

QUENCHING Rapid cooling. When applicable, the following more specific terms should be used: *direct quenching, fog quenching, hot quenching, interrupted quenching, selective quenching, spray quenching,* and *time quenching.*

RADIOGRAPH A photographic shadow image resulting from uneven absorption of radiation in the object being subjected to penetrating radiation.

RADIOGRAPHY A nondestructive method of internal examination in which metal or other objects are exposed to a beam of x-ray or gamma radiation. Differences in thickness, density, or absorption, caused by internal discontinuities, are apparent in the shadow image either on a fluorescent screen or on photographic film placed behind the object.

REDUCING FLAME A gas flame produced with excess fuel.

REFRACTORY (1) A material of very high melting point with properties that make it suitable for such uses as furnace linings and kiln construction. (2) The quality of resisting heat.

REFRACTORY ALLOY (1) A heat-resistant alloy. (2) An alloy having an extremely high melting point. See also *refractory metal.* (3) An alloy difficult to work at elevated temperatures.

REFRACTORY METAL A metal having an extremely high melting point. In the broad sense, it refers to metals having melting points above the range of iron, cobalt, and nickel.

REINFORCEMENT OF WELD (1) In a butt joint, weld metal on the face of the weld that extends out beyond a surface plane common to the members being welded. (2) In a fillet weld, weld metal that contributes to convexity. (3) In a flash, upset, or gas-pressure weld, the portion of the upset in excess of the original diameter or thickness.

RESISTANCE BRAZING Brazing by resistance heating, the joint being part of the electrical circuit.

RESISTANCE WELDING Welding with resistance heating and pressure, the work being part of the electrical circuit. Examples include resistance spot-welding, resistance seam-welding, projection welding, and flash-butt welding.

RESISTANCE-WELDING DIE The part of a resistance-welding machine, usually shaped to the work contour, with which the parts being welded are held and with which they can be discriminated.

ROCKWELL HARDNESS TEST A test for determining the hardness of a material based on the depth of penetration of a specified penetrator into the specimen under certain arbitrarily fixed test conditions.

ROLL WELDING Forge welding by heating in a furnace and applying pressure with rolls.

ROOT CRACK A crack in either the weld or the heat-affected zone at the root of a weld.

ROOT FACE The unbeveled portion of the groove face of a joint.

ROOT OF JOINT The location of closest approach between the parts of a joint to be welded.

ROOT OF WELD The points, as shown in cross-section, at which the bottom of the weld intersects the base-metal surfaces. It may be coincident with the *root of joint.*

ROOT OPENING The distance between the parts at the root of the joint.

ROOT PASS The first bead of a multiple-pass weld.

ROOT PENETRATION The depth to which weld metal extends into the root of a joint.

SCARF JOINT A butt joint in which the plane of the joint is inclined with respect to the main axis of the members.

SEAL WELD Any weld used primarily to obtain tightness and prevent leakage.

SEAM WELDING (1) Arc or resistance welding in which a series of overlapping spot welds is produced with rotating electrodes, rotating work, or both. (2) Making a longitudinal weld in sheet metal or tubing.

SEGREGATION Nonuniform distribution of alloying elements, impurities, or microphases.

SEMICONDUCTOR An electrical conductor whose resistivity at room temperature is in the range of 10^{-9} to 10^{-2} ohm-cm, and in which the conductivity increases with increasing temperature over some temperature range.

SERIES WELDING Making two or more resistance spot, seam, or projection welds simultaneously by a single welding transformer, with three or more electrodes forming a series circuit.

SHIELDED-ARC WELDING Arc welding in which the arc and the weld metal are protected by a gaseous atmosphere, the products of decomposition of the electrode covering, or a blanket of fusible flux.

SHIELDED METAL–ARC WELDING Arc welding in which the arc and the weld metal are protected by the decomposition products of the covering on a consumable metal electrode.

SIZE OF WELD (1) The joint penetration in a groove weld. (2) The lengths of the nominal legs of a fillet weld.

SLIP PLANE The crystallographic plane in which slip occurs in a crystal.

SLOT WELD Similar to *plug weld*, the difference being that the hole is elongated and may extend to the edge of a member without closing.

SOFT SOLDERING See *soldering*.

SOLDER EMBRITTLEMENT Reduction in mechanical properties of a metal as a result of local penetration of solder along grain boundaries.

SOLDERING Similar to *brazing*, with the filler metal having a melting-temperature range below an arbitrary value, generally 800°F. Soft solders are usually lead-tin alloys.

SOLID-STATE WELDING Any method of welding in which pressure, or heat and pressure, are used to consummate the weld without fusion.

SPHEROIDAL GRAPHITE CAST IRON Same as *nodular cast iron.*

SPHEROIDITE An aggregate of iron or alloy carbides of essentially spherical shape dispersed throughout a matrix of ferrite.

SPOT WELDING Welding of lapped parts in which fusion is confined to a relatively small, circular area. It is generally resistance welding, but may also be gas-shielded tungsten–arc, gas-shielded metal–arc, or submerged-arc welding.

SQUARE-GROOVE WELD A groove weld in which the abutting surfaces are square.

SQUEEZE TIME In resistance welding, the time between the initial applications of pressure and current.

STEEL An iron-base alloy, malleable in some temperature range as initially cast, containing manganese, usually carbon, and often other alloying elements. In carbon steel and low-alloy steel, the maximum carbon is about 2%; in high-alloy steel, about 2.5%. The dividing line between low-alloy and high-alloy steels is generally regarded as being at about 5% metallic alloying elements.

Steel is to be differentiated from two general classes of irons: the cast irons, on the high-carbon side; and the relatively pure irons such as ingot iron, carbonyl iron, and electrolytic iron, on the low-carbon side. In some steels containing very little carbon, the manganese content is the principal differentiating factor, steel usually containing at least 0.25%; ingot iron contains considerably less.

STORED-ENERGY WELDING Welding with electrical energy accumulated electrostatically, electromagnetically, or electrochemically at a relatively low rate and made available at the higher rate required in welding.

TACK WELDS Small, scattered welds made to hold parts of a weldment in proper alignment while the final welds are being made.

THROUGH WELD A weld of appreciable length made by either arc or gas welding through the unbroken surface of one member of a lap or T joint joining that member to the other.

T JOINT A joint in which the members are oriented in the form of a T.

TORCH A gas burner used to braze, cut, or weld. For brazing or welding, it has two gas-feed lines—one for fuel, such as acetylene or hydrogen, the other for oxygen. For cutting, there may be an additional feed line for oxygen. See also *oxygen cutting.*

TORCH BRAZING Brazing with a torch.

TRANSFORMATION TEMPERATURE The temperature at which a change in phase occurs. The term is sometimes used to denote the limiting temperature of a transformation range.

The following symbols are used for iron and steels:

- Ac_{cm}. In hypereutectoid steel, the temperature at which the solution of cementite in austenite is completed during heating.
- Ac_1. The temperature at which austenite begins to form during heating.
- Ac_3. The temperature at which transformation of ferrite to austenite is completed during heating.
- Ac_4. The temperature at which austenite transforms to delta ferrite during heating.
- Ae_{cm}, Ae_1, Ae_3, Ae_4. The temperatures of phase changes at equilibrium.
- Ar_{cm}. In hypereutectoid steel, the temperature at which precipitation of cementite starts during cooling.
- Ar_1. The temperature at which transformation of austenite to ferrite, or to ferrite plus cementite, is completed during cooling.
- Ar_3. The temperature at which austenite begins to transform to ferrite during cooling.
- Ar_4. The temperature at which delta ferrite transforms to austenite during cooling.
- M_s (or Ar''). The temperature at which transformation of austenite to martensite starts during cooling.
- M_f. The temperature at which martensite formation finishes during cooling.

Note: All these changes, except the formation of martensite, occur at lower temperatures during cooling than during heating, and depend on the rate of temperature change.

ULTIMATE STRENGTH The maximum conventional stress—tensile, compressive, or shear—that a material can withstand.

ULTRASONIC BEAM A beam of acoustical radiation with a higher frequency than the audible frequency range.

ULTRASONIC FREQUENCY A frequency, associated with elastic waves, that is greater than the highest audible frequency, generally regarded as being higher than 15 kHz.

UNDERBEAD CRACK A subsurface crack in the base metal near the weld.

UNDERCUT A groove melted into the base metal adjacent to the toe of a weld and left unfilled.

WEAVE WELD A weld made with oscillations transverse to the axis of the weld.

WELD A union made by welding.

WELDABILITY Suitability of a metal for welding under specific conditions.

WELD BEAD A deposit of filler metal from a single welding pass.

WELD CRACK A crack in weld metal.

WELD DELAY TIME In spot, seam, or projection welding, the time that current is delayed with respect to starting the forge delay timer in order to synchronize the forging pressure and the welding heat.

WELD GAGE A device for checking the shape and size of welds.

WELDING (1) Joining two or more pieces of material by applying heat, pressure, or both, with or without filler material, to produce a localized union through fusion or recrystallization across the interface. The thickness of the filler material is much greater than the capillary dimensions encountered in brazing. (2) May also be extended to include brazing.

WELDING CURRENT The current flowing through a welding circuit during the making of a weld. In resistance welding, the current used during preweld or postweld intervals is excluded.

WELDING CYCLE The complete series of events involved in making a resistance weld. Also applies to semiautomatic mechanized fusion welds.

WELDING LEAD (WELDING CABLE) A work lead or an electrode lead.

WELDING MACHINE Equipment used to perform the welding operation; for example, spot-welding machine, arc welding machine, and seam-welding machine.

WELDING PROCEDURE The detailed methods and practices, including joint-welding procedures, involved in the production of a *weldment.*

WELDING ROD Filler metal in rod or wire form used in welding.

WELDING TIP (1) A replaceable nozzle for a gas torch that is especially adapted for welding. (2) A spot welding or projection welding electrode.

WELDMENT An assembly whose component parts are joined by welding.

WELD METAL That portion of a weld which has been melted during welding.

WORK LEAD The electrical conductor connecting the source of arc-welding current to the work. Also called *welding ground*, *ground lead*, or *ground cable*.

X RAY Electromagnetic radiation, of wavelength less than about 500 angstrom units, emitted as the result of deceleration of fast-moving electrons (*Bremsstrahlung*, continuous spectrum) or decay of atomic electrons from excited orbital states (characteristic radiation). Specifically, the radiation produced when an electron beam of sufficient energy impinges upon a target of suitable material.

YIELD POINT The first stress in a material, usually less than the maximum attainable stress, at which an increase in strain occurs without an increase in stress. Only certain metals exhibit a yield point. If there is a decrease in stress after yielding, a distinction may be made between upper and lower yield points.

YIELD STRENGTH The stress at which a material exhibits a specified deviation from proportionality of stress and strain. An offset of 0.2% is used for many metals.

INDEX

ABCDEFGHIJ–H–798